PATTERN THE

PATTERN THEORY: FROM REPRESENTATION TO INFERENCE

Ulf Grenander and Michael I. Miller

OXFORD
UNIVERSITY PRESS

Great Clarendon Street, Oxford OX2 6DP

Oxford University Press is a department of the University of Oxford.
It furthers the University's objective of excellence in research, scholarship,
and education by publishing worldwide in

Oxford New York

Auckland Cape Town Dar es Salaam Hong Kong Karachi
Kuala Lumpur Madrid Melbourne Mexico City Nairobi
New Delhi Shanghai Taipei Toronto

With offices in

Argentina Austria Brazil Chile Czech Republic France Greece
Guatemala Hungary Italy Japan Poland Portugal Singapore
South Korea Switzerland Thailand Turkey Ukraine Vietnam

Oxford is a registered trade mark of Oxford University Press
in the UK and in certain other countries

Published in the United States
by Oxford University Press Inc., New York

© Oxford University Press, 2007

The moral rights of the authors have been asserted
Database right Oxford University Press (maker)

First published 2007

All rights reserved. No part of this publication may be reproduced,
stored in a retrieval system, or transmitted, in any form or by any means,
without the prior permission in writing of Oxford University Press,
or as expressly permitted by law, or under terms agreed with the appropriate
reprographics rights organization. Enquiries concerning reproduction
outside the scope of the above should be sent to the Rights Department,
Oxford University Press, at the address above

You must not circulate this book in any other binding or cover
and you must impose the same condition on any acquirer

British Library Cataloguing in Publication Data

Data available

Library of Congress Cataloging in Publication Data

Data available

Typeset by Newgen Imaging Systems (P) Ltd., Chennai, India
Printed in Great Britain
on acid-free paper by
CPI Bath

ISBN 0–19–850570–1 978–0–19–850570–9
ISBN 0–19–929706–1 978–0–19–929706–1

1 3 5 7 9 10 8 6 4 2

CONTENTS

1 Introduction — 1
    1.1 Organization — 3

2 The Bayes Paradigm, Estimation and Information Measures — 5
    2.1 Bayes Posterior Distribution — 5
        2.1.1 Minimum Risk Estimation — 6
        2.1.2 Information Measures — 7
    2.2 Mathematical Preliminaries — 8
        2.2.1 Probability Spaces, Random Variables, Distributions, Densities, and Expectation — 8
        2.2.2 Transformations of Variables — 10
        2.2.3 The Multivariate Normal Distribution — 10
        2.2.4 Characteristic Function — 11
    2.3 Minimum Risk Hypothesis Testing on Discrete Spaces — 12
        2.3.1 Minimum Probability of Error via Maximum A Posteriori Hypothesis Testing — 13
        2.3.2 Neyman–Pearson and the Optimality of the Likelihood Ratio Test — 14
    2.4 Minimum Mean-Squared Error Risk Estimation in Vector Spaces — 16
        2.4.1 Normed Linear and Hilbert Spaces — 17
        2.4.2 Least-Squares Estimation — 20
        2.4.3 Conditional Mean Estimation and Gaussian Processes — 22
    2.5 The Fisher Information of Estimators — 24
    2.6 Maximum-Likelihood and its consistency — 26
        2.6.1 Consistency via Uniform Convergence of Empirical Log-likelihood — 27
        2.6.2 Asymptotic Normality and $\sqrt{n}$ Convergence Rate of the MLE — 28
    2.7 Complete–Incomplete Data Problems and the EM Algorithm — 30
    2.8 Hypothesis Testing and Model Complexity — 38
        2.8.1 Model-Order Estimation and the $d/2 \log$ Sample-Size Complexity — 38
        2.8.2 The Gaussian Case is Special — 41
        2.8.3 Model Complexity and the Gaussian Case — 42
    2.9 Building Probability Models via the Principle of Maximum Entropy — 43
        2.9.1 Principle of Maximum Entropy — 44
        2.9.2 Maximum Entropy Models — 45
        2.9.3 Conditional Distributions are Maximum Entropy — 47

3 Probabilistic Directed Acyclic Graphs and Their Entropies — 49
    3.1 Directed Acyclic Graphs (DAGs) — 49
    3.2 Probabilities on Directed Acyclic Graphs (PDAGs) — 51
    3.3 Finite State Markov Chains — 54
    3.4 Multi-type Branching Processes — 56
        3.4.1 The Branching Matrix — 59
        3.4.2 The Moment-Generating Function — 60
    3.5 Extinction for Finite-State Markov Chains and Branching Processes — 62
        3.5.1 Extinction in Markov Chains — 62
        3.5.2 Extinction in Branching Processes — 63
    3.6 Entropies of Directed Acyclic Graphs — 64
    3.7 Combinatorics of Independent, Identically Distributed Strings via the Aymptotic Equipartition Theorem — 65

3.8	Entropy and Combinatorics of Markov Chains		66
3.9	Entropies of Branching Processes		68
	3.9.1	Tree Structure of Multi-Type Branching Processes	69
	3.9.2	Entropies of Sub-Critical, Critical, and Super-Critical Processes	70
	3.9.3	Typical Trees and the Equipartition Theorem	71
3.10	Formal Languages and Stochastic Grammars		74
3.11	DAGs for Natural Language Modelling		81
	3.11.1	Markov Chains and m-Grams	81
	3.11.2	Context-Free Models	82
	3.11.3	Hierarchical Directed Acyclic Graph Model	84
3.12	EM Algorithms for Parameter Estimation in Hidden Markov Models		87
	3.12.1	MAP Decoding of the Hidden State Sequence	88
	3.12.2	ML Estimation of HMM parameters via EM Forward/Backward Algorithm	89
3.13	EM Algorithms for Parameter Estimation in Natural Language Models		92
	3.13.1	EM Algorithm for Context-Free Chomsky Normal Form	93
	3.13.2	General Context-Free Grammars and the Trellis Algorithm of Kupiec	94

4 Markov Random Fields on Undirected Graphs — 95

4.1	Undirected Graphs		95
4.2	Markov Random Fields		96
4.3	Gibbs Random Fields		101
4.4	The Splitting Property of Gibbs Distributions		104
4.5	Bayesian Texture Segmentation: The log-Normalizer Problem		110
	4.5.1	The Gibbs Partition Function Problem	110
4.6	Maximum-Entropy Texture Representation		112
	4.6.1	Empirical Maximum Entropy Texture Coding	113
4.7	Stationary Gibbs Random Fields		116
	4.7.1	The Dobrushin/Lanford/Ruelle Definition	116
	4.7.2	Gibbs Distributions Exhibit Multiple Laws with the Same Interactions (Phase Transitions): The Ising Model at Low Temperature	117
4.8	1D Random Fields are Markov Chains		119
4.9	Markov Chains Have a Unique Gibbs Distribution		120
4.10	Entropy of Stationary Gibbs Fields		121

5 Gaussian Random Fields on Undirected Graphs — 123

5.1	Gaussian Random Fields		123
5.2	Difference Operators and Adjoints		124
5.3	Gaussian Fields Induced via Difference Operators		126
5.4	Stationary Gaussian Processes on $\mathbb{Z}^d$ and their Spectrum		133
5.5	Cyclo-Stationary Gaussian Processes and their Spectrum		134
5.6	The log-Determinant Covariance and the Asymptotic Normalizer		137
	5.6.1	Asymptotics of the Gaussian processes and their Covariance	138
	5.6.2	The Asymptotic Covariance and log-Normalizer	142
5.7	The Entropy Rates of the Stationary Process		142
	5.7.1	Burg's Maximum Entropy Auto-regressive Processes on $\mathbb{Z}^d$	143
5.8	Generalized Auto-Regressive Image Modelling via Maximum-Likelihood Estimation		144
	5.8.1	Anisotropic Textures	147

6	**The Canonical Representations of General Pattern Theory**	**154**
	6.1 The Generators, Configurations, and Regularity of Patterns	154
	6.2 The Generators of Formal Languages and Grammars	158
	6.3 Graph Transformations	162
	6.4 The Canonical Representation of Patterns: DAGs, MRFs, Gaussian Random Fields	166
	6.4.1 Directed Acyclic Graphs	167
	6.4.2 Markov Random Fields	169
	6.4.3 Gaussian Random Fields: Generators induced via difference operators	170
7	**Matrix Group Actions Transforming Patterns**	**174**
	7.1 Groups Transforming Configurations	174
	7.1.1 Similarity Groups	174
	7.1.2 Group Actions Defining Equivalence	175
	7.1.3 Groups Actions on Generators and Deformable Templates	177
	7.2 The Matrix Groups	177
	7.2.1 Linear Matrix and Affine Groups of Transformation	177
	7.2.2 Matrix groups acting on $\mathbb{R}^d$	179
	7.3 Transformations Constructed from Products of Groups	181
	7.4 Random Regularity on the Similarities	184
	7.5 Curves as Submanifolds and the Frenet Frame	190
	7.6 2D Surfaces in $\mathbb{R}^3$ and the Shape Operator	195
	7.6.1 The Shape Operator	196
	7.7 Fitting Quadratic Charts and Curvatures on Surfaces	198
	7.7.1 Gaussian and Mean Curvature	198
	7.7.2 Second Order Quadratic Charts	200
	7.7.3 Isosurface Algorithm	201
	7.8 Ridge Curves and Crest Lines	205
	7.8.1 Definition of Sulcus, Gyrus, and Geodesic Curves on Triangulated Graphs	205
	7.8.2 Dynamic Programming	207
	7.9 Bijections and Smooth Mappings for Coordinatizing Manifolds via Local Coordinates	210
8	**Manifolds, Active Models, and Deformable Templates**	**214**
	8.1 Manifolds as Generators, Tangent Spaces, and Vector Fields	214
	8.1.1 Manifolds	214
	8.1.2 Tangent Spaces	215
	8.1.3 Vector Fields on M	217
	8.1.4 Curves and the Tangent Space	218
	8.2 Smooth Mappings, the Jacobian, and Diffeomorphisms	219
	8.2.1 Smooth Mappings and the Jacobian	219
	8.2.2 The Jacobian and Local Diffeomorphic Properties	221
	8.3 Matrix Groups are Diffeomorphisms which are a Smooth Manifold	222
	8.3.1 Diffeomorphisms	222
	8.3.2 Matrix Group Actions are Diffeomorphisms on the Background Space	223
	8.3.3 The Matrix Groups are Smooth Manifolds (Lie Groups)	224
	8.4 Active Models and Deformable Templates as Immersions	226
	8.4.1 Snakes and Active Contours	226
	8.4.2 Deforming Closed Contours in the Plane	226
	8.4.3 Normal Deformable Surfaces	227
	8.5 Activating Shapes in Deformable Models	229
	8.5.1 Likelihood of Shapes Partitioning Image	229
	8.5.2 A General Calculus for Shape Activation	229

	8.5.3	Active Closed Contours in $\mathbb{R}^2$	232
	8.5.4	Active Unclosed Snakes and Roads	234
	8.5.5	Normal Deformation of Circles and Spheres	236
	8.5.6	Active Deformable Spheres	236
8.6		Level Set Active Contour Models	237
8.7		Gaussian Random Field Models for Active Shapes	240

9 Second Order and Gaussian Fields — 244

9.1		Second Order Processes (SOP) and the Hilbert Space of Random Variables	244
	9.1.1	Measurability, Separability, Continuity	244
	9.1.2	Hilbert space of random variables	247
	9.1.3	Covariance and Second Order Properties	249
	9.1.4	Quadratic Mean Continuity and Integration	251
9.2		Orthogonal Process Representations on Bounded Domains	252
	9.2.1	Compact Operators and Covariances	253
	9.2.2	Orthogonal Representations for Random Processes and Fields	257
	9.2.3	Stationary Periodic Processes and Fields on Bounded Domains	258
9.3		Gaussian Fields on the Continuum	262
9.4		Sobolev Spaces, Green's Functions, and Reproducing Kernel Hilbert Spaces	264
	9.4.1	Reproducing Kernel Hilbert Spaces	265
	9.4.2	Sobolev Normed Spaces	266
	9.4.3	Relation to Green's Functions	267
	9.4.4	Gradient and Laplacian Induced Green's Kernels	267
9.5		Gaussian Processes Induced via Linear Differential Operators	271
9.6		Gaussian Fields in the Unit Cube	274
	9.6.1	Maximum Likelihood Estimation of the Fields: Generalized ARMA Modelling	278
	9.6.2	Small Deformation Vector Fields Models in the Plane and Cube	280
9.7		Discrete Lattices and Reachability of Cyclo-Stationary Spectra	283
9.8		Stationary Processes on the Sphere	285
	9.8.1	Laplacian Operator Induced Gaussian Fields on the Sphere	289
9.9		Gaussian Random Fields on an Arbitrary Smooth Surface	293
	9.9.1	Laplace-Beltrami Operator with Neumann Boundary Conditions	293
	9.9.2	Smoothing an Arbitrary Function on Manifolds by Orthonormal Bases of the Laplace-Beltrami Operator	297
9.10		Sample Path Properties and Continuity	299
9.11		Gaussian Random Fields as Prior Distributions in Point Process Image Reconstruction	303
	9.11.1	The Need for Regularization in Image Reconstruction	304
	9.11.2	Smoothness and Gaussian Priors	304
	9.11.3	Good's Roughness as a Gaussian Prior	305
	9.11.4	Exponential Spline Smoothing via Good's Roughness	306
9.12		Non-Compact Operators and Orthogonal Representations	309
	9.12.1	Cramer Decomposition for Stationary Processes	311
	9.12.2	Orthogonal Scale Representation	312

10 Metrics Spaces for the Matrix Groups — 316

10.1		Riemannian Manifolds as Metric Spaces	316
	10.1.1	Metric Spaces and Smooth Manifolds	316
	10.1.2	Riemannian Manifold, Geodesic Metric, and Minimum Energy	317
10.2		Vector Spaces as Metric Spaces	319
10.3		Coordinate Frames on the Matrix Groups and the Exponential Map	320

	10.3.1	Left and Right Group Action	320
	10.3.2	The Coordinate Frames	321
	10.3.3	Local Optimization via Directional Derivatives and the Exponential Map	323
10.4		Metric Space Structure for the Linear Matrix Groups	324
	10.4.1	Geodesics in the Matrix Groups	324
10.5		Conservation of Momentum and Geodesic Evolution of the Matrix Groups via the Tangent at the Identity	326
10.6		Metrics in the Matrix Groups	327
10.7		Viewing the Matrix Groups in Extrinsic Euclidean Coordinates	329
	10.7.1	The Frobenius Metric	329
	10.7.2	Comparing intrinsic and extrinsic metrics in $\mathbf{SO}(2,3)$	330

11 Metrics Spaces for the Infinite Dimensional Diffeomorphisms 332

11.1 Lagrangian and Eulerian Generation of Diffeomorphisms 332
    11.1.1 On Conditions for Generating Flows of Diffeomorphisms 333
    11.1.2 Modeling via Differential Operators and the Reproducing Kernel Hilbert Space 335
11.2 The Metric on the Space of Diffeomorphisms 336
11.3 Momentum Conservation for Geodesics 338
11.4 Conservation of Momentum for Diffeomorphism Splines Specified on Sparse Landmark Points 340
    11.4.1 An ODE for Diffeomorphic Landmark Mapping 343

12 Metrics on Photometric and Geometric Deformable Templates 346

12.1 Metrics on Dense Deformable Templates: Geometric Groups Acting on Images 346
    12.1.1 Group Actions on the Images 346
    12.1.2 Invariant Metric Distances 347
12.2 The Diffeomorphism Metric for the Image Orbit 349
12.3 Normal Momentum Motion for Geodesic Connection Via Inexact Matching 350
12.4 Normal Momentum Motion for Temporal Sequences 354
12.5 Metric Distances Between Orbits Defined Through Invariance of the Metric 356
12.6 Finite Dimensional Landmarked Shape Spaces 357
    12.6.1 The Euclidean Metric 357
    12.6.2 Kendall's Similitude Invariant Distance 359
12.7 The Diffeomorphism Metric and Diffeomorphism Splines on Landmark Shapes 361
    12.7.1 Small Deformation Splines 361
12.8 The Deformable Template: Orbits of Photometric and Geometric Variation 365
    12.8.1 Metric Spaces for Photometric Variability 365
    12.8.2 The Metrics Induced via Photometric and Geometric Flow 366
12.9 The Euler Equations for Photometric and Geometric Variation 369
12.10 Metrics between Orbits of the Special Euclidean Group 373
12.11 The Matrix Groups (Euclidean and Affine Motions) 374
    12.11.1 Computing the Affine Motions 376

13 Estimation Bounds for Automated Object Recognition 378

13.1 The Communications Model for Image Transmission 378
    13.1.1 The Source Model: Objects Under Matrix Group Actions 379
    13.1.2 The Sensing Models: Projective Transformations in Noise 379
    13.1.3 The Likelihood and Posterior 379

		13.2	Conditional Mean Minimum Risk Estimation	381

13.2 Conditional Mean Minimum Risk Estimation 381
 13.2.1 Metrics (Risk) on the Matrix Groups 381
 13.2.2 Conditional Mean Minimum Risk Estimators 382
 13.2.3 Computation of the HSE for $\mathbf{SE}(2,3)$ 384
 13.2.4 Discrete integration on $\mathbf{SO}(3)$ 385
13.3 MMSE Estimators for Projective Imagery Models 385
 13.3.1 3D to 2D Projections in Gaussian Noise 385
 13.3.2 3D to 2D Synthetic Aperture Radar Imaging 389
 13.3.3 3D to 2D LADAR Imaging 392
 13.3.4 3D to 2D Poisson Projection Model 393
 13.3.5 3D to 1D Projections 395
 13.3.6 3D(2D) to 3D(2D) Medical Imaging Registration 397
13.4 Parameter Estimation and Fisher Information 398
13.5 Bayesian Fusion of Information 402
13.6 Asymptotic Consistency of Inference and Symmetry Groups 405
 13.6.1 Consistency 405
 13.6.2 Symmetry Groups and Sensor Symmetry 406
13.7 Hypothesis Testing and Asymptotic Error-Exponents 407
 13.7.1 Analytical Representations of the Error Probabilities and the Bayesian Information Criterion 408
 13.7.2 m-ary Multiple Hypotheses 412

14 Estimation on Metric Spaces with Photometric Variation 414
14.1 The Deformable Template: Orbits of Signature and Geometric Variation 414
 14.1.1 The Robust Deformable Templates 414
 14.1.2 The Metric Space of the Robust Deformable Template 415
14.2 Empirical Covariance of Photometric Variability via Principle Components 416
 14.2.1 Signatures as a Gaussian Random Field Constructed from Principle Components 417
 14.2.2 Algorithm for Empirical Construction of Bases 418
14.3 Estimation of Parameters on the Conditionally Gaussian Random Field Models 422
14.4 Estimation of Pose by Integrating Out EigenSignatures 424
 14.4.1 Bayes Integration 427
14.5 Multiple Modality Signature Registration 429
14.6 Models for Clutter: The Transported Generator Model 431
 14.6.1 Characteristic Functions and Cumulants 432
14.7 Robust Deformable Templates for Natural Clutter 438
 14.7.1 The Euclidean Metric 439
 14.7.2 Metric Space Norms for Clutter 439
 14.7.3 Computational Scheme 442
 14.7.4 Empirical Construction of the Metric from Rendered Images 444
14.8 Target detection/identification in EO imagery 445

15 Information Bounds for Automated Object Recognition 447
15.1 Mutual Information for Sensor Systems 447
 15.1.1 Quantifying Multiple-Sensor Information Gain Via Mutual Information 447
 15.1.2 Quantifying Information Loss with Model Uncertainty 449
 15.1.3 Asymptotic Approximation of Information Measures 452

15.2	Rate-Distortion Theory	456
	15.2.1 The Rate-Distortion Problem	456
15.3	The Blahut Algorithm	457
15.4	The Remote Rate Distortion Problem	459
	15.4.1 Blahut Algorithm extended	460
15.5	Output Symbol Distribution	465

16 Computational Anatomy: Shape, Growth and Atrophy Comparison via Diffeomorphisms — 468

- 16.1 Computational Anatomy — 468
  - 16.1.1 Diffeomorphic Study of Anatomical Submanifolds — 469
- 16.2 The Anatomical Source Model of CA — 470
  - 16.2.1 Group Actions for the Anatomical Source Model — 472
  - 16.2.2 The Data Channel Model — 473
- 16.3 Normal Momentum Motion for Large Deformation Metric Mapping (LDDMM) for Growth and Atrophy — 474
- 16.4 Christensen Non-Geodesic Mapping Algorithm — 478
- 16.5 Extrinsic Mapping of Surface and Volume Submanifolds — 480
  - 16.5.1 Diffeomorphic Mapping of the Face — 481
  - 16.5.2 Diffeomorphic Mapping of Brain Submanifolds — 481
  - 16.5.3 Extrinsic Mapping of Subvolumes for Automated Segmentation — 481
  - 16.5.4 Metric Mapping of Cortical Atlases — 483
- 16.6 Heart Mapping and Diffusion Tensor Magnetic Resonance Imaging — 484
- 16.7 Vector Fields for Growth — 488
  - 16.7.1 Growth from Landmarked Shape Spaces — 488

17 Computational Anatomy: Hypothesis Testing on Disease — 494

- 17.1 Statistics Analysis for Shape Spaces — 494
- 17.2 Gaussian Random Fields — 495
  - 17.2.1 Empirical Estimation of Random Variables — 496
- 17.3 Shape Representation of the Anatomical Orbit Under Large Deformation Diffeomorphisms — 496
  - 17.3.1 Principal Component Selection of the Basis from Empirical Observations — 497
- 17.4 The Momentum of Landmarked Shape Spaces — 498
  - 17.4.1 Geodesic evolution equations for landmarks — 498
  - 17.4.2 Small Deformation PCA Versus Large Deformation PCA — 499
- 17.5 The Small Deformation Setting — 502
- 17.6 Small Deformation Gaussian Fields on Surface Submanifolds — 502
- 17.7 Disease Testing of Automorphic Pathology — 503
  - 17.7.1 Hypothesis Testing on Disease in the Small Noise Limit — 503
  - 17.7.2 Statistical Testing — 505
- 17.8 Distribution Free Testing — 510
- 17.9 Heteromorphic Tumors — 511

18 Markov Processes and Random Sampling — 514

- 18.1 Markov Jump Processes — 514
  - 18.1.1 Jump Processes — 515
- 18.2 Random Sampling and Stochastic Inference — 516
  - 18.2.1 Stationary or Invariant Measures — 517
  - 18.2.2 Generator for Markov Jump Processes — 519
  - 18.2.3 Jump Process Simulation — 520
  - 18.2.4 Metropolis–Hastings Algorithm — 521

	18.3	Diffusion Processes for Simulation	523
		18.3.1 Generators of 1D Diffusions	525
		18.3.2 Diffusions and SDEs for Sampling	527
	18.4	Jump-Diffusion Inference on Countable Unions of Spaces	528
		18.4.1 The Basic Problem	529
19	**Jump Diffusion Inference in Complex Scenes**		**532**
	19.1	Recognition of Ground Vehicles	533
		19.1.1 CAD Models and the Parameter Space	533
		19.1.2 The FLIR Sensor Model	534
	19.2	Jump Diffusion for Sampling the Target Recognition Posterior	536
		19.2.1 The Posterior distribution	536
		19.2.2 The Jump Diffusion Algorithms	536
		19.2.3 Jumps via Gibbs' Sampling	539
		19.2.4 Jumps via Metropolis–Hastings Acceptance/Rejection	541
	19.3	Experimental Results for FLIR and LADAR	543
		19.3.1 Detection and Removal of Objects	543
		19.3.2 Identification	543
		19.3.3 Pose and Identification	544
		19.3.4 Identification and recognition via High Resolution Radar (HRR)	546
		19.3.5 The Dynamics of Pose Estimation via the Jump–Diffusion Process	546
		19.3.6 LADAR Recognition	548
	19.4	Powerful Prior Dynamics for Airplane Tracking	549
		19.4.1 The Euler-Equations Inducing the Prior on Airplane Dynamics	550
		19.4.2 Detection of Airframes	552
		19.4.3 Pruning via the Prior distribution	552
	19.5	Deformable Organelles: Mitochondria and Membranes	553
		19.5.1 The Parameter Space for Contour Models	553
		19.5.2 Stationary Gaussian Contour Model	554
		19.5.3 The Electron Micrograph Data Model: Conditional Gaussian Random Fields	555
	19.6	Jump–Diffusion for Mitochondria	556
		19.6.1 The jump parameters	557
		19.6.2 Computing gradients for the drifts	557
		19.6.3 Jump Diffusion for Mitochondria Detection and Deformation	558
		19.6.4 Pseudolikelihood for Deformation	560
	References		563
	Index		581

1 INTRODUCTION

This book is to be an accessible book on patterns, their representation, and inference. There are a small number of ideas and techniques that, when mastered, make the subject more accessible. This book has arisen from ten years of a research program which the authors have embarked upon, building on the more abstract developments of metric pattern theory developed by one of the authors during the 1970s and 1980s. The material has been taught over multiple semesters as part of a second year graduate-level course in pattern theory, essentially an introduction for students interested in the representation of patterns which are observed in the natural world. The course has attracted students studying biomedical engineering, computer science, electrical engineering, and applied mathematics interested in speech recognition and computational linguistics, as well as areas of image analysis, and computer vision.

Now the concept of *patterns* pervades the history of intellectual endeavor; it is one of the eternal followers in human thought. It appears again and again in science, taking on different forms in the various disciplines, and made rigorous through mathematical formalization. But the concept also lives in a less stringent form in the humanities, in novels and plays, even in everyday language. We use it all the time without attributing a formal meaning to it and yet with little risk of misunderstanding. So, what do we really mean by a pattern? Can we define it in strictly logical terms? And if we can, what use can we make of such a definition?

These questions were answered by *General Pattern Theory*, a discipline initiated by Ulf Grenander in the late 1960s [1–5]. It has been an ambitious effort with the only original sketchy program having few if any practical applications, growing in mathematical maturity with a multitude of applications having appeared in biology/medicine and in computer vision, in language theory and object recognition, to mention but a few. Pattern theory attempts to provide an algebraic framework for describing patterns as structures regulated by rules, essentially a finite number of both local and global combinatory operations. Pattern theory takes a *compositional* view of the world, building more and more complex structures starting from simple ones. The basic rules for combining and building complex patterns from simpler ones are encoded via graphs and rules on transformation of these graphs.

In contrast to other dominating scientific themes, in Pattern Theory we start from the belief that *real world patterns are complex*: Galielaen simplification that has been so successful in physics and other natural sciences will not suffice when it comes to explaining other regularities, for example in the life sciences. If one accepts this belief it follows that complexity must be allowed in the ensuing representations of knowledge. For this, probabilities naturally enter, superimposed on the graphs so as to express the variability of the real world by describing its fluctuations as randomness. Take as a goal the development of algorithms which assist in the ambitious task of *image understanding* or *recognition*. Imagine an expert studying a natural scene, trying to understand it in terms of the awesome body of knowledge that is informally available to humans about the context of the scene: identify components, relate them to each other, make statements about the fine structure as well as the overall appearance. If it is truly the goal to create algorithmic tools which assist experts in carrying out the time-consuming labor of pattern analysis, while leaving the final decision to their judgment, to arrive at more than ad hoc algorithms the subject matter knowledge must be expressed precisely and as compactly as possible.

This is the central focus of the book: 'How can such empirical knowledge be represented in mathematical form, including both structure and the all important variability?' This task of presenting an organized and coherent view of the field of Pattern theory seems bewildering at best. But what are today's challenges in signal, data and pattern analysis? With the advent of geometric increases in computational and storage resources, there has been a dramatic increase in the solution of highly complex pattern representation and recognition problems. Historically books on pattern recognition present a diverse set of problems with diverse methods for building recognition algorithms, each approach handcrafted to the particular task. The complexity and diversity of patterns in the world presents one of the most significant challenges to the pedagogical approach to the teaching of Pattern theory. Real world patterns are often the results of evolutionary change, and most times cannot be selected by the practitioner to have particular properties. The representations require models using mathematics which span multiple fields in *algebra, geometry, statistics and statistical communications.*

Contrasting this to the now classical field of statistical communications, it might appear that the task seems orders of magnitude bigger than modelling signal ensembles in the communication environment. Thinking historically of the now classical field of statistical communications, the discipline can be traced back far, to Helmholtz and earlier, but here we are thinking of its history in the twentieth century. For the latter a small number of parameters may be needed, means, covariances, for Gaussian noise, or the spectral density of a signal source, and so on. The development of communication engineering from the 1920s on consisted in part of formalizing the observed, more or less noisy, signals. Statistical signal processing is of course one of the great success stories of statistics/engineering. It is natural to ask why. We believe that it was because the pioneers in the field managed to construct representations of signal ensembles, models that were realistic and at the same time tractable both analytically and computationally (by analog devices at the time). The classical signalling models: choose $s_0(t), s_1(t)$ to be orthogonal elements in L^2, with the noise model additive stationary noise with covariance representation via a complete orthonormal basis. Such a beautiful story, utilizing ideas from Fourier analysis, stationary stochastic processes, Toeplitz forms, and Bayesian inference! Eventually this resulted in more or less automated procedures for the detection and understanding of noisy signals: matched filters, optimal detectors, and the like. Today these models are familiar, they look simple and natural, but in a historical perspective the phenomena must have appeared highly complex and bewildering.

We believe the same to be true for pattern theory. The central challenge is the formalization of a small set of ideas for constructing the *representations of the patterns themselves* which accommodate variability and structure simultaneously. This is the point of view from which this book is written. Even though the field of pattern theory has grown considerably over the past 30 years, we have striven to emphasize its coherence. There are essentially two overarching principles. The first is the representation of regularity via graphs which essentially encode the rules of combination which allow for the generation of complex structures from simpler ones. The second is the application of transformations to generate from the exemplars entire orbits. To represent typicality probabilistic structures are superimposed on the graphs and the rules of transformation. Naturally then the conditional probabilities encode the regularity of the patterns, and become the central tool for studying pattern formation.

We have been drawn to the field of pattern theory from backgrounds in communication theory, probability theory and statistics. The overall framework fits comfortably within the source-channel view of Shannon. The underlying deep regular structures are descriptions of the source, which are hidden via the sensing channel. We believe that the principle challenge is the representation of the source of patterns, and for that reason the majority of the book is focused precisely on this topic. A multiplicity of channels or sensor models will be used throughout the book, those appropriate for the pattern class being studied. They are however studied superficially, drawn from the engineering literature and taken as given, but certainly studied more deeply elsewhere. The channel sensor models of course shape the overall performance of the inference algorithms; but the major focus of our work is on the development of stochastic models for the structural understanding of the variabilities of the patterns at the source. This also explains the major deviation of this pattern theory from that which has come to be known as pattern recognition. Only in the final chapters will pattern recognition algorithms be studied, attempting to answer the question of how well the algorithm can estimate (recognize) the source when seen through the noisy sensor channel.

1.1 Organization

Throughout this book we use methods from estimation, stochastic processes and information theory. Chapter 2 includes the basic stalwarts of statistics and estimation theory which should be familiar to the reader, including minimum-risk estimation, Fisher Information, hypothesis testing and maximum-likelihood, consistency, model order estinmation, and entropies.

Chapters 3–6 bring into central prominence the role of representation of patterns via conditioning structure. Chapter 3 examines discrete patterns represented via probabilistic directed acyclic graphs (DAGs) emphasizing the pivoting properties and conditional factorizations of DAGs which are familiar for Markov chains and random branching processes. This provides ample opportunity to construct and study the syntactic theory of Chomsky Languages via the classical formulation of graphs and grammatical transformation. Chapter 4 relaxes away from the pivoting property of directed graphs to the conditioning structure of neighborhoods in Markov random fields. Chapter 5 brings the added structure of of the Gaussian law for representing real-valued patterns via Gaussian fields. In this context entropy and maximum entropy distributions are examined in these three chapters as a means of representing conditioning information for representing patterns of regularity for speech, language, and image analysis. Chapter 6 presents the abstract representation of patterns via generators and probabilistic structures on the generators. The generator representation is explored as it provides a unified way of dealing with DAGs and random fields.

Chapters 7 and 8 begin examining in their own right the second central component of pattern theory, groups of geometric transformation applied to the representation of geometric objects. The patterns and shapes are represented as submanifolds of $\mathbb{R}^n$, including points, curves, surfaces and subvolumes. They are enriched via the actions of the linear matrix groups, studying the patterns as orbits defined via the group actions. In this context active models and deformable templates are studied. The basic fundamentals of groups and matrix groups are explored assuming that the typical engineering graduate student will not be familiar with their structure.

Chapter 9 makes the first significant foray into probabilistic structures in the continuum, studying random processes and random fields indexed over subsets of $\mathbb{R}^n$. Classical topics are examined in some detail including second order processes, covariance representation, and Karhunen-Loeve transforms. This is the first chapter where more significant understanding is required for understanding signal and patterns as functions in a Hilbert space.

Chapters 10 and 11 continue the major thrust into transformations and patterns indexed over the continuum. In this context, the finite dimensional matrix groups are studied as diffeomorphic actions on $\mathbb{R}^n$, as well their infinite dimensional analogues are established. It is in these chapters in which the substantial bridge between Pattern theory, mechanics, and differential geometry is established. The links come through the study of flows of diffeomorphisms. Chapter 10 focuses on the study of the finite dimensional matrix groups, and Chapter 11 on the infinite dimensional diffeomorphisms acting on manifolds of $\mathbb{R}^n$ as a Riemannian metric space. The metric is induced by the geodesic length between elements in the space defined through the Riemannian length of the flow connecting one point to another. Herein the classical equations of motion for the geodesics in the finite dimensional case are expanded to include the Euler formulation of the infinite dimensional case.

Chapter 12 expands this view to examine the orbit of imagery as a deformable template under diffeomorphic action; the orbit is endowed with the metric structure through the length minimizing geodesics connecting images. In this chapter the photometric orbit is studied as well, adding notions from transport to define a metric on the product space of geometric and photometric variation.

Chapters 13–15 extend from the pure representations of shapes to the Bayes estimation of shapes and their parametric representation. Classical minimum-mean-squared error and maximum a-posteriori estimators of shapes are explored in these chapters as viewed through various remote sensing models. Chapter 13 focuses on estimating the pose of rigid objects; chapter 14 focuses on accommodating photometric variability superimposed on the geometric variability of rigid pose. Chapter 15 focuses on information bounds for quantifying estimation accuracy of the matrix group, comparing mean-squared error bounds with capacity and rate-distortion bounds for codebooks.

Chapters 16 and 17 turn from the estimation of finite dimensional matrix groups to the study of the estimation of infinite dimensional shape in the newly emergent field of Computational Anatomy. Chapter 15 focuses on estimating landmark and image based shape metrics in volumes, with Chapter 16 focusing on submanifolds and on the inference of disease and hypothesis testing in Computational Anatomy.

The last two Chapters 18 and 19 conclude on inference, exploring random sampling approaches for estimation of model order and parametric representing of shapes. Chapter 18 reviews jump and diffusion processes and their use in random sampling of discrete and continuum spaces. Chapter 19 examines a series of problems in object recognition.

We have made an attempt to keep the theory at a consistent level. The mathematical level is a reasonably high one, first-year graduate level, with a background of at least one good semester course in probability and a solid background in mathematics. We have, however, been able to avoid the use of measure theory.

Appendices outlining proofs, theorems and solutions to exercises together with a comphrehive list of figures, tables and plates are freely available on an accompanying website www.oup.com/uk/academic/companion/mathematics/patterntheory

In this book plates 1–16 appear between pages 180–181, plates 17–34 appear between pages 372–373, and plates 35–53 appear between 564–565.

2 THE BAYES PARADIGM, ESTIMATION AND INFORMATION MEASURES

ABSTRACT   The basic paradigm is the Bayesian setup, given is the source of parameters $X \in \mathcal{X}$ which are seen through a noisy channel giving observations $Y \in \mathcal{Y}$. The posterior distribution determines the bounds on estimation of X given Y, the risk associated with estimating it, as well as a characterization of the information in the observation in Y about X.

2.1 Bayes Posterior Distribution

The basic set up throughout is we are given a model of the *source* of possible objects $X \in \mathcal{X}$. These are observed through a *noisy channel* giving observations $Y \in \mathcal{Y}$. The source $X \in \mathcal{X}$ is modelled with distribution and density $P_X(dx) = p(x)dx$, $\int_{\mathcal{X}} p(x)dx = 1$. Generally the source can only be observed with loss of information due to observational noise or limited accuracy in the sensor. The mapping from the input source $X \in \mathcal{X}$ to the observed output $Y \in \mathcal{Y}$ expresses the physics of the sensing channels; the data $Y \in \mathcal{Y}$ will in general contain multiple components corresponding to several sensors $Y = (Y_1, Y_2, \dots)$. The observation process is characterized via a statistical transition law, transferring $X \to Y$ $P_{Y|X}(\cdot|\cdot) : \mathcal{X} \times \mathcal{Y} \to \mathbb{R}^+$, summarizing completely the transition law mapping the input model parameters X to the output Y, the likelihood of Y given X.

This Bayesian paradigm, separating the source from the channel is what has become the modern view of communications developed out of Shannon's theory of communications [6]. Figure 2.1, clearly delineates the separation of the source of information and the channel through which the messages are observed. To infer the transmitted message at the output of the channel, the observation Y must be processed optimally. The inference engine is a decoder, working to recover properties of the original message from the source.

Given such a communications *source/channel* decomposition, we shall be interested in both specifying optimal procedures for inferring properties of pattern generation systems given the observable measurements, and quantifying the information content and information gain of the observation system. Pattern deduction becomes an enterprise consisting of constructing the source and channel models, and essentially has several parts: (i) selection and fitting of parameters parametrizing the models representing the patterns, and (ii) construction of the family of probability models representing the knowledge about the pattern classes. For this we shall examine classical minimum-risk estimators, such as minimum-mean-squared-error (MMSE) estimators, maximum-aposteriori and likelihood (MAP, MLE). For constructing the models we shall examine various forms and principles of entropy and mutual information.

At the most fundamental level, the posterior distribution represents the information contained in the observables about the underlying imagery. All provably optimal structured methods of inference and information gathering fundamentally involve the posterior density or distribution of the random variables $X \in \mathcal{X}$ given the observed deformed image $Y \in \mathcal{Y}$.

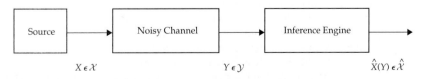

Figure 2.1 Shannon's source channel model for communications systems

Definition 2.1 *Define the* **prior distribution and density** *on the source of random variables* $P(dx) = p(x)dx$, *on* $\mathcal{X}$, $\int_{\mathcal{X}} p(x)dx$; *define the* **likelihood or transition density** *on the observations* $Y \in \mathcal{Y}$ *given the input* X *as* $P_{Y|X}(dy|x) = p(y|x)dy, Y \in \mathcal{Y}$. *Define the* **posterior** *as* $P_{X|Y}(dx|y) = p(x|y)dx$ *given by*

$$p(x|y) = \frac{p(x)p(y|x)}{p(y)}, \qquad \int_{\mathcal{X}} p(x|y)dx = 1. \tag{2.1}$$

If the random variables X and Y are discrete, then Bayes formula for the posterior is interpreted as a probability mass function.

2.1.1 Minimum Risk Estimation

Now in solving problems in pattern theory we will want to perform inference; how should this be done? It is not just a matter of calculating the posterior probabilities, we also have to operate on them to distill the essence of the knowledge they contain. This leads to various tasks in pattern inference:

1. *Synthesis*: to simulate the patterns from the probability distribution representing the source $X \in \mathcal{X}$ of patterns, $P(dX) = p(X)dx$.
2. *Restoration*: to recover the original source pattern X from the observed one Y suppressing noise and other measurement errors as far as possible.
3. *Recognition*: to determine to which pattern class X is likely to belong when Y has been observed.
4. *Extrapolation*: to extrapolate some part of the image that could not be observed because of obscuration or other reasons.
5. *Understanding*: both *intrinsic* understanding, assuming the knowledge representation to be essentially correct, and *extrinsic* understanding to spot misrepresentation and other model errors.

In solving problems in pattern theory a variety of techniques have to be developed and modified, examples including the estimation of unknown *nuisance* parameters in the representation, as well as model selection, i.e. choice of the actual representation being used. These can all be treated in a similar unified manner, all exploiting the basic properties of the likelihood function and prior distribution.

One of the most fundamental aspects is to construct estimators $X^\star$ that in some sense (to be made precise) are a reasonable approximation to the true but unknown objects X. In statistical terminology, generate a *point estimator*, a function $X^\star(Y)$, $X^\star : \mathcal{Y} \to \mathcal{X}$. But what should be meant by reasonable? Within the Bayesian paradigm, we could ask for an estimator which satisfies some optimality principle, or is of *minimum risk*. Fundamental to such an approach is the definition of a distortion measure, or distance between the estimator and the random source vector. If the source space $\mathcal{X}$ can be embedded in a vector space, $\mathbb{R}^d, \mathbb{C}^d$, then the concept of the squared-error metric can be introduced, and the conditional mean value could be computed. The risk associated with such metrics is often taken for granted, and has become part of the culture. The classic example is the Kalman filter, a fixture in modern estimation theory.

In such a setting, the choice to be made is the definition of the metric distance or risk function $R : \mathcal{X} \times \mathcal{X} \to \mathbb{R}^+$ quantifying the relative merit between the approximating estimator and the truth.

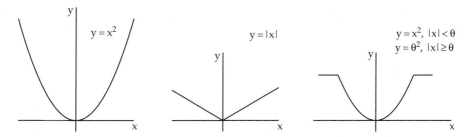

Figure 2.2 Left shows absolute value; middle shows squared error; right panel shows a thresholding function

Definition 2.2 *Then the* **positive-valued risk measure (loss)** $R : \mathcal{X} \times \mathcal{X} \to \mathbb{R}^+$ *has* **expected risk** $\mathcal{R} = E(R(X, \hat{X}(Y)))$ *with the* **minimum risk-estimator** *satisfying*

$$X^\star_\mathcal{R}(Y) = \arg\min_{\hat{X} \in \mathcal{X}} E\{R(X, \hat{X}(Y))\} \tag{2.2}$$

$$= \arg\min_{\hat{X} \in \mathcal{X}} \int_{\mathcal{X} \times \mathcal{Y}} R(x, \hat{x}(y)) p(x, y) dx\, dy. \tag{2.3}$$

Shown in Figure 2.2 are various examples.

2.1.2 Information Measures

Information measures quantify performance of the inference algorithms including accuracy of the estimators, exponential error bounds for decision, and combinatoric complexity of representation required. We introduce these information measures here following Cover and Thomas [7], where additional details may be found. Define the information measures for both discrete and continuous random variables.

Definition 2.3 *Given a discrete random variable* $X \in \mathcal{X}$ *with probability mass function* P_X *on* $\mathcal{X}$, *the* **entropy** *is*

$$H(X) = -E_{P_X} \log P_X = -\sum_{x \in \mathcal{X}} P_X(x) \log P_X(x),$$

where log *is base* e *giving nats or base* 2 *giving bits.*[1]

Given a second discrete random variable $Y \in \mathcal{Y}$ *jointly distributed with* X *according to the* $P_{X,Y}$ *on* $\mathcal{X} \times \mathcal{Y}$ *the* **joint entropy** *of* X *and* Y *and the* **conditional entropy** *of* X *given* Y *is*

$$H(X, Y) = -E_{P_{X,Y}} \log P_{X,Y} = -\sum_{x,y} P_{X,Y}(x,y) \log P_{X,Y}(x,y); \tag{2.4}$$

$$H(X|Y) = -E_{P_{X,Y}} \log P_{X|Y} = -\sum_{x,y} P_{X,Y}(x,y) \log P_{X|Y}(x|y). \tag{2.5}$$

---

[1] Throughout we shall use log with its base interpreted depending on the context, so that $\log e = 1$ when log is base e, and $\log 2 = 1$ when base 2.

Given two random variables X and Y with joint probability $P_{X,Y}$ on $\mathcal{X} \times \mathcal{Y}$ and marginals P_X on $\mathcal{X}$ and P_Y on $\mathcal{Y}$ the **mutual information** between X and Y is

$$\Delta H(X;Y) = H(X) - H(X|Y) = H(Y) - H(Y|X)$$
$$= E_{P_{X,Y}} \log \left(\frac{P_{X,Y}}{P_X P_Y} \right).$$

The **relative entropy or Kullback Leibler divergence** between the two probability mass functions P_X, P_Y, on $\mathcal{X}$ is defined as

$$D(P_X \| P_Y) = E_{P_X} \log \left(\frac{P_X}{P_Y} \right).$$

By manipulation $\Delta H(X;Y) = D(P_{X,Y} \| P_X P_Y)$.

For continuous random variables we use the convention of using small letters.

Definition 2.4 *Given a continuous random variable $X \in \mathcal{X}$ with probability density $P(dx) = p(x)dx$, the* **differential entropy** *is*

$$h(X) = -E_{P_X} \log p(X) = -\int_\mathcal{X} p(x) \log p(x) dx \, .$$

Given a second continuous random variable $Y \in \mathcal{Y}$ jointly distributed with density $P(dx, dy) = p(x,y)dx\,dy$ on $\mathcal{X} \times \mathcal{Y}$ the **joint differential entropy** *of X and Y and the* **conditional entropy** *of X given Y is*

$$h(X,Y) = -E_{P_{X,Y}} \log p(X,Y); \tag{2.6}$$
$$h(X|Y) = -E_{P_{X,Y}} \log p(X|Y). \tag{2.7}$$

Given two random variables X and Y with joint density $p(X,Y)$ on $\mathcal{X} \times \mathcal{Y}$ and marginals $p(X)$ on $\mathcal{X}$ and $p(Y)$ on $\mathcal{Y}$ the **mutual information** between X and Y is

$$\Delta h(X;Y) = h(X) - h(X|Y) = h(Y) - h(Y|X)$$
$$= E_{P_{X,Y}} \log \left(\frac{p(X,Y)}{p(X)p(Y)} \right).$$

The **relative entropy or Kullback-Leibler divergence** between two densities (assumed to be absolutely continuous) $P_X(dx) = p(x)dx, Q_Y(dx) = q(x)dx$ becomes

$$D(P\|Q) = E_{P_X} \log \left(\frac{p(X)}{q(X)} \right).$$

Similarly, $\Delta h(X;Y) = E_{P_{X,Y}} \log \left(p(X,Y)/p(X)p(Y) \right)$.

2.2 Mathematical Preliminaries

2.2.1 Probability Spaces, Random Variables, Distributions, Densities, and Expectation

First define the basic components of probabilities.

Definition 2.5 *A* **probability space** *$(\Omega, \mathcal{A}, \mu)$ is comprised of three quantities, a* **sample space** *Ω, a collection of subsets $A \in \mathcal{A}$ the* **field of events**, *and a* **probability measure** *μ.*

The field of events has the properties that (i) $\mathcal{A}$ is not empty, (ii) if $A \in \mathcal{A}$ then $A^c \in \mathcal{A}$, and (iii) if $A_1, A_2, \ldots$ are events then $\cup_i A_i \in \mathcal{A}$.

The probability measure $\mu : \mathcal{A} \to [0, 1]$ satisfies (i) $\mu(\Omega) = 1$ and (ii) if $A_1, A_2 \ldots$ are disjoint then $\mu(\cup_i A_i) = \sum_i \mu(A_i)$.

The sample space Ω with elements $\omega \in \Omega$ are possible outcomes of random experiments, and may be denumerable (outcomes of a dice throw) and nondenumerable (random events in the continuum such as the random time to failing of a system). As we shall see complicated spaces will be studied as well including function spaces.

A subset $A \subset \Omega$ termed an event is generally associated with describing the outcomes of the random experiment. The collection of subsets $\mathcal{A}$ will almost always contain all possible subsets for Ω finite. However, for nondenumerable sample spaces, $\mathcal{A}$ contains a subset of what we shall term the *measurable events*, those for which $\mu(\cdot)$ can be applied. Throughout the book, rather than working explicitly with the underlying probability space we will often refer to it through notation like $\Pr\{\omega \in A\}$. We shall also suppress the dependence on ω throughout.

Real-valued random variables and random vectors will be studied throughout.

Definition 2.6 *A function $X : \Omega \to \mathbb{R}$ is a* **real-valued random variable** *with respect to $(\Omega, \mathcal{A}, \mu)$ if every set $\{\omega : X(\omega) \leq x\} \in \mathcal{A}$.*

A function $(X_1, \ldots, X_n) : \Omega \to \mathbb{R}^n$ is an $\mathbb{R}^n$ **valued random variable** *with respect to $(\Omega, \mathcal{A}, \mu)$ if every set $\{\omega : (X_1(\omega), \ldots, X_n(\omega)) \leq (x_1, \ldots, x_n)\} \in \mathcal{A}$.*

Notice in the above definition intersections and unions of sets generated from such basic events $\cup_i \{\omega : X(\omega) \leq x_i\}, \cap_i \{\omega : X(\omega) \leq x_i\}$ are elements of $\mathcal{A}$ implying that sets generated from countable intersections and unions of intervals in $\mathbb{R}$ are legal sets as well to be asking for membership of random variables. Such sets we shall term *measurable*, meaning they are constructing from the basic intervals. The probability distribution associated with the random variable directly will be used throughout the entire book thereby suppressing the dependence on the underlying sample probability space. Densities, distributions, and joint distributions are defined as follows.

Definition 2.7 *The* **probability distribution function** *for the real-valued random variable $X(\omega)$ as the function such that for all sets $A \subset \mathbb{R}$ (Lebesgue measurable)*

$$P_X(A) = \mu\{\omega : X(\omega) \in A\}. \tag{2.8}$$

The random variable has **density** *$P_X(dx) = p(x)dx$ if for all such A,*

$$\Pr\{X(\omega) \in A\} = P_X(A) = \int_A p(x)dx. \tag{2.9}$$

The **joint probability distribution** *for the $\mathbb{R}^n$ valued $X_1(\omega), \ldots, X_n(\omega)$ is the function such that for all sets $A \subset \mathbb{R}^n$,*

$$P_X(A) = \mu\{\omega : (X_1(\omega), \ldots, X_n(\omega)) \in A\}. \tag{2.10}$$

Then $X = (X_1, \ldots, X_n)$ has **joint density** *$P_X(dx) = p(x_1, \ldots, x_n)dx$ if for all $A \in \mathcal{A}$,*

$$P_X(A) = \int_A p(x_1, \ldots, x_n)dx_1 \cdots dx_n. \tag{2.11}$$

In $\mathbb{R}^n$, the probability elements $p(x_1, \ldots, x_n)\Delta x_1 \cdots \Delta x_n$ is approximately $\Pr\{x_1 \leq X_1 \leq x_1 + \Delta x_1, \ldots, x_n \leq X_n \leq x_n + \Delta x_n\}$ if $p_{X_1, \ldots, X_n}$ is continuous.

Statistical independence and conditional probability are defined through relationships on the joint distribution.

Definition 2.8 *Random variables X, Y with joint distribution P_{XY} are* **mutually independent** *if*

$$P_{XY}(A, B) = P_X(A)P_Y(B), \tag{2.12}$$

where P_X, P_Y are the **marginal distributions**.

The **conditional distribution** $P_{X|Y}$ is given for all A, B (measurable),

$$P_{X|Y}(A|B) = \frac{P_{XY}(A,B)}{P_Y(B)} . \tag{2.13}$$

The moments are generated by taking the expected value with respect to the distribution.

Definition 2.9 *The **expected value** of a random function $f(X)$ with respect to distribution with density $P_X(dx) = p(x)dx$ will often be written $E_p f(X)$ given by*

$$E_p f(X) = \int_{\mathcal{X}} f(x) p(x) dx . \tag{2.14}$$

The nth moment is given by the expectation of the nth power of X:

$$E_p X^n = \int_{\mathcal{X}} x^n p(x) dx, \quad n = 1, 2, \ldots . \tag{2.15}$$

2.2.2 Transformations of Variables

Transforming random variables is something we will do all the time. Let $X_1, X_2, \ldots$ have joint density $P_X(dx) = p(x_1, x_2, \ldots) dx$, and consider the real-valued function $y_i = g_i(x_1, x_2, \ldots)$, $i = 1, \ldots, n$ with $g: \mathbb{R}^n \to \mathbb{R}^n$ one-to-one with inverse $x_i = g_i^{-1}(y_1, y_2, \ldots)$. Then $P_Y(dy)$ becomes

$$P_Y(dy) = p(g_1^{-1}(y_1, y_2, \ldots), \ldots, g_n^{-1}(y_1, y_2, \ldots)) \left| \det Dg^{-1}(y_1, y_2, \ldots) \right| dy, \tag{2.16}$$

where the $n \times n$ Jacobian matrix $Dg^{-1} = \left(\partial g_i^{-1} / \partial y_j \right)$.[2]

To prove this simply note the probability that

$$P_{Y_1, Y_2, \ldots}(A) = P_{X_1, X_2, \ldots}(g^{-1}(A)) = \int_{g^{-1}(A)} p(x_1, x_2, \ldots) dx \cdots \tag{2.17}$$

$$= \int_{g^{-1}(A)} p(g^{-1}(y_1, y_2, \ldots)) \left| \det Dg^{-1}(y) \right| dy. \tag{2.18}$$

Equating densities gives the result.

2.2.3 The Multivariate Normal Distribution

Normal random variables will be studied quite a bit. The univariate normal density with mean $EX = \mu$ and variance $E(X-\mu)^2$ is written as $P_X(dx) = \frac{1}{\sqrt{2\pi\sigma^2}} e^{-(1/2)(x-\mu)^2/\sigma^2} dx$. We shall of course be interested in the multivariate case. For this we define the $n \times 1$ and $n \times n$ vector notation[3]:

$$X = \begin{pmatrix} X_1 \\ X_2 \\ \vdots \end{pmatrix}, \quad \mu = \begin{pmatrix} EX_1 \\ EX_2 \\ \vdots \end{pmatrix}, \quad K_{XX} = E(X-\mu)(X-\mu)^* = \left(E(X_i - \mu_i)(X_j - \mu_j) \right). \tag{2.19}$$

---

[2] We use the notation $\left(A_{ij} \right)$ to denote the matrix with i, j entry A_{ij}; then $\left(\partial g_i^{-1} / \partial y_j \right)$ represents the $n \times n$ Jacobian matrix.

[3] Here $(A)^*$ denotes matrix transpose of real vectors and matrices A, and matrix hermitian transpose for the complex case.

2.2 MATHEMATICAL PRELIMINARIES

Definition 2.10 *Then we shall say that $X_1, X_2, \ldots$ is **multivariate normal** with mean vector $\mu = EX = (\mu_i)$ and covariance matrix K_{XX} having density $P(dX) = p(X_1, \ldots, X_n)dX$ with*

$$p(X_1, \ldots, X_n) = \det^{-1/2}(2\pi K_{XX})e^{-(1/2)(X-\mu)^*K_{XX}^{-1}(X-\mu)} . \quad (2.20)$$

The multivariate normal can be derived from the change of variables formula. Let $X = \begin{pmatrix} X_1 \\ X_2 \\ \vdots \end{pmatrix}$ be zero-mean, independent, normal, variance one with $p_X(x) = \frac{1}{(2\pi)^{n/2}} e^{-(1/2)x^*x}$. Define the random vector $Y = \begin{pmatrix} Y_1 \\ Y_2 \\ \vdots \end{pmatrix}$ by the linear transformation $Y = QX + \mu$, with Q an invertible $n \times n$ matrix. Defining $K = QQ^*$, then a change of the density formula on $P_Y(dy)$ gives

$$P_Y(dy) = p\left(Q^{-1}(y-\mu)\right) \left|\det Q^{-1}\right| dy \quad (2.21)$$

$$= \det^{-1/2}(2\pi K)e^{-(1/2)(y-\mu)^*K^{-1}(y-\mu)} dy. \quad (2.22)$$

This also provides the following result that invertible linear transformations of Gaussian random variables are Gaussian.

Theorem 2.11 *Let $X = \begin{pmatrix} X_1 \\ X_2 \\ \vdots \end{pmatrix}$ be jointly normal with mean vector μ and covariance K. Then invertible linear transformations $Y = AX + b$ with (A invertible) are normal with mean $A\mu + b$ and covariance AKA^*.*

The proof follows as above.

2.2.4 Characteristic Function

Random variables which are complex valued will be studied as well. For a complex variable to be well defined as a random variable, the real and imaginary parts must be measurable functions with respect to the underlying probability space, in which case the complex sum is also well defined. The characteristic function generated from a random variable is a complex random variable which we shall study a good bit.

Definition 2.12 *Given a random vector $X, v \in \mathbb{R}^n$ with distribution P_X then the **characteristic function** of X is*

$$M_X(jv) = E_{P_X} e^{jv^*X}, \quad v^*X = \sum_{i=1}^n v_i X_i . \quad (2.23)$$

Notice that moments can be determined from the characteristic function. Suppose that X is a scalar random variable, then the mth moment is generated by evaluating the mth derivative at $v = 0$:

$$\frac{d^m}{d(jv)^m} M_X(jv)|_{v=0} = E_{P_X} X^m . \quad (2.24)$$

Example 2.13 (The Gaussian) For random vectors with densities $P_X(dx) = p(x)dx$, then

$$M_X(jv) = \int_{\mathbb{R}^n} p(x)e^{jv^*x}dx \,. \tag{2.25}$$

Let X be a multivariate Gaussian with mean μ, covariance K, then $M_X(jv) = e^{jv^*\mu - (1/2)v^*Kv}$. To see this, define $Y = Q^{-1}(X - \mu)$ where $K = QQ^*$. The Y_i's are independent, univariate Gaussian, zero-mean random variables, with the characteristic function of the univariate zero-mean Gaussian given by

$$\begin{aligned} M_{Y_i}(jv) &= \int \frac{1}{\sqrt{2\pi}} e^{-(1/2)y^2} e^{jvy} dy \\ &= \int \frac{1}{\sqrt{2\pi}} e^{-(1/2)(y-jv)^2} e^{(1/2)(jv)^2} dy = e^{-(1/2)v^2} \,. \end{aligned} \tag{2.26}$$

The characteristic function of the Y vector becomes the product of characteristic functions since the Y_i's are independent:

$$M_Y(jv) = E_{P_Y} e^{jv^*Y} = \prod_{i=1}^n E_{P_{Y_i}} e^{jv_i Y_i} = \prod_{i=1}^n e^{-(1/2)v_i^2} = e^{-(1/2)v^*v} \,. \tag{2.27}$$

Now $X = QY + \mu$, so that

$$Ee^{jv^*X} = Ee^{jv^*QY} e^{jv^*\mu} = Ee^{j(Q^*v)^*Y} e^{jv^*\mu} \stackrel{(a)}{=} e^{-(1/2)v^*QQ^*v} e^{jv^*\mu} \,, \tag{2.28}$$

with (a) following from Eqn. 2.27, giving $M_X(jv) = e^{jv^*\mu - (1/2)v^*Kv}$.

Example 2.14 (The Poisson) Let N be Poisson with mean λ, so that

$$P_N(n) = e^{-\lambda} \frac{\lambda^n}{n!}, \quad n = 0, 1, \dots \,. \tag{2.29}$$

Then

$$M_N(jv) = \sum_{n=0}^\infty e^{-\lambda} \frac{\lambda^n}{n!} e^{jvn} = \sum_{n=0}^\infty e^{-\lambda} \frac{(\lambda e^{jv})^n}{n!} \tag{2.30}$$

$$= e^{\lambda(e^{jv} - 1)} \,. \tag{2.31}$$

2.3 Minimum Risk Hypothesis Testing on Discrete Spaces

Minimum-risk will take various forms for decision making on discrete and continuous parameter spaces. A good deal of inference in pattern theory requires deciding between discrete alternatives; object detection and identification is a fundamental example. The generalized hypothesis testing problem involves choosing between alternative hypotheses $X = 1, 2, \dots$, each with own probability law. Given a random sample $Y_1, Y_2, \dots, Y_n$, drawn from one of the populations, choose one of the hypotheses. Such hypothesis testing chooses between discrete hypotheses; thus the parameter spaces will be discrete and countable.

2.3.1 Minimum Probability of Error via Maximum A Posteriori Hypothesis Testing

For minimum probability of error testing on discrete spaces, it is therefore natural to identify the hypothesis spaces $\mathcal{X}$ with $\mathbb{Z}^+$, and work with the discrete metric, $R = \delta(\cdot,\cdot) : \mathbb{Z}^+ \times \mathbb{Z}^+ \to 0, 1$, with $\delta(X, X') = 1$ unless $X = X'$. This naturally reduces to a minimum probability of error estimation, the problem being given hypothesis X and random observation Y, generate estimators $\hat{X}(Y) : \mathcal{Y} \to \mathcal{X}$ which minimize the desired risk:

$$\mathcal{R} = E\delta(\hat{X}(Y), X) = \Pr(\hat{X} \neq X) . \tag{2.32}$$

Assume throughout the hypothesis space is finite, let $X = 1, \ldots, m$ be populations with densities $p(y|X = i), y \in \mathcal{Y}$, with prior probability $\pi(i), i = 1, \ldots, m$. We can view the optimum hypothesis testing as defining disjoint decision regions $\mathcal{D}_1, \ldots, \mathcal{D}_m$ on the space of observations $Y \in \cup_{i=1}^m \mathcal{D}_i = \mathcal{Y}$. Assuming identical unit costs for miscategorization corresponds to decision region design to minimize the probability of error:

$$\mathcal{R}(\mathcal{D}) = \Pr\{\hat{X}(Y) \neq X\} \tag{2.33}$$

$$= \sum_{i=1}^m \pi(i) \Pr\{\hat{X}(Y) \neq X | X = i\} \tag{2.34}$$

$$= \sum_{i=1}^m \pi(i) \sum_{j \neq i} \int_{\mathcal{D}_j} p(y|X = i) dy . \tag{2.35}$$

The goal is to choose the optimum regions $\mathcal{D}_1^\star, \ldots, \mathcal{D}_m^\star$ such that $\mathcal{R}(\mathcal{D}^\star)$ is the minimum risk. This gives the following theorem demonstrating maximum a posteriori estimation as the minimum risk estimator for identical costs (minimum probability of error).

Theorem 2.15 *Given are random observations $Y \in \mathcal{Y}$ from populations $X = i$ with densities $p(y|X = i)$ with a priori probabilities $\pi(i), i = 1, \ldots$ with equal costs of misclassification. Then the optimum decision regions $\mathcal{D}_1^\star, \ldots, \mathcal{D}_m^\star$ to minimize classification error are defined by assigning observation $Y = y$ to $\mathcal{D}_i^\star, i = 1, \ldots, m$ so that*

$$\mathcal{D}_i^\star = \{y \in \mathcal{Y} : i = \arg\max_{j=1,\ldots,m} \pi(j) p(y|X = j)\} . \tag{2.36}$$

Notice, if there are multiple indices which have identical posterior probability, assignment to either of the decision regions leaves the cost unchanged.

Proof To calculate such optimum decision regions rewrite the above cost Eqn. 2.35 according to

$$\sum_{i=1}^m \pi(i) \sum_{j \neq i} \int_{\mathcal{D}_j} p(y|X = i) dy = \sum_{j=1}^m \int_{\mathcal{D}_j} \sum_{i \neq j} \pi(i) p(y|X = i) dy , \tag{2.37}$$

and notice that minimizing total risk amounts to minimizing the risk for each measurement y since this is a convex combination. The goal becomes choosing the decision regions for measurement y so that the inner quantity is minimized:

$$j^\star = \arg\min_j \sum_{i \neq j} \pi(i) p(y|X = i) \tag{2.38}$$

$$= \arg\max_i \pi(i) p(y|X = i) , \tag{2.39}$$

where the quantity is made smallest by removing the largest contribution. Let $f_j(y) = \sum_{i \neq j} \pi(i) p(y|X = i)$ and defining $f(y|\mathcal{D}) = f_j(y), y \in \mathcal{D}_j$, then the expected loss for decision procedure $\mathcal{D}$ is

$$\sum_{j=1}^m \int_{\mathcal{D}_j} f_j(y) dy = \int_{\mathcal{Y}} f(y|\mathcal{D}) dy . \tag{2.40}$$

Taking two procedures $\mathcal{D}$ and optimum Bayes region $\mathcal{D}^\star$, then the difference in expected risk is

$$\int (f(y|\mathcal{D}) - f(y|\mathcal{D}^\star)) dy = \sum_{j=1}^m \int_{\mathcal{D}_j} (f_j(y) - \min_i f_i(y)) dy \geq 0 . \tag{2.41}$$

□

Remark 2.3.0 We point out that for unequal costs $C(j|i)$ of saying $X = j$ given $X = i$, then the risk of Eqn. 2.35 becomes

$$\mathcal{R}_C(\mathcal{D}) = \sum_{i=1}^m \pi(i) \sum_{j \neq i} C(j|i) \int_{\mathcal{D}_j} p(y|X = i) dy . \tag{2.42}$$

The decision regions are constructed by placing y in $\mathcal{D}_{j^\star}$ choosing $j^\star$ so that the sum is minimized:

$$j^\star = \arg\min_{j=1,\ldots,m} \sum_{i \neq j} \pi(i) p(y|X = i) C(j|i) . \tag{2.43}$$

2.3.2 Neyman–Pearson and the Optimality of the Likelihood Ratio Test

Almost without exception all hypothesis testing studied in this book will be based on the likelihood ratio because of its optimality associated with the Neyman–Pearson Lemma.

Definition 2.16 *Given $Y_1, Y_2, \ldots$ drawn independent and identically distributed from a model with density $p(\cdot|X = i), i = 1, \ldots, m$ corresponding to one of the hypotheses $X = i, i = 1, \ldots, m$. Then we shall say that* **likelihood-ratio testing** *is any decision strategy for choosing between hypotheses $X = i, X = j$ which has decision region determined by the test statistic of the form*

$$L_{ij}(Y_1, \ldots, Y_n) = \frac{p(Y_1, \ldots, Y_n | X = i)}{p(Y_1, \ldots, Y_n | X = j)} . \tag{2.44}$$

The fundamental role of likelihood-ratio testing is that it minimizes the risk of the error types.

Definition 2.17 *Define α_{ij} to be the* **risk** *of claiming $X = j$ given that $X = i$ is true.*
  In the **binary setting** *then α_{01} is usually called* **error of type I** *or* **probability of a false alarm**. *α_{10} is usually called* **error of type II** *or* **probability of a miss**.

Now for the optimality from Neyman and Pearson of such testing.

Theorem 2.18 (Neyman–Pearson) *Let Y be a random variable with density $P_Y(dy) = p(y) dy$. For choosing among hypotheses $X = 0, X = 1$ each from model with densities*

$p(\cdot|X=0), p(\cdot|X=1)$, respectively, define the decision region for choosing $X=0$ and $X=1$ to be defined, respectively, according to

$$\mathcal{D}(\theta) = \left\{ \frac{p(y|X=0)}{p(y|X=1)} \overset{X=0}{>} \theta \right\} \tag{2.45}$$

$$\mathcal{D}(\theta)^c = \left\{ \frac{p(y|X=0)}{p(y|X=1)} \overset{X=1}{\leq} \theta \right\}. \tag{2.46}$$

Given the error types

$$\alpha_{01}^\star = P(\mathcal{D}(\theta)^c | X=0), \tag{2.47}$$

then for any other decision region $B(\theta)$ with error probabilities α_{01}, α_{10}, then if $\alpha_{01} \leq \alpha_{01}^\star$ implies $\alpha_{10} \geq \alpha_{10}^\star$.

Proof Defining $1_\mathcal{D}(\cdot)$ as the indicator function on decision region $\mathcal{D} \subset \mathcal{Y}$,[4] then for B any other decision region, then

$$(1_\mathcal{D}(y) - 1_B(y))(p(y|X=0) - \theta p(y|X=1)) \geq 0. \tag{2.48}$$

Multiplying and integrating over $\mathcal{Y}$ gives

$$0 \leq \int_\mathcal{Y} \big(1_\mathcal{D}(y) p(y|X=0) - \theta 1_\mathcal{D}(y) p(y|X=1)$$
$$\quad - 1_B(y) p(y|X=0) + \theta 1_B(y) \theta p(y|X=1) \big) dy \tag{2.49}$$

$$= \int_\mathcal{D} (p(y|X=0) - \theta p(y|X=1)) dy$$
$$\quad - \int_B (p(y|X=0) - \theta p(y|X=1)) dy \tag{2.50}$$

$$= (1 - \alpha_{01}^\star) - \theta \alpha_{10}^\star - (1 - \alpha_{01}) + \theta \alpha_{10} \tag{2.51}$$

$$= \theta(\alpha_{10} - \alpha_{10}^\star) - (\alpha_{01}^\star - \alpha_{01}). \tag{2.52}$$

This implies

$$\alpha_{10} - \alpha_{10}^\star \geq \frac{\alpha_{01}^\star - \alpha_{01}}{\theta} \geq 0. \tag{2.53}$$

$\square$

Example 2.19 (Stein's Lemma and Error Exponents) Examine the role of entropy and cross entropy in quantifying the error types. Clearly the threshold trades one error for another. Examine the case when the error of the first kind is made to asymptotically go to zero, and the error of the second kind is quantified in terms of the best possible exponential decay rate. This is Stein's lemma, demonstrating that the Kullback distance controls the best possible exponential decay of the second error.

Theorem 2.20 *(Stein's Lemma) Let $Y_1, Y_2, \ldots$ be independent and identically distributed distribution P. Consider the hypotheses governed by two distributions with densities $P_0(dy) = p_0(y) dy$, $P_1(dy) = p_1(y) dy$ continuous with respect to each other with finite Kullback–Leibler distance*

$$D(P_0 \| P_1) = E_{P_0} \left(\log \frac{p_0}{p_1} \right) < \infty. \tag{2.54}$$

---

[4] The indicator function $1_A : \mathcal{Y} \to \{0,1\}$ on a set $A \subset \mathcal{Y}$ is defined as $1_A(y) = 1$ if $y \in A$, and 0 otherwise.

Define $\alpha_{01}(n), \alpha_{10}(n)$ to be the errors of type I,II indexed by sample size n.

Given that the probability of claiming $X = 1$ given $X = 0$ (type I error) goes to zero with sample size, $\lim_{n\to\infty} \alpha_{01}(n) = 0$, then the best error exponent for missing $X = 1$ and claiming $X = 0$ (type II error) is given by the Kullback–Leibler distance between the two probabilities:

$$\lim_{n\to\infty} \frac{1}{n} \log \alpha_{10}(n) = -D(P_0\|P_1). \tag{2.55}$$

Proof The proof follows by first defining an acceptance region, which satisfies the properties, and then showing you can not do better. Define the acceptance region

$$A_n = \Big\{(y_1,\ldots,y_n) \in \mathcal{Y}^n : n(D(P_0\|P_1) - \delta)$$

$$\leq \log \frac{p_0(y_1,\ldots,y_n)}{p_1(y_1,\ldots,y_n)} \leq n(D(P_0\|P_1) + \delta)\Big\}, \tag{2.56}$$

where

$$D(P_0\|P_1) = \int_{\mathcal{Y}} p_0(y) \log \frac{p_0(y)}{p_1(y)} dy. \tag{2.57}$$

Denote the probability that $(y_1,\ldots,y_n) \in A_n$ as $P(A_n)$, where P is associated with the density $p(y_1,\ldots,y_n) = p(y_1)p(y_2)\cdots p(y_n)$. Note $\alpha_{01} = 1 - P_0(A_n)$, and clearly $P_0(A_n) \to 1$ as $n \to \infty$ by the strong law of large numbers since $D(P_0\|P_1) = E_{P_0}(\log p_0(Y)/p_1(Y))$. So this decision region fits the first part of the bill. Now examine the error exponent for the second part using the definition of A_n to write

$$P_1(A_n) = \int_{A_n \subset \mathcal{Y}^n} p_1(y_1,\ldots,y_n) dy$$

$$\leq \int_{A_n \subset \mathcal{Y}^n} p_0(y_1,\ldots,y_n) dy\, 2^{-n(D(P_0\|P_1)-\delta)} \tag{2.58}$$

$$= 2^{-n(D(P_0\|P_1)-\delta)} P_0(A_n); \tag{2.59}$$

similar arguments give the upper bound $P_1(A_n) \geq 2^{-n(D(P_0\|P_1)+\delta)} P_0(A_n)$. Since $P_0(A_n) \to 1$, taking logarithms yields for all δ, there exists $n(\delta)$ large enough so that

$$-D(P_0\|P_1) - \delta \leq \frac{1}{n} \log \alpha_{10} \leq -D(P_0\|P_1) + \delta, \tag{2.60}$$

implying $\lim_{n\to\infty}(1/n) \log \alpha_{10} = -D(P_0\|P_1)$. □

2.4 Minimum Mean-Squared Error Risk Estimation in Vector Spaces

Minimum risk will take various forms for decision making on discrete and continuous parameter spaces. We now study continuous parameter spaces like $\mathbb{R}^n$. In vector spaces the familiar metrics will be the natural way to measure error for estimation. These metrics will become the risk functions for minimum-risk estimation. Begin with linear vector spaces and minimum-mean-squared error estimation.

2.4.1 Normed Linear and Hilbert Spaces

The objects being estimated are oftentimes vectors or elements in a normed vector space.

Definition 2.21 *A **normed linear vector space** is a vector space of elements $x \in \mathcal{X}$ on which there is defined a **norm** which is a function $\|\cdot\| : \mathcal{X} \to \mathbb{R}^+$ satisfying for all $x, y \in \mathcal{X}$:*

1. $\|x\| \geq 0$ *with* $\|x\| = 0$ *if and only if* $x = 0$;
2. $\|x + y\| \leq \|x\| + \|y\|$ *for each* $x, y \in \mathcal{X}$ *(triangle inequality);*
3. $\|\alpha x\| = |\alpha| \|x\|$ *for all scalars α and each $x \in \mathcal{X}$.*

Here are several examples of metric spaces which are familiar.

Example 2.22 (Euclidean) The Euclidean space of n-tuples $x = (x_1, \ldots x_n) \in \mathbb{R}^n$ is one of the principal normed spaces we are familiar with (and a Hilbert space as well, see below) with norm given by $\|x\| = \sqrt{\sum_{i=1}^{n} |x_i|^2}$.

Example 2.23 (Continuous functions) The normed linear space of continuous functions $C[0,1]$ on $[0,1]$ with the norm $\|x\| = \max_{t \in [0,1]} |x(t)|$. Clearly it is a vector space, and addition of continuity gives continuous functions. To see the triangle inequality is satisfied it follows

$$\max_{t \in [0,1]} |x(t) + y(t)| \leq \max_{t \in [0,1]} |x(t)| + \max_{t \in [0,1]} |y(t)|. \tag{2.61}$$

Oftentimes we require differentiability, then $C^1[0,1]$ denotes one time continuously differentiable functions with

$$\|x\| = \max_{t \in [0,1]} |x(t)| + \max_{t \in [0,1]} |\dot{x}(t)|. \tag{2.62}$$

Example 2.24 (Triangle Inequality in l^p) The square-summable and square-integrable spaces are used often. Let p be a real number, $p \geq 1$, the space l^p consisting of vectors $x = (x_1, x_2, \ldots)$ for which the positive quantity $\|\cdot\|_p$ is finite defined as $\|x\|_p = (\sum_{i=1}^{\infty} |x_i|^p)^{1/p} < \infty$. For p-arbitrary, see Luenberger [8]; let us do here $p = 1, 2$, showing the triangle inequality. For $p = 1$ it is straightforward:

$$\sum_{i=1}^{n} |x_i + y_i| \leq \sum_{i=1}^{n} |x_i| + \sum_{i=1}^{n} |y_i|. \tag{2.63}$$

Since this is true for every n, $\|x + y\|_1 \leq \|x\|_1 + \|y\|_1$. For $p = 2$, we use the Holder inequality (see proof below), for $1 \leq p, q \leq \infty$ with $x, y \in \mathbb{R}^n$, $\|x\|_p = \left(\sum_i |x_i|^p\right)^{1/p}$, then

$$\sum_{i=1}^{n} |x_i y_i| \leq \|x\|_p \|y\|_q \quad \text{with} \quad \frac{1}{p} + \frac{1}{q} = 1. \tag{2.64}$$

Using the Holder inequality[5] then,

$$\sum_{i=1}^{n}|x_i+y_i|^2 \le \sum_{i=1}^{n}|x_i+y_i||x_i| + \sum_{i=1}^{n}|x_i+y_i||y_i| \qquad (2.68)$$

$$\stackrel{(a)}{\le} \left(\sum_{i=1}^{n}|x_i+y_i|^2\right)^{1/2}\left(\left(\sum_{i=1}^{n}|x_i|^2\right)^{1/2} + \left(\sum_{i=1}^{n}|y_i|^2\right)^{1/2}\right), \qquad (2.69)$$

with (a) following for the Holder inequality with $p = q = 2$. Dividing through gives

$$\left(\sum_{i=1}^{n}|x_i+y_i|^2\right)^{1/2} \le \left(\sum_{i=1}^{n}|x_i|^2\right)^{1/2} + \left(\sum_{i=1}^{n}|y_i|^2\right)^{1/2}; \qquad (2.70)$$

since this is true for every n, we have the triangle inequality.

The square summable and integrable spaces l^2, L^2 will be used extensively for modeling. The extra structure afforded to them as a Hilbert space with inner product will be used extensively. The square-summable inner product associated with the inner product on l^2 will be $\langle x,y\rangle_2 = \sum_i x_i y_i$; the square-integrable space L^2 has an inner product associated with the norm which is $\langle x,y\rangle = \int_0^1 x(t) y^\star(t) dt$.

Definition 2.25 A **Hilbert space** *is a complete[6] linear vector space H together with the inner product producing a scalar $\langle \cdot, \cdot \rangle : H \times H \to \mathbb{C}$ satisfying for all $x, y \in H$, complex scalars*
 1. $\langle x,y\rangle = \langle y,x\rangle^*$;
 2. $\langle x+y, z\rangle = \langle x,z\rangle + \langle y,z\rangle$, $\langle \alpha x, y\rangle = \alpha \langle x, y\rangle$
 3. $\langle x,x\rangle \ge 0$ and equals 0 if and only if $x = 0$.

We note, that from the normed-distance to the origin the metric distances between elements is induced by the inner product. To see this, we only need prove the triangle inequality, which we do using the Cauchy–Schwartz inequality.

Theorem 2.26 (Cauchy–Schwartz) *In the Hilbert space H denotes $\|x\| = \langle x,x\rangle^{1/2}$, then for all $x, y \in H$,*

$$|\langle x,y\rangle| \le \|x\| \|y\|, \qquad (2.71)$$

---

[5] *Holder Inequality*: Consider $f(t) = t^\lambda - \lambda t + \lambda - 1, t \ge 1$ and $0 < \lambda < 1$, then $\dot{f}(t) = \lambda t^{\lambda-1} - 1$ giving $\dot{f}(t) > 0$ for $0 < t < 1$ and $\dot{f}(t) < 0$ for $t > 1$. Thus, for $t \ge 0, f(t) \le f(1) = 0$ implying $t^\lambda \le \lambda t + 1 - \lambda$. Substituting $t = (|x_i|/\|x\|_p)^p / (|y_i|/\|y\|_q)^q$ with $\lambda = 1/p, 1-\lambda = 1/q$ gives

$$\left(\frac{(|x_i|/\|x\|_p)^p}{(|y_i|/\|y\|_q)^q}\right)^{1/p} \le \frac{1}{p}\left(\frac{(|x_i|/\|x\|_p)^p}{(|y_i|/\|y\|_q)^q}\right) + \frac{1}{q}. \qquad (2.65)$$

Multiplying through by $(|y_i|/\|y\|_q)^q$ gives

$$\frac{|x_i y_i|}{\|x\|_p \|y\|_q} \le \frac{1}{p}\left(\frac{|x_i|}{\|x\|_p}\right)^p + \frac{1}{q}\left(\frac{|y_i|}{\|y\|_q}\right)^q \qquad (2.66)$$

and summing over i gives

$$\frac{\sum_i |x_i y_i|}{\|x\|_p \|y\|_q} \le \frac{1}{p} + \frac{1}{q} = 1. \qquad (2.67)$$

[6] A space for which Cauchy sequences converge within the space under the norm.

2.4 MINIMUM MEAN-SQUARED ERROR RISK ESTIMATION

and $\|\cdot\|$ is a norm satisfying the triangle inequality:

$$\|x+y\| \leq \|x\| + \|y\| \ . \tag{2.72}$$

Proof For all $x, y \in H$ then

$$0 \leq \left\langle x - \frac{\langle x,y \rangle}{\langle y,y \rangle} y, x - \frac{\langle x,y \rangle}{\langle y,y \rangle} y \right\rangle \tag{2.73}$$

$$= \langle x, x \rangle - \frac{|\langle x,y \rangle|^2}{\langle y,y \rangle} \ , \tag{2.74}$$

giving the Cauchy–Schwartz. The triangle inequality becomes

$$\|x+y\|^2 = \|x\|^2 + 2|\langle x,y \rangle| + \|y\|^2 \tag{2.75}$$

$$\overset{(a)}{\leq} \|x\|^2 + 2\|x\|\|y\| + \|y\|^2 \tag{2.76}$$

$$= (\|x\| + \|y\|)^2 \ , \tag{2.77}$$

(a) following from the Cauchy–Schwartz inequality. □

The added structure of the inner product in the Hilbert space with the squared-error metric provides the projection theorem for which errors from linear estimators are orthogonal to the approximating set of vectors.

Theorem 2.27 (Classical Projection Theorem) Let $M \subset H$ be the closed subspace of the Hilbert space generated from the linear independent basis $y_1, \ldots, y_n$. Then for all $x \in H$ there exists a unique vector $\hat{x} \in M$ of the form $\hat{x} = \sum_i \alpha_i y_i$ such that $\|x - \hat{x}\| \leq \|x - \tilde{x}\|$ for all $\tilde{x} \in M$. A necessary and sufficient condition for $\hat{x} \in M$ to be the unique minimizing vector is that $x - \hat{x}$ is orthogonal to M; that is, for $\hat{x} = \sum_{j=1}^n a_j y_j$ the normal equations $\langle x - \hat{x}, y_i \rangle = 0, i = 1, \ldots, n$:

$$\begin{pmatrix} \langle y_1,y_1 \rangle & \langle y_1,y_2 \rangle & \cdots & \langle y_1,y_n \rangle \\ \langle y_2,y_1 \rangle & \langle y_2,y_2 \rangle & \cdots & \langle y_2,y_n \rangle \\ \vdots & \vdots & \vdots & \vdots \\ \langle y_n,y_1 \rangle & \langle y_n,y_2 \rangle & \cdots & \langle y_n,y_n \rangle \end{pmatrix} \begin{pmatrix} a_1 \\ a_2 \\ \vdots \\ a_n \end{pmatrix} = \begin{pmatrix} \langle x,y_1 \rangle \\ \langle x,y_2 \rangle \\ \vdots \\ \langle x,y_n \rangle \end{pmatrix} \ . \tag{2.78}$$

Proof First, $x - \hat{x}$ must be orthogonal to the y's or any element $m \in M$. Suppose it isn't, so that $\langle x - \hat{x}, m \rangle = \delta \neq 0$, for some $m \in M$. Then defining $\tilde{x} = \hat{x} + \delta m$, with $\|m\| = 1$, then

$$\|x - \hat{x} - \delta m\|^2 = \|x - \hat{x}\|^2 - 2\delta \langle x - \hat{x}, m \rangle + \delta^2 \|m\|^2 \tag{2.79}$$

$$= \|x - \hat{x}\|^2 - \delta^2 < \|x - \hat{x}\|^2 \ ; \tag{2.80}$$

apparently this can not be since $\hat{x}$ is the minimum estimator. That $\hat{x}$ is unique follows from the property of the norm:

$$\|x - \tilde{x}\|^2 = \|x - \hat{x} + \hat{x} - \tilde{x}\|^2 \tag{2.81}$$

$$= \|x - \hat{x}\|^2 + \|\hat{x} - \tilde{x}\|^2 \ . \tag{2.82}$$

Thus $\|x - \tilde{x}\| > \|x - \hat{x}\|$ for $\tilde{x} \neq \hat{x}$. □

2.4.2 Least-Squares Estimation

We can construct linear least-square error (l.l.s.e.) estimator as a minimum norm problem in the Hilbert space of random vectors, in which the correlation plays the role of the inner product. In this way the geometric interpretation of the classical projection theorem will hold. Throughout we assume the finite dimensional setting; we return to this in Chapter 9 when we extend these notions to the infinite dimensional setting in which projection in the associated function spaces of L^2, l^2 becomes fundamental.

For now, construct the finite m-dimensional Hilbert space $H(Y_1, Y_2, \ldots, Y_m)$ of random variables generated from the random n-vectors $Y_1, Y_2, \ldots, Y_m \in \mathbb{R}^n$ of zero-mean random variables with finite second moment. Define the Hilbert space to be the space of random variables generated as linear combinations of the Y_is, with the inner product including the expectation.

Definition 2.28 *Define the **finite dimensional Hilbert space of random variables** generated from $Y_1, \ldots, Y_m$ as $H(Y_1, \ldots, Y_m)$ given by the set of all elements $Z = \sum_{i=1}^{m} a_i Y_i$, $a_i \in \mathbb{R}$ with inner product and norm*

$$\langle X, Y \rangle = \operatorname{tr} EXY^* = E\langle X, Y \rangle_{\mathbb{R}^n}, \qquad \|X\|^2 = \operatorname{tr} EXX^* = E\langle X, X \rangle_{\mathbb{R}^n}, \tag{2.83}$$

with $*$ denoting the transpose for real vectors so that $K_{XY} = EXY^*$, and $\langle X, Y \rangle_{\mathbb{R}^n}$ is the $\mathbb{R}^n$ inner product.

Now examine a classical problem illustrating least-squares estimation as an orthogonal projection.

Theorem 2.29 *Let $Y_i, i = 1, \ldots, m \in \mathbb{R}^n$ be zero-mean random vectors jointly distributed with covariances $K_{Y_i Y_j}$ with the optimum linear estimate $\hat{X} = \sum_{i=1}^{m} a_i Y_i$ of $X \in H(Y_1, \ldots, Y_m)$, has $a_1, \ldots, a_m$ satisfying*

$$\begin{pmatrix} E\langle Y_1, Y_1 \rangle_{\mathbb{R}^n} & \cdots & E\langle Y_1, Y_m \rangle_{\mathbb{R}^n} \\ \vdots & \vdots & \vdots \\ E\langle Y_m, Y_1 \rangle_{\mathbb{R}^n} & \cdots & E\langle Y_m, Y_m \rangle_{\mathbb{R}^n} \end{pmatrix} \begin{pmatrix} a_1 \\ \vdots \\ a_m \end{pmatrix} = \begin{pmatrix} E\langle X, Y_1 \rangle_{\mathbb{R}^n} \\ \vdots \\ E\langle X, Y_m \rangle_{\mathbb{R}^n} \end{pmatrix}. \tag{2.84}$$

Proof The projection theorem with inner product $\langle X, Y \rangle = \operatorname{tr} EXY^*$ implies orthogonality according to

$$\langle X - \hat{X}(Y), Y_j \rangle = \left\langle X - \sum_i a_i Y_i, Y_j \right\rangle = 0, \quad j = 1, \ldots, m. \tag{2.85}$$

The normal equations become

$$\begin{pmatrix} E\langle Y_1, Y_1 \rangle_{\mathbb{R}^n} & \cdots & E\langle Y_1, Y_m \rangle_{\mathbb{R}^n} \\ \vdots & \vdots & \vdots \\ E\langle Y_m, Y_1 \rangle_{\mathbb{R}^n} & \cdots & E\langle Y_m, Y_m \rangle_{\mathbb{R}^n} \end{pmatrix} \begin{pmatrix} a_1 \\ \vdots \\ a_m \end{pmatrix} = \begin{pmatrix} \langle X, Y_1 \rangle \\ \vdots \\ \langle X, Y_m \rangle \end{pmatrix}. \tag{2.86}$$

□

Corollary 2.30 *If $X, Y_1, Y_2, \cdots \in \mathbb{R}$ are scalar valued, then denoting $Y = \begin{pmatrix} Y_1 \\ \vdots \\ Y_m \end{pmatrix}$ with K_{YY}^{-1} existing, then the l.l.s.e and error covariance satisfies*

$$\hat{X} = K_{XY} K_{YY}^{-1} Y, \quad \text{with } K_{\hat{X}\hat{X}} = E|X - \hat{X}|^2 = K_{XX} - K_{XY} K_{YY}^{-1} K_{YX}. \tag{2.87}$$

2.4 MINIMUM MEAN-SQUARED ERROR RISK ESTIMATION

Proof For the $Y_i \in \mathbb{R}^1$ scalar case with $Y \in \mathbb{R}^m$ the m-vector of scalars, then each inner product reduces $\langle Y_i, Y_j \rangle = EY_iY_j$ so that the normal matrix becomes K_{YY}; with the inverse existing then $\hat{X} = K_{XY}K_{YY}^{-1}Y$. The covariance follows from computing

$$E(X - K_{XY}K_{YY}^{-1}Y)(X - K_{XY}K_{YY}^{-1}Y)^*$$
$$= K_{XX} - 2K_{XY}K_{YY}^{-1}K_{YX} + K_{XY}K_{YY}^{-1}K_{YX}, \qquad (2.88)$$

which gives the result. $\square$

Example 2.31 (Adding the Mean) Given random variable $X \in \mathbb{R}$ an n-vector $Y \in \mathbb{R}^n$, with means μ_X, μ_Y and cross covariance $K_{XY} = E(X - \mu_X)(Y - \mu_Y)^*$, $K_{YY} = E(Y - \mu_Y)(Y - \mu_Y)^*$, the linear-least-square-error estimator is adjusted by the mean values

$$\hat{X}(Y) = \mu_X + K_{XY}K_{YY}^{-1}(Y - \mu_Y). \qquad (2.89)$$

Notice the adjustment by the mean. To see this, linear estimators are of the form $\hat{X}(Y) = \sum_i a_i Y_i + b$. Using matrix notation, then define the vector $a^*Y = \sum_i a_i Y_i$, and the squared error gives

$$E|X - \hat{X}(Y)|^2 = E|X - a^*Y - b|^2 \qquad (2.90)$$

implying $b = \mu_X - a^*\mu_Y$. Thus

$$E|(X - \mu_X) - a^*(Y - \mu_Y)|^2 = K_{XX} + a^*K_{YY}a - 2a^*K_{YX} \qquad (2.91)$$

$$a = K_{YY}^{-1}K_{YX}. \qquad (2.92)$$

Example 2.32 Here is a classic problem in deconvolution. Let the observables Y be the matrix superposition via the linear operator P of the underlying unobserved random variables X with noise W:

$$Y = PX + W, \qquad (2.93)$$

with X and W zero-mean with covariance K_{XX} and K_{WW} and uncorrelated. Then the least-square estimator is given by

$$\hat{X} = K_{XX}P^*(PK_{XX}P^* + K_{WW})^{-1}Y. \qquad (2.94)$$

Example 2.33 (Recursive Estimation via Kalman Filtering) Examine the discrete-time dynamical system with state variables X_i and observables $Y_i, i = 0, 1, \ldots, n$:

$$X_{i+1} = \Phi_i X_i + U_i, \qquad (2.95)$$
$$Y_{i+1} = M_i X_i + W_i, \qquad (2.96)$$

with initial random vector X_0 with initial estimate $\hat{X}_0$ having error covariance $K_0 = E(\hat{X}_0 - X_0)(\hat{X}_0 - X_0)^*$, and input U_k and observation noise W_k zero-mean uncorrelated processes with covariances $EU_kU_l^* = Q_k\delta(k - l)$, $EW_kW_l^* = R_k\delta(k - l)$, with Q_k, R_k positive definite matrices.

Examine the estimation problem of obtaining an l.l.s.e. of the state X from the measurements Y. Introduce the special notation $\hat{X}_{k|j}$ to denote the l.l.s.e. of X_k from measurements $\{Y_i, j \geq i \geq 0\}$. For prediction we examine the case $k \geq j$.

Theorem 2.34 (Kalman) With initial conditions $\hat{X}_{0|-1} = \hat{X}_0$ with covariance K_0, the recursive algorithm which the l.l.s.e. and its covariance satisfies for $i \geq 1$ is given as follows:

$$\hat{X}_{i+1|i} = \Phi_i K_i M_i^*(M_i K_i M_i^* + R_i)^{-1}(y_i - M_i \hat{X}_{i|i-1}) + \Phi_i \hat{X}_{i|i-1}; \qquad (2.97)$$

$$K_{i+1} = \Phi_i K_i \left(I - M_i^*(M_i K_i M_i^* + R_i)^{-1} M_i K_i\right) \Phi_i^* + Q_i. \qquad (2.98)$$

Proof Suppose that $Y_0 = y_0, \ldots, Y_{i-1} = y_{i-1}$ have been observed with $\hat{X}_{i|i-1}$ the l.l.s.e. projected onto the space generated by the measurements $y_j, j \leq i - 1$ with covariance matrix K_i. Given observation $Y_i = y_i$ satisfying

$$y_i = M_i x_i + w_i, \qquad (2.99)$$

then $\hat{X}_{i|i}$ is the l.l.s.e. projected on the space generated to time i. Since M_i is a linear transformation, the updated estimator $\hat{X}_{i|i}$ is the old estimate $\hat{X}_{i|i-1}$ plus the best estimate of X_i in the subspace generated by $y_i - M_i \hat{X}_{i|i-1}$. From the above l.l.s.e. theorem the updated estimator to time i and error covariance becomes

$$\hat{X}_{i|i} = \hat{X}_{i|i-1} + K_i M_i^* (M_i K_i M_i^* + R_i)^{-1} (y_i - M_i \hat{X}_{i|i-1}), \qquad (2.100)$$

$$K_{i|i} = K_i - K_i M_i^* (M_i K_i M_i^* + R_i)^{-1} M_i K_i. \qquad (2.101)$$

Since Φ_i is a linear transformation, then the l.l.s.e. of $\Phi_i X_i$ given $y_j, j \leq i$ is $\Phi_i \hat{X}_{i|i}$. Since U_i is uncorrelated with $Y_j = y_j, j \leq i$ and $\hat{X}_{i|i}$, then we have the l.l.s.e. of X_{i+1} given by

$$\hat{X}_{i+1|i} = \Phi_i \hat{X}_{i|i}, \qquad (2.102)$$

with the error covariance given according to

$$K_{i+1} = \Phi_i K_{i|i} \Phi_i^* + Q_i. \qquad (2.103)$$

But substituting Eqns. 2.100 into 2.102 and Eqn. 2.101 into 2.103 gives the Kalman filter equations 2.97 and 2.98. □

2.4.3 Conditional Mean Estimation and Gaussian Processes

For elements of Hilbert spaces, the minimum-mean-squared error estimators have overwhelmingly become the most widely used estimators, principally because (i) they are firmly linked to conditional mean estimators, and (ii) for Gaussian random variables least-squared error estimators are linear functions of the data. For random variables and random vectors, the conditional mean is defined as should be expected, the average value using the conditional density.

Definition 2.35 *The **conditional mean** of X given Y is denoted by $E(X|Y)$.*

Theorem 2.36 *Given random vectors $X, Y \in \mathbb{R}^n$ with inner product $\langle X, Y \rangle = \sum_{i=1}^n X_i Y_i$ and associated norm $\|X\| = |\langle X, X \rangle|^{1/2}$, then the conditional mean $E(X|Y)$ has the particularly beautiful property that the correlation of the error $X - E(X|Y)$ with any other function (measurable) of the data $\psi(Y)$ is zero.*

The conditional mean also provides the minimum mean-squared error estimator over all estimators $\hat{X} : \mathcal{Y} \to \mathcal{X}$.

Proof To see this, examine

$$E\langle X - E(X|Y), \psi(Y) \rangle = E\langle X, \psi(Y) \rangle - E\langle E(X|Y), \psi(Y) \rangle \qquad (2.104)$$

$$= E\langle X, \psi(Y) \rangle - E\langle X, \psi(Y) \rangle = 0. \qquad (2.105)$$

2.4 MINIMUM MEAN-SQUARED ERROR RISK ESTIMATION

To show the minimum error property, let $\hat{X}(Y)$ be any other estimator. Then

$$\begin{aligned}
E\|X - \hat{X}(Y)\|^2 &= E\|X - E(X|Y) + E(X|Y) - \hat{X}(Y)\|^2 \\
&= E\|X - E(X|Y)\|^2 + E\|E(X|Y) - \hat{X}(Y)\|^2 \\
&\quad - 2E\langle X - E(X|Y), E(X|Y) - \hat{X}(Y)\rangle \\
&\stackrel{(a)}{=} E\|X - E(X|Y)\|^2 + E\|E(X|Y) - \hat{X}(Y)\|^2 \\
&\geq E\|X - E(X|Y(\cdot))\|^2,
\end{aligned}$$

with (a) following from the orthogonality property of the conditional mean. □

We now show that the conditional mean for Gaussian processes is precisely the least-squared error estimator.

Theorem 2.37 *Let the random vectors be zero-mean with components divided into two subvectors $[X, Y]$ with covariance matrix $K = \begin{pmatrix} K_{XX} & K_{XY} \\ K_{YX} & K_{YY} \end{pmatrix}$.*

Then if the distribution of X, Y is jointly normal, then the conditional distribution of X given Y is normal with mean $K_{XY} K_{YY}^{-1} Y$ and covariance $K_{XX} - K_{XY} K_{YY}^{-1} K_{YX}$.

Proof Define $Y^{(1)} = X - K_{XY} K_{YY}^{-1} Y$, $Y^{(2)} = Y$ then $Y^{(1)}, Y^{(2)}$ are Gaussian and independent with joint density

$$\begin{aligned}
p(Y^{(1)}, Y^{(2)}) &= \det{}^{(-1/2)} 2\pi (K_{XX} - K_{XY} K_{YY}^{-1} K_{YX}) e^{-(1/2) Y^{(1)*} (K_{XX} - K_{XY} K_{YY}^{-1} K_{YX})^{-1} Y^{(1)}} \quad (2.106) \\
&\quad \times \det{}^{(-1/2)} 2\pi K_{YY} e^{-(1/2) Y^{(2)*} K_{YY}^{-1} Y^{(2)}}. \quad (2.107)
\end{aligned}$$

The density on X, Y is obtained via the change of density formula using the transformation $Y^{(1)} = X - K_{XY} K_{YY}^{-1} Y$, $Y^{(2)} = Y$ with the Jacobian being one gives the joint density in X, Y according to

$$\begin{aligned}
p(X, Y) &= \det{}^{(-1/2)} 2\pi (K_{XX} - K_{XY} K_{YY}^{-1} K_{YX}) e^{-(1/2)(X - K_{XY} K_{YY}^{-1} Y)^* (K_{XX} - K_{XY} K_{YY}^{-1} K_{YX})^{-1} (X - K_{XY} K_{YY}^{-1} Y)} \\
&\quad \times \det{}^{-1/2} 2\pi K_{YY} e^{-(1/2) Y^* K_{YY}^{-1} Y}. \quad (2.108)
\end{aligned}$$

The conditional density divides by the Gaussian density on Y giving the result. □

Example 2.38 Let X and Y be jointly Gaussian random variables with the density

$$p(X, Y) = \frac{1}{2\pi \sigma_X \sigma_Y \sqrt{1 - \rho^2}}$$

$$\times \exp - \left(\frac{1}{2(1-\rho^2)} \frac{(X - \mu_X)^2}{\sigma_X^2} + \frac{(Y - \mu_Y)^2}{\sigma_Y^2} - \frac{2\rho (X - \mu_X)(Y - \mu_Y)}{\sigma_X \sigma_Y} \right) \quad (2.109)$$

also denoted as $N(\mu_X, \mu_Y, \sigma_X^2, \sigma_Y^2, \rho)$. A nice property of the bivariate Gaussian density is that normality is preserved under conditioning. The conditional density of random variable X given Y is

$$p(X|Y) = \frac{p(X,Y)}{p(Y)}$$

$$= \frac{\frac{1}{2\pi\sigma_X\sigma_Y\sqrt{(1-\rho^2)}} \exp -\left(\frac{1}{2(1-\rho^2)}\frac{(X-\mu_X)^2}{\sigma_X^2} + \frac{(y-y)^2}{\sigma_Y^2} - \frac{2\rho(X-\mu_X)(Y-\mu_Y)}{\sigma_X\sigma_Y}\right)}{\frac{1}{\sqrt{2\pi\sigma_Y^2}} \exp -1/2\left(\frac{(Y-\mu_Y)^2}{\sigma_Y^2}\right)}$$

(2.110)

$$= \frac{1}{\sigma_X\sqrt{2\pi(1-\rho^2)}} \exp -\frac{1}{2(1-\rho^2)}\left(\frac{(X-\mu_X)}{\sigma_X} - \frac{\rho(Y-\mu_Y)}{\sigma_Y}\right)^2 \quad (2.111)$$

which by observation is Gaussian distributed as $N(\mu_X+\rho(\sigma_X/\sigma_Y)(Y-\mu_Y), \sigma_X^2(1-\rho^2))$. The conditional mean $E(X|Y) = \mu_X + \rho(\sigma_X/\sigma_Y)(Y - \mu_Y)$, which is linear in Y.

For estimators which are linear combinations of the data, the least-square estimators are particularly simple involving only mean and covariances of the processes. Of course for the Gaussian process, conditional mean estimators are linear-least-squared error estimators.

Remark 2.4.1 That an estimator satisfies the orthogonality property is a natural way for defining the conditional mean which extends even to the case when the conditioning events are infinite. That is, given a random variable X and stochastic process $Y(\cdot)$, the **conditional expectation** of X given $Y(\cdot)$ is the unique random variable that is a functional of $Y(\cdot)$ satisfying the orthogonality condition $E(X - \hat{X}(Y(\cdot))\psi(Y(\cdot)) = 0$ for all measurable functions $\psi(\cdot)$.

2.5 The Fisher Information of Estimators

Thus far we have examined on discrete parameter spaces maximizing likelihood and posterior probability for minimum risk. For optimizing mean-squared error on the continuum the conditional mean fits the job. In general, however, there are many other point estimators as the conditional mean may be difficult to compute; maximum likelihood, MAP, method of moments estimators, and many others. In general they will not be optimal in the mean-square sense, although we would still like to know how good they can possibly be. For this reason Fisher information is extremely powerful as it allows us to bound the performance of any estimator in the mean-squared error sense.

Definition 2.39 Let X have density $P_X(dx) = p_\theta(x)dx$ indexed by parameter $\theta \in \Theta \subset \mathbb{R}^d$. An estimator $\hat{\theta} : \mathcal{X} \to \Theta$ of the real-valued parameter $\theta \in \Theta$ has **bias** defined by the true value θ and its average value:

$$\text{Bias } \theta = E\{\hat{\theta}(X) - \theta\}. \quad (2.112)$$

The **estimator is unbiased** if $\text{Bias}(\theta) = 0$.

It turns out that the mean-squared error for any unbiased estimator is lower bounded by the inverse of the Fisher information.

2.5 THE FISHER INFORMATION OF ESTIMATORS

Definition 2.40 Let $\theta \in \Theta$ be an m-dimensional real parameter, density $p_\theta(X)$. Then the $d \times d$ **Fisher information matrix** is defined as $F(\theta) = (F_{jk}(\theta))$, where

$$F_{jk}(\theta) = E\left(\frac{\partial \log p_\theta(X)}{\partial \theta_j} \frac{\partial \log p_\theta(X)}{\partial \theta_k}\right). \tag{2.113}$$

Theorem 2.41 Given sufficient smoothness of the density so that it is twice differentiable, then

$$F_{jk}(\theta) = E\left(-\frac{\partial^2 \log p_\theta(X)}{\partial \theta_j \partial \theta_k}\right). \tag{2.114}$$

Proof To see that the Fisher information is given by the negative second derivative follows from the following argument:

$$E\left(-\frac{\partial^2}{\partial \theta_j \partial \theta_k} \log p_\theta(X)\right)$$

$$= E\left(-\frac{(\partial^2/\partial \theta_j \partial \theta_k) p_\theta(X)}{p_\theta(X)} + \frac{(\partial/\partial \theta_j) p_\theta(X)(\partial/\partial \theta_k) p_\theta(X)}{(p_\theta(X))^2}\right) \tag{2.115}$$

$$\stackrel{(a)}{=} E\left(\frac{(\partial/\partial \theta_j) p_\theta(X)(\partial/\partial \theta_k) p_\theta(X)}{(p_\theta(X))^2}\right), \tag{2.116}$$

with (a) following from the interchange of differentiation[7] and integration in the first term with the integral a constant, $\int_{\mathcal{X}} (\partial^2/\partial \theta_j \partial \theta_k) p_\theta(x) dx = 0$. □

The mean-squared error performance is bounded by the inverse Fisher information. We shall sometimes denote the covariance of $\hat{\theta}$ as $\text{Cov}(\hat{\theta})$.

Theorem 2.42 (Cramer–Rao inequality) Let $\hat{\theta}(X) \in \Theta$ be an unbiased estimator of $\theta \in \Theta$. Then

$$\text{Cov}\,\hat{\theta}(X) \geq F(\theta)^{-1}. \tag{2.117}$$

Let $X_1, \ldots, X_n$ be i.i.d random variables. Then the Fisher information is n-times the single sample information with the covariance lower bound decreased by n:

$$\text{Cov}\,\hat{\theta}(X_1, \ldots, X_n) \geq (nF(\theta))^{-1}. \tag{2.118}$$

Proof Define the score function $V_\theta(X) = \nabla_\theta \log p_\theta(X)$, then $EV_\theta(X) = 0$ since

$$EV_\theta(X) = \int_{\mathcal{X}} \frac{\nabla_\theta p_\theta(x)}{p_\theta(x)} p_\theta(x) dx \tag{2.119}$$

$$= \nabla_\theta \int_{\mathcal{X}} p_\theta(x) dx = 0. \tag{2.120}$$

with the smoothness of Theorem 2.41 allowing interchange of differentiation for bounded convergence. Using the Cauchy–Schwartz inequality componentwise on the

---

[7] A sufficiently strong condition is if the second-derivative is continuous on Θ compact then it is bounded and so dominated convergence allows swapping of the derivative and integral.

entries in the matrix gives

$$\left(E(V_\theta(X) - EV_\theta(X))(\hat\theta - E\hat\theta)^*\right)^2$$
$$\leq E\left((V_\theta(X) - EV_\theta(X))(V_\theta(X) - EV_\theta(X))^*\right) E\left((\hat\theta - E\hat\theta)(\hat\theta - E\hat\theta)^*\right)$$
$$= F(\theta) \ \text{Cov} \ \hat\theta. \tag{2.121}$$

Note, this inequality is componentwise. Now use the fact that $EV_\theta(X) = 0$ according to Eqn. 2.120 simplifying the left-hand side according to

$$\left(E(V_\theta(X) - EV_\theta(X))(\hat\theta - E\hat\theta)^*\right)^2 = E(V_\theta(X)\hat\theta^*) \tag{2.122}$$

$$\stackrel{(a)}{=} \nabla_\theta \int_{\mathcal{X}} p_\theta(x) \hat\theta^*(x) dx \tag{2.123}$$

$$\stackrel{(b)}{=} \nabla_\theta \theta^* = \text{id}, \tag{2.124}$$

with (a) following from the smoothness as above and (b) following from unbiased assumption and where id is the $d \times d$ identity matrix.

To complete the proof, given $X_1, X_2, \ldots$ independent random variables, then

$$E\left(\frac{\partial}{\partial \theta_j} \log p_\theta(X_1, X_2, \ldots) \frac{\partial}{\partial \theta_k} \log p_\theta(X_1, X_2, \ldots)\right) = n F_{jk}(\theta). \tag{2.125}$$

□

Monotonicity of the Fisher information in sample size is important in multi-sensor fusion work; estimators get better when information is combined optimally.

Example 2.43 Let X be Poisson distributed, mean θ, then $p_\theta(X) = e^{-\theta + X \log \theta + \log X!}$, and $H(n, \theta) = (\theta - \bar{X} \log \theta)$, with the inverse Fisher information $F(\theta) = 1/\theta$.

2.6 Maximum-Likelihood and its consistency

The most commonly used estimation when not in a mean-squared error setting is one based on Bayesian inference which assumes that (i) *a priori* probabilities exist on the parameters, and (ii) they are known at least approximately. When this is not the case, maximum likelihood becomes the method of choice. If the data is strong, and the parameter space is of low-dimension the two perform similarly. In high dimensional settings the prior can play a fundamental role with special care required (see [9] for a detailed examination of such issues).

Let us begin with maximizing likelihood. Similar arguments hold for the maximum a posteriori formulation where the parameter is random with a priori distribution. Here is the basic setup. Let X be a random variable with density $P_\theta(dX) = p_\theta(X)dX$ parameterized by $\theta \in \Theta \subseteq \mathbb{R}^d$. Define ML estimators as follows.

Definition 2.44 *Given a sample $X_1, X_2, \ldots$ with joint density $P_\theta(dX) = p_\theta(X_1, X_2, \ldots)dX$ parameterized by $\theta \in \Theta \subseteq \mathbb{R}^d$. Then we shall say $\hat\theta$ is a* **maximum-likelihood (ML) estimator** *if $\hat\theta \in \mathcal{M}_n \subset \Theta$, $\mathcal{M}_n$ the set of maximizers of the likelihood:*

$$\mathcal{M}_n = \{\hat\theta \in \Theta : p_{\hat\theta}(x_1, \ldots, x_n) = \sup_{\theta \in \Theta} p_\theta(x_1, \ldots, x_n)\}. \tag{2.126}$$

2.6 MAXIMUM-LIKELIHOOD AND ITS CONSISTENCY

There are many classical proofs of optimality of likelihood estimators. In any case, what is the reason for maximizing likelihood? Why does it work? It is the law of large numbers and the fundamental role of the cross-entropy function which makes it work.

Consider the use of the normalized log-likelihood function defined on an independent identically distributed (i.i.d.) sample of n-terms with real-valued parameters $\theta \in \Theta$, $(1/n) \sum_{i=1}^{n} \log p_\theta(X_i)$. This forms an i.i.d. sample with expected value the negative cross-entropy with respect to the density parameterized by the true parameter $\theta^\star \in \Theta$. Thus if all goes well, and the log-likelihood function converges then $\lim_{n \to \infty} (1/n) \sum_{i=1}^{n} \log p_\theta(X_i) = E_{P_{\theta^\star}(X)} \log p_\theta(X)$. But we know that this has maximum for $\theta = \theta^\star$, so that

$$E_{P_{\theta^\star}(X)} \log p_\theta(X) \leq E_{P_{\theta^\star}(X)} \log p_{\theta^\star}(X) = \lim_{n \to \infty} \frac{1}{n} \sum_{i=1}^{n} \log p_\theta(X_i)|_{\theta=\theta^\star} . \quad (2.127)$$

We should then expect that maximizing with respect to $\theta \in \Theta$ should give a cross-entropy which is close to that given by the true parameter, and if the parameter space is well behaved should have maximum in the parameter space which is close to the true parameter $\theta^\star$. Essentially all proofs of consistency follow this line of argument, that the cross entropy is maximum when the true parameter is selected.

We shall now look at this a bit more carefully, first assuming a stronger form of uniform convergence for the empirical log-likelihood, and the second assuming a sufficient smoothness so that a rate of convergence can be established for the consistency.

2.6.1 Consistency via Uniform Convergence of Empirical Log-likelihood

We now examine this more precisely. Essentially we employ uniform convergence of the empirical log-likelihood function to establish consistency. We first define consistency as follows.

> **Definition 2.45** *Let $X_1, X_2, \ldots$ be a random sample jointly distributed with density $P_\theta(dX) = p_\theta(X_1, X_2, \ldots) dX, \theta \in \Theta$. Then we shall say that the MLE $\hat{\theta}(X_1, X_2, \ldots)$ is* **asymptotically consistent** *if it converges in probability to $\theta^\star$ the true parameter with $n \to \infty$.*

We shall establish uniform convergence for $H_n(\theta)$ using a Uniform Weak Law of Large numbers.

> **Theorem 2.46** *Let $X_1, X_2, \ldots$ be identically distributed with density $P_\theta(dX) = p_\theta(X_1, X_2, \ldots), \theta \in \Theta$, and define the sequence obtained from maximization of the log-likelihoods*
>
> $$\hat{\theta}_n = \arg\max_{\theta \in \Theta} \log p_\theta(X_1, X_2, \ldots) = \arg\min_{\theta \in \Theta} H_n(\theta) \quad (2.128)$$
>
> *where* $H_n(\theta) = -\frac{1}{n} \sum_{i=1}^{n} \log p_\theta(X_i)$. \quad (2.129)
>
> *Assume Θ is a compact subset of $\mathbb{R}^k$, and $H(\theta) = -E_{P_{\theta^\star}(X)} \log p_\theta(X)$ is continuously differentiable in $\theta \in \Theta$ so that $\{H_n(\theta)\}$ converges to $H(\theta)$ uniformly in probability.*[8]

[8] [The Uniform Weak Law [10]] Let $X_i \in \mathbb{R}^m$ be i.i.d. random samples, $f_\theta(X)$ a random variable on $\Theta \times \mathbb{R}^m$ which is continuous in θ for each X. Let Θ be a compact subset of $\mathbb{R}^m$ and $E \sup_{\theta \in \Theta} |f_\theta(X)| < \infty$. Then,

$$\text{for all } \epsilon > 0, \quad \theta \in \Theta, \quad \lim_{n \to \infty} \Pr\left\{ \left| \frac{1}{n} \sum_{i=1}^{n} f_\theta(X_i) - E f_\theta(X) \right| > \epsilon \right\} = 0. \quad (2.130)$$

Then if $H(\theta)$ has a unique minimum $\theta^\star \in \Theta$ with $H(\theta) > H(\theta^\star)$ for all $\theta \neq \theta^\star$, then the maximum-likelihood estimator is consistent in probability, also denoted $\hat{\theta}_n \xrightarrow{p} \theta^\star$.

Remark 2.6.2 We point out that $H(\theta)$ is only the true entropy of X (as in Eqn. 2.1.2) when $\theta = \theta^\star$, i.e.

$$-E_{P_{\theta^\star}(X)} \log p_{\theta^\star}(X) \leq -E_{P_{\theta^\star}(X)} \log p_\theta(X). \tag{2.131}$$

Proof Since $\theta^\star$ is the unique minimum, for an open neighborhood $N(\theta^\star) \subset \Theta$ of $\theta^\star$, then $H(\theta) > H(\theta^\star)$ for all $\theta \in N_\varepsilon(\theta^\star)$. Choose $\varepsilon = \min_{N_\varepsilon(\theta^\star)^c \cap \Theta} H(\theta) - H(\theta^\star) > 0$. Define the event Ω_n as the set

$$\Omega_n = \{\omega \in \Omega : |H_n(\theta) - H(\theta)| < \varepsilon/2, \forall \theta \in \Theta\}. \tag{2.132}$$

Then in Ω_n, we have

$$H(\hat{\theta}_n) - \frac{\varepsilon}{2} \stackrel{(a)}{<} H_n(\hat{\theta}_n) \tag{2.133}$$

$$\stackrel{(b)}{<} H_n(\theta^\star) \stackrel{(c)}{<} H(\theta^\star) + \frac{\varepsilon}{2}, \tag{2.134}$$

with (a,c) following from 2.132, and with (b) since $\hat{\theta}_n$ is an MLE of $\log p(X_1, \ldots, X_n)$ it minimizes $H_n(\theta)$. Thus we have $H(\hat{\theta}_n) < H(\theta^\star) + \varepsilon$, and apparently $\hat{\theta}_n \in N_\varepsilon(\theta^\star)$, with ε arbitrary. Thus $\hat{\theta}_n \to \theta^\star$ converges in probability. □

2.6.2 Asymptotic Normality and $\sqrt{n}$ Convergence Rate of the MLE

Here is another attack on consistency, in which by adding smoothness a convergence rate can be obtained. More restrictions on smoothness and definiteness of the Hessian actually gives a convergence rate.

Theorem 2.47 Let the parameter space Θ be as above with $H(\theta)$ twice continuously differentiable in $\theta \in \Theta$ with $H_n(\theta)$ converging to $H(\theta)$ uniformly in probability with $H(\theta)$ having a unique minimum $\theta^* \in \Theta$ as above.

Then if $F(\theta^\star) \stackrel{p}{=} \lim_{n\to\infty}((\partial^2 H_n(\theta))/(\partial \theta_j \partial \theta_k))_{\theta=\theta^\star}$ is a non-singular matrix, then the MLE is consistent converging in distribution to a Gaussian with covariance given by the inverse Fisher information:

$$\sqrt{n}\left(\hat{\theta}_n - \theta^\star\right) \xrightarrow{d} \mathcal{N}\left(0, F(\theta^\star)^{-1}\right). \tag{2.135}$$

Proof The mean value theorem implies that there exists $\lambda \in (0,1)$ with $\theta' = \theta^\star + \lambda(\hat{\theta}_n - \theta^\star)$ such that

$$\frac{\partial H_n(\hat{\theta}_n)}{\partial \theta} - \frac{\partial H_n(\theta^\star)}{\partial \theta} = \frac{\partial^2 H_n(\theta')}{\partial \theta_j \partial \theta_k}(\hat{\theta}_n - \theta^\star). \tag{2.136}$$

Since $\hat{\theta}_n$ is an MLE estimator, $(\partial H_n(\hat{\theta}_n))/\partial \theta = 0$, implying

$$\sqrt{n}\left(\hat{\theta}_n - \theta^\star\right) = -\sqrt{n}\left(\frac{\partial^2 H_n(\theta')}{\partial \theta_j \partial \theta_k}\right)^{-1} \frac{\partial H_n(\theta^\star)}{\partial \theta}. \tag{2.137}$$

2.6 MAXIMUM-LIKELIHOOD AND ITS CONSISTENCY

Examine the terms in the right-hand side of Eqn. 2.137. Asymptotic consistency of MLE implies $\hat{\theta}_n, \theta' \xrightarrow{P} \theta^\star$, which implies

$$\left(\frac{\partial^2 H_n(\theta')}{\partial \theta_j \partial \theta_k}\right)^{-1} \xrightarrow{P} \left(\frac{\partial^2 H_n(\theta^\star)}{\partial \theta_j \partial \theta_k}\right)^{-1} \xrightarrow{P} F(\theta^\star)^{-1} . \qquad (2.138)$$

Consider the second term in Eqn. 2.137,

$$-\sqrt{n}\frac{\partial H_n(\theta^\star)}{\partial \theta} = \sqrt{n}\frac{\partial}{\partial \theta}\left(\frac{1}{n}\sum_{i=1}^n \log p_\theta(X_i)\right)\bigg|_{\theta=\theta^\star} . \qquad (2.139)$$

The random variables $(\partial \log p_\theta(X_i))/\partial \theta|_{\theta=\theta^\star}$ are i.i.d. with zero mean and common covariance:

$$E\left(\frac{\partial \log p_\theta(X_i)}{\partial \theta}\right) = \int_{\mathcal{X}} \frac{\partial p_\theta(x)}{\partial \theta} \frac{p_\theta(x)}{p_\theta(x)} dx = \int_{\mathcal{X}} \frac{\partial p_\theta(x)}{\partial \theta} = \frac{\partial}{\partial \theta}\int_{\mathcal{X}} p_\theta(x) = 0, \qquad (2.140)$$

$$E\left(\frac{\partial \log p_{\theta^\star}(X_i)}{\partial \theta}\frac{\partial \log p_{\theta^\star}(X_i)'}{\partial \theta}\right) = E\left(-\frac{\partial^2 \log p_{\theta^\star}(X_i)}{\partial \theta_j \partial \theta_k}\right) = F(\theta^\star). \qquad (2.141)$$

where smoothness allows us to swap expectation and differentiation. The central limit theorem implies that and by the Uniform Weak Law theorem (see footnote 8) gives

$$-\sqrt{n}\frac{\partial H_n(\theta^\star)}{\partial \theta} \xrightarrow{d} \mathcal{N}(0, F(\theta^\star)) . \qquad (2.142)$$

Then, by Slutsky's theorem[9] applied to Eqn. 2.137 gives

$$\sqrt{n}\left(\hat{\theta}_n - \theta^\star\right) \xrightarrow{d} F(\theta^\star)^{-1} \mathcal{N}\left(0, F(\theta^\star)\right) \sim \mathcal{N}\left(0, F(\theta^\star)^{-1}\right). \qquad (2.144)$$

□

Example 2.49 (Gaussian) For the Gaussian setting then $p_\theta(X) = \frac{1}{\sqrt{2\pi\sigma^2}}e^{-\|X-\theta\|^2/2\sigma^2}$, with

$$H(n,\theta) = \frac{1}{n}\sum_{i=1}^n \frac{\|X_i - \theta\|^2}{2\sigma^2} , \qquad (2.145)$$

and $(\partial^2/\partial\theta^2)H(\theta) = 1/\sigma^2$. The MLE $(1/n)\sum_{i=1}^n X_i$ converges in probability to θ at rate $\sigma/\sqrt{n}$.

Example 2.50 Let $X_1, X_2, \ldots$ be drawn i.i.d with exponential density $p_\theta(X) = (1/\theta)e^{-X/\theta}$, then the MLE is unbiased:

$$\hat{\theta} = \arg\max_\lambda -n\log\theta - \frac{\sum_{i=1}^n X_i}{\theta} = \frac{\sum_{i=1}^n X_i}{n}. \qquad (2.146)$$

---

[9] **Theorem 2.48** *Slutsky's Theorem* [11] *Suppose a sequence of random vectors X_n converges in distribution to a random vector X, $X_n \xrightarrow{d} X$, and a sequence of random vectors Y_n converges in probability to a constant vector C, $Y_n \xrightarrow{P} C$. Then for any continuous function g,*

$$g(X_n, Y_n) \xrightarrow{d} g(X_n, C). \qquad (2.143)$$

2.7 Complete–Incomplete Data Problems and the EM Algorithm

We shall employ the expectation-maximization (EM) algorithm of Dempster, Laird and Rubin [12] to solve many problems in the estimation of parameters. Problems involving the EM algorithm are called complete–incomplete data problems, in which a function is estimated which parameterizes a known probability density; the actual data (denoted as the *complete-data*) described by the density are not observed. Rather, observations consist of data (denoted as *incomplete-data*) which nonuniquely specifies the complete-data via some set of many-to-one mappings. The motivation is that for many classic problems parameters are estimated from measurements which are both noisy, i.e. samples of a stochastic process, as well as incomplete.

The problem set up is to assume a prior density $f_\theta(X)$ describing the complete-data X, parameterized by some function θ, and observations $Y = h(X)$ where $h(\cdot)$ is a many-to-one vector mapping $h : \mathcal{X} \to \mathcal{Y}$ from the complete data observations. Then we define the *complete-data* random variable X as a measurable function with density $f(X)$ so that $\Pr(X \in B) = \int_B f(x)dx$ which is absolutely continuous with density parameterized by θ. The family of densities $f_\theta(X)$ parameterized by θ are termed the *complete-data* densities. We say that we are given *incomplete-data* if instead of observing X in $\mathcal{X}$, only the sample Y is available, where $Y = h(X)$ for some measurable m-dimensional vector mapping $h(\cdot)$; this mapping is, in general, many to one, so X is not uniquely specified by Y. Thus the incomplete data Y results in the existence of $m+1$ sample spaces, the complete data space $\mathcal{X}$ and the m incomplete data spaces $\mathcal{Y}_1, \mathcal{Y}_2, \ldots, \mathcal{Y}_m$. Denote the product space describing the incomplete data vector $\mathcal{Y} = \mathcal{Y}_1 \times \mathcal{Y}_2 \times \cdots \times \mathcal{Y}_m$. The subset $\chi(Y) \subset \mathcal{X}$ is given by $\chi(Y) = \{X : h(X) = Y\}$.

The family of densities $g_\theta(Y)$ describing the incomplete data are given by $g_\theta(Y) = \int_{\chi(Y)} f_\theta(x)dx$, with the conditional density on X given Y then

$$k_\theta(X|X \in \chi(Y)) = \frac{f_\theta(X)}{\int_{\chi(Y)} f_\theta(x)dx} = \frac{f_\theta(X)}{g_\theta(Y)}, \quad X \in \chi(Y). \tag{2.147}$$

Now it follows directly that maximum-likelihood estimation over parameterized families of log-likelihood density $g_\theta(\cdot)$ may be posed as a joint-entropy maximization.

Lemma 2.51 *Given is incomplete data $Y = h(X)$ with density parameterized by θ, $g_\theta(Y) = \int_{\chi(Y)} f_\theta(x)dx$. Then*

$$\arg\max_{\{\theta\}} \log g_\theta(Y) = \arg\max_{\{\theta\}} \max_{\{q:\int_{\chi(Y)} q(x)dx=1\}} -\int_{\chi(Y)} q(x) \log \frac{q(x)}{f_\theta(x)} dx. \tag{2.148}$$

Proof The proof results from the following equalities using Eqn. 2.147:

$$\log g_\theta(Y) = \log f_\theta(X) - \log k_\theta(X|X \in \chi(Y)) \tag{2.149}$$

$$= E_{k_\theta(X|X \in \chi(Y))}\{\log f_\theta(X) - \log k_\theta(X|X \in \chi(Y), \theta)\} \tag{2.150}$$

$$= -\int_{\chi(Y)} k_\theta(x|X \in \chi(Y)) \log \frac{k_\theta(x|X \in \chi(Y))}{f_\theta(x)} dx \tag{2.151}$$

$$= \max_{\{q:\int_{\chi(Y)} q(x)dx=1\}} -\int_{\chi(Y)} q(x) \log \frac{q(x)}{f_\theta(x)} dx. \tag{2.152}$$

□

Because of the equivalence between the conditional and maximum entropy densities, the incomplete-data log-likelihood is simply the joint-maximum with respect to q, θ of the entropy function. This results in the estimation problem being expanded to what appears to be a larger

2.7 COMPLETE–INCOMPLETE DATA PROBLEMS

problem in which both the parameters θ as well as density q must be estimated, which implies the following iterative algorithms of Dempster et al. [12] and Csiszar and Tusnady [13].

Theorem 2.52 *Given is incomplete data $Y = h(X)$ with density parametrized by θ, $g_\theta(Y) = \int_{\chi(Y)} f_\theta(X) dx$. Define the sequence of iterates $q^{(n)}, \theta^{(n)}; n = 1, 2, \ldots$ according to the following joint maximization:*

$$q^{(n+1)} = \underset{\{q: \int_{\chi(Y)} q(x) dx = 1\}}{\arg\max} -\int_{\chi(Y)} q(x) \log \frac{q(x)}{f_{\theta^{(n)}}(x)} dx \quad (2.153)$$

$$= \frac{f_{\theta^{(n)}}}{\int_{\chi(Y)} f_{\theta^{(n)}}(x) dx} = k_{\theta^{(n)}}; \quad (2.154)$$

EM – step
$$\theta^{(n+1)} = \underset{\{\theta\}}{\arg\max} -\int_{\chi(Y)} q^{(n+1)}(x) \log \frac{q^{(n+1)}(x)}{f_\theta(x)} dx \quad (2.155)$$

$$= \underset{\{\theta\}}{\arg\max} \, E_{k_{\theta^{(n)}}} \log f_\theta(X) . \quad (2.156)$$

Then this defines an expectation–maximization algorithm with iterates $\theta^{(n)}; n = 1, 2, \ldots$ which are monotone nondecreasing in the incomplete data likelihood sequence

$$\log g_{\theta^{(1)}}(Y) \leq \log g_{\theta^{(2)}}(Y) \leq \cdots . \quad (2.157)$$

Proof This iteration is a Csiszar and Tusnady alternating minimization of the cross entropy, where the parametrized family $f_\theta(\cdot)$ is varied through the parameters θ. The monotonicity of the log-likelihood for the EM algorithm is inherited from the fact that $q^{(n+1)}(X) = k_{\theta^{(n)}}(X|X \in \chi(Y))$ implying

$$\log g_{\theta^{(n-1)}}(Y) \stackrel{(a)}{=} -\int_{\chi(Y)} q^{(n)}(x) \log \frac{q^{(n)}(x)}{f_{\theta^{(n-1)}}(x)} dx \quad (2.158)$$

$$\stackrel{(b)}{\leq} -\int_{\chi(Y)} q^{(n)}(x) \log \frac{q^{(n)}(x)}{f_{\theta^{(n)}}(x)} dx \quad (2.159)$$

$$\leq -\int_{\chi(Y)} q^{(n+1)}(x) \log \frac{q^{(n+1)}(x)}{f_{\theta^{(n)}}(x)} dx \stackrel{(a)}{=} \log g_{\theta^{(n)}} , \quad (2.160)$$

with (a) using Eqns. 2.151 evaluated at $\theta^{(n-1)}, \theta^{(n)}$ and with (b) the EM-Step Eqn. 2.156.
□

Remark 2.7.3 Calculating the negative of the entropy function results in the alternating minimization of the K–L divergence of Csiszar and Tusnady [13] and Musicus [14]. The EM sequence is a particular example of an alternating minimization.

We now explicitly examine several maximum-likelihod estimation problems working with the conditional mean and the EM algorithm.

Example 2.53 (Maximum-Likelihood Estimation of Sinusoids in Gaussian Noise)
A classic problem in spectrum estimation involves stationary Gaussian noise in which the spectrum is unknown, and may contain time-varying means. This arises in many contexts in spectrum estimation [15] including direction of arrival processing [16–18], multi-dimensional magnetic resonance imaging spectroscopy [19–24], and radar imaging [25–28] to name a few. See Kay and Marple [29] for an extensive introduction to the problems. For the basic sinusoids in noise model, the measured field $Y(t), t \in [0, T]$ is

a superposition of exponentially decaying sinusoids in additive noise, with unknown amplitudes, frequencies and phases:

$$Y(t) = \sum_m a_m e^{-\lambda t} + W(t), \qquad (2.161)$$

and $W(t)$ is "white noise" with spectral density σ^2.

The log-likelihood is coupled across the sinusoids:

$$\log g_\theta(Y) = -\frac{2}{2\sigma^2} \int Y(t) \sum_m a_m e^{-\lambda_m t} dt + \frac{1}{2\sigma^2} \int \left| \sum_m a_m e^{-\lambda_m t} \right|^2 dt. \qquad (2.162)$$

Direct maximization involves coupling of parameters in the log-likelihood.

The EM algorithm [12] divides the problem into M independent maximizations. Define the complete data space to be a set of independent Gaussian processes $X_m(t), m = 1, 2, \ldots$, one for each signal component, each having mean $a_m e^{-\lambda_m t}$ with each noise component having spectral intensity σ_m^2 and $\sigma^2 = \sum_m \sigma_m^2$. The mapping between the complete and incomplete data is given by superposition $Y(t) = \sum_m X_m(t)$. Then the EM algorithm is given as follows.

Theorem 2.54 *Defining the complete-data $X_m(t) = a_m e^{-\lambda t} + W_m(t)$ with $W_m(t), m = 1, 2, \ldots$, independent white noise instantaneous variance $EW_m(t)^2 = \sigma_m^2$, with the many to one mapping $Y(t) = h(X_1(t), X_2(t), \ldots) = \sum_m X_m(t)$, then the sequence of iterates generated according to*

$$a_m^{\text{new}} = \frac{\int \bar{X}_m^{\text{old}}(t) e^{-\lambda_m^{\text{new}} t} dt}{\int |e^{-\lambda_m^{\text{new}} t}|^2}; \qquad \lambda_m^{\text{new}} = \arg\max_{\{\lambda_m\}} \frac{\left| \int \bar{X}_m^{\text{old}}(t) e^{-\lambda_m t} dt \right|^2}{\int |e^{-\lambda_m t}|^2}, \qquad (2.163)$$

$$\text{where } \bar{X}_m^{\text{old}}(t) = \frac{\sigma_m^2}{\sigma^2} \left(Y(t) - \sum_{m'} a_{m'}^{\text{old}} e^{-\lambda_{m'}^{\text{old}} t} \right) + a_m^{\text{old}} e^{-\lambda_m^{\text{old}} t}, \qquad (2.164)$$

are an instance of an EM algorithm implying that the likelihoods are monotonically increasing. In turn, stable points of the iteration satisfy the necessary maximizer conditions maximizing the likelihood of Eqn. 2.162.

Proof The expectation of the complete data log-likelihood becomes

$$E_{k_{\theta^{\text{old}}}} \log f_\theta(X_1, X_2 \ldots) = \sum_m 2 \int \bar{X}_m^{\text{old}}(t) a_m e^{-\lambda_m t} dt - \sum_m \int |a_m e^{-\lambda_m t}|^2 dt,$$

where $\bar{X}_m^{\text{old}} = E(X_m | \theta^{\text{old}}, Y)$. Now we have to prove Eqn. 2.164 for the conditional mean. The complete-data is Gaussian as is the incomplete data, implying that from Eqn. 2.31 of Example 2.31 the conditional mean becomes

$$\bar{X}_t^{\text{old}} = \mu_{X_t} + K_{X_t Y_t} K_{Y_t Y_t}^{-1} (Y - \mu_{Y_t}). \qquad (2.165)$$

Clearly $\mu_{X_t} = a_m e^{-\lambda_m^{\text{old}} t}$. Since $Y(t) = \sum_m X_m(t)$ and $X_1, X_2, \ldots$ are independent, we have $K_{X_t Y_t} = (\sigma_1^2, \sigma_2^2, \ldots)$, $K_{Y_t Y_t}^{-1} = \text{diag}(1/\sigma^2, 1/\sigma^2, \ldots)$ giving $\bar{X}_m^{\text{old}}$ in the form Eqn. 2.164. Using the conditional mean and maximizing gives the update EM iteration, Eqns. 2.163. □

Notice, the EM algorithm breaks the maximization into independent problems across the sinusoids.

2.7 COMPLETE–INCOMPLETE DATA PROBLEMS

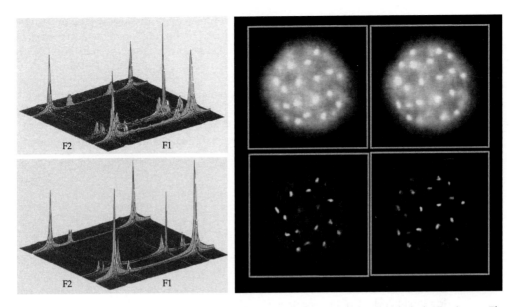

Figure 2.3 Left column: The figure shows the results from the N-butyl alchohol experiment. The top panel shows the original 2D spectrum of the N-butyl alcohol data. The bottom panel shows the spectrum reconstructed from the estimates of the EM-algorithm parameters. The data are taken from [23]. Right column: The figures show the EM algorithm reconstruction of the COSM data taken from Dr. Keeling of Washington University; the data reconstructions are from [30]. The top row shows X–Z sections through the COSM amoeba data. The bottom row shows the 200th EM algorithm iteration for sections through the COSM amoeba data (see also Plate 1).

In 1- and 2D nuclear magnetic resonance (NMR) spectroscopy time series are segmented into their constituent chemical shifts, modeled as sinusoids in noise [19–21] and modeled as exponentially decaying sinusoids signals in Gaussian noise [18, 22, 24]. In d-dimensional spectroscopy, the decaying sinusoids are indexed over $\mathbb{R}^d$ so that $t = (t_1, \ldots, t_d)$, and

$$Y(t) = \sum_m a_m e^{-\langle \lambda_m, t \rangle_d} dt + W(t), \quad \text{where } \langle \lambda_m, t \rangle_d = \sum_{i=1}^d \lambda_{mi} t_i . \tag{2.166}$$

Shown in Figure 2.3 are results from 2D magnetics resonance spectroscopy [22–24] for the analysis of N-butyl alcohol. In using the EM algorithm to determine the amplitude and signal parameters, twenty-two peaks were found. Column 1 of Figure 2.3 shows the original spectrum of the N-Butanol analysis data (top panel). The bottom panel shows the reconstructed spectrum from the maximum-likelihood fit of the peaks. The reconstructed spectrum closely resembles the original spectrum. There are minor differences observed in these two spectra, including small noise peaks present in the original spectrum not found in the reconstructed spectrum, and minor differences in the height and width of the peaks in the two spectra. Otherwise, the ML algorithm reconstructs the signal in the N-Butanol spectrum.

Example 2.55 (Image Restoration in Point Processes) Examine the image restoration problem via conditionally Poisson models. A variety of imaging applications involving measurement errors of various kinds have been studied on Poisson processes including forward looking infra-red imaging (FLIR) via CCD arrays [31, 32], computational optical sectioning microscopy (COSM) [30, 33], electron-microscopic-autoradiography [34], and emission tomography both positron (PET) [35–37] and single-photon (SPET) [38–40]. The image restoration must take into account two fundamental components characteristic of the imaging systems: (i) the number of measurement points are low and therefore dominated by Poisson statistics, and (ii) the physics of the measurement systems introduced in the creation of the observed data. The restoration is based on the model of Snyder [41] which hypothesizes the existence of two point-processes, hence the connection to complete–incomplete data.

Examine the discrete version of the counting process here. The photon counts are modeled as a Poisson counting process X_i the number of points with mean $EX_i = \lambda_i$ in discrete position i in the discrete lattice. The counts in subsets A are defined by the counting process $X(A) = \sum_{i \in A} X_i$, with mean $EX(A) = \sum_{i \in A} \lambda_i$. The Poisson law is given by $\prod_i (e^{-\lambda_i} \lambda_i^{X_i} / X_i!)$, with complete-data log-likelihood only a function of λ given by

$$\log f_\lambda(X) = -\sum_i \lambda_i + \sum_i X_i \log \lambda_i. \tag{2.167}$$

Generally, these are observed with measurement errors reflecting the uncertainty due to optical focus or detector uncertainty in optical sectioning and the line- and time-of-flight uncertainty in positron-emission tomography. The error vectors in measurement are assumed to be independent of the creation of the counting process, described via point-spread functions conditioned on voxel i, $p(\cdot|i)$, $\sum_j p(j|i) = 1$. The measurements Y_j are modeled as Poisson-distributed [42] with mean $EY_j = \sum_i p(j|i)\lambda_i$, and log-likelihood

$$\log g_\lambda(Y) = -\sum_i \lambda_i + \sum_j Y_j \log \left(\sum_i p(j|i)\lambda_i \right). \tag{2.168}$$

The EM iteration is as follows.

Theorem 2.56 *The sequence of iterates generated according to*

$$\lambda_i^{\text{new}} = \lambda_i^{\text{old}} \left(\sum_j \frac{p(j|i)}{\sum_{i'} p(j|i')\lambda_{i'}^{\text{old}}} Y_j \right), \tag{2.169}$$

are an instance of an EM algorithm with stable points of the iteration satisfying the necessary maximizer conditions for the maximizer of Eqn. 2.168.

Proof The new iterates maximize the conditional expectation of the complete-data log-likelihood

$$E_{\lambda^{\text{old}}}(X|Y) \log f_\lambda(X) = -\sum_i \lambda_i + \sum_i \bar{X}_i^{\text{old}} \log \lambda_i, \tag{2.170}$$

where $\bar{X}_i^{\text{old}} = E(X_i|Y, \lambda^{\text{old}}) = \lambda_i^{\text{old}} \left(\sum_j \frac{p(j|i)}{\sum_{i'} p(j|i')\lambda_{i'}^{\text{old}}} Y_j \right). \tag{2.171}$

Maximizing gives the iterates of Eqn. 2.169. The convergence point $\lambda^{\text{new}} = \lambda^{\text{old}}$ gives the fixed point condition[10]

$$1 = \left(\sum_j \frac{p(j|i)}{\sum_{i'} p(j|i')\lambda_{i'}} Y_j \right), \tag{2.172}$$

---

[10] [Convergence of the EM-algorithm to MLEs] Vardi et al. [43] proved that the discrete Eqn. 2.169 has global convergence properties; the initial estimate $\lambda_i^{(0)}$ can be any positive bounded function with the sequence converging to a an MLE satisfying the necessary and sufficient maximizer conditions. The neat proof of Vardi et al. breaks into two parts: (i) showing that if the iteration of 2.169 converges, the Kuhn–Tucker conditions are satisfied and therefore the convergence point of the algorithm maximizes the log-likelihood; and (ii) showing that every sequence converges. Proof of (i) follows since the log-likelihood is concave and Shepp and Vardi [36]

which is the necessary maximizer condition for interior points $\lambda_i > 0$. For the boundary term $\lambda_i = 0$, see [36]. □

The maximization of the discrete log-likelihood via the EM algorithm was concurrently derived and implemented by Shepp and Vardi [36] for positron-emission tomography and Lange and Carson [44] for transmission tomography, and subsequently for tomography systems with time-of-flight by Snyder and Politte [45]. Similar solutions have been derived for single-photon tomography and electron microscopic autoradiography, with the appropriate imaging models chosen in each [34, 46].

In the statistical models proposed for time-of-flight positron emission tomography [35] and computational optical sectioning microscopy (COSM) [30, 47, 48], the point-spread function $p(\cdot)$ reflects the 3D function in the source volume. In COSM the specimen fluoresces incoherently when illuminated with the photons detected only after they have undergone random translations as a result of the microscopes point spread function (see [30, 47, 48]). The point-spread function $p^{(k)}(j|i), k = 1, \ldots, K$, reflects the conditional probability-density that a photon incoherently flourescing at point x is detected at point y when the microscope is focused to plane k. Much work has been done on characterizing optical transfer (see [47, 48], e.g.).

Shown in the right column of Figure 2.3 are the results of the EM algorithm reconstruction in COSM of amoeba. The top row shows X–Z sections through the 3D data measurement of an amoeba collected using the optical sectioning microscope. Notice the blurriness of the measured data resulting from the optical sectioning point-spread function. Each measurement $Y^{(k)}$ has a mean corresponding to its point-spread function $EY_j^{(k)} = \sum_x p^{(k)}(j|i)\lambda_i$, with $\sum_k \sum_j p^{(k)}(j|i) = 1$. The set of measurements are Poisson distributed according to incomplete-data log-likelihood

$$\log g_\lambda(Y^{(1)}, Y^{(2)}, \ldots) = -\sum_i \lambda_i + \sum_{k=1}^{K} \sum_j Y_j^{(k)} \log \left(\sum_i p^{(k)}(j|i)\lambda_i \right). \quad (2.173)$$

The EM algorithm above is modified only slightly from Eqn. 2.169 adjusted to incorporate the multiple point-spread orientation terms according to

$$\lambda_i^{\text{new}} = \lambda_i^{\text{old}} \left(\sum_k \sum_j Y_j^{(k)} \frac{p_k(j|i)}{\sum_{i'} p_k(j|i')\lambda_{i'}^{\text{old}}} \right). \quad (2.174)$$

Shown in the right column of Figure 2.3 (bottom row) are the results of the reconstructed intensity of fluorescence resulting from 200 iterations of the EM algorithm for several sections. Notice the increased structure resulting from the deconvolution algorithm.

Example 2.57 (Segmentation of Brains) Gaussian mixture modeling has been ubiquitous in cortical brain analysis by several groups [49–57]. The brain is modelled as having multiple compartments including gray matter (G), white matter (W), and cerebrospinal fluid (CSF) with different tissue types having different mean and variance parameters.

Model the image as a Gaussian mixture of different regions each with its own parameters. Then the image $X_i, i = 1, \ldots, n$ of $n = |D|$ discrete voxels is modeled as a Gaussian density $f_{\mu, \sigma^2}(X_i) = \frac{1}{\sqrt{2\pi\sigma^2}} e^{-(X_i - \mu)^2 / 2\sigma^2}$ given the means and variances μ, σ^2. The data forming the histogram from the measured imagery is modeled

showed that the convergence point satisfies the necessary Kuhn–Tucker conditions. Proof of (ii) is more subtle, requiring the results of Csiszar and Tusnady [13]. For the particular alternating maximization of 2.169, the K–L divergence between any limit point of the sequence and successive iterates of the algorithm decreases. This coupled with the fact that every sequence has a set of subsequential limit points due to the compactness of the iteration set and the fact that the limit points are stable, implies global convergence for the full sequence of iterates.

as conditionally independent samples $X_1, \ldots, X_n$ from a Gaussian mixture density $g_\theta(X) = \sum_{m=1}^M a_m d(X; \theta_m)$, with $d(X; \theta_m)$ the Gaussian density with parameters $\theta_m = (\mu_m, \sigma_m^2, a_m)$.

Model the measured data forming the histogram as conditionally independent samples $Y_1, Y_2, \ldots$ from a Gaussian mixture density

$$g_\theta(Y_1, \ldots, Y_n) = \prod_{i=1}^n \sum_m a_m d_{\theta_m}(Y_i), \quad \text{with } d_{\theta_m}(Y) = \frac{1}{\sqrt{2\pi}\sigma_m} e^{\frac{-(Y-\mu_m)^2}{2\sigma_m^2}}.$$
(2.175)

The goal is to estimate the parameters making up the mixture $\theta_m = (a_m, \mu_m, \sigma_m^2)$, $m = 1, 2, \ldots$.

Model the complete data as the pairs $X_1 = (Y_1, M_1), X_2 = (Y_2, M_2) \ldots$, with M_i labeling which of the models the data arises from. The many to one mapping from complete to incomplete data discards the compartment tags. The complete data density becomes

$$f_\theta((Y_1, M_1), (Y_2, M_2), \ldots) = \prod_{i=1}^n \prod_m (a_m d_{\theta_m}(Y_i))^{\delta(M_i - m)}$$
(2.176)

with the indicator functions $\delta(M - m) = 1$ if $M = m$, and 0 otherwise. The EM algorithm is as follows.

Theorem 2.58 *The sequence of iterates generated according to*

$$a_m^{\text{new}} = a_m^{\text{old}} \sum_{i=1}^n \frac{d_{\theta_m^{\text{old}}}(Y_i)}{\sum_{\theta_{m'}} a_{m'}^{\text{old}} d_{\theta_{m'}^{\text{old}}}(Y_i)},$$

$$\mu_m^{\text{new}} = \frac{\sum_{i=1}^n Y_i (a_m^{\text{old}} d_{\theta_m^{\text{old}}}(Y_i) / \sum_{m'} a_{m'}^{\text{old}} d_{\theta_{m'}^{\text{old}}}(Y_i))}{\sum_{i=1}^n (d_{\theta_m^{\text{old}}}(Y_i) / \sum_{m'} a_{m'}^{\text{old}} d_{\theta_{m'}^{\text{old}}}(Y_i))},$$

$$(\sigma_m^2)^{\text{new}} = \frac{\sum_{i=1}^n (Y_i - \mu_m^{\text{new}})^2 (a_m^{\text{old}} d_{\theta_m^{\text{old}}}(v) / \sum_{m'} a_{m'}^{\text{old}} d_{\theta_{m'}^{\text{old}}}(Y_i))}{\sum_{i=1}^n (d_{\theta_m^{\text{old}}}(Y_i) / \sum_{m'} a_{m'}^{\text{old}} d_{\theta_{m'}^{\text{old}}}(Y_i))},$$
(2.177)

are an instance of an EM algorithm with nondecreasing likelihood. The stable points of the iterations satisfy the necessary maximizer conditions of the likelihood density of Eqn. 2.175.

Proof The expectation step of the EM algorithm requires the conditional expectation of the log-likelihood according to

$$E_{k_{\theta^{\text{old}}}} \log f_\theta(X_1, X_2, \ldots) = \sum_{i=1}^n \sum_m E\left\{\delta(M_i - m) | Y, \theta^{\text{old}}\right\} \log a_m d_{\theta_m^{\text{old}}}(Y_i) \quad (2.178)$$

where $E\{\delta(M - m) | Y, \theta\} = \dfrac{a_m d_{\theta_m^{\text{old}}}(Y)}{\sum_{m'} a_{m'} d_{\theta_{m'}^{\text{old}}}(Y)}.$
(2.179)

2.7 COMPLETE–INCOMPLETE DATA PROBLEMS

The maximization step is carried out with the constraint $\sum_m a_m = 1$ giving the new maximizers of the EM iteration of Eqn. 2.177:

$$\sum_{i=1}^n \frac{a_m^{\text{old}} d_{\theta_m^{\text{old}}}(Y_i)}{\sum_{m'} a_{m'}^{\text{old}} d_{\theta_{m'}^{\text{old}}}(Y_i)} \frac{1}{a_m} + \gamma = 0,$$

$$\sum_{i=1}^n \frac{a_m^{\text{old}} d_m^{\text{old}}(Y_i)}{\sum_{m'} a_{m'}^{\text{old}} d_{\theta_{m'}^{\text{old}}}(Y_i)} \left(\frac{\partial}{\partial \mu_m} d_{\theta_m}(Y_i) \right) \frac{1}{d_{\theta_m}(Y_i)} = 0,$$

$$\sum_{i=1}^n \frac{a_m^{\text{old}} d_m^{\text{old}}(Y_i)}{\sum_{m'} a_{m'}^{\text{old}} d_{\theta_{m'}^{\text{old}}}(Y_i)} \left(\frac{\partial}{\partial \sigma_m^2} d_{\theta_m}(Y_i) \right) \frac{1}{d_{\theta_m}(Y_i)} = 0. \quad (2.180)$$

To finish the proof, clearly at a fixed point $\theta^{\text{old}} = \theta^{\text{new}}$ then Eqns. 2.180 are the necessary maximizer conditions of the incomplete data log-likelihood $\sum_{i=1}^n \log \sum_m a_m d_{\theta_m}(Y_i; \theta)$ of Eqn. 2.175. □

Shown in Figure 2.4 are results from the cingulate and prefrontal gyrus. Panel 1 shows the EM algorithm fit of G,W,CSF, and partial volume compartments to a cingulate gyrus brain histogram to illustrate the parameter fitting. The top solid curve is the mixture model with fitted parameters from the EM algorithm. It superimposes the measured histogram data almost exactly. Shown below via the lower dashed lines are each of the compartment fits taken separately. Panel 2 shows a single MRI section through the cingulate gyrus. The minimum Bayes risk for segmentation (as in Section 2.3, Chapter 2) in the $m = 1, \ldots, M$ compartments selects for each i the model-type:

$$\hat{H}_i = \arg\max_{H_i \in \{1, \ldots, M\}} -\frac{1}{2} \log 2\pi \sigma^2(H_i) - \frac{1}{2} \frac{(X_i - \mu(H_i))^2}{\sigma^2(H_i)}. \quad (2.181)$$

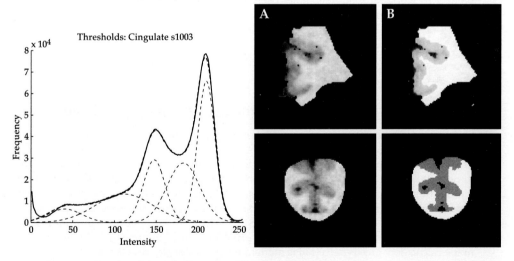

Figure 2.4 Panel 1 shows the EM algorithm fit of G,W,CSF and partial volume compartments to brain tissue histograms to illustrate the segmentation calibration. The top solid curve superimposes the measured histogram data almost exactly. The lower dashed lines depict each of the compartment fits taken separately. Panel 2 shows an MRI section of the cingulate; panel 3 shows the Bayes segmentation into G,W, CSF compartments of coronal sections of the cingulate gyrus; Panels 4 and 5 show the same as above (row 1) for the medial prefrontal gyrus. Data taken from the laboratories of Dr. John Csernansky and Dr. Kelly Botteron of Washington University (see also Plate 2).

Panel 3 shows the solution to the Bayes hypothesis testing problem for selecting G,W, and CSF in each image voxel X_i in the coronal section of the cingulate gyrus. The bottom row panels 4 and 5 show the same for the prefrontal gyrus.

2.8 Hypothesis Testing and Model Complexity

Consider the multihypothesis testing problem of determining model $m = 1, 2, \ldots$ from random observation $Y_1, Y_2, \ldots$. Proceeding in a Bayesian manner, the obvious solution is through minimum risk implying the fundamental role of Bayesian hypothesis testing and likehood ratio testing on the model given the observation. Given the sequence of conditional densities $p_m(Y)$ across models with priors $\pi_m, m = 1, 2, \ldots$ then Bayesian model selection is from Theorem 2.15 the minimum risk. Most often, however, the models involve random nuisance parameters with some prior density $\pi_m(X), X \in \mathcal{X}_m \subseteq \mathbb{R}^{d_m}$ of dimension d_m, requiring the calculation of the so-called *nuisance integral* of the conditional density on the random sample $Y_1, Y_2, \ldots$ for every m taking the form

$$\hat{m} = \arg\max_m \pi_m p_m(Y_1, \ldots, Y_n), \qquad (2.182)$$

where $p_m(Y_1, \ldots, Y_n) = \int_{\mathcal{X}_m} p_m(Y_1, \ldots, Y_n | x) \pi_m(x) dx$. $\qquad (2.183)$

Such celebrated integrals have received tremendous attention, and only in special cases can they be performed analytically. They have been focused on in the context of Bayesian integration of nuisance parameters via the Laplace's method (see Schwarz [58]) which is precisely the approach we shall follow. We return to this below in the context of model complexity and Rissanen's [59] pioneering work. Most influential has been the study of the information-theoretic asymptotics of Bayes methods by Barron and Clark [60].

2.8.1 Model-Order Estimation and the $d/2$ log Sample-Size Complexity

Since its introduction by Fisher, the method of maximum-likelihood has certainly proven to be one of the most ubiquitous and effective methods for vector parameter estimation when the dimension of the parameter space is fixed and finite. When the dimension of the parameter space itself needs to be estimated, maximum-likelihood techniques tend to be "greedy," consistently picking the models of greatest complexity to yield overly tight fits to the data. This is the so called **model order estimation problem**. The challenges of model order estimation were originally addressed by Akaike (1973), and subsequently Schwarz [58] and Rissanen [59]. Rissanen's **minimum description length principle** (MDL) seeks to remedy the problem by incorporating the complexity of the model in the calculation; greedy selection is moderated by the *complexity of the model*.

The exact formulation of the complexity for the Gaussian case results from the quadratic expansion. We should expect, if the posterior density is smooth then for large sample size the integrand can be approximated via Laplace's method (Polya & Szego [61], p. 96). The technical requirements for this are smoothness and identifiability.

Condition 2.59 (Smoothness Conditions) *Assume the parameter space from model $m = 1, 2, \ldots$ is closed and bounded with prior $\pi_m(x), x \in \mathcal{X}_m \subset \mathbb{R}^{d_m}$. Define the joint*

2.8 HYPOTHESIS TESTING AND MODEL COMPLEXITY

density on the i.i.d. sample $Y_1, \ldots, Y_n$ for all $x \in \mathcal{X}_m \subset \mathbb{R}^{d_m}$ to be of the form

$$p_m(Y_1, \ldots, Y_n, x) = e^{-nH_m(n,x)}, \quad H_m(n, x) = -\frac{1}{n} \sum_{i=1}^{n} \log p_m(Y_i|x) - \frac{1}{n} \log \pi_m(x). \tag{2.184}$$

Assume $H_m(n, x)$ satisfies the following smoothness and identifiability conditions for every n:

1. $H_m : x \mapsto H_m(x)$ are third derivative continuous, and
2. for all $(Y_1, \ldots, Y_n) \in \mathcal{Y}_0^n$, there exists a unique $\hat{x} \in \mathcal{X}_m$ such that

$$H_m(n, \hat{x}) < H_m(n, x) \quad \text{for all } x \in \mathcal{X}_m / \{\hat{x}\},$$

and for any closed set $G \subset \mathcal{X}_m$ not containing $\hat{x}$,

$$\inf_{x \in G} H_m(n, x) > H_m(n, \hat{x}); \text{ and}$$

3. positive definiteness of the $d_m \times d_m$ Hessian matrix at the MAP estimator so that

$$F_m(\hat{x}) = \left(\frac{\partial^2}{\partial x_j \partial x_k} H_m(n, \hat{x}) \right), \tag{2.185}$$

is non-singular in a neighborhood of $\hat{x} \in \mathcal{X}_m$.

Now the asymptotics on the nuisance integral follows. Laplace's approach employs a Taylor series expansion around the maximizer $\hat{x}$.

Theorem 2.60 (Nuisance Integral) *Assume models $m = 1, 2, \ldots$ with closed and bounded parameter spaces with priors $\pi_m(x), x \in \mathcal{X}_m \subset \mathbb{R}^{d_m}$. Define the joint density on $Y_1, \ldots, Y_n$ to be*

$$p_m(Y_1, \ldots, Y_n, x) = e^{-nH_m(n,x)}, \quad H_m(n, x) = -\frac{1}{n} \sum_{i=1}^{n} \log p_m(Y_i|x) - \frac{1}{n} \log \pi_m(x), \tag{2.186}$$

satisfying the smoothness conditions 2.59 with Hessian

$$F_m(x) = \left(\frac{\partial^2}{\partial x_j \partial x_k} H_m(n, x) \right). \tag{2.187}$$

Defining $A(n) \sim B(n)$ to mean $A(n)$ is asymptotically equal to $B(n)$, then as $n \to \infty$ with

$$p_m(Y_1, \ldots, Y_n) \sim p_m(Y_1, \ldots, Y_n, \hat{x}) \left(\frac{2\pi}{n} \right)^{d_m/2} \det^{-1/2} F_m(\hat{x}) \tag{2.188}$$

where $\hat{x} = \arg\max_{x \in \mathcal{X}_m} p_m(Y_1, \ldots, Y_n, x)$. \tag{2.189}

The asymptotic Bayes testing procedure is equivalent to

$$\max_m \pi_m p_m(Y_1, \ldots, Y_n) \sim \max_m \pi_m p_m(Y_1, \ldots, Y_n, \hat{x}) \left(\frac{2\pi}{n} \right)^{d_m/2} \det^{-1/2} F_m(\hat{x}). \tag{2.190}$$

Proof Expand H_m in a Taylor series attaining a minimum at $\hat{x}$ with $\nabla_x H_m(n,\hat{x}) = 0$ so that for some $x_1 \in B_\epsilon(\hat{x})$ a local neighborhood of $\hat{x} \in \mathcal{X}_m$,

$$H_m(n,x) = H_m(n,\hat{x}) + \frac{1}{2}(x-\hat{x})^* \left(\frac{\partial^2}{\partial x_j \partial x_k} H_m(n,\hat{x})\right)(x-\hat{x}). \tag{2.191}$$

Now use the uniqueness of the MLE of property 2 from 2.59 to get for all $x \in \mathcal{X} \setminus B_\epsilon(\hat{x})$ there exists a δ so that $H(n,x) > H(n,\hat{x}) + \delta$ giving

$$\int_{\mathcal{X}_m} e^{-nH_m(n,x)} dx = \int_{B_\epsilon(\hat{x})} e^{-nH_m(n,x)} dx + \int_{\mathcal{X}_m/B_\epsilon(\hat{x})} e^{-nH_m(n,x)} dx \tag{2.192}$$

$$= e^{-nH_m(n,\hat{x})} \left(\int_{B_\epsilon(\hat{x})} e^{-(n/2)(x-\hat{x})^* F_m(n,\hat{x})(x-\hat{x})} dx + O(e^{-\delta n})\right). \tag{2.193}$$

Third derivative continuity gives $k = \sup_{x \in B_\epsilon(\hat{x})}(\partial^3/\partial x_i x_j x_k) H_m(x)$ which with the intermediate value theorem implies for $x \in B_\epsilon(\hat{x})$,

$$|F_m(x)_{jk} - F_m(\hat{x})_{jk}| < k\|x - x^\star\| = O(\epsilon),$$

giving

$$\int_{\mathcal{X}_m} e^{-nH_m(n,x)} dx$$

$$= e^{-nH_m(n,\hat{x})} \left(\int_{B_\epsilon(\hat{x})} e^{-(n/2)(x-\hat{x})^* \left(F_m(\hat{x}) + O(\epsilon)\right)(x-\hat{x})} dx + O(e^{-\delta n})\right) \tag{2.194}$$

$$= e^{-nH_m(n,\hat{x})} \left(\int_{\mathbb{R}^{d_m}} e^{-(n/2)(x-\hat{x})^* \left(F_m(\hat{x}) + O(\epsilon)\right)(x-\hat{x})} dx + O\left(e^{-(n/2)c\|\epsilon^2\|}\right) + O(e^{-\delta n})\right). \tag{2.195}$$

Using property 3 then $F_m(\hat{x})$ is positive definite in the neighborhood of $\hat{x}$ to induce the density transformation of the Gaussian law. Fixing $\hat{x}$, substitute $u = (nF_m(\hat{x}) + O(\epsilon))^{1/2} x$ giving

$$\int_{\mathbb{R}^{d_m}} e^{(-1/2)u^* u} du \left(\frac{1}{n}\right)^{d_m/2} \det^{-1/2}\left(F_m(\hat{x}) + O(\epsilon)\right)$$

$$= \left(\frac{2\pi}{n}\right)^{d_m/2} \det^{-1/2}\left(F_m(\hat{x}) + O(\epsilon)\right), \tag{2.196}$$

since $\int e^{-(1/2)u^* u} du = (2\pi)^{d_m/2}$. Since this is true for arbitrarily small ϵ, this completes the proof. $\square$

Model selection by this asymptotic procedure is often called the Bayesian information criterion. It accomodates the prior distribution in a straightforward manner.

In i.i.d. large sample cases, as $n \to \infty$ then the model selection problem is dominated by the terms which are functions of n, yielding the classic approximation

$$\log p_m(Y_1, \ldots, Y_n) \approx \log p_m(Y_1, \ldots, Y_n | \hat{x}) - \frac{d_m}{2} \log n, \tag{2.197}$$

giving the complexity of the model is dimension $d_m \log sample - size$. We choose the model m which maximizes (2.197).

2.8.2 The Gaussian Case is Special

Examine the Gaussian case which can be handled analytically and demonstrates the fundamental role of the Fisher information in evaluating the integral. For the Gaussian case the nuisance integral of Eqn. 2.183 can be calculated directly. Examine the following time honored model, let $Y \in \mathbb{R}^n$ be a real-valued n-dimensional conditionally Gaussian process, $Y = \sum_{i=1}^d X_i \phi_i + W$, W white Gaussian noise variance σ^2, $\{\phi_i\}_{i=1}^d$ an orthonormal set of $n \times 1$ vectors spanning the d-dimensional subspace of $\mathbb{R}^n$. If the X_i's are Gaussian variates as well, then Y is Gaussian with a particularly interesting decomposition in the Fisher information. Then the MAP estimators $\hat{X}_i, i = 1, \ldots, d$ are given by

$$\hat{X}_i = \arg\max_{X_i \in \mathbb{R}^d} \log p(X_1, \ldots, X_d | Y) = \frac{\lambda_i^2}{\lambda_i^2 + \sigma^2} \langle Y, \phi_i \rangle_{\mathbb{R}^n}. \tag{2.198}$$

The $d \times d$ empirical Fisher information (including the prior) are given by

$$F = \left(-E \frac{\partial^2}{\partial x_i \partial x_j} \log p(X_1, \ldots, X_d | Y)\right) = \left(\frac{\lambda_i^2 + \sigma^2}{\lambda_i^2 \sigma^2} \delta(i-j)\right). \tag{2.199}$$

Theorem 2.61 *Given conditionally Gaussian process, $Y = \sum_{i=1}^d X_i \phi_i + W$, W white Gaussian noise variance σ^2, $\{\phi_i\}_{i=1}^d$ an orthonormal set of $n \times 1$ vectors. Let X_i be zero-mean independent Gaussian, variance λ_i^2, $i = 1, \ldots, d$. The density in terms of the Fisher information and MAP estimator are exact:*

$$p(Y) = p(Y, \hat{X}_1, \ldots, \hat{X}_d)(2\pi)^{d/2} \det{}^{-1/2} F. \tag{2.200}$$

As $\sigma \to 0$, the penalty goes as

$$\log \det{}^{-1/2} F \sim -d/2 \log \sigma^2 + O(1). \tag{2.201}$$

Proof The MAP estimator becomes

$$0 = \frac{\partial}{\partial x_j} \frac{\|Y - \sum_{i=1}^d X_i \phi_i\|_{\mathbb{R}^n}^2}{2\sigma^2} + \sum_{i=1}^d \frac{|X_i|^2}{2\lambda_i^2} = -\frac{2}{2\sigma^2} \left\langle Y - \sum_{i=1}^d X_i \phi_i, \phi_j \right\rangle_{\mathbb{R}^n} + \frac{2X_j}{2\lambda_j^2} \tag{2.202}$$

implying $\hat{X}_j = \frac{\lambda_j^2}{\lambda_j^2 + \sigma^2} \langle Y, \phi_j \rangle_{\mathbb{R}^n}$.

Defining

$$H(X_1, \ldots, X_d) = \frac{\|Y - \sum_{i=1}^d X_i \phi_i\|_{\mathbb{R}^n}^2}{2\sigma^2} + \sum_{i=1}^d \frac{|X_i|^2}{2\lambda_i^2}, \tag{2.203}$$

then rewriting the conditional density gives

$$p(Y) = \int_{\mathbb{R}^d} p(Y|x_1,\ldots,x_d)\pi(x_1,\ldots,x_d)dx_1\cdots dx_d \qquad (2.204)$$

$$= \int_{\mathbb{R}^d} \frac{1}{(2\pi\sigma)^{n/2}} e^{-\frac{\|Y-\sum_{i=1}^d x_i\phi_i\|_{\mathbb{R}^n}^2}{2\sigma^2}} \frac{1}{(2\pi)^{d/2}\prod_{i=1}^d \lambda_i} e^{-\sum_{i=1}^d \frac{|x_i|^2}{2\lambda_i^2}} dx_1\cdots dx_d \qquad (2.205)$$

$$= \int_{\mathbb{R}^d} \frac{1}{(2\pi\sigma)^{n/2}} \frac{1}{(2\pi)^{d/2}\prod_{i=1}^d \lambda_i} e^{-H(x_1,\ldots,x_d)} dx_1\cdots dx_d. \qquad (2.206)$$

Expand $H(x_1,\ldots,x_d)$ in a Taylor series around the MAP estimators $\hat{X}_1,\ldots$ using the fact that the gradient term is zero at the MAP estimator gives

$$H(x_1,\ldots,x_d) = H(\hat{X}_1,\ldots,\hat{X}_d) + \frac{1}{2}\sum_{i,j=1}^d (x_i - \hat{X}_i)(x_j - \hat{X}_j)F_{ij}. \qquad (2.207)$$

Substituting for the MAP estimator gives

$$p(Y) = p(Y,\hat{X}_1,\ldots,\hat{X}_d)\int_{\mathbb{R}^d} e^{-1/2\sum_{i,j=1}^d (x_i-\hat{X}_i)(x_j-\hat{X}_j)F_{ij}} dx_1\cdots dx_d$$

$$\stackrel{(a)}{=} p(Y,\hat{X}_1,\ldots,\hat{X}_d)(2\pi)^{d/2}\det{}^{-1/2} F = p(Y,\hat{X}_1,\ldots,\hat{X}_d)(2\pi\sigma^2)^{d/2}\left(\prod_{i=1}^d \frac{\lambda_i^2}{\lambda_i^2+\sigma^2}\right)^{1/2}, \qquad (2.208)$$

where (a) follows from the fact that the function $(2\pi)^{-d/2}\det{}^{1/2} Q e^{(-1/2)x^*Qx}$ integrates over $\mathbb{R}^d$ to 1 where Q is a positive definite $d\times d$ matrix. □

2.8.3 Model Complexity and the Gaussian Case

To introduce MDL, let us formulate it in a coding context as in Lanterman [62] in discrete and finite hypothesis space context. Suppose the data Y and parameters X are elements of discrete spaces, with the goal being to encode the data Y with a two-part message. The first part indicates the parameters X for that model, and the second encodes the data Y; the total message length becomes

$$\text{len}(Y,X) = \text{len}(Y|X) + \text{len}(X,m). \qquad (2.209)$$

The MDL principle selects the X and m which minimize (2.209) for the collected data Y. Shannon's theory dictates that for a given model m and parameter X, a uniquely decodable code exists for Y with codewords of length $\text{len}(Y|X,m) = \lceil -\log p(Y|X,m)\rceil$. (For convenience, drop the notation for "next largest integer" in the remaining discussion.) If X could somehow be transmitted cost-free, then the maximum-likelihood estimate would be selected for X since it would minimize $\text{len}(Y|X,m)$. However, the parameter X must be encoded as well as it indexes the law. In this data transmission viewpoint, the code for X must be a prefix (also called "self-punctuating") code. This means that the stream representing Y given X may follow the stream representing X without an additional "comma" symbol. This implies the code for X must satisfy the Kraft inequality (Cover and Thomas, 1991, section 5.2, p. 82 [7])

$$\sum_x e^{-\text{len}(X)} \leq 1. \qquad (2.210)$$

Hence, $\pi(X) \propto e^{-\text{len}(X)}$ gives a proper prior distribution on X. Similarly, if $\pi(X)$ is on hand then it can be used to find the code lengths $\text{len}(X)$.

To quantitatively measure the model complexity associated with parameters $(X_1, X_2, \ldots, X_d) \in \mathcal{X}_d$, define the complexity of model m as given by the difference between the information gained about the outputs of the model with and without knowledge of the parameters.

Definition 2.62 *Given model m outputting random sample Y parametrized by $X_1, X_2, \ldots$ with conditional density $p_m(Y|X_1, \ldots, X_{d_m})$, then the **complexity of model** m is given by the mutual information*

$$C(m) = h(Y) - h(Y|X_1, \ldots, X_{d_m}) \,. \tag{2.211}$$

For the quadratic (Gaussian) case the complexity can be calculated exactly.

Theorem 2.63 *Given is $Y \in \mathbb{R}^n$ a real-valued n-dimensional Gaussian vector with Gaussian mean $\sum_{i=1}^{d_m} X_i \phi_i$, $\phi \in \mathbb{R}^n$, in additive white-noise variance σ^2, with X_i zero-mean Gaussian variates with variances λ_i^2. The complexity of the d_m-parameter Gaussian model is*

$$C(m) = h(Y) - h(Y|X_1, \ldots, X_{d_m}) = \sum_{i=1}^{d_m} \frac{1}{2} \log \frac{\sigma^2 + \lambda_i^2}{\sigma^2} \,. \tag{2.212}$$

As $\sigma^2 \to 0$, $C(m) \sim \frac{d_m}{2} \log \sigma^2$.

Proof The set of $Y_i = \langle Y, \phi_i \rangle$, $i = 1, \ldots, d$ are zero-mean, Gaussian variance $\sigma^2 + \lambda_i^2$, with $Y_i, i = d+1, \ldots, n$ having variance σ^2. The entropy of Y becomes

$$h(Y) = \frac{1}{2} \sum_{i=1}^{d_m} \log 2\pi(\sigma^2 + \lambda_i^2) + \frac{n - d_m}{2} \log 2\pi \sigma^2 \,. \tag{2.213}$$

The conditional entropy $h(Y|X_1, \ldots, X_{d_m})$ becomes

$$h(Y|X_1, \ldots, X_{d_m}) = \frac{n}{2} \log 2\pi \sigma^2 \,. \tag{2.214}$$

Computing their difference gives the result. $\square$

2.9 Building Probability Models via the Principle of Maximum Entropy

As one proceeds in pattern theory we basically have two possible tools for solving inference problems. The first we have already seen, minimum-risk estimation which when given parametrized probability models delivers estimators of the random parametric functions. However, one might ask the fundamental question: *Where do the probability models over which the inference problems are constructed come from.* This involves our second basic tool, gathering context from the patterns and using it to construct probabilistic models which are of maximum entropy. Estimation and model building co-exist; almost without exception the estimation problems are solved in the probability models which are constructed to be of maximum entropy.

2.9.1 Principle of Maximum Entropy

There has, over the past several decades, been a tremendous increase in the application of maximum-entropy techniques to constraint problems with nonunique solutions. The rational has been most elequently formulated by Jaynes [63]: of all candidates consistent with a set of constraints the maximum-entropy (*maxent*) solution is the one which occurs with greatest multiplicity. The success of the entropy function is due to the property that the candidate solutions are concentrated strongly near the maxent one; solutions with appreciably lower entropy are atypical of those specified by the data [64].

Throughout, given is a prior density $p(x), x \in \mathcal{X}, \int_\mathcal{X} p(x)dx = 1$. In *Jaynes principle of maximum entropy*, probability densities $q(x), x \in \mathcal{X}, \int_\mathcal{X} q(x)dx = 1$ are the objects being estimated, and are chosen to maximize the entropy function $-D(q\|p) = -\int_\mathcal{X} q(x) \log (q(x))/(p(x))dx$. Incomplete data observations come in the form of mean values of known functions h with respect to the unknown density $H = \int_\mathcal{X} q(x)h(x)dx$. Jaynes principle is to choose the density which is consistent with the prior and satisfies the incomplete data observables.

Theorem 2.64 (Jaynes Principle of Maximum Entropy) *Given prior $p(\cdot)$ on $\mathcal{X}$, $\int_\mathcal{X} p(x)dx = 1$, with moment constraints via observation functions $h_m, m = 1, \ldots, M$ with expected values $H_m = E_q\{h_m\}, m = 1, \ldots, M$, then the unique density maximizing the entropy (minimizing Kullback–Leibler distance) $-D(q\|p)$ satisfies*

$$\hat{q}(\cdot) = \underset{q(\cdot):\int_\mathcal{X} q(x)dx=1}{\arg\max} -\int_\mathcal{X} q(x) \log \frac{q(x)}{p(x)} dx, \text{ subject to } H_m$$

$$= \int_\mathcal{X} q(x) h_m(x) dx, \quad m = 1, \ldots, M \qquad (2.215)$$

$$= e^{\lambda_0 + \sum_{m=1}^M \lambda_m h_m(\cdot)} p(\cdot). \qquad (2.216)$$

Proof For this use the proof of Cover and Thomas [7]. Take $\hat{q}(\cdot) = e^{\lambda_0 + \sum_{m=1}^M \lambda_m h_m} p(\cdot)$, and examine any other q satisfying the moment-constraints which satisfy the following set of inequalities:

$$-D(q\|p) = -\int_\mathcal{X} q(x) \log \frac{q(x)}{p(x)} dx = -\int_\mathcal{X} q(x) \log \frac{q(x)}{\hat{q}(x)} dx - \int_\mathcal{X} q(x) \log \frac{\hat{q}(x)}{p(x)} dx$$

$$(2.217)$$

$$= -D(q\|\hat{q}) - \int_\mathcal{X} q(x) \log \frac{\hat{q}(x)}{p(x)} dx \qquad (2.218)$$

$$\overset{(a)}{\leq} -\int_\mathcal{X} q(x) \left(\lambda_0 + \sum_{m=1}^M \lambda_m h_m\right) dx = -\int_\mathcal{X} \hat{q}(x) \left(\lambda_0 + \sum_{m=1}^M \lambda_m h_m\right) dx$$

$$(2.219)$$

$$= -\int_\mathcal{X} \hat{q}(x) \log \frac{\hat{q}(x)}{p(x)} dx = -D(\hat{q}\|p), \qquad (2.220)$$

where (a) follows from $\log x \geq 1 - (1/x)$, with equality attained if $q = \hat{q}$ almost everywhere in $\mathcal{X}$. □

2.9.2 Maximum Entropy Models

Almost all of the distributions we shall study will be of maximum-entropy. Maximum-entropy distributions are tilted, tilted by the expectations or moment constraints of the observable functions h.

The probability models we use are maximum entropy.

Theorem 2.65

1. *Independent processes:* Let $X_1, \ldots, X_N$, be finite valued, $X_i \in \{1, 2, \ldots, J\}$. For first order marginals which are stationary, $E\{1_j(X_n)\} = p_j$, $j = 1, \ldots, J$, the maxent distribution is the product law of independence

$$P(X_1, \ldots, X_N) = \prod_{n=1}^{N} P(X_n) ; \qquad (2.221)$$

and the sequence $X_1, \ldots, X_n$ is i.i.d.

2. *First order Markov:* Given joint stationary marginals, $E\{1_{ij}(X_n, X_{n+1})\} = Q_{i,j}$, $i, j = 1, \ldots, J$, the maxent distribution is 1st order Markov; with initial distribution $\Pr(X_1 = j) = \pi(j)$,

$$P(X_1, \ldots, X_N) = \prod_{n=2}^{N-1} P(X_{n+1}|X_n)\pi(X_1) . \qquad (2.222)$$

3. *Multivariate Gaussian:* Let X be real valued with first two moments $E[X] = \mu$, $EX^2 = \alpha$, the maximum entropy density is Gaussian with mean and variance $\mu, \alpha - \mu^2$.
   Let $X_1, \ldots, X_N$, be real valued, then given mean and covariance constraints $E\{X_n\} = \mu_n, E\{(X_m - \mu_m)(X_n - \mu_n)\} = K_{mn}$, the maximum entropy density is multivariate normal:

$$p(X_1, \ldots, X_N) = \frac{1}{(\sqrt{2\pi})^n} \det{}^{1/2} \Lambda e^{-(1/2) \sum_{m,n=1}^{N} X_m X_n \lambda_{mn}} \qquad (2.223)$$

$$\text{where } \Lambda = K^{-1} = \left(K_{ij}\right)^{-1} . \qquad (2.224)$$

4. *Whitening Model:* Let $X = X_1, \ldots, X_N$, be a real valued zero-mean process, $L = (L_{ij})$ an $n \times n$ nonsingular matrix. Then given second moments $E|LX_i|^2 = \sigma^2$ for all $i = 1, \ldots, n$, $LX_i = \sum_j L_{ij} X_j$, then X is a Gaussian vector process satisfying the equation

$$LX_i = W_i, \quad i = 1, \ldots, n, \qquad (2.225)$$

$W = $ "white Gaussian noise" variance σ^2.

5. *Exponential Interarrivals and Poisson density:* Let $N_{[0,t)}, t \geq 0$ be a counting process with arrival times $W_0 = 0, W_1, W_2, \ldots$ and interarrival times $T_1 = W_1, T_2 = W_2 - W_1, \ldots$. Then, if $E(T_i) = 1/\lambda$, then the maximum entropy density for waiting times are exponential

$$p(T_i) = \lambda e^{-\lambda T_i} . \qquad (2.226)$$

The counting process $N_{[0,t)}$ is Poisson with density

$$P(N_{[0,t)}) = \frac{e^{-\lambda t}(\lambda t)^{N_{[0,t)}}}{N_{[0,t)}!} . \qquad (2.227)$$

Proof *Marginal Independence:*

$$P(X_1,\ldots,X_N) = e^{\lambda_0 + \sum_{j=1}^{J} \sum_{n=1}^{N} \lambda_j(n) 1_j(X_n)} \qquad (2.228)$$

$$= e^{\lambda_0} \prod_{n=1}^{N} e^{\lambda_{X_n}(n)} . \qquad (2.229)$$

This is a product density, and choosing $e^{\lambda_j(n)} = P(j)$ satisfies the constraints.
Joint marginal Markov chain: To prove the Markov property,

$$P(X_N|X_1,\ldots X_{N-1}) = \frac{P(X_1,\ldots,X_N)}{\sum_{X_N=1}^{J} P(X_1,\ldots,X_N)}$$

$$= \frac{\prod_{n=1}^{N-1} e^{\lambda_0} e^{\lambda_{X_n,X_{n+1}}(n)} \pi(X_1)}{\sum_{X_N=1}^{J} \prod_{n=1}^{N-1} e^{\lambda_0} e^{\lambda_{X_n,X_{n+1}}(n)} \pi(X_1)} \qquad (2.230)$$

$$= \frac{e^{\lambda_{X_{N-1},X_N}(N-1)}}{\sum_{X_N=1}^{J} e^{\lambda_{X_{N-1},X_N}(N-1)}} . \qquad (2.231)$$

This is a function of X_{N-1}, X_N implying the first order Markov property.
Multivariate Gaussian:

$$p(X_1,\ldots,X_N) = e^{\lambda_0 + \sum_{n=1}^{N} \lambda_n X_n + \sum_{m=1}^{N} \sum_{n=1}^{N} \lambda_{mn}(X_m - \mu_m)(X_n - \mu_n)} . \qquad (2.232)$$

Whitening Process: The second moment constraints give

$$p(X_1,\ldots,X_n) = \frac{1}{Z(\lambda)} \prod_{i=1}^{n} e^{-\lambda_i (LX_i)^2} , \qquad (2.233)$$

and choosing the $\lambda_i = 1/2\sigma^2$ gives the Gaussian density. Choosing $W_i = LX_i$ gives the product law for W implying it is a white process, which is Gaussian (linear function of X).

Poisson Counting Process: The expected value on waiting times being given by the mean $E(T) = 1/\lambda$ implies the maximum entropy density is of the form $p(T) = \lambda e^{-\lambda T}$. Independence of interarrivals follows from the maximum entropy principle giving a product density

$$p(T_1,\ldots,T_n) = \lambda^n \prod_{i=1}^{n} e^{-\lambda T_i} . \qquad (2.234)$$

This is sufficient to imply the process is the number of arrivals is Poisson with parameter λ (see Snyder [65]). Defining the event times to be W_i, then

$$\Pr(N_{[0,t)} = n) = \Pr(W_n < t, W_{n+1} \geq t) = \int_0^t \int_t^\infty \lambda e^{-\lambda(y-x)} \frac{\lambda^n x^{n-1} e^{-\lambda x}}{(n-1)!} dx\, dy \qquad (2.235)$$

$$\stackrel{(a)}{=} \int_0^t \int_t^\infty \frac{\lambda^{n+1} x^{n-1} e^{-\lambda y}}{(n-1)!} dx\, dy = \frac{\lambda^n e^{-\lambda t}}{n!} , \qquad (2.236)$$

where (a) follows from the Gamma distribution on the n-th interarrival. □

2.9.3 Conditional Distributions are Maximum Entropy

Maximizing entropy has played an important role in the solution of problems in which the measurements correspond to moment constraints on some many-to-one mapping $h(\cdot)$. We now explore its role in estimation problems in which the measured data are statistical observations defined via many-to-one maps delimiting the domain of the density of inference, the so called *complete-incomplete* data problems. We conclude the density maximizing entropy is identical to the conditional density of the complete data given the incomplete data. The principle of maximum entropy is consistent with the rules of formal conditional probability for complete and incomplete data problems. Such a view is informative as it gives rise to the iterative approaches popularized by Dempster, Laird and Rubin [12] and Csiszar and Tusnady [13] already examined in the EM algorithm section. This equivalence results by viewing the measurements as specifying the domain over which the density is defined, rather than as a moment constraint on $h(\cdot)$.

Throughout, a function θ is estimated which parametrizes a known probability density; the actual data (denoted as the *complete-data*) described by the density are not observed. Rather, observations consist of data (denoted as *incomplete-data*) which nonuniquely specifies the complete-data via some set of many-to-one mappings.

Theorem 2.66 *Given the observable incomplete data $y \in \mathcal{Y}$ specified via the many-to-one mapping $y = h(x)$ on the complete data x a particular realization in $\mathcal{X}$, then the maximum-entropy density satisfying domain constraint $\int_{\chi(y)} q(x)dx$, $\chi(y) = \{x : h(x) = y\}$ is the conditional density of x given y:*

$$\hat{q}(\cdot) = \underset{\{q:\int_\mathcal{X} q(x)dx=1\}}{\arg\max} -\int_\mathcal{X} q(x) \log \frac{q(x)}{p(x)} dx \quad \text{subject to} \int_{\chi(y)} q(x)dx = 1. \quad (2.237)$$

$$\stackrel{(a)}{=} \frac{p_\theta(\cdot)}{\int_{\chi(y)} p_\theta(x)dx} \stackrel{(b)}{=} k_\theta(\cdot | x \in \chi(y)). \quad (2.238)$$

Proof Using the $\log x \geq 1 - 1/x$ inequality with equality $x = 1$ almost everywhere gives (a) with Bayes rules Eqn. 2.147 giving (b). □

Remark 2.9.4 [Conditional probability] The thesis work of Musicus [14] was the first to reveal the maximum entropy connection to one of the authors. The 1981 paper of Van Campenhout and Cover [66] is one of the more complete developments on the fundamental connection between conditional probability and the Jaynes principle. Viewing the measurements y as determining the domain over which the density is defined, rather than as a moment constraint, then the density closest to the prior f in the cross-entropy sense is the conditional density of x given y. By way of Van Campenhout and Cover, choose one particularly simple (their results are more general) many-to-one mapping for the incomplete data and apply their Theorems I and II. Assume $x_1, \ldots, x_N$ are the complete-data which are independent, identically distributed discrete random variables with mass function $p(x)$ on the range $x \in 1, 2, \ldots, m$. Given the incomplete data $y = \sum_{i=1}^N x_i$, the conditional probability of $x_1, x_2, \ldots, x_N$ given y is

$$p(x_1 = j_1, \ldots, x_N = j_N | y) = \frac{p(x_1 = j_1, \ldots, x_N = j_N)}{\sum_{(j_1, \ldots, j_N): j_1 + \cdots + j_N = y} p(x_1 = j_1, \ldots, x_N = j_N)},$$

$$(2.239)$$

and the independence of the x'_is it follows that

$$p(x_1 = j_1 | y) = p(x_1 = j_1) \left(\frac{\sum_{(j_2,\ldots,j_N): j_2+\cdots+j_N=y-j_1} \prod_{i=2}^{N} p(x_i = j_i)}{\sum_{(j_1,\ldots,j_N): j_1+\cdots+j_N=y} \prod_{i=1}^{N} p(x_i = j_i)} \right). \qquad (2.240)$$

For $N \to \infty$, by Theorems I and II of Van Campenhout and Cover, the above density converges to the *maxent* one of 2.216, with $h(x)$ replaced by x.

3 PROBABILISTIC DIRECTED ACYCLIC GRAPHS AND THEIR ENTROPIES

ABSTRACT Probabilistic structures on the representations allow for expressing the variation of natural patterns. In this chapter the structure imposed through probabilistic directed graphs is studied. The essential probabilistic structure enforced through the directedness of the graphs is **sites are conditionally independent of their nondescendants given their parents**. The entropies and combinatorics of these processes are examined as well. Focus is given to the classical Markov chain and the branching process examples to illustrate the fundamentals of variability descriptions through probability and entropy.

3.1 Directed Acyclic Graphs (DAGs)

Pattern theory builds complex structures from simpler ones. The structure is imposed via graph structures, directed and undirected, defining how the variables interact. Begin with discrete state spaces and for discrete graphs.[11]

The basic structures which glue the patterns together are the graphs, directed and undirected. Figure 3.1 depicts the contrast between directed (left panel) and undirected graphs (right panel). The patterns will be constructed via the graphs defined by their domain of sites D and edge system E. The graph structure is denoted using the shorthand notation $\sigma = \{D, E\}$.

We begin our study with the less general **directed graphs**, in which the arcs associated with the sites have a specific direction. For many of the examples, the finite state and tree graphs for languages, there is a natural root node associated with the start state or sentence state from which the direction is implicitly defined. Many graphs will be partially ordered (POSET graphs) in which the order is not total, but is transitive only. For this associate with the directed graph $\sigma = \{D, E\}$ an order defined by the set of edges, so that the edges $e \in E$ are directed. For directed graphs, assume at least a partial ordering, depicted via $<$ or $>$, so that two sites which have an edge between them are either $>$ or $<$ than one another. The ordering is denoted graphically via a directed arc, so that if there exists an edge $e \in E$ in the graph $\overset{i\ \ e\ \ j}{\bullet\text{---}\bullet}$ with direction of the arrow $\overset{i\ \ \ \ \ j}{\bullet\rightarrow\bullet}$ implying the ordering $i < j$, and $\overset{i\ \ \ \ \ j}{\bullet\leftarrow\bullet}$ implying the ordering $j < i$.

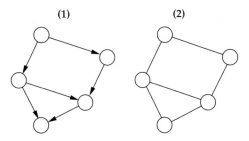

Figure 3.1 Panel 1 shows a directed graph consisting of sites D and edges E; panel 2 shows an undirected graph in which all edges have an orientation with no cycles.

---

[11] Standard measurability is assumed with respect to the underlying probability space $(\Omega, \mathcal{A}, P)$, with Ω being the set of sample points and $\mathcal{A}$ the sigma-algebra of events containing the P-measurable sets. The random processes on uncountable index sets will be studied in the following chapters and will require more careful examination of properties of measurablity, separability and continuity. Throughout the underlying measure theoretic issues are suppressed.

The left panel 1 of Figure 3.1 depicts a directed graph, with sites and edges with orientations which define a partial ordering. The right panel shows an undirected graph in which the edges have no direction.

Throughout these oriented sites, **paths** and **cycles** can be defined.

Definition 3.1 *A n-***path** $p_n(i,j)$ *on a directed graph connecting sites i,j is a list of $n+1$ sites $i_0, i, \ldots, i_n$, with $i_0 = i, i_n = j$ such that there exists oriented pairs $\overset{i_j}{\bullet} \to \overset{i_{j+1}}{\bullet}$ of sites $(i_j, i_{j+1}), j = 0, \ldots, n-1$.*

A **cycle** *in the directed graph σ is an n-path $p_n(i,j)$ for some n with $i = j$.*

Such paths induce the partial ordering $>$, then $j > i$ if there exists a path connecting i to j. This is a only a **partial ordering**, with the relation $>$ being transitive only, so that if $i > j$ and $j > k$ then this implies $i > k$. **Parents and descendents** can be defined as well.

Definition 3.2 *The* **parents** *of a site $i \in D$, denoted $\Pi_i = \{j \in D : \overset{j}{\bullet} \to \overset{i}{\bullet}\}$, are the sites connected via directed edges emanating from them. The* **descendents** *of a site are the set of nodes in the graph for which there exists finite length paths from the site to the node which respect the directed arcs:*

$$\Delta_i = \{j \in D : \exists p_n(i,j) \text{ for some } n \text{ finite}\}. \tag{3.1}$$

Probabilities will be put on acyclic graphs, implying the directed graphs will have no cycles.

Definition 3.3 *In summary, the graph $\sigma = \{D, E\}$, is a* **directed graph** *with* **parent system** $\Pi = \cup \Pi_i$ *and* **descendent system** $\Delta = \cup_i \Delta_i$, *if (1) $i \notin \Pi_i$, (2) $i \in \Pi_j \implies \exists e \in E$ with $\overset{i}{\bullet} \overset{e}{\to} \overset{j}{\bullet}$, (3) $i \in \Delta_j \implies \exists p_n(j,i)$ for some n finite.*

It is an **acyclic DAG** *if $\nexists p_n(i,j)$ for any n with $i = j$ for any $(i,j) \in D$.*

There are many examples of directed graphs, and here are three.

1. Panel 1 of Figure 3.2 shows a **linear graph**, $\sigma = LINEAR$, in which all the sites are ordered, with identically one parent for each internal site, and all sites are descendents and parents except for the special **root** and **leaf** nodes which have no parents and no children, respectively. For this reason, $\sigma = LINEAR$ has a total ordering. For all $i, j \in D$, either $i > j$ or $i < j$. We shall call a graph r-**linear** if internal sites have i-parents, $\Pi_i = \{i - r, i - r + 1, \ldots, i - 1\}$.

2. Panel 2 of Figure 3.2 shows a more general **tree graph** $\sigma = TREE$, in which all internal nodes have identically one parent. Each node internal to the tree has multiple children, the number called the **branching degree**. For this reason, $\sigma = TREE$ has only a partial ordering which is transitive.

3. Panel 3 of Figure 3.2 shows the most general direct acyclic graphs which we will be studying, termed a POSET, partially ordered set. Note that the nodes may have multiple parents.

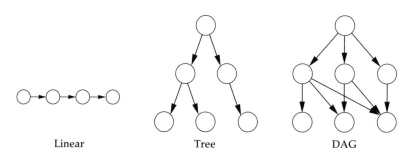

Linear   Tree   DAG

Figure 3.2 Left panel shows a directed graph LINEAR associated with a Markov chain; middle panel a graph type TREE associated with a random branching process; right panel most general DAG directed acyclic graph.

3.2 Probabilities on Directed Acyclic Graphs (PDAGs)

There is an exquisite connection of the picture associated with the graph showing arcs connecting the nodes and particular conditional independence structure providing natural and simple factorizations. This launches us into the notion of *directed acyclic graphs* and familiar probabilistic models such as *Markov chains, random branching processes* and most generally *probabilities on directed acyclic graphs (DAGs)*. For probabilities on DAGs we follow the influential paper of Dawid [67] and book by Pearl [68].

Begin assuming finite graphs $\sigma - \{D, E\}$ with a number of sites $n(\sigma) = n$. Assign to each point in the graph one of a finite set of random variables, $\{X_i, i \in D\}$ denoting the family of random variables indexed over D, each $X_i \in \mathcal{X}_0$ a finite state space. Define the set of all possible configurations $\mathcal{X} = \mathcal{X}_0^n$. To assign probability distributions to the set of all possible configurations $\mathcal{X}$, denote by X_G the configuration X restricted to the subgraph $G \subset D$:

$$X_G = \{X_i, i \in G \subset D\}. \tag{3.2}$$

For directed acyclic graphs, the full distribution P on $\mathcal{X}$ is determined by the conditional probability of a site i given its parents $P(X_i|X_j, j \in \Pi_i)$. Essentially, sites are conditionally independent of nondescendent nodes in the graph given their parents. This is the splitting property of directed graphs.

Definition 3.4 X *has* **realizations from a probabilistic directed acyclic graph (PDAG)** $\sigma = \{D, E\}$ *if for all nodes $i \in D$,* **parents split sites with their nondescendents***:

$$P(X_i, X_{D\setminus(i\cup\Delta_i)} | X_{\Pi_i}) = P(X_i | X_{\Pi_i})P(X_{D\setminus(i\cup\Delta_i)} | X_{\Pi_i}). \tag{3.3}$$

Notice, for all $i \in D$ conditional independence Eqn. 3.3 is equivalent to

$$P(X_i | X_j, j \notin (i \cup \Delta_i)) - P(X_i | X_j, j \in \Pi_i). \tag{3.4}$$

This provides us with a product of conditionals which is so familiar!

Theorem 3.5 *Given a directed acyclic graph $\sigma = \{D, E\}$ with parent system $\Pi = \cup_i \Pi_i$ and $n(\sigma) = n$ sites, then*

$$P(X_1, X_2, \ldots, X_n) = \prod_{i=1}^{n} P(X_i | X_{\Pi_i}). \tag{3.5}$$

Proof The proof involves recursively peeling the leaf nodes off the graph $\sigma = \{D, E\}$ by defining the sequence of graphs $\sigma_0 = \sigma, \sigma_1, \ldots$, with sites $D_0 = D \supset D_1 \supset \ldots$, the sites D_j from the graph σ_j with all of the leaf nodes removed. For any finite graph with $n(\sigma)$ sites, there are at most $j = 0, 1, \ldots, \max < n(\sigma)$ such peelings. See Figure 3.3.

Define the set of leaf nodes of each graph σ_j to be $L_j \subset D_j$, with $L_0 \subset D$ the leaf nodes for $D_0 = D$. Then, define the sites and extended parent system recursively according to

$$D_j = D_{j-1} \setminus L_{j-1}, \qquad D_0 = D, \tag{3.6}$$

$$\tilde{\Pi}_{L_{j-1}} = \cup_{i \in L_{j-1}} \Pi_i, \qquad \tilde{\Pi}_0 = \cup_{i \in L_0} \Pi_i, \tag{3.7}$$

for $j = 1, 2, \ldots, m = \max$, with $D_{\max+1} = \emptyset \implies \tilde{\Pi}_{L_{\max}} = \emptyset$.

This gives

$$P(X_D) = P(X_{L_0}|X_{D\setminus L_0})P(X_{D\setminus L_0}) \tag{3.8}$$

$$= P(X_{L_0}|X_{D_0\setminus L_0})P(X_{L_1}|X_{D_1\setminus L_1})P(X_{D_1\setminus L_1}) = \cdots \tag{3.9}$$

$$= \prod_{j=0}^{\max} P(X_{L_j}|X_{D_j\setminus L_j}). \tag{3.10}$$

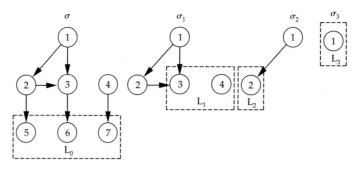

Figure 3.3 Showing a sequence of graphs generated by sequentially peeling the leaves (those sites with children).

Clearly $\tilde{\Pi}_{L_j} \subset D_j \setminus L_j$, and since the elements of the leaf set L_j are nondescendents of each other, they are conditionally independent given their parent set:

$$P(X_D) = \prod_{j=0}^{\max} P(X_{L_j}|X_{\tilde{\Pi}_{L_j}}) = \prod_{j=0}^{\max} \prod_{i \in L_j} P(X_i|X_{\Pi_i}). \tag{3.11}$$

□

Example 3.6 (Bayesian Belief Networks) A beautiful example is illustrated via Bayesian belief networks in which the causes and effects of events of interest in *expert systems* can be characterized probabilistically using *Bayesian belief networks* (see [69], e.g.). Belief networks are an example of general pattern theoretic graphs. The sites in the graph represent hypotheses. A directed arc from site X to site Y indicates that Y is probabilistically dependent on X. These dependences are characterized by conditional distributions. Sites with no in-bonds are characterized by their marginal distributions. Independence is assumed wherever it is consistent with the specified conditional distributions. For instance, if the graph indicates that X causes Y and Y causes Z, but there is no direct connection between X and Z, then X and Z are conditionally independent given Y. Note that the causal relationships may be positive, i.e. X true means Y is likely true, or negative, i.e. X true means Y is likely false.

Figure 3.4 shows an example from Cowell et al. [70] involving binary hypotheses. Suppose we wish to infer whether a patient has tuberculosis, lung cancer, or bronchitis, given the knowledge of possible causes of these conditions, such as smoking or a recent visit to Asia, and possible symptoms, such as a positive chest X-ray and dyspnoea. This would correspond to determining the probability distributions of sites T, L, and B, conditioned on the knowledge of A, S, X, and D. E enters as a nuisance variable. The joint probability of the configuration is given by

$$P(ASTLBEXD) = P(X|E)P(D|ESB)P(E|TL)P(T|A)P(L|S)P(A)P(S)P(B) \tag{3.12}$$

Remark 3.2.0 *Why restrict to acyclic graphs?* Essentially, cycles imply trivial independence. Let $X = (X_1, X_2, \ldots, X_n)$, since each site is a descendant.

Example 3.7 (Computational Biology) The advent of gene expression microchips and similar devices has recently opened a number of new avenues for investigators in medicine and biology. The elucidation of part of the intricacies of the intracellular mechanisms that lead to a given cellular state is one such example.

Biological information is stored as genes in the form of deoxyribonucleic acids (DNA). Genes from DNA can be read and copied (in a process termed *transcription*)

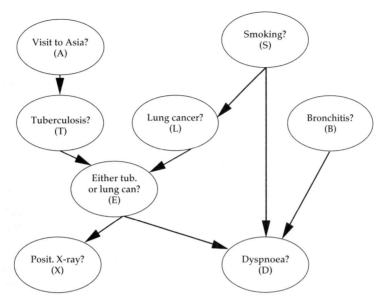

Figure 3.4 Bayesian belief network for medical diagnosis. From Cowell et al. [70].

into molecules called messenger ribonucleic acids (mRNA). mRNAs then serve as the blueprints to build proteins (in a process called *translation*), and these proteins are responsible for almost all biological functions.

The state of a cell can thus be thought of as the set of concentrations of all mRNAs and proteins as well as a variety of other molecules or ions described generally as metabolites. Now, genes and their expressed products (mRNAs, proteins) may interact with one another, either positively or negatively. That is, the expression of certain genes may activate the expression of certain genes while repressing that of others. The mapping of these interactions is referred to as the genetic regulatory network.

DNA microarray technologies enable one to acquire data related to the concentrations of the mRNAs for large numbers of genes in a given biological preparation. The challenge is then to find ways to interpret this information in the hope of identifying the nature of the interactions, if any, between genes and possibly even to quantify these. Among many other approaches that have been considered, directed acyclic graphs have been used. The motivation in using Bayesian networks rather than other approaches resides in their ability to integrate interactions patterns much more complex than pairwise, their robustness to noise (a major concern with gene expression data) and their ability to factor in latent or hidden variables.

In the case of galactose metabolism in *S. cerevisiae* (yeast), candidate DAGs of interactions between five mRNAs and two proteins were screened for the one that best accounts for the observed data. This work is by Hartemink et al. [71]. Two of the possible graphs of structures which are plausible candidates from a biological standpoint are shown in Figure 3.5. They differ by a single edge, that is whether Gal80p directly interacts with Gal4p or does so only via Gal4m. The graphs being directed and acyclic, the joint probability distribution can be factored into the product of the conditional probability distributions of each variable, given its parents.

$$P(X_D) = \prod_{i \in D} P(X_i | X_{\Pi_i}). \tag{3.13}$$

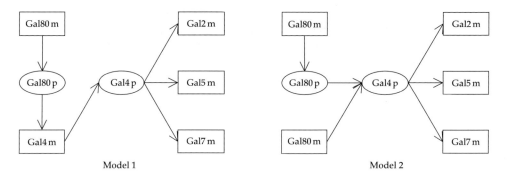

Figure 3.5 Genetic regulatory network candidate for yeast; taken from Hartemink et al. [71].

Now, each candidate graph X_D can be scored with the available data Y according to the conditional log-posterior

$$\log P(X_D|Y) \propto \log \pi(X_D) + \log P(Y|X_D), \tag{3.14}$$

where $\log \pi(X_D)$ and $\log P(Y|X_D)$ are the log prior distribution and log likelihood of realization Y given X_D. The evaluation of these terms and the choice of prior may vary, particularly with the size of the graph of interest. Hartemink et al. [71] have been able to correctly infer the most generally agreed upon candidate genetic regulatory networks (assessed through different biological means) using this method and yeast gene expression data obtained using Affymetrix's GeneChip technology.

3.3 Finite State Markov Chains

Finite state Markov chains are a classic example of probabilistic directed acyclic graphs which are of greatest familiarity. The special structure associated with *Markov chains* is that parents have at most one child, and children have only one parent in their parent system. Thus the Markov chain sites inherit a complete ordering $>$ on the site set $D \subset \mathbb{Z}$ the integers, with $\forall i,j \in \mathbb{Z}, i < j$ or $j < i$. When the sites are totally ordered, the global probability is rewritten as products of local probabilities in a familiar manner.

Begin by associating with $D \subset \mathbb{Z}$ the natural complete ordering $1 < 2 < \cdots < n \ldots$. Examine the left panel of Figure 3.2. This corresponds to Markov processes having a special structure implying that the *past* and *future* are conditionally independent given the present. This is a property which shall be exploited with the total ordering associated with the integers.

Definition 3.8 *A finite state $\mathcal{X}_0$-valued random process $X_1, X_2, \ldots$ is* **a Markov chain** *with initial distribution $\pi(\cdot)$ if for $i = 1, 2 \ldots$*

$$P(X_i|X_{i-1}, \ldots, X_0) = P(X_i|X_{i-1}) \tag{3.15}$$

with $P(X_0) = \pi(\cdot)$ for all $X_0, X_1, \ldots, X_n \in \mathcal{X}_0$.

The Markov chain is said to be **time-invariant** *with* **1-step state transition probabilities** $Q = \left(Q_{jk}\right)$, *with $\sum_k Q_{jk} = 1$, if for each j, and all times n,*

$$\Pr\{X_n = k | X_{n-1} = j\} = Q_{jk}. \tag{3.16}$$

With the conditional independence of Eq. 3.15 in hand, the familiar approach of rewriting joint probabilities as products of conditionals becomes available. The stationary Markov process is characterized by the initial distribution $\pi(\cdot)$ on X_0, and the transition distribution between states. Now lets connect to the splitting X_0, and the 1-step transition property.

Theorem 3.9 *Then $X_i, i = 0, 1, \ldots$ is a finite-state Markov chain with the joint probability given by*

$$P(X_0, \ldots, X_n) = \prod_{i=1}^{n} P(X_i|X_{i-1})\pi(X_0), \qquad (3.17)$$

if and only if it satisfies the **splitting property** *with the past and future conditional independence, i.e. for all ordered combinations of n-times $j = 0, \ldots, n$ and all j, then*

$$P(X_0, \ldots, X_{j-1}, X_{j+1}, \ldots, X_n|X_j) = P(X_0, \ldots, X_{j-1}|X_j)P(X_{j+1}, \ldots, X_n|X_j). \qquad (3.18)$$

Proof Using the splitting property gives

$$P(X_0, \ldots, X_{n-2}, X_n|X_{n-1}) = P(X_0, \ldots, X_{n-2}|X_{n-1})P(X_n|X_{n-1}) \qquad (3.19)$$

$$\frac{P(X_0, \ldots, X_n)}{P(X_{n-1})} = \frac{P(X_0, \ldots, X_{n-1})P(X_n|X_{n-1})}{P(X_{n-1})}. \qquad (3.20)$$

Continuing in this manner gives $P(X_0, \ldots, X_n) = \prod_{i=1}^{n} P(X_i|X_{i-1})\pi(X_0)$.

The product condition, Eqn. 3.17 implies the splitting property Eqn. 3.18 according to

$$P(X_0, \ldots, X_{j-1}, X_{j+1}, \ldots, X_n|X_j) = \frac{P(X_0, \ldots, X_n)}{P(X_j)} = \frac{\prod_{i=1}^{n} P(X_i|X_{i-1})\pi(X_0)}{P(X_j)}$$

$$= \frac{\prod_{i=1}^{j} P(X_i|X_{i-1})\pi(X_0)}{P(X_j)} \prod_{i=j+1}^{n} P(X_i|X_{i-1})$$

$$= \frac{P(X_0, \ldots, X_j)}{P(X_j)} P(X_{j+1}, \ldots, X_n|X_j)$$

$$= P(X_0, \ldots, X_{j-1}|X_j)P(X_{j+1}, \ldots, X_n|X_j).$$

$\square$

The factoring property results in the Chapman–Kolmogorov equations. In the time-invariant case, it is natural to introduce the matrix representing the multi-step transition probabilities. Define $(Q^i)_{jk}$ to be the j, k entry of the ith power.

Corollary 3.10 *The Chapman–Kolmogorov equations take the form, for all $1 \leq i \leq n$, then*

$$P(X_n|X_0) = \sum_{x \in \mathcal{X}_0} P(X_n|X_i = x)P(X_i = x|X_0). \qquad (3.21)$$

For the time invariant case, then

$$P(X_n|X_0) = (Q^n)_{X_0, X_n} = \sum_{x \in \mathcal{X}_0} (Q^{n-j})_{X_0, x} (Q^j)_{x, X_n}. \qquad (3.22)$$

Proof Conditional probability implies

$$P(X_n|X_0) = \sum_{x \in \mathcal{X}_0} P(X_n, X_i = x|X_0) = \sum_{x \in \mathcal{X}_0} P(X_n|X_i = x, X_0)P(X_i|X_0), \qquad (3.23)$$

from which using the splitting property of Markov chains gives the result.

For the time-invariant case use induction. For $n = 1$ it is clearly true, assume $n = k - 1$ is true, then for $n = k$ it follows by writing the probability of the event

$$P(X_k|X_0) = \sum_x P(X_k, X_{k-1} = x|X_0) = \sum_x P(X_k|X_{k-1} = x, X_0) P(X_{k-1} = x|X_0) \tag{3.24}$$

$$\stackrel{(a)}{=} \sum_x P(X_k|X_{k-1} = x) P(X_{k-1} = x|X_0) \tag{3.25}$$

$$\stackrel{(b)}{=} \sum_x P(X_1|X_0) P(X_{k-1} = x|X_0) = \sum_x (Q^{k-1})_{X_0 x} Q_{x X_1}, \tag{3.26}$$

with (a) following from the Markov property, and (b) following from time-invariance. Doing this j-times, gives the general results. □

An *m*-**memory Markov chain** simply has the splitting property of an m-linear graph.

Definition 3.11 *Let $D \subset \mathbb{Z}$ be finite subsets of the integers. Then a discrete time, finite state $\mathcal{X}_0$-valued random process $\{X_i, i = 0, 1, \ldots\}$ on the integers is an m-memory Markov chain if for all ordered combinations of n-times and m-conditioning times then*

$$P(X_0, \ldots, X_{j-m+1}, X_{j+1}, \ldots, X_k | X_j, \ldots X_{j-m})$$
$$= P(X_0, \ldots, X_{j-m+1} | X_j, \ldots X_{j-m}) P(X_{j+1}, \ldots, X_k | X_j, \ldots X_{j-m}), \tag{3.27}$$

with marginals adjusted at the boundary to the initial distribution $P(X_0, \ldots, X_{m-1}) = \pi(X_0, \ldots, X_{m-1})$.

Equivalently the product condition follows as in Theorem 3.9:

$$P(X_0, \ldots, X_n) = \prod_{i=m}^n P(X_i|X_{i-1}, \ldots, X_{i-m}) \pi(X_0, \ldots, X_{m-1}). \tag{3.28}$$

We can predict that the directed acyclic graphs with a complete ordering of the type shown in the left panel of Figure 3.2 are Markov chains, i.e. the splitting property of definition 3.8 holds since Theorem 3.5 gives the factored probability for the DAG.

Corollary 3.12 *Let $\{X_i, i \in D\}$ be a finite $\mathcal{X}_0$-valued random directed graph $\sigma = \{D, E\}$. Then with the complete order of the LINEAR graph of panel 1 of Fig. 3.2 with parent system $\Pi = \bigcup_{i=1}^n \{i - 1\}$, then*

$$P(X_i|X_{D \setminus (i \cup \Delta_i)}) = P(X_i|X_{i-1}), \tag{3.29}$$

and is statistically equivalent to a Markov chain satisfying the splitting property with the same transition probabilities.

3.4 Multi-type Branching Processes

We now construct the random branching process model historically termed the multi-type Galton Watson processes (for an extensive development see [72, 73]). Begin with the Markov chain view. Imagine that *particles* can generate additional objects of the same kind, and denote a particle type A. The initial set of particles is called the 0th generation, denoted by Z_0, which have children which are called the first generation, Z_1; their children, are the second generation Z_2, and so on. The simplest

3.4 MULTI-TYPE BRANCHING PROCESSES

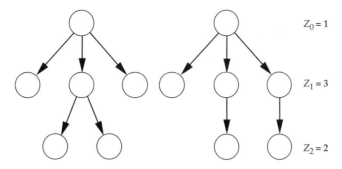

Figure 3.6 Panels showing two family generation trees from the 1-type process with particles A, evolving under different choices of birthing rules. Note that both families have the identical states $Z_0 = 1, Z_1 = 3, Z_2 = 2$.

model first proposed by Galton and Watson has indistinguishable particles with identical birthing laws; only the size of the successive generations are tracked. The more general model will have multiple particles. Shown in Figure 3.6 are two examples of family trees corresponding to particles generating offspring according to various offspring mechanisms, with both family trees having identical generations: $Z_0 = 1, Z_1 = 3, Z_2 = 2$.

We make the following critical assumptions:

1. If the size of the nth generation is known, then the probability law governing later generations does not depend on the size of generations preceeding the nth. Thus, the splitting property holds, and $Z_0, Z_1, \ldots$ forms a Markov chain on the positive integers.
2. Different particles within a generation do not interfere with one another. The number of children born to a particle, does not depend on how many other particles are present in that generation. Nodes are conditionally independent of their non-descendents given their parents.

The 1-step transition function of the Markov chain $P(Z_{n+1}|Z_n)$ which for the 1-type case is determined by a single probability law $p., \sum_{k=0}^{\infty} p_k = 1$, where p_k is the probability that a particle existing in the nth generation has k children in the $n + 1$st generation. The set of transformation laws to particle A takes the form

$$r_k : A \xrightarrow{p_k} \underbrace{A A \ldots A}_{k \text{ times}}, \quad k = 0, 1, \ldots. \tag{3.30}$$

The 1-type processes consist of indistinguishable particles. To accomodate V-distinguishable particles each with their own probabilistic behavior we introduce V−multi-type processes. Such processes arise as genetic types in animal populations, or bacterial types, or throughout as syntactic structures in computational linguistics: *noun-phrases, verb-phrases, determiners* etc.. Enumerate the V particle types $A_1, A_2, \ldots$ with their specific set of transformation (birthing rules) probability laws for generating offspring $p_v, v = 1, 2, \ldots$ the probability laws of transforming particles of type-v. The rules transforming each particle identify the specified number of the various children types. Defining $n_{vk}^{(j)}$ to be the number of type A_j created by substitution for particle v via rule k then

$$r_{vk} : A_v \xrightarrow{p_{vk}} \underbrace{A_1 A_2 A_3 \ldots}_{n_{vk}^{(1)} \text{ of type } A_1, n_{vk}^{(2)} \text{ of type } A_2, \ldots}, \quad \sum_k p_{vk} = 1, \quad v = 1, \ldots, V. \tag{3.31}$$

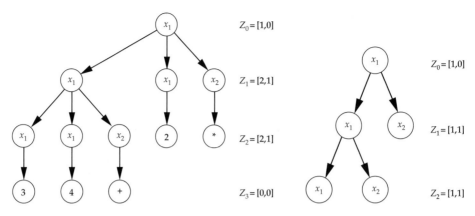

Figure 3.7 Left panel showing an example family tree from the 2-type *reverse polish*, with particles that birth according to the two rules $A_1 \to A_1\, A_1\, A_2$, $A_1 \to \mathbb{N}$, $A_2 \to \{+, *\}$. The derivation is $((3\ 4\ +)\ 2\ *)$ with state sequence $Z_0 = [1, 0], Z_1 = [2, 1], Z_2 = [2, 1]$. Right panel showing a 3-generation realization from a 2-type process, $p_{1,(1,1)} = 1, p_{2,(0,0)} = 1$, demonstrating the recurrence of the state $[1, 1]$ starting from $Z_0 = [1, 0]$.

The left panel of Figure 3.7 shows an example family tree from the 2-type *reverse polish* branching process, with particles birthing according to the rules $A_1 \to A_1\, A_1\, A_2$, $A_1 \to \mathbb{N}$, $A_2 \to \{+, *\}$. Notice, A_1 represents natural numbers, A_2 procedures. The derivation is $((3\ 4\ +)\ 2\ *)$; the state sequence $Z_0 = [1, 0], Z_1 = [2, 1], Z_3 = [2, 1]$.[12]

Associate with each generation the state vector $Z_n = [Z_n^{(1)}, \ldots, Z_n^{(V)}]$ the number of particles of each type in the nth generation. Since particles reproduce independently, the conditional distribution of Z_{n+1} given Z_n is distributed as the sum of independent random variables. The probability law becomes the convolution of probability laws. The state 0 is an absorbing state, so that if $Z_n = 0$, then $Z_{n+1} = 0$ with probability 1. Define the single particle state vectors

$$\mathbf{1}^{(v)} = [0, \ldots, 0, \ldots, \underbrace{1}_{v\text{th entry}}, \ldots, 0].$$

Let $p_{vk}, v = 1, \ldots, V$ be the probabilities that the particles type-v have $k = (k_1, k_2, \ldots, k_V)$ children type $j_1, j_2, \ldots, j_V$; $\sum_{k \in I^{+V}} p_{vk} = 1$. For branching processes, since there can be more than one particle born to any generation, the 1-step probability of transition from one state to another is given by the convolution of the basic single particle transitions.

Definition 3.13 *A* **Galton–Watson 1-type branching process** *is a Markov chain* $Z_n, n = 0, 1, \ldots$ *with infinite state space* $\mathcal{X}_0 = I^+$ *the nonnegative integers, with particle substitution law* $p., \sum_k p_k = 1$, *if for all times n,*

$$\Pr\{Z_{n+1} = k | Z_n = j\} = \begin{cases} (p^{*j})_k & \text{if } j \geq 1, \quad k \in I^+ \\ \delta(k) & \text{if } j = 0, \quad k \in I^+ \end{cases}, \quad (3.32)$$

with p^{*j} the j-fold convolution of the probability $p., \sum_k p_k = 1$.

A **V-multi-type (vector) branching process** $Z_n = [Z_n^{(1)}, \ldots, Z_n^{(V)}], n = 0, 1, \ldots$ *is a Markov chain with infinite state space* $\mathcal{X}_0 = I^{+V}$, *and particle substitution laws* p_v,

---

[12] Throughout this section the state vectors are taken in a row form.

3.4 MULTI-TYPE BRANCHING PROCESSES

$\sum_k p_{vk} = 1, v = 1, \ldots, V$, if for all times n,

$$\Pr\{Z_{n+1} = k | Z_n = j\} = \left((p_1)^{*j_1} * (p_2)^{*j_2} * \cdots * (p_V)^{*j_V}\right)_k \quad \text{if } j \neq 0, \ k \in I^{+V}$$

$$= \delta(k) \quad \text{if } j = 0, \ k \in I^{+V}, \quad (3.33)$$

with $p. * q.$ the convolution of distributions $p., q.$, and $p_.^{*i}$ is the i-fold convolution.[13]

Note that the definition of the transition law assumes the Markov structure of conditional independence of non-descendents given their parents, so that if $Z_n = (j_1, \ldots, j_V)$ then Z_{n+1} is the sum of $j_1 + j_2 + \cdots + j_V$ independent random vectors.

There are two tools for studying the branching processes, the branching rates given by the branching matrix and the moment generating function.

3.4.1 The Branching Matrix

The combinatorial expansion of the number of trees is given by the moments of generations of the process as determined by the mean branching matrix.

Definition 3.14 *The **mean branching matrix** is the positive $V \times V$ matrix $M = (M_{jk})$, defined to be the matrix of first moments:*

$$M_{jl} = E(Z_1^{(l)} | Z_0 = \mathbf{1}^{(j)}) = \sum_k p_{jk} n_{jk}^{(l)}. \quad (3.36)$$

*Define ρ to be the **largest eigenvalue** of M. For the **1-type**, the **mean** is the scalar $M = \rho = E(Z_1 | Z_0 = 1)$.*

If the process has exactly one offspring for each of its substitutions, then M is a stochastic matrix $\sum_l M_{vl} = 1, v = 1, \ldots, V$ corresponding to a Markov chain. Problem A.2 calculates the mean and variance of Z_n given $Z_0 = 1$ for the single type case. More generally, the branching matrix directly determines the expected generations.

Lemma 3.15 *Let $Z_0, Z_1, \ldots$ correspond to a branching process of the V-type. Then,*

$$E(Z_{n+m} | Z_m) = Z_m M^n, \quad n, m = 0, 1, 2, \ldots. \quad (3.37)$$

Proof Using conditioning and the Markov property gives

$$E(Z_{n+m} | Z_m) = E(E(Z_{n+m} | Z_{n+m-1}, Z_m) | Z_m) = E(E(Z_{n+m} | Z_{n+m-1}) | Z_m). \quad (3.38)$$

To compute the $V \times 1$ vector $E(Z_{n+m} | Z_{n+m-1})$ use the fact that for any n,

$$E(Z_n^{(v)} | Z_{n-1}) = \sum_{j=1}^{V} Z_{n-1}^{(j)} \underbrace{\sum_k n_{jk}^{(v)} p_{jk}}_{M_{jv}}, \quad (3.39)$$

implying $E(Z_{n+m} | Z_m) = E(Z_{n+m-1} M | Z_m)$. Doing this n-times completes the proof. □

---

[13] Let $p., q.$ have domains I^{+V}, then

$$(p. * q.)_k = \sum_{j \in I^{+V}} p_j q_{k-j}, \quad (3.34)$$

$$(p_.^{*j})_k = (p. * p_.^{*j-1})_k, \quad (p_.^{*0})_k = \delta(k). \quad (3.35)$$

The branching rates and entropies are determined by the critical eigenvalue of M being greater than 1.

Definition 3.16 *The random branching processes are called* **supercritical, critical, or subcritical** *according to the largest eigenvalue of M, $\rho >, =,$ or < 1, respectively.*

3.4.2 The Moment-Generating Function

Since the state vector involves convolutions of conditionally independent variates, it is natural that the moment or probability generating function plays an important role in studying branching processes.

Definition 3.17 *The* **1-single-type probability generating function** $f(s)$, s *complex, is given by*

$$f(s) = \sum_{k=0}^{\infty} p_k s^k, \quad |s| \leq 1, \tag{3.40}$$

with $f_n(s)$ the generating function of Z_n given $Z_0 = 1$:

$$f_n(s) = \sum_{k=0}^{\infty} \Pr\{Z_n = k | Z_0 = 1\} s^k. \tag{3.41}$$

For the V-multi-type, define the **vth probability generating function** associated with each particle type-$v = 1, \ldots, V$ as

$$f^{(v)}(s_1, s_2, \ldots, s_V) = \sum_{(k_1 \cdots k_V) \in I^{+V}} p_{v,(k_1,\ldots,k_V)} s^{k_1} s^{k_2} \cdots s^{k_V}. \tag{3.42}$$

The generating function of Z_n given $Z_0 = \mathbf{1}^{(v)}$ is defined as

$$f_n^{(v)}(s_1, \ldots, s_V) = \sum_{(k_1 \cdots k_V) \in I^{+V}} \Pr\{Z_n = (k_1, \ldots, k_V) | Z_0 = \mathbf{1}^{(v)}\} s_1^{k_1} \cdots s_V^{k_V}, \tag{3.43}$$

$$f_n(s_1, \ldots, s_V) = (f_n^{(1)}(s), f_n^{(2)}(s), \ldots, f_n^{(V)}(s)). \tag{3.44}$$

The shorthand notation $s^k = s_1^{k_1} s_2^{k_2} \cdots s_V^{k_V}$ shall be employed, so that $f^{(v)}(s) = \sum_k p_{vk} s^k$.

Generating functions provide a fundamental link to extinction as $\Pr\{Z_n = 0 | Z_0 = \mathbf{1}^{(v)}\} = f_{(n)}^{(v)}(0)$. As well the moments of the process are given by the generating functions. Defining $j = \sqrt{-1}$, the moments of the nth generation of the process are given by

$$E(Z_n^{(k_1)} Z_n^{(k_2)} \cdots Z_n^{(k_V)} | Z_0 = \mathbf{1}^{(v)}) = \frac{1}{j^r} \frac{\partial^r}{\partial \omega_{k_1} \cdots \partial \omega_{k_V}} f_{(n)}^{(v)} \left(s = \prod_{k=1}^{V} e^{(j\omega_k)} \right) \bigg|_{\omega=0}. \tag{3.45}$$

Such events as the probability that a random sequence goes to zero, i.e. the particle family goes extinct, and the behavior if the family does not become extinct are characterized by the moment or probability generating function. Watson discovered the basic result that for the single type the nth generating function is the nth iterate.

3.4 MULTI-TYPE BRANCHING PROCESSES

Theorem 3.18 *Defining the nth iterate recursively according to*

$$f_{(1)}(s) = f(s), \quad f_{(n)}(s) = f(f_{(n-1)}(s)), \tag{3.46}$$

then for the 1-type (Watson-1874) the generating function of Z_n given $Z_0 = 1$ becomes

$$f_{n+1}(s) = f(f_{(n)}(s)) = f_{(n+1)}(s), \quad n = 1, 2, \ldots. \tag{3.47}$$

For the V-multi-type, with $f_{(1)}^{(v)}(s) = f^{(v)}(s)$, then

$$f_{n+1}^{(v)}(s_1, \ldots, s_V) = f^{(v)}\left(f_{(n)}^{(1)}(s_1, \ldots, s_V), \ldots, f_{(n)}^{(V)}(s_1, \ldots, s_V)\right). \tag{3.48}$$

Proof The 1-step transition law $\Pr\{Z_n = k | Z_{n-1} = j\}$ is the j-fold convolution of the 1-particle transition probability, implying from the product rule for the transform (see Homework Problem F.2)

$$\sum_k \Pr\{Z_n = k | Z_{n-1} = j\} s^k = (f(s))^j. \tag{3.49}$$

Then

$$\sum_k \Pr\{Z_{n+1} = k | Z_0 = 1\} s^k \stackrel{(a)}{=} \sum_k \sum_m \Pr\{Z_{n+1} = k | Z_1 = m\} \Pr\{Z_1 = m | Z_0 = 1\} s^k$$

$$\stackrel{(b)}{=} \sum_m \Pr\{Z_1 = m | Z_0 = 1\} (f_n(s))^m = f(f_n(s)) \tag{3.50}$$

with (a) coming from Chapman–Kolmogorov and (b) from the m-fold convolution. Applying this recursively completes the first part of the proof.

For the V-multi-type, similar arguments follow:

$$f_{n+1}^{(v)}(s_1, \ldots, s_V)$$

$$= \sum_{k \in I^{+V}} \Pr\{Z_{n+1} = k | Z_0 = \mathbf{1}^{(v)}\} s_1^{k_1} s_2^{k_2} \cdots s_V^{k_V}$$

$$= \sum_{k \in I^{+V}} \sum_{m \in I^{+V}} \Pr\{Z_{n+1} = k | Z_1 = m\} \Pr\{Z_1 = m | Z_0 = \mathbf{1}^{(v)}\} s^k$$

$$= \sum_{m \in I^{+V}} \left(\sum_{k \in I^{+V}} \Pr\{Z_{n+1} = k | Z_1 = m\} s^k\right) \Pr\{Z_1 = m | Z_0 = \mathbf{1}^{(v)}\}$$

$$= \sum_{m \in I^{+V}} \left(f_n^{(1)}\right)^{m_1} \left(f_n^{(2)}\right)^{m_2} \cdots \left(f_{(n)}^{(V)}\right)^{m_V} \Pr\{Z_1 = m | Z_0 = \mathbf{1}^{(v)}\} = f^{(v)}(f_{(n)}(s)).$$

$$\tag{3.51}$$

□

Example 3.19 Let $V = 3$, $\{A_1, A_2, A_3\}$ with

$$\begin{array}{ll} A_1 \xrightarrow{4/7} A_1 A_1 & A_1 \xrightarrow{3/7} A_2 \\ A_2 \xrightarrow{1/35} \phi & A_2 \xrightarrow{16/35} A_1 A_3 \quad A_2 \xrightarrow{18/35} A_2 A_2, \\ A_3 \xrightarrow{3/7} A_2 & A_3 \xrightarrow{4/7} A_3 A_3 \end{array} \tag{3.52}$$

giving the probability generating function and branching matrix

$$f(s_1, s_2, s_3) = \left(f^{(1)}(s) = \frac{4}{7}s_1^2 + \frac{3}{7}s_2, \, f^{(2)}(s) = \frac{1}{35} + \frac{16}{35}s_1 s_3 + \frac{18}{35}s_2^2, \, f^{(3)}(s) = \frac{3}{7}s_2 + \frac{4}{7}s_3^2 \right),$$

$$M = \begin{pmatrix} \frac{8}{7} & \frac{3}{7} & 0 \\ \frac{16}{35} & \frac{36}{35} & \frac{16}{35} \\ 0 & \frac{3}{7} & \frac{8}{7} \end{pmatrix}.$$

3.5 Extinction for Finite-State Markov Chains and Branching Processes

We shall be interested in understanding the circumstances under which strings are finite, and when they are infinitely extended. For this we will have to understand the extinction of the finite length sequences. This has a particularly straightforward characterization for finite-state Markov chains; the typical sequences simply divide as probability 0 or 1. It will be more challenging for branching processes. Essentially, stationary Markov chains do not go extinct (or they would not be stationary); extinction for Markov chains will correspond to the transition matrix Q being sub-stochastic which will be characterized via Perron–Frobenius.

Definition 3.20 *Let $M = (M_{ij}) \geq 0$ be a nonnegative matrix, then it is said to be* **positively regular** *if there exists a finite integer n such that $M^n > 0$ is a strictly positive matrix.*

Clearly if $M = Q$ is the 1-step transition matrix of a Markov chain, then if M^n is strictly positive then there exists an n such that there is a nonzero probability of getting from any state to any other (the relationship to connectedness). Now we will use the well-known results from Perron (1909) and Frobenius (1912)

Theorem 3.21 (Perron–Frobenius) *Let M be positively regular. Then M has a simple, largest eigenvalue ρ, with right and left positive eigenvectors μ, ν, such that*

$$M^n = \rho^n \mu \nu + M_2^n, \quad \nu \mu = 1, \tag{3.53}$$

where $|M_2^n| = O(|\alpha|^n)$[14] with $0 \leq |\alpha| < \rho$, and μ is a column vector and ν a row vector.

See Gantmacher [74, volume 2] or Cinlar [75].

3.5.1 Extinction in Markov Chains

Let $X_1, X_2, \ldots$ be a time-invariant finite-state Markov chain with transitions $Q = (Q_{jk})$. Define extinction by introducing the *absorbing* or *dead* state $D = 0$. Stationarity of course corresponds to extinction probability 0, and the process never entering the terminating-state. For this, define $Z_n = 0$ ($X_n = 0$ the dead state); extinction is $Z_n = 0$ for some n.

Definition 3.22 *The* **probability of extinction** $\gamma = [\gamma^{(1)}, \gamma^{(2)}, \ldots]$ *is the probability that $Z_n = 0$ ($X_n = 0$ in dead-state) for any initial non-terminated state:*

$$\gamma^{(X_0)} = \Pr\{Z_n = 0 \text{ for some } n \mid X_0\}. \tag{3.54}$$

The extinction probability for finite-state Markov chains is either 0 or 1 depending upon whether any of the states are connected to the 0 dead-state; in this case the largest eigenvalue of

---

[14] A function f(n) is $O(\alpha^n)$ if $\lim_{n \to \infty} f(n)/\alpha^n < \infty$.

3.5 EXTINCTION FOR FINITE-STATE MARKOV CHAINS AND BRANCHING

the time-invariant transition out of states is $\rho < 1$. In the stationary case, $\rho = 1$, with transition matrix positively regular.

Theorem 3.23 *Let $X_1, X_2, \ldots$ be a finite state time-invariant Markov chain with dead-state 0 and transition matrix $\tilde{Q}$ of the form*

$$\tilde{Q} = \begin{pmatrix} 1 & 0 \\ \delta & Q \end{pmatrix}, \quad \mathbf{0} = [\underbrace{0\,0\,0\,\cdots\,0\,0}_{V \text{ times}}], \quad \delta = \begin{pmatrix} \delta_1 \\ \vdots \\ \delta_V \end{pmatrix}, \quad (3.55)$$

where $Q = (Q_{jk})$ is positively regular (states connected) with eigenelements $\nu Q = \rho \nu$, $Q\mu = \rho \mu$.

1. *If $\delta \neq 0$ then Q is sub-stochastic with $\rho < 1$ and extinction probability $\gamma = 1$.*
2. *If $\delta = 0$ then Q is stochastic with $\rho = 1$ with extinction probability $\gamma = 0$.*

Proof For $\delta \neq 0$, then

$$\lim_{n \to \infty} \tilde{Q}^n = \lim_{n \to \infty} \begin{pmatrix} 1 & 0 \\ \gamma & \rho^n \mu \nu \end{pmatrix}, \quad (3.56)$$

which if $\gamma \neq 0$ implies $\rho < 1$. Thus from the limit, $\gamma = 1$.
For $\delta = 0$, then

$$\lim_{n \to \infty} \tilde{Q}^n = \lim_{n \to \infty} \begin{pmatrix} 1 & 0 \\ 0 & Q^n \end{pmatrix}, \quad (3.57)$$

implying $\rho = 1$, Q is stochastic, $\gamma = 0$. □

3.5.2 Extinction in Branching Processes

Extinction in branching processes is far more subtle. Certain realizations branch quickly enough that the probability of termination is non-zero. The extinctions probability is determined by the critical maximal eigenvalue ρ of M and the moment generating function. Define the absorbing or terminating state $[0, 0, \ldots, 0]$, with the extinction probability determined by the generating function and its iterates.

Theorem 3.24

1. **1-single-type:** *The extinction probability γ is the limit $\gamma = \lim_{n \to \infty} f_n(0)$. If $\rho \leq 1 \implies \gamma = 1$. If $\rho > 1 \implies 0 \leq \gamma \leq 1$ and γ is the unique non-negative solution of the equation $\gamma = f(\gamma)$.*
2. **V-multi-type:** *If the process is positively regular and not a finite-state Markov chain, then $\rho \leq 1 \implies \gamma = 1$. If $\rho > 1 \implies 0 \leq \gamma \leq 1$, and γ satisfies the equation $\gamma = f(\gamma)$.*

Proof We only prove the 1-type case, see Harris [p. 72, 41] for the V-type vector case.

$$\gamma = \Pr\{Z_n = 0 \text{ for some } n | Z_0 = 1\} = \Pr\{\cup_{n=0}^{\infty} \{Z_n = 0\} | Z_0 = 1\}$$

$$\stackrel{(a)}{=} \lim_{n \to \infty} \Pr\{\cup_{i=1}^{n} \{Z_i = 0\} | Z_0 = 1\}$$

$$\stackrel{(b)}{=} \lim_{n \to \infty} \Pr\{Z_n = 0 | Z_0 = 1\} = \lim_{n \to \infty} f_n(0)$$

$$= \lim_{n \to \infty} f_{(n)}(0).$$

Equalities (a,b) follow from the semi-continuity of probability and the fact that $\{Z_1 = 0\} \subset \{Z_2 = 0\} \subset \cdots$. To see that $\gamma = f(\gamma)$, since $f_{(n+1)}(0) = f(f_{(n)}(0))$ and $\gamma = \lim_{n\to\infty} f_{(n)}(0) = \lim_{n\to\infty} f_{(n+1)}(0)$,

$$\lim_{n\to\infty} f(f_{(n)}(0)) = \lim_{n\to\infty} f_{(n)}(0) = \gamma \tag{3.58}$$

giving $f(\gamma) = \gamma$. That there is a unique solution $0 \leq \gamma \leq 1$ for $\rho > 1$ and that $\gamma = 1$ for $\rho \leq 1$ is proved in Harris [72]. □

Example 3.25 (Positive Regularity is Important!) Positive regularity is important for the various extinction and convergence theorems. Examine the $V = 2$ process A_1, A_2, with birthing probabilities

$$A_1 \xrightarrow{1} A_1 A_2, \qquad A_2 \xrightarrow{1} \phi, \tag{3.59}$$

with moment generating function and mean matrix

$$f(s_1, s_2) = (f^{(1)}(s_1, s_2) = s_1 s_2, f^{(2)}(s_1, s_2) = 1), \qquad M = \begin{pmatrix} 1 & 1 \\ 0 & 0 \end{pmatrix}. \tag{3.60}$$

This is not positively regular, $M = M^n$ for any n, and the state $Z = [1, 1]$ is recurrent, since

$$\Pr\{Z_n = [1,1] \text{ infinitely often } | Z_0 = [1,0]\} = 1. \tag{3.61}$$

One such three generation realization is shown in the right panel of Figure 3.7 demonstrating the recurrent state. Notice $\rho = 1$, yet $\gamma \neq 1$ contrary to Theorem 3.24 since it is not positively regular.

Example 3.26 (Reverse Polish) Here is the basis for the Reverse Polish language, with two rewriting rules

$$A \xrightarrow{p} a \quad A \xrightarrow{1-p} AAb, \quad a \in \{\mathbb{Z}, \mathbb{R}\}, \quad b \in \{+, -, \times\} \tag{3.62}$$

with moment generating function and mean

$$f(s) = p + (1-p)s^2, \qquad \rho = m = \left.\frac{\partial f(s)}{\partial s}\right|_{s=1} = 2(1-p). \tag{3.63}$$

The extinction probability is the root $0 \leq s \leq 1$ solving $f(s) = s$ giving

$$s = p + (1-p)s^2, \tag{3.64}$$

then γ is the unique positive value

$$\gamma = \frac{1}{2}\frac{1}{1-p} \pm \sqrt{\frac{1}{(1-p)^2} - \frac{4p}{1-p}} = \frac{p}{1-p} \quad \text{for } p < \frac{1}{2} \tag{3.65}$$

$$= 1 \quad \text{for } p \geq \frac{1}{2}. \tag{3.66}$$

3.6 Entropies of Directed Acyclic Graphs

As Shannon and Weaver [6,76] established, entropy characterizes the combinatoric number of patterns generated by random sources generating from a random family of strings. Directed graphs have particularly simple forms for their entropies; with this comes the classical theorems on numbers of strings. Begin with standard definitions of entropies and conditional entropies of families of random variables.

Definition 3.27 *Let X be a $\mathcal{X}_0$-valued process on a probabilistic directed acyclic graph $\sigma = \{D, E\}$ with $n(\sigma) = n$ sites with joint distribution P on $\mathcal{X}_0^n$. The* **joint entropy** *of X is defined to be*

$$H(X_1, X_2, \ldots, X_n) = -E_{P(X_1,\ldots,X_n)} \log P(X_1, \ldots, X_n), \qquad (3.67)$$

with the **joint entropy rate** *normalized by the number of random variables.*

For two jointly distributed random variables X, Y, the entropy of the conditional is defined by

$$H(X|y) = -E_{P(X|Y=y)} \log P(X|Y = y). \qquad (3.68)$$

Then the **conditional entropy** *of X given Y is defined to be*

$$H(X|Y) = -E_{P(X,Y)} \log P(X|Y), \qquad (3.69)$$

$$= \sum_y P(y) H(X|y). \qquad (3.70)$$

Then the realizations from the directed graphs inherit the factoring property from their parent systems.

Theorem 3.28 *Given the directed acyclic graph $\sigma = \{D, E\}$ with parent system $\Pi = \cup_i \Pi_i$ and $n(\sigma) = n$ sites, then the entropy is given by*

$$H(X) = \sum_{i=1}^n H(X_i | X_{\Pi_i}). \qquad (3.71)$$

Proof Write the joint probability via its factorization,

$$P(X_1, \ldots, X_n) = \prod_{i=1}^n P(X_i | X_{\Pi_i}), \qquad (3.72)$$

use the additive property of logarithms, and take the negative expectation. □

3.7 Combinatorics of Independent, Identically Distributed Strings via the Asymptotic Equipartition Theorem

We will use the entropy measures in various ways, one of the most prominent being to use it to count the typical strings which are likely to occur. For independent and identically distributed (i.i.d.) samples $X_1, X_2, \ldots$, the law of large numbers states that under reasonable conditions empirical averages of functions of the random variables $1/n \sum_{i=1}^n f(X_i)$ converge to their expectation $E_{P(X)} f(X)$. The asymptotic equipartition theorem (AEP) exploits this for the special $-\log P(\cdot)$ function, implying $-1/n \sum_{i=1}^n \log P(X_i)$ converge to the entropy constant. This being the case implies that the number of such sequences whose empirical log-probability is close to the entropy must be exponential in the entropy thus providing a method for counting these sequences. Let us define typical sets.

Definition 3.29 Let $X_1, X_2, \ldots$, be i.i.d. with $P(X)$ on $\mathcal{X}_0$. The **typical set** $\mathcal{T}_\epsilon^{(n)} \subset \mathcal{X}_0^n$ with respect to $P(X)$ with entropy $H(X)$ is the set of sequences $(X_1, \ldots, X_n) \in \mathcal{X}_0^n$ with the property

$$2^{-n(H(X)+\epsilon)} \leq P(X_1, \ldots, X_n) \leq 2^{-n(H(X)-\epsilon)}. \tag{3.73}$$

The AEP and its implications are stated as follows.

Theorem 3.30
1. For all $\epsilon > 0$, $\Pr\{\mathcal{T}_\epsilon^{(n)}\} > 1 - \epsilon$ for n sufficiently large.
2. For all n, $2^{n(H(X)+\epsilon)} \geq |\mathcal{T}_\epsilon^{(n)}|$ and for n sufficiently large, $|\mathcal{T}_\epsilon^{(n)}| > (1-\epsilon)2^{n(H(X)-\epsilon)}$.

Proof From the law of large numbers,

$$\lim_{n \to \infty} \Pr\left\{ \left| -\frac{1}{n} \log P(X_1, \ldots, X_n) - H(X) \right| < \epsilon \right\} = 1, \tag{3.74}$$

implying part 1. Part 2 follows from

$$1 \geq \sum_{\mathcal{T}_\epsilon^{(n)}} P(X_1, \ldots, X_n) \geq \sum_{\mathcal{T}_\epsilon^{(n)}} 2^{-n(H(X)+\epsilon)} \tag{3.75}$$

$$= 2^{-n(H(X)+\epsilon)} |\mathcal{T}_\epsilon^{(n)}|. \tag{3.76}$$

The second part of the inequality follows from $\lim_{n \to \infty} \Pr\{|-(1/n) \log P(X_1, \ldots, X_n) - H(X)| < \epsilon\} = 1$ giving for n sufficiently large,

$$1 - \delta < \Pr\{\mathcal{T}_\epsilon^{(n)}\} \tag{3.77}$$

$$\leq \sum_{\mathcal{T}_\epsilon^{(n)}} 2^{-n(H(X)-\epsilon)} = 2^{-n(H(X)-\epsilon)} |\mathcal{T}_\epsilon^{(n)}|. \tag{3.78}$$

Choose $\delta = \epsilon$. □

3.8 Entropy and Combinatorics of Markov Chains

Let us now examine the combinatorics of large graphs using entropy and the AEP beginning with Markov chains. First begin with the entropy of n-length strings. The entropy of the process will be a mixture of the visitation probability with the entropy of the conditional transition probability from each state in the transition matrix Q. For this, reintroduce the state occupancy vector

$$Z_n = \mathbf{1}^{(X_n)} = [0, \ldots, 0, \ldots, \underbrace{1}_{X_n \text{ entry}}, \ldots, 0]. \tag{3.79}$$

It will be convenient to exploit the matrix property that $(Z_n Q)_j = P(X_{n+1} = j | X_n)$.

Lemma 3.31 Let $X_0, X_1, \ldots$ be a time-invariant finite-state Markov chain with matrix of transition probabilities $Q = \left(Q_{jk}\right)$, $\sum_k Q_{jk} = 1$ and conditional transition entropy

$$h_j = -\sum_k Q_{jk} \log Q_{jk}, \quad j \in \mathcal{X}_0. \tag{3.80}$$

3.8 ENTROPY AND COMBINATORICS OF MARKOV CHAINS

The entropy conditioned on the initial state Z_0 (X_0) is given by

$$H_{Z_0}(X_1,\ldots,X_n) = E(-\log P(X_1,\ldots,X_n|X_0)|X_0) \tag{3.81}$$

$$= \sum_{i=0}^{n-1} Z_0 Q^i h. \tag{3.82}$$

Proof Computing the entropies using the conditional expectation

$$E(-\log P(X_1,\ldots,X_n|X_0)|X_0) = \sum_{i=1}^{n} E(E(-\log P(X_i|X_{i-1},X_0))|X_0) \tag{3.83}$$

$$= \sum_{i=1}^{n} E(E(-\log P(X_i|X_{i-1}))|X_0) \tag{3.84}$$

$$= \sum_{i=1}^{n} E(Z_{i-1}|Z_0)h) = \sum_{i=1}^{n} Z_0 Q^{i-1} h. \tag{3.85}$$

□

Examining Theorem 3.30 it is clear that the typical sets for which the entropy rate describes the exponential growth rate $\log |\mathcal{T}_\epsilon^{(n)}| \sim nH(X)$ are realizations which are infinitely long, so that subsequences for arbitrarily large n can always be found for the limit. The complement of the typical set will certainly contain those strings which terminate and of finite length. Thus understanding the extinction of the finite length sequences is essential. As we have seen this has a particularly straightforward characterization for Markov chains; the typical sequences simply divide as probability 0 or 1. Stationary Markov chains do not go extinct (or they would not be stationary); extinction for Markov chains corresponds to the transition matrix Q being sub-stochastic.

Let $X_1, X_2, \ldots$ be a time-invariant finite-state Markov chain with transitions $Q = \left(Q_{jk}\right)$. Stationarity corresponds to the extinction probability 0, and the process never entering the terminating-state, i.e. extinction is $Z_n = 0$ for some n.

For positively regular, stationary finite-state Markov chains, the entropy rate is given by the familiar mixture of conditional entropies in each state mixed against the stationary distribution.

Theorem 3.32 *Let $X_1, X_2,\ldots$ be a finite state time-invariant Markov chain with $Q = \left(Q_{jk}\right)$ positively regular and eigenelements $\nu Q = \rho \nu, Q\mu = \rho\mu$. With **1** defined as the all 1's vector, then the normalized entropy rate becomes*

$$H_{Z_0}(X) = \lim_{n\to\infty} \frac{-E(\log P(X_1,\ldots,X_n|Z_0)|Z_0)}{E(\sum_{i=1}^n Z_i \mathbf{1}|Z_0)}. \tag{3.86}$$

Then

1. *if $\gamma = 1$ and Q is sub-stochastic with $\rho < 1$, then the normalized entropy rate is a function of the initial state:*

$$H_{X_0}(X) = \frac{Z_0(I-Q)^{-1}h}{Z_0(I-Q)^{-1}\mathbf{1}}; \tag{3.87}$$

2. *if $\gamma = 0$ and Q is stochastic, $\rho = 1$, $1/n \sum_{i=1}^n Z_i \to \nu$, $\tilde{Q}^n \to \begin{pmatrix} 1 & 0 \\ 0 & \mu\nu \end{pmatrix}$, and the entropy rate is independent of initial state:*

$$H(X) = \nu h. \tag{3.88}$$

Proof *Proof for $q \neq 0$ ($\rho < 1$):* The geometric series in Q of Eqn. 3.82 gives the first entropy rate since the numerator and the denominator each have terms

$$\sum_{i=0}^{n-1} Q^i = (I-Q)^{-1}(I-Q^n) \to (I-Q)^{-1} \quad \text{as } n \to \infty. \tag{3.89}$$

Proof for $\gamma = 0$ ($\rho = 1$): Since $Q^n \to \mu v$ as $n \to \infty$ with μ the all 1's vector, this implies

$$(Q^n)_{X_0, j} \to v_j \quad \text{as } n \to \infty, \tag{3.90}$$

which gives the second entropy rate. □

Proving the normalized empirical negative log-probability converges to the entropy rate requires the convergence of the occupancy rate of states to the stationary distribution.

Theorem 3.33 *Let $X_1, X_2, \ldots,$ be a finite-state positively regular Markov chain with transition $Q = \left(Q_{jk}\right)$ and unique stationary distribution $v = vQ$, entropy rate $H(X) = \sum_j v_j h_j$. The **typical set** $\mathcal{T}_\epsilon^{(n)} \subset \mathcal{X}_0^n$ defined as*

$$2^{-n(H(X)+\epsilon)} \leq P(X_1, \ldots, X_n | X_0) \leq 2^{-n(H(X)-\epsilon)}, \tag{3.91}$$

satisfies the AEP Theorem 3.30.

Proof Now to prove that $\Pr\{\mathcal{T}_\epsilon^{(n)}\} > 1 - \epsilon$ use the fact that the process is ergodic (unique stationary probability, [p. [77], 136]) so that the state occupancy defined as $N_j(n) = \sum_{i=1}^n Z_{i-1}^{(j)}$ when normalized converges $(N_j(n)/n) \to v_j$ with probability one. As well, defining the subsequence $X_{i_1}, X_{i_2}, \ldots, X_{i_{N_j(n)}}$ of states immediately following state j in the sequence $X_1, X_2, \ldots, X_n$, then

$$\sum_{l=1}^{N_j(n)} f(X_{i_l}) \to \sum_k Q_{jk} f(k) \quad \text{w.p. 1}, \tag{3.92}$$

implying

$$\lim_{n \to \infty} -\frac{1}{n} \log P(X_1, \ldots, X_n | X_0) = \lim_{n \to \infty} -\frac{1}{n} \sum_{i=1}^n \log P(X_i | X_{i-1}) \tag{3.93}$$

$$= -\sum_j \frac{N_j(n)}{n} \frac{\sum_{l=1}^{N_j(n)} \log Q_{j X_{i_l}}}{N_j(n)} \tag{3.94}$$

$$= \sum_j v_j h_j. \tag{3.95}$$

□

3.9 Entropies of Branching Processes

While entropies of the Markov chains have been calculated since the time of Shannon, the entropies of the branching processes have been more recently developed [78–80]. Thus far in studying

3.9 ENTROPIES OF BRANCHING PROCESSES

the branching processes, the role of $Z_n, n = 0, 1, \ldots$ have been emphasized as a Markov chain. These vector processes do not completely characterize the complete properties of the random realizations, for example their entropies. For this, the random processes as manifest by their description as tree graphs are required. Throughout particles A_i are assumed to have a finite set of particle generation probabilities, $\{p_{ij}, j = 1, \ldots J\}$, J finite. Figure 3.6 illustrates the basic issue. Note that the two realizations are different trees although the state vectors are identical, $Z_0 = 1, Z_1 = 3, Z_2 = 2$, irrespective of the fact that the trees were generated with different rule choices. To characterize realizations of the entire family tree and its underlying probability space we follow Harris by defining each tree by a sequence of finite sequences which specify the generation numbers including the offspring in each level of the tree.

3.9.1 Tree Structure of Multi-Type Branching Processes

Identify with the random n-generation base $T_n = X_1, \ldots, X_n$ of a perhaps infinite random tree $T(\omega)$ consisting of a sequence of sequences of rules specifying each of the generations $X_1 = (X_{11}, X_{12}, \ldots), X_2 = (X_{21}, X_{22}, \ldots)$, with X_{il} defined as a substitution rule-valued random variable generating offspring for the lth node in the ith generation of $T(\omega)$. The random n-generation base is identified with the sequence of random variables

$$T_n = \underbrace{X_1}_{X_1 \text{ consisting of 1 rewrite rule of } Z_0}, \underbrace{X_{2,1}, X_{2,2}, \ldots,}_{X_2 \text{ consisting of } Z_1^{(1)}, Z_1^{(2)}, \ldots, \text{ rewrite rules}}, \ldots, \underbrace{X_{n,1}, X_{n,2}, \ldots,}_{X_n \text{ consisting of } Z_{n-1}^{(1)}, Z_{n-1}^{(2)}, \ldots, \text{ rewrite rules}}. \quad (3.96)$$

Random realizations are generated by successively applying the generation rules to the variables in the tree which can have children, until a tree is generated consisting solely of particles at the leaves which are infertile. The conditional independence structure of the random branching processes is imposed via the fact that offspring rules are chosen conditionally independent of their siblings given their parents. Consistency in the trees is maintained by choosing the probability law which is appropriate for the particular node of the tree. This directed graph structure is depicted in the middle panel of Figure 3.2.

The probability of an n generation base T_n given the start node Z_0 is given by

$$P(T_n|Z_0) = P(X_1, \ldots, X_n|Z_0) \quad (3.97)$$

$$= \prod_{i=1}^{n} \prod_{v=1}^{V} \prod_{k=1}^{J_v} (p_{vk})^{\sum_{l=1}^{\sum_j z_i^{(j)}} \delta_{r_{vk}}(X_{il}(\omega))}. \quad (3.98)$$

These n-generation bases form the natural cylinders from which the probability law on the infinite realizations are constructed. This extends naturally to a distribution on infinitely extended trees as shown in [2]. Let $\mathcal{T}$ be the space of all realizations of n-generation trees, for all n, including infinitely extended trees. Then the cylinder set $C(T_n) \subset \mathcal{T}$ is the set of all infinitely extended trees with base T_n. The probability distribution is extended to the sigma-algebra generated by all cylinder sets by defining the probability of cylinders $C(T_n) \subset \mathcal{T}$ according to Eqn. 3.98 applied to the first n generation base

$$P(T_n) = \Pr\{T \in C(T_n)\}. \quad (3.99)$$

3.9.2 Entropies of Sub-Critical, Critical, and Super-Critical Processes

Examine the entropies of the V-type branching processes. Let T_n be the base of a perhaps infinite random tree generated by successively applying the generation rules to the variables in the tree which can have children. Then $T_n = X_1, \ldots, X_n$ is a sequence of sequences $X_1 = (X_{1l}, l = 1, \ldots), X_2 = (X_{2l}, l = 1, \ldots), \ldots$ with X_{il} the rule used to generate offspring at node l of generation i. The probability of an n generation base conditioned on Z_0 is given by Eqn. 3.98.

Calculation of the entropies is essentially determined by the rate at which the tree branches. Roughly speaking, each path to a leaf node is a Markov chain; the entropy of the tree is the number of paths multiplied by the Markov chain-like entropy of each path. This of course links the branching rates and eigenvalues ρ of M the mean branching matrix and the entropies as defined originally in [78, 79].

Theorem 3.34 *Letting the entropy of the set of rewrite rules of the v-type, $v = 1, \ldots, V$, be defined as*

$$h_v = -\sum_{k=1}^{J_v} p_{vk} \log p_{vk}, \qquad h = \begin{pmatrix} h_1 \\ \vdots \\ h_V \end{pmatrix}, \tag{3.100}$$

then the entropy of n generation trees conditioned on the initial state Z_0 of the random branching process with mean matrix M is given by

$$H_{Z_0}(T_n) = -E(\log P(T_n|Z_0)|Z_0) = \sum_{i=0}^{n-1} Z_0 M^i h. \tag{3.101}$$

Proof Evaluating the conditional expectation of $-\log P(T_n|Z_0)$ yields

$$H_{Z_0}(T_n) = E(-\log P(T_n|Z_0)|Z_0) = E(-\log P(X_1, \ldots, X_n|Z_0)|Z_0) \tag{3.102}$$

$$= \sum_{i=1}^{n} E(E(-\log P(X_i|Z_{i-1}, Z_0))|Z_0) \tag{3.103}$$

$$= \sum_{i=1}^{n} E(E(-\log P(X_i|Z_{i-1}))|Z_0) \tag{3.104}$$

$$= \sum_{i=1}^{n} E(Z_{i-1}|Z_0)h = \sum_{i=0}^{n-1} Z_0 M^i h. \tag{3.105}$$

□

Since these directed graphs have random numbers of nodes in n-generation graphs defined by the **normalized entropy per parent in the tree** H. As in the Markov chain case, there are two limits.

Theorem 3.35 *Given a branching process with positively regular mean matrix M, then define*

$$H_{Z_0}(T) = \lim_{n \to \infty} \frac{-E(\log P(T_n|Z_0)|Z_0)}{E(\sum_{i=0}^{n-1} Z_i \mathbf{1}|Z_0)}. \tag{3.106}$$

The normalized entropy per branching variable has different forms for the sub-critical, and the critical, super-critical branching processes:

(1) *for M having largest positive eigenvalue $\rho < 1$ then $H_{Z_0}(T) = ((Z_0 (I - M)^{-1}h)/ (Z_0 (I - M)^{-1} \mathbf{1}))$;*

(2) *for M having largest positive eigenvalue $\rho \geq 1$ then $H(T) = vh/v\mathbf{1}$.*

Proof *Proof for $\rho < 1$*: The denominator in the last expression of Eqn. 3.106 results from Eqn. 3.37. Since the largest eigenvalue of M is less than 1, $I - M$ is invertible. Using the geometric series on M in 3.106 yields the result that

$$H = \lim_{n\to\infty} \frac{Z_0\,(I-M)^{-1}(I-M^n)h}{Z_0\,(I-M)^{-1}(I-M^n)\mathbf{1}}. \qquad (3.107)$$

Examining the limit as $n \to \infty$ with the largest eigenvalue $\rho < 1$, yields the first part.
Proof for $\rho \geq 1$: For $\rho > 1$, substituting into Eqn. 3.106 yields

$$H = \lim_{n\to\infty} \frac{\sum_{i=0}^{n-1}\left(\rho^i Z_0\,\mu\,v\,h + Z_0\,M_2^i\,h\right)}{\sum_{i=0}^{n-1}\left(\rho^i Z_0\,\mu v\,\mathbf{1} + Z_0\,M_2^i\,\mathbf{1}\right)}$$

$$= \lim_{n\to\infty} \frac{((\rho^L-1)/(\rho-1))Z_0\,\mu\,vh + O(\alpha^n)}{((\rho^n-1)/(\rho-1))Z_0\,\mu\,v\mathbf{1} + O(\alpha^n)} = \lim_{n\to\infty} \frac{Z_0\,\mu\,vh + O(\alpha^n/\rho^n)}{Z_0\mu v\mathbf{1} + O(\alpha^n/\rho^n)} = \frac{vh}{v\mathbf{1}}. \qquad (3.108)$$

For $\rho = 1$, $(\rho^n - 1)/(\rho - 1)$ is replaced by n and $O(\alpha^n)$ is replaced by $O(1)$ in 3.108, with the conclusion identical as for $\rho > 1$. □

3.9.3 Typical Trees and the Equipartition Theorem

The typical strings correspond to trees which have infinite extent. Characterizing the combinatoric properties of such processes requires an understanding of extinction. Do the family trees die out? This is more subtle for branching processes than for Markov chains where if there was a connected terminating state, then all the connected states are transient and with probability one the process terminates. For branching processes the possibility of birthing multiple offspring implies that extinction can occur with probability between zero and 1. By extinction we mean the event that the random sequence $Z_0, Z_1, \ldots$ consists of zero for all but a finite number of values of n.

Definition 3.36 *For the **V-multi-type extinction probability** define $\gamma = [\gamma^{(1)}, \gamma^{(2)}, \ldots, \gamma^{(V)}] \in [0,1]^V$ to be the extinction probability corresponding to the initial state $Z_0 = \mathbf{1}^{(v)}$:*

$$\gamma^{(v)} = \Pr\{Z_n = 0 \text{ for some } n \mid Z_0 = \mathbf{1}^{(v)}\}, \quad v = 1, \ldots, V. \qquad (3.109)$$

*For the **1-type extinction probability**, γ is a scalar.*

The moments of generations of the process given by the mean matrix M will determine the extinction probability and entropy properties. The mean matrix M is a positive matrix (all entries ≥ 0). In order to obtain standard limiting forms we restrict to the class of positively regular branching processes. Clearly if M^n is strictly positive, then there exists an n generation birthing route of nonzero probability from each parent type to every other parent type.

Clearly Z_n has the potential to grow. For $\rho \leq 1$, the limiting behavior has $Z_n \to 0$ with probability 1 ($\gamma = 1$). Surprisingly, Z_n in fact converges with probability 1 for $\rho > 1$. This will be apparent in the asymptotics section below.

Now an equipartition theorem is proved showing in essence that the set of trees are cut into two sets, the **typical trees** corresponding to the non-extinct set of infinite, non-terminated trees of probability $1 - q$, and the **a-typical trees** of all trees terminating in a finite number of derivations of probability q. For typical infinitely extended trees, the state vectors Z_i converge, with typicality corresponding to the relative node occurrences following the rule birthing probabilities.

Definition 3.37 *Let Ω_I be the set of* **non-terminating, or infinitely extended trees** *given according to* $\Omega_I = \{T(\omega) : Z_n \neq 0 \text{ for any } n\}$.

For Markov chains which are stationary, the empirical averages of state occupancy equal the stationary probability given by the left eigenvector of the transition probability. This is true as well for branching processes even though the number of variables at every generation is a random number determined by the branching rate.

Lemma 3.38 *Given a random branching process with positively regular mean branching matrix M, with largest eigenvalue $\rho \geq 1$ and associated left eigenvector v, then the occupancy vectors have ratios given by the left eigenvector of the branching matrix:*

$$\lim_{n\to\infty} \frac{\sum_{i=0}^{n-1} Z_i}{\sum_{i=0}^{n-1} \rho^i} = c(\omega)v, \quad c(\omega) \geq 0, \text{ a.e. } \omega \in \Omega \text{ (w.p. 1)}, \qquad (3.110)$$

with $c(\omega) > 0$ for infinitely extended trees implying

$$\lim_{n\to\infty} \frac{\sum_{i=0}^{n-1} Z_i^{(j)}}{\sum_j \sum_{i=0}^{n-1} Z_i^{(j)}} = v_j, \quad \text{a.e. } \omega \in \Omega_I \text{(w.p. } 1-\gamma), \qquad (3.111)$$

Proof This follows by using Theorem 9.2 of Harris [72]:

$$\lim_{n\to\infty} \frac{Z_n}{\rho^n} = c(\omega)v, \quad \text{a.e. } \omega \in \Omega \text{(w.p. 1)}, \qquad (3.112)$$

Then the triangle inequality gives almost everywhere $\omega \in \Omega$, for every $\epsilon > 0$, $\exists n(\epsilon, \omega)$ such that with $i > n(\epsilon, \omega)$, $|Z_i - \rho^i c(\omega)v| < \epsilon/2$ implying with $N > n(\epsilon, \omega)$ chosen large enough and the triangle inequality:

$$\left| \frac{\sum_{i=0}^{N-1} Z_i}{\sum_{i=0}^{N-1} \rho^i} - c(\omega)v \right| \leq \underbrace{\left| \frac{\sum_{i=0}^{N-1} Z_i}{\sum_{i=0}^{N-1} \rho^i} - \frac{\sum_{i=n(\epsilon,\omega)}^{N-1} Z_i}{\sum_{i=n(\epsilon,\omega)}^{N-1} \rho^i} \right|}_{\epsilon/2} + \underbrace{\left| \frac{\sum_{i=n(\epsilon,\omega)}^{N-1} Z_i}{\sum_{i=n(\epsilon,\omega)}^{N-1} \rho^i} - c(\omega)v \right|}_{\epsilon/2} = \epsilon.$$

(3.113)

Since ϵ is arbitrary the first part is completed.

To show that $c(\omega)$ is strictly positive a.e. Ω_I follows from its connection to the extinction probability (Theorem 7.1 of Harris [72]) implying

$$\Pr\{c(\omega) = 0 | Z_0 = \mathbf{1}^{(1)},$$
$$T(\omega) \in \Omega_I\} = \Pr\{Z_i(\omega) = 0 \text{ for some } i | Z_0, Z_i \neq 0 \text{ for any } i\} = 0. \qquad \square$$

Theorem 3.39 (AEP) *Given is a random branching process with positively regular mean branching matrix M, with largest eigenvalue $\rho \geq 1$ and associated left eigenvector v. Define the infinitely extended trees with* **typical set** $\mathcal{T}_\epsilon^{(n)}$ *with respect to $P(\cdot)$ with entropy rate $H(T)$ as the trees T_n with the property*

$$2^{-(\sum_{i=0}^{n-1} Z_i \mathbf{1})(H(T)+\epsilon)} \leq P(T_n) \leq 2^{-(\sum_{i=0}^{n-1} Z_i \mathbf{1})(H(T)-\epsilon)}. \qquad (3.114)$$

Then, for all $\epsilon > 0$, $P(\mathcal{T}_\epsilon^{(n)}) > 1 - \epsilon$.

3.9 ENTROPIES OF BRANCHING PROCESSES

Proof Factoring the probability of an n-generation tree gives

$$\log P(T_n|Z_0) = \sum_{i=1}^{n} \log P(X_1,\ldots,X_n|Z_0) = \sum_{i=1}^{n} \log P(X_i|Z_{i-1},Z_0) \quad (3.115)$$

$$= \sum_{i=1}^{n} \log P(X_{i1}, X_{i2},\ldots|Z_{i-1}). \quad (3.116)$$

Defining $N_j(n) = \sum_{i=1}^{n} Z_{i-1}^{(j)}$, $N(n) = \sum_j N_j(n)$, then from Lemma 3.38, Eqn. 3.111 $N_j(n)/N(n) \to v_j$ a.e. Ω_I. As well on the subsequence of substitutions of particle type j, $X_{il_1},\ldots,X_{i_{Z_{i-1}^{(j)}}}$,

$$\sum_{m=1}^{Z_{i-1}^{(j)}} f(X_{il_m}) \to \sum_k p_{jk} f(k) \text{ a.e. } \Omega_I, \quad (3.117)$$

giving almost everywhere in infinitely extended trees (probability $1-q$),

$$\lim_{n\to\infty} -\frac{1}{N(n)} \log P(T_n|Z_0) = -\lim_{n\to\infty} \sum_{i=1}^{n} \sum_{j} \sum_{l=1}^{Z_{i-1}^{(j)}} \log P(X_{il}|Z_{i-1}^{(j)}) \quad (3.118)$$

$$= -\lim_{n\to\infty} \sum_{j} \frac{N_j(n)}{N(n)} \frac{1}{N_j(n)} \sum_{i=1}^{n} \sum_{l=1}^{Z_{i-1}^{(j)}} \log p_{jX_{il}} \quad (3.119)$$

$$= \sum_{j} v_j h_j. \quad (3.120)$$

$\square$

The branching rate and entropy are the definitive quantity describing the combinatorics.

Theorem 3.40 *Given is a random branching process with positively regular mean branching matrix.*

1. *If each birthing law has identically $\rho \geq 1$ children, then for all $\epsilon > 0$, for all n, $2^{(\sum_{i=0}^{n-1} \rho^i)(H(T)+\epsilon)} \geq |T_\epsilon^{(n)}|$ and for n sufficiently large, $|T_\epsilon^{(n)}| > (1-\epsilon) 2^{(\sum_{i=0}^{n-1})(H(T)-\epsilon)}$.*

2. *If $\rho > 1$, then typical trees grow at super-geometric rate ρ: for all $\epsilon > 0$, for all n, $(\sum_{i=0}^{n-1}(\rho+\epsilon)^i) \geq \log |T_\epsilon^{(n)}|$ and for n sufficiently large, $\log |T_\epsilon^{(n)}| > (\sum_{i=0}^{n-1}(\rho-\epsilon)^i)$.*

Proof (Part 1): From the fact that $\sum_{i=0}^{n-1} Z_i \mathbf{1} = \sum_{i=0}^{n-1} \rho^i$ then the identical argument as above Theorem 3.30 in which $P(T_\epsilon^{(n)})$ is upper bounded by 1, and lower bounded by $1-\epsilon$ in the limit $n \to \infty$ for any $\epsilon > 0$.

(Part 2): For the combinatoric statement, the definition of $\mathcal{T}_\epsilon^{(n)}$ implies that for all $T_n(\omega) \in \mathcal{T}_\epsilon^{(n)}$, $\log P(T_n(\omega)) \geq -(\rho + \delta)^n$, and since $P(\mathcal{T}_\epsilon^{(n)}) \leq 1$,

$$1 \geq \sum_{T(\omega) \in \mathcal{T}_\epsilon^{(n)}} P(T_n) \geq |\mathcal{T}_\epsilon^{(n)}| \min_{T(\omega) \in \mathcal{T}_\epsilon^{(n)}} P(T_n) \qquad (3.121)$$

$$\geq |\mathcal{T}_\epsilon^{(n)}| 2^{-(\rho+\delta)^n},$$

$$\implies 2^{(\rho+\delta)^n} \geq |\mathcal{T}_\epsilon^{(n)}|. \qquad (3.122)$$

The lower bound follows from the convergence in probability implying for all $\epsilon > 0$, $\exists n(\epsilon)$ with $n > n(\epsilon)$,

$$1 - \gamma - \epsilon < P(\mathcal{T}_\epsilon^{(n)}) < |\mathcal{T}_n^\delta| \max_{T(\omega) \in \mathcal{T}_\epsilon^{(n)}} P(T_n) \qquad (3.123)$$

$$< |\mathcal{T}_\epsilon^{(n)}| 2^{-(\rho-\epsilon)^n}. \qquad (3.124)$$

$\square$

Remark 3.9.1 Part (1) of the theorem is stronger than part (2) for the family of special cases in which the number of descendents below every node is identically ρ, the branching rate, such as for the singular Markov case $\rho = 1$, and the binary case $\rho = 2$ [80] having exactly two non-terminal variables for every rule. For these cases, $\nu\mathbf{1} = 1$ and $c(\omega) = 1$ for all $\omega \in \Omega$. The combinatorics resulting is strengthened simply because the number of nodes in an n depth tree is a deterministic quantity growing at a geometric rate.

3.10 Formal Languages and Stochastic Grammars

The formal languages of Chomsky regular, context-free and context-sensitive [81–83] are an exquisite example of directed acyclic graphs; they should be studied in their own right. The regular and context-free correspond to finite-state and multi-type random branching processes; context-sensitive correspond to more general directed graphs.

Following Chomsky, define the class of strings and the languages or configurations associated with a *regular* or *finite-state grammar*,[15] *context-free grammar*, and *context-sensitive grammar*.

Definition 3.41 A **formal grammar** G *is a quadruple,* $G = \langle V_N, V_T, R, S \rangle$, *where* V_N *is a finite set of non-terminal symbols (also called states in the finite-state case),* V_T *is a finite set of terminal symbols, R is a finite set of rewrite rules, $S \in V_N$ is a sentence start symbol. The finite set of non-terminal symbols and combinations of them* $V_N = \{A_1, A_2, \ldots\}$ *are the variables which can be rewritten, with the terminal symbols* $V_T = \{w_1, w_2, \ldots\}$ *the words forming the strings in the language.*

The **grammatical rules** R *take the following form:*

1. *For a* **finite-state grammar** *the rules R rewrite the non-terminals (these are states in the finite-state case) and are independent of the context in which the substitutions to the nonterminal symbols are applied, of the form*

$$A \to wB, \quad A \to w \quad \text{where } w \in V_T, \ A, B \in V_N. \qquad (3.125)$$

---

[15] Generally these grammars are called regular, as well as their languages. We prefer to use finite-state as all the configurations which we shall deal with are *regular* in the sense of general pattern theory associated with the first structure formula Eqn. 6.2 of Chapter 6.1.

3.10 FORMAL LANGUAGES AND STOCHASTIC GRAMMARS

2. *For a **context-free grammar** rules R extend the form of the finite state rules to include combinations of non-terminals on their right-hand side, taking the form*

$$A \to \psi, \quad A \in V_N, \quad \psi \in (V_N \cup V_T)^*, \tag{3.126}$$

  ψ being a collection of terminal and non-terminal symbols.

3. *For a **context-sensitive grammar** the rules R extend the form of the context-free rules to include context around the non-terminal being written on the left hand side, giving rules of the form*

$$\alpha A \beta \to \alpha \psi \beta, \quad A \in V_N, \alpha, \beta, \psi \in (V_N \cup V_T)^*. \tag{3.127}$$

*The **language** L(G) generated by the grammar G is the set of all sequences consisting of terminal symbols that can be generated by starting in S and recursively applying production rules to the nonterminating expressions until the expressions contain only terminal symbols.*

Now examine the probabilistic versions of the grammars and associated stochastic languages. The languages correspond to the set of strings generated by probabilistic application of the rules. For the finite-state (regular) and context-free languages, these are Markov chains and multi-type random branching processes.

Definition 3.42 *A **stochastic formal grammar** $G(P)$ is a formal grammar $G = \langle V_N, V_T, R, S \rangle$ with a family of probability laws $P = \{p_v, \sum_k p_{vk} = 1, v = 1, \ldots, V\}$ on the production rules $R = \{r_{vk}, v = 1, \ldots, k = 1, \ldots\}$:*

$$\alpha A \beta \stackrel{p_{(\alpha A \beta), k}}{\to} \alpha \psi \beta, \quad A \in V_N, \alpha, \beta, \psi \in (V_N \cup V_T)^*, \quad \sum_k p_{(\alpha A \beta), k} = 1. \tag{3.128}$$

*The **stochastic language** $L(G(P))$ generated by the stochastic grammar $G(P)$ is the set of all sequences consisting of terminal symbols that can be generated by starting in S and recursively applying production rules applied with probability P to the nonterminating expressions until the expressions contain only terminal symbols.*

Formal grammars are an exquisite illustration of the pattern theory. They directly provide a mechanism for building the space of directed graphs recursively: *the grammar specifies the legal rules of transformation of one graph type to another*. Naturally, finite-state grammars provide transformations which build the space of linear graphs, the context-free grammars trees and finally context sensitive grammars provide transformations through the partially ordered set graphs.

Example 3.43 (Run-Length and Parity Finite-State Languages) Strings corresponding to finite-state languages can all be accepted by finite-state machines [84], and therefore correspond to paths through finite state graphs. For finite-state grammars, the finite set of non-terminal symbols $V_N = \{A^{(1)}, A^{(2)}, \ldots\}$ are the *states* of the equivalent finite-state graph which can be used to generate the strings.

Examine the run-length and parity languages. From the alphabet $\mathcal{X}_0 = \{0, 1\}$, the state graphs are depicted in Figure 3.8.

Begin with the 1-1 constraint language, consisting of strings containing no more than a single 1 symbol in a row. The grammar $G = \langle V_T = \{0, 1\}, V_N = \{\text{ZERO}, \text{ONE}\}, S = \text{ZERO} \rangle$, with the rule set (generators) simply the transitions in the state graph:

$$R = \{\text{ONE} \stackrel{r_1}{\to} 0\text{ZERO}, \text{ZERO} \stackrel{r_2}{\to} 0\text{ZERO}, \text{ZERO} \stackrel{r_3}{\to} 1\text{ONE}\}. \tag{3.129}$$

Elements in the language can end in either states, augment the rule set with the terminating rules $\text{ZERO} \stackrel{r_4}{\to} 1$, $\text{ZERO} \stackrel{r_5}{\to} 0$, $\text{ONE} \stackrel{r_6}{\to} 0$.

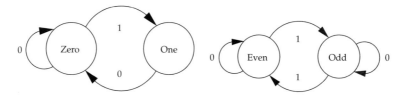

Figure 3.8 Showing two finite-state graphs corresponding to run-length (left panel) and parity (right panel) languages. For the 1-1 languages, both states are accepting; for even parity the even state is accepting.

For parity languages, $V_N = \{\text{EVEN}, \text{ODD}\}$ with $S = \text{EVEN}$, and the rules

$$R = \{\text{EVEN} \xrightarrow{r_1} 0\text{EVEN}, \text{EVEN} \xrightarrow{r_2} 1\text{ODD}, \text{ODD} \xrightarrow{r_3} 0\text{ODD}, \text{ODD} \xrightarrow{r_4} 1\text{EVEN}\}. \tag{3.130}$$

Since only strings ending in EVEN parity are in the language, augment with the terminating rules $\text{EVEN} \xrightarrow{r_5} 0, \text{ODD} \xrightarrow{r_6} 1$.

Example 3.44 (M-Gram Language Models) Consider **bigram** and **trigram** Markov models for natural language as in Shannon [85]. The m-gram model treats a string of words as a realization of an mth-order Markov chain in which the transition probability is the conditional probability of a word in the string given by the previous $m-1$ words. The probability of the word string is then the product of the transition probabilities for each word, $P(W) = \prod_{i=1}^{n} p(W_i | W_{i-1}, \ldots, W_{i-m+1})$.

The grammatical rules for the m-ary processes are

$$\underbrace{w_{i-m}, w_{i-(m-1)}, \ldots, w_i}_{A} \to \underbrace{w_{i-(m-1)}, w_{i-(m-2)}, \ldots, w_{i+1}}_{w_{i+1} \, A'}$$

$$\underbrace{w_{i-m}, w_{i-(m-1)}, \ldots, w_i}_{A} \to \underbrace{w_{i+1-m}, w_{i+1-(m-1)}, w_{i+1-(m-2)}, \ldots, w_i, \phi)}_{w_{i+1}}$$

where $\phi = null$ (punctuation). For each such transition there are then two rule types

$$A \to wA', \quad w \in V_T, \quad A, A' \in V_N, \quad A \to w \quad w \in V_T, \quad A \in V_N. \tag{3.131}$$

The regular grammar rewrites the non-terminals which are states in the finite-state case independently of the context in which the substitutions to the nonterminal symbols are applied.

Example 3.45 (Finite-State Languages) Strings corresponding to finite-state languages can all be accepted by finite-state machines [84], and therefore correspond to paths through finite state graphs. For finite-state grammars, the finite set of non-terminal symbols $V_N = \{A_1, A_2, \ldots\}$ are the "states" of the equivalent finite-state graph which can be used to generate the strings.

Examine run-length and parity languages of previous examples from the alphabet $\Lambda = \{0, 1\}$. The state graphs are depicted in Figure 3.8; strings are generated as paths through the finite-state graph shown in Figure 3.8.

Begin with the 1-1 constraint language, strings containing no more than a single 1 symbol in a row. The grammar $G = \langle V_T = \{0, 1\}, V_N = \{\text{ZERO}, \text{ONE}\}, S = \text{ZERO} \rangle$, with the rule set (generators) simply the transitions in the state graph:

$$R = \{\text{ONE} \xrightarrow{r_1} 0\text{ZERO}, \text{ZERO} \xrightarrow{r_2} 0\text{ZERO}, \text{ZERO} \xrightarrow{r_3} 1\text{ONE}\}. \tag{3.132}$$

3.10 FORMAL LANGUAGES AND STOCHASTIC GRAMMARS

Elements in the language can end in either states, augment the rule set with the terminating rules ZERO $\xrightarrow{r_4}$ 1, ZERO $\xrightarrow{r_5}$ 0, ONE $\xrightarrow{r_6}$ 0.

For parity languages, $V_N = \{\text{EVEN}, \text{ODD}\}$ with $S = \text{EVEN}$, and the rules (generators)

$$R = \{\text{EVEN} \xrightarrow{r_1} 0\text{EVEN}, \text{EVEN} \xrightarrow{r_2} 1\text{ODD}, \text{ODD} \xrightarrow{r_3} 0\text{ODD}, \text{ODD} \xrightarrow{r_4} 1\text{EVEN}\}. \tag{3.133}$$

Since only strings ending in EVEN parity are in the language, augment with the terminating rules EVEN $\xrightarrow{r_5}$ 0, ODD $\xrightarrow{r_6}$ 1.

In terms of combinatoric constraints placed by the grammars, notice $L(G) \subset 2^n$ shrinks at an exponential rate. Using a standard argument dating back to at least Shannon [76], define $N_n = \begin{pmatrix} N_n(\text{ZERO}) \\ N_n(\text{ONE}) \end{pmatrix}$ to be the 2×1 vector of the number of strings ending in states ZERO, ONE, then

$$\underbrace{\begin{pmatrix} N_n(\text{ZERO}) \\ N_n(\text{ONE}) \end{pmatrix}}_{N_n} = \underbrace{\begin{pmatrix} 1 & 1 \\ 1 & 0 \end{pmatrix}}_{A} \underbrace{\begin{pmatrix} N_n(\text{ZERO}) \\ N_n(\text{ONE}) \end{pmatrix}}_{N_{n-1}}. \tag{3.134}$$

The exponential growth rate is given by the largest eigenvalue $\rho = ((1 + \sqrt{5})/2)$ of A since by Perron–Frobenius A is strictly positive. This $\log_2$-exponential growth rate is in Shannon's words termed the **capacity** of the language of strings.

Note that the EVEN-parity language of strings ending only in the EVEN state has the capacity $\rho = 1$.

Example 3.46 (Finite-State Markov Chain: 3-1's Language) Let the $V = $ 4-type, $\{A_1, A_2, A_3, A_4\}$, and birthing laws

$$\begin{array}{llll} A_1 \xrightarrow{0.382} 0A_2 & A_2 \xrightarrow{1} 1A_3 & A_3 \to 1A_4 & A_4 \xrightarrow{1} 0A_1 \\ A_1 \xrightarrow{0.618} 1A_3 & & A_3 \to 0A_1 & \end{array}. \tag{3.135}$$

Define N_n interpreted as the total number of derivation trees. The branching matrices M are stochastic with largest eigenvalue $\rho = 1$, $\nu M = \nu$, since every rewrite rule gives rise to one syntactic variable in the strongly connected set, $\sum_k n_{\nu k}^{(l)} = 1$:

$$M = \begin{pmatrix} 0 & 0.382 & 0.618 & 0 \\ 0 & 0 & 1 & 0 \\ 0.618 & 0 & 0 & 0.618 \\ 1 & 0 & 0 & 0 \end{pmatrix}. \tag{3.136}$$

To find the growth rates of the start state A_1, define the initial vector $Z_0 = [1, 0, 0, 0]$, then

$$\lim_{n \to \infty} \frac{\log N_n(A_1)}{n} = \lim_{n \to \infty} \frac{Z_0 \log N_n}{n} = \nu h = 0.4812, \tag{3.137}$$

where $H = \nu h$ is the entropy of the Markov chain. Since M is stochastic, ν is the limit distribution of the Markov chain induced by the particular choice of probabilities on the rewrite rules.

Example 3.47 (Adding Closure: Sentences to Paragraphs) Extend the regular grammar example forbidding three consecutive ones or zeros by adding rules which generate the language consisting of the concatenation of all sentences in the run-length language, commonly called the closure of the language. Define $G_1^* = \langle V_T, V_N^*, R^*, A_0 \rangle$, with terminal symbols as above, non-terminals $V_N^* = V_N \cup A_0$, and rewrite rules

$R^* = R \cup \{A_0 \to A_1 A_0, A_0 \to A_1\}$. G_1^* is not strongly connected since there is no derivation originating in A_1 which results in a A_0. It is, however, context-free and contains one rewrite rule with more than one non-terminal variable. To calculate the combinatorics note that the number of derivations starting in A_k for $k \in \{1,2,3,4\}$ is the same as in the first regular grammar example and are given above, with $\rho = 1$. Using these, proceed by examining $\log N_n(A_0)$ given by

$$\log N_n(A_0) = \log N_{n-1}(A_1) + \log(N_{n-1}(A_0) + 1). \tag{3.138}$$

By ignoring the "+1", this equation may be used to get a lower bound on $\log N_n(A_0)$ which is asymptotically tight. With increasing n, $N_n(A_0) \to \infty$ so $\log(N_n(A_0)+1) \to \log N_n(A_0)$. Since, $\log(N_n(A_1))/n \to H$ as $n \to \infty$, this implies that $\log(N_n(A_1) + 1)/n \to H$ with n, yielding $(\log(N_n(A_0)) - \log(N_{n-1}(A_0)))/n - 1 \to H$. From this it follows that the branching rate $\rho = 1$ and that

$$g_n(A_0) - g_{n-1}(A_0) \to (n-1)H. \tag{3.139}$$

Thus $g_n(A_0)$ is asymptotically a polynomial in n of degree 2.

Example 3.48 (Pseudolinear Grammars) Now examine the set of pseudolinear languages [86, 87] which demonstrate the important role of the strong connectedness property. In these examples the grammars have more than one non-terminal on their right-hand side, yet $\rho = 1$, but have higher growth rates than regular in that $\log N_n$ are polynomial in n. This language lies between the context-free languages having branching rate $\rho > 1$, and the regular languages for which their log-rate of growth is linear in n, and illustrates properties of the entire family of pseudolinear grammars characterized by Kuich [87].

A grammar G is called pseudolinear if it is never the case that for any finite set of substitutions to any non-terminal A_v can a derivation $A_v \to \phi_1 A_v \phi_2 A_v \phi_3$ be generated, where ϕ's are strings of terminals and non-terminals, possibly empty. As shown by Kuich such a grammar can be decomposed into a set of strongly connected subcomponents each of which contains rewrite rules which have at most one non-terminal on their right-hand side. For each of these subcomponents, the branching rates are at most 1, and the combinatorics can be calculated directly as has been done for the regular component in the above example. To determine the branching rate of all rewrite rules having more than one non-terminal on its right side, recursive equations for the log-rates of growth are derived, where the homogeneous growth parameter is 1 (since there can be at most one syntactic variable equal to itself on the right side of any of these rewrite rules or it would not be pseudolinear). The additive driving function on this linear equation is n-varying and given by the log-rate of growth of the linear, strongly connected subcomponents. For example, in the above the homogeneous equation $\log N_n(A_0) = \log N_{n-1}(A_0)$ has $\rho = 1$, and driving function $\log N_{n-1}(\sigma) = (n-1)H$, with the driving function growing with n giving the polynomial properties for $\log N_n$.

To illustrate this approach more comprehensively, now examine a more complex example in the class of pseudolinear grammars from Kuich [86]. Define $G_4 = \langle V_T, V_N, R, A_1 \rangle$ with terminal symbols $V_T = \{a,b,c,d,e\}$, syntactic variables $V_N = \{A_1, A_2, A_3, A_4, A_5, A_6\}$, and production rules

$$R = \{A_1 \to A_1 A_2 A_5 \quad A_2 \to a A_2 a \quad A_2 \to A_3 A_6 \quad A_3 \to A_5 A_4 \tag{3.140}$$

$$A_4 \to A_3 A_6 \quad A_5 \to cA_5c \quad A_5 \to d \quad A_6 \to bA_6 b \quad A_6 \to \}. \tag{3.141}$$

The combinatorics for A_5, A_6 are given by linear equations $N_n(A_5) = N_{n-1}(A_5) + 1$, $N_n(A_6) = N_{n-1}(A_6) + 1$, giving $N_n(A_5) = N_n(A_6) = n$. Next, for A_3,

$$N_n(A_3) = N_{n-1}(A_5) N_{n-1}(A_4) = (n-1)(n-2) N_{n-2}(A_3). \tag{3.142}$$

3.10 FORMAL LANGUAGES AND STOCHASTIC GRAMMARS

This is a linear equation in the log with solution $\log N_n(A_3) = \sum_{i=2}^{n-1} \log i$, yielding from Stirling's formula for large n

$$(n - \tfrac{1}{2})\log(n-1) - (n-1) + k_2 \ < \ \log N_n(A_3) \ < \ (n - \tfrac{1}{2})\log(n-1) - (n-1) + k_1 \tag{3.143}$$

implying that $\rho = 1$ and that $\log N_n(A_3)$ grows as $(n - \tfrac{1}{2})\log(n-1) - (n-1)$. Now for A_2,

$$N_n(A_2) = N_{n-1}(A_2) + N_{n-1}(A_3)N_{n-1}(A_6) = N_{n-1}(A_2) + (n-2)!(n-1), \tag{3.144}$$

$$= \sum_{i=1}^{n}(i-1)! = (n-1)!(1 + O(n^{-1})). \tag{3.145}$$

Clearly $\rho = 1$ and $\log N_n(A_2)$ is similar to $\log N_n(A_3)$. Finally, the start state has combinatorics determined by

$$\frac{N_n(A_1)}{N_{n-1}(A_1)} = N_{n-1}(A_2)N_{n-1}(A_5) = \left(\sum_{i=1}^{n-1}(i-1)!\right)(n-1). \tag{3.146}$$

giving

$$N_n(A_1) = \prod_{k=1}^{n}\left((k-1)\sum_{i=1}^{n}k - 1(i-1)!\right) = (n-1)!\sum_{i=1}^{n-1}((i-1)!)^{n-l}. \tag{3.147}$$

Asymptotically, this implies $\log N_n(A_1)$ grows at least as a quadratic in n times $\log n$.

Example 3.49 (Binary Trees) Let $V = 2, \{A_1, A_2\}$, with birthing laws

$$\begin{array}{ll} A_1 \stackrel{p}{\to} A_1 A_1 & A_1 \stackrel{1-p}{\to} A_2 A_2 \\ A_2 \stackrel{1}{\to} A_1 A_2 & \end{array}. \tag{3.148}$$

Now we can calculate the probabilities to maximize the entropy H. Start with the branching matrix

$$M = \begin{pmatrix} 2p & 2-2p \\ 1 & 1 \end{pmatrix}, \tag{3.149}$$

having largest eigenvalue $\rho = 2$ irrespective of p.

Example 3.50 (Arithmetic Expression Grammar) Now the branching corresponding to the arithmetic expression language [84, 88] is examined and is demonstrated to have branching rate $\rho = 1.75488$. Let $V = 4, \{A_1, A_2, A_3, A_4\}$ with $A_1 =$ roman expression, $A_2 =$ roman term, $A_3 =$ roman factor, $A_4 =$ roman variable, with birthing rules

$$\begin{array}{llll} A_1 \stackrel{p_{11}}{\to} A_1 + A_2 & A_2 \stackrel{p_{21}}{\to} A_2 * A_3 & A_3 \stackrel{p_{31}}{\to} (A_1) & A_4 \stackrel{p_{41}}{\to} a \\ A_1 \stackrel{p_{12}}{\to} A_1 - A_2 & A_2 \stackrel{p_{22}}{\to} A_2/A_3 & A_3 \stackrel{p_{32}}{\to} A_4 & A_4 \stackrel{p_{42}}{\to} b \\ A_1 \stackrel{p_{13}}{\to} A_2 & A_2 \stackrel{p_{23}}{\to} A_3 & & A_4 \stackrel{p_{43}}{\to} c \\ & & & A_4 \stackrel{p_{44}}{\to} d \end{array}. \tag{3.150}$$

A derivation of the expression $a * b$ takes the form

$$A_1 \to A_2 \to A_2 * A_3 \to A_3 * A_3 \to \text{roman} * A_3$$
$$\to \text{roman} * \text{roman} \to a * \text{roman} \to a * b. \tag{3.151}$$

Associate with the first two rewrite rules of A_1 probabilities p_{11} and p_{12} with the third rule probability $p_3^{(1)} = 1 - p_{11} - p_{12}$. Let the probabilities for rewriting A_2 be given by p_{21}, p_{22} and $1 - p_{21} - p_{22}$, respectively. The mean matrix becomes

$$M = \begin{pmatrix} p_{11} + p_{12} & 1 & 0 \\ 0 & p_{21} + p_{22} & 1 \\ 1 & 0 & 0 \end{pmatrix}. \tag{3.152}$$

Clearly, choosing $p_{11} + p_{12} = 1$ and $p_{21} + p_{22} = 1$ gives the matrix $\bar{M}$ with largest eigenvalue, implying the branching rate is the largest eigenvalue of

$$\begin{pmatrix} 1 & 1 & 0 \\ 0 & 1 & 1 \\ 1 & 0 & 0 \end{pmatrix}. \tag{3.153}$$

The branching rate is the largest root of the characteristic equation $\lambda^3 - 2\lambda^2 + \lambda - 1 = 0$ which is $\rho = 1.75488$.

To emphasize the significance of the branching parameter, shown in Figure 3.9 are the results of counting arithmetic expression programs as a function of generation level n. Panel 1 shows a plot of $\log(\log(N_n(1)))$ versus n, with N_n the number of terminated arithmetic expressions. This plot is linear in n for large n (note that $n = 12$ is as far as could be calculated using exponents less than 300). Panel 2 shows $\log N_n(1) / \log N_{n-1}(1)$ converging to $\rho = 1.75488$, demonstrating that the branching rate calculated for the branching process equals the branching rate for the context-free language as predicted by the theory. Panel 3 demonstrates a deterministic large deviation result showing that for $\delta = 0.08$, $(\rho \pm \delta)^n$ bounds $\log N_n(1)$ for large n.

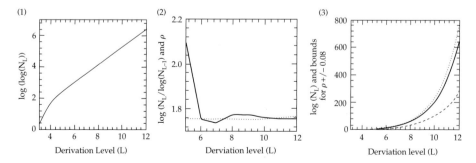

Figure 3.9 Figure shows super-exponential growth rates of the arithmetic expression language. Panel 1 shows log–log growth rate with derivation depth n. Panel 2 shows the near line ratio of the logarithms of successive generations. Panel 3 shows branching rate upper and lower bounds for ρ. Results taken from O'Sullivan [79].

3.11 DAGs for Natural Language Modelling

There are two fundamental approaches to language modeling: knowledge-based approaches and statistical approaches. In recent years, the success of statistical approaches has renewed interest in the statistical analysis of text [89]. In particular, the success of statistical part-of-speech taggers [90, 91] and of trigram models in speech recognition systems [92] has promoted the usefulness of statistical models. The two fundamental statistical language models are the m-gram model and the stochastic context-free model. Both models have their distinct advantages and disadvantages which will be discussed after presenting each model.

3.11.1 Markov Chains and m-Grams

The m-gram language models of Shannon [6] model the language as an mth-order Markov chain on the word strings. A drawback of this model is its failure to capture the hierarchical structure of language. For applications such as message understanding or machine translation, the syntactic structure of a sentence is useful and desired. In order to address the problem of incorporating syntactic structure, recent research [93, 94] has focused on stochastic context-free language models.

Figure 3.10 depicts several of these language models. Panel 1 shows the Markov chain models, and panel 2 shows the context-free random branching process models.

Considering the dependencies between words as Shannon did in his discussion of successive approximations to English [85] results in the Markov chain graph. Since language is produced as a time sequence of words, a first approach towards modeling it is a stationary Markov chain. The m-gram model treats a string of words as a realization of an mth-order Markov chain. The joint probability is then the product of the conditional transition probabilities according to

$$P(W) = \prod_{i=1}^{n} P(W_i | W_{i-1}, \ldots, W_{i-m+1}). \tag{3.154}$$

For fitting model parameters to the statistics of natural language, standard ML estimators are used. For bigrams, define the counting function $N_{ij}(W) = \sum_{k=2}^{g} 1_{ij}(W_{k-1}, W_k)$ the number of occurrences of the sequence ij in the string $W_1, W_2, \ldots$. The MLE of the model transition

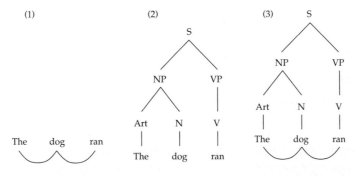

Figure 3.10 Three graph structures for natural language: Panel 1 Markov chain graph, panel 2 tree graph, and panel 3 directed acyclic graph.

Table 3.1 Data support for bigram and trigram estimation

Data Support	Bigrams With Support	Trigrams With Support
1	250,146	587,339
2–9	83,129	83,043
10–99	11,393	6,334
100–999	612	222
>1000	32	7
Total	345,312	676,945

probabilities maximizes

$$\log P(W_1, \ldots, W_n) = \log P(W_1) + \sum_i \sum_j N_{ij}(W) \log Q_{ij}, \qquad (3.155)$$

and adding the summability condition $\sum_j Q_{ij} = 1$ gives the ML solution $\hat{Q}_{ij} = ((N_{ij}(W))/\left(\sum_j N_{ij}(W)\right)$. Similarly, the trigram MLEs are given by $\hat{Q}_{(ij)k} = ((N_{(ij)k}(W))/\left(\sum_k N_{(ij)k}(W)\right)$.

While these estimates are asymptotically unbiased, it is often the case that there is not enough data to adequately support an estimate. The data support for trigrams and bigrams taken from the Penn TreeBank are shown in Table 3.1. Of particular interest is the fact that for over 72% of the bigrams and 86% of the trigrams there is only one occurrence in our dataset of a million words. In fact in a small held-out test set of 43,000 words, 30% of the bigram types did not appear in the training set and 60% of the trigram types did not appear in the training set. For sentences containing these missing bigrams or trigrams, zero probability would be assigned to them by the corresponding Markov chain model.

Several approaches have been taken to resolve this problem including deleted estimation and the use of the Good–Turing method [95, 96]. For this the deleted interpolation method of the IBM group is used to interpolate the parameters.[16] This method is used for the interpolation of the bigram and trigram models from the Penn TreeBank shown below. In the part-of-speech tagging problem, trigram models ($m = 3$) have achieved high performance for estimating the parts-of-speech of a word string with >95% accuracy [90]. The trigram model has also proven to be valuable in speech recognition systems because of its accuracy. However, this accuracy comes at a cost. The number of parameters in the trigram model is proportional to N_w^3 where N_w is the number of distinct words in the domain of interest. For example, the number of possible parameters in a typical 20,000 word vocabulary is $20,000^3 = 8 \times 10^{12}$! Although a large number of these parameters are zero, there is still a large number of parameters. In the Penn Treebank data used, there are 744,000 nonzero trigrams out of a possible $90,000^3$ parameters.

3.11.2 Context-Free Models

The m-gram models have large numbers of parameters and fail to capture the hierarchical structure of language. For applications such as message understanding or machine translation, the syntactic

---

[16] The deleted interpolation method [97, 98] makes use of lower order models to smooth our parameters. In this case, the interpolated probabilities are given by

$$\hat{p}(W_3|W_1, W_2) = \lambda_1(W_1, W_2)p(W_3) + \lambda_2(W_1, W_2)p(W_3|W_2) + (1 - \lambda_1 - \lambda_2)p(W_3|W_1, W_2)$$

The interpolated probability of a trigram is given in terms of weighted unigram, bigram and trigram terms. The estimation of the interpolation weights is done by using the forward-backward algorithm for hidden Markov models on held-out training data [99].

structure of a sentence is useful and desired. In order to address the problem of incorporating syntactic structure, recent research [93,94] has focused on stochastic context-free language models.

This shortcoming has serious implications for higher level speech processing such as message understanding or machine translation. What is needed for these applications in the context-free model based on a stochastic context-free grammar which is a quintuple, $\langle V_N, V_T, R, S, P \rangle$, where V_N is a finite set of non-terminal symbols, V_T a finite set of terminal symbols, R the set of rewrite rules, S the start symbol ($S \in V_N$), and P the parameter vector such that if $r \in R$, then P_r is the probability of using the rewrite rule r. Strings in the language defined by grammar are derived by successive applications of the rewrite rules to nonterminals beginning with the start symbol S. Associated with each terminal sequence is a labeled tree corresponding to the hierarchical application of the rewrite rules.

Figure 3.11 shows a small grammar written by Steve Abney of BellCore. This core grammar contains rules that form a foundation for most context-free grammars. Abney has also provided a more substantial grammar containing 411 rules. The probability associated with each rule is shown to the left of the rule. Note that the probabilities for rules rewriting a given left-hand side symbol, such as "Matrix", add up to one.

The tree $T = (R_1, R_2, \ldots, R_{n_T})$ is a sequence of rules applied in order to the leftmost non-terminal which derives the preterminal string $\gamma_1, \gamma_2, \ldots, \gamma_n$ where $\gamma_i \in V_P$ and the word string $W = W_1, W_2, \ldots, W_n$ are the terminal symbols. The probability of a derivation tree for a given tree T, W

$$\pi(T, W) = \pi(T) \prod_{i=1}^{n} \pi(W_i | \gamma_i) \tag{3.156}$$

$$= \prod_{i=1}^{n_T} P_{R_i(T)} \prod_{i=1}^{n} \pi(W_i | \gamma_i), \tag{3.157}$$

where n_T is the number of rules in tree T and $R_i(T)$ is the ith rule used in tree T.

For the tree in Figure 3.10,

$$\pi(T, W) = P_{S \to \text{NP VP}} P_{\text{NP} \to \text{Art N}} P_{\text{VP} \to \text{V}}$$
$$\pi(\text{'The'}|\text{Art}) \pi(\text{'dog'}|N) \pi(\text{'ran'}|V).$$

0.85	Matrix → S		0.05	VP → VP adv
0.05	Matrix → Wh-Question		0.10	VP → VP PP
0.05	Matrix → Yes-No-Question		0.30	VP → v
0.05	Matrix → Imperative		0.20	NP → NP-Core
0.60	Wh-Question → Wh-NP aux NP VP		0.20	NP → det NP-Core
0.40	Wh-Question → Wh-NP VP		0.20	NP → pron
1.00	Yes-No-Question → aux NP VP		0.20	NP → NP PP
1.00	Imperative → VP		0.20	NP → n
1.00	S → NP VP		0.30	NP-Core → adj NP-Core
1.00	That-Clause → that S		0.70	NP-Core → n
1.00	Infinitive → to VP		0.30	Wh-NP → wh det NP-Core
0.30	VP → v NP		0.60	Wh-NP → whpron
0.10	VP → v NP PP		0.10	Wh-NP → Wh-NP PP
0.10	VP → v That-Clause		1.00	PP → p NP
0.05	VP → v Infinitive			

Figure 3.11 Stochastic context-free grammar constructed by Steven Abney.

The probability of a word string $W_{1,n} = W_1 W_2 \ldots W_n$ is given by

$$P(W_1,\ldots,W_n) = \sum_{T \in \text{Parses}(W_1,\ldots,W_n)} \pi(T) \tag{3.158}$$

where Parses(W) is the set of parse trees for the given word string. For an unambiguous grammar, Parses(W) consists of a single parse.

3.11.3 Hierarchical Directed Acyclic Graph Model

Examine the class of stochastic language models introduced by Mark [100, 101] incorporating both the lexical dependence in context-free natural language models with the Markov chain statistical dependences of word relations in sentences. This is depicted in panel 3 of Figure 3.10. Such a model is a hierarchical, layered directed acyclic graph in which word-tree derivations are given by a stochastic context-free prior on trees down to the preterminal (part-of-speech) level and word attachment is made by non-homogeneous Markov chain models. The hierarchical digraph model of Mark [100, 101] addresses the inadequacies of the two basic language models providing the hierarchical structure of context-free languages while maintaining the

The Mark hierarchical digraph language model with conditional word dependencies is a directed acyclic graph. The strengths of both the m-gram model and the stochastic context-free model are combined by adding bigram/trigram relative frequencies as constraints on the stochastic context-free trees. The directed acyclic graph structure is depicted in panel 3 of Figure 3.10. The word string with its bigram relationships and its underlying parse tree structure is depicted.

In Mark's digraphs [101] the preterminals or parts-of-speech in the word-tree configuration are considered to be the boundary between two Markov processes, one the random branching process which generates the preterminal sequence and the other the Markov chain which attaches words to the generated preterminal sequence. The stochastic context free grammar generates the parts-of-speech, Art N V P Art N. Now, given this preterminal sequence, words are attached according to a Markov chain.

The stochastic context free grammar. An element $T \in \mathcal{T}$ is a tree with part-of-speech leaves, $\gamma_i, i = 1, \ldots, n$. The probability of the word sequence given the tree, $P(W|T)$ is considered here to be an inhomogeneous Markov chain in the words given the preterminal sequence, i.e. $P(W|T) = \prod_{i=1}^{n} P(W_i|\gamma_i, W_{i-1})$ where γ_i is the ith preterminal. The resulting DAG is neither context-free nor strictly a Markov chain, with probability of (W, T) written

$$P(W,T) = \frac{1}{K} \exp\left(\sum_{k=1}^{N} \log P(W_k|\gamma_k, W_{k-1}) + \sum_{i=1}^{N_t} \log P_{r_i(T)} \right). \tag{3.159}$$

The normalizer is calculated by exploiting the fact that the sum of the probabilities $P(W|T)$ over all word strings may be either 0 or 1. If, for a given tree T, there does not exist a word string that may be attached to it, then $P(W|T) = 0$ for all $W \in \mathcal{W}$. We shall say that for this case "the tree is non-parseable". Hence, partitioning the space of trees $\mathcal{T} = \tau_P \cup \tau_P^c$ where τ_P is the set of parseable-trees (exists at least one non-zero probability word string attachment), then $\sum_{W \in \mathcal{W}} P(W|T) = 1_{\tau_P}(T)$, so for a complete word-tree configuration (w, t), the probability is given by the probability of all consistent parses that have word strings $K = \pi(\tau_P)$ for $\tau_P = \{t : t \in \text{Parse}(W) \text{ for some } W \in \mathcal{W}\}$. In order to gain a better understanding of these two models we consider the following example. First, we are given a simple noun phrase grammar:

0.8	NP → DET N	1.0	ADJ → red
0.2	NP → DET ADJ N	0.5	N → car
1.0	DET → the	0.5	N → dog

3.11 DAGS FOR NATURAL LANGUAGE MODELLING

Suppose have the following observations and parameters,

Noun Phrase	Number of Occurrences	$\sigma_a\sigma_b$	$\hat{H}_{\sigma_a,\sigma_b}$	$\alpha_{\sigma_a,\sigma_b}$
The car	40	The car	0.4	0.0
The dog	40	The dog	0.4	0.0
The red car	18	The red	0.1	0.0
The red dog	2	Red car	0.09	1.176
		red dog	0.01	−3.219

With these parameters we have the following tilted distributions Note that under the prior,

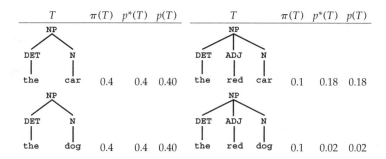

the two trees for "the red car" and "the red dog" are equally likely. However, when we add the bigram constraints, the likelihood of "the red dog" is greatly reduced.

The entropies of natural languages have been looked at historically in the information theory community by many, including originally Shannon [85], then later by Cover and King [102], and [103]. Mark [101, 104] was the first to calculate the entropies of the directed acyclic graph model.

Theorem 3.51 *The entropy of the DAG model given by the joint entropy $H(W,T) = H(T) + H(W|T)$ where τ_P is the set of parseable trees is given by*

1.

$$H(T) = \frac{1}{\pi(\tau_P)} H_\pi(T) + \frac{1}{\pi(\tau_P)} \sum_{t \in \tau_P^c} \pi(t) \log \pi(t) + \log \pi(\tau_P), \qquad (3.160)$$

where $H_\pi(T)$ is the entropy of the random branching process Theorem 3.34, and

2.

$$H(W|T) = \sum_{t \in \tau_P} \frac{\pi(t)}{\pi(\tau_P)} H(W|T = t), \qquad (3.161)$$

where $H(W|T = t)$ is the entropy of the non-stationary Markov chain.

Proof The second term on the right-hand side is given by the Markov chain. In order to compute this entropy, $H(T)$, the marginal probability of a tree t is given by

$$p(T) = \frac{\pi(T) 1_{\tau_P}(T)}{\pi(\tau_P)}. \qquad (3.162)$$

The entropy of the trees which can be parsed into legal bigram/trigram leaf strings

$$H(T) = E\{-\log P(T)\} = -\sum_{t \in \mathcal{T}} \frac{\pi(t) 1_{\tau_P}(t)}{\pi(\tau_P)} \log \frac{\pi(t) 1_{\tau_P}(t)}{\pi(\tau_P)} \quad (3.163)$$

$$= -\frac{1}{\pi(\tau_P)} \sum_{t \in \tau_P} \pi(t) \log \pi(t) + \log \pi(\tau_P) \quad (3.164)$$

$$= \frac{1}{\pi(\tau_P)} H_\pi(T) + \frac{1}{\pi(\tau_P)} \sum_{t \in \tau_P^c} \pi(t) \log \pi(t) + \log \pi(\tau_P), \quad (3.165)$$

where $H_\pi(T)$ is the entropy of the random branching process Theorem 3.34, Eqn. 3.101. The second term is dependent on the partition τ_P of the tree space $\mathcal{T}$ and must be computed numerically. □

Such an entropy measure can be used to compare the entropies of the four language models: bigrams, trigrams, context-free, and the layered Markov model. The parameters for these models were estimated from a subset of Dow Jones newswire articles from the Penn TreeBank corpus. This data has 1,013,789 words in 42,254 sentences which have been machine-parsed and hand-corrected. The number of rules in the underlying context-free grammar are 24,111 down to the preterminal level and 78,929 down to the word level. There are 389,440 distinct bigrams in the data set and 744,162 distinct trigrams.

The performance of such models for limiting the uncertainty of the language can be studied by calculating the empirical entropy of the probability law generated from empirical representations of the underlying statistics. This was done by Mark for the bigram model with parameters estimated from a subset of Dow Jones newswire articles from the Penn TreeBank corpus. The entropy of the purely context-free model can be calculated from the same corpus as for the bigram model. Using the parameters derived from the corpus, (see Table 3.1). Figure 3.12 shows the model entropies for the purely bigram/trigram ($m = 2, 3$) models and the context-free model. The model entropy has the interpretation of the number of bits required to code a configuration (word string, tree, or word-tree) under the given language model. The entropy of the

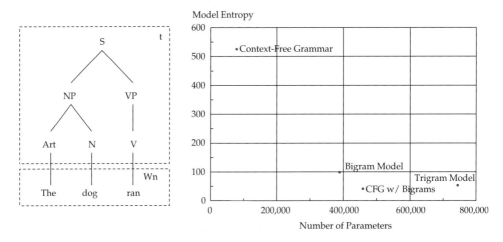

Figure 3.12 Left panel shows two components of a derivation tree T: the tree t deriving the preterminal string "Art N V" and the word string W_n = "The dog ran". Right panel shows comparison of the model entropy of four language models, bigrams, trigram, context-free, and directed acyclic graph.

Markov chain bigram model is given according to Lemma 3.31, Eqn. 3.82. The state transition parameters of the model were estimated using the ML estimates from the bigram counts and trigram counts using the deleted interpolation method to interpolate the transitions for which there was insufficient data. The bigram model has entropy of 99 bits; the trigram model has entropy of 53 bits.

The entropy of the purely context-free model can be calculated from the same corpus as for the bigram model. Using the conditional probabilities generated via the ML estimates from the empirically estimated Penn Treebank training corpus, Figure 3.12 shows the model entropy for the context-free model $H_\pi(T)$ given by Theorem 3.34, Eqn. 3.101. The context-free model has entropy $H_\pi(T) = 525$ bits.

Figure 3.12 also shows the model entropy for the layered digraph model compared to the context-free and bigram models alone plotted versus the number of parameters in the corresponding model. The model entropy has the interpretation of the number of bits required to code a configuration (word string, tree, or word-tree) under the given language model. A dramatic reduction in the model entropy is seen in the layered digraph model with entropy of 41 bits. This order-of-magnitude reduction in entropy from that of the SCFG demonstrates the effect of the bigram constraints. These constraints place probability only on configurations that satisfy the lexical preferences in the training corpus which rules out a very large number of trees which violate these preferences.

This reduction in entropy does come at the cost of additional parameters since approximately 400,000 parameters have been added in the form of bigram constraints. However, the entropy of the trigram model is 53 bits which is greater than that of the layered Markov model. So, even though it has almost twice as many parameters as the layered Markov model, the trigram model has higher entropy. The number of additional parameters added via context-free rules to the bigram model is a fraction of those added via the trigram model, yet a comparable reduction in entropy results.

3.12 EM Algorithms for Parameter Estimation in Hidden Markov Models

Typical speech recognition systems use hidden Markov models to estimate strings of phonemes from the acoustic signal (see the review [105]). At a higher level of processing models are used which describe the way in which words or subwords (phonemes) are put together to form sentences. Hidden Markov models (HMMs) with directed graph structures often provide a computationally efficient framework for computing probabilities of sequences. Methods based on such computationally efficient structures have been given many names, dynamic programming, Dijkstra's algorithm, Viterbi algorithm [106]. For a beautiful paper on this topic, see Forney's 1973 incisive paper [107]. The particular model descriptions presented here follow the developments of V. Goel and S. Kumar of the Center for Language and Speech Processing at Johns Hopkins University. Examine the HMM DAG depicted in Figure 3.13 showing the hidden process a Markov chain $X_1, X_2, \ldots$ of states in $\mathcal{X}_0$ with transition probabilities $Q_{X_{i-1}X_i} = P(X_i|X_{i-1})$, and the output or observation process $Y_1, Y_2, \ldots$ taking values in $\mathcal{Y}$ with output probability stationary $o_{X_i \to Y_i} = P(Y_i|X_i)$, and initial state distribution $\pi(X_1)$.

The practical usability of HMMs arise from their efficient solution to the following two basic problems: (i) how to choose a state sequence $(X_1, X_2, \ldots)$ that is maximum *a posteriori* probability given the observation sequence $(Y_1, Y_2, \ldots)$, (ii) how to efficiently compute $P_\phi(Y_1, Y_2, \ldots)$ and adjust the parameters ϕ of the model so as to maximize $P_\phi(Y_1, Y_2, \ldots)$ given the observation sequence $(Y_1, Y_2, \ldots)$.

3 PDAGS AND THEIR ENTROPIES

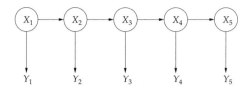

Figure 3.13 The Hidden Markov model with hidden state sequence $X_1, X_2, \ldots$ and conditionally independent outputs $Y_1, Y_2, \ldots$ given the state-sequence.

3.12.1 MAP Decoding of the Hidden State Sequence

Examine the graph structure dictating the conditional independence of the time evolution of the process. This is best represented by a structure called *trellis*, as shown in Figure 3.14.

Examine the problem of deciding on a hidden state sequence $X_1, X_2, \ldots$ that best explains the observation $Y_1, Y_2, \ldots$. The MAP state sequence is given by

$$(\bar{X}_1, \bar{X}_2, \ldots) = \arg\max_{(X_1, \ldots, X_n) \in \mathcal{X}_0^n} P(X_1, \ldots, X_n, Y_1, \ldots, Y_n). \tag{3.166}$$

Brute force calculation of the MAP estimator is $O(|\mathcal{X}_0|^n)$. Alternatively, dynamic programming based solutions based on the *Viterbi* algorithm [107] computes the MAP sequence in $O(n|\mathcal{X}_0|^2)$ computations. Examine Figure 3.14 illustrating a three-state graph for an $n = 5$ length state sequence $X_1, \ldots X_5$. The most likely state sequences are scored. The Viterbi algorithm exploits the fact that given the highest scoring paths to time k ending in all of the states, then to compute the solution to time $k+1$ one has to examine all of $|\mathcal{X}_0|$ states sequentially trying $|\mathcal{X}_0|$ new costs for each state, hence producing an order $|\mathcal{X}_0|^2$ algorithm per sequence length, and $n|\mathcal{X}_0|^2$ complexity algorithm for n-length strings.

Lemma 3.52 (Viterbi Algorithm) *Denote the score of the most likely state sequences up to time k ending in state $X_k = x$ as $S_k(x)$, defined as*

$$S_k(x) = \max_{X_1, \ldots, X_{k-1} \in \mathcal{X}_0^{k-1}} P(X_1, \ldots, X_{k-1}, X_k = x, Y_1, \ldots, Y_k). \tag{3.167}$$

The maximizing state sequence is generated recursively in order $O(|\mathcal{X}_0|^2)$ via the recursion with $S_1(x) = \pi(x) o_{x \to Y_1}$, then $S_n = \max_{x \in \mathcal{X}_0} S_n(x)$ with

$$S_{k+1}(x) = \max_{X_k \in \mathcal{X}_0} Q_{X_k x} o_{x \to Y_{k+1}} S_k(X_k), \quad x \in \mathcal{X}_0. \tag{3.168}$$

Proof

$$S_{k+1}(x) = \max_{(X_1, \ldots, X_k) \in \mathcal{X}_0^k} P(X_1, \ldots, X_k, X_{k+1} = x, Y_1, \ldots, Y_k, Y_{k+1}) \tag{3.169}$$

$$= \max_{(X_1, \ldots, X_k) \in \mathcal{X}_0^k} P(X_{k+1} = x, Y_{k+1} | X_1, \ldots, X_k, Y_1, \ldots, Y_k) P(X_1, \ldots, X_k, Y_1, \ldots, Y_k) \tag{3.170}$$

$$= \max_{X_k \in \mathcal{X}_0} P(X_{k+1} = x | X_k) P(Y_{k+1} | x) S_k(X_k) = \max_{X_k \in \mathcal{X}_0} Q_{X_k x} o_{x \to Y_{k+1}} S_k(X_k). \tag{3.171}$$

□

3.12 EM ALGORITHMS FOR PARAMETER ESTIMATION IN HMMS

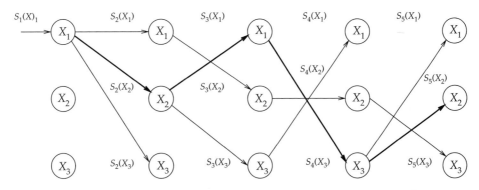

Figure 3.14 Representation of the Viterbi algorithm on a three-state state space. The arcs in bold show the path which obtains the state sequence with the highest output probability.

3.12.2 ML Estimation of HMM parameters via EM Forward/Backward Algorithm

Examine ML estimation of the model parameters $\pi(x), Q_{xx'}, o_{x \to y}$. For estimating the parameters of the hidden state sequences, procedures based on the EM algorithm have been used extensively (see Section 2.7). To see the difficulty, a direct assault on maximizing likelihood with respect to the parameters $\phi = (\pi, Q_{xx'}, o_{x \to y})$ requires the computation of the observation sequence

$$P_\phi(Y_1, \ldots, Y_n) = \sum_{(X_1, \ldots, X_n) \in \mathcal{X}_0^n} P_\phi(X_1, \ldots, X_n, Y_1, \ldots, Y_n) \tag{3.172}$$

$$= \sum_{(X_1, \ldots, X_n) \in \mathcal{X}_0^n} \pi(X_1) o_{X_1 \to Y_1} \prod_{i=2}^n Q_{X_{i-1} X_i} o_{X_i \to Y_i}, \tag{3.173}$$

where $\mathcal{X}_0^n$ is the set of all n length state sequences. Because of the hidden variates, direct calculation of maximizers is difficult.

Here is the Baum–Welch algorithm [108] commonly used in speech recognition. It is an EM algorithm benefiting from the properties of monotonicity of likelihood.

Algorithm 3.53 (Baum–Welch) *Defining $\phi^m = (\pi^m, Q_{xx'}^m, o_{x \to y}^m)$ and the sufficient statistics $N_{xx'}(X)$ the number of times state x' follows state x and $N_{x \to y}(X, Y)$ is the number of times state x outputs observation y, then the sequence of iterates $\phi^1, \phi^2, \ldots$ defined by the following iteration are an EM algorithm:*

$$Q_{xx'}^{new} = E(N_{xx'} | \phi^{old}, Y) = \frac{\sum_{k=1}^{n-1} P_{\phi^{old}}(X_k = x, X_{k+1} = x' | Y)}{\sum_{k=1}^{n-1} P_{\phi^{old}}(X_k = x | Y)} \tag{3.174}$$

$$o_{x \to y}^{new} = E(N_{x \to y} | \phi^{old}, Y) = \frac{\sum_{k=1}^{n} P_{\phi^{old}}(X_k = x | Y) 1_{\{y\}}(y_k)}{\sum_{k=1}^{n} P_{\phi^{old}}(X_k = x | Y)}, \tag{3.175}$$

with initial state estimate $\pi(X)^{new} = P_{\phi^{old}}(X_1 | Y)$.

Proof That this is an EM algorithm (Theorem 2.52) follows with complete data $(X, Y) = (X_1, \ldots, X_n, Y_1, \ldots, Y_n)$ and complete data likelihood written in terms of the sufficient statistics

$$P(X,Y) = \pi(X_1) \prod_{k=2}^{n} P(X_k|X_{k-1}) \prod_{k=1}^{n} P(Y_k|X_k)$$

$$= \pi(X_1) \prod_{x \in \mathcal{X}_0} \prod_{x' \in \mathcal{X}_0} Q_{xx'}^{N_{xx'}(X,Y)} \prod_{x \in \mathcal{X}_0} \prod_{y \in \mathcal{Y}} o_{x \to y}^{N_{x \to y}(X,Y)}.$$

The expectation E-step in the EM algorithm (Eqn. 2.154) computes the expected value of the complete data log-likelihood given the data $Y = Y_1, Y_2, \ldots$ and the previous estimates of the parameters $\phi = \{\pi, Q_{xx'}, o_{x \to y}\}$ according to

$$E\big(\log P(X,Y)|Y,\phi\big) = \sum_{x \in \mathcal{X}_0} \sum_{x' \in \mathcal{X}_0} E\big(N_{xx'}(X)|Y,\phi\big) \log Q_{xx'}$$

$$+ \sum_{x \in \mathcal{X}_0} \sum_{y \in \mathcal{Y}} E\big(N_{x \to y}(X,Y)|Y,\phi\big) \log o_{x \to y}$$

$$+ \sum_{x \in \mathcal{X}_0} E\big(1_{\{x\}}(X_1)|Y,\phi\big) \log \pi(x).$$

For the maximization M-step (Eqn. 2.156), the conditional expectation is maximized subject to normalization constraints on the probabilities. For the state transition parameter $Q_{xx'}$, maximize the first term of Eqn. 3.176 with the constraint $\sum_{x' \in \mathcal{X}_0} Q_{xx'} = 1$ giving

$$Q_{xx'} = \frac{E\big(N_{xx'}(X)|Y,\phi\big)}{\sum_{x' \in \mathcal{X}_0} E\big(N_{xx'}(X)|Y,\phi\big)} = \frac{\sum_{k=1}^{n-1} P_\phi(X_k = x, X_{k+1} = x'|Y)}{\sum_{k=1}^{n-1} P_\phi(X_k = x|Y)}.$$

For the output distribution parameter $o_{x \to y}$, maximize the second term of Eqn. 3.176 with constraint $\sum_{y \in \mathcal{Y}} o_{x \to y} = 1$ giving

$$o_{x \to y} = \frac{E\big(N_{x \to y}(X,Y)|Y,\phi\big)}{\sum_{x \in \mathcal{X}_0} E\big(N_{x \to y}(X,Y)|Y,\phi\big)} = \frac{\sum_{k=1}^{n} P_\phi(X_k = x|Y) 1_{\{y\}}(y_k)}{\sum_{k=1}^{n} P_\phi(X_k = x|Y)}.$$

For the initial state parameter $\pi(x)$, $\hat{\pi}(x) = ((E\{1_{\{x\}}(X_1)|Y,\phi\})/(\sum_{x \in \mathcal{X}_0} E\{1_{\{x\}}(X_1)|Y,\phi\})) = P_\phi(X_1 = x|Y)$. □

To implement the EM algorithm, there is a core computation, the conditional probability of occupying a pair of states, $P_\phi(X_k = x, X_{k+1} = x'|Y)$. From this the single state conditional probability $P_\phi(X_k = x|Y)$ can be directly calculated. Brute force calculation would involve summing out the hidden variables of Eqn. 3.173, a prohibitive $O(|\mathcal{X}_0|^n)$ calculation. The recursive structure exploited by the Viterbi algorithm associated with the trellis diagram is commonly used. This is the so-called *Forward/Backward algorithm* commonly used in speech recognition.

Lemma 3.54 (Forward/Backward) *Defining the forward and backward scores*

$$\alpha_k(x) = P(Y_1, \ldots, Y_k, X_k = x), \qquad \beta_k(x) = P(Y_{k+1}, \ldots, Y_n|X_k = x) \qquad (3.176)$$

the joint state probabilities are computed order $O(n|\mathcal{X}_0|^2)$ according to

$$\xi_k(x,x') = P_\phi(X_k = x, X_{k+1} = x'|Y) \qquad (3.177)$$

$$= \frac{\alpha_k(x) Q_{xx'} o_{x' Y_{k+1}} \beta_{k+1}(x')}{\sum_{x \in \mathcal{X}_0} \alpha_n(x)}. \qquad (3.178)$$

3.12 EM ALGORITHMS FOR PARAMETER ESTIMATION IN HMMS

The EM algorithm parameters are computed according to

$$Q_{xx'} = \frac{\sum_{k=1}^{n-1} P_\phi(X_k = x, X_{k+1} = x'|Y)}{\sum_{k=1}^{n-1} P_\phi(X_k = x|Y)} = \frac{\sum_{k=1}^{n-1} \xi_k(x,x')}{\sum_{k=1}^{n-1} \sum_{x' \in \mathcal{X}_0} \xi_k(x,x')}$$

$$o_{x \to y} = \frac{\sum_{k=1}^{n} P_\phi(X_k = x|Y) 1_{\{y\}}(y_k)}{\sum_{k=1}^{n} P_\phi(X_k = x|Y)} = \frac{\sum_{k=1}^{n} \sum_{x' \in \mathcal{X}_0} \xi_k(x,x') 1_{\{y\}}(y_k)}{\sum_{k=1}^{n} \sum_{x' \in \mathcal{X}_0} \xi_k(x,x')}.$$

with initial state estimate $\pi(X_1) = P_\phi(X_1|Y) = \sum_{x' \in \mathcal{X}_0} \xi_1(X_1, x')$.

Proof Using the forward and the backward scores, then the joint state probability becomes

$$\xi_k(x, x') = P(X_k = x, X_{k+1} = x'|Y)$$

$$= \frac{P(Y_1, \ldots, Y_k, X_k = x, X_{k+1} = x', Y_{k+1}, \ldots, Y_n)}{P(Y)} \quad (3.179)$$

$$= \frac{\alpha_k(x) Q_{xx'} o_{x'Y_{k+1}} \beta_{k+1}(x')}{\sum_{x \in \mathcal{X}_0} \alpha_n(x)}. \quad (3.180)$$

To demonstrate efficient procedures $O(n|\mathcal{X}_0|^2)$ for computing *forward* and *backward* scores exploit the conditioning structure of the graph. For $\alpha_k(x_k)$ with $\alpha_1(x) = \pi(x)P(Y_1|x)$, then

$$\alpha_{k+1}(x_{k+1}) = \sum_{x_k \in \mathcal{X}_0} P(Y_1, \ldots, Y_k, Y_{k+1}, x_k, x_{k+1})$$

$$= \sum_{x_k \in \mathcal{X}_0} P(Y_{k+1}, x_{k+1}|x_k, Y_1, \ldots, Y_k) P(Y_1, \ldots, Y_k, x_k)$$

$$= \sum_{x_k \in \mathcal{X}_0} o_{x_{k+1} Y_{k+1}} Q_{x_k x_{k+1}} \alpha_k(x_k) \quad \text{with } P(Y_1, \ldots, Y_n) = \sum_{x \in \mathcal{X}_0} \alpha_n(x). \quad (3.181)$$

Similarly for the *backward score* initializing with $\beta_n(x) = 1$, then

$$\beta_k(x) = P(Y_{k+1}, Y_{k+2}, \ldots, Y_n | X_k = x) \quad (3.182)$$

$$= \sum_{x' \in \mathcal{X}_0} P(Y_{k+1}, Y_{k+2}, \ldots, Y_n, X_{k+1} = x' | X_k = x), \quad (3.183)$$

$$= \sum_{x' \in \mathcal{X}_0} P(Y_{k+2}, \ldots, Y_n, | X_{k+1}, X_k = x, Y_{k+1},) P(Y_{k+1}, X_{k+1} = x' | X_k = x), \quad (3.184)$$

$$= \sum_{x' \in \mathcal{X}_0} P(X_{k+1} = x | X_k = x') P(Y_{k+1} | X_{k+1} = x) \beta_{k+1}(x). \quad (3.185)$$

□

Example 3.55 (**HMM Based System for Large Vocabulary Conversational Speech Recognition**) An automatic speech recognition(ASR) system was designed at the Center for Language and Speech Processing (CLSP) summer workshop [109] with the goal of producing a word-transcription of an acoustic input signal. We now describe a HMM-based ASR system for the Switchboard corpus [110]. Switchboard is a large vocabulary conversational speech recognition corpus where speech data is collected

Table 3.2 A summary of the automatic speech recognition system from a training corpus of 45 h with 2.2 million words from the 1997 CLSP Workshop [109] on a test corpus of 93 minutes with the dimension of the features 39

	Switchboard Corpus
Front-End Features	PLP Cepstral Coefficients [111]
HMM Topology	Three-state, Left-to-Right
Number of Gaussians/HMM State	12
Number of HMM states	7461
Number of Gaussian Mixtures	89,544
Word Error Rate	40.4%

over the telephone lines. The acoustic signal is processed by a front-end and converted into a sequence of acoustic feature vectors, typically 100 vectors or frames per second. Given this sequence of acoustic feature vectors O, the system seeks from amongst all word sequences W, the sequence $\hat{W}$ satisfying:

$$\hat{W} = \arg\max_{W \in \mathcal{W}} P(W|O) = \arg\max_{W \in \mathcal{W}} P(O|W)P(W), \tag{3.186}$$

where $P(W)$ is the *a priori* probability of a word sequence W and $P(O|W)$ is the conditional probability of observing O when the word sequence W is uttered by the speaker. The estimates of these probabilities are computed using parametric models and the parameters of these models are estimated from data. An acoustic model gives the estimate of $P(O|W)$ while a language model gives an estimate of $P(W)$.

The state-of-the-art ASR systems use a phonetic HMM as the acoustic model. Context dependent versions of phones are modeled by concatenating HMMs. In our system, each HMM state corresponds to a subphonetic unit with a multivariate Gaussian mixture as the output distribution. These Gaussian mixtures have diagonal covariances. A decision tree-based procedure was used to cluster the HMM states based on phonetic and word boundary information (Table 3.2).

The Language Model used in the CLSP system is a trigram model where the probability of the word string W is given by

$$P(W) = \prod_{i=3}^{n} P(W_i|W_{i-1}, W_{i-2}) \pi(W_1, W_2). \tag{3.187}$$

The classification performance in this task is measured by a weighted Levenshtein distance [112] between speech recognizer transcript and the manual transcription.

3.13 EM Algorithms for Parameter Estimation in Natural Language Models

An important problem in both of the proposed language models is the estimation of the model parameters. The *Inside-Outside* algorithm, first established by Baker [113], is used to estimate the rule probabilities in the stochastic context-free model (see Lari and Young [93] and Jelinek for reviews [94]). This estimation algorithm is a parallel of the forward-backward algorithm for hidden Markov models. It can be shown that the Inside-Outside algorithm is an EM algorithm.

These models clearly take care of the problem of linguistic structure. Primary among the problems is the failure to model lexical preferences such as those expressed by bigrams and trigrams.

3.13.1 EM Algorithm for Context-Free Chomsky Normal Form

For a grammar in Chomsky Normal Form, the familiar Inside/Outside Algorithm is used to estimate the stochastic grammar substitution rule probabilities. The Inside/Outside algorithm is an extension of the Baum–Welch re-estimation algorithm for hidden Markov models in which the complete data is the word string with its underlying parse tree, and the incomplete data is the observed word string.

For the estimation of rule probabilities from a word string $W_1, \ldots, W_n$, examine ML estimation of the model parameters $p_{\sigma \to r_0 r_1}$ the probabilities of rewriting non-terminal σ as $r_0 r_1$. Assume the context-free grammar comes in Chomsky Normal form [84,94], with each non-terminal rewriting into either two non-terminals $\sigma \to \sigma_1 \sigma_2$ or into a terminal or word $\sigma \to w$. Again the EM algorithm is used, appealing to the *Inside/Outside Algorithm* as first introduced by Baker [113] generalizing the *Forward/Backward* Baum–Welch algorithm for Markov chains (see 3.12.2).

Here the hidden state sequences correspond to the underlying parse of the word string; for nonambiguous grammars there would be a unique parse for each word string, although context-free grammars are generally ambiguous. Maximizing likelihood with respect to the parameters $\theta = (p_{\sigma \to \sigma_1 \sigma_2}, p_{\sigma \to w})$ uses the *Inside/Outside Algorithm* commonly used in language parsing which is an EM algorithm benefiting from the properties of monotonicity of likelihood.

Algorithm 3.56 *Defining* $\theta^m = (p^m_{\sigma \to \sigma_1 \sigma_2}, p^m_{\sigma \to w})$, *the sufficient statistics in the complete-data are the number of instantiations of the rewrites rules* $N_{\sigma \to \sigma_0 \sigma_1}(T), N_{\sigma \to w}(T)$ *in the underlying tree T. The sequence of iterates* $\theta^1, \theta^2, \ldots$ *defined by the following iteration are an EM algorithm:*

$$p^{\text{new}}_{\sigma \to \sigma_1 \sigma_2} = \frac{E\big(N_{\sigma \to \sigma_1 \sigma_2}(T)|W_1, \ldots W_n, \theta^{\text{old}}\big)}{E\big(N_\sigma(T)|W_1, \ldots, W_n, \theta^{\text{old}}\big)}, \tag{3.188}$$

$$p^{\text{new}}_{\sigma \to w} = \frac{E\big(N_{\sigma \to w}(T)|W_1, \ldots, W_n, \theta^{\text{old}}\big)}{E\big(N_\sigma(T)|W_1, \ldots, W_n, \theta^{\text{old}}\big)}. \tag{3.189}$$

Proof That this is an EM algorithm (Theorem 2.52) follows with complete data the words and tree parse (W, T) and complete data probability given by

$$P(W, T) = \prod_{\sigma \to \sigma_0 \sigma_1} p_{\sigma \to \sigma_0 \sigma_1}^{N_{\sigma \to \sigma_0 \sigma_1}(T)} \prod_{\sigma \to w} p_{\sigma \to w}^{N_{\sigma \to w}(T)}.$$

The E-step in the EM algorithm (Eqn. 2.154) computes the expected value of the complete data log-likelihood given the incomplete data W and the previous estimates of the parameters θ^{old}

$$E\big(\log P(W, T)|W_1, \ldots, W_n, \theta^{\text{old}}\big)$$
$$= \sum_{\sigma \to \sigma_0 \sigma_0} E\big(N_{\sigma \to \sigma_0 \sigma_1}(W)|W_1, \ldots, W_n, \theta^{\text{old}}\big) \log p_{\sigma \to \sigma_0 \sigma_1}$$
$$+ \sum_{\sigma \to w} E\big(N_{\sigma \to w}(T)|W_1, \ldots, W_n, \theta^{\text{old}}\big) \log p_{\sigma \to w}$$

For the maximization M-step (Eqn. 2.156), the conditional expectation is maximized subject to normalization constraints on the probabilities, $\sum_{\sigma \to \sigma_0 \sigma_1} p_{\sigma \to \sigma_0 \sigma_1} + \sum_{\sigma \to w} p_{\sigma \to w} = 1$. Maximizing subject to the constraint gives the theorem statement. □

3.13.2 General Context-Free Grammars and the Trellis Algorithm of Kupiec

However, most grammars are not in this normal form. Although the grammar could be easily converted to CNF, maintaining its original form is necessary for linguistic relevance. Hence, we need an algorithm that can estimate the probabilities of rules in the more general form given above.

The algorithm derived by Mark [101] is a specific case of Kupiec's trellis-based algorithm [114]. Kupiec's algorithm estimates parameters for general recursive transition networks. In our case, we only have rules of the following two types:

1. $H \to G_1 G_2 \cdots G_k$ where $H, G_i \in V_N$ and $k = 1, 2, \ldots$;
2. $H \to T$ where $H \in V_N$ and $T \in V_T$.

For this particular topology, we derived the following trellis-based algorithm.

Trellis-based Algorithm

1. Compute inner probabilities $\alpha(i, j, \sigma) = \Pr[\sigma \text{ derives } W_{ij}]$ where $\sigma \in V_N$ the set of nonterminals and W_{ij} denotes the substring $W_i \ldots W_j$.

$$\alpha(i,i,\sigma) = P^{old}_{\sigma \to w_i} + \sum_{\sigma_1 : \sigma \to \sigma_1} P^{old}_{\sigma \to \sigma_1} \alpha(i,i,\sigma_1),$$

$$\alpha(i,j,\sigma) = \sum_{\sigma_n : \sigma \to \ldots \sigma_n} \alpha_{nte}(i,j,\sigma_n,\sigma)$$

$$\alpha_{nte}(i,j,\sigma_m,\sigma) = \begin{cases} P^{old}_{\sigma \to \sigma_m \ldots} \alpha(i,j,\sigma_m) & \text{if } \sigma \to \sigma_m \ldots \text{ or } m = 1 \\ \sum_{k=i+1}^{j-1} \alpha_{nte}(i,k,\sigma_{m-1},\sigma) \alpha(k,j,\sigma_m) & \text{if } \sigma \to \ldots \sigma_{m-1} \sigma_m \ldots \end{cases}$$

2. Compute outer probabilities $\beta(i,j,\sigma) = \Pr[S \overset{*}{\Rightarrow} W_{1,i-1} \sigma W_{j+1,n}]$ where $\sigma \in V_N$. Choose $\beta(1,n,S) = 1.0$,

$$\beta(i,j,\sigma) = \sum_{n \to \sigma \ldots} P^{old}_{n \to \sigma \ldots} \beta_{nte}(i,j,\sigma,n) + \sum_{n \to \ldots p\sigma \ldots} \sum_{k=0}^{i-1} \alpha_{nte}(k,i,p,n) \beta_{nte}(k,j,\sigma,n)$$

$$\beta_{nte}(i,j,\sigma_m,\sigma) = \begin{cases} \beta(i,j,\sigma) & \text{if } \sigma \to \ldots \sigma_m \\ \sum_{k=j+1}^{L} \alpha(j,k,\sigma_{m+1}) \beta_{nte}(i,k,\sigma_{m+1},\sigma) & \text{if } \sigma \to \ldots \sigma_m \sigma_{m+1} \ldots \end{cases}$$

3. Re-estimate P.

$$P^{new}_{\sigma \to \sigma_1 \sigma_2 \ldots \sigma_n} = \frac{\sum_{i=1}^{N-1} \sum_{j=i+1}^{N} \alpha_{nte}(i,j,\sigma_n,\sigma) \beta(i,j,\sigma)}{\sum_{i=1}^{N} \sum_{j=i}^{N} \alpha(i,j,\sigma) \beta(i,j,\sigma)}$$

$$P^{new}_{\sigma \to T} = \frac{\sum_{i : w_i = T} \alpha(i,i,\sigma) \beta(i,i,\sigma)}{\sum_{i=1}^{N} \sum_{j=i}^{N} \alpha(i,j,\sigma) \beta(i,j,\sigma)}.$$

Note that for CNF grammars, the trellis-based algorithm reduces to the Inside-Outside algorithm. As an EM algorithm, the trellis-based algorithm has the important property that the sequence of likelihood values converges monotonically, that is, the likelihood increases after each iteration unless $P^{new} = P^{old}$ which indicates convergence of the sequence of estimates.

4 MARKOV RANDOM FIELDS ON UNDIRECTED GRAPHS

ABSTRACT This chapter focuses on random fields on lattices and undirected graphs. Discrete finite state spaces are examined in the context of Markov and Gibbs fields. The subsequent chapter studies state spaces on the continuum through Gaussian fields.

This chapter examines texture representation and segmentation, exploring the Gibbs random fields. For dealing with the partition function asymptotics are derived allowing for the explicit calculation of the asymptotic approximation to the log-normalizer required for the Bayes solution. Zhu's maximum-entropy model is explored for representing random field textures.

4.1 Undirected Graphs

Thus far we have examined only directed graphs. Now examine the most general undirected graphs such as shown in the left panel of Figure 4.1. The patterns are constructed via the graphs $\sigma = \{D, E\}$ defined by their domain of sites D and edge system E. For undirected graphs, the edges of the graph will have no orientation, but play the role of defining the *neighborhood* and *clique* structure of the graph. To begin with, two points which are sites are neighbors if there is an edge connecting them. A set of points form a clique if every pair of points in the set are neighbors.

Definition 4.1 *Undirected graph $\sigma = \{D, E\}$ has **neighborhood system** $\mathcal{N} = \cup_i N_i$ with neighborhoods if (1) $j \notin N_j$, (2) $j \in N_i \Leftrightarrow i \in N_j$, and (3) $j \in N_i \implies \exists e \in E$ with $\overset{j}{\bullet}\overset{e}{-}\overset{i}{\bullet}$.*

*Define $\mathcal{C} = \cup C$ as the **clique system** of $\sigma = \{D, E\}$ with subsets $C \subset D$ the **cliques** if every two distinct sites in C are neighbors: $i, j \in C \implies i \in N_j, j \in N_i$.*

There are many examples of patterns with undirected graph structures; a few are depicted in Figure 4.1.

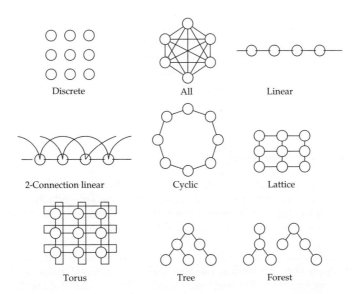

Figure 4.1 Figure showing various undirected graphs.

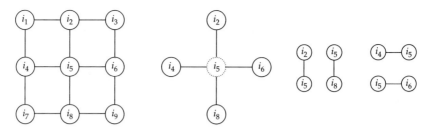

Figure 4.2 Left panel shows a 3 × 3 lattice. Middle panel shows the neighborhood system $N_5 = \{i_2, i_4, i_6, i_8\}$ for i_5. Right panel shows the clique system of i_5, $\{i_2, i_5\} \cup \{i_5, i_8\} \cup \{i_4, i_5\} \cup \{i_5, i_6\}$.

1. If there are no edges or bonds connecting sites in the graph then, σ = DISCRETE. If, on the other hand all sites are connected with segments then, σ = ALL.

2. Begin with the integer lattice, nearest neighbor connected, σ = LINEAR. Let $D = \{i : 1 \leq i \leq m\}$ be the integer lattice. Let the neighborhood system be of the form

$$\mathcal{N} = \cup_{1 \leq i \leq m} N_i ; N_i = \{j \in D : 0 < |j - i| \leq c\}. \tag{4.1}$$

   with $c = 1$, these are neighborhoods corresponding to a linear graph, with cliques subsets of the form $\{i\}$, $\{i, i+1\}$. Notice sites at the boundary have 1 less neighbor. With $c = 2$, it is the 2-connection linear structure shown. Notice the boundary sites have 2 less neighbors, and 1 site in from the boundary has 1 less neighbor.

3. The cyclic graph σ = CYCLIC is similar to the linear one, with the first and last sites connected, so that neighborhoods are identical throughout the graph.

4. Let σ = LATTICE, then let $D = \{1 \leq i_1, i_2 \leq m\}$ be the $m \times m$ integer lattice. Let the neighborhood system be of the form

$$\mathcal{N} = \cup_{1 \leq i_1, i_2 \leq m} N_i ; N_i = \{j \in D : \|j - i\| \leq c\}. \tag{4.2}$$

   for $c = 1$, cliques are subsets of the form $\{(i_1, i_2)\}$, $\{(i_1, i_2), (i_1 + 1, i_2)\}$, or $\{(i_1, i_2), (i_1, i_2 + 1)\}$. Changing to the σ = TORUS, the neighborhoods only change along the boundary.

5. Multiple trees correspond to σ = FOREST also as depicted. Notice, there are multiple root nodes.

To illustrate the neighborhood and clique structures for a regular lattice, shown in Figure 4.2 is a 3 × 3 set of sites $D = \{i_1, \ldots, i_9\}$, with its edges. The middle panel shows the neighbors of i_5, $N_5 = \{i_2, i_4, i_6, i_8\}$; the right panel shows the cliques. Now examine the placement of probabilities on the graphs to respect the directed edge structure.

4.2 Markov Random Fields

Thus far conditional independence and splitting properties of directed graphs have been examined. We now examine random fields on lattices and arbitrary graphs which are not directed.

Since graphs play a fundamental role in the description of patterns, Markov random fields provide an important probabilistic structure. Of course Markov random fields extend Markov processes to settings in which there are no natural ordering supporting finite, fixed size boundaries for arbitrary cylinders during the peeling process. This allows us to study multidimensional lattices. The earliest applications of Markov random fields are contained in the works of Ising (1925) [115] and later Onsager on the now classic Ising random field models for the characterization of magnetic domains. Of course much of the formal mathematics on the existence of probability measures on the infinite graph can be credited to Dobrushin [116–119] and Ruelle [120]. These mathematical treatments notwithstanding, much of the popularization of MRFs in the 1980s can be credited to Besag's early 1974 paper [121], Geman and Geman's subsequent influential paper [122] on image processing, and the *must-read* monographs of Kinderman and Snell [123] and Geman [124].

Now examine the placement of probabilities on the graphs to respect the neighborhood and edge structure.

MRFs are an important probabilistic tool for the configurations of pattern theory. Begin assuming finite graphs $\sigma = \{D, E\}$ with number of sites $n(\sigma) = n$. Assign to each point in the graph one of a finite set of random variables, $\{X_i, i \in D\}$ denoting the family of random variables indexed over D, each random variable taking values in $\mathcal{X}_0$ a finite state space. The set of possible configurations becomes $\mathcal{X} = \mathcal{X}_0^n$. To assign probability distributions to the set of all possible configurations of $X \in \mathcal{X}$, begin with the *local characteristics* of the probability distribution on $\mathcal{X}$ which are the conditional probabilities of the form

$$P(X_i | X_j, j \in D/i). \tag{4.3}$$

Then P will be said to define a *Markov random field* with respect to a graph with neighborhood and cliques if the local characteristic depends only on the neighbors.

Definition 4.2 X *is a* $\mathcal{X}_0 = \{0, 1, \ldots, m-1\}$ *valued* **Markov random field** *on* $\sigma = \{D, E\}$ *with* $n = |D|$ *and neighborhoods* $\mathcal{N} = \cup_i N_i$, *if for all* $X \in \mathcal{X} = \mathcal{X}_0^n$, *and* $X_i, i \in D$,

$$P(X) > 0, \tag{4.4}$$

$$P(X_i | X_j, j \neq i) = P(X_i | X_j, j \in N_i).$$

The conditional probabilities of an MRF globally determine the joint distribution.

Theorem 4.3 *The distribution on X is determined by its local characteristics.*

Proof We will show that for all $x, y \in \mathcal{X}$,

$$\frac{\Pr\{X = x\}}{\Pr\{X = y\}}$$

$$= \frac{\prod_{i=1}^n \Pr\{X_i = x_i | X_1 = x_1, \ldots, X_{i-1} = x_{i-1}, X_{i+1} = y_{i+1}, \ldots, X_n = y_n\}}{\prod_{i=1}^n \Pr\{X_i = y_i | X_1 = x_1, \ldots, X_{i-1} = x_{i-1}, X_{i+1} = y_{i+1}, \ldots, X_n = y_n\}}. \tag{4.5}$$

This determines the distribution $P(X)$ since taking another candidate $P'(X)$ satisfying $P(x)/P(y) = P'(x)/P'(y)$ implies

$$\sum_{x \in \mathcal{X}} \frac{P(x)}{P(y)} = \sum_{x \in \mathcal{X}} \frac{P'(x)}{P'(y)} \implies P(y) = P'(y), \quad \forall y \in \mathcal{X}. \tag{4.6}$$

Now to $\Pr\{X = x\}/\Pr\{X = y\}$:

$$\Pr\{X = x\} = \frac{\Pr\{X_1 = x_1, \ldots, X_n = x_n\}}{\Pr\{X_1 = x_1, \ldots, X_{n-1} = x_{n-1}\}}$$

$$\times \frac{\Pr\{X_1 = x_1, \ldots, X_{n-1} = x_{n-1}\}}{\Pr\{X_1 = x_1, \ldots, X_n = y_n\}} \Pr\{X_1 = x_1, \ldots, X_n = y_n\}$$

$$= \frac{\Pr\{X_n = x_n | X_1 = x_1, \ldots, X_{n-1} = x_{n-1}\}}{\Pr\{X_n = y_n | X_1 = x_1, \ldots, X_{n-1} = x_{n-1}\}} \Pr\{X_1 = x_1, \ldots, X_n = y_n\}. \tag{4.7}$$

Do it again with

$$\Pr\{X_1 = x_1, \ldots, X_n = y_n\} = \frac{\Pr\{X_{n-1} = x_{n-1} | X_1 = x_1, \ldots, X_n = y_n\}}{\Pr\{X_{n=1} = y_{n-1} | X_1 = x_1, \ldots, X_n = y_n\}}$$
$$\times \Pr\{X_1 = x_1, \ldots, X_{n-1} = y_{n-1} \cdot X_n = y_n\}. \quad (4.8)$$

Continuing in this manner n-times gives

$$\Pr\{X_1 = x_1, \ldots, X_n = x_n\}$$
$$= \frac{\prod_{i=1}^n \Pr\{X_i = x_i | X_1 = x_1, \ldots, X_{i-1} = x_{i-1}, X_{i+1} = y_{i+1}, \ldots, X_n = y_n\}}{\prod_{i=1}^n \Pr\{X_i = y_i | X_1 = x_1, \ldots, X_{i-1} = x_{i-1}, X_{i+1} = y_{i+1}, \ldots, X_n = y_n\}}$$
$$\times \Pr\{X_1 = y_1, \ldots, X_n = y_n\}. \quad (4.9)$$

Dividing by $\Pr\{X_1 = y_1, \ldots, X_n = y_n\}$ gives the result. □

Example 4.4 (Markov Chains) Let $X_1, X_2, \ldots$ be a Markov chain, then the neighborhoods are the parents and children for $N_i = \{i - 1, i + 1\}, i = 2, \ldots$.

$$P(X_i | X_j, j \neq i) = \frac{P(X)}{\sum_{X_i} P(X)} = \frac{\prod_{k=2}^n P(X_k | X_{k-1}) \pi(X_1)}{\sum_{X_i} \prod_{k=1}^n P(X_k | X_{k-1}) \pi(X_1)}$$
$$= \frac{P(X_i | X_{i-1}) P(X_{i+1} | X_i)}{\sum_x P(X_i = x | X_{i-1}) P(X_{i+1} | X_i = x)}. \quad (4.10)$$

Example 4.5 (MRF Representation of Textured Images) Random field models have been used extensively in image processing. The number of applications are extensive; here is one. Examine the local texture representations via Markov Random Fields described in[125–128]Exploit the local MRF characterization by associating with each region type in an image one of M-models $\theta_1, \theta_2, \ldots, \theta_M$. Then X is modeled as a Markov random field with local probabilities

$$P_{\theta_k}(X_i | X_j, j \neq i) = P_{\theta_k}(X_i | X_j, j \in N_i), \quad i \in D.$$

The neighborhood system $\mathcal{N} = \cup_{i \in D} N_i$ is taken to be nearest neighbors on the regular square lattice.

The left two panels 1 and 2 of Figure 4.3 are several electron micrograph images at a magnification of approximately 30,000 used in an experiment for representing the textures of mitochondria and cytoplasm in subcellular regions viewed via electron microscopy. The conditional probabilities $P(X_i | X_j, j \in N_i)$ are estimated from a set

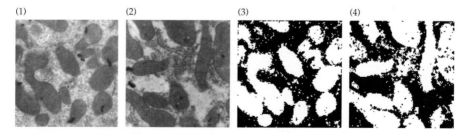

Figure 4.3 Panels 1 and 2 show electron micrographs at 30,000 magnification containing mitochondria and cytoplasm. Panels 3 and 4 show the Bayes segmentation via an MRF model of nearest-neighbors with zero boundary and four gray levels. Results from [127, 128]; data from Jeffrey Saffitz of the Department of Pathology at Washington University.

of training micrographs labeled into mitochondria and cytoplasm based on the local average gray-level feature and a four-gray level texture feature, $\mathcal{X}_0 = \{0, 1, 2, 3\}$ with nearest-neighbor structure [127, 128]. Estimating the local conditional probabilities for the organelles is straight forward. The Von–Mises estimators for the conditional probabilities are computed by counting the relative frequency of occurrence of $X_i \in \{0, 1, 2, 3\}$, given its neighboring configuration. This allows for the direct estimation of the essential characteristics as encoded via the local conditional probabilities of the various organelles.

The right two panels 3 and 4 of Figure 4.3 show the pixel based optimal Bayesian hypothesis test performed pixel by pixel solving

$$\theta_i = \underset{\theta \in \{\text{mito,cyto}\}}{\arg\max} \; \log P_\theta(X_i | X_j, j \in N_i). \tag{4.11}$$

Each shows the segmentation into two regions; white corresponds to mitochondria and black corresponds to cytoplasm.

Example 4.6 (Hidden Markov Chains) Hidden processes are very important as they change the graph structures. Consider the hidden Markov model used extensively in speech recognition in which there is a hidden state sequence $X = (X_1, X_2, \ldots, X_n)$ assumed generated by a Markov chain and a sequence of observables $Y = (Y_1, Y_2, \ldots, Y_n)$ generated independently conditioned on the state sequence. The probability of the state sequence and the probability of the observation sequence conditioned on the states become

$$P(X) = \prod_{k=1}^{n} P(X_k | X_{k-1}), \quad P(Y) = \prod_{k=1}^{n} P(Y_k | X_k). \tag{4.12}$$

The joint probability is then defined as

$$P(X, Y) = P(X)P(Y|X) = \prod_{k=1}^{n} P(X_k | X_{k-1}) P(Y_k | X_k).$$

If the state sequence is known, the neighborhood structure is simple, that is,

$$P(Y_i | (X, Y)/Y_i) = P(Y_i | X_i) \quad \text{and} \quad P(X_i | (X, Y)/X_i) = P(X_i | X_{i-1}, X_{i+1}, Y_i).$$

Shown in the left panel of Figure 4.4 is the directed graph structure for the joint X, Y process.

In general, however, the state sequence is hidden and only the observation sequence is measured with marginal probability $P(Y) = \sum_X P(X, Y)$ where the sum is over all possible state sequences.

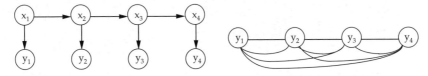

Figure 4.4 Left panel shows the directed graph structure corresponding to the joint X, Y process. Right panel shows the completely connected undirected graph corresponding to the marginal process Y only.

Under the marginal probability, the neighborhood structure is revealed by considering

$$P(Y_i|Y/Y_i) = \frac{P(Y)}{\sum_{Y_i} P(Y)}$$

$$= \frac{\sum_{X_1} \sum_{X_2} \cdots \sum_{X_n} \prod_{k=1}^{n} P(X_k|X_{k-1})P(Y_k|X_k)}{\sum_{Y_i} \sum_{X_1} \sum_{X_2} \cdots \sum_{X_n} \prod_{k=1}^{n} P(X_k|X_{k-1})P(Y_k|X_k)}$$

$$= P(Y_i|Y_1, \ldots, Y_{i-1}, Y_{i+1}, \ldots, Y_n).$$

Shown in the right panel of Figure 4.4 is the graph structure for the marginal of Y of the joint X, Y process, a fully connected graph.

Example 4.7 (Hidden Random Branching Processes) Now consider a hidden branching process model in which there is a branching process specified by a sequence of rules $T = (R_1, R_2, \ldots, R_N), R_i \in R$, a set of context-free rules, and a sequence of observed words $W = (W_1, W_2, \ldots, W_n)$ generated conditioned on the preterminating leaves of the tree $\gamma_1, \ldots, \gamma_n$ derived by the rules above. The probability of the rule sequence and the probability of the word sequence conditioned on the rule sequence is

$$P(T) = \prod_{j=1}^{N} P(R_j), \quad P(W|T) = \prod_{j=1}^{n} P(W_j|\gamma_j),$$

with the joint probability

$$P(W, T) = P(T)P(W|T) = \prod_{j=1}^{N} P(R_j) \prod_{k=1}^{n} P(W_k|\gamma_k).$$

If the underlying branching process T is known, the neighborhood structure is

$$P(W_i|(W, T)/W_i) = P(W_i|\gamma_i). \tag{4.13}$$

Shown in the left panel of Figure 4.5 is the directed graph structure for the marginal on Y of the joint X, Y process.

If the underlying tree is not observed but only the word sequence, then the marginal probability on the words W becomes, $P(W) = \sum_T P(W, T)$ where the sum is

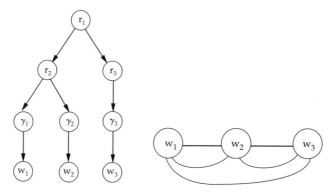

Figure 4.5 Left panel shows the directed graph branching process structure for the joint $T = R, W$ process. The $\gamma_1, \gamma_2, \ldots, \gamma_n$ denote the pre-terminal symbols. Right panel shows resulting completely connected undirected graph corresponding to the marginal process on the word W.

over all possible trees. In the case of an unambiguous context-free grammar, such as for programming languages, there is only one tree associated with each word sequence. In general, however, context-free languages are ambiguous [84].

Under the marginal probability, the neighborhood structure is shown by considering

$$P(W_i|W/W_i) = \frac{P(W)}{\sum_{w_i} P(W)}$$

$$= \frac{\sum_t \prod_{j=1}^N P(R_j) \prod_{k=1}^n P(W_k|\gamma_k)}{\sum_{w_i} \sum_t \prod_{j=1}^N P(R_j) \prod_{k=1}^n P(W_k|\gamma_k)}$$

$$= P(W_i|W_1,\ldots,W_{i-1},W_{i+1},\ldots,W_n).$$

So, the underlying graph structure is fully connected. Shown in the right panel of Figure 4.5 is the graph structure for marginal process on the words W for the joint T, W process, a fully connected graph.

4.3 Gibbs Random Fields

Now examine the equivalence between Gibbs distributions and Markov random fields as proved first for lattices in [129], and extended to arbitrary graphs in [130–132]. To connect Gibbs representations with probabilities of the form $P(X) = e^{-U(X)}/Z$ having $U(X)$ the potential function to the graph $\sigma = \{D, E\}$ it must be established that the neighborhood and clique structure of the MRF determines the form of the potential in the Gibbs representation. Proceeding first descriptively, the potential $U : \mathcal{X} \to \mathbb{R}$ will be constructed from locally defined potentials which associates values to every configuration of sites $X_G, G \subset D$. In particular, the Gibbs probability on the entire graph having potential U is

$$P(X) = \frac{e^{-U(X)}}{Z} \quad \text{where } Z = \sum_{X \in \mathcal{X}} e^{-U(X)}. \tag{4.14}$$

The exquisite connection between the probabilistic structure of the Gibbs field with potential $U(\cdot)$ and the picture of a graph with neighborhood and clique system is available. Most importantly, as first shown in [129-132], a Gibbs random field with potential $U(\cdot)$ must involve all cliques in assigning energy for the Gibbs distribution to have the neighborhood structure and local characteristics of the MRF.

Definition 4.8 *A* **Gibbs distribution** *is a probability distribution on $\mathcal{X}$ with*

$$P(X) = \frac{1}{Z}e^{-U(X)} \quad \text{with } Z = \sum_{X \in \mathcal{X}} e^{-U(X)}, \tag{4.15}$$

with the normalizing constant Z called the **partition function** *and $U : \mathcal{X} \to \mathbb{R}$ the* **energy function**.

We shall say that $P(X)$ is a **Gibbs distribution respecting the graph** $\sigma = \{D, E\}$ *with neighborhood and cliques $C \in \mathcal{C}$ if the potential $U : \mathcal{X} \to \mathbb{R}$ is of the form*

$$U(X) = \sum_{C \in \mathcal{C}} \Phi_C(X) \quad \text{with } \Phi_C(X) = \Phi_C(X_C), \tag{4.16}$$

where $\Phi_C : \mathcal{X} \to \mathbb{R}$ depends only on those coordinates in $C \subset D$.

Naturally this desire to have finite potentials links to the non-zero probabilities of Eqn. 4.4.

Theorem 4.9 X is an MRF with respect to graph $\sigma = \{D, E\}$ with neighborhood and clique structure $\mathcal{N}, \mathcal{C}$ if and only if $P(X)$ is a Gibbs distribution with respect to $\mathcal{N}, \mathcal{C}$.

Proof If $P(X), X = (X_1, \ldots, X_n) \in \mathcal{X} = \mathcal{X}_0^n$ is a $\mathcal{X}_0 = \{0, 1, \ldots, m-1\}$-valued Gibbs distribution with respect to neighborhood and clique system $\mathcal{N}, \mathcal{C}$ then

$$P(X) = \frac{1}{Z} e^{-U(X)}, \qquad Z = \sum_{X \in \mathcal{X}_0^n} e^{-U(X)}, \tag{4.17}$$

and $U(X)$ is the energy function. Fix $i \in D$ and X_i, then by rules of conditional probability

$$P(X_i | X_j, j \neq i) = \frac{P(X)}{\sum_{X_i \in \mathcal{X}_0} P(X)} = \frac{e^{-\sum_{C \in \mathcal{C}} \Phi_C(X)}}{\sum_{X_i \in \mathcal{X}_0} e^{-\sum_{C \in \mathcal{C}} \Phi_C(X)}}$$

$$= \frac{e^{-(\sum_{C \in \mathcal{C}: i \in C} \Phi_C(X) + \sum_{C \in \mathcal{C}: i \notin C} \Phi_C(X))}}{\sum_{X_i \in \mathcal{X}_0} e^{-(\sum_{C \in \mathcal{C}: i \in C} \Phi_C(X) + \sum_{C \in \mathcal{C}: i \notin C} \Phi_C(X))}}. \tag{4.18}$$

Notice the right-hand side of Eqn. 4.18 depends only on X_i and on X_j, where $j \in N_i$, since any site in a clique containing i must be a neighbor of i. Hence,

$$P(X_i | X_j, j \neq i) = P(X_i | X_j, j \in N_i). \tag{4.19}$$

Converse: Define $x_i = \mathbf{0}$ to be an entire realization of 0s, and $X^{(i)}$ the realization $(X_1, \ldots, X_{i-1}, 0, X_{i+1}, \ldots, X_n)$. Then with the assumption that $P(\mathbf{0}) > 0$, define $Q(X) = \log\left(P(X)/P(\mathbf{0})\right)$. Introduce all subsets of $\mathcal{X}$ of the form

$$E_{i,j,\ldots,i_r} = \{X \in \mathcal{X} | X_i > 0, X_j > 0, \ldots, X_{i_r} > 0, \text{all other } X_i = 0\}, \quad r = 1, 2, \ldots, n.$$

It is convenient to assume the ordering $i < j < \cdots < i_r$. Denote the indicator function of $E_{i,j,\ldots,i_r}$ by $1_{i,j,\ldots,i_r}(\cdot)$. In an E-set the x-components that are non-zero can be divided out, and since the E-sets are disjoint and cover $\mathcal{X}$ this gives the identity

$$Q(X) = \sum_{r=1}^{n} \sum_{i: i < j \ldots i_r} Q(X) 1_{i < j \ldots i_r}(X) = \sum_{r=1}^{n} \sum_{i: i < j \ldots i_r} X_i \cdots X_{i_r} F(X), \tag{4.20}$$

where $F(X) = (Q(X)/(X_i \cdots X_{i_r}))$. Reordering the variables gives

$$Q(X) = \sum_{1 \leq i \leq n} X_i F_i(X_i) + \sum_{1 \leq i < j \leq n} X_i X_j F_{i,j}(X_i, X_j) + \cdots$$

$$+ X_1 X_2 \ldots X_n F_{1,2,\ldots,n}(X_1, X_2, \ldots, X_n). \tag{4.21}$$

Since

$$\exp\left(Q(X) - Q(X^{(i)})\right) = \frac{P(X)}{P(X^{(i)})} = \frac{P(X_i | X_1, \ldots, X_{i-1}, X_{i+1}, \ldots, X_n)}{P(X_i = 0 | X_1, \ldots, X_{i-1}, X_{i+1}, \ldots, X_n)}, \tag{4.22}$$

and this is a Markov random field, the last equation suggests that $\exp\left(Q(X) - Q(X^{(i)})\right)$ in Eqn. 4.22 can depend only on X_i and its neighborhood. Without loss of generality assume $i = 1$, then

$$Q(X) - Q(X^{(1)}) = X_1 \Big(F_1(X_1) + \sum_{2 \leq j \leq n} X_j F_{1,j}(X_1, X_j)$$

$$+ \sum_s 2 \leq j < \sum_{k \leq n} X_j X_k F_{1,j,k}(X_1, X_j, X_k)$$

$$+ X_2 X_3 \cdots X_n F_{1,2,\ldots,n}(X_1, X_2, \ldots, X_n) \Big).$$

Now suppose that site $l \neq 1$ is not a neighbor of site 1. Then $Q(X) - Q(X^{(1)})$ must be independent of X_l for all $X \in \mathcal{X}$. Putting $X_i = 0$ for $i \neq 1$ or l gives $F_{1,l}(X_1, X_l) = 0$ on $\mathcal{X}$. Similarly by other suitable choices of X it is deduced that all $3-, 4-, \ldots, n-$ variable F-functions involving X_1 and X_l must be null. The analogous result holds for any pair of sites which are not neighbors of each other and hence, in general, $F_{i,j,\ldots,s}$ can only be non-null if the sites $i, j, \ldots, s$ form a clique. Thus, $Q(X)$ has the form $Q(X) = \sum_{C \in \mathcal{C}} \Phi_C(X)$ and is hence Gibbsian. $\square$

Given the global Gibbs probability it is straightforward to calculate the local conditionals of the MRF.

Example 4.10 (The Ising Model on Regular Lattices) The Ising model is a celebrated example. Consider an $n \times n$ lattice of sites, $D = \{1 \leq i, j \leq n\}$. At each site place a dipole, "up" or "down" element of the state space $\mathcal{X}_0 = \{+1, -1\}$, with $x_{ij} \in \{+1, -1\}$, $(i, j) \in D$, with configuration space $\mathcal{X} = \{+1, -1\}^{n^2}$ of n^2 dipole orientations. Shown in the left panel of Figure 4.6 is such a configuration of dipoles. Ising's probability on $\mathcal{X}$ is constructed from the local potential

$$\Phi_{ij}(X) = -JX_{i,j}(X_{i,j-1} + X_{i-1,j}) - mHX_{ij}. \tag{4.23}$$

Then to each configuration $X \in \mathcal{X}_0^{n^2}$ assign the potential

$$U(X) = -J \sum_{1 \leq i,j \leq n} X_{i,j}(X_{i,j-1} + X_{i-1,j}) - mH \sum_{1 \leq i,j \leq n} X_{i,j}. \tag{4.24}$$

with $J > 0$ this is termed the attractive case encouraging neighboring spins to be identically aligned. H encourages spins to be in the same direction associated

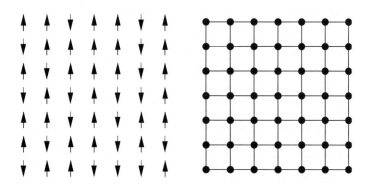

Figure 4.6 Left panel shows a 2D configuration of dipoles; right panel shows the graph structure for the Ising model.

with an external field resulting in net magnetization. The probability assigned to a configuration becomes

$$P(X) = \frac{1}{Z} e^{\frac{-U(X)}{KT}}, \quad \text{where } Z = \sum_{X \in \mathcal{X}} e^{\frac{-U(X)}{KT}}. \quad (4.25)$$

Notice the Φ_{ij} for internal sites are a function of only sites in the cliques of X_{ij}:

$$\{(ij), (i+1, j)\} \cup \{(ij), (i-1, j)\} \cup \{(ij), (i, j+1)\} \cup \{(ij), (i, j-1)\} \cup \{(ij)\}. \quad (4.26)$$

The neighborhood structure for internal sites is straightforwardly calculated using the conditional probabilities

$$P(X_{ij}|X_{kl}, (kl) \neq (ij)) = \frac{P(X)}{\sum_{X_{ij} \in \{-1,+1\}} P(X)}$$

$$= \frac{e^{-JX_{ij}(X_{i+1,j}+X_{i-1,j}+X_{i,j+1}+X_{i,j-1})}}{\sum_{X_{i,j} \in \{-1,+1\}} e^{-JX_{ij}(X_{i+1,j}+X_{i-1,j}+X_{i,j+1}+X_{i,j-1})}}. \quad (4.27)$$

4.4 The Splitting Property of Gibbs Distributions

The crucial aspect of examining Gibbs distribution is of course understanding the role of the boundary in determining probabilities of event on cylinders in the interior. This is of course the significant departure in multiple dimensions from the 1D Markov process case.

For this take a more careful look at Gibbs distributions on finite state spaces associated with infinite lattices. Choose a regular lattice $D = \mathbb{Z}^d$ with state space $\mathcal{X} = \mathcal{X}_0^{\mathbb{Z}^d}$, X taking values of d-dimensional array of elements from $\mathcal{X}_0$. As before, for $G \subset \mathbb{Z}^d$, take the configuration associated with the subset of sites $G \subset \mathbb{Z}^d$ as $X_G = \{X_i, i \in G\}$.

For the stationary setting the full Gibbs potential is constructed by shifting the local Φ potential around the subgraphs of $G \subset \mathbb{Z}^d$, with interactions at the boundaries. The potentials will have finite support; we quantify the range of support of the Gibbs potential as R as follows.

Definition 4.11 *Define the* **support set** *$S \subset \mathbb{Z}^d$ positioned to the left of the origin of size R^d:*

$$S \subset \{i \in \mathbb{Z}^d : (-R+1, -R+1, \dots) \leq (i_1, i_2, \dots) \leq (0, 0, \dots)\}. \quad (4.28)$$

Then R is defined to be the **range of the interaction** *if the potential is constructed from locally X_S supported functions, and $\Phi : \mathcal{X} \to \mathbb{R}$ having the property that Φ depends only on the random subconfiguration X_S, so that $\Phi(X) = \Phi(X_S)$.*

Given set G, then define its **interior** $G^0 = \{i \in G \subset \mathbb{Z}^d : (i + S) \subset G\}$, its **closure** $\bar{G} = \cup_{i \in G}(i + S)$, and **interior boundary** $\partial G = G/G^0$ and **exterior boundary** $\partial \bar{G} = \bar{G}/\bar{G}^0$.

Examine Figure 4.7. Notice the interior boundary defined ∂G here is the internal boundary including only points inside the set.

The configuration determining the potential associated with the finite set G with the infinitely extended boundary configuration $y \in \mathcal{X}$ is defined to be $X_G \vee y$:

$$(X_G \vee y)_i = \begin{cases} X_i & i \in G \\ y_i & i \notin G \end{cases}. \quad (4.29)$$

Examine distributions generated from fixed boundaries.

4.4 THE SPLITTING PROPERTY OF GIBBS DISTRIBUTIONS

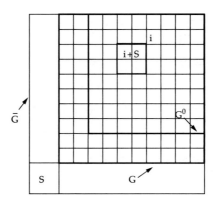

Figure 4.7 Showing the interior and closure of the sets $G^0, \bar{G}$. Double bold line denotes the set G, with G^0 the interior square, and $\bar{G}$ the exterior square.

Definition 4.12 *Define τ the **shift operator** so that $\tau_i(X_G), i \in G$ is the configuration obtained by shifting the coordinates to the origin:*

$$\tau_i(X_G) = \{X_{i'} : i' = j - i,\, j \in G\}. \tag{4.30}$$

*The **finite volume Gibbs distribution** $P(X_G|y)$ **with interaction** Φ **and fixed boundary** y on finite cylinders X_G is defined to be*

$$P(X_G|y) = \frac{e^{-U_G^y(X_G)}}{Z_G^y},$$

$$\text{where } U_G^y(X_G) = \sum_{i \in G^0} \Phi(\tau_i(X_G)) + \sum_{i \in \partial G} \Phi(\tau_i(X_G \vee y)), \tag{4.31}$$

with the normalizer $Z_G^y = \sum_{X_G \in \mathcal{X}_0^{|G|}} e^{-U_G^y(X_G)}$.

In the context of conditional probability, the range R of the local support of Φ can be given a precise interpretation as the critical distance[17] between subsets G, G' so that events on these subsets are conditionally independent.

This provides the basic splitting property for Markov random fields, split by the external boundaries. $\partial \bar{G}, \partial \bar{G}'$.

Theorem 4.13 *Let P be constructed from locally supported potentials $\Phi(x) = \Phi(x_S)$ with range R, then the probabilities on sets $G, G' \subset \mathbb{Z}^d$ with $d(G, G') \geq R$ are split by conditioning on the boundary:*

$$P(X_G, X_{G'}|X_{(G \cup G')^c}) = P(X_G|X_{\partial \bar{G}})P(X_{G'}|X_{\partial \bar{G}'}) \tag{4.32}$$

in particular, the probability of a configuration on set G is determined by its external boundary:

$$P(X_G|X_{G^c}) = P(X_G|X_{\partial \bar{G}}). \tag{4.33}$$

---

[17] The distance between sets $G, G' \subset \mathbb{Z}^d$ is defined as the minimum distance between coordinates: $d(G, G') = \min_{k=1}^d \min_{i \in G, j \in G'} |i_k - j_k|$.

Proof By definition

$$P(X_G, X_{G'}|X_{(G \cup G')^c}) = \frac{1}{Z^{X_{(G \cup G')^c}}_{G \cup G'}} \exp -U^{X_{(G \cup G')^c}}_{G \cup G'}(X_G, X_{G'}) \quad (4.34)$$

$$= \frac{1}{Z^{X_{(G \cup G')^c}}_{G \cup G'}} \exp \left(- \sum_{i \in (G \cup G')^0} \Phi(\tau_i(X_{G \cup G'})) \right.$$

$$\left. - \sum_{i \in \partial(G \cup G')} \Phi(\tau_i(X_{(G \cup G')} \vee X_{(G \cup G')^c})) \right).$$

Since $d(G, G') > R$ then for all $i \in \partial(G \cup G')$, then either

$$\begin{cases} (i+S) \cap G = \emptyset & \text{if } i \in \partial G' \\ (i+S) \cap G' = \emptyset & \text{if } i \in \partial G' \end{cases} \quad (4.35)$$

implying the separation of the potentials:

$$\sum_{i \in (G \cup G')^0} \Phi(\tau_i(X_{(G \cup G')})) = \sum_{i \in G^0} \Phi(\tau_i(X_G)) + \sum_{i \in G'^0} \Phi(\tau_i(X_{G'})),$$

$$\sum_{i \in \partial(G \cup G')} \Phi(\tau_i(X_{(G \cup G')} \vee X_{(G \cup G')^c})) = \sum_{i \in \partial G} \Phi(\tau_i(X_G \vee X_{(G \cup G')^c}))$$

$$+ \sum_{i \in \partial G'} \Phi(\tau_i(X_{G'} \vee X_{(G \cup G')^c})).$$

Substituting into 4.34 gives

$$P(X_G, X_{G'}|X_{(G \cup G')^c})$$

$$= \frac{1}{Z^{X_{(G \cup G')^c}}_{G \cup G'}} \exp \left(- \sum_{i \in G^0} \Phi(\tau_i(X_G)) - \sum_{i \in \partial G} \Phi(\tau_i(X_G \vee X_{(G \cup G')^c})) \right)$$

$$\exp \left(- \sum_{i \in G'^0} \Phi(\tau_i(X_{G'})) - \sum_{i \in \partial G'} \Phi(\tau_i(X_{G'} \vee X_{(G \cup G')^c})) \right). \quad (4.36)$$

The first multiplier terms are only a function $i \in \bar{G}$, the second term a function $i \in \bar{G}'$. Therefore, the partition function factors as $Z^{X_{\bar{G}}}_G Z^{X_{\bar{G}'}}_{G'}$ completing the proof giving the factorization and the splitting property of Eqn. 4.32:

$$P(X_G|X_{(G \cup G')^c}) = P(X_G|X_{\partial \bar{G}}), \quad P(X_{G'}|X_{(G \cup G')^c}) = P(X_{G'}|X_{\partial \bar{G}'}). \quad (4.37)$$

$\square$

Corollary 4.14 *The Markov chain splitting takes the form for Markov random fields*

$$P(X_G, X_{G'}|X_{\partial \bar{G}}) = P(X_G|X_{\partial \bar{G}})P(X_{G'}|X_{\partial \bar{G}}). \quad (4.38)$$

4.4 THE SPLITTING PROPERTY OF GIBBS DISTRIBUTIONS

Proof To see this form of the splitting property

$$P(X_G|X_{G'}, X_{\partial \bar{G}}) = \frac{\exp(-U_G^{(X_{G'}, X_{\partial \bar{G}})}(X_G))}{Z_G^{(X_{G'}, X_{\partial \bar{G}})}}$$

$$= \frac{\exp(-\sum_{i \in G^0} \Phi(\tau_i(X_G)) - \sum_{i \in \partial G} \Phi(\tau_i(X_G \vee (X_{G'}, X_{\partial \bar{G}}))))}{Z_G^{(X_{G'}, X_{\partial \bar{G}})}}$$

$$= \frac{\exp(-U_G^{X_{\partial \bar{G}}})}{Z_G^{X_{\partial \bar{G}}}}. \qquad (4.39)$$

This gives the splitting property of Eqn. 4.38 according to

$$P(X_G, X_{G'}|X_{\partial \bar{G}}) = P(X_G|X_{G'}, X_{\partial \bar{G}})P(X_{G'}|X_{\partial \bar{G}}) \stackrel{(a)}{=} P(X_G|X_{\partial \bar{G}})P(X_{G'}|X_{\partial \bar{G}}), \qquad (4.40)$$

with the final equality (a) following from Eqn. 4.39 above. □

The identical argument gives $P(X_G, X_{G'}|X_{\partial \bar{G}'}) = P(X_G|X_{\partial \bar{G}'})P(X_{G'}|X_{\partial \bar{G}'})$.

Example 4.15 (Markov chains) Choose $D = \{0 \leq i \leq n\}$, then the 1-memory Markov chain is a range $R = 2$ random field. The energy $U : \mathcal{X} \to \mathbb{R}$ is constructed from $S = \{-1, 0\}$ with $\Phi(X) = \Phi(X_{-1}, X_0)$, $\Phi(\tau_i(X)) = \Phi(X_{i-1}, X_i)$.

The m-memory Markov chain is a range $R = m + 1$ random field; the energy $U : \mathcal{X} \to \mathbb{R}$ constructed from locally $S = \{-m, \ldots, 0\}$ supported functions

$$\Phi(X) = \Phi(X_{-m}, X_{-m+1}, \ldots, X_0), \quad \Phi(\tau_i(X)) = \Phi(X_{i-m}, X_{i-m+1}, \ldots, X_i). \qquad (4.41)$$

The open sets and closed sets become

$$D^0 = \{m, \ldots, n\}, \quad \bar{D} = \{-m, \ldots, 0, \ldots, n\}, \qquad (4.42)$$

with internal boundary $\partial D^0 = \{0, \ldots, m-1\}$, $\partial \bar{D} = \{-m, \ldots, -1\}$.

Example 4.16 (Segmentation via the Ising Model [133]) The Ising model has been studied extensively, not only for the description of magnetic spins but also in image analysis for capturing low-level aggregation features in segmentation. As proposed by Geman and Geman [122], the Ising model can be used for segmentation to treat the case where the image has 2 parts, one light and one darker, so that the segmentation variables will be $+1$ on the light part, and -1 on the darker. An example is illustrated in Figure 4.8 from Mumford [133].

Take the graph to be on $D = \{1 \leq i, j \leq m\}$ the $m \times m$ square-lattice with two layers, the bottom layer observable random variables $Y_{ij} \in \mathbb{R}, 1 \leq i, j \leq m$ associated to the image measurement, and the top layer hidden random variable $X_{ij} \in \{+1, -1\}$ associated to the segmentation. Connect by edges each $Y_{i,j}$ vertex to the X_{ij} vertex above and to $X_{i,j}$ its 4 neighbors $X_{i \pm 1, j}, X_{i, j \pm 1}$ in the X–graph (adjusted to the boundary) and no others. The cliques are just the pairs of vertices connected by edges. The clique potentials become

$$\Phi_{ij}^{(1)}(X) = X_{i,j} \cdot X_{i',j'} \quad \text{for } (i,j), (i',j') \text{ for two adjacent vertices}, \qquad (4.43)$$

$$\Phi_{ij}^{(2)}(X) = Y_{i,j} \cdot X_{i,j} \quad \text{for } (i,j). \qquad (4.44)$$

Figure 4.8 Figure taken from Mumford [133] showing the solutions from the Ising model following the cooling schedule.

The potentials become

$$U(X,Y) = \alpha \sum_{i,j} X_{i,j}(X_{i-1,j} + X_{i,j-1}) + \sum_{ij} Y_{i,j} X_{i,j}, \tag{4.45}$$

with probability $P(X,Y) = \frac{1}{Z} e^{-U(X,Y)}$. Notice the observed image Y is being applied as the external field locally to each site X of the hidden field.

The modes of $P(X|Y)$ seek to make adjacent X vertices equal, and have the same sign as the corresponding Ys. These are in conflict with rapidly varying sign changes of the external image field. The probable values of X will align with the large areas where Y is consistently of one sign.

Shown in Figure 4.8 are results from Mumford [133] showing the solutions from the Ising model. Panel 1 shows the original image which was scaled to make the dark and light areas have opposite signs. Panel 2 shows the final segmentation resulting from the sequence of Ising segmentations. Panels 3–8 show successive realizations representing local minimum from the potential $P(X,Y)^{1/T}$, where T = temperature in a cooling schedule. Panel 3 shows T large, and panels 4–8 show the realizations coming from successive decreasing of T on the probability law. The decreasing T cooling schedule forces the probability to concentrate on fewer and fewer realizations each occupying higher and higher probability.

Example 4.17 (Hierarchical Fields via Line Processes) We have seen that hidden processes change the graph structures. As emphasized by Geman and Geman [122], this same notion holds for the regular lattice random field case. Define the $n \times n$ pixel lattice $D^{(1)} = \{1 \leq i, j \leq n\}$ and "dual" lattice $D^{(2)} = \{1 \leq i, j \leq n\}$, the dual sites placed midway between each vertical or horizontal pair of pixels. The dual lattice represents the possible locations of *edge elements*. Associate with each lattice site the joint MRF (X, L) consisting of the pixel intensity field $X = \{X_{i,j}, 1 \leq i,j \leq n\}$ with eight gray level values and the line site field $L = \{L_{i,j}, 1 \leq i,j \leq n\}$ which is $\{0,1\}$ valued denoting the existence or non-existence of an edge at site (i,j). Associate nearest neighborhoods with the pixel lattice and dual lattice, respectively, $\mathcal{N}^{(1)} = \cup_{(i,j) \in D^{(1)}} N_i^{(1)}$, $\mathcal{N}^{(2)} = \cup_{(i,j) \in D^{(2)}} N_{i,j}^{(2)}$, as depicted in Figure 4.9. The neighborhood

4.4 THE SPLITTING PROPERTY OF GIBBS DISTRIBUTIONS

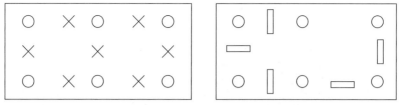

Figure 4.9 Left panel shows pixel sites marked with circles and edge sites marked with crosses; right panel shows the dual lattice with edges and pixels.

system is constructed from the basic neighborhoods for the $k = 1, 2$ lattices:

$$N_{i,j}^{(k)} = \{(i-1, j)^{(k)}, (i+1, j)^{(k)}, (i, j-1)^{(k)}, (i, j+1)^{(k)}\} \subset D^{(k)}. \quad (4.46)$$

The set of cliques becomes

$$\mathcal{C}^{(1)} = \{(i, j)^{(1)}\}, \quad \mathcal{C}^{(2)} = \{((i, j)^{(1)}, (i+1, j)^{(1)}), ((i, j)^{(1)}, (i, j+1)^{(1)})\}, \quad (4.47)$$

$$\mathcal{C}^{(3)} = \{((i, j)^{(1)}, (i-1, j)^{(2)}), ((i, j)^{(1)}, (i+1, j)^{(2)}), ((i, j)^{(1)}, (i, j-1)^{(2)}),$$
$$\times ((i, j)^{(1)}, (i, j+1)^{(2)})\}$$

with the potential

$$U(X) = \alpha \sum_{i,j} X_{i,j}(X_{i+1,j} + X_{i,j+1})$$
$$+ \beta \sum_{i,j} X_{i,j}(L_{i-1,j} + L_{i+1,j} + L_{i,j-1} + L_{i,j+1}) + \gamma \sum_{i,j} X_{i,j}. \quad (4.48)$$

The joint probability takes the form

$$P(X, L) = \frac{1}{Z} e^{\alpha \sum_{i,j} X_{i,j}(X_{i+1,j}+X_{i,j+1})} e^{\beta \sum_{i,j} X_{i,j}(L_{i-1,j}+L_{i+1,j}+L_{i,j-1}+L_{i,j+1})} e^{\gamma \sum_{i,j} X_{i,j}}, \quad (4.49)$$

with $L \in \{0, 1\}^{m^2}$ taking $0, 1$-values corresponding to the absence or presence of an edge.

Marginalizing with respect to $X = x$ corresponds to summing the density over all values of line sites. Note that summation of a term of the form $\exp(xl)$ over all values of l gives $(1 + \exp(x))$. Hence we have to only sum the exponent term in the probability density which contains y. The other two exponents factor out. Summation over all l gives

$$\sum_{l_{i,j}, 1 \leq i, j \leq n} e^{(\beta \sum_{i,j} x_{i,j}(l_{i-1,j}+l_{i+1,j}+l_{i,j-1}+l_{i,j+1}))}$$
$$= \prod_{i,j} (1 + e^{x_{i-1,j}+x_{i,j}})(1 + e^{x_{i+1,j}+x_{i,j}})(1 + e^{x_{i,j-1}+x_{i,j}})(1 + e^{x_{i,j+1}+x_{i,j}}).$$

Hence the marginal density has the form

$$P(X_{i,j}, 1 \leq i,j \leq n)$$
$$= \frac{1}{Z} \left[\prod_{i,j} \{(1 + e^{X_{i-1,j}+X_{i,j}})(1 + e^{X_{i+1,j}+X_{i,j}})(1 + e^{X_{i,j-1}+X_{i,j}})(1 + e^{X_{i,j+1}+X_{i,j}})\} \right.$$
$$\left. \times e^{(\gamma \sum_{i,j} X_{i,j})} e^{(\alpha \sum_{i,j} X_{i,j}(X_{i+1,j}+X_{i,j+1}))} \right].$$

Note the density is no longer of Gibbs form with local neighborhood structure. In fact it is a sum of exponents now. If we take the conditional expectation with respect to a particular lattice site the density instead of having a local neighborhood structure shows a fully connected graph.

4.5 Bayesian Texture Segmentation: The log-Normalizer Problem

Bayesian approaches have enjoyed increasing prominence throughout the spectrum of computer vision paradigms, ranging from traditional bottom-up "early vision" algorithms to systems exploiting high-level deformable shape models [134–136]. Statistical characterizations of the available data are fundamental to any Bayesian approach. In ideal cases, different objects may be effectively identified by their intensities; more commonly, they will be differentiated via their textures. A tremendous amount of research has been performed and published in the area of texture analysis, much of which seeks to characterize the relationships between pixels in a textured object. Clearly, in the independent pixel models, the basic object being modeled is the pixel. Under such circumstances the log-normalizer will present no difficulty. However, for textures of extended objects having interiors forming a disjoint partition of the image this changes dramatically.

Model the interiors of the shapes of real-valued images as scalar random fields defined on the index sets of integer lattices. We take the viewpoint of Bayesian testing calculating the likelihood of shapes from Gaussian and Gibbs fields. Because of splitting, the partitions are conditionally independent given their boundaries. The goal of taking a Bayesian approach presents a substantial conceptual hurdle. Since the normalizing partition function must be calculated for each partition $D^{(m)} \subset \mathbb{Z}^d, d = 2, 3, m = 1, \ldots, M$ the number of random fields depending on the boundaries of the random shapes. To avoid the difficulty with the partition function, Besag suggested replacing the likelihood with a *pseudolikelihood*, a product of conditional probabilities, yielding a modified Bayesian paradigm as described in [137]. In this chapter we examine direct assaults on the normalization constants using asymptotics thus providing a purely Bayesian formulation.

4.5.1 The Gibbs Partition Function Problem

To illustrate, examine the goal of understanding the partition function over randomly shaped objects, roads, lakes, etc. Model their interiors as scalar real-valued random fields. The Bayesian probability of accepting one shape over another is obtained by evaluating the likelihood of a particular disjoint partition of the scene, each region a random field. Conditioned on the boundaries, the subregion Markov fields are assumed independent.

4.5 BAYESIAN TEXTURE SEGMENTATION: THE LOG-NORMALIZER PROBLEM

For the Gibbs distribution let X on D have state space $\mathcal{X}_0^{|D|}$, taking the form

$$P(X) = \frac{e^{-U(X)}}{Z_D}, \quad \text{where } Z_D = \sum_{x \in \mathcal{X}_0^{|D|}} e^{-U(x)}. \quad (4.50)$$

Choosing model type α from a group of Gibbs fields with potentials U^α requires computation of $\log Z_D^\alpha$:

$$\hat{\alpha} \leftarrow \arg\max_\alpha -U^\alpha(x) - \log Z_D^\alpha. \quad (4.51)$$

Notice the role the partition function plays. Clearly, if the region corresponds to the interior of an unknown shape which is to be inferred and is evolving during the inference procedure, calculation of the partition function over the sub-graph becomes impossible.

Nicely enough, the normalized partition functions are relatively independent of the boundary condition as proved in [138, 139]. This is essentially the same as the convergence of the log-determinant of the covariance for Gaussian fields in the large graph limit which we show in the next chapter (see Corollary 5.21, Chapter 5).

Theorem 4.18 *Let X correspond to a* **Gibbs distribution** *P with potential $U^\Phi = \sum_i \Phi \circ \tau_i$. Then there exists a constant c depending on the range R and the interaction $\|\Phi\| = \sup_{x \in \mathcal{X}} \Phi(x)$ such that for any $D \subset \mathbb{Z}^d$, the normalized log-partition functions have the following boundary to volume dependence:*

$$\frac{1}{|D|}\left|\log \frac{Z_D^y}{Z_D^z}\right| \leq c\frac{|\partial D|}{|D|}. \quad (4.52)$$

The distributions under the various boundary conditions have the form

$$\frac{1}{|D|}\left|\log \frac{P_D^y}{P_D^z}\right| \leq c\frac{|\partial D|}{|D|}, \quad \frac{1}{|D|}\left|\log \frac{P_D}{P_D^z}\right| \leq c\frac{|\partial D|}{|D|}. \quad (4.53)$$

Proof Begin with the inequality $|U_D^y - U_D^z| \leq 2\|\Phi\||\partial D|$ on the arbitrary boundaries $y, z \in \mathcal{X}_0^{\mathbb{Z}^d/D}$ giving

$$e^{-2\|\Phi\||\partial D|} \leq \frac{e^{-U_D^y}}{e^{-U_D^z}} \leq e^{+2\|\Phi\||\partial D|}; \quad (4.54)$$

$$e^{-U_D^z}e^{-2\|\Phi\||\partial D|} \leq e^{-U_D^y} \leq e^{-U_D^z}e^{+2\|\Phi\||\partial D|}. \quad (4.55)$$

Using Eqn. 4.55 gives

$$Z_D^y = \sum_{x_D} e^{-U_D^y(x_D)} \stackrel{(a)}{\leq} e^{+2\|\Phi\||\partial D|} \sum_{x_D} e^{-U_D^z(x_D)} = e^{+2\|\Phi\||\partial D|}Z_D^z \quad (4.56)$$

$$= \sum_{x_D} e^{-U_D^y(x_D)} \stackrel{(b)}{\geq} e^{-2\|\Phi\||\partial D|}Z_D^z, \quad (4.57)$$

where (a) follows from R.H.S. Eqn. 4.55 and (b) follows from L.H.S. Eqn. 4.55. This gives $|\log(Z_D^y)/(Z_D^z)| \leq 2\|\Phi\||\partial D|$, completing the first part of the Theorem with the constant $c = 2\|\Phi\|$.

From this,

$$\frac{e^{-U_D^y}}{Z_D^y} \stackrel{(a)}{\leq} \frac{e^{-U_D^z}e^{2\|\Phi\||\partial D|}}{Z_D^y} \stackrel{(b)}{\leq} \frac{e^{-U_D^z}e^{2\|\Phi\||\partial D|}}{Z_D^z e^{-2\|\Phi\||\partial D|}} = P_D^z e^{4\|\Phi\||\partial D|}, \quad (4.58)$$

where (a) follows from the R.H.S. Eqn. 4.55 and (b) follows from Eqn. 4.57. This gives the second part

$$\left|\log \frac{P_D^y}{P_D^z}\right| \leq 4\|\Phi\||\partial D|. \tag{4.59}$$

To finish, then

$$P_D^z e^{-4\|\Phi\||\partial D|} \leq P_D^y \leq P_D^z e^{4\|\Phi\||\partial D|}, \tag{4.60}$$

$$P_D^z e^{-4\|\Phi\||\partial D|} \leq E\{P_D^y\} = P_D \leq P_D^z e^{4\|\Phi\||\partial D|} \tag{4.61}$$

giving $|\log(P_D/P_D^z)| \leq 4\|\Phi\||\partial D|$, completing the proof. □

4.6 Maximum-Entropy Texture Representation

Markov Random field models are all maximum entropy models (see Chapter 2, Section 2.9). As discussed in Section 2.9 maximum-entropy techniques provide methods for constructing representations of the world. As Jaynes [63] eloquently formulated, given empirical observations, the maximum-entropy model occurs with greater multiplicity than any other choice. Zhu and Mumford have exploited this for texture representation in natural imagery. They address the following problem in pattern representation and texture modeling: Assume there exists a joint probability density $p_X(\cdot)$ on $\mathcal{X}$ the space of images $X \in \mathcal{X}$ with $X = X_i, i \in D$ the discrete graph D, and the objective is to estimate $p(\cdot)$ via some density $q(\cdot)$. Assume throughout that the only direct observation of $p(\cdot)$ is through realizations of it, or samples $x^{(s)} = \{x_i^{(s)}, i \in D\}, s = 1, \ldots, S$ the size of the empirical sample. The basic strategy is to construct $\hat{p}(\cdot)$ so that it reproduces features of the empirical samples. Defining the M-features $h_m(X), m = 1, \ldots, M$, then the model $\hat{p}(\cdot)$ approximating $p(\cdot)$ is constructed so that the expectation under the approximating model equals the empirical average of the feature under the sample set. Then the maximum entropy density takes the familiar form from Jaynes theorem on moment constraints.

How should the features be chosen? Zhu et al. [140] exploit the fact that probability densities can be represented through their marginals. This is given by the following theorem characterizing any smooth density via its characteristic functional.

Theorem 4.19 *Let $X = X_1, \ldots, X_n$ associated with discrete lattice D of size $|D| = n$ have joint density $p_X(\cdot)$ on $\mathcal{X} = \mathbb{R}^n$. Then given filters $f_i, i = 1, \ldots, n$ and filtered versions of the process,*

$$X_f = \sum_{i=1}^{n} X_i f_i \tag{4.62}$$

have marginal densities

$$p_{X_f}(r) = E\{\delta(r - X_f)\}, \qquad r \in \mathbb{R},$$

which determine $p_X(\cdot)$.

Proof Let $|D| = n$, then relate $p_X(\cdot)$ to its Fourier transform $\hat{p}_X(\cdot)$ according to

$$p_X(x_1, \ldots, x_n) = \int_{\mathbb{R}^n} e^{j2\pi \sum_{i=1}^n x_i f_i} \hat{p}_X(f_1, \ldots, f_n) \, df \tag{4.63}$$

4.6 MAXIMUM-ENTROPY TEXTURE REPRESENTATION

with

$$\hat{p}_X(f_1,\ldots,f_n) = \int_{\mathbb{R}^n} e^{-j2\pi \sum_{i=1}^{n} x_i f_i} p_X(x_1,\ldots,x_n)\, dx \qquad (4.64)$$

$$= \int_{\mathbb{R}} e^{-j2\pi r}\, dr \int_{\mathbb{R}^n} \delta(r - \sum_{i=1}^{n} x_i f_i) p_X(x_1,\ldots,x_n)\, dx \qquad (4.65)$$

$$= \int_{\mathbb{R}} e^{-j2\pi r} p_{X_f}(r)\, dr, \qquad (4.66)$$

where $p_{X_f}(r) = \int_{\mathbb{R}^n} \delta(r - \sum_{i=1}^{n} x_i f_i) p_X(x_1,\ldots,x_n)\, dx \qquad (4.67)$

$$= E\{\delta(r - X_f)\}. \qquad (4.68)$$

□

Zhu takes these filtered versions as the empirical test functions in the maximum entropy formalism.

Theorem 4.20 *Given are moment constraints on marginal densities of filtered versions of the process:*

$$H_m(r) = E\{\delta(r - X_{f^{(m)}})\}, \quad m = 1,\ldots M,\ r \in \mathbb{R} \qquad (4.69)$$

$$= \int_{\mathbb{R}^n} \delta(r - \sum_{i=1}^{n} x_i f_i^{(m)}) p_X(x_1,\ldots,x_n)\, dx. \qquad (4.70)$$

Then the density maximizing entropy with moment equalling the observation functions takes the form

$$p_X(x_1,\ldots,x_n) = \frac{1}{Z_m(\Lambda)} e^{-\sum_{m=1}^{M} \lambda^{(m)}(\sum_{i=1}^{n} x_i f_i^{(m)})}, \qquad (4.71)$$

where $\lambda^{(m)}(r), r \in \mathbb{R}$ are real-valued functions.

4.6.1 Empirical Maximum Entropy Texture Coding

For constructing texture fields Zhu exploits ergodic properties of stationary random fields to model textures from individual realizations by generating empirical averages of the marginal densities from realizations of the single pictures. For this, given the random process $X_D = \{x_i, i \in D\}$ a realization of the stationary random field X on the infinite lattice, with $D \subset \mathbb{Z}^k$ assumed large. Then take as the empirical estimator $\hat{H}_m(r)$ of $H_m(r) = E\{\delta(r - X_{f^{(m)}})\}$ the function

$$\hat{H}_m(r) = \frac{1}{|D|} \sum_{j \in D} \delta\left(r - \sum_{i \in D} x_{i+j} f_i^{(m)}\right). \qquad (4.72)$$

While this is a powerful theorem demonstrating the characterization of processes through filtered versions, it requires knowledge of an infinite number of such filter functions to completely specify the inverse transform. For texture representations in visual coding Zhu et al. exploit the fact that small numbers of selective filtering functions may be adequate for explaining large numbers

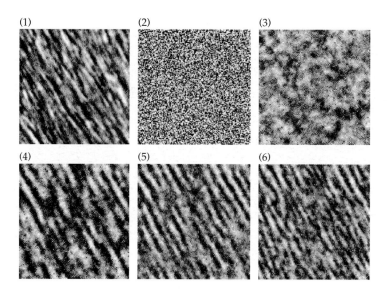

Figure 4.10 Panel 1 shows animal fur texture, panel 2 shows white noise, panel 3 shows maximum-entropy with 5 pixel Laplacian of a Gaussian filter f_1, panel 4 shows added filters $f_{6,120°}$, $f_{2,30°}$, then panel 5 adding the filter $f_{12,60°}$, and panel 6 adds the filter $f_{10,120°}$ and DC component.

of textures in the world. Examples of filters include the frequency and orientation selective Gabor filters [141] modeling cells in the mammalian visual cortex. Zhu et al. define a general filter bank including (i) a DC component $f(\cdot) = \delta(\cdot)$, (ii) the Laplacian of Gaussian filters which are isotropic center-surrounded and are used to model retinal cells with impulse response functions

$$f_\sigma(x,y) \propto (x^2 + y^2 - \sigma^2)e^{-(x^2+y^2)/2\sigma^2}, \tag{4.73}$$

where σ controls the scales of the filters, and (iii) the Gabor filters modeling frequency and orientation sensitivity:

$$f_{\sigma,\theta}(x,y) \propto e^{-1/(2\sigma^2)4(x\cos\theta+y\sin\theta)^2+(-x\sin\theta+y\cos\theta)^2}e^{-j(2\pi/\sigma)(x\cos\theta+y\sin\theta)}, \tag{4.74}$$

where σ controls the scale and θ controls the orientation.

Examine texture fields from Zhu et al. Figure 4.10 shows various examples of animal fur and its synthesis. Panel 1 shows the fur texture. Panel 2 shows a white noise field. Panel 3 shows a texture field synthesized from the maximum entropy distribution resulting from the five pixel Laplacian of a Gaussian filter f_1 with scale $\sigma = 1.0$. Panels 4, 5 and 6 show synthesized textures with the dc component added and the Gabor filters added, panel 4 adding two $f_{6,120°}, f_{2,30°}$, then panel 5 adding the filter $f_{12,60°}$, and panel 6 adding $f_{10,120°}$. With more filters added, the synthesized texture image moves closer to the observed one.

When model complexity is fixed, the filters can be chosen based on the MDL principle. Given a maximum entropy distribution $p_X(x_1,\ldots,x_n)$, which meets moment constraints on marginal densities of filtered versions of the process, the goodness of this distribution is given by the Kullback–Leibler distance between the true density $f(x_1,\ldots,x_n)$ and $p_X(x_1,\ldots,x_n)$:

$$D(f\|p_X) = E_{f(x_1,\ldots,x_n)}\log f(x_1,\ldots,x_n) - E_{f(x_1,\ldots,x_n)}\log p_X(x_1,\ldots,x_n). \tag{4.75}$$

Thus, minimizing the Kullback–Leibler distance between $f(x_1,\ldots,x_n)$ and $p_X(x_1,\ldots,x_n)$ is equivalent to minimizing $-E_{f(x_1,\ldots,x_n)}\log p_X(x_1,\ldots,x_n)$ which is the expected coding length of the data. In other words, one should make use of all the information to specify $p_X(x_1,\ldots,x_n)$. Let F be

4.6 MAXIMUM-ENTROPY TEXTURE REPRESENTATION

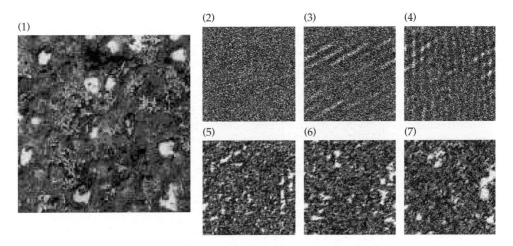

Figure 4.11 Panel 1: observed texture of mud. Panels 2–7 show synthesized versions from Zhu model.

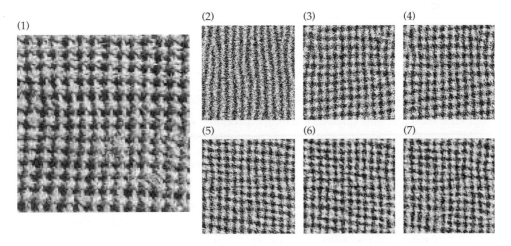

Figure 4.12 Panel 1: observed texture of fabric. Panels 2–7 show synthesized versions from Zhu model.

the set of all possible filters. Then the optimal set $F^*(K)$ of K filters should be chosen according to

$$F^*(K) = \arg\min_{|F(K)|=K} -E_{f(x_1,\ldots,x_n)} \log p_X(x_1,\ldots,x_n). \tag{4.76}$$

Figure 4.11 shows an example of mud texture and the results of its synthesis. Panel 2 shows a white noise field. Panel 3 shows a texture field synthesized from the maximum entropy distribution resulting from adding the filter $f_{8,30}$. Panels 4, 5, 6, 7 and 8 show synthesized textures with the filters cosine component of $f_{7,90}$, Laplacian of a Gaussian filter f_2, Laplacian filter with mask size 5x5, cosine component of the Gabor filter $f_{3,90}$ and cosine component of $f_{3,0}$, respectively.

Figure 4.12 shows similar results for fabric textures.

4.7 Stationary Gibbs Random Fields

Thus far the specification of the Gibbs fields has been in terms of an assumed boundary; the probabilities on finite graphs $G \subset \mathbb{Z}^d$ are determined by the value of the field on the boundary of the graph. Now define the underlying Gibbs distribution P associated with the infinite lattice $\mathbb{Z}^d$. This is far more than just an interesting theoretical indulgence. If we hope to be able to understand what is happening "in the large graph limit," that is for large patterns understanding the infinite distribution and its properties, the entropy and partition function will determine the properties of the finite graph patterns.

Let $D \subset \mathbb{Z}^d$, with X defined on D with state space $\mathcal{X}_0$. The Gibbs distribution takes the form

$$P(X) = \frac{e^{-U(X)}}{Z_D}, \quad \text{where } Z_D = \sum_{x \in \mathcal{X}_0^{|D|}} e^{-U(x)}. \tag{4.77}$$

The partition function is determined by values of the fields on the boundary. Similar issues apply as above for the Gaussian case. The probabilities on finite graphs $D \subset \mathbb{Z}^d$ are determined by the value of the field on the boundary ∂D. Fortunately, in the large graph limit there is a diminishing effect of the boundary on the partition function.

4.7.1 The Dobrushin/Lanford/Ruelle Definition

Define the underlying Gibbs distribution P associated with the infinite lattice $\mathbb{Z}^d$. To define probabilities on the cylinders for any cylinder, the infinite distribution must be defined consistently with the finite cylinder distributions of the same potentials. The most natural definition is to construct it from the fixed boundary distributions of the same potentials:

$$P_D(x_D) = E_{P_Y}\{E\{1_{x_D}(X_D)|Y\}\}. \tag{4.78}$$

Definition 4.21 (Dobrushin, Lanford, Ruelle) *Then P is the **Gibbs distribution** corresponding to interaction Φ on $\mathbb{Z}^d$ if for any $D \subset \mathbb{Z}^d$ the marginal P_D of P on D is given by*

$$P_D(X_D) = \sum_{y \in \mathcal{X}_0^{\mathbb{Z}^d/D}} P_D(X_D|y) P_{\mathbb{Z}^d/D}(y). \tag{4.79}$$

Denote the set of **stationary Gibbs probabilities with interaction** Φ *as* $\mathcal{G}_s^\Phi$.

This Eqn. 4.79 is known as the Dobrushin, Lanford, and Ruelle equation. One of the remarkable things about Gibbs distributions, which is unlike irreducible Markov chains, is that there may be multiple distributions in $\mathcal{G}_s^\Phi$ with precisely the same interactions Φ; *the set may contain more than one element*. In particular, the definition implies

$$P_D(X_D|y_{\mathbb{Z}^d/D}) = P_D(X_D|y_{\partial \bar{D}}), \quad \text{for all } P \in \mathcal{G}_s^\Phi. \tag{4.80}$$

All of the Gibbs distributions with the same interaction attach the same conditional probabilities to finite sets [120].

4.7.2 Gibbs Distributions Exhibit Multiple Laws with the Same Interactions (Phase Transitions): The Ising Model at Low Temperature

Thus far, as we have defined it, the finite volume Gibbs distributions are determined by their boundaries. Perhaps the most critical difference between 1D random fields which are essentially Markov chains and two or higher dimensional fields is the fact that on the infinite lattice there can be multiple Gibbs distributions with exactly the same local conditionals. We would be delinquent not to highlight this fundamental departure. Since this is an advanced topic, we do it by illustrating the issue on the Ising model.

Examine the proof originally constructed by Peierls [142] following Kinderman and Snell [123] showing that the 2D Ising model does have a phase transition—that is, it exhibits spontaneous magnetization at sufficiently low temperatures. Examine the pure +1 boundary, setting $X_i = +1$ on the boundary—then letting the boundary move off to infinity, or equivalently, letting the size of the lattice go to infinity. For low temperatures the probability that $X_0 = +1$ (or -1) for any site O is no longer $\frac{1}{2}$ but dependent on the boundary condition, no matter how far away the site O is from the boundary. In other words, the boundary condition has a strong effect on internal sites if the temperature is sufficiently low. Recall that in the 2D Ising model, the probability of being in a particular configuration $X = (X_1, \ldots, X_N)$ is given by

$$P(X) = \frac{1}{Z} e^{1/KT} \left(J \sum_{\langle i,j \rangle} X_i X_j + mH \sum_i X_i \right), \tag{4.81}$$

with H the strength of the external field, $J > 0, m > 0$ are weights of the internal and external energies, with K Boltzmann's constant and T the temperature. The first summation is taken over all pairs of sites that are connected by a bond (i.e. nearest neighbors), and the second summation is over all sites in the lattice. Defining the number of even bonds $n_e(X) = X_i X_j > 0$, and the number of odd bonds $n_o(X) = X_i X_j < 0$ and $n_b = n_e + n_o$, with $H = 0$ Eqn. 4.81 is simplified to

$$P(X) = \frac{1}{Z} e^{J/KT (n_e(X) - n_o(X))} = \frac{1}{Z} e^{J/KT (n_b(X) - 2n_o(X))} \tag{4.82}$$

$$= \frac{1}{Z'} e^{-b n_o(X)}. \tag{4.83}$$

Z' is a new normalization factor and $b = 2J/KT$. In the following, we consider a positive boundary condition, and show that at low enough temperatures $\Pr\{X_0 = -1\} < \frac{1}{2}$ and $\Pr\{X_0 = +1\} > \Pr\{X_0 = -1\}$. Similar reasoning shows that $\Pr\{X_0 = +1\} < \Pr\{X_0 = -1\}$ under a negative boundary condition. We consider O to be in the middle of the lattice, that is, the site deepest in the interior. The left panel of Figure 4.13 illustrates one configuration that site O takes a value of -1. In the figure, we have drawn lines separating sites of opposite signs, or odd bonds. Because of the boundary conditions, the lines must form closed curves. It is clear that the number of odd bonds $n_o(X)$ is just the total length of all the curves in Figure 4.13. This was the essential device introduced by Peierls to study the probability distribution. Since $X_0 = -1$, there must be one closed curve that encloses O. Let's call this curve a circuit and denote it by S, and say that its length is $L(S)$. Suppose we fix S, and consider the set Ω_S of configurations having S as the circuit that encloses O. Then for any configuration $\bar{X} \in \Omega_S$, we associate a new configuration X' which agrees with $\bar{X}$ except that all sites inside S are changed to $+1$. This has the effect of removing the circuit S and decreasing the number of odd bonds by $L(S)$. That is, $n_o(X') = n_o(\bar{X}) - L(S)$. The

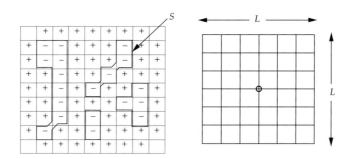

Figure 4.13 Left panel shows a configuration with positive boundary condition. Right panel shows the bounding box for the circuit of length L.

mapping $\bar{X} \to X'$ is one-to-one. Then the probability that the circuit S around O occurs is

$$P(S) = \frac{\sum_{\bar{X}} e^{-bn_o(\bar{X})}}{\sum_X e^{-bn_o(X)}} \leq \frac{\sum_{\bar{X}} e^{-bn_o(\bar{X})}}{\sum_{X'} e^{-bn_o(X')}} \qquad (4.84)$$

$$= \frac{\sum_{\bar{X}} e^{-bn_o(\bar{X})}}{\sum_{\bar{X}} e^{-bn_o(\bar{X})+bL(S)}} = e^{-bL}. \qquad (4.85)$$

Now consider the set Ψ of all possible circuits which surround the lattice site O, and we get

$$\Pr\{X_0 = -1\} = \sum_{S \in \Psi} P(S) \leq \sum_{S \in \Psi} e^{-bL}$$

$$= \sum_{L=4,6,\ldots} r(L) e^{-bL}, \qquad (4.86)$$

where $r(L)$ is the number of circuits of length L which surround the lattice site O. Consider such a circuit. Since it is required to surround the lattice site 0, this circuit must have each point at a distance no more than $L/2$ from the site O. See panel 2 of Figure 4.13. There are at most L^2 ways to choose a starting point for the circuit. Having chosen a starting point, there are at most four ways to choose the next point and three ways to choose each succeeding point moving around the circuit. Thus there are at most $\frac{1}{L} 4L^2 3^{L-1}$ circuits of length L surrounding O. (The factor $1/L$ comes from the fact that each circuit gets counted n times since any point along it can be considered as the starting point). Thus $r(L) \leq 4L3^{L-1} \leq L4^L$ implying

$$\Pr\{X_0 = -1\} \leq \sum_{L=4,6,\ldots} L4^L e^{-bL} \leq \sum_{L=1}^{\infty} L(4e^{-b})^L \qquad (4.87)$$

$$\stackrel{(a)}{=} \frac{4e^{-b}}{(1-e^{-b})^2} \qquad (4.88)$$

with (a) coming from

$$\frac{x}{(1-x)^2} = x(1 + 2x + 3x^2 + \cdots) = \sum_{n=1}^{\infty} nx^n, \quad \text{if } |x| < 1.$$

since $b = 2J/KT$, we can make the R.H.S. of the inequality (4.88) arbitrarily small (strictly less than $\frac{1}{2}$) by letting the temperature $T \to 0$. Thus, phase transition is guaranteed at low temperature.

4.8 1D Random Fields are Markov Chains

The DAG models are a special case of Gibbs random fields, and Markov chains are a 1D (linear) Gibbs random field with transition matrix determined by the potentials.

Theorem 4.22 *Let $X_1, X_2, \ldots$ be a finite-valued m-range Gibbs random field with locally supported potential Φ. Then there is a statistically equivalent Markov chain with transition matrix determined by the locally supported clique function Φ.*

Proof The energy $U : \mathcal{X} \to \mathbb{R}$ is constructed from locally supported functions $\Phi(x) = \Phi(x_S), S \subset \mathbb{Z}$ with range $R = m$, $\max_{\{i,j \in S\}} |i - j| \leq m$, implying

$$P(X) = \frac{1}{Z}e^{-U(X)} = \frac{1}{Z}e^{-\sum_{j=m}^n \Phi(\tau_j(X))}\pi(X_0,\ldots,X_{m-1}) \quad (4.89)$$

$$= \frac{1}{Z}e^{-\sum_{j=m}^n \Phi(X_{j-m},\ldots,X_j)}\pi(X_0,\ldots,X_{m-1}), \quad (4.90)$$

with the Φ function corrected at the boundary by the initial distribution $\pi(\cdot)$. The Markov property follows from

$$P(X_n|X_{n-1}, X_{n-2},\ldots) = \frac{P(X_n, X_{n-1}, X_{n-2},\ldots)}{\sum_{X_n \in \mathcal{X}_0} P(X_n, X_{n-1}, X_{n-2},\ldots)} \quad (4.91)$$

$$= \frac{e^{\Phi(X_{n-m},\ldots,X_n)}}{\sum_{X_n \in \mathcal{X}_0} e^{\Phi(X_{n-m},\ldots,X_n)}}. \quad (4.92)$$

□

Example 4.23 (Ising Model) The locally supported potentials $\Phi : \mathcal{X} \to \mathbb{R}$ determine the transition matrix for the chain. Let us do it explicitly for the 1D Ising case of Section 4.10 following Kinderman and Snell [123]. Let $D = \{0 \leq i \leq n\}, \mathcal{X}_0 = \{-1, +1\}$ and associate no external field. Then

$$P(X) = \frac{e^{\frac{-J}{KT}\sum_{i=1}^n X_i X_{i-1} + f(X_0)}}{Z_{n+1}}. \quad (4.93)$$

Define $n_e(X), n_o(X)$ to be the number of *even* and *odd* bonds; a bond even corresponds to either $(X_i = 1) \wedge (X_{i-1} = 1)$ or $(X_i = -1) \wedge (X_{i-1} = -1)$, with $n_e + n_o = n$. Then

$$P(X) = e^{-2(J/KT)n_e(X)}e^{(J/KT)n+f(X_0)} \quad (4.94)$$

$$= \frac{1}{Z_{n+1}}e^{(J/KT)n+f(X_0)}e^{(-2J/KT)n_e(X)}.$$

Define a two state Markov chain with 2×2 transition matrix $Q = \begin{pmatrix} p & 1-p \\ 1-p & p \end{pmatrix}$, then

$$P(X) = \pi(X_0)p^{n_e(X)}(1-p)^{n_o(X)} = \pi(X_0)p^{n_e(X)}(1-p)^{(n-n_e(X))}$$

$$= \pi(X_0)(1-p)^n \left(\frac{p}{1-p}\right)^{n_e(X)} = \pi(X_0)(1-p)^n e^{n_e(X)(\log \frac{p}{1-p})}. \quad (4.95)$$

Choose $-2J/KT = \log(p)/(1-p)$.

Apparently, the partition function may be calculated for this example. Equating the normalizers gives

$$\frac{1}{Z_{n+1}}e^{(J/KT)n+f(X_0)} = \pi(X_0)(1-p)^n. \tag{4.96}$$

Choosing $\pi(X_0) = e^{f(X_0)}$ as the distribution on the initial state, and substituting for $J/KT = \frac{1}{2}\log(1-p)/(p)$ gives

$$\frac{1}{Z_{n+1}}\left(\frac{1-p}{p}\right)^{n/2} = (1-p)^n \implies Z_{n+1} = \frac{1}{[(1-p)p]^{n/2}}. \tag{4.97}$$

Using the fact that $e^{2J/KT} = (1-p)/p$ implies $p = 1/(1+e^{2J/KT})$, $1-p = e^{2J/KT}/(1+e^{2J/KT})$ giving

$$Z_{n+1} = \frac{(1+e^{2J/KT})^n}{(e^{2J/KT})^{n/2}}. \tag{4.98}$$

Notice, with $n \to \infty$, $(\log Z_n)/n \to \log Z = \frac{1}{2}\log(1+e^{2J/KT})/(e^{2J/KT})$.

4.9 Markov Chains Have a Unique Gibbs Distribution

Clearly this claim that there can be multiple stationary probabilities with the same interactions is a departure from what we usually think about the Markov chain setting. This has something to do with the dimension of the boundary for random fields in $\mathbb{Z}^d, d > 1$. Let us now prove that for irreducible Markov chains there is only one stationary probability with the interactions corresponding to the transition law of the chain.

Theorem 4.24 *Given a Markov chain on finite states space $\mathcal{X}_0$ with positively regular transition matrix $Q = (q_{kj})$, there is a unique stationary probability P^S satisfying the Dobrushin, Lanford, Ruelle condition Eqn. 4.79, where P_n^S on the finite cylinders is given by*

$$P_n^S(X_1, \ldots, X_n) = \pi(X_1) \prod_{i=1}^{n-1} Q_{X_i, X_{i+1}}, \tag{4.99}$$

with $\pi = \pi Q$ the left eigenvector of Q.

Proof Let us show the DLR condition is satisfied by P^S; defining $\mathbb{Z}_n$ to be the first n-integers, then

$$\sum_{y \in \mathcal{X}_0^{\mathbb{Z}/\mathbb{Z}_n}} P^S(X_1, \ldots, X_n|y)P^S(y) = \sum_{X_0 \in \mathcal{X}_0} P^S(X_1, \ldots, X_n|x_0)P^S(X_0) \tag{4.100}$$

$$= \sum_{X_0 \in \mathcal{X}_0} \prod_{i=0}^{n-1} Q_{X_i, X_{i+1}} \pi(x_0) \tag{4.101}$$

$$= \pi(X_1) \prod_{i=1}^{n-1} Q_{X_i, X_{i+1}} = P_n^S(X_1, \ldots, X_n). \tag{4.102}$$

Uniqueness follows from the fact that π is the unique left eigenvector (Perron–Frobenius) for the irreducible case. □

4.10 Entropy of Stationary Gibbs Fields

For stationary fields the entropy rates on large graphs are well defined.

Theorem 4.25 *Let $X_i, i \in \mathbb{Z}^d$ be a stationary Gibbs random field. The **entropy rate** defined as*

$$H(X) = \lim_{|D| \to \infty} \frac{H(X_i, i \in D)}{|D|} \tag{4.103}$$

exists and letting $D_n \subset \mathbb{Z}^d$ be an increasing family of cylinders size $|D_n| = n_1 \times \cdots \times n_d$ with $D_n \uparrow D_\infty$, then the limiting entropy rate $H(X)$ is given by

$$H(X) = \lim_{n \to \infty} H(X_1|X_{D_n/1}) = \lim_{n \to \infty} H(X_n|X_{D_n/n}) \tag{4.104}$$

$$= \lim_{n \to \infty} \frac{1}{|D_n|} H(X_{D_n}). \tag{4.105}$$

Proof The proof generalizes the 1D proof in [7]. The first limit follows since conditioning reduces entropy, $H(X|Y,Z) \leq H(X|Y)$, implying

$$H(X_1|X_{D_n/1}) \leq H(X_1|X_{D_{n-1}/1}). \tag{4.106}$$

Thus, $H(X_1|X_{D_n/1}), H(X_1|X_{D_{n+1}/1})$ is a decreasing sequence of non-negative numbers implying it has a limit, call it $H^{(1)}(X)$. Also, $H(X_n|X_{D_n/n}), H(X_{n-1}|X_{D_{n-1}/n-1})$ is a decreasing positive sequence which has a limit as well, call it $H^{(2)}(X)$, since

$$H(X_n|X_{D_n/n}) \geq H(X_n|X_{1+D_{n-1}/n}) \stackrel{(a)}{=} H(X_{n-1}|X_{D_{n-1}/n-1}), \tag{4.107}$$

with the equality (a) using the stationarity of the field. To prove that these limits are the same and equal to the normalized entropy of the cylinders, factor the entropy using the chain rule:

$$H(X_{D_n}) = H(X_{1,1,\ldots,1}|X_{D_n/1}) + H(X_{2,1,\ldots,1}|X_{D_n/\{1,(2,1,\ldots,1)\}}) + \cdots$$

$$= \sum_{i_d=1}^{n_d} \cdots \sum_{i_1=1}^{n_1} H(X_{(i_1,\ldots,i_d)}|X_{D_n/\cup_{j=1}^i \{(j_1,\ldots,j_d)\}}). \tag{4.108}$$

Choosing $m(\epsilon)$ large enough so that $|H(X_1|X_{D_m/1}) - H^{(1)}| < \epsilon$, then construct cylinder D_n with $n > m(\epsilon)$ so that for all $i \in D_{n-m(\epsilon)} \subset D_n$,

$$H^{(1)}(X) + \epsilon \geq H(X_1|X_{D_m/1}) \tag{4.109}$$

$$\stackrel{(a)}{=} H(X_{i+1}|X_{i+D_m/i+1}) \tag{4.110}$$

$$\stackrel{(b)}{\geq} H(X_{i+1}|X_{D_n/\cup_{j=1}^{i+1} j}) \geq H^{(1)}(X), \tag{4.111}$$

where (*a*) follows from stationarity, and (*b*) follows since conditioning decreases entropy. Since $\lim_{n\to\infty}(|D_{n-m(\epsilon)}|)/|D_n| = 1$,

$$H^{(1)}(X) + \epsilon \geq \lim_{n\to\infty} \frac{1}{|D_n|} H(X_{D_n}) \qquad (4.112)$$

$$= \lim_{n\to\infty} \frac{\sum_{i\in D_n/D_{n-m(\epsilon)}} H(X_i|X_{D_n/\cup_{j=1}^{i} j})}{|D_n|}$$

$$+ \frac{\sum_{i\in D_{n-m(\epsilon)}} H(X_i|X_{D_n/\cup_{j=1}^{i} j})}{|D_n|} \qquad (4.113)$$

$$\geq H^{(1)}(X). \qquad (4.114)$$

Since ϵ is arbitrary, the limit is $H^{(1)}(X)$. The identical limit is obtained by factoring the joint entropy using $H(X_n|X_{D_n/n})$, implying $H^{(1)}(X) = H^{(2)}(X) = H(X)$. □

5 GAUSSIAN RANDOM FIELDS ON UNDIRECTED GRAPHS

ABSTRACT This chapter focuses on state spaces of the continuum studying Gaussian random fields on discrete lattices. Covariances are induced via difference operators and the associated neighborhood structure of the resulting random field is explored. Texture representation and segmentation are studied via the general Gaussian random field structures. For dealing with the partition function determined by the log-determinant of the covariance asymptotics are derived connecting the eigenvalues of the finite covariance fields to the spectrum of the infinite stationary extension.

5.1 Gaussian Random Fields

In pattern theoretic representations the graphs often arise in the process of performing discrete computation associated with continuum representations associated with differential operators expressing interactions in the continuum. The differential operators induce the graphs, the random fields with covariance structure induced via differential operators arising in the continuum. Naturally the order of the differential operator determines the neighborhood and graph structure.

Inducing Gaussian time series via differential operators is familiar through auto-regressive processes. In a similar manner this can be done for generalizations to fields on more general lattices $D \subset \mathbb{Z}^d$. For this, Gaussian fields on regular lattices play an important role. Gaussian fields in general correspond to index spaces perhaps on the continuum $D \subset \mathbb{R}^d$, and state spaces vector valued, $\mathcal{X}_0 = \mathbb{R}^m$. For now, examine a discrete lattice $D \subset \mathbb{Z}^d$, scalar valued $\mathcal{X}_0 = \mathbb{R}$. The most familiar view of the specification of Gaussian processes and Gaussian random fields is through arbitrary linear combinations of samples of the field that are Gaussian. The Gaussian fields studied will be associated with Hilbert spaces $H = L^2, l^2$; therefore it will be powerful to define the Gaussian fields through their properties in inner products with elements of the Hilbert space.

Definition 5.1 *A scalar valued stochastic process or random field $\{X_i(\omega), i \in D \subseteq \mathbb{Z}(\mathbb{Z}^d)\}$ is a **Gaussian process (field)** with mean $m_X : D \to \mathbb{R}$, and **covariance** $K_X : D \times D \to \mathbb{C}$ if finite linear combinations are normally distributed; for any integer n and any n-tuple $a_1, \ldots, a_n$ and points $i_1, \ldots, i_n \in D$, $Y(\omega) = \sum_{k=1}^n a_k X_{i_k}(\omega)$ is Gaussian with mean and variance*

$$\sum_{k=1}^n a_k m_X(i_k), \quad \sum_{k=1}^n \sum_{j=1}^n a_k K_X(i_k, i_j) a_j. \tag{5.1}$$

We shall say X is a **Gaussian process (field)** with respect to the Hilbert space of square summable sequences $l^2(\mathbb{Z}^d), l^2(D \subset \mathbb{Z}^d)$ with inner product $\langle f, g \rangle = \sum_{i \in D} f_i g_i$ with mean $m_X \in H$ and covariance operator K_X, if for all $f \in l^2$, $\langle f, X \rangle$ is normally distributed with mean and variance

$$m_f = \langle f, m_X \rangle, \quad \sigma_f^2 = \langle f, K_X f \rangle. \tag{5.2}$$

This illustrates the more general case which is specified according to arbitrary operators on lattices $D \subset \mathbb{Z}^d$. If the operators are linear, they induce Gaussian fields with particular graph structures dependent on the differential order.

Remark 5.1.1 It is usual to specify a Gaussian process as having all marginals which are Gaussian, so that the process is zero-mean Gaussian if for any n variates $X_1, \ldots, X_n$ with covariance K has density

$$p(X_1, \ldots, X_n) = \det^{-(1/2)} 2\pi K e^{-(1/2) \sum_{ij} X_i K_{ij}^{-1} X_j}. \tag{5.3}$$

That the above definition is equivalent is sufficient to examine the characteristic function, using the fact that if Y is Gaussian with variance σ_Y^2 then $E e^{i\gamma Y} = e^{-(1/2)\gamma^2 \sigma_Y^2}$ (see Example 2.13, Chapter 2). Use the fact that for any coefficient $\sum_{i=1}^n a_i X_i$ is Gaussian with variance $\sum_{ij} a_i a_j K_X(i,j)$, so choosing $\omega = (\omega_1, \ldots, \omega_n)$, then the characteristic function becomes

$$E e^{i \sum_{i=1}^n \omega_i X_i} = e^{-(1/2) \sum_{ij} \omega_i \omega_j K_X(i,j)}, \tag{5.4}$$

since $\sum_{i=1}^n \omega_i X_i$ is Gaussian with variance $\sum_{ij} \omega_i \omega_j K_X(i,j)$. This is the joint characteristic function of a Gaussian vector with covariance K_X (see [143], p. 168), which is a unique identification of the density, so $X_1, \ldots, X_n$ have joint density given by 5.3.

5.2 Difference Operators and Adjoints

The Gaussian fields are studied induced by stationary linear difference operators adjusted at the boundary to reflect boundary conditions. Let L be a finite difference, bounded support stationary operator with its action on functions $f \in l^2(\mathbb{Z}^d)$ of the form

$$(Lf)_i = \sum_{j \in \mathbb{Z}^d} L(i,j) f_j = \sum_{j \in \mathbb{Z}^d} a_{j-i} f_j \tag{5.5}$$

$$= \sum_{s \in S} a_s f_{i+s}, \quad i \in \mathbb{Z}^d, \tag{5.6}$$

with bounded support implied by $|S|$ finite.

To induce Gaussian processes on finite domains $D \subset \mathbb{Z}^d$, follow the approach as above for Gibbs fields in which boundaries are specified determining the form of the operator at the boundary configuration. The finite domain operator L_D^g extended with infinite boundary $g \in l^2(\mathbb{Z}^d)$ is constructed from L according to

$$(L_D^g f)_i = (L f^g)_i = \sum_{s \in S} a_s (f_{i+s} \vee g_{i+s}), \tag{5.7}$$

where $f^g = (f \vee g)_i = \begin{cases} f_i & i \in D \\ g_i & i \notin D \end{cases}. \tag{5.8}$

As before define the interior and boundary associated with the operator determined by the support set S:

$$D^0 = \{i \in D : i + S \subset D\}, \quad \bar{D} = \cup_{i \in D}(i+S), \quad \partial D = D/D^0. \tag{5.9}$$

In what follows the 0-boundary will be used extensively, and will be denoted for convenience as L_D corresponding to the operator L on the interior of D extended with the zero boundary. The covariance of the fields generated will be defined by the operator and its adjoint. The adjoint operator is defined through the inner product and its action on functions (see Luenberger [8]).

5.2 DIFFERENCE OPERATORS AND ADJOINTS

Definition 5.2 *Let H_1, H_2 be two Hilbert spaces of functions with inner products $\langle \cdot, \cdot \rangle_{H_1}$, $\langle \cdot, \cdot \rangle_{H_2}$, respectively. Then the **adjoint** L^* of $L : H_1 \to H_2$ is a mapping $L^* : H_2 \to H_1$ satisfying*

$$\text{for all } f \in H_1, \; g \in H_2, \quad \langle f, L^*g \rangle_{H_1} = \langle Lf, g \rangle_{H_2}. \tag{5.10}$$

We study difference operators, therefore $H_1 = H_2 = l^2(\mathbb{Z}^d)$, and with the same inner product denoted $\langle \cdot, \cdot \rangle_{l^2}$. The adjoint operators for the infinite and finite processes become as follows:

Theorem 5.3 *Given are L on $\mathbb{Z}^d$ and L_D L extended with 0-boundary on D defined according to*

$$(Lf)_i = \sum_{s \in S} a_s f_{i+s}, \quad i \in \mathbb{Z}^d, \tag{5.11}$$

$$(L_D f)_i = \sum_{s \in S} a_s f^0_{i+s}, \quad i \in D. \tag{5.12}$$

1. *The adjoint operators L^* on $\mathbb{Z}^d$ and L_D^* on D are given by*

$$(L^* f)_i = \sum_{s \in S} a_s f_{i-s}, \quad i \in \mathbb{Z}^d, \tag{5.13}$$

$$(L_D^* f)_i = \sum_{s \in S} a_s f^0_{i-s}, \quad i \in D. \tag{5.14}$$

2. *The operators L^*L on $\mathbb{Z}^d$ and $L_D^* L_D$ on D are given by*

$$(L^* L f)_i = \sum_{s' \in S} \sum_{s \in S} a_s a_{s'} f_{i+s-s'}, \quad i \in \mathbb{Z}^d, \tag{5.15}$$

$$(L_D^* L_D f)_i = \sum_{s \in S} \sum_{s' \in S} a_s a_{s'} f^0_{i-s+s'}, \quad i \in D. \tag{5.16}$$

Proof First part follows from the identity

$$\langle g, L^* f \rangle_{l^2} = \sum_{i \in \mathbb{Z}^d} g_i (L^* f)_i = \langle Lg, f \rangle_{l^2} = \sum_{i' \in \mathbb{Z}^d} (Lg)_{i'} f_{i'} = \sum_{i' \in \mathbb{Z}^d} \sum_{s \in S} a_s g_{s+i'} f_{i'}$$

$$= \sum_{i \in \mathbb{Z}^d} g_i \underbrace{\sum_{s \in S} a_s f_{i-s}}_{(L^* f)_i} \quad (i = s + i'). \tag{5.17}$$

The adjoint operator L_D^{0*} is constructed similarly. For all $f, g \in l^2(D)$ with f^0, g^0 extended to $\mathbb{Z}^d$ with the 0-boundary, then

$$\langle g, L_D^{0*} f \rangle = \sum_{i \in D} g_i (L_D^{0*} f)_i = \langle L_D g, f \rangle = \sum_{i \in D} \underbrace{\sum_{s \in S} a_s g^0_{i+s}}_{(L_D g)_i} f_i = \sum_{i \in \mathbb{Z}^d} \sum_{s \in S} a_s g^0_{i+s} f^0_i \quad (i = s + i')$$

$$= \sum_{i' \in \mathbb{Z}^d} \sum_{s \in S} a_s f^0_{i'-s} g^0_{i'} = \sum_{i' \in D} \underbrace{\sum_{s \in S} a_s f^0_{i'-s}}_{(L_D^* f)_{i'}} g_{i'}. \tag{5.18}$$

Then L^*L is given by

$$(L^*Lf)_i = \sum_{s \in S} a_s L f_{i-s} = \sum_{s \in S} a_s \sum_{s' \in S} a_{s'} f_{s'+i-s}.$$

Similarly $L_D^* L_D$ is derived as above. □

5.3 Gaussian Fields Induced via Difference Operators

We shall look only at the 0-boundary case, so again it is implicitly assumed L_D is the L operator defined on D with 0-boundary. Begin by defining the Minkowski set difference.

Definition 5.4 *Given a set $S \subset D \subset \mathbb{Z}^d$, then* **the Minkowski set difference and addition** *is*

$$S \ominus S = \{s - s'; \forall s \neq s' \in S\} \subset \mathbb{Z}^d, \tag{5.19}$$

$$S \oplus S = \{s + s'; \forall s \neq s' \in S\} \subset \mathbb{Z}^d. \tag{5.20}$$

Theorem 5.5 *Assume L_D is an invertible operator on $l^2(D)$ and let X satisfy the stochastic difference equation*

$$(L_D X)_i = W_i, \tag{5.21}$$

with $W_i, i \in D$ a white noise zero mean process of independent, identically distributed Gaussian variables, variance σ^2.

Then $\{X_i(\omega), i \in D \subset \mathbb{Z}^d\}$ is a real-valued Gaussian random field having density

$$p(X) = \frac{1}{(2\pi)^{|D|/2} \det^{1/2} K_X} \prod_{i \in D} e^{\frac{-|(L_D X)_i|^2}{2\sigma^2}}, \tag{5.22}$$

with the inverse covariance the product of the operator with its adjoint $K_X^{-1} = (1/\sigma^2) L_D^* L_D$, with Markov structure on graph $\sigma = \{D, E\}$ with neighborhoods $N_i = D \cap (i + S \ominus S)$.

Proof First we must prove for any vector of coefficients $f_i, i \in D$, then the linear combination $\sum_{i \in D} f_i X_i$ is Gaussian distributed. Use the fact that $L_D X = W$ defines a bijection with inverse L_D^{-1} implying

$$\sum_{i \in D} f_i X_i = \sum_{i \in D} f_i (L_D^{-1} W)_i$$

$$= \langle f, L_D^{-1} W \rangle = \langle \underbrace{L_D^{-*} f}_{g}, W \rangle, \tag{5.23}$$

which is Gaussian distributed since by definition $\langle g, W \rangle$ is Gaussian for all g. Thus $\langle f, X \rangle$ is Gaussian, with covariance following from Eqn. 5.23:

$$E \langle f, X \rangle^2 = \langle f, K_X f \rangle \tag{5.24}$$

$$= \langle L_D^{-*} f, K_W L_D^{-*} f \rangle = \sigma^2 \langle f, L_D^{-1} L_D^{-*} f \rangle. \tag{5.25}$$

This is true for all a, implying $K_X^{-1} = L_D^* L_D / \sigma^2$.

5.3 GAUSSIAN FIELDS INDUCED VIA DIFFERENCE OPERATORS

To calculate the graph structure defined by the neighborhood system, examine the case that $i + S \ominus S \subset D$. Use the adjoint and write the density according to

$$\frac{1}{Z} e^{-\langle L_D X, L_D X \rangle / 2\sigma^2} = \frac{1}{Z} e^{-\langle X, L_D^* L_D X \rangle / 2\sigma^2} = \frac{1}{Z} \prod_{i \in D} e^{-X_i (L_D^* L_D X)_i / 2\sigma^2}. \quad (5.26)$$

Applying the definitions gives $L_D^* L_D$ as

$$(L_D^* L_D X)_i = \sum_{s \in S} \sum_{s' \in S} a_s a_{s'} X^0_{i-s+s'}. \quad (5.27)$$

The number of edges in the graph emanating from internal sites is $|S|^2 - |S|$. Only terms in the quadratic form involving X_i are those sites $j \in i + S \ominus S$ giving the neighborhood structure for $i \in D^0$ according to

$$p(X_i | X_j, j \neq i) = \frac{e^{-(1/2\sigma^2) \sum_{j \in D} X_j \sum_{s \in S} \sum_{s' \in S} a_s a_{s'} X^0_{j-s+s'}}}{\int_{\mathbb{R}} e^{-(1/2\sigma^2) \sum_{j \in D} X_j \sum_{s \in S} \sum_{s' \in S} a_s a_{s'} X^0_{j-s+s'}} dx_i}$$

$$= \frac{e^{-(1/2\sigma^2) X_i \sum_{s \in S} \sum_{s' \in S} a_s a_{s'} X_{j-s+s'}}}{\int_{\mathbb{R}} e^{-(1/2\sigma^2) X \sum_{s \in S} \sum_{s' \in S} a_s a_{s'} X_{j-s+s'}} dx}. \quad (5.28)$$

The edges internal to the graph are constructed by taking a site $i \in D^0$ and equipping it with bonds enumerated by $j = 1, 2, \ldots |S \ominus S|$. Two sites, i, i' are connected by an edge in the graph if $i = i' + k$ for some $k = s - s' \in S \ominus S$. □

Remark 5.3.1 The same result can be obtained using the standard change of distribution formula. Since $W_i, i \in D$ is independent Gaussian with product density having the form $p(W) = \prod_{i \in D} p(W_i)$, then

$$p(X) = \frac{1}{Z} \prod_{i \in D} e^{-|(L_D X)_i|^2 / 2\sigma^2}, \quad (5.29)$$

with the constant $1/Z$ determined by the Jacobian of the transformation. The Gaussian case is a special case of a linear operator with $p(W_i) = e^{f(W_i)}$ where $f(\cdot) = \| \cdot \|^2$. Let $p(W_i) = e^{f(W_i)}$ with L_D a linear or non-linear difference operator. Then

$$p(X) = \frac{1}{Z} \prod_{i \in D} e^{f((L_D X)_i)}, \quad (5.30)$$

with Z given by the Jacobian of the transformation L_D.

Example 5.6 (Auto-Regression: 1 Derivative, 0-boundary) In order to motivate the general theorem on graphs and generator structure induced via difference operators, examine the simplest auto-regressive model which is familiar.

Take the index set on the real line (time), $D = [0, \infty)$, $L = (d/dt) + a$ with zero boundary $X(0) = 0$, associated with the differential equations $\dot{X}(\omega, t) + aX(\omega, t) = W(\omega, t)$, and $W(\cdot)$ "white noise". This induces a Gaussian process with particular covariance; we will return to these continuum processes in some detail in Chapter 9. For now, focus on the associated difference equation and the resulting nearest neighbor Markov chain graph. Then, taking backward differences with zero boundary $X_0 = 0$ gives

$$(1 + a) X_i - X_{i-1} = W_i, \quad i \geq 1, \quad (5.31)$$

implying

$$X_i = \sum_{k=1}^{i} \left(\frac{1}{1+a}\right)^{i+1-k} W_k \qquad i \geq 1, \ X_0 = 0. \tag{5.32}$$

This is a Gaussian process, which is first order Markov; its directed graph structure is as shown in the left panel of Figure 3.2, with parents of internal sites $\pi_i = \{i-1\}$. The linear graph structure is induced by the first order operator $L = (d/dt) + a$. If W is zero-mean, covariance $EW_n W_j = \delta(n-j)$, then

$$K_X(i,j) = EX_i X_j = \sum_{k=0}^{i-1} \sum_{k'=0}^{j-1} \left(\frac{1}{1+a}\right)^{i-k} \left(\frac{1}{1+a}\right)^{j-k'} EW_k W_{k'} \tag{5.33}$$

$$= \sum_{k=0}^{\min(i,j)-1} \left(\frac{1}{1+a}\right)^{i+j-2k}. \tag{5.34}$$

Examine the operator as inducing a Gaussian field $\{X_i, 1 \leq in\}$ with backward differences giving the stochastic equation

$$(LX)_i = \sum_{s \in S} a_s X_{s+i} = X_i - X_{i-1} + aX_i = W_i, \quad X_0 = 0.$$

Applying Theorem 5.5

$$S = \{0, -1\}, \qquad S \ominus S = \{-1, 1\}, \tag{5.35}$$

$$\mathcal{N} = \cup_{i=2}^{n-1} \underbrace{\{i-1, i+1\}}_{i+S\ominus S} \cup \{2\} \cup \{n-1\}. \tag{5.36}$$

Viewing $L_n : \mathbb{R}^n \to \mathbb{R}^n$ as a matrix operator in equation $L_n X = W$, the L_n is invertible with

$$L_n^0 = \begin{pmatrix} 1+a & 0 & 0 & 0 & \cdots \\ -1 & 1+a & 0 & 0 & \cdots \\ 0 & -1 & 1+a & 0 & \cdots \\ \vdots & \vdots & \vdots & \vdots & \vdots \\ 0 & 0 & \cdots & -1 & 1+a \end{pmatrix},$$

$$L_n^{0*} = \begin{pmatrix} 1+a & -1 & 0 & 0 & \cdots \\ 0 & 1+a & -1 & 0 & \cdots \\ 0 & 0 & 1+a & -1 & \cdots \\ \vdots & \vdots & \vdots & \vdots & \vdots \\ 0 & 0 & \cdots & 0 & 1+a \end{pmatrix}. \tag{5.37}$$

Using the standard density transformation with zero boundary gives

$$p(X_1, \ldots, X_n) = \frac{1}{Z} \prod_{i=1}^{n} e^{(-|(L_n^0 X)_i|^2/2\sigma^2)}$$

$$= \frac{1}{Z} e^{(-|(1+a)X_1|^2/2\sigma^2)} \prod_{i=2}^{n} e^{(-|(1+a)X_i - X_{i-1}|^2/2\sigma^2)}, \tag{5.38}$$

Z absorbing the normalization given by the Jacobian of the coordinate transformation, which is the determinant of L_n. In this case, L_n is lower triangular and the determinant is just $(1+a)^n$, giving $Z = (2\pi\sigma^2)^{n/2}(1+a)^{-n}$.

Example 5.7 (Derivative operator: Periodic boundary) Reconsider the same basic operator $L = (\partial/\partial t) + a$, discretized with backward differences as before, but with periodic boundary conditions on $1 \leq i \leq n$:

$$(LX)_i = \sum_{s \in S} a_s X_{s+i} = X_i - X_{i-1} + aX_i = W_i, \quad X_0 = X_n.$$

This is the cyclic torus graph structure shown in Figure 4.1. Then

$$L_n = \begin{pmatrix} 1+a & 0 & 0 & \cdots & 0 & -1 \\ -1 & 1+a & 0 & \cdots & 0 & 0 \\ 0 & -1 & 1+a & 0 & \cdots & 0 \\ \vdots & & & & & \\ 0 & 0 & 0 & \cdots & -1 & 1+a \end{pmatrix},$$

$$L_n^* = \begin{pmatrix} 1+a & -1 & 0 & \cdots & 0 & 0 \\ 0 & 1+a & -1 & 0 & \cdots & 0 \\ 0 & 0 & 1+a & -1 & \cdots & 0 \\ \vdots & & & & & \\ -1 & 0 & 0 & 0 & \cdots & 1+a \end{pmatrix}, \quad (5.39)$$

with eigenfunctions for L_n, L_n^* being $\phi_i(\omega_k = (2\pi/n)k) = e^{(-j2\pi/n)ki}$ and eigenvalues $\lambda_k = 1 - e^{(+j2\pi k/n)} + a = -2je^{j\pi k/n}\sin(\pi k/n) + a$ and eigenvalues for L_n^* given by $\lambda_k^* = 2je^{(-j\pi k)/n}\sin(\pi k/n) + a$. The inverse covariance is

$$K_n^{-1} = \frac{L_n^* L_n}{\sigma^2} \quad (5.40)$$

$$= \frac{1}{\sigma^2}\begin{pmatrix} (1+a)^2+1 & -(1+a) & 0 & \cdots & 0 & -(1+a) \\ -(1+a) & (1+a)^2+1 & -(1+a) & 0 & \cdots & 0 \\ 0 & -(1+a) & (1+a)^2+1 & -(1+a) & \cdots & 0 \\ \vdots & & & & & \\ -(1+a) & 0 & 0 & \cdots & -(1+a) & (1+a)^2+1 \end{pmatrix},$$

which is symmetric having real eigenvalues

$$\lambda_k \lambda_k^* = (1+a)^2 + 1 - (1+a)e^{-j2\pi k/n} - (1+a)e^{-j2\pi(n-1)k/n} \quad (5.41)$$

$$= 2 + 2a + a^2 - 2(1+a)\cos\frac{2\pi}{n}k = \lambda_k \lambda_k^*. \quad (5.42)$$

These illustrate 1D cases. Now to operators on lattices $D \subset \mathbb{Z}^2$.

Example 5.8 (2D Derivative Operator) Examine the operator $L = (\partial/\partial x_1) + (\partial/\partial x_2) + a$, with backward differences with 0-boundary $X_{0,j} = X_{i,0}$, inducing the random field $X_{(i,j)}, (i,j) \in D = \{1 \leq i, j \leq n\}$ according to

$$-X_{i-1,j} + (2+a)X_{i,j} - X_{i,j-1} = W_{i,j}, \quad (5.43)$$

and $W_{i,j}$ a zero-mean white noise field, independent identically distributed Gaussian variables, variance σ^2. Thus $(L_m X)_i = \sum_{s \in S} a_s X_{s+i}$, and applying Theorem 5.5, gives

$$S = \{(0,0), (0,-1), (-1,0)\}, \quad S \ominus S = \{(0,-1), (-1,0), (0,1), (-1,1), (1,0), (1,-1)\},$$

$$N_i = D \cap i + S \ominus S = \{(i,j-1), (i-1,j), (i,j+1), (i-1,j+1), (i+1,j), (i+1,j-1)\}.$$

Then the interior neighborhoods are of size $|S \ominus S| = 6$.

The operator equation is invertible, since there is a bijection between the X and W vectors in $(L_m X)_{i,j} = W_{i,j}$, $(i,j) \in D$, as evidenced by

$$\begin{aligned}
(i=1, j=1) \quad & (2+a)X_{1,1} = W_{1,1} \\
(i=1, j=2) \quad & (2+a)X_{1,2} - X_{1,1} = W_{1,2} \\
& \vdots \\
(i=1, j=m) \quad & (2+a)X_{1,m} - X_{1,m-1} = W_{1,1} \\
(i=2, j=1) \quad & (2+a)X_{2,1} - X_{1,1} = W_{2,1} \\
(i=2, j=2) \quad & (2+a)X_{2,2} - X_{1,1} - X_{2,1} = W_{2,2} \\
& \vdots
\end{aligned} \tag{5.44}$$

Thus L_m is invertible, and using the change of density formula $p(X_{i,j}, (i,j) \in D)$ the density becomes

$$1/Z \prod_{i \in D^0} e^{-1/2\sigma^2 (X_{i-1,j} - (2+a)X_{i,j} + X_{i,j-1})^2} \prod_{j \in D} e^{-1/2\sigma^2 (X_{1,j})^2} \prod_{i \in \mathbb{Z}_m} e^{-1/2\sigma^2 (X_{i,1})^2}.$$
(5.45)

The left panel in Figure 5.1 shows the induced graph structure for the 2D derivative operator.

Example 5.9 (1D Self-adjoint Laplacian: Periodic boundary) Examine the Laplacian $L = -(\partial^2/\partial x^2) + a$ with periodic boundary conditions. Assume the derivative is approximated according to $((\partial/\partial x)x)_i = x_{i+(1/2)} - x_{i-(1/2)}$ giving $L = L^*$ with

$$(LX)_i = -X_{i-1} + (a+2)X_i - X_{i+1} = W_i \tag{5.46}$$

giving

$$L_n = \begin{pmatrix} (a+2) & -1 & 0 & \cdots & 0 & -1 \\ -1 & (a+2) & -1 & 0 & \cdots & 0 \\ \vdots & \vdots & \vdots & \vdots & \vdots & \\ -1 & 0 & 0 & \cdots & -1 & a+2 \end{pmatrix} \tag{5.47}$$

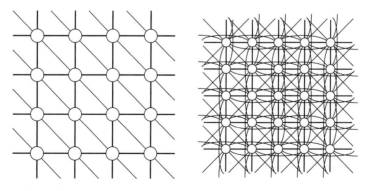

Figure 5.1 Left panel shows the graph induced by the first order difference operator; right panel shows the graph associated with the second order Laplacian difference operator.

5.3 GAUSSIAN FIELDS INDUCED VIA DIFFERENCE OPERATORS

with eigenvalues $\lambda_k = a + 2 - 2\cos(2\pi k/n)$. Notice, a makes the operator positive definite.

Since this is symmetric, $L_n = L_n^*$, the inverse covariance is given by

$$K_n^{-1} = \frac{1}{\sigma^2} L_n^2$$

$$= \frac{1}{\sigma^2} \begin{pmatrix} (a+2)^2+2 & -2(a+2) & 1 & 0 & \cdots & 0 & 1 & -2(a+2) \\ -2(a+2) & (a+2)^2+2 & -2(a+2) & 1 & 0 & \cdots & 0 & 1 \\ \vdots & & & & & & & \\ -2(a+2) & 1 & 0 & \cdots & 0 & 1 & -2(a+2) & (a+2)^2+2 \end{pmatrix}$$
(5.48)

The normalizer is given by

$$\det^{-1/2} K_n = \det^{-1/2} K_n = (\det L^*L)^{1/2}$$

$$= \det L = \prod_{k=1}^{n} \lambda_k = \prod_{k=1}^{n} \left(a + 2 - 2\cos\frac{2\pi k}{n}\right).$$

Example 5.10 (2D Laplacian: 0-boundary) Examine the Laplacian operator $L = -\Delta + a = -(\partial^2/\partial x_1^2) - (\partial^2/\partial x_2^2) + a$, with 0-boundary, inducing the random field $X_{i,j}, (i,j) \in D = \{1 \le i, j \le n\}$ according to

$$-(X_{i+1,j} + X_{i-1,j} + X_{i,j+1} + X_{i,j-1}) + (4+a)X_{i,j} = W_{i,j},$$
(5.49)

$X_{0,j} = X_{i,0} = X_{m,j} = X_{i,m} = 0$, and $W_{i,j}, (i,j) \in D$ a white noise field. The density $p(X_{i,j}, (i,j) \in D)$ becomes

$$\frac{1}{Z} \prod_{(i,j) \in D^0} e^{-1/2\sigma^2(-(X_{i+1,j}+X_{i-1,j}+X_{i,j+1}+X_{i,j-1})+(4+a)X_{i,j})^2}$$

$$\times \prod_{i \in \{1,n\}, 1 \le j \le n} e^{-1/2\sigma^2(X_{i,j})^2} \prod_{j \in \{1,n\}, 1 \le i \le n} e^{-1/2\sigma^2(X_{i,j})^2},$$

giving

$$S = \{(-1,0), (1,0), (0,1), (0,-1), (0,0)\}$$
(5.50)

$$S \ominus S = \{(-1,0), (-2,0), (1,0), (2,0), (-1,-1), (1,1), (0,-1), (0,-2),$$
$$(0,1), (0,2), (-1,1), (1,-1)\},$$

with interior neighborhood $N_i = i + S \ominus S$ of size $|S \ominus S| = 12$. The graph for the Laplacian induced random field is shown in the right panel of Figure 5.1. Figure 5.2 illustrates the result of synthesizing such random fields. For this the noise fields were generated as a set of independent and identically distributed Gaussian random variables with mean μ and standard deviation σ. The random field was generated by solving Eqn. 5.49. The top row, panel 1 shows a realization of the white noise fields with $\sigma = 100, \mu = 128$. Panels 2–4 show the random field solved from the stochastic differential equation with $a = 1$ (panel 2), $a = 4$ (panel 3), and $a = 12$ (panel 4). The independent nature of the noise-image pixels is clearly evident by the high degree of granularity (left panel). The inter-pixel dependence induced by the Laplacian is also clear. The bottom row of Figure 5.2 shows the histograms of the statistics generated from the fields.

Figure 5.3 illustrates the effect that varying the noise standard deviation and restoring force have on the resulting textures for fixed noise mean $\mu = 128$. The columns 1,2

132   5 GAUSSIAN RANDOM FIELDS ON UNDIRECTED GRAPHS

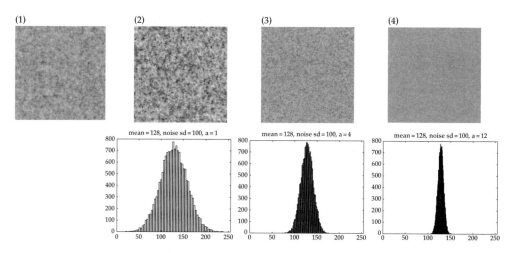

Figure 5.2 Top row: Panel 1 shows a noise realization driving the stochastic difference equation with $\sigma = 100$, $\mu = 128$. Panels 2–4 show the solution of the stochastic difference equation with $a = 1$ (panel 2), $a = 4$ (panel 3), and $a = 12$ (panel 4). Bottom row: Histograms of statistics of the images.

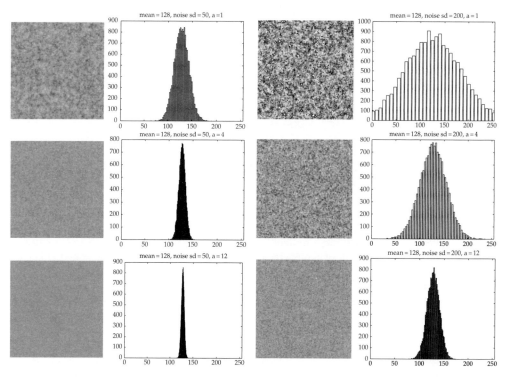

Figure 5.3 Random fields and histograms resulting from different sets of parameters. Rows have restoring forces $a = 1$, $a = 4$ and $a = 12$ (top to bottom) for fixed noise mean $\mu = 128$, with noise standard deviations of $\sigma = 50$ (columns 1, 2) and $\sigma = 200$ (columns 3, 4).

show noise standard deviations of $\sigma = 50$ and columns 3,4 show standard deviations of $\sigma = 200$ with the three rows showing restoring forces $a = 1$ (left column), $a = 4$ (middle column), and $a = 12$ (right column). Columns 2 and 4 of Figure 5.3 present the histograms of the images. Note how different sets of parameters result in similar histograms, but the images are distinguished by their different textures.

5.4 Stationary Gaussian Processes on $\mathbb{Z}^d$ and their Spectrum

For the independent pixel model (e.g. Example 2.57, Chapter 2), the normalizing variance may be calculated for each pixel separately. There is no difficulty. As clusters of pixels that are correlated are examined, the closed form calculation of the normalizer becomes far more challenging. The basic paradigm now examined provides the techniques for handling the normalizing covariance of stationary Gaussian random fields on arbitrary but large finite subgraphs. Clearly the boundaries of the random fields determine the covariance; the question of how significant the boundary is in determining the log-determinant of the normalizer can be answered definitively. We do this by demonstrating asymptotics via the fully stationary case.

The asymptotics will correspond to an approximation of the finite processes by their stationary analogue. Stationary Gaussian random fields are characterized through their covariance and spectrum.

Definition 5.11 *Given a stationary real-valued zero mean stochastic process* $\{X_i, i \in \mathbb{Z}^d\}$, *then define the* **autocovariance function** *for lag* $\tau \in \mathbb{Z}^d$ *as*

$$K_\tau = EX_i X_{i+\tau}. \tag{5.51}$$

Let $\omega = (\omega_1, \dots, \omega_d) \in [-\pi, \pi]^d$ and $\langle \omega, k \rangle = \sum_{i=1}^d \omega_i k_i$ then the **power spectral density** is defined as the Fourier transform of the covariance:

$$S(\omega) = \sum_{\tau \in \mathbb{Z}^d} K_\tau e^{-j\langle \omega, \tau \rangle}, \quad \omega \in [-\pi, \pi]^d. \tag{5.52}$$

Induce the stationary Gaussian fields via the translation invariant operator $L : l^2(\mathbb{Z}^d) \to l^2(\mathbb{Z}^d)$, with its action on functions defined to be invariant to shift where $L(i,j) = a_{j-i}, j-i \in S \subset \mathbb{Z}^d$. The calculation of the eigenelements $\{\phi(\omega), \sigma(\omega)\}$ of the operator L associated with $\mathbb{Z}^d$ (infinite boundary) is determined by the Fourier transform, since the operator is shift invariant.

Theorem 5.12 *Let* $X_i, i \in \mathbb{Z}^d$ *satisfy the stochastic difference equation* $LX_i = W_i$ *with* L *the shift invariant invertible operator* $L(i,j) = a_{j-i}$ *given by*

$$\sum_{j \in \mathbb{Z}^d} L(i,j) X_j = \sum_{j \in \mathbb{Z}^d} a_{j-i} X_j = W_i, \tag{5.53}$$

W an i.i.d Gaussian, zero-mean, variance 1 process. Then X is a Gaussian process (inverse covariance LL^*) with spectral density the inverse of the spectrum of LL^*:

$$S(\omega) = \frac{1}{\sigma^2(\omega)}$$

$$= \frac{1}{|\sum_{s \in S} a_s e^{-j\langle \omega, s \rangle}|^2}, \quad \text{where } \sigma(\omega) = \sum_{s \in S} a_s e^{j\langle \omega, s \rangle}. \tag{5.54}$$

Proof The Gaussianity is proven as in Theorem 5.5, extended to the stationary case on $\mathbb{Z}^d$. Using the fact that L is invertible, then for all $f \in l^2(\mathbb{Z}^d)$,

$$\langle f, X \rangle = \langle f, L^{-1}W \rangle = \langle L^{-*}f, W \rangle. \tag{5.55}$$

Thus $\langle f, X \rangle$ is Gaussian with inverse covariance operator $K_X^{-1} = L^*L$.

The eigenfunctions are the complex exponentials $\phi(\omega) = e^{j\langle \omega, \cdot \rangle}$, from the discrete Fourier transform given from the definition of the operator and its adjoint Eqn. 5.15

$$(L\phi(\omega))_i = \sum_{s \in S} a_s e^{j\langle \omega, (i+s) \rangle} = \sum_{s \in S} a_s e^{j\langle \omega, s \rangle} e^{j\langle \omega, i \rangle} \tag{5.56}$$

$$= \sigma(\omega) e^{j\langle \omega, i \rangle}$$

$$(L^*L\phi(\omega))_i = \sum_{s \in S} a_s (L\phi(\omega))_{i-s} = \sum_{s \in S} a_s \sigma(\omega) e^{j\langle \omega, i-s \rangle} \tag{5.57}$$

$$= |\sigma(\omega)|^2 e^{j\langle \omega, i \rangle}.$$

Defining the covariance of the X process to be $K_X(i,j) = EX_i X_j$, then we have

$$EW_i W_j = ELX_i LX_j = \sum_{h'} \sum_h a_{h'} a_h EX_{i+h'} X_{j+h}$$

$$= \sum_{h'} \sum_h a_{h'} a_h K_X(i+h', j+h) \tag{5.58}$$

$$= \sum_r \sum_h a_{r+h} a_h K_X(i+r+h, j+h). \tag{5.59}$$

Now, $EW_i W_j$ is independent of i, j so define $j = i + \tau$ implying $K_X(i,j) = K_X(0,\tau)$ is a function of the difference of its arguments, so the process is Gaussian and stationary. Calling the covariance $K_X(i, i+\tau) = K_\tau, \tau \in \mathbb{Z}^d$, then we have

$$\delta(\tau) \stackrel{(a)}{=} ELX_i LX_{i+\tau} = \sum_r \sum_h a_{r+h} a_h K_{\tau-r}, \tag{5.60}$$

with (a) following from the independent properties of W. Taking the Fourier transform completes the proof:

$$\left| \sum_h a_h e^{j\langle \omega, h \rangle} \right|^2 S(\omega) = 1, \quad \omega \in [-\pi, \pi]^d. \tag{5.61}$$

□

5.5 Cyclo-Stationary Gaussian Processes and their Spectrum

For cyclic conditions on cubic lattices $D \subset \mathbb{Z}_N^d$ there is no difficulty with the normalizer. The spectrum of the process and the eigenvalues of the covariance are given by the discrete periodic Fourier transform.

Cyclo-stationary Gaussian random fields are characterized through periodic covariances and spectra.

5.5 CYCLO-STATIONARY GAUSSIAN PROCESSES

Definition 5.13 *Given a cyclo-stationary real-valued zero mean stochastic process $\{X_i, i \in \mathbb{Z}_N^d\}$ with $X_i = X_{i+mN}, m$ integer, then define the **autocovariance function** for lag $\tau \in \mathbb{Z}_N^d$ as*

$$K_\tau = EX_i X_{i+\tau} = K_{\tau+mN}, \quad m \text{ integer}. \tag{5.62}$$

*Let $\omega_k = (2\pi k_1/N_1, \ldots, 2\pi k_d/N_d), k \in \mathbb{Z}_N^d$ and $\langle \omega_k, \tau \rangle = \sum_{i=1}^d (2\pi k_i/N_i)\tau_i$ then the **power spectral density** is defined as the periodic Fourier transform of the covariance:*

$$S\left(\frac{2\pi k}{N}\right) = \sum_{\tau \in \mathbb{Z}_N^d} K_\tau e^{-j\langle 2\pi k/N, \tau\rangle}, \quad k \in \mathbb{Z}_N^d. \tag{5.63}$$

Induce the stationary Gaussian fields via the square-root of the inverse covariance which is cyclo-stationary to shift, $L(i,j) = a_{j-i}, j-i \in \mathbb{Z}_N^d$. The calculation of the eigenelements $\{\phi(2\pi k/N), \sigma(2\pi k/N)\}$ of L associated with $\mathbb{Z}^d$ (infinite boundary) is determined by the periodic Fourier transform. This is a direct corollary of Theorem 5.12.

Corollary 5.14 (Theorem 5.12) *Let $X_i, i \in \mathbb{Z}_N^d$ satisfy the stochastic difference equation $LX_i = W_i$ with L the cyclo-shift invariant invertible operator $L(i,j) = a_{j-i}$ given by*

$$\sum_{j \in \mathbb{Z}_N^d} L(i,j) X_j = \sum_{j \in \mathbb{Z}_N^d} a_{j-i} X_j = W_i, \tag{5.64}$$

W an i.i.d Gaussian, zero-mean, variance 1 process. Then X is a Gaussian process (inverse covariance LL^*) with spectral density the inverse of the spectrum of LL^*:

$$S\left(\frac{2\pi k}{N}\right) = \frac{1}{\sigma^2(2\pi k/N)}$$

$$= \frac{1}{|\sum_{s \in \mathbb{Z}_N^d} a_s e^{-j\langle 2\pi k/N, s\rangle}|^2}, \quad \text{where } \sigma\left(\frac{2\pi k}{N}\right) = \sum_{s \in \mathbb{Z}_N^d} a_s e^{j\langle 2\pi k/N, s\rangle}. \tag{5.65}$$

Example 5.15 (Cyclo-Stationary Gaussian Texture Modelling) Examine models of the textured images in which the covariances are unconstrained other than to full cyclo-stationarity, then $K(i,j), i,j \in D \subset \mathbb{Z}^d$ a cyclic $K(i+mN, j+nN) = K(i,j)$ and toeplitz $K(i,j) = K(i-j, 0)$ having square root factorization $K = (LL^*)^{-1}$. Then the eigenelements are

$$K(i,j) = \sum_k \lambda_k^2 e^{j\langle \frac{2\pi k}{N}, (i-j)\rangle}, \tag{5.66}$$

implying $L(i,j) = \sum_k (1/\lambda_k) e^{j\langle \frac{2\pi k}{N}, (i-j)\rangle}$. With this model in mind, let X solve $LX = W$ with L generated as the inverse square root of the cyclo-stationary covariance $L^{-1} = \sqrt{K}$. The Toeplitz, cyclic covariance satisfies $K(i,j) = K(i-j,0) = EX_i X_j$ with the added property that $X_i = X_{i+N}$. The Toeplitz covariance $K(i,j)$ is estimated from the training data according to $K(i,0) = 1/|D| \sum_{j \in D} X_{i+j} X_j$.

Due to cyclo-stationarity of the process $X(\cdot)$, the eigenvalues of the Toeplitz covariance are given by complex exponentials. The random process is generated from the i.i.d. Gaussian white process transformed by $L^{-1} = \sqrt{K}$ so that the random Gaussian process satisfies $LX = W$ with $X = \sqrt{K}W = L^{-1}W$. Shown in Figure 5.4 are results of cyclo-stationary Gaussian process modelling of various textures. In each panel a

136   5 GAUSSIAN RANDOM FIELDS ON UNDIRECTED GRAPHS

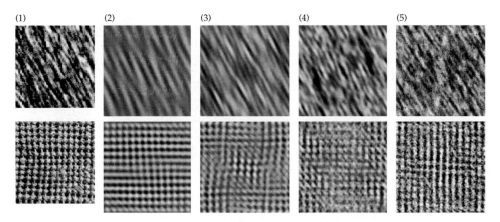

Figure 5.4 Panels show textures being synthesized with stationary Gaussian process. Top row: Panel 1 shows the original fur synthesis, panel 2 shows synthesis with the largest 8 eigenvalues, panel 3 shows synthesis with the largest 128 eigenvalues, panel 4 shows synthesis with the largest 512 eigenvalues, and panel 5 shows synthesis with all of the eigenvalues. Bottom row is similar but for fabric. Taken from Dimitri Bitouk.

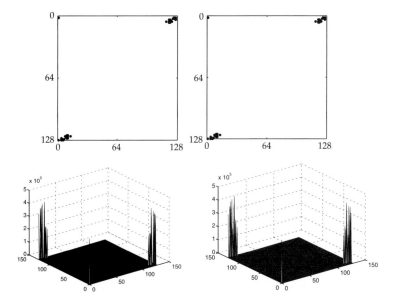

Figure 5.5 Figure presents the animal spectrum and the spectrum estimated from the synthesized textures. The two panels on the left show sparsity pattern of the animal fur texture eigenvalues and the eigenvalues estimated from the synthesized texture images; the two panels on the right show empirical versus synthesized spectra. Taken from Dimitri Bitouk.

different number of eigenvalues were used for generating the empirical covariance. Since the complex exponentials are the eigenfunctions, the covariances were generated from increasing numbers of empirically generated eigenvalues. Panels 1, 6 show the original fur and fabric textures. Panels 2, 7 show synthesis using the largest 8 eigenvalues. Panels 3, 8 show synthesis with the largest 128 eigenvalues. Panels 4, 9 show synthesis via the largest 512 eigenvalues. Panels 5, 10 show synthesis via all of the eigenvalues.

(1) 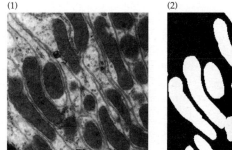 (2)  (3)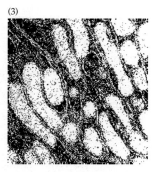

Figure 5.6 Panel 1 shows the micrograph data including mitochondria and cytoplasmic structures. Panel 2 shows a decomposition into two region types, mitochondria (white) and background (black). Panel 3 shows the pixel-by-pixel segmentation of the image based on the Bayesian hypothesis test under the two models. Data taken from Dr. Jeffrey Saffitz of the Department of Pathology at Washington University.

Figure 5.5 presents the animal spectrum and the spectrum estimated from the synthesized textures. Left two panels show sparsity pattern of the animal fur texture eigenvalues and the eigenvalues estimated from the synthesized texture images; right two panels show empirical versus synthesized spectra.

5.6 The log-Determinant Covariance and the Asymptotic Normalizer

For segmentation of textured objects, model the image as a collection of objects with textured random fields of specific type over the interior. Define the background space $D \subset \mathbb{Z}^d$ is made up of a collection of textured interiors forming a disjoint partition of the image, $D = \cup_j D_j$. The goal is to segment the interiors D_j forming a covering of the complete image. Shown in Figure 5.6 are examples of micrographs showing textured mitochondria and cytoplasm. The data are from rabbit-kidney cells at 20,000 times magnification. Panels 1–3 show the micrograph data, with the two region types, mitochondria (white) and background (black).

Generally, the interiors of the shapes are modelled as real-valued scalar random fields. We take the viewpoint of Bayesian testing for calculating the likelihood of the shapes and construct the regions $D_j, j = 1, 2, \ldots$ so that the probability of the observed imagery $P(X_D)$ is a maximum. Panel 2 of Figure 5.6 shows such a decomposition of the micrograph image into a disjoint partition. For Gaussian random fields under the independent pixel model the global probability over the shape is simply the product of the probabilities over the interior, with the normalizer just the product of the variances:

$$p(X_D) = \prod_i \frac{1}{2\pi\sigma^2} e^{-|X_i - \mu|^2/2\sigma^2}. \tag{5.67}$$

This is straightforwardly calculated. Panel 3 of Figure 5.6 shows the segmentation performed by fitting the mean and variance statistics to the mitochondria and membrane. For this an independent pixel to pixel hypothesis test is performed selecting the two model types. Notice the granularity of the segmentation; there is no notion of the global structure associated with the connected regions. The deformable active models attempt to model the global structure.

We model the interiors of the shapes of real-valued images as scalar random fields defined on the index sets of integer lattices. Because of splitting, the partitions are conditionally independent given their boundaries. The Bayesian approach presents a substantial conceptual hurdle. Since the normalizing partition function must be calculated for each partition $D_j \subset \mathbb{Z}^d, d = 2,3, j = 1,\ldots,J$ the number of object regions. The partition function and log-normalizer must be calculated for each shape. Let X_D be a real-valued Gaussian field on $D \subset \mathbb{Z}^d$, with mean and covariance μ, K_D. We model the inverse covariance via the difference operator which is zero at the boundary of the shape so that $L_D L_D^* = K_D^{-1}$; the probability density requires the computation of the determinant of the covariance corresponding to the log-partition function for Gaussian fields:

$$p(X_D) = \det{}^{-1/2}(2\pi K_D) e^{-1/2 \|L_D(X_D - \mu)\|^2}, \tag{5.68}$$

with $\|L_D(X-\mu)\|^2$ the integral-square of the field over the domain D. Notice the role the partition functions play; segmentation requires computation of log K_D for the random shape. Construct the background space $D \subset \mathbb{Z}^d$ as a disjoint partition of simply connected regions $D(g_j)$ with random image field $X_{D(g_j)}$. Conditioned on the parametric models, $g_1, g_2, \ldots$, the subregion Markov random fields are assumed conditionally independent. Write the potential via energy densities $E_j(x), x \in D(g_j)$ representing the log-normalizer and quadratic data statistic according to

$$p(X_{D_1}, \ldots, X_{D_J}) = \prod_{j=1}^{J} \det{}^{-(1/2)}(2\pi K_{D_j}) e^{-1/2\sigma^2 \|L(X_{D_j} - \mu_j)\|^2}. \tag{5.69}$$

5.6.1 Asymptotics of the Gaussian processes and their Covariance

Let X_D be a real-valued Gaussian field on $D \subset \mathbb{Z}^k$, with mean and covariance μ, K_D. let $D \subset \mathbb{Z}^k$, with X real-valued on D since the determinant of the covariance replaces the log-partition function. The probability density requires the computation of the determinant of the covariance corresponding to the log-partition function for Gaussian fields:

$$p(X_D) = \det{}^{-(1/2)}(2\pi K_D) e^{-(1/2)\langle X - \mu, K_D^{-1} X - \mu \rangle}. \tag{5.70}$$

As we noted in Chapter 4 the problem isn't any better for Gibbs distributions. Now examine Gaussian random fields $\{X_i(\omega), i \in D_n \subset \mathbb{Z}^d\}$ defined on a sequence of the finite increasing graphs $D_n \subset D_{n+1} \subset \cdots$, with $\lim_{n \to \infty} D_n = \mathbb{Z}^d$. Associate with $D_n \subset \mathbb{Z}^d$ the operator $L_n : l^2(D_n) \to l^2(D_n)$ constructed from the stationary operator L restricted to D_n according to $L_n X = L X^0$, where $X_i^0 = X_i, i \in D_n$ and 0 otherwise. Then X on D_n solves the stochastic difference equation

$$(L_n X)_i = W_i, \quad i \in D_n, \tag{5.71}$$

with W_i white noise with variance 1. As proven in Theorem 5.5, the operator L_n determines the covariance structure with $K_{X_D} = (L_n^* L_n)^{-1}$. Of course the fundamental issue is that for every domain shape D_n, the boundary of the operator ∂D_n determines the normalizing determinant of the covariance.

The asymptotics are a direct assault on the influence of the boundary on the covariance, appealing to results familiar to the Toeplitz aficionados, similar in spirit to Weyl's results. For this, crucially the operators defining the fields are stationary with respect to shift over the interior sites. This allows for the relationship of the determinant of the covariance on D_n of the finite operator to the Fourier transform of the stationary operator associated with $\mathbb{Z}^d$. Assuming the boundaries are thin relative to their internal volumes, so that $\lim_{n \to \infty} |\partial D_n|/|D_n| = 0$, then the asymptotic moments of the eigenvalue distribution hold irrespective of the boundary conditions. It is natural to study

5.6 THE LOG-DETERMINANT COVARIANCE

the 0-boundary. Now we prove a theorem on the trace and log-trace of the operator which provides the normalizer for the Gaussian case.

Definition 5.16 *Let the operator L_n on D_n have eigenvalues $\lambda_m, m = 1, \ldots, |D_n|$, with the pth **power operator** defined as*

$$L_n^p(i_0, i_p) = \sum_{i_1 \in D_n} \cdots \sum_{i_{p-1} \in D_n} L_n(i_0, i_1) L_n(i_1, i_2) \cdots L_n(i_{p-1}, i_p). \tag{5.72}$$

*The **operator trace** is defined as the sum of the eigenvalues $\operatorname{tr} L_n^p = \sum_{m=1}^{|D_n|} \lambda_m^p$; the **boundary** of the pth power of L_n^p as*

$$\partial D_n^{(p)} = D_n / D_n^{(p)o} \quad \text{where } D_n^{(p)o} = \{i \in D_n : i + \underbrace{(S \ominus S \ominus \cdots S)}_{p \text{ times}} \subset D_n\}. \tag{5.73}$$

We now show that the normalized pth moment of eigenvalues of the finite-domain operator converges to the pth moment of the Fourier transform of the infinite domain operator.

Theorem 5.17 (Dedicated to the Memory of Gabor Szego) *Let $L_n : l^2(D_n) \to l^2(D_n)$ according to $L_n f = L f^0$ where $(L f^0)_i = \sum_{s \in S} a_s f_{i+s}^0$, $i \in D_n$ with eigenvalues of L_n given by $\lambda_m, m = 1, \ldots, |D_n|$, and the Fourier transform of L given by $\sigma(\omega) = \sum_s a_s e^{j\langle \omega, s \rangle}$.*

Assuming the boundary is thin so that $\lim_{n \to \infty} |\partial D_n^{(p)}|/|D_n| = 0$ for all p, then the asymptotic trace is given by

$$\lim_{n \to \infty} \frac{\operatorname{tr} L_n^p}{|D_n|} = \lim_{n \to \infty} \frac{\sum_{m=1}^n \lambda_m^p}{|D_n|} = \frac{1}{(2\pi)^d} \int_{[-\pi, \pi]^d} \sigma^p(\omega) \, d\omega. \tag{5.74}$$

Assuming that the domain of eigenvalues does not include 0, then

$$\lim_{n \to \infty} \frac{\sum_{m=1}^{|D_n|} \log \lambda_m}{|D_n|} = \frac{1}{(2\pi)^d} \int_{[-\pi, \pi]^d} \log \sigma(\omega) \, d\omega. \tag{5.75}$$

Proof

$$\lim_{n \to \infty} \frac{\operatorname{tr} L_n^p}{|D_n|} = \lim_{n \to \infty} \frac{\sum_{i=k \in D_n} L_n^p(i, k)}{|D_n|} \tag{5.76}$$

$$= \lim_{n \to \infty} \left(\frac{\sum_{i=k \in D_n^{(p)o}} L_n^p(i, k)}{|D_n|} + \frac{O(|\partial D_n^{(p)}|)}{|D_n|} \right). \tag{5.77}$$

Now clearly, $L^p(i, \cdot)$ and $\sigma^p(\cdot)$ form a Fourier transform pair, since the complex exponentials are eigenfunctions of the stationary operator L:

$$L^p(i, k) = \frac{1}{(2\pi)^d} \int_{[-\pi, \pi]^d} \sigma^p(\omega) e^{j\langle \omega, (k-i) \rangle} \, d\omega, \tag{5.78}$$

and substituting in Eqn. 5.77 completes the first part of the theorem:

$$\lim_{n \to \infty} \frac{\operatorname{tr} L_n^p}{|D_n|} = \lim_{n \to \infty} \left(\frac{|D_n^{(p)o}|}{|D_n|} \frac{1}{(2\pi)^d} \int_{[-\pi, \pi]^d} \sigma^p(\omega) \, d\omega + \frac{O(|\partial D_n^{(p)}|)}{|D_n|} \right) \tag{5.79}$$

$$= \frac{1}{(2\pi)^d} \int_{[-\pi, \pi]^d} \sigma^p(\omega) \, d\omega. \tag{5.80}$$

The second part follows since the eigenvalues λ are contained in the convex hull C_W of the set $W = \{\sigma(\omega); \omega \in [-\pi,\pi]^d\} \subset \mathbb{C}$ a closed subset of the complex plane. This follows since if λ is an eigenvalue of L_n with eigenvector $\phi \in l^2(D_n)$ of norm one, then

$$\lambda = \lambda \|\phi\|^2 = \sum_{i \in D_n} (L_n\phi)_i \phi_i^\star \tag{5.81}$$

$$= \sum_{i \in D_n} \left(\sum_{k \in D_n} a_{k-i} \phi_k \right) \phi_i^\star$$

$$= \sum_{i \in D_n} \left(\int_{[-\pi,\pi]^d} \frac{1}{(2\pi)^d} \sum_{k \in D_n} \phi_k e^{j\langle \omega, k-i \rangle} \sigma(\omega)\, d\omega \right) \phi_i^\star \tag{5.82}$$

$$= \frac{1}{(2\pi)^d} \int_{[-\pi,\pi]^d} \sigma(\omega) \left| \sum_{i \in D_n} \phi_i e^{-j\langle \omega, i \rangle} \right|^2 d\omega. \tag{5.83}$$

But the function $g(\omega) = 1/(2\pi)^d |\sum_{i \in D_n} \phi_i e^{-j\langle \omega,i\rangle}|^2$ is non-negative and has integral one $\int_{[-\pi,\pi]^d} g(\omega)\, d\omega = 1$ since the ϕ-vector had l_2 norm one. Hence $\lambda \in C_W$ the convex hull.

Defining the eigenvalue distribution according to $E^n(d\lambda) = 1/|D_n| \sum_{m=1}^{|D_n|} \delta_{\lambda_m}(d\lambda)$, on C_W, then rewrite the pth moment of the eigenvalues according to

$$\lim_{n \to \infty} \frac{\sum_{m=1}^{|D_n|} \lambda_m^p}{|D_n|} = \lim_{n \to \infty} \int_{C_W} \lambda^p E^n(d\lambda) \tag{5.84}$$

$$\stackrel{(a)}{=} \lim_{n \to \infty} \frac{1}{(2\pi)^d} \int_{[-\pi,\pi]^d} \sigma^p(\omega)\, d\omega, \tag{5.85}$$

where (a) follows from Eqn. 5.74. This is true for all p, implying for any polynomial $P(\cdot)$ on C_W,

$$\lim_{n \to \infty} \int_{C_W} P(\lambda) E^n(d\lambda) = \frac{1}{(2\pi)^d} \int_{[-\pi,\pi]^d} P(\sigma(\omega))\, d\omega. \tag{5.86}$$

From the Stone–Weierstrass theorem, choose a sequence of polynomials $P_j, j = 1, \ldots$ converging uniformly to h a continuous function on the compact domain C_W implying

$$\lim_{n \to \infty} \frac{\sum_{m=1}^{|D_n|} h(\lambda_m)}{|D_n|} = \frac{1}{(2\pi)^d} \int_{[-\pi,\pi]^d} h(\sigma(\omega))\, d\omega. \tag{5.87}$$

This is true for all continuous h, and since $0 \notin C_W$, then choose $h(\cdot) = \log(\cdot)$ as the particular continuous function in the complex plane. Since $\log \lambda_k$ is unambiguously defined by systematically taking the main branch of the log-function for $0 \notin C_W$, the function $\log z$ is analytic in C_W and from Chebychev the function can then be uniformly approximated by polynomials in C_W completing the proof of the theorem. □

Remark 5.6.1 If L is self-adjoint (symmetric) $L = L^*$, the Fourier transform $\sigma(\cdot)$ is real. In the above theorem it requires the condition $0 \notin C_W$ which is satisfied, for example, when the values of $\sigma(\omega)$ fall in a half plane not including imaginary axis, for example when the spectrum is never identically zero. Note, if the operator is self-adjoint, then this is equivalent to saying it is bounded away from 0, so that $\lambda \in [a, B], a > 0$, for

5.6 THE LOG-DETERMINANT COVARIANCE

all eigenvalues, and as long as the eigenvalues are bounded away from 0 lying in a closed set, log is a particular continuous function.

Example 5.18 (Periodic Boundary) The claim is that, for example, the discrete coefficients for the periodic Fourier transform converge into the infinite transform spectrum. This would correspond, for example, to the periodic boundary condition most often studied for cyclo-stationary and the fully stationary limit. The periodic boundary condition on a rectangular domain is generated via shift on the domain $D = \{i = (i_1, \ldots, i_d) \in \mathbb{Z}^d : (0, \ldots, 0) \leq i \leq (n_1 - 1, \ldots, n_d - 1)\}$ according to

$$x_D^P = \{x_i^P = x_{(i \bmod n)}, i \in \mathbb{Z}^d\} \in \Lambda^{\mathbb{Z}^d}. \tag{5.88}$$

Remark 5.6.2 The conclusion is qualitatively the same as the classical Toeplitz theorem in a rectangle: the shape does not matter as long as the discretized boundaries are thin. It is also related to Weyl's result on the asymptotics of the eigenvalues associated with an elliptic differential operator.

Example 5.19 Let us examine the following 1D example to understand the asymptotics. Let L_n be an $n \times n$ operator matrix with i, j entries $L_n(i, j) = a_{j-i}, j - i \in S$. Then, L_n^2 is given by

$$L_n^2(i, j) = \sum_{k=1}^n L_n(i, k) L_n(k, j) \quad \text{implying} \quad \text{tr}(L_n^2) = \sum_{i=1}^n \sum_{k=1}^n a_{i-k} a_{k-i}. \tag{5.89}$$

The trace reduces to

$$\lim_{n \to \infty} \frac{1}{n} \text{tr}(L_n^2) = \lim_{n \to \infty} a_0^2 + 2\left(1 - \frac{1}{n}\right) a_1 a_{-1} + 2\left(1 - \frac{2}{n}\right) a_2 a_{-2} \ldots$$

$$+ 2\left(1 - \frac{n-1}{n}\right) a_{n-1} a_{-(n-1)}, \tag{5.90}$$

$$= a_0^2 + 2 \sum_{j \in S}^{\infty} a_j a_{-j} = (a * a)_0 = \frac{1}{2\pi} \int_0^{2\pi} \sigma^2(\omega) \, d\omega. \tag{5.91}$$

This is Eqn. 5.74 for $p = 2$.

Example 5.20 Examine the Laplacian Example 5.9 in 1D $-(\partial^2/\partial x^2) + a$ with periodic boundary conditions:

$$(L_n X)_i = -X_{i-1} + (a + 2) X_i - X_{i+1} \tag{5.92}$$

with eigenvalues $\lambda_k = a + 2 - 2\cos(2\pi k/n)$ and therefore covariance normalizer

$$\det^{-1/2} K_n = \det L_n = \prod_{k=1}^n \left(a + 2 - 2\cos\frac{2\pi k}{n}\right). \tag{5.93}$$

Theorem 5.17 states that since $\sigma(\omega) = (a + 2) - 2\cos\omega$,

$$\lim_{n \to \infty} \frac{1}{n} \log \det K_n^{-1} = \lim_{n \to \infty} \frac{2 \sum_{k=1}^n \log(a + 2 - 2\cos(2\pi k/n))}{n}$$

$$= \frac{1}{2\pi} \int_0^{2\pi} \log(a + 2 - 2\cos\omega) \, d\omega. \tag{5.94}$$

5.6.2 The Asymptotic Covariance and log-Normalizer

We shall now examine the log-determinant of the covariance (log-normalizer) of the stationary Gaussian fields generated from the finite graphs. We shall exploit the fact that for stationary fields limits on increasing cylinders given a log-determinant are given by the integral of the spectrum independent of the boundary.

Corollary 5.21 *Let X be a Gaussian process induced by operator L_n on $l^2(D_n)$ satisfying Theorem 5.17 solving the difference equation $L_n X = W$ on D_n, W white Gaussian noise. Then the normalized log-determinant of the covariance of the process converges to the integral of the log-spectrum, so that*

$$\lim_{n \to \infty} \frac{1}{|D_n|} \log \det K_{X_D} = \frac{1}{(2\pi)^d} \int_{(-\pi,\pi)^d} \log S(\omega)\, d\omega, \qquad (5.95)$$

$$\text{where } S(\omega) = \frac{1}{|\sigma(\omega)|^2} = \frac{1}{|\sum_h a_h e^{j\langle \omega, h \rangle}|^2}. \qquad (5.96)$$

Proof From the definition of the operator Eqn. 5.15 $K_{X_{D_n}} = (L_{D_n}^* L_{D_n})^{-1}$, and from Theorem 5.17 (Eqn. 5.75), the normalizing asymptotic trace is given by the log spectrum of the stationary operator $(LL^*)^{-1}$. Then $S(\omega) = 1/|\sigma(\omega)|^2$ is the eigenvalue of this operator satisfying

$$(L^* L \phi(\omega))_i = \sum_{s \in S} a_s (L\phi(\omega))_{i-s} = \sum_{s \in S} a_s \sigma(\omega) e^{j\langle \omega, i-s \rangle} = |\sigma(\omega)|^2 e^{j\langle \omega, i \rangle}.$$

□

5.7 The Entropy Rates of the Stationary Process

We shall now examine the entropy of stationary Gaussian fields generated from the finite graphs exploiting the fact that for stationary fields limits on increasing cylinders of the normalized entropies converge to the so-called entropy rate. Denote the differential entropy rates for the continuum-valued Gaussian fields in small letters $h(X)$. The entropy of the real-valued Gaussian process $X_i, i \in D$ is given by

$$\frac{1}{|D_n|} h(X_i, i \in D) = \frac{1}{2} \left(\frac{\log \det 2\pi e K_{X_D}}{|D_n|} \right). \qquad (5.97)$$

As we now show, there is a limiting entropy rate for the large graphs of the stationary processes generated via shift invariant operators. Theorem 5.17 tells us that the limiting entropy is given by the logarithm of the spectrum.

Corollary 5.22 *Let X be a Gaussian process induced by operator L_n on $l^2(D_n)$ satisfying Theorem 5.17 solving the difference equation $L_n X = W$ on D_n, W white Gaussian noise. Then the normalized log-determinant of the covariance of the process converges to the integral of the log-spectrum, and the differential entropy rate becomes*

$$\lim_{n \to \infty} \frac{1}{|D_n|} h(X_i, i \in D_n) = \frac{1}{2} \left(\frac{1}{(2\pi)^d} \int \log 2\pi e S(\omega)\, d\omega \right). \qquad (5.98)$$

Proof Substituting into Eqn. 5.97 the result for the log-determinant covariance from Corollary 5.21 given by the log-spectral density gives the result. □

5.7.1 Burg's Maximum Entropy Auto-regressive Processes on $\mathbb{Z}^d$

We will now show that stationary processes induced through auto-regressive operators are maximum entropy. This is Burg's maximum-entropy principle for Gaussian stationary processes.

Theorem 5.23 *Let $X_i, i \in \mathbb{Z}^d$ be a real-valued stationary Gaussian process with covariances $K_\tau, \tau \in S \subset \mathbb{Z}^d$, with $K_\tau = K_{-\tau}$. Then the process maximizing entropy rate $1/(2\pi)^d \int_{[-\pi,\pi]^d} \log 2\pi e S(\omega) \, d\omega$ with covariance constraints*

$$K_\tau = \frac{1}{(2\pi)^d} \int_{[-\pi,\pi]^d} S(\omega) e^{j\langle \omega, \tau \rangle} \, d\omega, \quad \tau \in S - S \tag{5.99}$$

satisfies the difference equation

$$LX = W, \tag{5.100}$$

with the difference operator $L(i,j) = a_{j-i}, j - i \in S$ where

$$a_s = \int \frac{1}{\sqrt{S(\omega)}} e^{-j\langle \omega, h \rangle} \, d\omega, \quad s \in S. \tag{5.101}$$

Proof Maximizing the entropy subject to the constraints gives the equation which the maximum entropy solution must satisfy

$$\max_{S(\cdot)} \frac{1}{(2\pi)^d} \int_{[-\pi,\pi]^d} \log S(\omega) \, d\omega + \sum_{\tau \in S-S} \lambda_\tau \frac{1}{(2\pi)^d} \int_{[-\pi,\pi]^d} S(\omega) e^{j\langle \omega, \tau \rangle} \, d\omega \tag{5.102}$$

giving

$$\frac{1}{S(\omega)} + \sum_{\tau \in S-S} \lambda_\tau e^{j\langle \omega, \tau \rangle} = 0, \tag{5.103}$$

implying the maximum entropy spectrum is of the form

$$S(\omega) = \frac{1}{\sum_{\tau \in S-S} \lambda_\tau e^{j\langle \omega, \tau \rangle}}. \tag{5.104}$$

The spectrum of the finite-difference process is

$$S(\omega) = \frac{1}{|\sum_{s \in S} a_s e^{j\langle \omega, s \rangle}|^2} = \frac{1}{\sum_{s \in S} \sum_{s' \in S} a_s a_{s'} e^{j\langle \omega, s-s' \rangle}} \tag{5.105}$$

$$= \frac{1}{\sum_{\tau \in S-S} \sum_{s \in S} a_s a_{\tau+s} e^{j\langle \omega, \tau \rangle}}. \tag{5.106}$$

Choosing

$$\lambda_\tau = \sum_{s \in S} a_s a_{\tau+s}, \quad \tau \in S \tag{5.107}$$

demonstrates the maximum entropy form and clearly $a_s \leftrightarrow 1/\sqrt{S(\omega)}$ form a Fourier transform pair. □

5.8 Generalized Auto-Regressive Image Modelling via Maximum-Likelihood Estimation

With the above asymptotic results we have turned texture modelling into spectral analysis analogous to ARMA modelling for stationary time series. The integral of the log-spectrum is precisely what is needed for the normalizer in the likelihood function of any kind of estimation. For this examine the maximum-likelihood estimation. Let $X_D = \{X_i(\omega), i \in D\}$ be a realization of mean μ-Gaussian random field with covariance $K_D = (L_D L_D^*)^{-1}$ solving $L_D X(\omega) = W(\omega)$, $W(\omega)$ white noise and L_D the stationary difference operator interior to the graph D determined by the coefficients $a_s, s \in S$:

$$(L_D f)_i = \sum_{s \in S} a_s f_{i+s}, \quad i \in D/\partial D, \tag{5.108}$$

and with L_D having boundary conditions on ∂D.

Parametric maximum-likelihood estimates of the operator representation of the field are defined in the standard way.

Definition 5.24 *Then the **maximum-likelihood estimate** L^{MLE} determining the covariance K^{MLE} maximizing likelihood are given by the parameters*

$$\{a_s^{\text{MLE}}, s \in S\} = \arg\max_{\{a_s, s \in S\}} -\log \det K_D - \|L_D X\|^2. \tag{5.109}$$

Then we have the following.

Theorem 5.25 *The asymptotic maximum likelihood estimators*

$$\{a_s^{\text{AMLE}}, s \in S\} = \arg\max_{\{a_s, s \in S\}} -\int_{[-\pi,\pi]^d} \log |\sum_{s \in S} a_s e^{j\langle \omega, s\rangle}| \, d\omega - \frac{\|L_D X\|^2}{|D|}, \tag{5.110}$$

are asymptotic MLE's in the sense that

$$\left| -\int_{[-\pi,\pi]^d} \log |\sum_{s \in S} a_s^{\text{AMLE}} e^{j\langle \omega, s\rangle}| \, d\omega - \frac{\|L_D^{\text{AMLE}} X\|^2}{|D|} \right.$$
$$\left. + \frac{\log \det K_D^{\text{MLE}}}{|D|} + \frac{\|L_D^{\text{MLE}} X\|^2}{|D|} \right| \leq \frac{O(|\partial D|)}{|D|}. \tag{5.111}$$

Example 5.26 (Isotropic Laplacian) Examine as the first example the Laplacian in $\mathbb{R}^2$ made non-singular $-\Delta + a$, where $a > 0$ is the constant in Hooke's law corresponding to a restoring force returning the texture field to its original average displacement, and $\Delta = (\partial^2/\partial x^2) + (\partial^2/\partial y^2)$. Model regions in the image $\{X_i, i \in \mathbb{Z}_{m^2}\}$ as Gaussian random fields with mean μ and covariance structures induced via the operator, the discrete Laplacian operator inducing nearest-neighbor structure:

$$L X_i = (-\Delta + a) X_i \quad i \in \mathbb{Z}^2, \tag{5.112}$$

where $\Delta X_i = X_{i_1-1, i_2} + X_{i_1+1, i_2} + X_{i_1, i_2-1} + X_{i_1, i_2+1} - 4 X_{i_1, i_2}$, \tag{5.113}

The noise process is taken to be white Gaussian noise with mean μ and variance σ^2. The three parameters, a and σ^2, completely specify the model. With 0-boundary, L induces the random field according to

$$-(X_{i_1+1, i_2} + X_{i_1-1, i_2} + X_{i_1, i_2+1} + X_{i_1, i_2-1}) + (4 + a) X_{i_1, i_2} = W_{i_1, i_2}, \tag{5.114}$$

5.8 GENERALIZED AUTO-REGRESSIVE IMAGE MODELLING

with $X_{0,i_2} = X_{i_1,0} = X_{n,i_2} = X_{i_1,n} = 0$.
For $L = -\Delta + a$, the spectral density $\sigma(\omega), \omega = (\omega_1, \omega_2)$ of L is

$$\sigma(\omega) = 2\sum_{k=1}^{2}(1 - \cos\omega_k) + a, \tag{5.115}$$

yielding the normalization used in the large size limit of the shape with $n = m^2$ pixels:

$$\frac{\log \det^{1/2} K_D}{n} = \frac{1}{n}\sum_{m=1}^{n}\log \lambda_m \tag{5.116}$$

$$= \frac{1}{4\pi^2}\int_{[-\pi,\pi]^2}\log\left(2\sum_{k=1}^{2}(1-\cos\omega_k) + a\right)d\omega + \frac{O(|\partial D|)}{n}. \tag{5.117}$$

Apply ML-estimation to the normalized log-likelihood function

$$(\hat{\sigma}^2, \hat{a}) = \arg\max_{\{\sigma^2, a\}} -\frac{1}{2}\log\sigma^2 + \frac{1}{2n}\sum_{m=1}^{n}\log\lambda_m^2 - \frac{1}{2n\sigma^2}\|LX\|^2. \tag{5.118}$$

Since the log-partition function is independent of the variance, the maximum likelihood estimate of σ^2 becomes

$$\hat{\sigma}^2 = \arg\max_{\sigma^2} -\frac{1}{2}\log\sigma^2 + \frac{1}{2n}\sum_{m}\log\lambda_m^2 - \frac{1}{2n\sigma^2}\|LX\|^2 \tag{5.119}$$

$$= \frac{\|LX\|^2}{n}. \tag{5.120}$$

Upon substitution, then estimating the restoring force and using the asymptotic eigenvalue expression gives

$$\hat{a} = \arg\max_{a} -\frac{1}{2}\log\frac{\|LX\|^2}{n} + \frac{1}{2n}\sum_{m=1}^{n}\log\lambda_m^2 - \frac{1}{2}$$

$$= \arg\max_{a} -\frac{1}{2}\log\frac{\|LX\|^2}{n} + \frac{1}{4\pi^2}\int_{[-\pi,\pi)^2}\log|\sigma(\omega)|d\omega - \frac{1}{2}$$

$$= \arg\max_{a}\left(-\frac{1}{2}\log\frac{\sum_{i\in D}|-(\Delta X)_i + aX_i|^2}{n}\right.$$

$$\left.+\frac{1}{4\pi^2}\int_{[-\pi,\pi]^2}\log\left(2\sum_{k=1}^{2}(1-\cos\omega_k)+a\right)d\omega\right). \tag{5.121}$$

One of the integrations can be done in closed form, yielding

$$\frac{1}{4\pi^2}\int_{[-\pi,\pi]^2}\log\left(2\sum_{k=1}^{2}(1-\cos\omega_k)+a\right)d\omega$$

$$= \frac{1}{2\pi}\int_{[-\pi,\pi]}\log\frac{4+a-2\cos\omega_1+\sqrt{(4+a-2\cos\omega_1)^2-4}}{2}d\omega_1 \tag{5.122}$$

with the remaining integration performed by numerical quadrature.

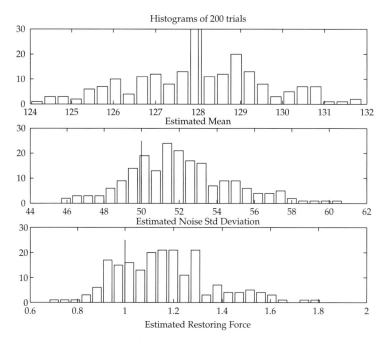

Figure 5.7 Showing maximum-likelihood estimates of μ, σ, a from an experiment of 200 random fields generated with $\mu = 128$, $\sigma = 50$, and $a = 1$. Results courtesy of David Maffit, Washington University at St. Louis.

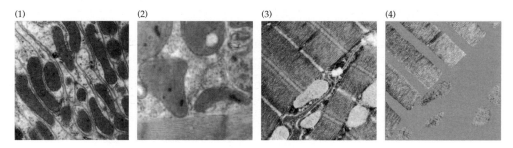

Figure 5.8 Panels 1–3 show example micrographs containing mitochondria and actin–myosin complexes; panel 4 illustrates the hand segmentations of the mitochondria and actin–myosin shapes used to estimate the oriented texture parameters. Everything not labelled in gray is not mitochondria or actin–myosin. Data taken from Dr. Jeffrey Saffitz of the Department of Pathology at Washington University.

Results of experiments measuring how well various model parameters can be estimated from random fields with known parameters are presented in Figure 5.7. Experiments were performed corresponding to the data shown in Figures 5.2 and 5.3 illustrating the effect of varying the noise standard deviations and the various parameters of the differential operator. For this the noise fields were generated as a set of independent and identically distributed Gaussian random variables with mean μ and standard deviation σ, and solved for the random field as a fixed point of Eqn. 5.112. Figure 5.7 shows the distributions of the estimates from random field realizations generated with $\mu = 128$, $\sigma = 50$, and $a = 1$. Shown in Figure 5.8 are examples of estimating the parameters of Eqn. 5.112 with two different model types: Mitochondria and

5.8 GENERALIZED AUTO-REGRESSIVE IMAGE MODELLING

Table 5.1a Showing the maximum-likelihood estimates of the parameters μ, variance σ^2, and restoring force a in the isotropic model.

	Mean Gray Level	Variance of White Noise σ^2	Restoring Force a
Mitochondria	58.7	847.3	0.62
Cytoplasm	130.3	2809.9	0.15

Table 5.1b Estimates of the texture parameters a_{11}; a_{12}; a_{22} for the actin–myosin complex and the mitochondria for the orientation dependent model.

Texture	Mean	a_{11}	a_{12}	a_{22}	Standard Deviations
Actin–myosin	87.70	0.87	−0.68	0.82	64.16
Mitochondria	179.56	0.98	0.02	1.03	55.45

cytoplasm. The data are from rabbit-kidney cells at 20,000 times magnification. Panels 1–3 show the micrograph data, with the two region types, mitochondria (white) and background (black). Panel 4 shows a pixel-by-pixel segmentation of the image based on the likelihood of each pixel under the two models solving the Bayesian hypothesis test for each pixel separately. The parameters estimated from the micrograph data for these regions are shown in Table 5.1 assuming the Gaussian model $(-\Delta + a)X = W$ has parameters the noise mean μ, noise power σ^2, and restoring force a.

Remark 5.8.3 The benefits of maximum-likelihood applied directly are obvious. Reasoning heuristically for a moment, it appears to be common sense that a should be estimated by the following ratio of inner products $a^* = (\Delta X, X)/(X, X)$ that minimizes the quadratic form $\|-\Delta X + aX\|^2$. After all, this is just the least squares estimate which ought to perform fairly well. This, however, is not the case and we must be a bit more careful. Now notice that $-\Delta$ is a negative operator (with the proper boundary conditions) so that this will give a negative value for the estimate, but we know that a is positive. Therefore this estimate must be discarded; the MLE has to be used instead.

5.8.1 Anisotropic Textures

Thus far Gaussian fields which are isotropic and homogeneous have been described with textures which are rotationally invariant (no preferred direction). However, there are clearly applications for which this is not appropriate. Shown in the bottom row of Figure 5.8 are exquisite examples of actin–myosin complexes with an orientation to the texture field. A great deal of work has been done on the representation of oriented textured fields, such as those evidenced in wood grains and or finger prints (e.g., see [144]). Now we extend to the more general operator formulation $L = -\sum_{i,j=1}^2 a_{ij}(\partial^2/(\partial x_i \partial x_j)) + a_0$, modelling the texture field $X_i(\omega), i \in D_n = \{1 \le i_1, i_2 \le n\}$ and solve the stochastic equation $LX = W$ with the discrete operator becoming

$$LX_{i_1,i_2} = -a_{11}(X_{i_1+1,i_2} - 2X_{i_1,i_2} + X_{i_1-1,i_2}) - a_{22}(X_{i_1,i_2+1} - 2X_{i_1,i_2} + X_{i_1,i_2-1})$$
$$- \tfrac{1}{2}a_{12}(X_{i_1+1,i_2+1} - X_{i_1+1,i_2-1} - X_{i_1-1,i_2} + X_{i_1-1,i_2-1}) + a_0 X_{i_1,i_2}. \tag{5.123}$$

For maximum-likelihood fitting, treat the constant a_0 and the 2×2 symmetric matrix $L = \begin{pmatrix} a_{11} & a_{12} \\ a_{21} & a_{22} \end{pmatrix}$ as unknowns to be estimated. It is convenient to interchange the role of L as an operator, and L the matrix representation of the operator. The log-determinant normalizer is given by the Fourier transform

$$\sigma(\omega) = a_0 + 4a_{11} \sin^2 \frac{\omega_1}{2} + 2a_{12} \sin \omega_1 \sin \omega_2 + 4a_{22} \sin^2 \frac{\omega_2}{2}. \tag{5.124}$$

Theorem 5.17 obtains

$$\frac{1}{n^2} \sum_{m=1}^{n^2} \log \lambda_m = \frac{1}{(2\pi)^2} \int_{[-\pi,\pi]^2} \log \sigma(\omega) \, d\omega + \frac{O(|\partial D_n|)}{n^2}. \tag{5.125}$$

Using maximum likelihood the matrix of parameters a_0, L can be estimated directly from the data. The eigenvectors of the estimated matrix $\hat{L}$ give the principle directions of the texture; the ratio between the eigenvalues determine how anisotropic it is.

Consider the micrographs shown in panels 4–6 of Figure 5.8. The hand segmentation used to extract the regions for estimation of the mitochondria and actin–myosin are shown on the right, with the region means subtracted out. Note it could be assumed that σ is fixed, with the a parameters then estimated. Since the ML estimate for σ can be computed in closed form, $a_0 = 1$ is set; this reduces the number of parameters to be searched over. As well, the maximum-likelihood estimate of the average pixel value $\hat{\mu}$ is subtracted from all the object pixels before analysis.

The bottom row of Table 5.1 lists the maximum-likelihood estimates of the texture parameters for this image. For the mitochondria, $a_{11} \approx a_{22}$ and $a_{12} \approx 0$, suggesting the isotropic models are appropriate. The parameters for the actin–myosin complex suggest a highly oriented texture; the ratio of eigenvalues is $1.5278/0.1670 = 9.1485$, with the principle eigenvector pointing in the direction $(-0.7193, 0.6947)$, which is nearly diagonal as can be easily observed in the micrograph.

It is reasonable to synthesize the actin–myosin complexes under the model to determine the appropriateness via visual inspection. For this a white random field W is generated, with the Gaussian random field solving $LX = W$. The random field X is solved for using a standard fixed-point iteration with "checkerboard" interchange, updating the red and black squares on alternate iterations. Updating all pixel values simultaneously can result in divergence.

The top row of Figure 5.9 shows results of such solutions. Panel 1 shows a 128×128 random field synthesized with zero mean and the texture parameters given in Table 5.1. Panel 2 uses the same values for a_0 and the noise variance a_0 but employs $a_{11} = 1.5278, a_{22} = 0.1670, a_{12} = 0$, which is analogous to rotating the actin–myosin texture to align its principal directions along the axis. The MRF model effectively captures the concept of directionality, which here is the essential property that differentiates the different textures. Panel 3 has $a_0 = 1$ and the same noise variance as the actin–myosin complex but employed $a_{11} = 2, a_{22} = 6$, and $a_{12} = 0$, resulting in a thicker-grained appearance. Panel 4 has the thicker-grained texture operator rotated by $60°$ (as described in the next section), yielding $a_{11} = 5, a_{12} = 1.7321$, and $a_{22} = 3$. The maximum-likelihood estimates of the texture parameters deduced from these simulations are shown in Table 5.2.

Example 5.27 (Anisotropic Filters: 3-Dimensions) In 3-dimensions, extend to the more general operator formulation $L = \sum_{i,j=1}^{3} a_{ij} (\partial^2/(\partial x_i \partial x_j)) + a_0$. Treat the constant a_0 and the 3×3 symmetric matrix L as unknowns

$$L = \begin{pmatrix} a_{11} & a_{12} & a_{13} \\ a_{21} & a_{22} & a_{23} \\ a_{31} & a_{32} & a_{33} \end{pmatrix}. \tag{5.126}$$

Given the texture field $X_i(\omega), (i_1, i_2, i_3) \in D_n = \{1 \leq i_1, i_2, i_3 \leq n\}$ with boundary conditions, ∂X, the unknown parameters are estimated by maximum likelihood. The

5.8 GENERALIZED AUTO-REGRESSIVE IMAGE MODELLING

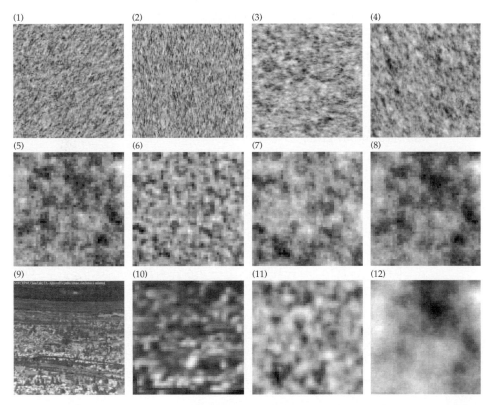

Figure 5.9 Top row: Synthetic Gaussian texture images with directional components. Panels 1–4 show simulated random fields with varying parameter choices. Panel 1 shows fields based on parameters estimated from micrographs of the actin–myosin complexes; panel 2 shows similar parameters rotated to align with the Y-axis; panel 3 shows the thicker grain parameters; panel 4 shows these rotated by $60°$. Middle row: Simulated mitochondria textures; panels 5 and 6 show the preferable first-order model and the less-likely second order model, with panels 7 and 8 showing the mixed and first-order models, respectively. Bottom row: Panel 9 shows cityscape observed via an infrared imager. Panel 10 shows a subimage of urban clutter extracted from the left portion of the cityscape. (Data courtesy Howard McCauley, Naval Air Warfare Center Weapons Division, NAWCWPNS, China Lake, CA. Approved for public release; distribution is unlimited.) Panels 11,12 show simulated infrared urban clutter; panel 11 shows the preferable second-order model; panel 12 shows the less-likely first order model.

Table 5.2 Maximum-likelihood estimates derived from the simulated images of the top row of Figure 5.9. Left 3 columns show the parameters used for the simulations; right 3 columns show the estimates.

a_{11}	a_{12}	a_{22}	a_{11}	a_{12}	a_{22}	σ
0.87	−0.68	0.82	0.87	−0.69	0.82	63.57
1.53	0	0.17	1.50	−0.02	0.17	63.35
2	0	6	1.89	0.05	5.73	61.51
5	1.73	3	4.90	1.76	2.99	61.51

log-eigenvalue normalizer given by Theorem 5.17 is given according to

$$\sigma(\omega) = \frac{1}{8\pi^3}\left\{a_0 + \left(2j\sin\frac{\omega_1}{2}, 2j\sin\frac{\omega_2}{2}, 2j\sin\frac{\omega_3}{2}\right)\right.$$
$$\left.\times \begin{pmatrix} a_{11} & a_{12} & a_{13} \\ a_{21} & a_{22} & a_{23} \\ a_{31} & a_{32} & a_{33} \end{pmatrix} \begin{pmatrix} 2j\sin(\omega_1/2) \\ 2j\sin(\omega_2/2) \\ 2j\sin(\omega_3/2) \end{pmatrix}\right\}, \quad (5.127)$$

with Theorem 5.17 giving

$$\frac{1}{n^2}\sum_m \log \lambda_m = \frac{1}{8\pi^3}\int_{[-\pi,\pi]^3} \log \sigma(\omega)\,d\omega + \frac{O(|\partial D_n|)}{n^2}. \quad (5.128)$$

Example 5.28 (Higher-Order Isotropic Texture Models) Examine sums of powers of the Laplacian of the form

$$L = \sum_{m=0}^{M}(-1)^m \Delta^m = (-1)^M a_M + \cdots + a_2\Delta^2 - a_1\Delta + a_0, \quad (5.129)$$

where Δ is the Laplacian, Δ_0 is defined to be the identity operator, and $\Delta^m = \Delta\Delta^{(m-1)}$ are powers of the Laplacian. Then the Fourier–Bessel transform of the Green's function of (5.129) is

$$H(\omega) = \frac{1}{\sum_{m=0}^{M} a_m\omega^{2m}} = \frac{1}{a_M\omega^{2M} + \cdots + a_2\omega^4 + a_1\omega^2 + a_0}. \quad (5.130)$$

This is a low-pass filter with DC gain $1/a_0$. The a's must be non-negative for H to be well-behaved. If $a_m = 0$ for $0 < m < M$, then the resulting filter

$$H(\omega) = \frac{1}{a_M\omega^{2M} + a_0} \quad (5.131)$$

is commonly referred to in the medical imaging field as an Mth order Butterworth filter with cutoff frequency $\omega_c = (a_0/a_M)^{1/M}$. Eqn. (5.131) is actually the *power* versus frequency response of the 1D *causal* filters, with associated phase shift, originally studied by Butterworth, so (5.131) is not strictly a Butterworth filter in the traditional sense.

Discretize the second order operator $L = a_2\Delta^2 - a_1\Delta + a_0$ according to

$$LX_{i_1,i_2} = a_0 X_{i_1,i_2} - a_1(X_{i_1,i_2} + X_{i_1,i_2} + X_{i_1,i_2} + X_{i_1,i_2} - 4X_{i_1,i_2})$$
$$+ a_2[X_{i_1+2,i_2} + X_{i_1-2,i_2} + X_{i_1,i_2+2} + X_{i_1,i_2-2}$$
$$- 8(X_{i_1+1,i_2} + X_{i_1-1,i_2} + X_{i_1,i_2+2} + X_{i_1,i_2-2})$$
$$+ 2(X_{i_1+1,i_2+1} + X_{i_1-1,i_2+1} + X_{i_1-1,i_2+1} + X_{i_1-1,i_2-1}) + 20X_{i_1,i_2}]. \quad (5.132)$$

The corresponding Fourier representation used to compute the log-eigenvalue sum normalizer is

$$\sigma(\omega) = a_0 - a_1(2(\cos\omega_1 + \cos\omega_2) - 4) + a_2(2(\cos 2\omega_1 + \cos 2\omega_2) \quad (5.133)$$
$$- 16(\cos\omega_1 + \cos\omega_2) + 8(\cos\omega_1 \cos\omega_2) + 20).$$

5.8 GENERALIZED AUTO-REGRESSIVE IMAGE MODELLING

Table 5.3 Maximum-likelihood parameter estimates derived from the simulated images

Experiment	a_1	a_2	Standard Deviation	a_1	a_2	Standard Deviation
Mitochondria	2.77	0	63.55	2.64	0	61.06
Mitochondria $w/a_1 = 0$	0	0.23	29.92	0	0.22	29.91
Cytoplasm	5.91	0.16	340.01	5.24	0.17	311.93
Cytoplasm $w/a_2 = 0$	7.67	0	395.25	7.16	0	371.36
Cytoplasm $w/a_2 = 0$	7.67	0	395.25	6.54	0.16 (set)	372.72
Urban clutter	0	1.73	136.86	0	1.72	136.27
Urban clutter $w/a_2 = 0$	105.25	0	2913	82.60	0	2294.3

The (set) tag denotes that a parameter was set to the specified value in order to explore different modeling choices. Means chosen for mitochondria $\mu_1 = 57.2922$, cytoplasm $\mu_2 = 135.2851$ and urban clutter $\mu_3 = 106.6847$.

Returning to the mitochondria example shown in Figure 5.8, the first line of Table 5.3 shows ML values $a_1 = 2.77$, $a_2 = 0$ supporting the first-order Laplacian models explored previously. The second line shows the results of artificially setting $a_1 = 0$. Shown in panels 5 and 6 of the middle row of Figure 5.9 are random fields from these parameters. As in the previous section on nonisotropic operators, we set $a_0 = 1$ and use the closed-form solution of the ML estimate for σ, so we only have two parameters to maximize over. Again we subtract the average pixel value before analysis.

The second set of three lines of Table 5.3 show the results of analysing the cytoplasm background in the left panel of Figure 5.8. This is the first example of a truly mixed model in which a_1 and a_2 are both deduced to be nonzero. The last line shows the result of employing just a first-order model. Synthesized textures with the mixed and first-order models are shown in panels 7 and 8 of Figure 5.9. The fifth line illustrates the deduction from the first-order cytoplasm simulation which results if a_2 is artificially set to the value deduced from the original mixed model. As might be expected, a_1 is estimated to be at a lower value, closer to that found in the original mixed model.

Consider the forward-looking infrared (FLIR) image displayed in the bottom row panel 9 of Figure 5.9. A 50×50 subimage of urban clutter extracted from the left side of the image is shown in panel 10. Interestingly, the ML parameter estimates for the urban clutter sample yield $a_1 = 0$, $a_2 = 1.7278$, as listed in the bottom two lines of Table 5.3, suggesting a second-order Butterworth filter model; this is a special case of the "Model A" clutter proposed by US Army Missile Command [145]. As shown in the last line of the table, if the coefficient $a_2 = 0$ is artificially restricted and a first-order model is enforced, maximum-likelihood chooses an extremely large a_1 value. Simulated IR clutter using the higher-likelihood second order model and the lower-likelihood first order are shown in the panels 11 and 12 of Figure 5.9. Even at first glance, the clutter synthesized from the second-order model appears to more closely capture the characteristics of the real clutter than the first-order model.

Example 5.29 (Anisotropic Textures with Variable Orientation) Examine the case where the principal directions of the texture are variable. Parameterize the matrix $L = L(O)$ via the unknown rotation $O \in \mathbf{SO}(3)$ describing the orientation of the texture for the particular instance. The standard way in which differentiation depends upon the choice of coordinate system choose $y = Ox$, O an orthogonal matrix with

$$O = \begin{pmatrix} o_{11} & o_{12} & o_{13} \\ o_{21} & o_{22} & o_{23} \\ o_{31} & o_{32} & o_{33} \end{pmatrix}, \tag{5.134}$$

then $(\partial/\partial X_i) = \sum_{j=1}^{3}(\partial/\partial y_j)o_{ji}$ or in shorthand notation $(\partial/\partial x) = O^*(\partial/\partial y)$. Iterating gives $(\partial^2/\partial x_i \partial x_j) = \sum_{k,l}(\partial^2/(\partial y_k \partial y_l))o_{ki}o_{lj}$, therefore

$$\sum_{i,j=1}^{3} a_{ij} \frac{\partial^2}{\partial x_i \partial x_j} = \sum_{k,l=1}^{3} b_{kl} \frac{\partial^2}{\partial y_k \partial y_l}$$

with $b_{kl} = \sum_{ij} a_{ij} o_{ki} o_{lj}$. Then

$$B = \begin{pmatrix} b_{11} & b_{12} & b_{13} \\ b_{21} & b_{22} & b_{23} \\ b_{31} & b_{32} & b_{33} \end{pmatrix} = OLO^*, \tag{5.135}$$

dictating the relationship between $L(O)$ and the rotation O in the continuum $\mathbb{R}^3$. This transfers directly to finite differences. Without loss of generality assume L is of diagonal form and introduce the nuisance parameter O to estimate the matrix $L'(O) = OLO^*$. To determine the rotation that makes the estimate L', use the Hilbert–Schmidt norm $\|M\|^2 = \mathrm{tr}(M^2)$ for symmetric matrices M corresponding to the inner product $\langle M_1, M_2 \rangle = \mathrm{tr}(M_1 M_2)$. Then

$$\arg\min_{O} \|L' - OLO^*\|^2 = \arg\max_{O} \mathrm{tr}(L'OLO^*). \tag{5.136}$$

To solve the maximum problem express L and L' in their spectral representation

$$L = \sum_{\nu=1}^{3} \lambda_\nu E_\nu, \qquad L' = \sum_{\nu=1}^{3} \lambda'_\nu E'_\nu, \tag{5.137}$$

with the E's projection operators corresponding to respective resolutions of the identity. Then

$$\mathrm{tr}(L'OLO^*) = \sum_{\nu,\mu=1}^{3} \lambda_\nu \lambda'_\mu c_{\nu\mu} \tag{5.138}$$

with the coefficients $c_{\nu\mu} = \mathrm{tr}(E_\nu O E'_\mu O^*)$.

Express the projections E_ν, $OE'_\mu O^*$ expressed in terms of the unit vectors $e = (e_i; i = 1, 2, 3)$ and $e' = (e'_i; i = 1, 2, 3)$, we have

$$c_{\nu\mu} = \mathrm{tr}(E_\nu E'_\mu) = \sum_{i,j} e_i e_j e'_i e'_j = \left(\sum_i e_i e'_i\right)^2 \geq 0, \tag{5.139}$$

so that the c-coefficients are non-negative. On the other hand

$$\sum_{\mu=1}^{3} c_{\nu\mu} = \sum_\mu \mathrm{tr}(E_\nu O E'_\mu O^*) = \mathrm{tr}(E_\nu) = 1,$$

since $\sum_\mu E'_\mu = $ *Identity* and any projection operator to a 1D sub-space has trace 1. Similarly $\sum_\nu c_{\nu\mu} = 1$, $\forall \mu$.

Hence matrix C belongs to the set $\mathcal{D}$ of doubly stochastic matrices. Since $\mathcal{D}$ is convex and compact any of its elements can be expressed as a convex combination of the extreme points of $\mathcal{D}$, i.e. the permutation matrices Π_k, so that

$$C = \sum_{k=1}^{6} w_k \Pi_k; \quad w_k \geq 0, \quad \sum w_k = 1.$$

5.8 GENERALIZED AUTO-REGRESSIVE IMAGE MODELLING

The maximum is attained by solving $\max_k \sum_v \lambda_v^L \lambda_{(\Pi_k)_v}^B$, where the vector Π_k means the reordering of $(1,2,3)$ by the permutation matrix Π_k. This is a classical problem (see Chapter 10 [146]). The answer is that if we have ordered the eigenvalues λ_i^L in non-decreasing order then we should choose the Π_k that orders the λ also in increasing order.

Now we have the solution. After having estimated the L-matrix solve for its eigenelements. Order the eigen-vectors so that the corresponding eigen-values form a nondecreasing sequence and use them to form an orthogonal matrix (in the same order): This rotation is the solution. Since eigen-vectors are determined up to its sign, there will typically be 8 solutions, it does not matter which one we choose.

Extending to differential operators of order greater than two presents some new difficulties. Indeed, instead of quadratic forms higher degree forms will be encountered with various complications.

6 THE CANONICAL REPRESENTATIONS OF GENERAL PATTERN THEORY

ABSTRACT Pattern theory is combinatory in spirit or, to use a fashionable term, connectionist: complex structures are built from simpler ones. To construct more general patterns, we will generalize from combinations of sites to combinations of primitives, termed generators, which are structured sets. The interactions between generators is imposed via the directed and undirected graph structures, defining how the variables at the sites of the graph interact with their neighbors in the graph. Probabilistic structures on the representations allow for expressing the variation of natural patterns. Canonical representations are established demonstrating a unified manner for viewing DAGs, MRFs, Gaussian random fields and probabilistic formal languages.

6.1 The Generators, Configurations, and Regularity of Patterns

To construct more general patterns, the random variables and state spaces are generalized via the introduction of primitives called generators which are structured sets. The generators become the random variables at the nodes of the graphs; the structure is imposed via the edge relations in the graphs constraining how the sets at the vertices of the graph interact with their neighbors. The graphs imply the conditional probabilities associated with the directed and undirected random fields.

When moving to the more general patterns on arbitrary graphs $\sigma = \{D, E\}$ it is clear that building structure requires the aggregation of the field variables. This is precisely what the generators do! Examine the goal of moving from unstructured representations of pixelated images to those containing *edge* or *line* sites and continuing to *object* elements and highly structured sets. In an attempt to add structure, more complex abstractions are defined; the edge vertices which are aggregations of the pixels, and line and boundary vertices which are in turn aggregations of the edge and line vertices, respectively. The generators arise as aggregations of the field elements.

In an atomistic spirit we build the representations by combining simple primitives, *generators*. To begin with the generators will be treated as abstract mathematical entities whose only structure will be expressed in terms of *bonds* defining how they communicate on the graph. The mathematical objects, the *generators*, are from the *generator space* $\mathcal{G}$ and will appear in many forms: they can be positive pixel values in an image, states in a Markov chain, geometric objects such as vectors and surface elements, or rewriting rules in language theory.

As before, begin with the *connector graph* σ with $n = |\sigma|$ the number of vertices in σ. At each vertex place one of the *generators* $g \in \mathcal{G}$ the *generator space*. A pattern or configuration is denoted $c(\sigma) = \sigma(g_1, \ldots, g_n)$. Associate with the structured generators *bonds*, information which is used to communicate to other neighbors in the graph. The bonds establish the local rules of interaction; an edge in the graph corresponds to a bond between generators. As the generators are stitched together, only certain patterns will have the structure regularity of the pattern class. To enforce rigid structure, a *bond function* $\beta : \mathcal{G} \to \mathcal{B}$ is associated to each generator and its target defined by edges $e = (i, j)$ between pairs of generators which are neighbors in the graph. If two vertices $i, j \in D$ of the graph are neighbors, $j \in N_i, i \in N_j$, then the generator g_i has a bond $\beta_j(g_i)$ which must agree with the bond $\beta_i(g_j)$ from g_j. For sites i, j which have no edge, we assume the null bonds, with agreement trivially satisfied. The value of the bond couples determine how the generators interact and therefore how the patterns form.

The *configurations* are illustrated as in Figure 6.1. Panel 1 shows two generators in the graph with their bonds. Panel 2 shows a graph with the set of 6 vertices in D with associated edges dictating where the bonds must form; the right panel, the set of generators with their bonds.

6.1 GENERATORS, CONFIGURATIONS AND REGULARITY OF PATTERNS

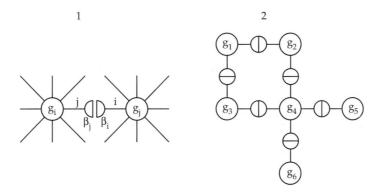

Figure 6.1 Panel 1 shows two generators with their bonds; panel 2 shows a configuration with the set of generators, bonds, and edges in the graph.

The bonds constrain the configuration defining which ones are regular; the space of regular configurations will generally be a subset of the full configuration space $\mathcal{G}^n$.

Definition 6.1 *Define the* **unconstrained configurations** *on graph* $\sigma = \{D, E\}$ *as a collection of n-generators,*

$$\mathcal{C}(\sigma) = \{c(\sigma) = \sigma(g_1, g_2, \ldots, g_n) : g_i \in \mathcal{G}^{(i)}, i = 1, \ldots, n\}. \tag{6.1}$$

A configuration is determined by its **content**, *the set of generators in* $\mathcal{G}$ *making it up*, $\text{content}(c) = (g_1, g_2, \ldots, g_n)$, *and its combinatory structure determined by the graph with the* **internal bonds** *of the configuration denoted* $\text{int}(c)$, *and the set of the remaining ones*, **the external bonds**, *denoted by* $\text{ext}(c)$.

Defining the **bond function** $\rho(\cdot, \cdot) : \mathcal{B} \times \mathcal{B} \rightarrow (\text{TRUE}, \text{FALSE})$, *then a configuration* $c(\sigma) = \sigma(g_1, g_2, \cdots, g_n) \in \mathcal{C}$ *is said to be* **regular** *when the bond relation is satisfied over the graph* $\sigma = (D, E)$:

$$\bigwedge_{e=(i,j) \in E} \rho(\beta_j(g_i), \beta_i(g_j)) = \text{TRUE}. \tag{6.2}$$

The **space of regular configurations**, *a subset of the full configuration space, is denoted as*

$$\mathcal{C}_\mathcal{R} \subseteq \mathcal{C}(\sigma) = \mathcal{G}^{(1)} \times \cdots \times \mathcal{G}^{(n)}.$$

For many problems, we will also be interested in configuration spaces associated with the collection of graphs $\mathcal{C}_\mathcal{R}(\Sigma) = \cup_{\sigma \in \Sigma} \mathcal{C}_\mathcal{R}(\sigma)$.

Formula 6.2 is a structural formula expressing relations between the generators at the nodes of the graph. Sometimes **local regularity** may be distinguishable in which all of the bonds in the graph type σ are satisfied, although the graph type itself may not be in the set of allowed graphs. The connected bonds are essentially the internal bonds of the configuration, with the set of the remaining ones, the external bonds.

In general, connector graphs may be sparse, so that many of the sites do not have an edge. The right graph in Figure 6.1 is such a case. It is then natural to explicitly index the bonds at each generator $\beta_k(g), k = 1, \ldots, \omega(g)$, with total arity $\omega(g)$ designating the number of bonds attached to g.

Example 6.2 (Regular Lattice Magnetic Spins) In the magnetic spin models such as the Ising model from statistical physics, $\sigma = \text{LATTICE}$, the generators are the magnetic dipoles in plus or minus orientation $\mathcal{G} = \{+, -\}$, and the bond values are the terms which multiply to form the energy function: $\mathcal{B} = \{-1, +1\}$ with $\omega(g) = 2, 4, 6$ for the

6 THE CANONICAL REPRESENTATIONS OF GENERAL PATTERN THEORY

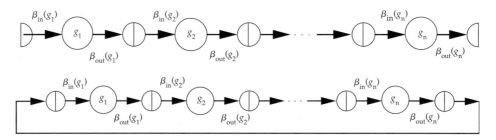

Figure 6.2 Top and bottom rows show configurations $LINEAR(n, g_1, \ldots, g_n)$ and $CYCLIC(n, g_1, \ldots, g_n)$.

1,2,3D models, respectively. For the Ising case, $\rho(\beta, \beta') = $ TRUE for all $(\beta, \beta') \in \mathcal{B}^2$; all $+1, -1$ configurations are allowed.

In 1D let $D = \{1 \leq i \leq n\}$ and choose the two bond values to be $\beta_1(g) = \beta_2(g)$ defined by $\beta(+) = 1, \beta(-) = -1$, with the truth table taking the form for the generators

$$\rho : \mathcal{B} \times \mathcal{B} = \begin{array}{c|cc} & +1 & -1 \\ \hline +1 & \rho = T & \rho = T \\ -1 & \rho = T & \rho = T \end{array}, \quad \text{and} \quad \mathcal{C}_\mathcal{R} = \mathcal{C} = \{+,-\}^n. \qquad (6.3)$$

Example 6.3 (Unclosed and Closed Contours) Examine unclosed contours in the plane, and generate the boundaries from line segments. Then $\sigma = $ LINEAR, $\omega(g) = 2$, generators are arcs in the plane $g = (z_1, z_2); z_1, z_2 \in \mathbb{R}^2, \mathcal{G} = \mathbb{R}^4$, and bond-values are start and end-points of the generators, $\beta_{\text{in}}(g) = z_1; \beta_{\text{out}}(g) = z_2$. As depicted in Figure 6.2, notice that the boundary vertex generators have one less in and out bond for LINEAR. For cyclic, the last generator interacts with the first generator.

The continuity constraints mean that consecutive line segments are joined to each other and enforced via the regularity of bond consistency $\mathcal{R}$(LINEAR) so that the outbond of the ith generator equals the inbond of the $i + 1$st:

$$\rho(\beta_{\text{out}}(g_i), \beta_{\text{in}}(g_{i+1})) = \text{TRUE} \quad \text{if and only if} \quad \beta_{\text{out}}(g_i) = \beta_{\text{in}}(g_{i+1}). \qquad (6.4)$$

For closed contours, $\sigma = $ CYCLIC, with the first and last vertices having an arc in the graph, adding the cyclic regularity enforced by the added bond relation $\rho(\beta_{\text{out}}(g_n), \beta_{\text{in}}(g_1)) = $ TRUE (bottom row of Figure 6.2).

Here the generators are chosen as directed line segments, which could be generalized to other arc elements from conic sections or other curve families. For curves in $\mathbb{R}^3, z_1, z_2 \in \mathbb{R}^3, \mathcal{G} = \mathbb{R}^6$.

Example 6.4 (Triangulated Graphs) In cortical surface generation the surfaces are generated from vertices using triangulations of the sphere. Panel 1 of Figure 6.3 depicts a triangulated sphere, the generators being

$$g_i = (v_i^{(1)}, v_i^{(2)}, v_i^{(3)}) \in (S^2)^3, \qquad (6.5)$$

with $\omega(g) = 3$, and $\beta_j(g_i) = v_i^{(j)}; j = 1, 2, 3$. Thus $\mathcal{G} = \mathbb{R}^9$. The graph family $\Sigma = \cup_n$ TRIANG(n) expresses the topology corresponding to patches which are topologically connected according to the triangulated sphere topology. A complete triangulation with congruent triangles is only possible for $n = 4, 6, 8, 12, 20$, the Platonic solids. Generally large n-values are used, so that noncongruent triangles form the patterns. Panels 2 and 3 show the triangulated graphs associated with the bounding closed surface of the hippocampus in the human brain and the macaque cortex.

6.1 GENERATORS, CONFIGURATIONS AND REGULARITY OF PATTERNS

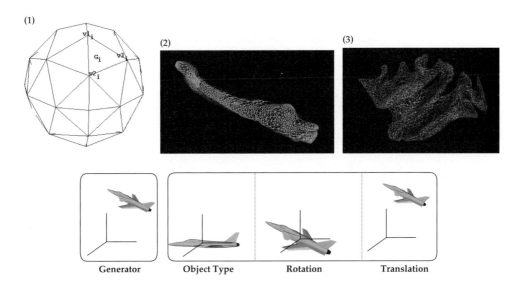

Figure 6.3 Top row: Panel 1 shows the triangulated graph for the template representing amoeba corresponding to spherical closed surfaces. Generators are $g_i = (v_1, v_2, v_3)$, elements of $\mathbb{R}^9$. Panel 2 shows the triangulated for the closed surface representing the bounding closed surface of the hippocampus in the human brain. Panel 3 shows the surface representation of a macaque brain from the laboratory of David Van Essen. Bottom row: Shows a generator shown in standard position, orientation under rotation and translation transformation. (see plate 3).

Alternatively, choose the generators to be the vertices in $\mathcal{G} = \mathbb{R}^3$, and the generating points are the n-points sampled around the sphere:

$$g_{i,j} = \left(\sin \frac{2\pi i}{m} \cos \frac{2\pi j}{m}, \sin \frac{2\pi i}{m} \sin \frac{2\pi j}{m}, \cos \frac{2\pi i}{m} \right), \quad (i,j) \in \mathbb{Z}_{m^2}. \tag{6.6}$$

Example 6.5 In the setup for observing moving bodies, say for tracking an aircraft, define each generator to an aircraft, $\mathcal{G} = \{\text{AIRCRAFTS}\}$ of specified types in specified positions and orientation. The center of gravity is located at some $(x_1, x_2, x_3) \in \mathbb{R}^3$; the attitude is the natural coordinate axes of the aircraft forming given angles with those of the inertial frame. Then $\dim(\mathcal{G}) = 6$. A convenient choice of generator space is to let each generator consist of an arc, for example a line segment, in location and orientation space.

A system of moving bodies need not be constrained by local regularity unless they do not occupy common volume. Then a generator has an indefinite number of bonds, all of whose values should be the set occupied by the body. The bond relation takes the form

$$\beta_1 \cap \beta_2 = \emptyset. \tag{6.7}$$

A configuration for tracking is a linear graph LINEAR(m_1) of m_1 generators (airplanes) specified via the special Euclidean motion. Multiple airplanes correspond to unions of linear graphs, MULT(m_2, LINEAR). Examine the set of graph transformations associated with discovering tracks. These consist of such transformations as depicted in the Figure in the right panel of Figure 6.7 and in Figure 6.4 for airplane tracking.

Example 6.6 (Finite State Graphs) Let $X_i, i = 1, 2, \ldots$ be a realization from the binary 1-1s languages $\mathcal{X}_0 = \{0, 1\}$ of binary strings not containing 2 consecutive 1 symbols in a row. Let the generators be the binary values, $\mathcal{G} = \{0, 1\}$.

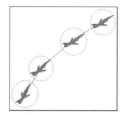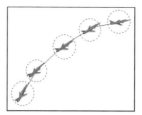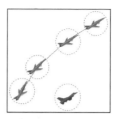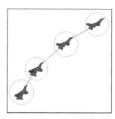

Figure 6.4 Figure showing graph transformations forming and deleting tracks and track segments.

With $\Sigma = \text{LINEAR}, \omega(g) = 2$ with $\beta_1(g) = \beta_2(g) = g \in \mathcal{B} = \mathcal{G}$, choose the bond-relation truth table to respect the 1-1 constraint:

$$\rho : \mathcal{B} \times \mathcal{B} = \begin{array}{c|cc} & 0 & 1 \\ \hline 0 & \rho = T & \rho = T \\ 1 & \rho = T & \rho = F \end{array} . \quad (6.8)$$

The regular configurations become $c(\text{LINEAR}) \in \mathcal{C}_\mathcal{R}(\text{LINEAR})$ if and only if $\beta_{\text{out}}(g_i) = \beta_{\text{in}}(g_{i+1}), i = 1, \ldots, n - 1$.

Example 6.7 (Model-Order, Sinusoid estimation) In a sinusoid estimation such as for NMR [22], signals are constructed from sinusoids $x(t) = \sum_{i=1}^{n} a_i \sin \omega_i t e^{-\lambda_i t}$, generators are sinusoids. Then a configuration $c(\text{LINEAR}) = \text{LINEAR}(a_1, \ldots, a_n)$, linear graph denoting order so that a_i is associated with $\sin \omega_i t$, and $\mathcal{C}_\mathcal{R}(n) = \mathcal{C}(n) = \mathbb{R}^n$. All choices of amplitudes are allowed: it is a vector space. The dimension of the model will not generally be known, so that $\mathcal{C}(\Sigma) = \cup_{n=0}^{\infty} \mathbb{R}^n$.

6.2 The Generators of Formal Languages and Grammars

The formal grammars of Chapter 3, Section 3.10 are an exquisite illustration of the pattern theory. They directly provide a mechanism for building the space of directed graphs Σ recursively: *the grammar specifies the legal rules of transformation of one graph type to another.* Naturally, finite-state grammars provide transformations which build the space of linear graphs $\Sigma = \text{LINEAR}$, the context-free grammars the space of trees $\Sigma = \text{TREE}$ and finally context-sensitive grammars provide transformations through the partially ordered set graphs $\Sigma = \text{POSET}$.

What is so beautiful about the formal language theory of Chomsky is that the grammars precisely specify the generators for the configurations and the rules of transformation. They also illustrate one more non-trivial movement to more abstract generators. The patterns (strings) are built by connecting the generators in the various graph structures to satisfy the consistency relationship defined by the bond structure: LINEAR, TREE, DAG. The connection of the Chomsky transformation rules is that the rules of transformation are the generators. These are the formal grammars! The transformations allowed by the rules are constrained by the consistency placed via the bonds between the generating rules.

The generators are the grammatical rules themselves, and the bond-values are subsets of the non-terminals and terminals on the right and left hand sides of the rules, elements of the terminal and non-terminal set. To see the forms for the rules of the three grammar types, refer to Figure 6.5 which shows various examples. Left panel shows a finite-state grammar production rule $A \to B$, $\omega_{\text{in}} = \omega_{\text{out}} = 1$. The middle panel shows a production rule (generator) $S \to NP\ VP$ for a context-free grammar; $\omega_{\text{in}}(g) = 1, \omega_{\text{out}}(g) = 2, \beta_{\text{in}}(g) = S, \beta_{\text{out1}}(g) = NP$, and $\beta_{\text{out2}}(g) = VP$. The right panel shows a context-sensitive production rule $A\ B \to C\ D$.

6.2 THE GENERATORS OF FORMAL LANGUAGES AND GRAMMARS

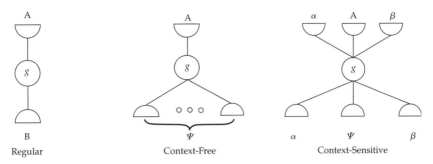

Figure 6.5 Rewriting rules corresponding to the finite-state, context-free, and context-sensitive generators.

The local regularity comes from the fact that for two bonds β, β' joined in the graph, then

$$\text{outbond } \beta = \text{inbond } \beta'. \tag{6.9}$$

In other words, when a rewriting rule has produced certain non-terminal syntactic variables as their daughters, then this rule can only be connected to a rule with those syntactic variables on their left hand side: no jumps are allowed. In languages, we shall be interested in the set of strings having only terminals.

Definition 6.8 *Associated with every configuration define the* **closed or internal bonds** *of the configuration, denoted $int(c)$, and the set of the remaining ones, the* **external bonds** *denoted as $ext(c)$.*

Theorem 6.9 *Given is a formal grammar $G(P) = <V_N, V_T, R, S>$ with language $L(G)$. Then let the regular configurations be constructed from generators the rules $\mathcal{G} = R$, with the bonds the non-terminals and terminals on the right and left hand sides of the rules in $\mathcal{B} = V_N \cup V_T$:*

1. *for finite-state, $\sigma = $ LINEAR with generators and bonds*

$$A \xrightarrow{g} wB, \ \beta_{\text{in}}(g) = A, \ \beta_{\text{out}}(g) = A', \ \omega(g) = 2; \tag{6.10}$$

2. *for context-free, $\sigma = $ TREE with generators and bonds*

$$A \xrightarrow{g} \psi, \ \beta_{\text{in}}(g) = A, \ \beta_{\text{out}}(g) = \psi, \ \omega(g) = 1 + |\psi|; \tag{6.11}$$

3. *for context-sensitive, $\sigma = $ DAG with generators and bonds*

$$\alpha A \beta \xrightarrow{g} \alpha \psi \beta, \ \beta_{\text{in}}(g) = \alpha A \beta, \ \beta_{\text{out}}(g) = \psi, \ \omega(g) = |\alpha A \beta| + |\psi|. \tag{6.12}$$

The regular configurations $c(\sigma) = \sigma(n, g_1, g_2, \ldots, g_n) \in \mathcal{C}_\mathcal{R}(\sigma)$ satisfy

$$\bigwedge_{e=(i,i') \in E} \rho(\beta_{j(e)}(g_i), \beta_{j'(e)}(g_{i'})) = \text{TRUE}, \text{ where } \rho(\beta, \beta') = \text{TRUE} \iff \beta = \beta'.$$

$$\tag{6.13}$$

Then the regular configurations $\bar{\mathcal{C}}_\mathcal{R}(\sigma) \subset \mathcal{C}_\mathcal{R}(\sigma), \sigma \in \Sigma = \{\text{LINEAR}, \text{TREE}, \text{DAG}\}$ having no unclosed external bonds is the formal language:

$$\bar{\mathcal{C}}_\mathcal{R}(\sigma) = \{c(\sigma) = \sigma(n, g_1, \ldots, g_n) \in \mathcal{C}_\mathcal{R}(\sigma) : ext(c) = \phi\} = L(G). \tag{6.14}$$

Figure 6.5 depicts the various forms of the generators for the three grammar types. The special rewrite rules involving the sentence symbol, $S \to \psi$ have $\omega_{\text{in}}(g) = 0$, and the rules $\alpha A \beta \xrightarrow{g} \psi \in V_T^*$ with only terminal symbols on the right hand side, $\beta_{\text{out}}(g) = 0$.

6 THE CANONICAL REPRESENTATIONS OF GENERAL PATTERN THEORY

The probabilistic versions of formal languages, the stochastic languages, the set of strings generated by probabilistic application of the rules. For the finite-state (regular) and context-free languages, these are Markov chains and multi-type random branching processes.

Example 6.10 (Finite-State Languages) Examine the finite-state languages of Example 3.43. Begin with the 1-1 constraint language, $V_N = \{\text{ZERO}, \text{ONE}\}$ with $S = \text{ZERO} >$, with the generators the transitions in the state graph:

$$\mathcal{G} = R = \{\text{ONE} \xrightarrow{r_1} 0\text{ZERO}, \text{ZERO} \xrightarrow{r_2} 0\text{ZERO}, \text{ZERO} \xrightarrow{r_3} 1\text{ONE}\}. \tag{6.15}$$

Elements in the language can end in either states, augment the rule set with the terminating rules $\text{ZERO} \xrightarrow{r_4} 1$, $\text{ZERO} \xrightarrow{r_5} 0$, $\text{ONE} \xrightarrow{r_6} 0$.

For parity languages, $V_N = \{\text{EVEN}, \text{ODD}\}$ with $S = \text{EVEN}$, and the rules (generators)

$$\mathcal{G} = R = \{\text{EVEN} \xrightarrow{r_1} 0\text{EVEN}, \text{EVEN} \xrightarrow{r_2} 1\text{ODD}, \text{ODD} \xrightarrow{r_3} 0\text{ODD}, \text{ODD} \xrightarrow{r_4} 1\text{EVEN}\}. \tag{6.16}$$

Since only strings ending in EVEN parity are in the language, augment with the terminating rules $\text{EVEN} \xrightarrow{r_5} 0, \text{ODD} \xrightarrow{r_6} 1$. The graph is $\sigma = \text{LINEAR}$, arity $\omega(g) = 2$ with in- and out-bond values

$$\beta_{\text{in}}(g) = \text{LHS}(g), \quad \beta_{\text{out}}(g) = \text{RHS}(g), \tag{6.17}$$

with

$$\rho(\beta_{\text{out}}(g_i), \beta_{\text{in}}(g_{i+1})) = \text{TRUE} \quad \text{if and only if } \text{RHS}(g_i) = \text{LHS}(g_{i+1}).$$

The string 001001 is regular in the 1-1 language with generator representation $\text{LINEAR}(r_2, r_2, r_3, r_1, r_2, r_4)$. The parity string 001001 in the EVEN parity language is regular with generator representation $\text{LINEAR}(r_1, r_1, r_2, r_3, r_3, r_6)$. They both satisfy the structure relations $\beta_{\text{out}}(g_n) = \phi$ with

$$\beta_{\text{out}}(g_i) = \beta_{\text{in}}(g_{i+1}) \quad \text{for } i = 1, \ldots, n-1. \tag{6.18}$$

For the **bigram** and **trigram** language models the m-gram model treats a string of words as a realization of an mth-order Markov chain in which the transition probability is the conditional probability of a word in the string given by the previous $m-1$ words. The generators for the m-ary processes are

$$\underbrace{w_{i-m}, w_{i-(m-1)}, \ldots, w_i}_{A} \rightarrow \underbrace{w_{i-(m-1)}, w_{i-(m-2)}, \ldots, w_{i+1}}_{w_{i+1}\, A'}$$

$$\underbrace{w_{i-m}, w_{i-(m-1)}, \ldots, w_i}_{A} \rightarrow \underbrace{w_{i+1-m}, w_{i+1-(m-1)}, w_{i+1-(m-2)}, \ldots, w_i, \phi)}_{w_{i+1}}$$

where $\phi = \text{null}$ (punctuation). Notice, for $m = 1, m = 2$ these rules are precisely the generators of node i, (X_i, X_{π_i}) in Theorem 6.19, with in-arity 1 over the entire graph.

An element of the *configuration space* $\mathcal{C}(\text{LINEAR})$ consists of a set of generators which are the rewrite rules $\mathcal{G} = R$ placed at the vertices of the directed graph. A configuration consists of n nodes in a linear graph $c = \text{LINEAR}(n, r_1, \ldots, r_n)$; rule $r_i \in R$ transforms node i.

Example 6.11 (Phrase Structure Grammars) Examine the structures associated with context free languages and their associated tree graphs.

6.2 THE GENERATORS OF FORMAL LANGUAGES AND GRAMMARS

The generators are the rewrite rules, and for each **generator** $g \in \mathcal{G}$, the arity of the in-bonds $\omega_{in}(g) = 1$ and the arity of the out-bonds $\omega_{out}(g) \geq 1$, [18] the in-bonds corresponding to the left hand side of the production rule, the out-bonds the right hand side of the production rule determined by the number of non-terminals. Context-free means $\Sigma = \text{TREE}$, as depicted in 6.6. An element of the configuration space $c(\text{TREE}) = \text{TREE}(n, g_1, g_2, \ldots, g_n) \in \mathcal{C}$ consists of a set of n-vertices at which the rewriting rule generators are placed. The in- and out-bonds of a generator are $\beta_1(g)$, the in-bond value the LHS of the production rule, and $\beta_2(g), \ldots, \beta_{\omega(g)}(g)$, the out-bond values which are elements of the set of bond values $\mathcal{B} = V_N \cup V_T$, where V_N is the set of non-terminal symbols and V_T is the set of terminal symbols.

Examine the simple phrase-structure grammar with seven basic rules

$$R = \left\{ \begin{array}{ccc} S \xrightarrow{r_1} NP\ VP & NP \xrightarrow{r_2} ART\ N & \\ VP \xrightarrow{r_3} V & N \xrightarrow{r_4, r_5} \text{frogs} \vee \text{eggs} & \\ ART \xrightarrow{r_6} \text{the} & V \xrightarrow{r_7} \text{jump} & \end{array} \right\}. \tag{6.19}$$

Define the *sentence–language* as the set of all context-free trees rooted in the syntactic variable S. Panel 1 of Figure 6.6 shows a sentence $\text{TREE}(6, r_1, r_2, r_3, r_6, r_4, r_7)$. Notice, it is a single tree rooted in the sentence root node $S = sentence$ with all

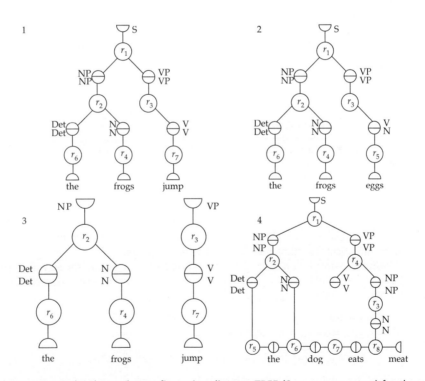

Figure 6.6 Panel 1 shows the configuration diagram TREE $(6, r_1, r_2, r_3, r_6, r_4, r_7)$ for the sentence *The frogs jump*; panel 2 shows the configuration diagram TREE $(6, r_1, r_2, r_3, r_6, r_4, r_5)$ for the sentence *The frogs eggs*, which is not locally regular. Panel 3 shows the configuration PHRASE–FOREST $(2, c_1, c_2)$ of two phrases PHRASE $(3, r_2, r_6, r_4)$, PHRASE $(2, r_3, r_7)$ each locally regular. Panel 4 shows the configuration DIGRAPH $(8, r_1, r_2, r_4, r_5, r_6, r_7, r_3, r_8)$ which is context-sensitive and is a DIGRAPH.

---

[18] As proven by Chomsky a context-free language is fundamentally not finite-state, if at least one rule in the connected set has more than one non-terminal in the right-hand side ($\omega_{out}(g) \geq 2$), the so called self-embedding property.

bonds agreeing (local regularity). Panel 2 of Figure 6.6 shows a configuration $c = \text{TREE}(6, r_1, r_2, r_3, r_6, r_4, r_5)$ which is not regular. Notice the disagreement of the $VP \to V, N \to jump$ rules. Notice, $\beta_{\text{out}}(r_3) = V \neq \beta_{\text{in}}(r_5) = N$. Panel 3 of Figure 6.6 shows a configuration consisting of two trees each of which are phrases. The graph type PHRASE–FOREST, consists of all tree derivations rooted in any syntactic variable with leaves as the terminal strings. The configuration $c=\text{PHRASE–FOREST}(2, c_1, c_2)$ where

$$c_1 = \text{PHRASE}(3, r_2, r_6, r_4), \quad c_2 = \text{PHRASE}(2, r_3, r_7), \quad (6.20)$$

is locally and globally regular.

Example 6.12 (Context-Sensitive Languages and POSET Graphs) An example of a context-sensitive grammar is $V_N = \{S, NP, VP, V, DET, N\}, V_T = \{\text{The, dog, eat, meat}\}$ with rules

$$R = \left\{ \begin{array}{lll} S \xrightarrow{r_1} NP\ VP, & NP \xrightarrow{r_2} DETN, & NP \xrightarrow{r_3} N, \\ VP \xrightarrow{r_4} V\ NP, & DET \xrightarrow{r_5} \text{the}, & \text{the}\ N \xrightarrow{r_6} \text{the dog}, \\ \text{dog}\ V \xrightarrow{r_7} \text{dog eats}, & \text{eats}\ N \xrightarrow{r_8} \text{eats meat} \end{array} \right\}. \tag{6.21}$$

The context sensitive rules are those that rewrite a part of speech in the context of the previous word. Context-sensitive grammars derive configurations that are partially ordered sets. Examine panel 4 of Figure 6.6 showing the corresponding graph. Notice the choice of grammar places context at the terminal leaves resulting in loss of the conditional independence structure at the bottom of the tree.

6.3 Graph Transformations

Now explore *graph transformations* as the mechanism for constructing more complex structures from simpler ones. The construction of complex patterns requires rules of transformation so that more complex patterns may be constructed from simpler ones. This is certainly one of the most fundamental aspects of pattern theory, building more complex graphs from simpler ones. Graph transformations are familiar, as in linguistic parsing, where the role of the *grammatical rules* define the structure of the transformation. Such grammatical transformations are chosen to be rich enough to represent valid structures in the language, while at the same time restrictive enough to enforce regular grammatical structure associated with the language.

During transformation of graphs, bonds will be left unclosed, the so called external bonds, $(ext(c))$, and bonds will be closed during combination. For such changes, introduce the set of graph transformations $\mathcal{T} : \mathcal{C_R} \to \mathcal{C_R}$ of either birth or death type.

Definition 6.13 *Define the discrete graph transformation $T \in \mathcal{T}$ to be either of the **birth** or **death** type according to the following:*

$$T \in \mathcal{T}(c) : c(\sigma) \stackrel{\text{birth}}{\mapsto} c'(\sigma') = \sigma(c, c''); \quad c, c'', c' \in \mathcal{C_R}; \tag{6.22}$$

$$T \in \mathcal{T}(c) : c = \sigma(c', c'') \stackrel{\text{death}}{\mapsto} c'(\sigma'); \quad c, c', c'' \in \mathcal{C_R}. \tag{6.23}$$

*For any regular configuration c then the **upper neighborhood** $\mathcal{N}_+(c)$ consists of all configurations obtained from c by a **birthing** and the **lower neighborhood** $\mathcal{N}_-(c)$ of all configurations obtained via a **death** transformation to c, with the full neighborhood*

$$\mathcal{N}(c) = \mathcal{N}_+(c) \cup \mathcal{N}_-(c).$$

*Then the set of graphs will be traversed by the **family of graph transformations** $T \in \mathcal{T} = \cup_{c \in \mathcal{C}} \mathcal{T}(c)$ of simple moves consisting of either the **birth** or **death** types.*

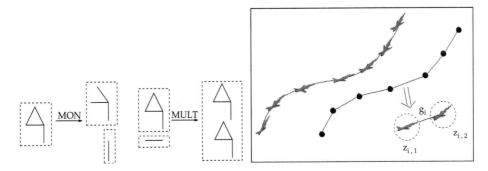

Figure 6.7 Left panel depicts MON and MULT extensions. Right panel shows figure tracking airplanes as LINEAR graphs with generators.

We shall see in defining inference algorithms with desirable properties that it is helpful to require that the transformations $\mathcal{T}$ give the graph space a *reversibility* and *connectedness* property.

Definition 6.14 $\mathcal{T} : \mathcal{C}_\mathcal{R} \to \mathcal{C}_\mathcal{R}$ shall be **reversible** *in the sense that*

$$T \in \mathcal{T}(c) : c \mapsto c' \in \mathcal{C}_\mathcal{R} \implies \exists T \in \mathcal{T}(c) \subset \mathcal{T} : c' \mapsto c = T(c') \in \mathcal{C}_\mathcal{R}$$

$\mathcal{T}$ shall **act transitively** on $\mathcal{C}_\mathcal{R}$ *in the sense that for any* $c, c' \in \mathcal{C}_\mathcal{R}$ *there shall exist a sequence of* $T_i \in \mathcal{T}; i = 1, 2, \ldots$ *such that* $c_{i+1} = T_i c_i$ *with* $c_1 = c$ *and* $c_n = c'$.

As the graphs are traversed, they are transformed into unions of simpler and more complex graphs.

Definition 6.15 *It is called* **monotonic** *if* $\sigma \in \Sigma$ *implies that any subgraph of* σ *also belongs to* Σ; *it is then also monatomic.*

For any given connection type Σ, *it can be extended by adding all subgraphs of the connectors it contains, as well as repetitions containing multiple bodies. Define the* **monotonic extension**, *denoted by* $\mathcal{MON}(\Sigma)$, *to be the extension of* Σ *generated by adding all subgraphs of the connectors it contains.*

Define the **multiple extension** *of* Σ, *denoted by* $\mathcal{MULT}(\Sigma)$, *to be the family of graphs generated by unions with a finite and arbitrary number of repetitions, for all* $\sigma_i \in \Sigma$,

$$\sigma_1 \cup \sigma_2 \cup \sigma_3 \cdots . \tag{6.24}$$

Figure 6.7 shows various examples to illustrate transformations which give rise to MON (Σ) and MULT(Σ). Panel 1 shows a monotonic extension transformation; panel 2 shows a multiple extension transformation. A configuration for tracking is a linear graph LINEAR(m_1) of m_1.

Example 6.16 (Multiple Object Discovery) For multiple object recognition where the number of shapes and objects are unknown it is natural to define a *hierarchy* or *nested family* of graphs corresponding to unions of multiple graphs. The generators in the higher level graphs become the single objects themselves.

Begin with the object graphs $\sigma \in$ OBJ = {CYCLIC, LINEAR}. Define m to be the number of objects present in the scene. Extend the graph space to include MULT(m_1, LINEAR) and MULT(m_2, CYCLIC), the disconnected union of m_1, m_2 LINEAR and CYCLIC graphs. Then a scene $\sigma \in$ SCENE is an element of the graph set

$$\text{SCENE} = \cup_{m_1 \geq 0} \text{MULT}(m_1, \text{LINEAR}) \times \cup_{m_2 \geq 0} \text{MULT}(m_2, \text{CYCLIC}) \tag{6.25}$$

consisting of all graphs that are finite disconnected unions of graphs from OBJ. The entire configuration space becomes the union of configuration spaces $\mathcal{C} = \cup_{\sigma \in \text{SCENE}} \mathcal{C}(\sigma)$.

164   6 THE CANONICAL REPRESENTATIONS OF GENERAL PATTERN THEORY

A configuration $c(\text{MULT}(m)) \in \mathcal{C}(\text{MULT}(m))$ associated with an m-object graph is just a collection of single object configurations making up the scene.

The multiple object scenes are synthesized by transformation of new objects in the multiple object graph or dropping already existing objects. The graph transformations become

$$\text{MULT}(m_1) \rightarrow \text{MULT}(m_1 + 1)$$
$$\text{MULT}(m_2) \rightarrow \text{MULT}(m_2 - 1)$$
(6.26)

where the first adds a new object from OBJ, and the second deletes.

Example 6.17 (Parsing and Language Synthesis) Examine standard English parsing using rules from the phrase structure grammar. Then Σ = PHRASE–FOREST and is constructed recursively as the set of all phrases in the language which terminate in a given English sentence and can be generated by parsing the English string. Parsing is the process of transforming graph types $\sigma \rightarrow \sigma'$ so that the parse *pushes towards* a tree rooted in a complete sentence. A set of parse trees resulting from such transformations are depicted in Figure 6.8.

The set Σ =PHRASE–FOREST is generated recursively through the family $T \in \mathcal{T} = \cup_{c \in \mathcal{C}} \mathcal{T}(c)$ of simple moves, beginning with the starting configuration, a sequence of isolated words. This is depicted in the left panel of Figure 6.8 showing the discrete graph corresponding to the three words *The frogs jump*. This is the starting configuration during the parsing process. Simple moves are allowed that add or delete exactly one generator at the boundary $\partial \sigma$ of the connector. From c only configurations in the neighborhood $\mathcal{N}_{\text{add}}(c) = \{c' | Tc = c', T \in \mathcal{T}(c)\} \subset \mathcal{C}$ where

$$\mathcal{N}_{\text{add}}(c) = \text{set of all } c' \text{ such that } c' \text{ is obtained}$$
$$\text{from } c \text{ by adding a new generator.}$$

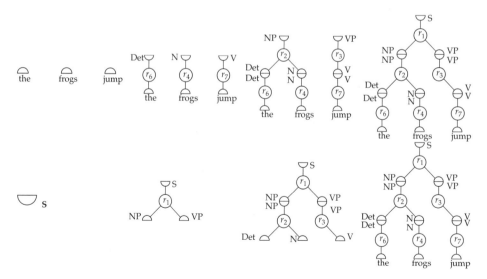

Figure 6.8 Top row shows graphs from Σ = PHRASE-FOREST corresponding to a sequence of graph transformations generated during the parsing process of the sentence *The frogs jump*. Left panel shows the starting configuration; successive panels show parsing transformations from the phrase–structure context-free grammar. Bottom row shows a sequence of transformations through Σ = PHRASE-FOREST to the root node S resulting in the synthesis of the sentence *The frogs jump*.

6.3 GRAPH TRANSFORMATIONS

Notice, during parsing, external$(c) \neq \phi$; upon completion of the parse to the root node external$(c) = \phi$.

Language synthesis involves sampling elements in the language. For this, start with the root node S as the sole configuration. Using identical grammar generators are added to external bonds, with $\mathcal{N}_{add}(c)$ as above. See Figure 6.8 illustrating synthesis of the string in the language *The frogs jump*.

Example 6.18 (Computational Biology) Learning genetic regulatory networks from microarray data using Bayesian Networks is an area of active research in the Computational Biology community. The use of DAGs is an active area of research. Friedman et al. [2000]. use a Bayesian approach for the scoring of learned graphical structures based on the use of a conditional log-posterior and the choice of Dirichlet distributions for the prior. Once an appropriate scoring function has been defined, the space of all possible directed acyclic graph is explored using graph transformations in order to find the structure that best fits the data. For this, simple moves or operations resulting

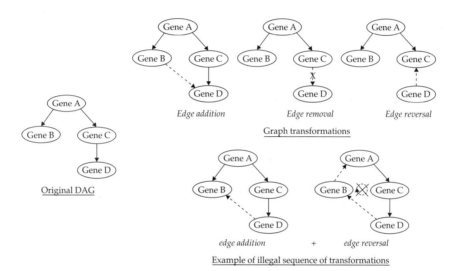

Figure 6.9 Top panel shows the set of elementary graph transformations over a sample DAG. Lower panel depicts a sequence of transformations resulting in an illegal structure.

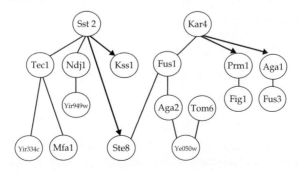

Figure 6.10 Gene subnetwork for mating response in *Saccharomyces cerevisiae*, from [D. Pe'er: From Gene Expression to Molecular Pathways Ph D dissertation Hebrew University]. The widths of the arcs correspond to feature confidence and they are oriented only when there is high confidence in their orientation.

in so-called neighboring structures are used (namely the addition, removal or reversal of an edge, see figure 6.9 for some examples). At each step of the learning process, only operations leading to "legal" structures (i.e. structures with no directed cycles) are permitted.

Friedman et al. [2000] and Pe'er [D. Pe'er: From Gene Expression to Molecular Pathways Ph D dissertation Hebrew University]. have applied these ideas to the study of gene regulatory networks in Yeast (*Saccharomyces cerevisiae*) and they have been able to correctly identify some generally agreed upon relations, as shown in figure 6.10.

6.4 The Canonical Representation of Patterns: DAGs, MRFs, Gaussian Random Fields

In the *pattern theory*, the first job in constructing the solution to a problem is the construction of the generators and the graphs on which they sit. This is representation. It would be nice if for each problem there were a unique definition of generators, bonds and their respective graphs. *This will not be the case*; there are no unique choices in general. There is, however, a *canonical representation* in which the generators are defined from the field variables in such a way as to accommodate the arbitrary nature of cliques in at most a pairwise interaction on the graph. This is analogous to Gaussian fields in which there is at most pairwise interactions of the cliques via the pairwise product in the potentials. The challenge is that the regular structures associated with the variability of Markov random fields on an arbitrary graph can in general have binary, ternary, arbitrary dependencies. The order depends upon the locally supported potential functions and the size of the cliques in the graph. To construct a canonical representation in pairwise interactions only, the generators must aggregate the field variables so that arbitrary dependency structure of the cliques is accommodated. This is a very general idea: essentially that the potentials of the MRFs of regular patterns can always be realized as the products of pairs of generators. To this end an explicit construction is established incorporating all order dependencies reduced to binary interactions by the introduction of the more abstract generators.

Thus far, rigid regularity has been introduced through the bond function. To accommodate variability over the regular configurations we introduce a Gibbs probability $P(\cdot)$ on the configuration space, $\mathcal{C}_\mathcal{R}$, the regular configurations. To place probabilities on the patterns assume that the generator spaces $\mathcal{G}$ are discrete collections making them directly connected to Gibbs distributions. The probabilities are replaced with densities once the generators move to the continuum.

Definition 6.19 *Define the* **local potential functions** *expressing interaction between generators* $\Phi_{ij} : \mathcal{B}_i \times \mathcal{B}_j \to R^+$ *governing the probabilistic behavior of the configurations as strictly finite so that in exponential form* $e^{-\Phi(\beta,\beta')}$.

Then the Gibbs probability of a configuration $P(C)$ restricted to $\mathcal{C}_\mathcal{R}$ for discrete generator spaces $\mathcal{G}$ can be written as the product of potentials over pairwise interactions in the graph:

$$P(c) = \frac{1}{Z} \prod_{e=(i,j) \in E} e^{-\Phi_{ij}(\beta_j(g_i), \beta_i(g_j))}, \qquad (6.27)$$

with Z normalizing to $\mathcal{C}_\mathcal{R}$ so that $\sum_{c \in \mathcal{C}_\mathcal{R}} P(c) = 1$.

It is often useful to highlight Eqn. 6.27 via a modification which distinguishes the coupling between generators and their marginal frequency of occurence in the graph:

$$\frac{1}{Z} \prod_{e=(i,j) \in E} e^{-\Phi_0(\beta_j(g_i), \beta_i(g_j))} \prod_{i=1}^{n(\sigma)} Q(g_i). \qquad (6.28)$$

The role of $e^{-\Phi_0}$ is to control the couplings between generators and the weight function Q governs the frequencies of occurrence of different generators. In many applications, the A, A_0, Q functions are node dependent, depending upon the generator index such as for non-stationary. The above formulas 6.27, 6.28 specify that the random structures on the generators can be constructed to involve at most pairwise interactions on the graph.

For generator spaces $\mathcal{G}$ on the continuum, interpret the left-hand side as a density $p(\cdot)$ with $q(\cdot)$ a density on $\mathcal{G}$:

$$p(c) = \frac{1}{Z} \prod_{e=(i,j) \in E} e^{-\Phi_0(\beta_j(g_i), \beta_i(g_j))} \prod_{i=1}^{n(\sigma)} q(g_i). \tag{6.29}$$

Let us now establish that the question "is binary enough?" can be answered in the affirmative for all of the probabilistic models discussed thus far. The probabilistic models of directed acyclic graphs, Markov random fields, and Gaussian random fields all have a canonical representation via pairwise generator cliques. Our claim is that we can view all of the so far discussed random fields on graphs in a unified manner, in at most pairwise interactions. It is not a matter of principle but of convenience what form is preferred, binary or higher order. This is an old idea well known to systems theorists using state variable descriptions to study ODEs where one can always assume that the order of the equation is one. Any nth order differential equation can be studied as a first order differential equation, simply via expansion of the state space. As we show in the general digraph case, and in particular for Markov chains, this is familiar to the reader as the standard way in which the state space is enlarged to accommodate higher order dependencies. For Markov chains this corresponds to changing the graph structure from n-nearest neighbor to 1-neighbor with the expanded state space (nth power). This is reminiscent of the principle expressed by the great probabilist William Feller that any process is a Markov process with a sufficiently large state space.

6.4.1 Directed Acyclic Graphs

The conditional probability formulas Eqns. 6.27, 6.28 provide a common representation that only involves nearest neighbor interactions. For directed acyclic graphs, the factoring as a product supports a straightforward representation in products of probabilities (independent) with conditional dependence determined through the regularity $\mathcal{R}$ of the graph.

Theorem 6.20 *Let X be a realization from a finite $\mathcal{X}_0$-valued directed acyclic graph $\sigma = \{D, E\}$ with parent system Π with probability*

$$P(X) = \prod_{i=1}^{n(\sigma)} P(X_i | X_{\pi_i}). \tag{6.30}$$

The probability law $P(X)$ written in the pairwise formula Eqn. 6.27 has generators of the form determined by the parent system, $g_i = (X_i, X_{\pi_i})$, with arity $\omega(g_i) = 1 + |\pi_i|$, $\beta_{\text{in}}(g_i) = X_i, \beta_{\text{out}} = X_{\pi_i}, \rho(\beta, \beta') = \text{TRUE}$ defining local regularity.

The conditioned probability on configurations $c(DAG) = DAG(g_1, \ldots, g_n)$ becomes

$$P(c) = \prod_{i=1}^{n(\sigma)} Q(g_i), \quad \text{where } Q(g_i) = P(X_i | X_{\pi_i}). \tag{6.31}$$

Proof *Proof by construction:* Divide the sites of the DAG into classes

$$D = \{1, 2, \ldots, n\} = \cup_{k \geq 0} D_k$$

where the sites in subset $D_k \subset D$ have in-arity k; that is $i \in D_k \Rightarrow |\pi_i| = k$. Define the maximum parent set size $k_{\max} \leq n - 1$. Then the joint probability of the random vector $X = (X_1, X_2, \ldots, X_n)$ can be written as

$$P(X) = \prod_{i_0 \in D_0} P(X_{i_0}) \times \prod_{i_1 \in D_1} P(X_{i_1}|X_{\pi_{i_1}}) \cdots \times \prod_{i_{k_{\max}} \in D_{k_{\max}}} P(X_{i_{k_{\max}}}|X_{\pi_{i_{k_{\max}}}})$$

where the functions $P(\cdot|\cdot)$ depend upon l parents, $\pi_{i_l} = \{j_1, j_2, \ldots, j_l\}$ and the site i. The factorization suggests the form of the generators. If $l > 1$ construct a new generator g carrying the information $X_{i_l}; X_{j_1}, X_{j_2}, \ldots, X_{j_l}$ and associate it with the Q-function:

$$Q(g_i) = P(X_{i_l}|X_{j_1}, X_{j_2}, \ldots, X_{j_l}). \tag{6.32}$$

Consistency requires that if two new generators contain the same X_j add a segment to the graph linking the sites with a bond relation $\rho = $ EQUAL. Also link g with any other site to which one of the xs constituting g was already connected. Then the probability density can be written as in the product structure formula with the new graph and generators.

Notice that the regularity constrained configurations are strictly contained in the full configuration space except for the independent DAG case:

$$\mathcal{C}_\mathcal{R} = \mathcal{X}_0^n \subset \mathcal{C} = \prod_{j=0}^{k_{\max}} (\mathcal{X}_0^{j+1})^{|D_j|}. \tag{6.33}$$

$\square$

Example 6.21 (m-memory Markov Chain) The $\mathcal{X}_0$-valued m-memory Markov chains have canonical representation in the regular configurations $\mathcal{C}_\mathcal{R}$ defined by the $\sigma =$ LINEAR graph structures. Let $X_i, i = 1, 2, \ldots$ be an m-memory Markov chain,

$$P(X_1, \ldots, X_n) = \prod_{i=1}^{n} P(X_i|X_{i-1}, X_{i-2}, \ldots, X_{i-m}), \tag{6.34}$$

with transition law $Q_{(X_{i-m}, \ldots, X_{i-1}), X_i}$.

For the LINEAR graph with two in and out bonds with values $\beta_{\text{in}}(g) = (y_j, j = 1, \ldots, m)$, $\beta_{\text{out}}(g) = (X_j, j = 2, \ldots, m+1)$, with regularity $\mathcal{R}$ if and only if $\beta_{\text{out}}(g_i) = \beta_{\text{in}}(g_{i+1})$, $i = 1, \ldots, n$. Figure 6.11 shows the 1-memory and 2-memory cases, 1-memory having generators $g = (x, y) \in \mathcal{G} = \mathcal{X}_0^2$, and 2-memory $g = (x, y, z) \in \mathcal{G} = \mathcal{X}_0^3$. Regularity implies the in- and out-bond relations are satisfied reducing the space

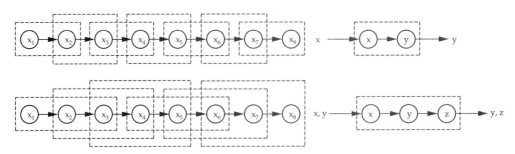

Figure 6.11 Top row shows $m = $ 1-memory Markov chain with generators $g = (x, y)$ the pairs in the boxes and bond values $\beta_{\text{in}}(g) = x$, $\beta_{\text{out}}(g) = y$. Bottom row shows for $m = $ 2-memory, with $g = (x, y, z)$ and $\beta_{\text{in}}(g) = (x, y)$, $\beta_{\text{out}}(g) = (y, z)$.

of allowable strings to $\mathcal{C}_\mathcal{R}(n) = \mathcal{X}_0^n \subset \mathcal{C} = (\mathcal{X}_0^{m+1})^n$. Such a string of generators in $\mathcal{C}_\mathcal{R}$ has probability

$$P(\text{LINEAR}(g_1, \ldots, g_n)) = \prod_{i=1}^{n} Q_{(X_{i-m}, \ldots, X_{i-1}), X_i}. \qquad (6.35)$$

6.4.2 Markov Random Fields

The conditional probability formulas Eqns. 6.27 and 6.28 provide a representation involving pairwise interaction of cliques. This must interface to the Hammersley Clifford result which allows arbitrary interactions.

Theorem 6.22 *Consider a finite $\mathcal{X}_0$-valued Markov random field $\{X_i, i \in D\}$ with respect to graph $\sigma = \{D, E\}$ size $n(\sigma)$, with Gibbs form*

$$P(X) = \frac{e^{-\sum_{C \in \{\text{set of cliques}\}} \Phi_C(x)}}{Z}, \quad x \in \mathcal{X}. \qquad (6.36)$$

Then $P(X)$ is of the form of the pairwise probability formula, Eqn. 6.27 with graph σ', generators $g \in \mathcal{G}$, and acceptors $A(\cdot, \cdot)$ to be defined within the construction below.

Proof Define the sets of cliques of size κ to be $\mathcal{C}_\kappa \subset 2^{n(\sigma)}, \kappa = 1, 2, \ldots n$, with the complete set of cliques $\cup_\kappa \mathcal{C}_\kappa$. The general form of the probability of a configuration becomes

$$P(X) = \frac{1}{Z} \prod_{\kappa=1}^{n} \prod_{\{i_1, i_2, \ldots, i_\kappa\} \in \mathcal{C}_\kappa} e^{-\Phi_{i_1, i_2, \ldots, i_\kappa}(X_i, X_j, \ldots, X_{i_\kappa})}. \qquad (6.37)$$

Now define max $\leq n$ to be the largest clique size in the graph σ. If max ≤ 2, then the interactions are in the pairwise form and nothing need be done. Otherwise, for any clique $(i_1, i_2, \ldots, i_{\max}) \in \mathcal{C}_\kappa$ replace the sites (generators) in the clique by a new generator $g = (X_{i_1}, X_{i_2} \ldots, X_{i_{\max}})$ and connect g with bond value

$$\beta(g) = (X_{i_1}, \ldots, X_{i_{\max}}) \in \mathcal{G}_{\max} = \mathcal{X}_0^{\max} \qquad (6.38)$$

to each site value $i \in D$ that was connected to any of the $X_{i_1}, X_{i_2} \ldots, X_{i_{\max}}$, in the original graph σ. The factors in the clique probabilities are replaced by functions $Q(g_i), i = 1, \ldots, |\mathcal{C}_{\max}|$ depending upon the single (new) generators

$$Q(g_i) = e^{-\Phi_{i_1, i_2, \ldots i_{\max}}(X_{i_1}, X_{i_2}, \ldots, X_{i_{\max}})}, \quad i = 1, \ldots, |\mathcal{C}_{\max}|.$$

Figure 6.12 illustrates the reduction of the various largest cliques of size four and size three. Do this recursively; this is guaranteed to terminate since max $< n$. At this point σ' has largest cliques of size $\leq$ max -1. Now repeat this process replacing max by max -1. This procedure is repeated until all cliques are of size 2 or less.

The resulting graph σ' is binary and will consist of some old and m new generators. The reformulated graph is shown on the right. For the other segments in σ' the associated factor in the product depends upon at most two generators, old or new. Hence the probability of a configuration becomes a mixture of the original random field elements, call them $(X_i, X_j, \ldots, X_{i_m}), i_j \in \sigma$, all entering into the

170   6 THE CANONICAL REPRESENTATIONS OF GENERAL PATTERN THEORY

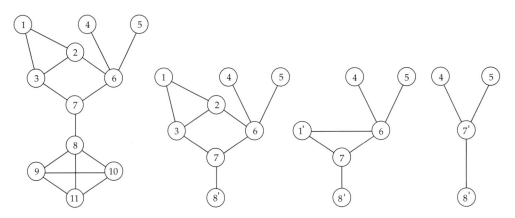

Figure 6.12 Left panel showing various cliques of size three and four transformed to the binary graph case (right panel).

cliques of size $\kappa \leq 2$, as well as the m'-newly defined generators, $g_1, g_2, \ldots, g_{m'}$. Then $c(\sigma') = \sigma'(X_{i_1}, \ldots, X_{i_m}, g_1, \ldots, g_{m'})$ is of a form

$$P(c) = \prod_{i \in \mathcal{C}_1} e^{-\Phi_i(X_i)} \prod_{\{i,i'\} \in \mathcal{C}_2} e^{-\Phi_{(i,i')}(X_i, X_{i'})} \prod_{i=1}^{m'} Q(g_i).$$

Figure 6.12 shows the new graph σ' for the creation of m' new generators. The density has been reformulated into binary form as claimed.   □

There are many ways of reducing the representation to binary form. Another way is to introduce new generators so that each clique consists of at most two new generators. This is more efficient in the sense that the multiplicity of the new generators can be made smaller. Apparently, the question "is binary enough?" can be answered in the affirmative.

6.4.3 Gaussian Random Fields: Generators induced via difference operators

Returning to Gaussian random fields, we see directly that the difference operators specify the generators in the representations. Here the generators and bond values $\mathcal{G}, \mathcal{B}$ are continua for Gaussian random fields, with the conditional probability formulas Eqns. 6.27, 6.28 interpreted as probability densites with respect to Lebesgue measure.

Theorem 6.23 *Given is a Gaussian random field $\{X_i, i \in D \subset \mathbb{Z}^d\}$ satisfying the stochastic difference equation*

$$L_D X_i = W_i \quad \text{with } L_D X_i = \sum_{s \in S} a_s X_{i+s}, \tag{6.39}$$

$S \subset \mathbb{Z}^d$, W a white noise process of independent, identically distributed Gaussian variables, variance σ^2, with L_D with associated boundary conditions, an invertible difference operator. The probability distribution can be represented as pairwise cliques with generators $g_i = \{X_{i+s}, s \in S\}$, $g \in \mathcal{G} = \mathbb{R}^{|S|}$ having bond structure as defined below within the proof and determined by the Minkowski addition set $K = S \ominus S \subset \mathbb{Z}^d$ of Eqn. 5.19 with arity $\omega(g) = |K|$.

6.4 THE CANONICAL REPRESENTATION OF PATTERNS

Proof Define the new generator at site $i \in \mathbb{Z}^d$ as the vector of length $|S|$

$$g_i = (X_{i+s_1}, X_{i+s_2}, \ldots, X_{i+s_{|S|}}),$$

implying that the generator defines the field on a patch of $\mathbb{Z}^d$. With $Q(g_i) = \exp -1/2\tau^2 \left[\sum_{s \in S} a_s X_{i+s} \right]^2$, the desired probability density is obtained as a product of Q's. Consistency is established in the sense that if two patches have a non-empty intersection the X-fields defined by the two generators must coincide on this intersection. The bond structure of a generator is constructed by placing it at the origin of the lattice $\mathbb{Z}^d$ and equipping it with $|K|$ bonds enumerated by $j, j = 1, 2, \ldots |K|$. Define the bonds as the vector

$$\beta_j(g_i) = (X_{i+l}; l \in K \cap \{K + k_j\}), k_j \in K, j = 1,2 \ldots |K|.$$

Two generators, $g_i, g_{i'}$ situated at i, i' are connected by a segment if $i = i' + k$ for some $k = h - h' \in K$.

With the bond relation $\rho = \text{TRUE}$ when $\beta_j(g_i) = \beta_{j'}(g_{i'})$ implies consistency, the X-fields defined on patches for the two generators coincide on the intersection of the patches. $\square$

Example 6.24 (The derivative operator) Examine the non-self-adjoint differential operator $L = (\partial/\partial x_1) + (\partial/\partial x_2)$, with 0-boundary conditions $X_{0,j} = X_{i,0} = 0$ with operator $LX_i = \sum_{s \in S} a_s X_{i+s}$, given in Eqn. 5.43 inducing the random field $\{X_{i,j}, (i,j) \in \mathbb{Z}_{n^2}\}$. Then

$$S = \{(0,0), (0,1)(1,0)\}, \tag{6.40}$$

$$K = S \ominus S = \{(0,-1), (-1,0), (0,1), (-1,1), (1,0), (1,-1)\}. \tag{6.41}$$

Theorem 5.5 dictates the neighborhood structure of the random field interior to the graph as $N_i = i + S \ominus S$. The generators carry information in the form of 3-vectors

$$g_i = X_{i+s}, s \in S = (X_{i,j}, X_{i+1,j}, X_{i,j+1})$$

according to panel 1 in Figure 6.13 where the dashed triangle represents a generator. Then with

$$Q(g_i) = e^{-\frac{1}{2\sigma^2}(X_{i+1,j} - (2-c)X_{i,j} + X_{i,j+1})^2}$$

the probability density is obtained as the product $\prod_i Q(g_i)$. To enforce consistency, choose the $|K| = 6$ bond values as illustrated in the right panel of Figure 6.13, the bonds given by

$$\beta_1(g_i) = X_{i,j}; \beta_2(g) = X_{i+1,j}; \beta_3(g) = X_{i+1,j}; \beta_4(g) = X_{i,j+1}; \beta_5(g) = X_{i,j+1};$$
$$\beta_6(g) = X_{i,j}.$$

The bond relation will be EQUAL as consistency demands that the 6th bond of $g_{i+1,j}$ is equal to the 3rd bond of $g_{i,j}$, so that $\beta_3(g_{i,j}) = \beta_6(g_{i+1,j})$, and so on for the other relations between bonds.

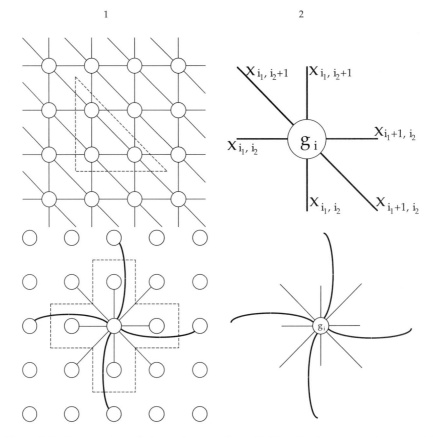

Figure 6.13 Top row: Panel 1 shows the grouping of lattice sites into generators; panel 2 shows the bond values for generator $g_{i,j} = (X_{i,j}, X_{i+1,j}, X_{i,j+1})$ for the derivative operator. Bottom row shows the same for the Laplacian operator.

Example 6.25 (The Laplacian operator) Examine the Laplacian operator $L = -\Delta + a = -\partial^2/\partial x_1^2 - \partial^2/\partial x_2^2 + a$, with 0-boundary according to Eqn. 5.49. Then

$$S = \{(1,0), (-1,0), (0,0), (0,1), (0,-1)\},$$
$$K = S \ominus S = \{(-1,0), (-2,0), (1,0), (2,0), (-1,-1), (-1,1) \quad (6.42)$$
$$(0,-1), (0,-2), (0,1), (0,2), (-1,1), (1,-1)\},$$

with generators $g = \{X_{i+s}, s \in S\}$ given by

$$g_{i_1,i_2} = (X_{i_1,i_2}, X_{i_1-1,i_2}, X_{i_1+1,i_2}, X_{i_1,i_2+1}, X_{i_1,i_2-1}), \quad (6.43)$$

and $\omega(g) = |K| = 12$ the number of bonds. Use a generator coordinate $i = (i_1, i_2)$ and a bond coordinate $j = 1, 2, 3, \ldots, 12$ which will be treated modulo 4 enumerating the directions S,E,N,W. Let

$$L = -\Delta + a$$

with the discrete 4-neighbor Laplacian Δ. Introduce generators of the form

$$g_i = (x_i, y_i^0, y_i^1, y_i^2, y_i^3) \in \mathbb{R}^5,$$

6.4 THE CANONICAL REPRESENTATION OF PATTERNS

and with arity $\omega(g_i) = 4$. Associate the bond values

$$\beta_j(g_i) = (x_i, y_i^j; j = 0, 1, 2, 3).$$

Consider two neighboring generators $g_i; i = (i_1, i_2)$ and $g_{i'}; i' = (i'_1, i'_2)$, for example $i' = i+1, j = j'$; the other three possibilities are treated in the same way. Then the bond relation ρ is of the following form:

$$\rho(\beta_1(g_i), \beta_3(g'_i)) = \{y_{i'}^3 = x_i\} \wedge \{y_i^1 = x_{i'}\}. \tag{6.44}$$

Then with

$$Q(g_i) = e^{-|(1+a)x_i - \frac{1}{4}(y_i^0 + y_i^1 + y_i^2 + y_i^3)|^2}, \tag{6.45}$$

the probability density obtained is $1/Z \prod_i Q(g_i)$ which is a Gaussian process.

7 MATRIX GROUP ACTIONS TRANSFORMING PATTERNS

ABSTRACT Thus far Pattern theory has been combinatore constructing complex patterns by connecting simpler ones via graphs. Patterns typically occurring in nature may be extremely complex and exhibit invariances. For example, spatial patterns may live in a space where the choice of coordinate system is irrelevant; temporal patterns may exist independently of where time is counted from, and so on. For this matrix groups as transformations are introduced, these transformations often forming groups which act on the generators.

7.1 Groups Transforming Configurations

Pattern theory is transformational. One of the most fundamental transformation types for studying shape is defined via groups and group actions acting on the generators, providing a vehicle for representing the natural invariances of the real world. The fundamental role of groups as transformations on a background space what we term the generators is familiar and is at the very core of much of the geometry which is familiar to us. Even though the concept of groups did not appear in Euclid's original axioms, congruence did. For Euclid, congruence was defined through the equivalence defined by the rigid motions carrying one figure into another. Klein [147] discovered the central role of groups in all the classical geometries (see Boothby [148]). The *Erlangen* program was essentially that. Geometry becomes the study of the groups which leave geometric properties invariant. The emphasis is transferred from the things being acted upon to the things doing the action. This is familiar to the Kolmogoroff complexity model. With each geometry is associated a group of transformations on the background space in which the properties of the geometry and its theorems are invariant. In Euclid's geometry the subgroup of rigid motions of the affine group leaves distances invariant and preserves the congruence relation. Two figures are congruent if they are equal modulo an element of the rigid motion group.

7.1.1 Similarity Groups

In Pattern theory, patterns or shapes are understood through *exemplars* or *representers* of *equivalence classes*. Elements in the *equivalence classes* will often be called *congruent*, with the examplar congruent to all elements of its congruence class. The formal notion of pattern is based on congruence modulo the similarity group. This section follows Artin [149] and Boothby [148].

> **Definition 7.1** *A* **group** *is a set S together with a law of co(S, ∘)mposition $\circ : S \times S \to S$ which is associative and has an identity element $e \in S$ and such that every element $s \in S$ has an inverse element $s^{-1} \in S$ satisfying $s \circ s^{-1} = e$.*
>
> *A subset H of a group $S, H \subset S$, is called a* **subgroup** *if*
>
> 1. *it is closed: $a_\circ \in H, b \in H \implies a \circ b \in H$,*
> 2. *it has an identity $e \in H$ and*
> 3. *the inverse is in H: $a \in H \implies a^{-1} \in H$.*

As a notational shorthand, a group will be represented by a set, leaving the law of composition implicit, when it can be done so without ambiguity. Subgroups will be important. *Direct products* of groups will allow for increasing dimensions.

Definition 7.2 *The* **direct product** *group*

$$(\mathcal{S}, \circ) = (\mathcal{S}_1, \circ_1) \times (\mathcal{S}_2, \circ_2) \times \cdots \times (\mathcal{S}_n, \circ_n) \tag{7.1}$$

has law of composition $s = (s_1, \ldots, s_n), s' = (s'_1, \ldots, s'_n) \in \mathcal{S}_1 \times \mathcal{S}_2 \times \cdots \times \mathcal{S}_n$ *given by*

$$s \circ s' = (s_1 \circ_1 s'_1, \ldots, s_n \circ_n s'_n) \tag{7.2}$$

with $\circ_i$ the law of composition for group $\mathcal{S}_i$.

It will be helpful to partition groups according to their subgroups and associated cosets.

Definition 7.3 *A* **left (right) coset** *of subgroup $H \subset \mathcal{S}$ is a subset of the form*

$$aH = \{ah | h \in H\} \quad (Ha = \{ha | h \in H\}). \tag{7.3}$$

Example 7.4 (Familiar groups and subgroups.) Groups and subgroups are very familiar.

1. The group $(\mathbb{Z}, +), (\mathbb{R}, +), (\mathbb{C}, +)$, the integers, reals and complex numbers with addition, $e = 0$. A subgroup of the integers is $H = m\mathbb{Z}$, the subset of integers of multiples of m, $m\mathbb{Z} = \{n \in (\mathbb{Z}, +) : n = mk, k \in (\mathbb{Z}, +)\}$.

2. The groups $(\mathbb{R} - \{0\}, \times), (\mathbb{C} - \{0\}, \times)$, the nonzero real and complex numbers, with multiplication, $e = 1$. Obvious subgroups include $H = \{c \in \mathbb{C}^\times : |c| = 1\}$, the points on the unit circle in the complex plane, or the discrete subgroup of equally spaced points on the unit circle $H = \{e^{i(2\pi k/n)}, k = 0, 1, \ldots, n-1\}$.

Unlike the matrix groups we study below, all these commute giving them the special name of **Abelian groups**.

7.1.2 Group Actions Defining Equivalence

Congruence or sameness is an equivalence relation.

Definition 7.5 *Let X be a set. An equivalence relation on X is a relation which holds between certain elements of X, written as $x \sim y$, having the properties*

1. *transitivity: if $x \sim y$ and $y \sim z$ then $x \sim z$,*
2. *symmetric: if $x \sim y$ then $y \sim x$ and*
3. *reflexive: $x \sim x$ for all $x \in X$.*

Equivalence classes, i.e. subsets of X which contain equivalent elements, form a disjoint covering, or **partition** of X. This is an important property which we now prove.

Theorem 7.6 *Let $\sim$ be an equivalence relation on set X. Then the equivalence classes partition X.*

Proof For $x \in X$, let $[x]_\sim = \{y \in X : y \sim x\}$ denote the equivalence class containing a. By the reflexive property $\forall x \in X, x \sim x$, so $x \in [x]_\sim$. Thus, the family or collection of equivalence classes covers X. We need only show that two equivalence classes are disjoint or equal. Suppose we have two equivalence classes identified by $[x]_\sim$ and $[y]_\sim$, respectively, and that there exists a shared element $z \in [x]_\sim, z \in [y]_\sim$. By definition, $x \sim z$ and $y \sim z$ so by commutativity $z \sim y$, and by transitivity, $x \sim y$. Now suppose $w \in [x]_\sim$, then by the definition of equivalence class, $w \sim x$, but by transitivity $w \sim y$, so $w \in [y]_\sim$. Thus, $[y]_\sim$ contains $[x]_\sim$. Similarly, $[x]_\sim$ contains $[y]_\sim$. So, $[x]_\sim$ and $[y]_\sim$ are equal, i.e. they represent the same equivalence class. □

A classic example are the EVEN,ODD integers, represented by $\bar{0}, \bar{1}$, equivalent mod2. Through the group action on sets, particularly beautiful equivalence relationships emerge.

Definition 7.7 *Let S be a group with group operation $\circ$, and X be a set. Then define a **group action** Φ on X which is a mapping $\Phi : S \times X \to X$, $\Phi(s, x) = s \cdot x, s \in S, x \in X$, with the properties that*

1. *if e is the identity element of S then*

$$\Phi(e, x) = x \quad \forall x \in X, \tag{7.4}$$

2. *if $s_1, s_2 \in S$ then the associative law holds according to*

$$\Phi(s_1, \Phi(s_2, x)) = \Phi(s_1 \circ s_2, x). \tag{7.5}$$

A shorthand notation used throughout for the group action will be $\Phi(s, x) = s \cdot x, s \in S, x \in X$. This is extremely familiar for matrices.

The group action S allows the decomposition of the set X into its *orbits*.

Definition 7.8 *The **orbit** $Sx \subset X, x \in X$ of the group S is just the set of all images of x under arbitrary group action of $s \in S$:*

$$Sx = \{y \in X : y = sx \text{ for some } s \in S\} \tag{7.6}$$

Theorem 7.9 *Let S denote a group, X a set and $\Phi : S \times X \to X$ a group action. Define the relation*

$$x \sim y \quad \text{if } \exists s \in S : \Phi(s, x) = y. \tag{7.7}$$

With the equivalence classes under this relation denoting $[x]_S$, the orbits are the equivalence classes, $[x]_S = Sx$, and the orbits $\{Sx\}$ of the group S partition X.

Proof We need to show that this is an equivalence relation. First, $x \sim x$ since $x = ex, e \in S$ the identity. $x \sim y$ implies $y \sim x$ (reflexivity) according to $x \sim y \Rightarrow y = sx$ implying $x = s^{-1}y, s^{-1} \in S$. Finally, $x \sim y, y \sim z \Rightarrow y = sx, z = s'y$ giving $z = (s's)x, (s's) \in S \Rightarrow x \sim z$.

That the equivalence classes and orbits are equal, $[x]_S = Sx$ follows since $x \sim y$ implies $y = sx$ and thus $y \in Sx$ giving $[x]_S \subset Sx$. Conversely, $y \in Sx$ implies $x \sim y$ so $Sx \subset [x]_S$. □

Definition 7.10 *Define the **set of equivalence classes** as X/S called the orbits of the action.*

In pattern theory, the set of equivalence classes X/S will arise often, S expressing the natural invariances in which the pattern is viewed.

Certain sets are essentially the same from the point of view of the group actions, i.e. if we look for example at X/S there is but one equivalence class $X = [x]_S$. This is formalized through the notion transitive action and homogeneous spaces.

Definition 7.11 *Let $\Phi : S \times X \to X$ be a group action. Then Φ is **transitive** if for all $x, y \in X$ there exists $s \in S$ such that $\Phi(s, x) = y$.*

*In turn, X is said to be a **homogeneous space** of the group S if there exists a transitive group action on X.*

Example 7.12 Equivalence relations are fundamental to many things that we do in image analysis. Examine an oriented and piecewise linear closed curve in the plane. Assume that it does not intersect itself so that Jordan's curve theorem dictates that it divides the plane into two parts, an inside and an outside. Define $set(c) = $ *inside of the curve* and an equivalence relation $\sim$ by

$$c \sim c' \text{ implies } set(c) = set(c'). \tag{7.8}$$

This defines the equivalence relation, with $[c]_\sim$ the orbit representing the image.

Example 7.13 (Null Space of a Linear Operator) Let $A : \mathcal{H} \to \mathcal{H}$, A a linear operator $A : x \in \mathcal{H} \mapsto Ax$, $\mathcal{H}$ a Hilbert space, either finite or infinite dimensional. Two elements x, x' can be identified via the identification rule if $x = x' + null$, where $null$ is an element from the null space of A.

In particular, let $A : \mathbb{R}^n \to \mathbb{R}^n$, with $A = \sum_{i=1}^{m} \alpha_i \phi_i \langle \phi_i, \cdot \rangle_{\mathbb{R}^n}$, with $m < n$ and $\{\phi_i\}_{i=1}^{n}$ a C.O.N. basis for $\mathbb{R}^n$. Then $x \sim x'$ if and only if $Ax = Ax'$. Divide $\mathbb{R}^n$ into the null space of the operator and its orthogonal complement, $\mathbb{R}^n = \mathcal{N} \oplus \mathcal{N}^\perp$. To construct a disjoint partition via equivalence classes define $x^\perp = \sum_{i=1}^{m} \phi_i \langle \phi_i, x \rangle_{\mathbb{R}^n}$, and $x^\perp$ is the projection onto the orthogonal complement of the null space, $x^\perp \in \mathcal{N}^\perp$. The equivalence classes or images $= [x]_\sim$ become

$$[x]_\sim = x^\perp \oplus \mathcal{N} = \left\{ x' \in \mathbb{R}^n : x' = x^\perp + \sum_{i=m+1}^{n} \alpha_i \phi_i, \ (\alpha_{m+1}, \ldots, \alpha_n) \in \mathbb{R}^{n-m} \right\}.$$

7.1.3 Groups Actions on Generators and Deformable Templates

The patterns and shapes will be represented via the *exemplars* or *generators*; the natural invariances are expressed through the groups acting on the generators. Equivalence being defined through the orbits and partition of the pattern space into the orbits.

For this, let the similarity group be of some fixed dimension acting on the generator space. We shall use the shorthand notation to define the group action $\Phi(s, g) = sg$.

Definition 7.14 *The **group action on a single generator** $\Phi : \mathcal{S} \times \mathcal{G} \to \mathcal{G}$ is defined as*

$$\Phi(s, \Phi(s', g)) = \Phi(s \circ s', g) = (s \circ s', g). \tag{7.9}$$

7.2 The Matrix Groups

Shapes and structures will be studied using the low-dimensional matrix groups translations, rotations, rigid motions of translations and rotations acting on the finite dimensional $X = \mathbb{R}^d$ background spaces.

7.2.1 Linear Matrix and Affine Groups of Transformation

For the geometry of shape, the finite dimensional groups and their subgroups generated from the group of matrices, the *generalized linear group* is used throughout.

Define the matrix groups explicitly as follows.

Definition 7.15 (Generalized Linear Group) *The $d \times d$ **general linear group**, denoted as $\mathbf{GL}(d)$, is the group of all $d \times d$ matrices*

$$\mathbf{GL}(d) = \{d \times d \text{ real matrices } A \text{ with } \det A \neq 0\}, \qquad (7.10)$$

with non-zero determinant (invertible) and with law of composition matrix multiplication,

$$A \circ B = AB = \left(\sum_j A_{ij} B_{jk} \right), \qquad (7.11)$$

with the identity element $I = \operatorname{diag}[1, \ldots, 1]$.

That this is a group follows from the fact that the identity $I \in \mathbf{GL}(d)$, the product of two matrices, is in the group $A \circ B = AB \in \mathbf{GL}(d)$ and the inverse is in the group as well $A^{-1} \in \mathbf{GL}(d)$.

Subgroups of $\mathbf{GL}(d)$ are used as well.

1. *Define the **special linear group** $\mathbf{SL}(d) \subset \mathbf{GL}(d)$ to be the subgroup of volume preserving transformations:*

$$\mathbf{SL}(d) = \{A \in \mathbf{GL}(d) : \det A = 1\}. \qquad (7.12)$$

2. *Define the **orthogonal group** $\mathbf{O}(d) \subset \mathbf{GL}(d)$ to be the orthogonal subgroup of matrices and $\mathbf{SO}(d) \subset \mathbf{O}(d) \subset \mathbf{GL}(d)$ to be the special orthogonal subgroup with determinant 1:*

$$\mathbf{O}(d) = \{A \in \mathbf{GL}(d) : A^* A = I\}, \qquad (7.13)$$

$$\mathbf{SO}(d) = \{A \in \mathbf{O}(d) : \det A = 1\} = \mathbf{O}(d) \cap \mathbf{SL}(d). \qquad (7.14)$$

3. *Define the **uniform scale group** $\mathbf{US}(d) \subset \mathbf{GL}(d)$ of diagonal matrices:*

$$\mathbf{US}(d) = \{A \in \mathbf{GL}(d) : A = \rho I, \rho > 0\}. \qquad (7.15)$$

Notice, the group operation for $\mathbf{GL}(d)$ does not commute. There are various groups which are generated as products of the subgroups.

Definition 7.16 *The **affine group** $\mathbf{A}(d)$ is the semi-direct product of groups $\mathbf{GL}(d) \otimes \mathbb{R}^d$ with elements $\{(A, a) : A \in \mathbf{GL}(d), a \in \mathbb{R}^d\}$ and law of composition semi-direct product*

$$\mathbf{A}(d) = \mathbf{GL}(d) \otimes \mathbb{R}^d, \quad \text{with} \quad (A, a) \circ (B, b) = (AB, Ab + a). \qquad (7.16)$$

*The **Euclidean** and **special Euclidean** groups $\mathbf{E}(d)$, $\mathbf{SE}(d)$, respectively, are the subgroups of the Affine Group consisting of the rigid motions generated from orthogonal and special orthogonal $\mathbf{O}(d), \mathbf{SO}(d)$ motions with group operation, the semi-direct product:*

$$\mathbf{E}(d) = \mathbf{O}(d) \otimes \mathbb{R}^d, \quad \mathbf{SE}(d) = \mathbf{SO}(d) \otimes \mathbb{R}^d. \qquad (7.17)$$

*The **similitudes** are given by the direct product of uniform scale with orthogonal motions*

$$\mathbf{Sim}(d) = \mathbf{US}(d) \times \mathbf{SO}(d). \qquad (7.18)$$

7.2.2 Matrix groups acting on $\mathbb{R}^d$

When the matrix groups act on the infinite background space $X = \mathbb{R}^d$ according to the usual convention of identifying points $x \in X = \mathbb{R}^d$ with column vectors, the action is matrix multiplication on the vectors.

Definition 7.17 *Let X be the background space and the **affine group** $A \in \mathbf{GL}(d)$ acts on the background space according to*

$$\Phi(A,x) = Ax = \begin{pmatrix} \sum_j A_{1j}x_j \\ \vdots \\ \sum_j A_{dj}x_j \end{pmatrix} \in X. \tag{7.19}$$

Let $(A,a) \in \mathcal{S} \subset \mathbf{A}(d)$, then

$$\Phi((A,a),x) = Ax + a. \tag{7.20}$$

*For **homogeneous coordinates**, then represent*

$$\bar{A} = \begin{pmatrix} A & a \\ 0 & 1 \end{pmatrix} \in \bar{\mathbf{A}}, \quad A \in \mathbf{GL}(d),\ a \in \mathbb{R}^d, \quad \bar{x} = \begin{pmatrix} x \\ 1 \end{pmatrix}, \tag{7.21}$$

with group action $\Phi(\bar{A}, \bar{x}) = \bar{A}\bar{x}$.

We emphasize that the semi-direct product of Eqn. (7.16) matches $\Phi((A,a),x)$ of Eqn. 7.20 a group action. We leave it to the reader to verify this.

Example 7.18 (Polar Decomposition) Polar decomposition is useful for clarifying the degrees of freedom inherent in the class of operators. Any invertible matrix $A \in \mathbf{GL}(d)$ can be written as

$$A = PO = \underbrace{O'^{*}\mathrm{diag}(\lambda_1,\ldots,\lambda_d)O'}_{P}\ O, \tag{7.22}$$

P being a positive definite symmetric matrix, with $O, O' \in \mathbf{O}(d)$ the orthogonal matrices. With that said, rewrite elements of $\mathbf{GL}(2)$ to have 4 parameters, noticing $O, O' \in \mathbf{SO}(2)$ have one each, with two for the diagonal entries. Similarly elements of $\mathbf{GL}(3)$ have 9 free paramaters, 3 each for the orthogonal groups and 3 diagonal entries.

Example 7.19 (Tracking via Euclidean motions) To accomodate arbitrary position and pose of objects appearing in a scene introduce the subgroups of the generalized linear group $\mathbf{GL}(3) : X \leftrightarrow X$. The generators $\mathcal{G}$ are the CAD models of various types as depicted in Figure 7.1 showing 3D renderings of sample templates. In this case each template consists of a set of polygonal patches covering the surface, the material description (texture and reflectivity), and surface colors.

The Euclidean group $\mathbf{E}$ including rigid translation and rotation operate on the generators 2D surface manifolds in $\mathbb{R}^3$. The Euclidean group of rigid motions denoted $\mathbf{E}(n) = \mathbf{O}(n) \otimes \mathbb{R}^n$, where $O \in \mathbf{O}(n)$ are $n \times n$ orthogonal matrices; the group action on $X = \mathbb{R}^n$ is $\Phi((O,a),\cdot) : \mathbf{O}(n) \otimes \mathbb{R}^n \to \mathbb{R}^n$ is $\Phi((O,a),x) = Ox + a$, with the law of composition given by the semi-direct product: That Euclidean distance is preserved $d(\Phi((O,a),x), \Phi((O,a),y)) = d(x,y)$ where $d(x,y) = \sum_{i=1}^{n} |x_i - y_i|^2$ for rotation around the origin followed by translation is clear, hence the name rigid motions.

Example 7.20 (Flip Group for Symmetry) Biological shapes exhibit often symmetries. The right panel of Figure 7.1 illustrates symmetry seen in the human brain. The *cosets of*

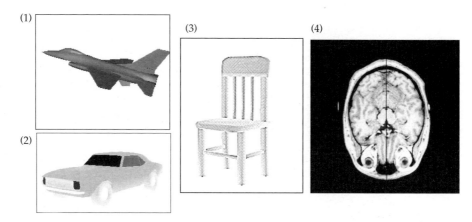

Figure 7.1 The panels 1–3 show various CAD models for objects. Panel 4 shows the human brain section depicting symmetry.

the orthogonal group defined by the flips which are appropriate for studying symmetry, $\mathbf{D}(2,3) = \{R, I\} \subset \mathbf{O}(2,3)$, given by the $2 \times 2, 3 \times 3$ matrices

$$\mathbf{D} = \left\{ I = \begin{pmatrix} 1 & 0 \\ 0 & 1 \end{pmatrix}, \quad R = \begin{pmatrix} -1 & 0 \\ 0 & 1 \end{pmatrix} \right\}, \tag{7.23}$$

$$\mathbf{D} = \left\{ I = \begin{pmatrix} 1 & 0 & 0 \\ 0 & 1 & 0 \\ 0 & 0 & 1 \end{pmatrix}, \quad R = \begin{pmatrix} -1 & 0 & 0 \\ 0 & 1 & 0 \\ 0 & 0 & 1 \end{pmatrix} \right\}. \tag{7.24}$$

with the cosets $I\{\mathbf{SO}(2,3), R\mathbf{SO}(2)(3)\}$.

Example 7.21 (Similar and Congruent Triangles) *Congruence:* Let $\mathcal{S} = \mathbf{E}(2)$ be the set of Euclidean rigid motions in the plane, X the set of points, lines, or triangles in the plane. For $X =$ TRIANGLES, then $[\triangle]_\mathcal{S}$ is the subset of TRIANGLES congruent to $\triangle$. Let a nondegenerate $\triangle = (v_1, v_2, v_3) \in \mathbb{R}^6$ be identified with a 2×3 matrix constructed from the vertices, the triangle nondegenerate so that $\triangle$ is a rank 2 matrix. Position the triangle $\triangle_0$ at the origin so that $v_1 = (0,0) \in \mathbb{R}^2$.

Let $\mathcal{G} =$ TRIANGLES $\subset \mathbb{R}^6$ be generated by the affine motions on $\triangle_0$:

$$\text{TRIANGLES} = \{(A, a)\triangle_0, (A, a) \in \mathbf{A}(2)\}, \tag{7.25}$$

where the group action applied to the 2×3 matrix generators becomes

$$(A, a)\triangle = A\triangle + \begin{pmatrix} a_1 & a_1 & a_1 \\ a_2 & a_2 & a_2 \end{pmatrix}. \tag{7.26}$$

Define equivalence classes by the orbits of $\mathbf{E}$, $[\triangle_0]_\mathbf{E}$. Then two triangles $\triangle_1 \sim \triangle_2$ if both are in the same orbit: $\triangle_2 = s\triangle_1$ for some $s \in \mathbf{E}$. Since orbits form disjoint partitions we have essentially established the Euclidean partition of TRIANGLES into congruence classes.

Notice in this example a triangle $\triangle$ is not a picture but rather a point in the manifold of 2×3 matrices.

Example 7.22 (Similar Triangles) Now return to Euclid's more general notion of similarity and notice that the subgroups are normal allowing us to use quotient group ideas. Examine the orbit of $\triangle_0$ under the scale-rotation group matrices of the type

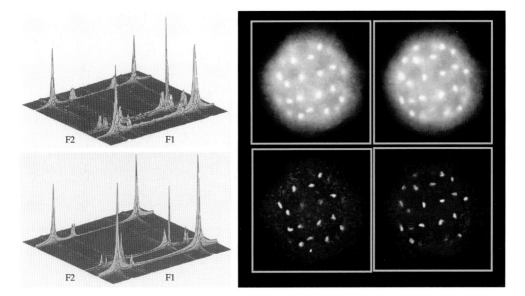

Plate 1 See page 33.

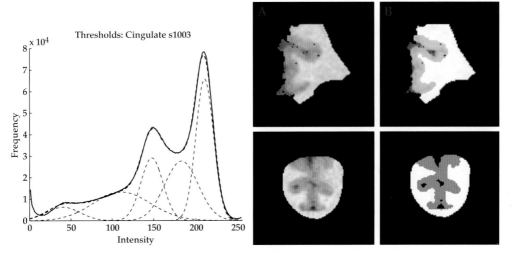

Plate 2 See page 37.

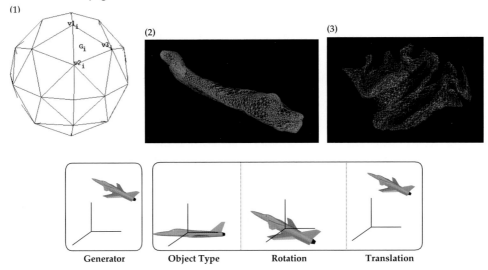

Plate 3 See page 157.

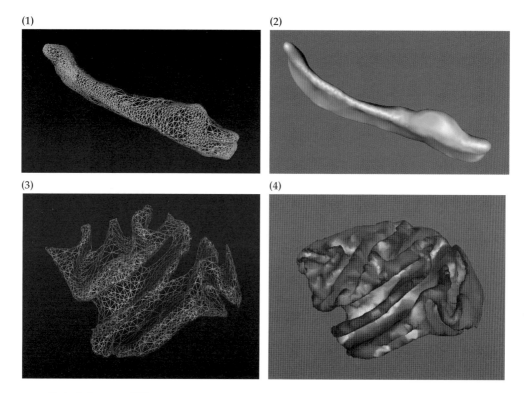

Plate 4 See page 203.

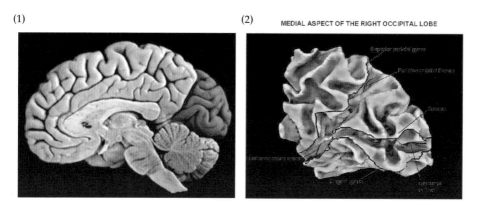

Plate 5 See page 204.

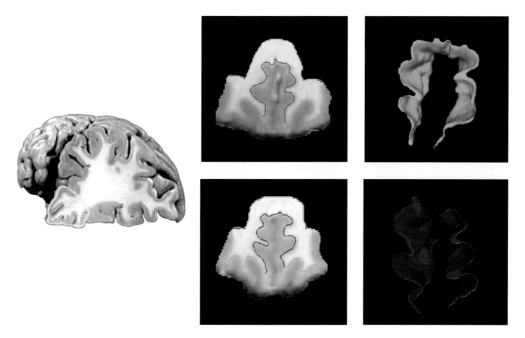

Plate 6 See page 204.

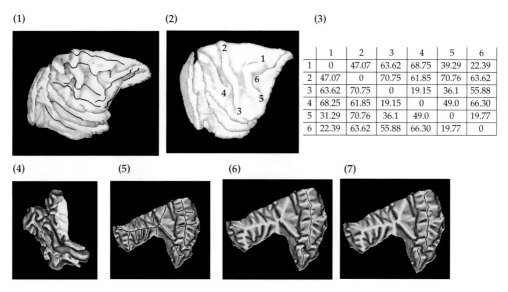

Plate 7 See page 209.

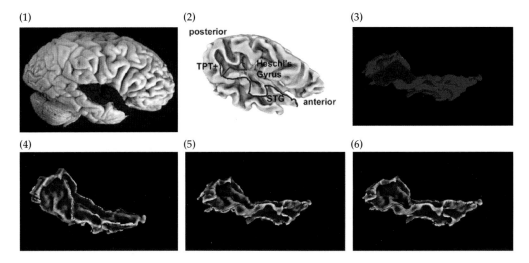

Plate 8 See page 209.

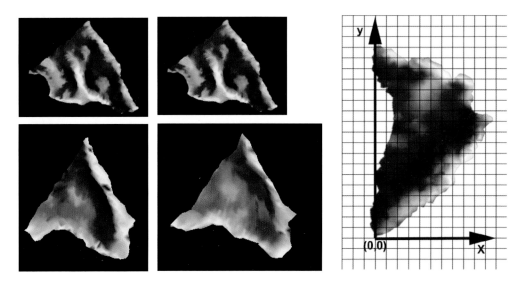

Plate 9 See page 211.

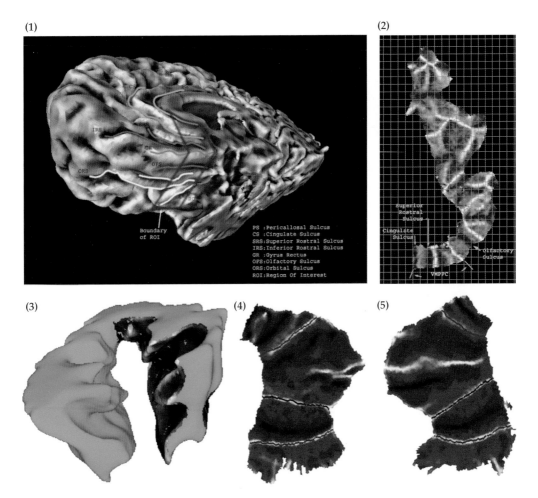

Plate 10 See page 212.

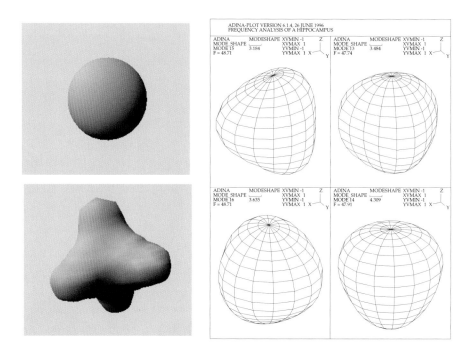

Plate 11 See page 228.

Plate 12 See page 239.

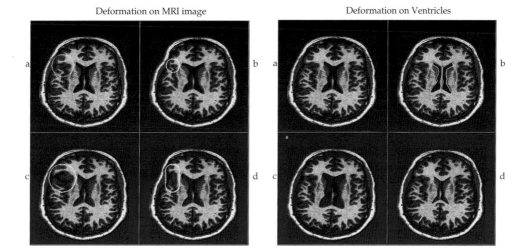

Plate 13 See page 281.

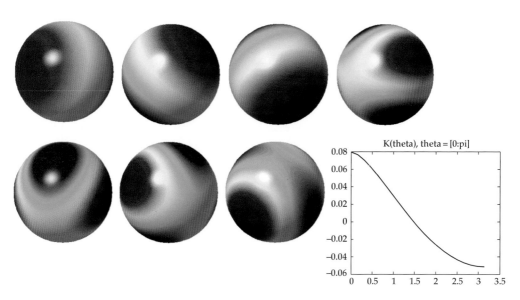

Plate 14 See page 287.

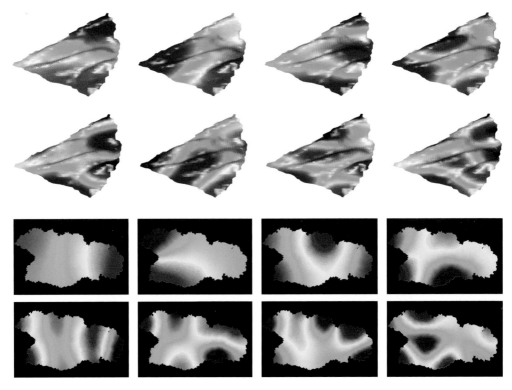

Plate 15 See page 297.

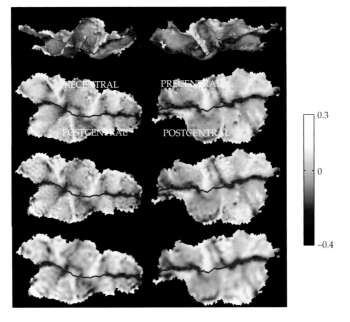

Plate 16 See page 299.

$s = \begin{pmatrix} u & v \\ -v & u \end{pmatrix}, (u,v) \neq (0,0)$. Then, both **SO**(2), **US**(2) are normal subgroups implying, for example, **SR**(2)/**SO**(2) is the quotient group. This is the union of all congruence classes (left cosets of **SO**(2)) in the sense of Euclid:

$$[\triangle_o]_{\mathbf{SR}(2)/\mathbf{SO}(2)} = \{s\triangle_o : s \in \rho\mathbf{SO}(2), \rho > 0\}. \tag{7.27}$$

These are the congruence classes of triangles rooted at the origin. Each class is similar to each other in the extended sense of Euclid but not congruent.

Example 7.23 (Target and Object Symmetry) Examine target and object symmetry, the generators $\mathcal{G}$ the targets. Man-made objects are characterized by well defined shapes with symmetries. The standard notion of symmetry of an object is studied through the subgroup for the object $\alpha \in \mathcal{A}$,

$$\mathcal{S}(\alpha) = \{s \in \mathcal{S} : sg^\alpha = g^\alpha\} \subset \mathcal{S}.$$

Due to this equivalence, the inference set has to be reduced to the quotient space $\mathcal{S}/\mathcal{S}(\alpha)$. For this define the identification rule corresponding to the set of symmetries:

$$[g]_\sim = \{g' : g' = sg, s \in \mathcal{S}(g)\}. \tag{7.28}$$

Let $\mathcal{S} = \mathbf{SE}(3)$ the special Euclidean group, for $\alpha =$ cylinder, then $\mathcal{S}(\text{cylinder}) = \mathbf{SO}(2)$ and the inference is performed over the 2-sphere, $\mathbf{SO}(3)/\mathbf{SO}(2) = S^2$.

If on the other hand more structure is added to the cylinder such as for a missile which also has four tail fins separated by the angle 90° then the relative symmetry set instead includes the discrete 4-torus.

7.3 Transformations Constructed from Products of Groups

Since complex patterns are often constructed by combining simpler ones, it will be necessary to work with the products of groups, each one applied to a single generator in the scene, thereby representing the variability of multiple subconfigurations. To extend the domain of the group transformations to the full configurations combining multiple generators, define the product group acting on the multiple generators making up the configurations. We now generalize the transformations so that they can be applied locally throughout the space required. The product groups are extended to act on the regular configurations $c(\sigma) = \sigma(g_1, \ldots, g_n) \in \mathcal{C}_\mathcal{R}$ in the obvious way. The graph type σ will remain unaffected by the transformations; however, the configuration may not be regular after the similarities $\mathcal{S} = \mathcal{S}_0^n$ act on them.

For this, let the similarity group be of dimension d acting on the ith generator $\mathcal{S}_0 : \mathcal{G}_0 \leftrightarrow \mathcal{G}_0$, with the product $\mathcal{S} = \mathcal{S}_0 \times \mathcal{S}_0 \times \cdots$ of dimension d times the number of generators.

Definition 7.24 The **group action on multiple generators** $\Phi^n : \mathcal{S}_0^n \times \mathcal{G}^n \to \mathcal{G}^n$ becomes

$$\Phi^n((s_1,s_2,\ldots), \Phi^n((s_1',s_2',\ldots),(g_1,g_2,\ldots))) = ((s_1 \circ s_1')g_1, (s_2 \circ s_2')g_2, \ldots). \tag{7.29}$$

Define the **action on the configurations** $\mathcal{S}_0^n : \mathcal{C}_\mathcal{R} \to \mathcal{C}$ as

$$s\sigma(g_1, g_2, \ldots) = \sigma(s_1 g_1, s_2 g_2, \ldots,). \tag{7.30}$$

Denote a particular regular configuration within the configuration space as the **template**, one fixed element in the orbit:

$$c^0 = \sigma(g_1^0, g_2^0, \ldots, g_n^0). \tag{7.31}$$

The **deformable template** becomes the orbit $[c^0]_\mathcal{S}$.

The significant extension being made here is to accommodate the glueing together of the structured generating sets, with the transformations acting locally upon them. Thus far the group transformations have acted globally on the entire generating background space. This presents several significant difficulties for the pattern theory; clearly with multiple groups acting regular configurations may not be regular. Examine the application to the representation of a semi-rigid chair swiveling around its base. Imagine that a translation and axis-fixed rotation group is applied globally via the Euclidean matrix $\mathbf{E}(2)$. However, the chair should rotate independently around the same axis fixed orientation. Should the cartesian product $\mathbf{E}(2) \times \mathbf{E}(2)$ unconstrained be applied to the rigid substructure? Apparently not, *otherwise the chair would tear apart*. It is natural to wonder why this difficulty does not arise in the application of single group transformations. For the matrix groups acting globally local structure is mapped with its topology preserved. This is the property of the matrix groups when viewed as smooth 1–1 and onto transformations (diffeomorphisms). We will return to this later.

It will be necessary to define the subspace of transformations for which the regular configurations stay regular.

Definition 7.25 *Then the* **regularity constrained subset of similarities** $\mathcal{S}_\mathcal{R} \subset \mathcal{S} = \mathcal{S}_0^n$ *become*

$$\mathcal{S}_\mathcal{R} = \{s \in \mathcal{S}_0^n : sc \in \mathcal{C}_\mathcal{R}, \; \forall c \in \mathcal{C}_\mathcal{R}\} \tag{7.32}$$

$$= \cap_{c \in \mathcal{C}_\mathcal{R}} \mathcal{S}_\mathcal{R}(c) \quad \text{where} \quad \mathcal{S}_\mathcal{R}(c) = \{s \in \mathcal{S}_0^n : sc \in \mathcal{C}_\mathcal{R}\}. \tag{7.33}$$

From the configurations which have acted upon it is natural to denote a particular regular configuration within the configuration space of the *template*.

Definition 7.26 *The* **template** *is one fixed element in the orbit:*

$$c^0 = \sigma(g_1^0, g_2^0, \cdots, g_n^0). \tag{7.34}$$

The **deformable template** *is the full orbit* $[c^0]_\mathcal{S}$.

The template expresses typical structure, the similarities variability around it. These subsets play an important role in pattern theory, and in general they will not be subgroups of $\mathcal{S}$! Typically they are of lower dimension than that of $\mathcal{S}$, but equality can hold:

$$\dim(\mathcal{S}_\mathcal{R}(c)), \dim(\mathcal{S}_\mathcal{R}) \leq \dim(\mathcal{S}). \tag{7.35}$$

Although they are not generally a subgroup, $\mathcal{S}_\mathcal{R} \subset \mathcal{S} : \mathcal{C}_\mathcal{R} \to \mathcal{C}_\mathcal{R}$ is a semi-group with identity. Clearly $e \in \mathcal{S}_\mathcal{R}(c)$ for all c. If $s, s' \in \mathcal{S}_\mathcal{R}$ then this implies

$$(s \circ s')c = s(s'c) = sc', \quad c' \in \mathcal{C}_\mathcal{R} \tag{7.36}$$

and since $s \in \mathcal{S}_\mathcal{R}$ then $sc' \in \mathcal{C}_\mathcal{R}$. It is sometimes the case that $\mathcal{S}_\mathcal{R}$ is a subgroup.

Example 7.27 (TANKS) Examine the almost rigid body case corresponding to the representation of tanks with movable turrets atop tractor assemblies. Let $\mathcal{G} = \{\text{tractors, turrets}\}$ the set of tractors and turrets at all possible orientations and positions, $\mathcal{G}^0 = \{g_1^0, g_2^0\} \subset \mathcal{G}$ the tractors and turrets at the origin at $0°$ orientation. The graph type DIMER consists of bonds between the mounting support of the tractor and the pivot point of the turret: the configurations $c = \text{DIMER}(g_1, g_2)$ are regular if $\beta_{\text{out}}(g_1) = \beta_{\text{in}}(g_2)$ meaning the turret is mounted on the tractors. The basic similarity group $\mathbf{E}(2) = \mathbf{SO}(2) \otimes \mathbf{R}^2 : \mathcal{G} \leftrightarrow \mathcal{G}$ rotates the tractors and turrets around the body reference frame, and translates. The full product similarity $\mathcal{S} = \mathbf{E}(2)^2$ does not generate regular configurations; the similarity constrained subset $\mathcal{S}_\mathcal{R} = \mathbf{E}(2) \times \mathbf{SO}(2)$, with its action rotating the turret around the center point of contact of the tractor according to

$$s = ((O_1, a_1), (O_2, a_1)) \in \mathcal{S}_\mathcal{R} : \text{DIMER}(g_1, g_2) \mapsto \text{DIMER}((O_1, a_1)g_1, (O_2, a_1)g_2).$$

7.3 TRANSFORMATIONS CONSTRUCTED FROM PRODUCTS OF GROUPS

Example 7.28 (Unclosed Contours: SNAKES and ROADS on the Plane) Examine the representation such as in HANDS [150] for unclosed ROAD contours in which the generators are vectors $\mathcal{G} = \mathbb{R}^2$, essentially directed line segments from the origin, $g_i^0 = \begin{pmatrix} g_{i1}^0 \\ g_{i2}^0 \end{pmatrix}$, with graph type $\sigma = \text{LINEAR}(n)$ denoting the order of generators. The polygonal representations are generated by adding the generators sequentially; hence the ordering is required in the configuration. To generate n-length curves in the plane, attach the basic similarities rotations $\mathcal{S}_0 = \mathbf{SO}(2)$ rotating each of the generators:

$$O(\theta) : g_k \mapsto \begin{pmatrix} \cos\theta & \sin\theta \\ -\sin\theta & \cos\theta \end{pmatrix} \begin{pmatrix} g_{k1} \\ g_{k2} \end{pmatrix}. \quad (7.37)$$

The regularity constrained set is $\mathcal{S}_\mathcal{R} = \mathbf{SO}(2)^n$, and the configuration space $\mathcal{C}_\mathcal{R}$ is the space of all n-length piece-wise linear curves rooted at the origin. This is a homogeneous space under the group action $\mathcal{S}_\mathcal{R}$ on $\mathcal{C}_\mathcal{R}$; the identity are copies of the 0-degree rotations, with law of composition

$$s \circ s' = (\theta_1 + \theta_1', \ldots, \theta_n + \theta_n') \in \mathcal{S}_\mathcal{R}.$$

To generate all n-length, piece-wise linear curves, add the global translation group $\mathcal{S}_\mathcal{R} = \mathbf{SO}(2)^n \times \mathbb{R}^2$. An obvious template is a straight line along the x-axis of length n so that the configuration $c^0 = \text{LINEAR}(g_1^0, g_2^0, \ldots)$, $g_k^0 = \begin{pmatrix} 1 \\ 0 \end{pmatrix}$.

Example 7.29 (Closed Contours) Adding the closure condition, the regularity constrained product transformation $\mathcal{S}_\mathcal{R}$ on $\mathcal{C}_\mathcal{R}$ is no longer a subgroup, just a linear submanifold. Take as generators $\mathcal{G} = \mathbb{R}^2$ chords from the circle $g_k^0 = \begin{pmatrix} \cos(2\pi k/n) - \cos(2\pi(k-1)/n) \\ \sin(2\pi k/n) - \sin(2\pi(k-1)/n) \end{pmatrix}$; with these similarity groups $\mathcal{S}_0 = \mathbf{US}(2) \times \mathbf{O}(2)$ the scales/rotations $s = \begin{pmatrix} u_1 & u_2 \\ -u_2 & u_1 \end{pmatrix}$, where $u_1 = \rho\cos\phi$, $u_2 = \rho\sin\phi$, $u_1, u_2 \in \mathbb{R}^2$, ρ the scale parameters, and ϕ the rotation parameter. The full space of generators is generated via the transitive action $s \in \mathcal{S}_0 : g^0 \mapsto g = sg^0$ according to

$$g_k = \begin{pmatrix} g_{1k} \\ g_{2k} \end{pmatrix} = \begin{pmatrix} u_{1k} & u_{2k} \\ -u_{2k} & u_{1k} \end{pmatrix} \begin{pmatrix} g_{1k}^0 \\ g_{2k}^0 \end{pmatrix}.$$

The full space of transformations (regular and irregular configurations) becomes $\mathcal{S} = \mathcal{S}_0^n = (\mathbf{US}(2) \times \mathbf{SO}(2))^{2n} = \mathbb{R}^{2n}$. Figure 7.2 shows the circular closed contour templates and an example deformation via the scale-rotation groups.

The template becomes $c^0 = (g_1^0, g_2^0, \ldots, g_n^0)$. The closure condition means the endpoint of the nth generator must be 0 giving $\sum_{i=1}^n g_i = 0$ implying $\mathcal{S}_\mathcal{R} = \mathbb{R}^{2n-2} \subset \mathbb{R}^{2n} = \mathcal{S}_0^n$, and $\mathcal{S}_\mathcal{R}$ is not a subgroup but is a submanifold! The loss of two-dimensions from the closure condition is a linear manifold constraint given by the discrete Fourier transform (DFT) condition, for $j = \sqrt{-1}$, then

$$\sum_{k=1}^n \begin{pmatrix} u_{1k} & -u_{2k} \\ u_{2k} & u_{1k} \end{pmatrix} \begin{pmatrix} \cos(2\pi k/n) - \cos(2\pi(k-1)/n) \\ \sin(2\pi k/n) - \sin(2\pi(k-1)/n) \end{pmatrix}$$

$$= \sum_{k=1}^n (u_{1k} - ju_{2k})(e^{j(2\pi k/n)} - e^{j(2\pi(k-1)/n)}). \quad (7.38)$$

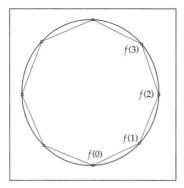 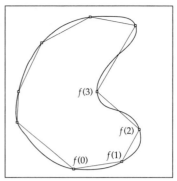

Figure 7.2 Showing the circular closed contour template and its deformation via the scale-rotation groups.

This implies the highest frequency DFT component is zero:

$$\sum_{k=1}^{n}(u_{1k} - ju_{2k})e^{-j(2\pi k(n-1)/n)} = e^{-j(2\pi/n)}\sum_{k=1}^{n}(u_{1k} - ju_{2k})e^{-j(2\pi k(n-1)/n)} = 0. \tag{7.39}$$

This $\mathcal{S}_\mathcal{R}$ is not a subgroup acting on $\mathcal{C}_\mathcal{R}$, hence $s \circ s' \notin \mathcal{S}(c, \mathcal{R})$ since the closure condition is a linear constraint on the scale/rotation elements according to Eqn. 7.39 which is not satisfied. Addition of the vector of similarities would maintain closure since this would maintain the highest frequency DFT coefficient being zero. However, composition is not addition.

7.4 Random Regularity on the Similarities

A fundamental interplay in pattern theory is the interaction betweeen probability laws on the generators and probability laws on the transformations or similarities. The probability law on transformations may be induced directly via the probabilities on the similarities $\mathcal{S}$ and the regularity constrained similarities $\mathcal{S}_\mathcal{R}$. We are reminded of the work by Kolmogoroff in defining Kolmogoroff complexity, transferring the focus from the program (template) to the input to the program (transformation).

Assume that the configuration spaces can be identified with Euclidean vector spaces, $\mathcal{C}(\sigma) = \mathbb{R}^m$, with $\mathcal{S}: \mathbb{R}^m \to \mathbb{R}^m$ which is a bijection. Then the two probability densities are related via the standard *transformation of probability formulas*.

Since the roles of the generators and the similarities are so often interchanged in so many of the examples in pattern theory, it is instructive to be more explicit about how these densities are related. For this, assume there is a bijection between generators and similarities, so that the equation $g = sg^0, g, g^0 \in \mathcal{G}$ has a unique solution in $s \in \mathcal{S}$. Then the density on the similarities induces a density on the configurations.

Theorem 7.30 *Assume $c = sc^0 \in \mathbb{R}^n$ with a bijection between configurations and similarities. The two densities on the similarities $p_S(s)$ on $\mathcal{S}$ and configurations $p_C(c)$ on $\mathcal{C}(\sigma)$ are related via the standard probability transformation formula with $m(ds)$ the measure on similarities:*

$$p_C(c|_{sc^0})dc = p_S(s) |\det D_s c| m(ds), \tag{7.40}$$

with $D_s c$ the Jacobian matrix of the transformation $c = sc^0$.

Here are several examples.

7.4 RANDOM REGULARITY ON THE SIMILARITIES

Example 7.31 (Closed Contours with Directed Arcs as Generators) Examine closed contours, the graph $\sigma = \text{CYCLIC}(n)$ with generators the directed arcs $g_k = (g_{1k}, g_{2k}) \in \mathcal{G} = \mathbb{R}^4$, $g_{1k} = \text{start point}$, $g_{2k} = \text{end point}$, with bond values $\beta_{\text{in}}(g) = g_1, \beta_{\text{out}}(g) = g_2$ start and end points of the generators in the plane and bond function

$$\beta_{\text{out}}(g_k) = \beta_{\text{in}}(g_{k+1})) \text{ so that } g_{2k} = g_{1k+1}.$$

The start point of the $k+1$st arc equals the endpoint of the kth arc. For the graph type CYCLIC, unlike LINEAR, the in-bond of the first generator g_1 is connected to the out-bond of the last generator g_n as well.

Apply the similarity of scales-rotations-translations S_0 extending the template generators $S_0 : \mathcal{G}^0 \to \mathcal{G} = \mathbb{R}^4$ rotating and scaling the generators around their origin, followed by translation, so that $s = (A, a) \in S_0 : g'_k \mapsto g_k = s_k g'_k$,

$$g_k = s g'_k = (g'_{1k} + a_k), \left(\begin{pmatrix} u_{1k} & u_{2k} \\ -u_{2k} & u_{1k} \end{pmatrix} (g'_{2k} - g'_{1k}) + g'_{1k} + a_k \right). \quad (7.41)$$

This is a transitive action on $\mathbb{R}^4$. Choose as generators the chords from the circle and similarities so that

$$g_k^0 = \left[\begin{pmatrix} \cos(2\pi(k-1)/n) \\ \sin(2\pi(k-1)/n) \end{pmatrix}, \begin{pmatrix} \cos(2\pi k/n) \\ \sin(2\pi k/n) \end{pmatrix} \right], \quad s_k^0 = \left(I, \begin{pmatrix} \cos(2\pi(k-1)/n) \\ \sin(2\pi(k-1)/n) \end{pmatrix} \right);$$

then the identity $I = \begin{pmatrix} 1 & 0 \\ 0 & 1 \end{pmatrix}$ and the template is a connected circle $c^0 = \text{CYCLIC}(s_1^0 g_1^0, s_2^0 g_2^0, \ldots, s_n^0 g_n^0)$. The graph type CYCLIC associated with the closure condition implies g_1 given the first translation, all others are fixed giving the set of $2n$ rotation/scale elements implying $S_\mathcal{R} = \mathbb{R}^2 \times \mathbb{R}^{2n-2} \subset S = \underbrace{(\mathbf{US}(2) \times \mathbf{SO}(2)}_{\mathbb{R}^2} \times \mathbb{R}^2)^n$

which is not a group. The loss of two-dimensions follows from the closure condition on the sum of the generators, which is a linear manifold constraint.

The probability on the transformations induces the probability on the generators. Define the generators as 4×4 matrices operating on the similarity according to

$$g_k = s_k g_k^0 = \begin{pmatrix} g_{21,k}^0 & g_{22,k}^0 & 1 & 0 \\ g_{22,k}^0 & -g_{21,k}^0 & 0 & 1 \\ 0 & 0 & 1 & 0 \\ 0 & 0 & 0 & 1 \end{pmatrix} \begin{pmatrix} u_{1k} \\ u_{2k} \\ a_{1k} \\ a_{2k} \end{pmatrix}; \quad (7.42)$$

$p_S(s_1, s_2, \ldots, s_n) m(ds)$ induces a probability on the generators according to

$$p(g_1, \ldots, g_n) dg_1, \ldots, dg_n = p_S(s_1|_{s_1 g_1^0 = g_1}, \ldots, s_n|_{s_n g_n^0 = g_n}) \prod_{k=1}^{n} |\det D_{s_k} g_k| \, ds_k. \quad (7.43)$$

Since

$$|\det D_{s_k} g_k| = \left| \det \begin{pmatrix} g_{21,k}^0 & g_{22,k}^0 & 1 & 0 \\ g_{22,k}^0 & -g_{21,k}^0 & 0 & 1 \\ & & 1 & 0 \\ & & 0 & 1 \end{pmatrix} \right| = (g_{21,k}^0)^2 + (g_{22,k}^0)^2, \quad (7.44)$$

then the determinant of the Jacobian matrix is the product of the lengths of the generators squared, $\prod_{k=1}^{n} (g_{1,k}^0)^2 + (g_{2,k}^0)^2$.

Example 7.32 (Unclosed Contours: Vectors as generators) Now take as the generators vectors $g = (g_1, g_2) \in \mathbb{R}^2$ and $\mathcal{S}_0$ the two-parameter scale rotation group plane $\mathcal{S}_0 = \mathbf{US}(2) \times \mathbf{SO}(2)$, each subset $[g]_{\tilde{\mathcal{S}}_0}$ identified with a vector in $\mathbb{R}^2$. Thus a generator is an equivalence class. The elements of $\mathcal{S}_0$ are 2×2 matrices $s = \begin{pmatrix} u_1 & u_2 \\ -u_2 & u_1 \end{pmatrix}$, $u_1, u_2 \in \mathbb{R}^2$. Two vectors (generators) are similar if one can be brought into the other by rotating and stretching it.

Defining the template generators to be $\begin{pmatrix} g^0_{1,k} \\ g^0_{2,k} \end{pmatrix}$ notice that

$$g_k = s_k g^0_k = \begin{pmatrix} u_{1k} & u_{2k} \\ -u_{2k} & -u_{1k} \end{pmatrix} \begin{pmatrix} g^0_{1k} \\ g^0_{2k} \end{pmatrix}; \quad (7.45)$$

then for the template generators nondegenerate, g_k, g^0_k uniquely determine s_k. The density on the transformations induces the density on the configuration same as above:

$$p_C(c(\sigma)) dc(\sigma) = p_S(s_1|_{g_1 = s_1 g^0_1}, \ldots, s_n|_{g_n = s_n g^0_n}) \left| \det \prod_{i=1}^{n} D_{s_i} g_i \right| ds_1, \ldots, ds_n. \quad (7.46)$$

Then the determinant of the Jacobian is the product of determinants of matrices

$$D_{s_k} g_k = \begin{pmatrix} \partial g_{1k}/\partial u_{1k} & \partial g_{1k}/\partial u_{2k} \\ \partial g_{2k}/\partial u_{1k} & \partial g_{2k}/\partial u_{2k} \end{pmatrix} = \begin{pmatrix} g^0_{1k} & g^0_{2k} \\ g^0_{2k} & -g^0_{1k} \end{pmatrix} \quad (7.47)$$

and the absolute value of the determinant of the Jacobian is the product of the lengths of the generators squared as above.

Example 7.33 (Triangulated Graphs) Consider triangulations T_n of the 2-sphere $S^2 \subset \mathbb{R}^3$ and let the connection type Σ express the topology of T_n. Generators are patches which are topologically connected according to the triangulated sphere topology TRIANG$(n) \in \Sigma$ shown in Figure. 6.3, Chapter 6.1. Let the n generators represent nondegenerate triangles with vertices v_{ji}, and identify the generators g with 3×3 matrices which are nonsingular if the vertices are not on a great circle, so that $g_i = (v_{1i}, v_{2i}, v_{3i}); v_{ji} \in S_2; i = 1, 2, \ldots, n$ with arity $\omega(g) = 3$ and the bonds $\beta_j(g_i) = v_{ji}; j = 1, 2, 3$. The connector σ connects $\beta_j(g_i)$ with $\beta_{j'}(g_{i'})$ if the two triangles g_i and $g_{i'}$ are contiguous with $v_{ji} = v_{j'i'}$ which shall also be the bond relation. The same generator sometimes appears in more than one of the closed loops in the system.

Choose the general linear group as the similarity group, $S = \mathbf{GL}(3)$. Then, for any two generators g_1 and g_2 none of which is on a great circle, the equation $g_1 = sg_2$ has a unique solution in s. Denote the number of edges and vertices by n_e and n_v so that Euler's relation can be written as

$$n_v - n_e + n = 2.$$

But each triangle has three edges, each of which appears in two triangles, so that $n_e = \frac{3}{2}n$ and $n_v = (n/2) + 2$. Each vertex has 3 degrees of freedom so that for fixed n the manifold $\mathcal{C}_\mathcal{R}$ will have dimension $\frac{3}{2}n + 6$. The dimension of the similarity group we have chosen is $\dim(S) = 9n$, implying that $\frac{15}{2}n - 6$ independent constraints on the group elements are required.

7.4 RANDOM REGULARITY ON THE SIMILARITIES

The template becomes $c^0 = \text{TRIANG}(n, g_1^0, g_2^0, \ldots, g_n^0)$; the deformed template becomes

$$c = \text{TRIANG}(n, g_1, g_2, \ldots, g_n) = \text{TRIANG}(n, s_1 g_1^0, s_2 g_2^0, \ldots, s_n g_n^0). \quad (7.48)$$

Solving for the similarities in matrix form gives $s_i = g_i(g_i^0)^{-1}$, establishing a linear relation with constant Jacobian between the vertices and the similarities, so far unconstrained.

A density $p(s_1, s_2, \ldots, s_n)$ on the similarities induces a density on the generators,

$$p(g_1, \ldots, g_n) dg_1, \ldots, dg_n = p_S(s_1|_{g_1 = s_1 g_1^0}, \ldots, s_n|_{g_n = s_n g_n^0}) ds_1, \ldots, ds_n,$$

where the vertices in the g_is are not constrained.

Example 7.34 (LEAVES) Real world systems can have topological structure with complex connection types. A moderately complex connector is shown in panel 1 of Figure 7.3 showing a template for stylized maple leaves. The generators are directed line segments of arity $\omega(g) = 2$, the bond relation $\rho = \text{EQUAL}$, and the basic similarity group $\mathcal{S} = \text{US}(2) \times \text{O}(2)$. The connector σ is the directed graph in the figure, where $n(\sigma) = 24$ has been chosen small for descriptive purposes.

Note that the same generator appears in more than one of the closed loops in the system. To express closure of loops we obtain linear vector relations generalizations of the closure condition in example 7.29.

Introduce the adjacency matrix of length $n = 24$ with entries $0, 1, -1$, for each loop that we are considering. The first loop gives a vector with zeros everywhere

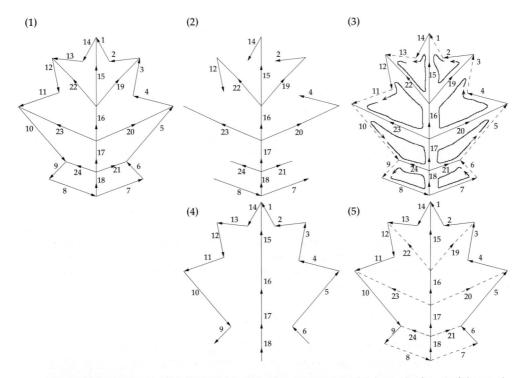

Figure 7.3 Top row: Panel 1 shows the basic leaf connector; panel 2 shows the forest of the maple leaf with panel 3 showing the corresponding independent loops. Bottom row: Panel 4 shows a second forest with panel 5 showing its associated independent loops.

except at locations 13, 14, 15, where the entry is 1, and location 22 with entry -1. In the same way the second loop would have zeros except at locations 11,12,16,22 and location 23 with the entry -1. The closure of loops are encapsulated via the following sets of linear vector relations:

$$0 = g_{15} + g_{14} + g_{13} - g_{22},$$
$$0 = g_{16} + g_{22} + g_{12} + g_{11} - g_{23},$$
$$0 = g_{17} + g_{23} + g_{10} - g_{24},$$
$$0 = g_{18} + g_{24} + g_9 + g_8,$$
$$0 = g_7 + g_6 - g_{21} - g_{18},$$
$$0 = g_{21} + g_5 - g_{20} - g_{17},$$
$$0 = g_{20} + g_4 + g_3 - g_{19} - g_{16},$$
$$0 = g_{19} + g_2 + g_1 - g_{15},$$
$$0 = g_{14} + g_{13} + g_{12} + g_{11} + g_{10} + g_9 + g_8 + g_7 + g_6 + g_5 + g_4 + g_3 + g_2 + g_1. \tag{7.49}$$

Now form matrix A which is a 9×24 matrix, a composition of the rows denoting the nine coefficient vectors, and the columns denoting the generators. $A = (a_{ij})$ is the matrix whose elements have the following values:

$$a_{ij} = 1 \quad \text{if generator } j \text{ is in vector } i \text{ and their orientations concide,} \tag{7.50}$$

$$a_{ij} = -1 \quad \text{if generator } j \text{ is in vector } i \text{ and their orientations do not concide,} \tag{7.51}$$

$$a_{ij} = 0 \quad \text{if generator } j \text{ is not in vector } i. \tag{7.52}$$

The A matrix for the leaf connector becomes

$$A = \begin{pmatrix} 0 & 0 & 0 & 0 & 0 & 0 & 0 & 0 & 0 & 0 & 0 & 0 & 1 & 1 & 1 & 0 & 0 & 0 & 0 & 0 & 0 & -1 & 0 & 0 \\ 0 & 0 & 0 & 0 & 0 & 0 & 0 & 0 & 0 & 0 & 1 & 1 & 0 & 0 & 0 & 1 & 0 & 0 & 0 & 0 & 0 & 1 & -1 & 0 \\ 0 & 0 & 0 & 0 & 0 & 0 & 0 & 0 & 0 & 1 & 0 & 0 & 0 & 0 & 0 & 0 & 1 & 0 & 0 & 0 & 0 & 0 & 1 & -1 \\ 0 & 0 & 0 & 0 & 0 & 0 & 0 & 1 & 1 & 0 & 0 & 0 & 0 & 0 & 0 & 0 & 0 & 1 & 0 & 0 & 0 & 0 & 0 & 1 \\ 0 & 0 & 0 & 0 & 0 & 1 & 1 & 0 & 0 & 0 & 0 & 0 & 0 & 0 & 0 & -1 & 0 & 0 & -1 & 0 & 0 & 0 & 0 & 0 \\ 0 & 0 & 0 & 0 & 1 & 0 & 0 & 0 & 0 & 0 & 0 & 0 & 0 & 0 & -1 & 0 & 0 & -1 & 1 & 0 & 0 & 0 & 0 & 0 \\ 0 & 0 & 1 & 1 & 0 & 0 & 0 & 0 & 0 & 0 & 0 & 0 & 0 & -1 & 0 & 0 & -1 & 1 & 0 & 0 & 0 & 0 & 0 & 0 \\ 1 & 1 & 0 & 0 & 0 & 0 & 0 & 0 & 0 & 0 & 0 & 0 & -1 & 0 & 0 & 0 & 1 & 0 & 0 & 0 & 0 & 0 & 0 & 0 \\ 1 & 1 & 1 & 1 & 1 & 1 & 1 & 1 & 1 & 1 & 1 & 1 & 0 & 0 & 0 & 0 & 0 & 0 & 0 & 0 & 0 & 0 & 0 & 0 \end{pmatrix}. \tag{7.53}$$

To find the rank of the matrix A we have to find the number of linearly independent rows. Here all the rows are linearly independent except row 9, which is just a linear addition of the first eight rows. So the rank of the matrix A is 8. There are of course other loops such as

$$g_{24} + g_9 + g_8 + g_7 + g_6 - g_{21} = 0. \tag{7.54}$$

The probability on the similarities is induced as follows. The generators are directed line segments of arity $\omega(g) = 2$, with the bond relation $\rho = \text{EQUAL}$, and the basic similarity group $\mathcal{S}_0 = \mathbf{US}(2) \times \mathbf{O}(2)$ with elements $\begin{pmatrix} u_1 & u_2 \\ -u_2 & u_1 \end{pmatrix}$. The connector σ is the directed graph as shown in Figure 7.4.

7.4 RANDOM REGULARITY ON THE SIMILARITIES

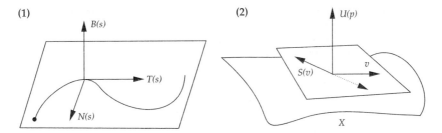

Figure 7.4 (i) Panel 1 shows the three orthonormal vector fields T, N, B of the curve x. Panel 2 shows a surface with the normal vector field and the corresponding shape operator $S(v)$.

To construct the probability on the full similarity group $\mathcal{S} = \mathcal{S}_0^{24}$, begin with the potential function on $s_i = (u_{i1} u_{i2})$ with

$$e^{-\Phi_{ij}(s_i, s_j)} = e^{-(1/2\tau^2)((u_{i1}-1)-a(u_{j1}-1))^2 - (1/2\tau^2)(u_{i2}-au_{j2})^2}, \quad (i,j) \in \sigma, \quad a \neq 1.$$

Let s denote the 48-vector of similarities $s = (u_{11}, u_{12}, u_{21}, u_{22}, \dots)$ and introduce the quadratic form

$$\sum_{ij} s_i^* K_{ij}^{-1} s_j = \sum_{(i,j) \in \sigma} \left(((u_{i1} - 1) - a(u_{j1} - 1))^2 + (u_{i2} - au_{j2})^2 \right), \quad (7.55)$$

and the density is Gaussian proportional to $e^{-(1/2\tau^2)Q(s)}$, $Q(s) = \sum_{ij} s_i^* K_{ij}^{-1} s_j$ with similarity $s \in \mathcal{S} = \mathbb{R}^{48}$. Now to condition the closure constraint express the closed cycle conditions in matrix vector form via the constraint matrix C of size $n_{\text{cyclo}} \times n$, where n_{cyclo} is the so-called *cyclomatic number*, and C is of full rank:

$$C \begin{pmatrix} g_1 \\ \vdots \\ g_n \end{pmatrix} = 0, \quad (7.56)$$

with the generator matrix of size $n \times 2$. Now apply known properties of conditioned Gaussian distributions implying that the s-vector, conditioned as above, will have the modified covariance matrix

$$K - KC(C^*KC)^{-1}C^*K. \quad (7.57)$$

This implies that pattern synthesis can be achieved by a fast direct algorithm without using iterative methods. Indeed, if we choose the Cholesky decomposition of the covariance matrix, $K = LL^*$, in terms of a lower triangular matrix L, the above expression can be written as

$$L[I - L^*C(C^*KC)^{-1}C^*L]L^* = LPL^*. \quad (7.58)$$

Note that P is symmetric and idempotent so that the whole expression takes the form of a "square" $K = (LP)(LP)^*$ so that $s = LPx$ where x is a Gaussian vector with independent components.

Example 7.35 (Graph Theory) To determine the set of independent loops, recall some definitions in graph theory. A *spanning tree* is a connected subgraph of a connected graph containing all the nodes of the graph but containing no loops. The branches of a tree are called *twigs* and those branches not in a tree are called *links*. If a graph is not connected, the object corresponding to a tree is a *forest*.

Now let us write the procedure for determining the *independent or fundamental loops*. Given a graph, a forest is selected. Then, one at a time, the union of a link with the forest is considered. Each resulting subgraph contains one loop. In each case, the loop will be characterized by the fact that all but one of its branches are twigs of the chosen forest. The orientation of the loop is chosen to coincide with that of its defining link. Each of the loops is an independent one and the total number of such loops which will be equal to the number of links is $b - n - 1$. The rank of the matrix is then $rank = b - n - 1$ where b is the number of branches and n is the number of nodes in the graph generated by our generators. In the case above $b = 24$ and $n = 17$; the rank is 8.

Returning to the problem, we first consider the first forest shown in the panels 2 and 3 of the top row of Figure 7.3. One at a time, take the union of links 13,11,10, 9,6,5,3 and 1 with the forest to get the eight independent loops corresponding to each link. The eight independent loops with their orientations are {13,22,15,14} {11,23,16,22,12}, {10,24,17,23}, {9,8,18,24}, {6,21,18,7}, {5,20,17,21}, {3,19,16,20,4}, and {1,15,19,2}. They are shown in panel 3 of Figure 7.3.

Although the total number of independent loops is always fixed for a given graph, the independent loops are not unique. Depending on the forest we choose we obtain different independent loops corresponding to the different links of the chosen forest. Consider the forest shown in the bottom row panels of Figure 7.3. Like the previous steps, one at a time take the unions of links 19,20,21,7,8,24,23, and 22 with the forest obtaining the eight independent loops corresponding to each link. The set of these eight loops are {19,2,1,15}, {20,4,3,19,16}, {21,5,20,17}, {7,6,21,18}, {8,18,24,9}, {24,10,23,17}, {23,11,12,22,16}, and {22,13,14,15}.

7.5 Curves as Submanifolds and the Frenet Frame

In the following two sections let us examine curves and surfaces which often provide generators for the patterns we study in shape. Almost all the geometric work which we do in pattern theory concentrates on submanifolds of $\mathbb{R}^d, d = 1, 2, 3$, corresponding to points, lines, surfaces, and subvolumes. Begin with the geometry of curves defined in the continuum based on their definition via curvature, torsion through the Frenet representation of curves.

Definition 7.36 *A* **regular curve** $x : D = [0, 1] \to \mathbb{R}^3$ *is a differentiable function*

$$x : s \in D = [0, 1] \subset \mathbb{R} \mapsto x(s) \in \mathbb{R}^3 \tag{7.59}$$

assumed to be regular so that $\dot{x}(s) \neq 0, s \in [0, 1]$ *with* **velocity vector** $\dot{x}(s)$. *and* **speed** *the norm* $\|\dot{x}(s)\|$.

Curves will be defined by their **speed, curvature, torsion**.

Definition 7.37 *Let* $x(\cdot)$ *be a regular smooth curve,* $x(s), s \in [0, 1]$, *then the* **speed, curvature, and torsion** *are given by*

$$\alpha(s) = \|\dot{x}(s)\|_{\mathbb{R}^3}, \quad \kappa(s) = \frac{\|\dot{x}(s) \times \ddot{x}(s)\|_{\mathbb{R}^3}}{\alpha(s)^3}, \quad \tau(s) = \frac{\langle (\dot{x}(s) \times \ddot{x}(s)), \dddot{x}(s) \rangle_{\mathbb{R}^3}}{\|\dot{x}(s) \times \ddot{x}(s)\|}. \tag{7.60}$$

7.5 CURVES AS SUBMANIFOLDS AND THE FRENET FRAME

For **unit speed curves** $\alpha = 1$ with

$$\kappa(s) = \|\ddot{x}(s)\|_{\mathbb{R}^3}, \quad \tau(s) = \frac{1}{\kappa^2(s)} \det \begin{pmatrix} \dot{x}_1(s) & \dot{x}_2(s) & \dot{x}_3(s) \\ \ddot{x}_1(s) & \ddot{x}_2(s) & \ddot{x}_3(s) \\ \dddot{x}_1(s) & \dddot{x}_2(s) & \dddot{x}_3(s) \end{pmatrix}. \tag{7.61}$$

The arc length of x from $s = a$ to $s = b$ is given by $l(a,b) = \int_a^b \|\dot{x}(s)\| \, ds$. Length does not change under reparameterization, and since the curvature and torsion are simpler we will work in the arc-length parameterization.

Lemma 7.38 *Let $\phi : [0,1] \to [0,1]$ be any diffeomorphism, $\phi_0 = 0, \phi_1 = 1, \dot{\phi} > 0$, then the length of any curve x in $\mathbb{R}^d$ is invariant to all such ϕ, so that*

$$\int_0^1 \left\| \frac{d}{dt} x \circ \phi_t \right\|_{\mathbb{R}^d} dt = \int_0^1 \|\dot{x}_t\|_{\mathbb{R}^d} \, dt. \tag{7.62}$$

*In particular on defining $\phi_t = (\int_0^t \|\dot{x}_t\| \, dt)/(\int_0^1 \|\dot{x}_t\| \, dt)$, $x(\cdot)$ in **arc-length parameterization** $x \circ \phi_t^{-1}, t \in [0,1]$ is constant speed:*

$$\left\| \frac{d}{dt} x \circ \phi_t^{-1} \right\|_{\mathbb{R}^d} = \int_0^1 \|\dot{x}_t\|_{\mathbb{R}^d} \, dt, \ t \in [0\ 1]. \tag{7.63}$$

Proof Since ϕ is a diffeomorphism, $\dot{\phi} > 0$, then

$$\int_0^1 \left\| \frac{d}{dt} x \circ \phi_t \right\|_{\mathbb{R}^d} dt = \int_0^1 \|\dot{x} \circ \phi_t\|_{\mathbb{R}^d} |\dot{\phi}_t| \, dt \stackrel{(a)}{=} \int_0^1 \|\dot{x}_s\|_{\mathbb{R}^d} \, ds, \tag{7.64}$$

with (a) following from the substitution of variables $s = \phi_t$. Let $\psi_t = \phi_t^{-1}$, then $\dot{\psi}_t = 1/(\dot{\phi}(\phi_t^{-1}))$, giving $|\dot{\psi}_t| = (\int_0^1 \|\dot{x}_t\|_{\mathbb{R}^d} \, dt)/(\|\dot{x} \circ \psi_t\|_{\mathbb{R}^d})$ implying

$$\left\| \frac{d}{dt} x \circ \phi_t^{-1} \right\|_{\mathbb{R}^d} = \|\dot{x} \circ \psi_t\|_{\mathbb{R}^d} |\dot{\psi}_t| \, dt = \int_0^1 \|\dot{x}_t\|_{\mathbb{R}^d} \, dt. \tag{7.65}$$

$\square$

Now examine the Frenet representation following [151, 152] describing the mathematical measurements of the turning (curvature) and twisting (torsion) curves in $\mathbb{R}^3$. It follows that in the Frenet representation, the derivatives of the tangents, curvature, and torsion form an orthogonal frame.

Theorem 7.39 (Frenet) *Let $x : D \subset \mathbb{R}^3 \to \mathbb{R}^3$ be a unit speed curve, so that $\|\dot{x}_s\|_{\mathbb{R}^3} = 1$ for each s in D. Then the three vector fields associated with the curve x, (i) the **unit tangent vector field** $T = \dot{x}_s$, (ii) the **unit normal curvature vector field** $N = \dot{T}/\|\dot{T}\|_{\mathbb{R}^3}$, and (iii) the **binormal vector field** $B = T \times N$, provide an orthonormal frame $T(s), N(s), B(s)$, $s \in [0,1]$,*

$$\dot{T} = \kappa N, \quad \dot{N} = -\kappa T + \tau B, \quad \dot{B} = -\tau N, \tag{7.66}$$

Proof First for the orthonormality of T, N, B. Since T has constant length 1, differentiation of $\langle T, T \rangle_{\mathbb{R}^3}$ gives $2\langle \dot{T}, T \rangle_{\mathbb{R}^3} = 0$, so $\dot{T}$ is always orthogonal to T, that is, normal to x. Thus the unit normal field $N = \dot{T}/\kappa$ is normal to T and tells the direction in which x is turning. B is orthonormal to T, N by definition of the cross product.

Now for the three equations. The first equation is the definition of curvature fields $\dot{T} = \kappa N$ with $\kappa = \|\ddot{x}\|$. The third equation follows since $\langle \dot{B}, B \rangle_{\mathbb{R}^3} = 0$ as B is unit length, and since $\langle B, T \rangle_{\mathbb{R}^3} = 0$ differentiating gives

$$0 = \langle \dot{B}, T \rangle_{\mathbb{R}^3} + \langle B, \dot{T} \rangle_{\mathbb{R}^3} = \langle \dot{B}, T \rangle_{\mathbb{R}^3} + \langle B, \kappa N \rangle_{\mathbb{R}^3} = \langle \dot{B}, T \rangle_{\mathbb{R}^3}. \tag{7.67}$$

Thus $\dot{B}$ is orthogonal to B, T implying $\dot{B} = \alpha N$ with $\alpha = -\tau$.

For the second equation expand $\dot{N}$ in terms of the orthonormal vectors T, N, B:

$$\dot{N} = \langle \dot{N}, T \rangle_{\mathbb{R}^3} T + \langle \dot{N}, N \rangle_{\mathbb{R}^3} N + \langle \dot{N}, B \rangle_{\mathbb{R}^3} B. \tag{7.68}$$

The first term is $-\kappa T$ since differentiation of $\langle N, T \rangle_{\mathbb{R}^3} = 0$ gives

$$\langle \dot{N}, T \rangle_{\mathbb{R}^3} + \langle N, \dot{T} \rangle_{\mathbb{R}^3} = 0 \Rightarrow \langle \dot{N}, T \rangle_{\mathbb{R}^3} = \langle -N, \dot{T} \rangle_{\mathbb{R}^3} = \langle -N, \kappa N \rangle_{\mathbb{R}^3} = -\kappa. \tag{7.69}$$

The second term is 0 since $\langle \dot{N}, N \rangle_{\mathbb{R}^3} = 0$ since N is a unit vector field. The third term follows differentiating $\langle N, B \rangle_{\mathbb{R}^3} = 0$ giving

$$\langle \dot{N}, B \rangle_{\mathbb{R}^3} = \langle -N, \dot{B} \rangle_{\mathbb{R}^3} = \langle -N, -\tau N \rangle_{\mathbb{R}^3} = \tau. \tag{7.70}$$

$\square$

The fields T, N, B are depicted in panel 1 of Figure 7.4. The Frenet equations show that the rotating orthonormal frame fields T, N, B, and therefore the curves themselves, are completely determined by the torsion and curvature functions up to rigid motions. If $\kappa = 0$ the curve is a straight line; if $\tau = 0$, the curve is contained in a plane. The values of κ are always nonnegative; those of τ can take both signs.

Example 7.40 (Frenet representation of curves in the cortex [153, 154]) External features of great importance in the brain are the curves depicting the regions of folding in the neocortex called the sulcal curves. These have been well characterized in the macaque [155], and are actively being studied in humans by various groups [156]. Despite their anatomic and functional significance, the sulcus folding of the neocortex in mammals appear consistently and exhibit pronounced variability in size and configuration. Khaneja [153, 154] has examined representations of sulcal curve variation via the Frenet representation of curves. Figure 7.5 shows three macaque brains (panels 1–3) depicting curves delineating the folds of the neocortical surface termed the sulcal curves, including the Arcuate Sulcus, Lateral Sulcus, Central Sulcus, and others. Let the sulcal curve in parametric form be $x(s), s \in [0, L]$ with arclength parameter s. A discrete representation supporting curvature and torsion is obtained from the $x_i = (x_{i1}, x_{i2}, x_{i3}), i = 1, 2, \ldots, N$, equidistant points spaced some distance $\delta = 1$ apart, from which first, second, and third difference vectors are generated, $\dot{x}_i^\delta, \ddot{x}_i^\delta, \dddot{x}_i^\delta$. The orthogonal moving frame $T_i, N_i, B_i, i = 1, \ldots$ are obtained via Gram-Schmidt orthogonalization. The discrete curvatures and torsions are expressed in terms of the frame vectors via the discretized Frenet relations to solve

$$(T_{i+1} - T_i) = \kappa_i N_i, \quad (N_{i+1} - N_i) = -\kappa_i T_i + \tau_i B_i, \quad (B_{i+1} - B_i) = -\tau_i N_i, \tag{7.71}$$

with torsions and curvatures generated using discrete versions of Eqn. 7.61.

To associate a distribution, define the sequence of triples curvatures, torsions, and lengths as a Gaussian process $\{(l_i, \kappa_i, \tau_i), i = 1, \ldots, N\}$, mean process

Figure 7.5 Panels 1–3 show the three macaque brain sulcal curves used for estimating the random sulcal model. Panel 4 shows the mean sulcal curves computed from the three brains displayed on the average brain generated from the 3 in the left panels; panel 5 shows samples of the sulcal curves from the distribution on curvature and torsion. Data courtesy of David Van Essen, Washington University.

$\bar{l}_i = (1/M) \sum_j l_i^j$, $\bar{\kappa}_i = (1/M) \sum_j \kappa_i^j$, $\bar{\tau}_i = (1/M) \sum_j \tau_i^j$, and independent variances. This gives the potential for the log-prior distribution of the form

$$\log \pi(l_i, \kappa_i, \tau_i, i=1,\dots,N) = \eta \sum_i (l_i - \bar{l}_i)^2 + \beta \sum_i (\kappa_i - \bar{\kappa}_i)^2 + \gamma \sum_i (\tau_i - \bar{\tau}_i)^2. \tag{7.72}$$

Panel 1–3 of Figure 7.5 shows the loci of sulcal curves on the three macaque brains depicting the locations of the deep folds. From these the mean and variability of the length, curvature and torsion functions of the sulcus are empirically computed using the standard empirical mean formulas across the hand traced sulci.

To synthesize the sulcus curves associated with this distribution generate the random variables l_i, κ_i, τ_i $i = 1, 2, \dots, N$ from the Gaussian distribution and then solve the discretized Frenet equations sequentially. Note the starting vectors v_1^1, v_1^2 are undetermined expressing the fact that the patterns are in the equivalence set modulo the special Euclidean group $SE(3)$. Each principal sulcal curve was sampled into 50 points. Panel 4 shows the mean sulcus curves displayed on the mean brain. Panel 5 shows the sulcal curves synthesized from the Frenet equations with random curvature and torsion functions sampled from the empirically estimated means and variances.

Example 7.41 (Representations in Pattern Theory are Not Unique) The representations in pattern theory are not unique. Alternative to the Frenet representation of the motion of the orthogonal frame through the instantaneous skew-symmetric matrix acting on it, the orthogonal frame can be represented directly as vectors. Let $x_i \in \mathbb{R}^3; i = 1, 2, \dots$ denote equidistant points on the space curve representing the sulcus curves assumed to be equi-spaced some distance l apart. The transformations are defined from the generalized linear group $\mathbf{GL}(3)$ and constructed as a Gauss-Markov process. Then generators $g \in \mathcal{G} = \mathbb{R}^9$ are defined to be the triplet of vectors

$$g_i = (v_i^{(1)} = x_{i+1} - x_i, \; v_i^{(2)} = x_{i+2} - x_{i+1}, \; v_i^{(3)} = x_{i+3} - x_{i+2}). \tag{7.73}$$

Then the similarity transformation $s_i \in \mathbf{GL}(3)$ takes its action

$$g_i' = s_i g_i, \text{ with } g_i = (v_i^{(1)}, v_i^{(2)}, v_i^{(3)}), \quad g_i' = (v_i^{(1)\prime}, v_i^{(2)\prime}, v_i^{(3)\prime}). \tag{7.74}$$

Attach to a generator g_i an inbond as $(\rho_{\text{in}}(g_i) = x_i)$ and outbond as $(\rho_{\text{out}}(g_i) = x_{i+3})$ with relation ρ=EQUAL sticking generators together.

Define $S_0(\omega), S_1(\omega), \ldots$ to be a 9×1 Gaussian vector process with means $\mu_0, \mu_1, \ldots$ and covariance induced by the first order autoregression

$$(S_j - \mu_j) = \alpha(S_{j-1} - \mu_{j-1}) + W_j, \tag{7.75}$$

where μ_j is the mean of the jth transformation group $ES_j = \mu_j$, and W_j represents a zero-mean 3×1 white noise process with covariance $EW_i W_j^* = \delta(i-j)\sigma^2 I$, I the 9×9 identity matrix. Reparameterizing in $Y_j = S_j - \mu_j$ gives

$$Y_{i-1} = \alpha Y_{i-2} + W_{i-1}, \quad \text{with } Y_0 = W_0. \tag{7.76}$$

Drawing $Y_0 = W_0$ from the stationary distribution so that $EY_0 Y_0^* = EW_0 W_0^* = \sigma^2 I$, then the Y_i are stationary, implying $EY_i Y_i^* = E(\alpha Y_{i-1} + W_i)(\alpha Y_{i-1} + W_i)^* = EY_{i-1} Y_{i-1}^*$, giving

$$K_{ii} = EY_i Y_i^* = \frac{\sigma^2}{1-\alpha^2} I, \quad K_{i,i+j} = EY_{i+j} Y_i^* = \alpha^j K_{ii}, \tag{7.77}$$

with the $9n \times 9n$ covariance matrix $K = (K_{ij})$, and the I the 9×9 identity matrices representing assumed independence between the similarities. The inverse covariance has special tri-diagonal structure

$$K^{-1} = \frac{1}{\sigma^2} \underbrace{\begin{pmatrix} I & -\alpha I & & & \\ -\alpha I & (1+\alpha^2)I & -\alpha I & & \\ & \ddots & \ddots & \ddots & \\ & & -\alpha I & (1+\alpha^2)I & -\alpha I \\ & & & -\alpha I & I \end{pmatrix}}_{J} = \frac{1}{\sigma^2} J. \tag{7.78}$$

Given M realizations of the above process each of length N, the joint density has the form of M independent Gaussian vectors with the log-likelihood given by

$$\log p(Y_1, Y_2, \ldots, Y_M; K) = -\frac{M}{2} \log \det 2\pi K - \frac{1}{2} \sum_{m=1}^{M} Y_m^* K^{-1} Y_m, \tag{7.79}$$

$$= -\frac{M}{2} \log \det 2\pi K - \frac{M}{2} \operatorname{tr}(K^{-1} \Sigma), \tag{7.80}$$

where $\operatorname{tr}(\cdot)$ is the matrix trace with sample covariance matrix $\Sigma = (1/M) \sum_{m=1}^{M} y_m y_m^*$. The maximum likelihood estimate of the parameters $\hat{\alpha}, \hat{\sigma} \leftarrow \arg\max_{(\alpha,\sigma)} \log p(\Sigma; K)$, is given by

$$\hat{\sigma}^2 = \arg\max_{\sigma^2} \log p(\Sigma; K) = \arg\max_{\sigma^2} -9n \log(\sigma^2) + \log |J| - \frac{\operatorname{tr}[J\Sigma]}{\sigma^2} \tag{7.81}$$

$$= \frac{\operatorname{tr}[J\Sigma]}{9n} = \frac{\operatorname{tr}(\Sigma) + \hat{\alpha}^2 \Upsilon(\Sigma) - \hat{\alpha}\Theta(\Sigma)}{9n}, \tag{7.82}$$

where $\Upsilon(B) = \sum_{i=2}^{i=N-1} \operatorname{tr}(B_{i,i}), \quad \Theta(B) = \sum_{i=1}^{i=N-1} \operatorname{tr}(B_{i,i+1}) + \sum_{i=2}^{i=N} \operatorname{tr}(A_{i,i-1})$.

$$\tag{7.83}$$

To derive an equation only in σ^2, define the block matrices $B = (B_{ij})$; then the second equation for calculating $\hat{\alpha}$ arises computing the variation with respect to α.

The variation of $\log |J|$ can be written as $\delta \log |J| = \text{tr}(J^{-1}\delta J)$, implying that maximizing with respect to α and keeping σ^2 constant gives

$$\left.\frac{\partial \log p(\Sigma; K)}{\partial \alpha}\right|_{\hat{\alpha}} = \text{tr}\left[(J^{-1} - \Sigma/\hat{\sigma}^2)\frac{\partial J}{\partial \alpha}\right] = 0 \quad \text{where}$$

$$\frac{\delta J}{\delta \alpha} = \begin{pmatrix} 0 & -1 & & & \\ -1 & 2\alpha & -1 & & \\ & \ddots & \ddots & \ddots & \\ & & -1 & 2\alpha & -1 \\ & & & -1 & 0 \end{pmatrix}. \tag{7.84}$$

This gives $2\hat{\alpha}\Upsilon(J^{-1} - (\Sigma/\hat{\sigma})) = \Theta(J^{-1} - (\Sigma/\hat{\sigma}^2))$. Using $\Upsilon(J^{-1}) = (9(n-2))/(1-\hat{\alpha}^2)$, $\Theta(J^{-1}) = (2\hat{\alpha}9(n-1))/(1-\hat{\alpha}^2)$ yields a second equation which the MLE $\hat{\alpha}, \hat{\sigma}^2$ must satisfy:

$$\hat{\sigma}^2 = \frac{1-\hat{\alpha}^2}{18\hat{\alpha}}\Theta(\Sigma) - \frac{(1-\hat{\alpha}^2)}{9}\Upsilon(\Sigma). \tag{7.85}$$

Solving Eqns. 7.82 and 7.85 simultaneously in $\hat{\sigma}^2, \hat{\alpha}$ gives the maximum-likelihood estimates.

Figure 7.5 shows three brains which were used for estimating sulcal curve variability (panels 1–3). Using these three macaque brain maps the mean sulcal curves were computed, with the $\hat{\alpha}, \hat{\sigma}^2$ estimates generated by maximizing the log-likelihood. Panel 3 shows different realizations of sulcal curves superimposed over the template. The parameters of the Markov process $\hat{\alpha}, \hat{\sigma}^2$ were estimated from the population of three brains for the Gaussian distributions for each sulcus on the space $\mathbf{GL}(3)^M \times \mathbb{R}^3$. Each sulcus was named following the nomenclature in Felleman and Van Essen [155]. The values found for the sulci were Superior Temporal Sulcus $\alpha = 0.25, \sigma = 0.01$, Arcuate Sulcus $\alpha = 0.07, \sigma = 0.02$, Central Sulcus $\alpha = 0.32, \sigma = 0.01$, Intra-Parietal Sulcus $\alpha = 0.19, \sigma = 0.02$, Inferior Occipital Sulcus $\alpha = 0.30, \sigma = 0.01$.

7.6 2D Surfaces in $\mathbb{R}^3$ and the Shape Operator

Curves and surfaces in $\mathbb{R}^3$ provide a rich source of patterns, and provide concrete examples of manifolds, in particular 2D manifolds, which are familiar. Now it is natural that for a surface $M \subset \mathbb{R}^3$ there exists a tangent space at each of its points which resembles a region in the plane R^2.

Definition 7.42 *A* **2D surface** *is a subset $M \subset \mathbb{R}^3$ having the property that for each point $p \in M$ there exists a neighborhood $O \subset M$ of p, an open set $D \subset \mathbb{R}^2$, and a* **smooth coordinate patch** *(infinitely differentiable) $x : D \to O$ which is a diffeomorphism*

$$x : (u,v) \in D \subset R^2 \mapsto x(u,v) = (x_1(u,v), x_2(u,v), x_3(u,v)). \tag{7.86}$$

Define the **tangent space** $T_p(M)$ *of the surface at $p \in M$ as the set of all vectors generated by the 2D span of tangent vectors $(\partial x(u,v))/\partial u, (\partial x(u,v))/\partial v$. These are often called the* **coordinate frames** *spanning the tangent space.*

7.6.1 The Shape Operator

Just as the shape of a curve in $\mathbb{R}^3$ can be measured via its curvature and torsion functions, similarly the shape of a surface M in $\mathbb{R}^3$ is described infinitesimally by a feature associated with the tangent plane called the shape operator S, a 2×2 linear operator describing how the tangent plane changes as one follows particular curves on the surface. The algebraic invariants (determinant, trace, etc) of the shape operator encode the geometric meaning for the surface M.

Now natural local coordinates which we shall use to understand the surface are developed as follows. Define the partial derivative notation of a function of two variables $f(u,v)$ as $f_u = \partial f/\partial u, f_v = \partial f/\partial v$.

Theorem 7.43 *Given is smooth surface submanifold $M \subset \mathbb{R}^3$, with local coordinate patch $x: D \subset \mathbb{R}^2 \to O \subset M$ with orthogonal frame E_1, E_2, E_3 at $p \in M$ with E_1, E_2 spanning the tangent plane $T_p(M)$. Then the local coordinate representation representing the surface locally around $p \in O$ given by*

$$x(u,v) = p + uE_1 + vE_2 + E_3 f(u,v) \tag{7.87}$$

has tangent vectors and surface normal

$$\frac{\partial x(u,v)}{\partial u} = E_1 + E_3 f_u(u,v), \quad \frac{\partial x(u,v)}{\partial v} = E_2 + E_3 f_v(u,v), \tag{7.88}$$

$$n = \frac{-f_u E_1 - f_v E_2 + E_3}{\sqrt{1 + f_u^2 + f_v^2}}. \tag{7.89}$$

Proof Over the patch the tangent space $T_x(M)$ is clearly spanned by the natural coordinate vectors $E_1 + f_u E_3, E_2 + f_v E_3$. The normal vector to the surface n is given by the cross product

$$n = \det \begin{pmatrix} E_1 & -E_2 & E_3 \\ 1 & 0 & f_u \\ 0 & 1 & f_v \end{pmatrix} = \frac{-f_u E_1 - f_v E_2 + E_3}{\sqrt{1 + f_u^2 + f_v^2}}. \tag{7.90}$$

□

Following the usual conventions the coordinate frame dependence is suppressed with local coordinate representation $x(u,v) = (u, v, f(u,v))$ and orthogonal frame on $(u,v) \in D$ written as

$$\frac{\partial x}{\partial u} = \begin{pmatrix} 1 \\ 0 \\ f_u \end{pmatrix}, \quad \frac{\partial x}{\partial v} = \begin{pmatrix} 0 \\ 1 \\ f_v \end{pmatrix}, \quad n = \frac{1}{\sqrt{f_u^2 + f_v^2 + 1}} \begin{pmatrix} -f_u \\ -f_v \\ 1 \end{pmatrix}. \tag{7.91}$$

Notice, in this local coordinate system $p = x(0,0)$ and $f(0,0) = 0$, and $f_u(0,0) = f_v(0,0) = 0$, since the tangent vectors along the u and v curves (keeping first v fixed, and alternately u fixed) must be in the tangent space given by the E_1, E_2 plane. Since understanding the shape operator which involves curvature will only be quadratic terms, so that f can be approximated near $(0,0)$ via the quadratic Taylor series

$$f(u,v) = f_{uu}(0)\frac{u^2}{2} + 2f_{uv}(0)\frac{uv}{2} + f_{vv}(0)\frac{v^2}{2} + \sum_{k=0}^{2} o(u^k v^{2-k}). \tag{7.92}$$

7.6 2D SURFACES IN $\mathbb{R}^3$ AND THE SHAPE OPERATOR

Definition 7.44 The **shape-operator** of M at p is a linear operator $S_p : T_p(M) \to T_p(M)$ attaching to each tangent vector $v_p \in T_p(M)$, the vector $S_p(v_p)$ given by the negative of the directional derivative along v_p:

$$S_p(v_p) = -\frac{d}{dt} n(p + tv_p)|_{t=0} . \tag{7.93}$$

The **directional derivative** of the vector field n in the direction of v_p according to $(d/dt)n(p + tv_p)|_{t=0}$.

The right panel of Figure 7.4 shows the shape operator.
With this the following can be established.

Lemma 7.45 Let the surface M have local coordinate patch $x : D \subset \mathbb{R}^2 \to O \subset M$ with orthogonal frame E_{1p}, E_{2p}, E_{3p} at $p \in O$ and E_{1p}, E_{2p} spanning the tangent space $T_p(M)$. Then with $n = (-f_u E_1 - f_v E_2 + E_3)/\sqrt{1 + f_u^2 + f_v^2}$ the unit normal vector field on M over the patch, the shape operator at $x(0,0) = p$ the 2×2 symmetric operator is

$$S_{x(0,0)} = \begin{pmatrix} f_{uu}(0,0) & f_{uv}(0,0) \\ f_{vu}(0,0) & f_{vv}(0,0) \end{pmatrix} . \tag{7.94}$$

The shape operator over the local coordinate patch takes the form

$$S_{x(u,v)} = \frac{1}{\sqrt{1+f_u^2+f_v^2}} \begin{pmatrix} 1+f_u^2 & f_u f_v \\ f_u f_v & 1+f_v^2 \end{pmatrix}^{-1} \begin{pmatrix} f_{uu} & f_{uv} \\ f_{vu} & f_{vv} \end{pmatrix}, \quad u, v \in D. \tag{7.95}$$

Proof To compute the shape operator at the origin $(0,0)$, take perturbations along tangent vectors of the form $v_0 = \alpha_1 E_1 + \alpha_2 E_2$ giving

$$S_0(\alpha_1 E_1 + \alpha_2 E_2) = -\frac{d}{dt} n(t\alpha_1 E_1 + t\alpha_2 E_2)|_{t=0} \tag{7.96}$$

$$= (\alpha_1 f_{uu} + \alpha_2 f_{uv})E_1 + (\alpha_1 f_{vu} + \alpha_2 f_{vv})E_2. \tag{7.97}$$

Therefore, identifying the shape operator with 2×2 matrices, it is a symmetric linear operator uniquely represented by the 2×2 symmetric matrix of the form S_0 of Eqn. 7.94.
$\square$

Example 7.46 (The Sphere) Return to the sphere Example 8.8, $S^2 = \{y \in \mathbb{R}^3 : \|y\| = \sqrt{y_1^2 + y_2^2 + y_3^2} = r\}$, and examine the coordinate patch $x : (u,v) \in D \mapsto x(u,v) = (u, v, \sqrt{r - u^2 - v^2})$ at the north pole $p = (0,0,1)$ with coordinate frames $E_1 = (1,0,0), E_2 = (0,1,0)$. The tangents and curvatures correspond to

$$x_u(u,v) = \left(1, 0, \frac{-u}{\sqrt{r^2 - u^2 - v^2}}\right), \quad x_v(u,v) = \left(0, 1, \frac{-v}{\sqrt{r^2 - u^2 - v^2}}\right), \tag{7.98}$$

$$x_{uu}(u,v) = \left(0, 0, \frac{-1}{\sqrt{r^2 - u^2 - v^2}} + \frac{u^2}{(r^2 - u^2 - v^2)^{3/2}}\right)$$

$$= \left(0, 0, -\frac{r^2 - v^2}{(r^2 - u^2 - v^2)^{3/2}}\right),$$

$$0 x_{vv}(u,v) = \left(0, 0, -\frac{r^2 - u^2}{(r^2 - u^2 - v^2)^{3/2}}\right), \quad x_{uv}(u,v) = x_{vu}(u,v) = 0. \tag{7.99}$$

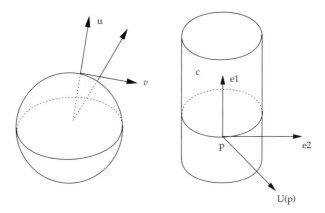

Figure 7.6 Shape operator on surfaces.

Then n the "outward normal" over the disk D orthogonal to the span of the tangent space x_u, x_v of Eqn. 7.98. becomes

$$n = \frac{1}{r}(u, v, \sqrt{r^2 - u^2 - v^2}). \qquad (7.100)$$

The shape operator $S(v)$ in the direction of the tangent vector $v = (v_x, v_y, v_z)$ becomes

$$\nabla_v n = \frac{d}{dt}(u + tv_x, v + tv_y, z + tv_z)|_{t=0} = \frac{(v_x, v_y, v_z)}{r}, \qquad (7.101)$$

where $z = \sqrt{r^2 - u^2 - v^2}$. The outward normal is shown in the left panel of Figure 7.6. Then $S(v) = -v/r$ for all v, so the shape operator is merely scalar multiplication by $-1/r$. This uniformity reflects that a sphere bends equally in all the directions.

Example 7.47 (The Cylinder) Examine the circular cylinder $C = \{y : y_1^2 + y_2^2 = r^2\}$ in $\mathbb{R}^3$. At any point p of C, let E_{1p}, E_{2p} be unit orthonormal tangent vectors to the surface at p, E_1 tangent to the ruling of the cylinder through p, and E_2 tangent to the cross-sectional circle. The outward normal is shown in the right panel of Figure 7.6. Now, when n moves from p in the E_1 direction, it stays parallel to itself just as on the plane; hence $S(E_1) = 0$. When n moves forward in the E_2 direction it topples forward exactly as in case of a sphere of radius r; hence $S(E_2) = -E_2/r$. In this way S describes the half flat, half-round shape of the cylinder.

7.7 Fitting Quadratic Charts and Curvatures on Surfaces

7.7.1 Gaussian and Mean Curvature

For fitting curvatures, we write the local coordinate patch up to second order in the Taylor series expansion, so that

$$f(u,v) = \frac{1}{2}(s_{2,0}u^2 + 2s_{1,1}uv + s_{0,2}v^2). \qquad (7.102)$$

The shape operator eigenvalues and eigenvectors, its trace and determinant all have geometric meaning of first importance for the surface $M \subset \mathbb{R}^3$. The shape operator and curvature at a point

7.7 FITTING QUADRATIC CHARTS AND CURVATURES ON SURFACES

$p \subset M$ define the surface shape locally around p. To construct the local coordinate patches around points $p \in M$ define the following.

Definition 7.48 *Let T_p be the **tangent space** of M at the point p with the **orthonormal basis of the tangent plane** at p $E_{1p}, E_{2p} \in T_p(M)$, with $E_{3p} = E_{1p} \times E_{2p}$ the unit normal.*

*The **quadratic patch** approximating the surface passing through the point p is written as*

$$x(u,v) = p + uE_{1p} + vE_{2p} + \left((u,v)S_p \begin{pmatrix} u \\ v \end{pmatrix}\right) E_{3p}, \qquad (7.103)$$

*where the **shape operator** S_p is the symmetric 2×2 matrix of the form $S_p = \begin{pmatrix} s_{20} & s_{11} \\ s_{11} & s_{02} \end{pmatrix}$.*

*Let $u = (u_1, u_2)^*$ be the 2×1 vector of expansion coefficients of the unit vector tangent to $M \subset \mathbb{R}^3$ at p representing the expansion coefficients expanding the tangent in the coordinate basis E_{1p}, E_{2p}. Then the **normal curvature in the u direction** is defined to be the number $\kappa(u) = u^* S u$.*

*The **principal curvatures** κ_1, κ_2 of M at p are the maximum and minimum values of the normal curvature $\kappa(u)$ of M at p. The directions in which these extreme values occur are the **principal directions** denoted by $\vec{t}_1$ and $\vec{t}_2$.*

The maximum and minimum values of the normal curvature are the eigenvalues of the matrix S, and the principal vectors are the corresponding eigenvectors, i.e. $S\vec{t}_i = \kappa_i \vec{t}_i$, $i = 1, 2$. The sign of the normal curvature has a geometric meaning. If x is a curve in M with initial velocity $\dot{x}(0) = u$ and N is the normal to the curve as shown in Figure 7.7 then if $\kappa(u) > 0$ then $N(0) = U(p)$, so that the surface is bending towards $U(p)$. If $\kappa(u) < 0$ then $N(0) = -U(p)$, so that surface is bending away from $U(p)$ as shown in the Figure.

Return to the cylinder example Fig 7.6. It is clear that along the ruling $\kappa_1 = 0$ and $\kappa_2 < 0$ occur in the direction tangent to the cross section. A point p of $M \subset \mathbb{R}^3$ is umbilic provided the normal curvature $\kappa(u)$ is constant on all unit tangent vectors u at p, as is the case of a sphere where $\kappa_1 = \kappa_2 = -1/r$.

Definition 7.49 *The **gaussian and mean curvature** of $M \subset \mathbb{R}^3$ are real-valued functions of the eigenvalues, the determinant and arithmetic mean, respectively.*

*The **gaussian curvature** is the product of the two principal curvatures which is the determinant given by the product of the eigevalues,*

$$K = \kappa_1 \kappa_2 = s_{02} s_{20} - s_{11} s_{11}.$$

*The **mean curvature** of $M \subset \mathbb{R}^3$ is the trace of S given by the mean of the eigenvalues.*

$$H = \frac{1}{2} \operatorname{tr} S = \frac{s_{20} + s_{02}}{2} = (\kappa_1 + \kappa_2)/2. \qquad (7.104)$$

The Gaussian curvature $K(p) > 0$ implies the principal curvatures $\kappa_1(p)$ and $\kappa_2(p)$ have the same sign. Thus $\kappa(u) > 0$ for all tangent directions or $\kappa(u) < 0$. Thus M is either bending towards or away in all directions. The surface locally looks like that shown in panel 2 of Figure 7.7. The Gaussian curvature $K(p) < 0$, then the principal curvatures $\kappa_1(p)$ and $\kappa_2(p)$ have opposite signs. Thus the quadratic approximation of M near p is a hyperboloid, so M also is saddle shaped near p, as shown in panel 3 of Figure 7.7.

If the Gaussian curvature $K(p) = 0$ then there are two cases. (i) If one principal curvature is zero, the shape of the surface then looks as shown in panel 4 of Figure 7.7. (ii) If both principal curvatures are zero, the surface is planar.

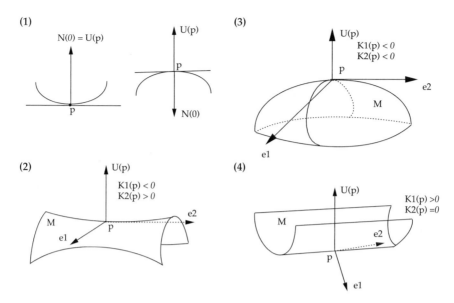

Figure 7.7 Panel 1 shows the positive and negative normal curvatures. Panel 2 shows the case of $K(p) > 0$. Panel 3 shows the case of $K(p) < 0$; panel 4 shows the case $K(p) = 0$.

7.7.2 Second Order Quadratic Charts

Upon these triangulated graphs calculus is performed, surface areas are computed, torsions and curvatures of various curves on the surface are generated. For this, local quadratic charts associated with the tangent spaces are generated. The surface manifold M representing the structure of interest is assumed to be a smooth 2D differentiable submanifold of $\mathbb{R}^3$.

The problem of fitting a second order local coordinate chart is to estimate the symmetric 2×2 matrix S_p at each point $p \in M$. Based on the neighborhood N_p of the point p, a minimum mean squared error estimate can be generated [157].

Algorithm 7.50 (Hamann [157]) Let $\{x_i \in M \subset \mathbb{R}^3, i = 1, \ldots, M\}$ be the vertices of the triangulated graph $M_\triangle$ with neighborhoods $\cup_{i=1}^M N_i$. The coordinate frames and neighborhoods N_p of point p are depicted in Figure 7.8.

1. Let T_p be the tangent plane at p, with coordinate frames E_{1p}, E_{2p} and unit normal $E_{3p} = E_{1p} \times E_{2p}$. Then, $T_p = \{x \in \mathbb{R}^3 : \langle E_{3p}, (x-p) \rangle_{\mathbb{R}^3} = 0\}$.

2. The orthonormal basis vectors are chosen based on the outward unit normal vector E_{3p}. Choose a vector v perpendicular to E_{3p}, that is, $\langle v, E_{3p} \rangle = 0$.

   The orthonormal basis vectors E_{1p}, E_{2p} are given by $E_{1p} = v/\|v\|$, $E_{2p} = E_{3p} \times E_{1p}$, where $\times$ is the cross product in $\mathbb{R}^3$. Figure 7.8 shows the neighborhood of a point p and the three vectors E_{3p}, E_{1p}, E_{2p}.

3. Let $N_p = \{x_j \in M_\triangle, j = 1, \ldots, n\}$ be the neighborhood of the point p. For each point $x_j \in N_p$ define h_j to be the distance of the point x_j to the tangent plane T_p. Define a set of $n \times 1$ vector of distances, $H = [h_1, \ldots, h_n]^*$.

4. Let x_j^p be the projection of the point $x_j \in N_p$ on to the tangent plane T_p. Let u_j and v_j be the local coordinates of the points x_j^p in the tangent plane given by

$$(u_j, v_j) = (\langle (x_j^p - p), E_{1p} \rangle_{\mathbb{R}^3}, \langle (x_j^p - p), E_{2p} \rangle_{\mathbb{R}^3}). \tag{7.105}$$

7.7 FITTING QUADRATIC CHARTS AND CURVATURES ON SURFACES

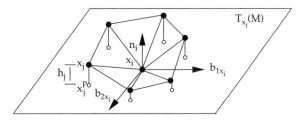

Figure 7.8 Figure depicts the neighborhood of a point p and the three vectors E_{3p}, E_{1p}, E_{2p} used for fitting the local quadratic charts; taken from Joshi et al. [158].

5. The minimum mean squared estimate of the matrix S_p in it is the solution of the minimization

$$S_p = \arg \min_{\{S: 2 \times 2 \text{ symmetric matrices}\}} \sum_{x_j \in N_p} \|x_j$$
$$- (p + u_j E_{1p} + v_j E_{2p} + E_{3p}[u_j, v_j] S_p[u_j, v_j]^*)\|^2 \qquad (7.106)$$

Identifying the symmetric 2×2 shape matrix S_p with a 3×1 column vector $S = \begin{pmatrix} s_{11} \\ s_{02} \\ s_{20} \end{pmatrix}$, and defining the $n \times 3$ matrix of the local coordinates $A = \begin{pmatrix} u_1^2 & 2u_1v_1 & v_1^2 \\ \vdots & \vdots & \vdots \\ u_n^2 & 2u_nv_n & v_n^2 \end{pmatrix}$, then the minimization of Eqn. 7.106 yields

$$\hat{S} = \arg \min_S \|AS - H\|^2 = (A^*A)^{-1} A^* H.$$

7.7.3 Isosurface Algorithm

Numerous computer scientists are studying the folding and tangent structures of Biologically related surfaces and manifolds. The basic idea behind isosurface generation in volumetric data is to determine where and how the surface representing the image value of constancy will intersect a given voxel. A popular approach to this problem is called the "Marching Cubes" algorithm [159] which was later refined [160] to handle ambiguities in the previous. The algorithm details follow.

The space in which the volume lies is subdivided into a series of small cubes, each the size of a voxel and centered at the intersection of every $2 \times 2 \times 2$ group of voxels (thus for voxel neighbors (1,1,1), (2,1,1), (1,2,1), ..., (2,2,2) the cube center will be at (1.5,1.5,1.5)). The eight vertices of each cube are assigned the voxel value of the volume data they lie in. By comparing each vertex of the cube with a user-defined isovalue, the vertex can be considered "in" or "out" of the surface. There are 2^8 or 256 different arrangements of in/out labeled vertices that each cube can have in this manner. For each of the 256 cases, a lookup table is consulted to see how the surface would intersect a particular cube arrangement. This lookup table pre-defines a set of polygons that will be located within that cube. Each of the polygons have vertices that lie at the midpoint of an edge between "in" and "out" vertices of a cube centered at the origin. The most trivial cases — no vertices in or all vertices in — have zero polygons that will contribute to the surface. A simple case where 1 vertex is in and all others are out (or vice versa) results in a single triangle whose vertices lie along each of the three edges emanating from the "in" vertex of the cube (Figure 7.9). The other cases are more complex and contain between 1 and 30 triangles.

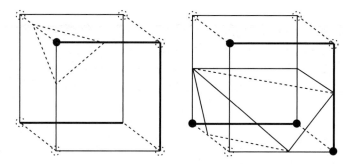

Figure 7.9 Panel 1 shows the resulting single triangle (shown as thick lines) from a cube where only 1 vertex (solid circle) is considered inside the surface. Panel 2 shows a slightly more complicated example where 3 vertices are considered inside the surface.

For each cube the contributing polygons from the lookup table are added to the final surface first by shifting polygon vertices along the edge of the cube corresponding to how close each edge vertex value is to the isovalue (using interpolation) and then by adding a translation using the center of the cube. Since neighboring cubes will have the same interpolated polygon vertices along each edge this results in a continuous surface crossing the cubes.

Algorithm 7.51 (Isosurface Algorithm) *Here is the pseudocode algorithm. Define I as volume data, S is the new (empty) surface, nx is the number of voxels in the X dimension of I, ny is the number of voxels in the Y dimension of I. nz is the number of voxels in the Z dimension of I. θ = isovalue threshold for surface (within range of I values); table is the lookup table with 256 entries, each defining some number of polygons (see [159, 160] for a sample table).*

```
for k = 1...nz-1 ; for j = 1...ny-1 ; for i = 1...nx-1
  vertex[0] = I(i-1,j-1,k-1), vertex[1] = I(i,j-1, k-1), vertex[2] = I(i,j,k-1),
  vertex[3] = I(i-1,j,k-1), vertex[4] = I(i-1,j-1,k), vertex[5] = I(i,j-1,k),
  vertex[6] = I(i,j,k), vertex[7] = I(i-1,j,k)
  index = 8 bit bitmask where bit i is 1 iff vertex[i] ≤ Θ (threshold), else 0
  for each polygon P in table[index]
    for each vertex V of P
      E = edge of cube V lies on
      a = cube index of first vertex in E (0..7)
      b = cube index of second vertex in E (0..7)
      [ by definition of table one of vertex[a] or vertex[b] will be
       'in', the other will be 'out']
      shift V along E by interpolating beween vertex[a] and vertex[b]
      using Θ
    endfor
    translate P by (i-0.5,j-0.5,k-0.5)
    add P to S
  endfor
endfor; endfor; endfor
```

Example 7.52 (Human Hippocampus Surface and Macaque Cerebral Cortex) The folding structure of the neocortex and other subvolumes of the brain is being studied by constructing triangulated graphs representing such surface submanifolds. To construct a triangulated graph, $M_\triangle$, surface contours are generated defining the boundaries of interest, which are pieced together, via isocontouring algorithms (e.g. Marching Cubes [161]) or hand tracing in the tissue sections. Shown in Figure 7.10 are the triangulated graphs representing the human hippocampus (top row, panel 1) and macaque cortex (bottom row, panel 3).

Shown in Figure 7.10 are the mean curvature maps superimposed on the triangulated graphs representing the hippocampus in the human (top row) and the macaque neocortex (bottom row). The left column shows the triangulated meshes

7.7 FITTING QUADRATIC CHARTS AND CURVATURES ON SURFACES

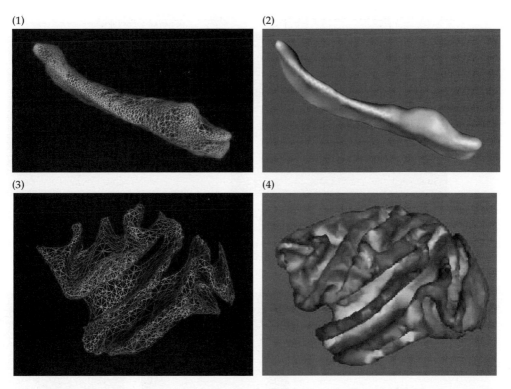

Figure 7.10 Top row: Panel 1 shows the triangulated graph representing the hippocampus generated by isocontouring; panel 2 shows the mean curvature map superimposed on the hippocampus surface of the human. Data taken from Haller et al. [162]. Bottom row: Panels 3 and 4 show the same as above for macaque cortical surface reconstructed from cryosection data from the laboratory of Dr. David Van Essen. Bright areas represent areas of high positive mean curvature; dark areas represent areas of high negative mean curvature (see Plate 4).

representing the surfaces; the right column shows the mean curvature representing the curving nature of the surfaces. To compute the mean curvature map, the shape matrix S_p is estimated at each point $p \in M_\triangle$ on the triangulated graph. With the local charts established at each vertex, then the mean curvature defined as the mean of the principle curvatures given by the eigenvalues of the shape matrix $S_p, p \in M_\triangle$ can be computed. The mean curvature maps (right column) were generated by fitting with the least-squares Algorithm 7.50. The mean curvature is depicted through a intensity scale. Note that the brightly lit areas correspond to areas of high curvature (valleys) and dark regions are areas of negative curvature (ridges). Notice in the cortex the places of deep folding.

Example 7.53 (Occipital Cortex) The mammalian cerebral cortex has the form of a layered, highly convoluted thin shell of grey matter surrounding white matter, with its folding pattern a current topic of intense scientific investigation. Coordinatizing the local gyral and sulcal submanifolds of the human brain is a significant area of research in the study of human shape. Figure 7.11 depicts reconstructions of the occipital cortex. Panel 1 shows the medial wall of the cortex depicting the occipital cortex (back right). Panel 2 shows the surface reconstruction (panel 2) of the occipital cortex depicting the major sulcal and gyral principal curves generated via dynamic programming including the inferior Calcarine sulcus and Lingual and Parietal sulci. The heat scale color map depicts the Gaussian curvature profiles. Data taken from Dr. Steven Yantis of the Johns Hopkins University.

(1)   (2)

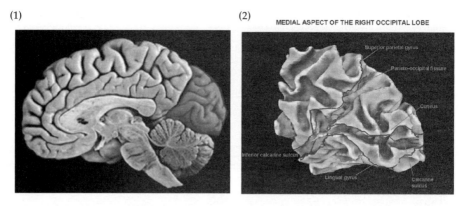

Figure 7.11 Top row: Panel 1 shows the atlas depiction of the occipital cortex. Panel 2 shows the reconstruction of the occipital cortex depicting the major sulcal and gyral principal curves including the inferior Calcarine sulcus and Lingual and Parietal gyri. Data taken from Dr. Steven Yantis of the Johns Hopkins University (see Plate 5).

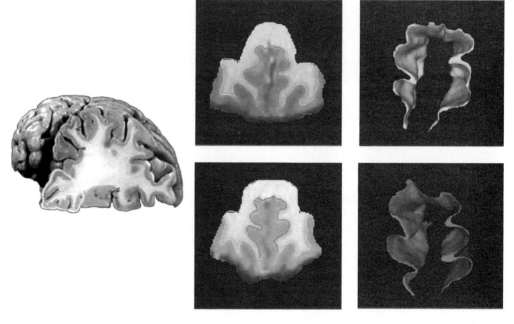

Figure 7.12 Left column shows the medial prefrontal cortex section from the Duvernoy atlas. Middle column top and bottom panels show the isosurface reconstruction of the prefrontal medial cortex. Sections through the two different MRI brains show the embedded surfaces. Right column, top and bottom panels, shows the medial prefrontal cortex reconstructions. Data taken from Dr. Kelly Botteron of Washington University (see Plate 6).

Example 7.54 (Medial Prefrontal Cortex Reconstruction) Figure 7.12 shows results from the segmentation and iscontour triangulated graph construction for the medial prefrontal cortex (MPFC) taken from Dr. Kelly Botteron of Washington University. The left column shows the medial prefrontal cortex section from the Duvernoy atlas. The middle column top and bottom panels show the isosurface reconstruction of the prefrontal medial cortex. Sections through the two different MRI brains show the embedded surfaces. The right column top and bottom panels shows the medial prefrontal cortex reconstructions.

7.8 Ridge Curves and Crest Lines

Ridge curves are important features in computational vision. Intuitively ridge curves on a surface are where the curvature is changing most rapidly. For a detailed description of ridge curves see [151, 163]. Ridge curves can be defined in terms of the shape operator. Begin with the local surface representation in $\mathbb{R}^3$ by the equation $z = f(u, v)$. The tangent space of the point $p(u, v) = (u, v, f(u, v))$ in local coordinates is given by the two vectors over the patch $E_1 = (1, 0, f_u(u, v))^*$, $E_2 = (0, 1, f_v(u, v))^*$. The principal curvature and directions are the eigenvalues and eigenvectors of the shape operator S, respectively, the shape operator taking the form $S = A^{-1} B$, where

$$A = \begin{pmatrix} 1 + f_u^2 & f_u f_v \\ f_u f_v & 1 + f_v^2 \end{pmatrix}, \quad B = (1 + f_u^2 + f_v^2)^{-1/2} \begin{pmatrix} f_{uu} & f_{uv} \\ f_{vu} & f_{vv} \end{pmatrix}.$$

Principal curves on the surface are the curves whose tangent directions always point in the principal direction. There are two principal curves passing through a point corresponding to the larger and smaller eigenvalues of the shape operator at that point. Along a principal curve the value of the principal curvature changes and at certain points has local extrema; these points are called the ridge points, and set of such points form the collection of curves called the ridge curves. There are two set of curves one for the larger, and one for the smaller eigenvalue. Each ridge point can be classified according to whether the curvature is maximal or minimal. We will deal with a special subset of the ridge curves called the crest lines, as described by [164]. Let us define an algebraic condition for a curve to be ridge curve.

Definition 7.55 *If $\vec{t} = (t_1, t_2)$ is a principal direction for the surface in $\mathbb{R}^3$ represented by $z = f(u, v)$, with non-zero principal curvature κ_1, then the point $p = (u, v, f(u, v))$ is on a* **ridge curve** *if and only if*

$$\begin{aligned} R(u, v) =& (t_1^3 f_{uuu} + 3 t_1^2 t_2 f_{uuv} + 3 t_1 t_2^2 f_{uvv} + t_2^3 f_{vvv}) \\ & - 3(1 + f_u^2 + f_v^2)^{1/2} (t_1^2 f_{uu} + 2 t_1 t_2 f_{uv} + t_2^2 f_{vv})(t_1 f_u + t_2 f_v) \kappa_1 = 0. \end{aligned} \quad (7.107)$$

Crest lines *are the loci of points whose maximal (in absolute value) principal curvature is a local extremum along a corresponding principal curve, expressed as*

$$\langle \nabla \kappa_m, t_m \rangle_{\mathbb{R}^3} = 0, \quad (7.108)$$

where t_m is the principal direction corresponding to maximal principal curvature κ_m.

7.8.1 Definition of Sulcus, Gyrus, and Geodesic Curves on Triangulated Graphs

Modern whole brain cryosection imaging provides excellent high resolution data required for the study of such anatomical features of cortex [155, 165] such as the arrangement of the sulcal fissures visible throughout the cortical surface of a mammalian brain with major sulci and gyri now being cataloged in human atlases which are becoming available [166, 167]. Computational metrics defined by cortical geometry such as geodesic length is drawing the attention of the neuroscience community in terms of the role of wiring length in the general layout of the nervous system [168]. Despite their anatomic and functional significance, the gyri, sulci, and many stable cortical pathways consistently appearing in all normal anatomies exhibit pronounced variability in size and configuration [169]. Methods are beginning to appear for characterizing their variation [156]. The sulci and gyri exhibit strong features associated with their extrema of bending. The deepest beds of the sulci are called the fundus beds; associated with the gyri are the crowns. Empirical evidence [156, 166, 167] suggests that in most part of their length fundus beds resemble crest lines (see [151, 170]) corresponding to points where the maximal absolute principal curvature has, a local maximum. There exist algorithms for extracting ridge and crest lines from surface geometries [151, 163, 170] defined as the loci of points $x \in S$ where the maximal absolute principal curvature $\kappa_{\max}(x)$ has a maximum. At these points, $\langle \nabla \kappa_{\max}, t_{\max} \rangle_{\mathbb{R}^3} = 0$ for $t_{\max}$ the principal unit direction corresponding to the maximal principal curvature, $\kappa_{\max}$. Such zero tracing methods are sensitive to noise. An alternate approach based on dynamic programming for tracking optimal trajectories on surfaces has noise immunity. Instead of finding the extremum of principal curvature using higher derivatives of curvature, we define a sequentially additive energy associated with candidate curves and use dynamic programming for its minimization. For this define the cost of a candidate curve $\alpha(s,t)$ to be $\int_{\alpha(s,t)} (\kappa_{\max}(x) - \mathcal{K})^2 \, d\alpha$, with $\mathcal{K}$ assigned the largest maximal curvature on the surface, and minimize over all such paths on the triangulated graph representation of the surface. For a surface symmetrical about a crest line, i.e. one where $t_{\max}$ is perpendicular to the crest line, minimizing gives $(\kappa_{\max}(x) - \mathcal{K}) \langle \nabla \kappa_{\max}, t_{\max} \rangle_{\mathbb{R}^3} = 0$ implying $\langle \nabla \kappa_{\max}, t_{\max} \rangle_{\mathbb{R}^3} = 0$ which is precisely the equation for the crest line. For regions of the sulcus where the basin is flat, $\kappa_{\max}$ is constant, and the minimizer of the above functional produces shortest paths through these regions.

For generating *geodesics*, we adapt the continuum definition associated with length on the surface as measured by the integral of the norm of the tangent vector along the curve. If s and t are points of a smooth connected surface $M \subset \mathbb{R}^3$, the intrinsic distance $\rho(s,t)$ from s to t in M is the lower bound of the lengths of these curves. The curve $\hat{\alpha}$ for which the minimum length is achieved is called a geodesic. Throughout we denote the triangulation of points on the surface as i, j, and the set of coordinates in $\mathbb{R}^3$ taken by the sites of the graph and or positions of the triangle vertices as $x_i, x_j \in \mathbb{R}^3$.

Definition 7.56 *Given a 2D triangulation of the surface* $\{x_i \in M\}$ *define the* **platelet** $\mathcal{P}_i$ *of point i as the set of triangles (with index-triples (j_1, j_2, j_3) specifying their vertices) sharing x_i as a common vertex* $\mathcal{P}_i = \cup \{(j_1, j_2, j_3) | x_i = x_{j_1} \text{ or } x_i = x_{j_2} \text{ or } x_i = x_{j_3}\}$.

Define a path on the surface $\alpha(s,t)$ routed and terminated, respectively, in nodes s on the surface as

$$\alpha(s,t) = (s = j_1, j_2), (j_2, j_3), \ldots, (j_{k-1}, j_k), \ldots, (j_{N-1}, t = j_N), \quad \text{such that}$$

$$j_k \in \mathcal{P}_{j_{k-1}}, \forall k,$$

and the collection of all paths connecting (s,t) as $\alpha(s,t) \in P_{s,t}(S)$.

Define the N-length discrete geodesic and discrete fundus bed as cost minimizing paths given by

$$\text{fundus}(s,t) = \arg \min_{\alpha(s,t) \in P_{s,t}(S)} \sum_{k=1}^{N} d_{\text{fundus}}(j_k, j_{k+1}), \text{ where} \qquad (7.109)$$

$$d_{\text{fundus}}(j_k, j_{k+1})$$

$$= \left(\frac{(\kappa_{\max}(x_{j_k}) + \kappa_{\max}(x_{j_{k+1}}))}{2} - \mathcal{K}^2 \right) \|x_{j_k} - x_{j_{k+1}}\|; \qquad (7.110)$$

$$\text{geodesic}(s,t) = \arg \min_{\alpha(s,t) \in P_{s,t}(S)} \sum_{k=1}^{N} d_{\text{geo}}(j_k, j_{k+1}), \text{ where} \tag{7.111}$$

$$d_{\text{geo}}(j_k, j_{k+1}) = \sqrt{(x_{1,j_k} - x_{1,j_{k+1}})^2 + (x_{2,j_k} - x_{2,j_{k+1}})^2 + (x_{3,j_k} - x_{3,j_{k+1}})^2}.$$

Notice the cost of a discrete curve $\alpha(s)$ connecting (s,t) is defined by assuming piecewise constant function between successive nodes for the curvature integral $\int_{\alpha(s,t)} (\kappa_{\max}(x) - \mathcal{K})^2 \, dx$.

7.8.2 Dynamic Programming

Examine an approach based on dynamic programming for tracking optimal trajectories on surfaces similar to that done in boundary and artery tracking [171, 172]. We search for curves that pass through regions of highest maximal curvature joining the prespecified start and end points in the surface. For generating these curves and geodesics we follow Khaneja et al. [153, 154] using dynamic programming adapted to optimization on triangulated surfaces. Denote the finite state space S of size $\|S\| = N$; on these triangulated graphs the positions of the nodes of the surface itself are the states. The goal is to compute optimal shortest paths between the specified initial states s and the final state t. Assuming that the optimal path has no more than K nodes, the total number of paths of length K between points s and t are of the order N^K. If the cost is additive over the length of the path, dynamic programming reduces the complexity of the search algorithm to order of KN^2. Let $c^k(x_k, x_{k+1})$ denote the cost incurred for the transition from state $x_k \in S$ to $x_{k+1} \in S$ at each time k. Suppression of k dependence in $c(i,j)$ means the cost is independent of time. We shall assume that $c(i,j) \geq 0$, and arcs of infinite cost $c(i,j) = \infty$ signify that there is no arc from node i to node j. An optimal path need not have more than N arcs (number of nodes in the graph) and hence take no more than N moves. We formulate the problem as one of finding the optimal path in exactly N moves allowing degenerate moves from a node i to itself with cost $c(i,i) = 0$. The degenerate moves signify that the length of the path may be less than N.

The efficiency of dynamic programming on the triangulated graphs representing the surface is that the states spaces $S_k \subset S$ can be dynamically defined and of reduced complexity. Curves passing through a point on the graph must pass through one of it neighbors (analogos to being in the tangent space for the continuum representation).

Algorithm 7.57 (Dynamic Programming Algorithm) *Denote the optimal cost for getting from node i to node t in $(N - k)$ moves as $J_k(i), i \in S, k = 0, 1, \ldots, N - 1$. Then the optimal N-length path $J_0(i)$ from i to t is given by the final step of the following algorithm,*

$$\text{with } J_{N-1}(i) = c^{N-1}(i,t), \text{ and} \tag{7.112}$$

$$J_k(i) = \min_{j=1,\ldots,N} \{c^k(i,j) + J_{k+1}(j)\}, \quad k = 0, 1, \ldots, N - 2, \quad i \in M. \tag{7.113}$$

Define the state spaces dynamically $S_{N-1} = \{i | i \in \mathcal{P}_t\}$, $S_k = \{i | i \in \mathcal{P}_j, j \in S_{k+1}\}$, and implement the algorithm according to Initialize: $J_k(i) \leftarrow \infty$ $i \neq t$, for all k, $S_N \leftarrow t$, $J_k(t) \leftarrow 0$;
  For $k \leftarrow N - 1$ down to 0 do

$$S_k \leftarrow \{i \mid i \in \mathcal{P}_j, j \in S_{k+1}\}, \quad \text{set } c^k(i,j), j \in S_{k+1}, i \in S_k, \tag{7.114}$$

with $J_{N-1}(i) = c^{N-1}(i,t), i \in S_{N-1}$, (7.115)

$$J_k(i) = \min_{j \in \{S_{k+1} \cap \mathcal{P}_i\}} \{c^k(i,j) + J_{k+1}(j)\}, \quad i \in S_k, \quad k = 0, 1, \ldots, N-2. \quad (7.116)$$

Theorem 7.58

1. **Geodesic generation.** *Given the costs for transition*

$$c^k(i,j) = d_{\text{geo}}(i,j), \quad j \in \mathcal{P}_i, \quad c^k(i,j) = \infty \text{ for } j \notin \mathcal{P}_i,$$

then the length $J_0(s)$ in the algorithm is a a geodesic (not necessarily unique) between nodes s and t:

$$\text{geodesic}(s,t) = \arg \min_{\alpha(s,t) \in P_{s,t}(S)} \sum_k d_{\text{geo}}(k, k+1).$$

2. **Fundus curve generation.** *Given the costs for transition*

$$c^k(i,j) = d_{\text{fundus}}(i,j), \quad j \in \mathcal{P}_i, \quad c^k(i,j) = \infty, \quad j \notin \mathcal{P}_i,$$

then the length $J_0(s)$ in the algorithm is an optimal principal curve from nodes s to t:

$$\hat{\alpha}(s,t) = \arg \min_{\alpha(s,t) \in P_{s,t}(S)} \sum_k d_{\text{fundus}}(k, k+1).$$

Proof The original proof was provided by Khaneja et al. [154] which follows since the costs are additive over the curves. For $k = N-1$, $J_{N-1}(i) = c(i,t), i = 1, 2, \ldots, N$, with $c(i,t) = \infty$ for $i \notin \mathcal{P}_t$, implying that there is no one arc path from i to t. Therefore the only values of J that get updated in the first step are for $S_{N-1} = \{i \in \mathcal{P}_t\}$. Writing the kth step in the original algorithm $J_k(i) = \min_{j=1,\ldots,N} \{c(i,j) + J_{k+1}(j)\}, k = 0, 1, \ldots, N-2$, observe that by definition S_k consists of all j from which the terminal node t can be reached via a curve with less than or equal to $N-k$ arcs or $N-k$ moves (degenerate moves allowed). Therefore if $j \notin S_{k+1}$ then $J_{k+1}(j) = \infty$, and as in previous case if $i \notin \mathcal{P}_j$, $c(i,j) = \infty$, in either case $J_k(i) = \infty$. Hence only $i \in S_k$ need be considered $J_k(i) = \min_{j \in \{S_{k+1} \cap \mathcal{P}_i\}} \{c^k(i,j) + J_{k+1}(j)\}, i \in S_k, k = 0, 1, \ldots, N-1$. □

By using the Riemannian length function for the minimizing cost then the shortest length geodesic paths on the surface can be generated. Shown in Figure 7.13 are examples of such shortest length paths on the macaque cortex. The bottom row of Figure 7.13 depicts results on the optimality of DP on the superior temporal gyrus (STG) and also shows a reconstruction of the STG. Panels 4–7 show the DP generation of principal curves on the superior temporal gyrus of the Visible Human. Panel 4 shows various cortical gyri depicted in different colors by Van Essen in his reconstruction of the Visible Human cryosection. Panel 5 shows the temporal gyrus extracted from the Visible Human in $\mathbb{R}^3$ with dynamic programming generated fundus curves. Panel 6 shows principal curves choosing multiple start and end points for the dynamic programming solution finding common flows on the surface. The paths are illustrated on the planar representation of the STG illustrating robustness of the solution. The DP algorithm was run for multiple starting and end points. Notice how the trajectories initiated at different starting points merge into a common trajectory, illustrating the optimality of the path. Panel 7 shows the dynamic programming generation of the superior temporal sulcus jumping across the break connecting the start and end points which were manually selected.

Example 7.59 (Cutting Surfaces with Dynamic Programming) Figure 7.14 shows a depiction of the surface of the superior temporal gyrus and generation of its boundaries via dynamic programming. The triangulated graphs represent the gray/white boundary of the STG in MRI data. Panels 1 and 2 show the whole brain and STG

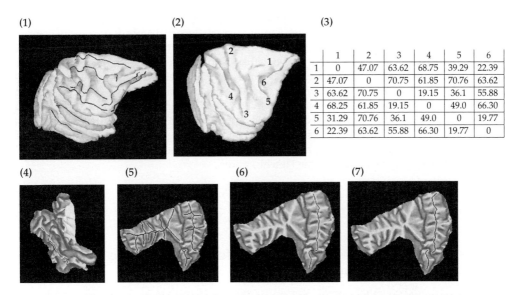

Figure 7.13 Top row: Panel 1 shows eight geodesics generated on the neocortex by picking the start and end points manually. Panel 2 depicts geographical landmarks on the macaque cortex; labels 1, 2, 3, 4, 5, 6. Panel 3 shows a table of Riemannian distances in millimeters between the predefined points. Data taken from the laboratory of David Van Essen, Washington University. Bottom row: Figure shows optimality of dynamic programming. Panel 4 shows the Visible Human cortex extracted by David Van Essen; panel 5 shows choosing multiple terminal points for the DP solution; panels 6, 7 show the DP generation of the superior temporal sulcus jumping across the break connecting the start and end points which were manually selected (see Plate 7).

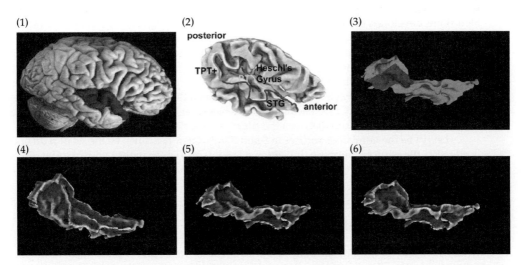

Figure 7.14 Top row panel 1 shows the external view of the Superior Temporal Gyrus (STG). Panel 2 shows Heschl's gyrus and the posterior boundary of the plenum temporale (PT) defined via dynamic programming. Panel 3 shows the delineation of the STG surface into two with the PT as the blue region which is extracted. Bottom row shows the application of dynamic programming to extract the PT from STG surface. Panel 4 tracks Heschl's gyrus; panel 5 tracks the STG as far as the posterior ascending (or descending) ramus; panel 6 tracks the geodesic from the end of the STG to the retro-insular end of the Heschl's gyrus. Data taken from the laboratory of Drs. Godfrey Pearlson and Patrick Barta, reconstructions from Dr. Tilak Ratnanather (see Plate 8).

reconstructions. To extract the planum temporale (PT) from the reconstructed STG surface, we use dynamic programming to track the gyral and sulcal principal curves defining the boundaries of the submanifold graphs. The PT is delineated by the HG and the STG as far as at the start of either the posterior ascending ramus or posterior descending ramus. The posterior boundary of PT is defined by the geodesic from the end of the STG and the retro-insula point of the HG. Panel 3 shows the delineation of the STG surface into two with the PT as the blue region that is extracted. The bottom row shows the application of dynamic programming to extract the PT from STG surface. Panel 4 tracks Heschl's gyrus with panel 5 tracking the STG as far as the posterior ascending (or descending) ramus. Panel 6 tracks the geodesic from the end of the STG to the retro-insular end of the Heschl's gyrus.

7.9 Bijections and Smooth Mappings for Coordinatizing Manifolds via Local Coordinates

Various investigators have been studying methods for mapping the surface manifold to spherical and planar coordinates [166–168, 173–177]. The approach being taken is to define local diffeomorphisms between the submanifolds and their locally Euclidean representation. To construct the local diffeomorphisms $\phi : S \to D \subset \mathbb{R}^2$ between the 2D manifold $S \subset \mathbb{R}^3$ of the cortical surface to the plane we describe methods based on quasi-conformal flat mapping which is a computational procedure for implementing the Riemann mapping theorem developed [178, 179]. The computational approach for generating ϕ relies on the "flattening" of the triangulated graphs representation $S(\Delta)$ of the original surface S. Once a "flattened graph" has been generated then the diffeomorphism is associated with the bijection between the vertices in the original triangulate surface and the discrete points in $D \subset \mathbb{R}^2$. The quasi-conformal flat mapping is computed via the circle packing algorithms developed on triangulated graphs, which control angular distortion [180, 181].

Let the triangulated graph be denoted by $S(\Delta) = (N, V)$ where N is the neighborhood structure or connectivity array of the graph vertices and V is the set of coordinates of the graph vertices. The goal is to generate an equivalent graph $S' = (N, V')$ such that the curvature at each vertex of V' is zero, i.e. a flattened graph with a one-to-one correspondence between vertices of V and V'. This is achieved by circle packing for S' in which the vertices are the center of circles that are tangent to each other. Circle packing [180] is based on the Riemann Mapping Theorem [181] in which the angular measure between the edges is preserved.

Let $|V|$ be the number of vertices in S and $R = \{r_v, v = 1, \ldots, |V|\}$ be the set of radii of circles centered at the vertices. Suppose the vertex v has k faces (i.e. triangles) with associated vertices $F_v = \{v : v_1, \ldots, v_{k+1} \text{ s.t. } v_{k+1} = v_1\}$, then the angle sum at v is

$$\theta(v; R) = \sum_{\langle v,u,w \rangle : u,w \in F_v} \alpha(r_v; r_u, r_w),$$

where the triple $\langle v, u, w \rangle$, denotes the face with vertices v, u, w, and α is the angle of the face at v subtended by the edge joining u and w readily computed via the cosine rule

$$\alpha(r_v; r_u, r_w) = \arccos\left(\frac{(r_v + r_w)^2 + (r_v + r_u)^2 - (r_u + r_w)^2}{2(r_v + r_w)(r_v + r_u)}\right).$$

The angle sum at the vertex is a measure of the curvature that is concentrated at the vertex. Thus for flatness, it is required that $\theta(v; R) = 2\pi$. If $\theta(v; R) < 2\pi$, curvature at v is positive like

7.9 BIJECTIONS AND SMOOTH MAPPINGS

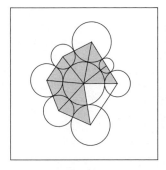 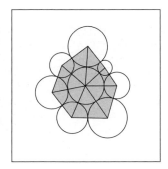 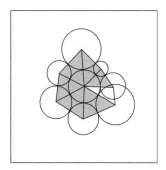

Figure 7.15 Panel 1 (Panel 3) show the radius of circle at vertex with positive (negative) curvature must be reduced (increased) to ensure that circles at all vertices are tangent to each other as shown in Panel 2. Figure from Ratnanather et al. (2003).

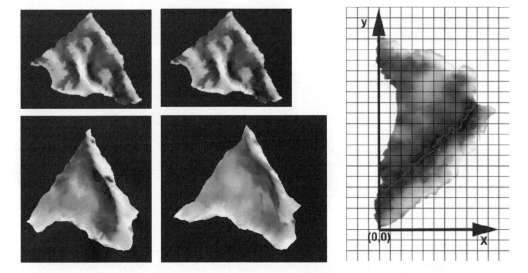

Figure 7.16 Left column: Top two panels show automatically and hand generated PT surfaces from one MRI brain; bottom two panels show a second PT generated automatically and by hand contouring. Right column shows the surface of the left PT from the STG shown superimposed with the mean-curvature map drawn over the planar coordinates at the location defined by the bijection $\phi : S(\triangle) \to D$. The retro-insular end of the HG and the positive y-axis passes through the posterior STG where the ramus begins. Also superimposed is the Heschl's sulcus in blue generated by dynamic programming tracking on the original surface. Data taken from the laboratory of Drs. Godfrey Pearlson and Patrick Barta; reconstructions from Dr. Tilak Ratnanather (see Plate 9).

a cone point alongwith a gap between circles around v as in panel 1 of Figure 7.15 which is rectified by reducing the radii of the circles around v until tangency is attained as in panel 2 of Figure 7.15. If $\theta(v; R) > 2\pi$, curvature at v is negative like a saddle point with overlapping circles as in panel 3 of Figure 7.15 which is rectified by expanding the radii of the circles around v until all circles are tangent. For our computational procedure we use the methods of Hurdal et al. [178, 182].

Thus configuration $S' = (N, V')$ is achieved by iteratively computing radii of the circles at interior vertices, with fixed radius at the boundary vertices equal to the average of half of the lengths of the two edges of the vertex. In practice, instead of repeated refinement of triangulation giving zero angular distortion, angular distortion is minimized. The iteration is initialized by assigning arbitrary radii at interior vertices and convergence is accelerated based on the algorithm developed by Collins and Stephenson [180]. The result is an Euclidean flat map which preserves Euclidean

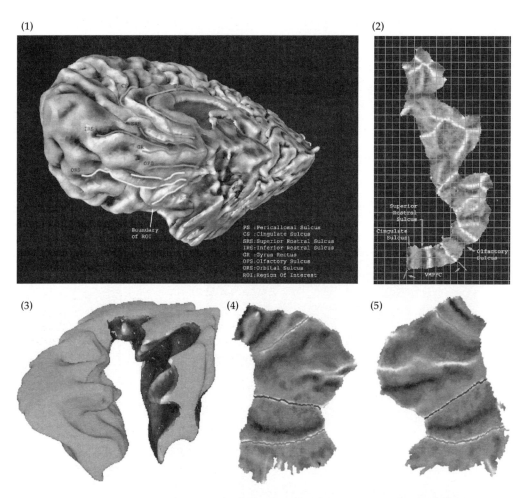

Figure 7.17 Top row: Panel 1 shows the reconstruction of the cortical surface with the curvature map superimposed. Shown depicted are various sulcal principal curves generated via dynamic programming. Panel 2 shows the planar representation of the medial cortex. Bottom row: Panel 3 shows the reconstruction of the left and right MPFC from Dr. Kelly Botteron. Panels 4 and 5 show the planar maps of the MPFC reconstructions superimposed curvature profiles. Data taken from Dr. Kelly Botteron of Washington University (see Plate 10).

lengths on the edges. The quasi-conformal flat map is normalized by assigning a landmark to an origin and another vertically above the origin.

Example 7.60 (Planum Temporale in the Temporal Gyrus) Figure 7.16 depicts such results on the planum temporal in the superior temporal gyrus from Ratnanather et al. [183]. The left column depicts automatically (panels 1 and 3) and hand contour generated (panels 2 and 4) PT surfaces from two MRI brains. The middle column shows the left PT surface from the STG with the superimposed mean-curvature map. The blue line depicts Heschl's sulcus in blue generated by dynamic programming. The right column shows the same curvature map drawn over the planar coordinates at the location defined by the diffeomorphism $\phi : S(\Delta) \to D \subset \mathbb{R}^2$. The retro-insular end of the HG and the positive y-axis passes through the posterior STG where the ramus begins. Also superimposed is the Heschl's sulcus in blue generated by dynamic programming tracking on the original surface.

The quasi-conformal mapping algorithm of Hurdal permits the definition of a coordinate system on the PT manifold. The origin must be unambiguously present. Thus the retro-insular end of the HG where the HG meets the SF is defined as the origin. The y-axis is aligned such that it passes through the point on the posterior STG where the ramus begins.

Example 7.61 (Medial Prefrontal Cortex Reconstruction) Shown in Figure 7.17 are the global reconsructions of the Medial Prefrontal cortex from Dr. Kelly Botteron. Shown in the top row panel 1 is the reconstructions of the MPFC with superimposed heat scale color map showing the Gaussian curvature profiles superimposed on the medial walls defining the MPFC. Panel 2 shows the same curvature maps shown mapped to planar coordinates $D \subset \mathbb{R}^2$ via the diffeomorphisms $\phi : S \to D$ generated via the circle packing algorithm of Hurdal applied to the left and right MPFC. Shown superimposed are dynamic programming generated contours depicting the various sulcal principle curves. The color map depicts the Gaussian curvature profiles of the MPFC. Shown in panels 4 and 5 are the curvature maps shown mapped to planar coordinates $D \subset \mathbb{R}^2$ via the diffeomorphisms $\phi : S \to D$ generated via the circle packing algorithm applied to the left and right MPFC. Shown superimposed are dynamic programming generated principal curves.

8 MANIFOLDS, ACTIVE MODELS, AND DEFORMABLE TEMPLATES

ABSTRACT To study shape we introduce manifolds and submanifolds examined in the continuum as the generators. Transformations are constructed which are built from the matrix groups and infinite products. This gives rise to many of the widely used structural models in image analysis often termed active models, essentially the deformable templates. These deformations are studied as both diffeomorphisms as well as immersions. A calculus is introduced based on transport theory for activating these deformable shapes by taking variations with respect to the matrix groups parameterizing them. Segmentation based on activating these manifolds is examined based on Gaussian random fields and variations with respect to the parameterizations.

8.1 Manifolds as Generators, Tangent Spaces, and Vector Fields

From the previous two Chapters 6, 7 it is clear that for studying shape the generators are often the structured sets of line segments, curves, quadrics, subvolumes, to name some. Thus the quantitive study of shape and variability behoves us to study quantitatively manifolds in $\mathbb{R}^n$, and their transformation.

8.1.1 Manifolds

Roughly speaking, an n-dimensional smooth manifold will locally look like $\mathbb{R}^n$ with the property that it has local coordinates which overlap smoothly.

> **Definition 8.1** *A* **manifold** *M of dimension m is a set with a set of open sets (topology)[19] which is locally Euclidean of dimension m, i.e. for each $p \in M$, there exists a smooth local* **coordinate chart** *which is a diffeomorphism $\phi = (\phi_1, \ldots, \phi_m) : O \to D \subset \mathbb{R}^m$ (a 1-1, onto bijection with both ϕ, ϕ^{-1} continuous and differentiable) from an open neighborhood $O \subset M$ containing p to an open set $D \subset \mathbb{R}^m$.*
>
> *A manifold is called a C^∞* **smooth manifold** *if there exists a family of coordinate charts (O_α, ϕ_α) covering the manifold, pairwise compatible, so that if $O_\alpha \cap O_\beta$ nonempty implies that the maps $\phi_\alpha \circ \phi_\beta^{-1} : \phi_\beta(O_\alpha \cap O_\beta) \to \phi_\alpha(O_\alpha \cap O_\beta)$ giving the change of smooth coordinates.*

This is depicted in the following equation:

$$p \in O \subset M \tag{8.1}$$

$$\Downarrow \phi \tag{8.2}$$

$$(x_1, \ldots, x_m) = \phi(p) \in D \subset R^m \tag{8.3}$$

---

[19] A topology $\mathcal{T}$ on a set X is a collection of open subsets of X satisfying the following axioms: unions of open sets are open, finite intersections are open, and $X, \emptyset$ are both open (closed). See Boothby [148] for additional properties of Hausdorff and countability of basis in the definition of the manifold. We will take these as given.

8.1 MANIFOLDS AS GENERATORS, TANGENT SPACES, AND VECTOR FIELDS

Example 8.2 (The Circle) The circle S^1 is $M = \{(x,y) \in \mathbb{R}^2 : x^2 + y^2 = 1\}$, choose for the point $p = (1,0)$ the chart $O \subset N, \phi$ with O the right half circle given by the open set $\{(x,y) \in \mathbb{R}^2 : x > 0\} \cap S^1$, $\phi : O \to D \subset \mathbb{R}, \phi^{-1} : D \to O \subset N$ to be

$$\phi^{-1}(x) = (\cos x, \sin x) \qquad x \in D = \left(-\frac{\pi}{2}, \frac{\pi}{2}\right),$$

$$\phi(p_1, p_2) = \arcsin(p_2) \qquad (p_1, p_2) \in O = \phi^{-1}\left(\frac{-\pi}{2}, \frac{\pi}{2}\right).$$

Example 8.3 (Linear Manifolds and Equality Constraints) Linear manifolds are well known through equality constraints in vector spaces. Let $X = \mathbb{R}^n$, and $M \subset X$ defined by

$$M = \left\{ y \in \mathbb{R}^n : \sum_i \alpha_i y_i = 0 \right\}. \tag{8.4}$$

Then M is an $m = n - 1$ dimensional manifold with coordinate chart $(O = M, \phi)$ with $\phi : M \to \mathbb{R}^{n-1}$ given by

$$\phi^{-1} : (x_1, \ldots, x_{n-1}) \in \mathbb{R}^{n-1} \mapsto \sum_{i=1}^{n-1} x_i b_i \tag{8.5}$$

where $\{b_i \in \mathbb{R}^n, i = 1, \ldots, n-1\}$ span the orthogonal complement of the 1D vector space $(\alpha_1, \ldots, \alpha_n)$. Assuming the b_i's are orthogonal, then $\phi(y) = (\langle b_1, y \rangle_{\mathbb{R}^n}, \ldots, \langle b_{n-1}, y \rangle_{\mathbb{R}^n}) \in \mathbb{R}^{n-1}$ for all $y \in M$.

8.1.2 Tangent Spaces

Roughly speaking, a *tangent vector* to a manifold M at point $p \in M$ will correspond to tangents to curves on the manifold through the point p. The collection of all tangents generated from tangent curves on M through p will be denoted $T_p(M)$ the tangent space which will be a vector space. Figure 8.1 shows an example of a 2-D manifold M with an associated curve.

Tangent vectors to M at $p \in M$, and the tangent vector space $T_p(M)$ is defined as the vector space generated from a basis $\{E_i, i = 1, \ldots, m\}$ of the tangent space.

Definition 8.4 *Define the **local coordinate frames** for $M = \mathbb{R}^m$ to be $E_i = \partial/\partial x_i$, $i = 1, \ldots, m$, with action on smooth functions*

$$E_i f = \frac{\partial f}{\partial x_i}, \qquad i = 1, \ldots, m. \tag{8.6}$$

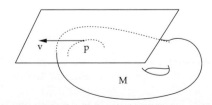

Figure 8.1 Panel shows a curved submanifold $M \subset \mathbb{R}^3$ with tangent plane $T_p(M)$ at point $p \in M$.

For manifold M curved with local coordinates $(O \subset M, \phi)$ containing p, then define the *m*-**local coordinate frames** at $p \in M$ as $E_{ip}, i = 1, \ldots, m$ taking action

$$E_{ip}f = \frac{\partial}{\partial x_i}(f \circ \phi^{-1})\Big|_{x=\phi(p)}, \quad i = 1, \ldots, m. \tag{8.7}$$

The coordinate frames form a basis for the **tangent space** $T_p(M)$ of M at p which is a vector space given by the set of all vectors generated from the basis:

$$V_p = \sum_{i=1}^{m} \alpha_i E_{ip}. \tag{8.8}$$

The most important properties of the basis are that they are an *m*-dimensional span, and they satisfy the Liebnitz rule of ordinary calculus. [20]

Theorem 8.5 *The coordinate frames $E_{1p}, \ldots, E_{mp}$ of $T_p(M)$ in Definition 8.4 satisfy (i) independence implying $\dim T_p(M) = m$, and (ii) Liebnitz rule*

$$E_i(fg) = gE_if + fE_ig. \tag{8.9}$$

Proof To see that the $E_{ip}, i = 1, \ldots m$ are independent assume that they are not. This would imply that for some $i \neq j$, then for all $f \in C^\infty(p)$ $E_{ip}f = \sum_{j \neq i} \alpha_j E_{jp}f$, and choose $f = \phi_i$ the ith local coordinate function of $\phi = (\phi_1, \ldots, \phi_m)$ implying

$$E_{ip}f = \frac{\partial}{\partial x_i}(\phi_i \circ \phi^{-1})\Big|_{\phi(p)} = \frac{\partial}{\partial x_i}x_i = 1 \tag{8.10}$$

$$= \sum_{j \neq i} \alpha_j E_{jp}f = \sum_{j \neq i} \alpha_j \frac{\partial}{\partial x_j}(\phi_i \circ \phi^{-1})\Big|_{\phi(p)} = 0. \tag{8.11}$$

Contradiction, so the E_{ip} must be independent. For Liebnitz rule,

$$E_{ip}(fg) = \frac{\partial}{\partial x_i}(fg) \circ \phi^{-1} = \frac{\partial}{\partial x_i}\left(f \circ \phi^{-1}\right)\left(g \circ \phi^{-1}\right)$$

$$= g \circ \phi^{-1}\Big|_{\phi(p)} \frac{\partial}{\partial x_i}f \circ \phi^{-1}\Big|_{\phi(p)} + f \circ \phi^{-1}\Big|_{\phi(p)} \frac{\partial}{\partial x_i}g \circ \phi^{-1}\Big|_{\phi(p)}$$

$$= gE_{ip}f + fE_{ip}g. \tag{8.12}$$

$\square$

Example 8.6 Of course, then any $V_p \in T_p(M)$ can be written $V_p = \sum_{i=1}^{m} \alpha_i E_{ip}$, with the α_i calculated by operating the tangent vector on the ith local coordinate function ϕ_i of $\phi = (\phi_1, \ldots, \phi_m)$ according to

$$V_p\phi_i = \sum_{j=1}^{m} \alpha_j E_{jp}\phi_i = \sum_{j=1}^{m} \alpha_j \frac{\partial}{\partial x_j}(\phi_i \circ \phi^{-1})\Big|_{\phi(p)} = \sum_{j=1}^{m} \alpha_j \frac{\partial}{\partial x_j}x_i\Big|_{\phi(p)} = \alpha_i. \tag{8.13}$$

Example 8.7 (Directional Derivatives) These are familiar as directional derivatives used all the time in optimization via Lagrange multipliers to constrain the optimization

---

[20] A standard way to define the tangent space of M at p abstractly (as in Boothby [148]) is to define it as the space all tangent vectors V_p at $p \in M$ is a map $V_p : C^\infty(p) \mapsto \mathbb{R}$ satisfying linearity and Liebnitz rule, so that for all f, g smooth, then (i) $V_p(af + bg) = aV_pf + bV_pg$, $a, b \in \mathbb{R}$, and (ii) $V_p(fg) = f(p)V_pg + g(p)V_pf$.

8.1 MANIFOLDS AS GENERATORS, TANGENT SPACES, AND VECTOR FIELDS

to a constraint space (manifold). Let $M = \mathbb{R}^m$, with tangent space $\mathbb{R}^m$ itself so that $\phi = $ id the identity map, and the global coordinate frames (independent of p) become $E_{ip} = \partial/\partial x_i, i = 1,\ldots,m$. Then for $V_p = \sum_{i=1}^{m} \alpha_i(\partial/\partial x_i)$ and $V_p f$ is just the directional derivative in the direction of α:

$$V_p f = \langle \alpha, \nabla f \rangle_{\mathbb{R}^m}. \tag{8.14}$$

Note that the basis is independent of p, the position at which the tangent space $T_p(\mathbb{R}^m)$ is attached: the special property of Euclidean spaces.

Example 8.8 (Circle and Sphere) Return to Example 8.2. The natural coordinate frame E_{1p} becomes $E_{1p} f = (\partial/\partial x) f \circ \phi^{-1}$ giving

$$E_{1p} f = \left. \frac{\partial}{\partial x} f \circ \phi^{-1} \right|_{\phi(p)} = \left. \frac{\partial f}{\partial y_1} \frac{d \cos x}{dx} \right|_{\arcsin p_2} + \left. \frac{\partial f}{\partial y_2} \frac{d \sin x}{dx} \right|_{\arcsin p_2}. \tag{8.15}$$

The coordinate frame becomes $E_p = -p_2(\partial/\partial y_1) + p_1(\partial/\partial y_2)$ which is a globally defined, dimension 1 vector field on all of S^1.

For surfaces, then for each point $p \in M$ there exists a local coordinate system $\phi^{-1} : x \in \mathbb{R}^2 \to M$ given by

$$\phi^{-1}(x_1, x_2) = (\phi_1^{-1}(x_1, x_2), \phi_2^{-1}(x_1, x_2), \phi_3^{-1}(x_1, x_2)). \tag{8.16}$$

For the 2-sphere $S^2 = \{(y_1, y_2, y_3) : y_1^2 + y_2^2 + y_3^2 = 1\}$, use the azimuth-elevation coordinates

$$\phi^{-1}(x_1, x_2) = (\cos x_1 \sin x_2, \cos x_1 \cos x_2, \sin x_1).$$

The two coordinate frames applied to smooth functions $E_{ip} f = (\partial/\partial x_i) f \circ \phi^{-1}|_{\phi(p)}$ are

$$E_{1p} = -\sin x_1 \sin x_2 \frac{\partial}{\partial y_1} - \sin x_1 \cos x_2 \frac{\partial}{\partial y_2} + \cos x_1 \frac{\partial}{\partial y_3},$$

$$= -\frac{p_3 p_1}{\sqrt{1-p_3^2}} \frac{\partial}{\partial y_1} - \frac{p_3 p_2}{\sqrt{1-p_3^2}} \frac{\partial}{\partial y_2} + \sqrt{1-p_3^2} \frac{\partial}{\partial y_3}, \tag{8.17}$$

$$E_{2p} = \cos x_1 \cos x_2 \frac{\partial}{\partial y_1} - \cos x_1 \sin x_2 \frac{\partial}{\partial y_2}$$

$$= \frac{p_2}{\sqrt{1-p_3^2}} \frac{\partial}{\partial y_1} - \frac{p_1}{\sqrt{1-p_3^2}} \frac{\partial}{\partial y_2}. \tag{8.18}$$

Notice, this is not a globally defined basis of dimension 2; at $x_1 = \pi/2$, $E_{2p} = 0$.

8.1.3 Vector Fields on M

Associating to every point of the manifold a vector which varies smoothly across the manifold is a vector field.

Definition 8.9 *A smooth vector field V on M assigns to each $p \in M$ a tangent vector $V_p \in T_p(M)$ with components in the coordinate frames which are smooth.*

The coordinate frames being E_{ip}, then the function $V_p f = \sum_{i=1}^{m} \alpha_i(p) E_{ip} f$ is a smooth function.

Vector fields will be important for us in their role in defining a basis for the tangent spaces. A set of m-vector fields on a manifold of dimension m which are linearly independent at every point in the manifold form a basis for the tangent spaces at each point of the manifold. In general it is not possible to find a set of m independent vector fields on an arbitrary manifold. For $\mathbb{R}^m$ it is straightforward, $\partial/\partial x_i, i = 1, \ldots, m$ are such an example; for matrix groups it will be possible as well.

This property is so significant that it is given a name.

Definition 8.10 *A manifold M of dimension m with the property that there exists a full set of m coordinate frames is called* **parallelizable**.

8.1.4 Curves and the Tangent Space

The tangent space can be defined via the curves as the set of all tangent vectors of curves $x(\cdot)$ on M. Depicted in Figure 8.1 is an example of a curve with tangent vector $v_{x(t)=p}$.

Definition 8.11 *A* **curve** *on the manifold M of dimension m is a smooth mapping from an open real interval to the manifold M according to* $x : D \subset \mathbb{R}^1 \to M$.

The tangents to these curves will be elements of the tangent space of the manifold.

Theorem 8.12 *Given the curve $\Phi_t \in M$ on the manifold M rooted in $p \in M$ so that $\Phi_0(p) = p$, then the vectors $V_{\Phi_{t_0}(p)}$ for $p \in M$ defined by tangents to curves through the manifold*

$$V_{\Phi_{t_0}(p)} f = \left. \frac{d}{dt} f \circ \Phi_t(p) \right|_{t_0}, \tag{8.19}$$

are elements of the tangent space of the manifold $V_{\Phi_{t_0}(p)} \in T_{\Phi_{t_0}(p)}(M)$.

Proof Applying the definition assuming the coordinate frames $E_{i\Phi_{t_0}(p)}$ spanning the tangent space at $\Phi_{t_0}(p)$ gives

$$\left. \frac{d}{dt} f \circ \Phi_t \right|_{t_0} = \left. \frac{d}{dt} f \circ \phi^{-1} \circ \phi \circ \Phi_t \right|_{t_0} = \sum_{i=1}^{m} \left. \frac{\partial f \circ \phi^{-1}}{\partial x_i} \right|_{\phi \circ \Phi_{t_0}(p)} \dot{x}_i(t_0) \tag{8.20}$$

$$= \sum_{i=1}^{m} \dot{x}_i(t_0) E_{i\Phi_{t_0}(p)} f, \tag{8.21}$$

with $\dot{x} = (d/dt) \phi \circ \Phi$ the derivative of the curve in local coordinates. □

Example 8.13 (Translation) For illustration examine the simple version of translation in the plane, then $\Phi_t : p \in \mathbb{R}^n \to p + at \in \mathbb{R}^n$; then

$$\frac{d}{dt} f \circ \Phi_t|_p = \frac{d}{dt} f(p_1 + a_1 t, \ldots, p_n + a_n t) = \sum_{i=1}^{n} \frac{\partial f}{\partial x_i} |_{\Phi_t(p)} (\dot{\Phi}_t(p))_i, \tag{8.22}$$

with $V_{\Phi_{t_0}(p)} = \sum_{i=1}^{n} a_i \frac{\partial}{\partial x_i}. \tag{8.23}$

Notice that it is independent of $p \in \mathbb{R}^n$.

8.2 Smooth Mappings, the Jacobian, and Diffeomorphisms

We have already started looking at shapes corresponding to generators which are sub-manifolds sewn together; the tranformations were built from groups acting locally on submanifolds of the whole. This is our general construction of deformable templates and active models. The graph based regularity on the transformations constrains the local action of these transformations so that the total manifold stays regular (connected) (see Chapter 7, Section 7.3). In the continuum this corresponds to smoothness of the mapping, and local bijective properties. For computation, the number of submanifolds (generators) are chosen to be finite; to understand them analytically we now study these transformations as smooth mappings on the continuum in which the number of generating manifolds goes to infinity as well as the groups acting on them. This takes us into the study of homeomorphisms and diffeomorphisms.

8.2.1 Smooth Mappings and the Jacobian

The basic model to be studied is pictured in Figure 8.2 below in which manifolds are studied via mappings of one to the other, $F : M \subset \mathbb{R}^m \to N \subset \mathbb{R}^n$ with local coordinates $(\phi, M), (\psi, N)$.

$$p \in M \quad \stackrel{F}{\Longrightarrow} \quad F(p) \in N \tag{8.24}$$

$$\Downarrow \phi \qquad\qquad \Downarrow \psi \tag{8.25}$$

$$\phi(p) \in R^m \quad \psi \circ F(p) \in R^n \tag{8.26}$$

Figure 8.2 The basic model of transforming manifold M with local coordinates ϕ under smooth mapping $F : M \to N$ with local coordinates ψ.

Definition 8.14 *Let $F : M \to N$ with **coordinate neighborhoods** (O, ϕ) and (P, ψ) with $F(O) \subset P$. Then F **in local coordinates**, denoted $\hat{F} : \phi(O) \to \psi(P)$ is*

$$\hat{F}(p) = (y_1, \ldots, y_n) = \psi \circ F \circ \phi^{-1}|_{\phi(p)=(x_1,\ldots,x_m)} . \tag{8.27}$$

There are various properties to be understood corresponding to smoothness, continuity, and bijective properties which we shall examine again below more carefully.

Definition 8.15 *A **mapping** $F : M \to N$ is C^k **smooth** if for each $p \in M$ there exists local coordinate (O, ϕ) of M and (P, ψ) of N with $p \in O$, $F(p) \in P$ such that $\hat{F}$ is C^k. For X_1, X_2 are topological spaces, with $F : X_1 \to X_2$, then various properties of F are as follows:*

   (i) *F is **continuous** if the inverse image of every open set in X_2 is an open set in X_1;*

   (ii) *F is an **open mapping** if F takes open sets in X_1 to open sets in X_2;*

   (iii) *F is a **homeomorphism** if F and F^{-1} are both 1-1 and onto (bijections), continuous and open;*

   (iv) *F is a **smooth** C^k **map** from $\mathbb{R}^n$ to $\mathbb{R}^m$ if $F : x \in \mathbb{R}^n \mapsto F(x) = (F_1(x), \ldots, F_m(x)) \in \mathbb{R}^m$ has each component C^k, k-times continuously differentiable;*

   (v) *a smooth C^k mapping $F : \mathbb{R}^n \to \mathbb{R}^m$ is a **diffeomorphism** iff it is a homeomorphism (1-1 and onto and continuous) with F^{-1} smooth C^k.*

Note that F, a homeomorphism, implies that F, F^{-1} are both open mappings. To see that F^{-1} is open, for example, take an open set $O_2 \in \mathcal{T}_2$, since F is continuous, $F^{-1}(O_2)$ is open, thus F^{-1} is an open mapping. The identical argument shows that F is open.

The C^k-differentiability of each component is defined in the standard iterative way from vector calculus. Let $F : O \subset \mathbb{R}^m \to \mathbb{R}$ be in C^k and let $D^k F$ represent the $m \times m \times \cdots \times m$ (k times) vector of its kth partial derivatives. Then $F \in C^{k+1}$ on O if and only if for each component of that vector, say $D^k F_i$, there exists an m-vector of functions, A_i and an m-tuple $R_i(x,a)$ of functions defined on $O \times O$ such that $\|R_i(x,a)\| \to 0$ as $x \to a$ and for each $x \in O$ we have

$$D^k F_i(x) = D^k F_i(a) + A_i(x-a) + \|x-a\| R_i(x,a).$$

Then $F : \mathbb{R}^m \to \mathbb{R}^n$ is smooth if each component is smooth. Clearly the Jacobian of the transformation will play an important role in determining *at least local* 1-1 and onto characteristics of the maps. For this the rank of the mapping as defined through the Jacobian in local coordinates is important.

Definition 8.16 *The rank of F at p is defined to be the rank of $\hat{F} = (y_1, \ldots, y_n)$ at $\phi(p) = (x_1, \ldots, x_m)$ of the Jacobian matrix of the transformation in local coordinates:*

$$D\hat{F}(x_1, \ldots, x_m) = \begin{pmatrix} \frac{\partial y_1}{\partial x_1} & \cdots & \frac{\partial y_1}{\partial x_m} \\ \vdots & \vdots & \vdots \\ \frac{\partial y_n}{\partial x_1} & \cdots & \frac{\partial y_n}{\partial x_m} \end{pmatrix}. \quad (8.28)$$

We will be interested in understanding how the tangent spaces transform under smooth mappings $F : M \to N$, M, N smooth manifolds of dimension m, n. For this we define the *differential* of the map F.

Definition 8.17 *Let M, N be smooth manifolds with $F : M \to N$ a smooth mapping. Then the **differential** of F at $p \in M$ is the linear map $F_\star : T_p(M) \to T_{F(p)}(N)$ defined as, for all $V_p \in T_p(M)$ and f smooth,*

$$F_\star(V_p)f = V_p(f \circ F). \quad (8.29)$$

Let us in fact verify that the map $F_\star$ is a linear map into the tangent space $T_{F(p)}$. Assume coordinate charts (O, ϕ) and (P, ψ) on M, N, with $p \in O, F(p) \in P$. For this assume coordinate frames

$$\tilde{E}_{jF(p)} = \psi_\star^{-1} \frac{\partial}{\partial y_j}, \quad j = 1, \ldots, n. \quad (8.30)$$

Let us establish that $F_\star E_{ip}$ is an element of the tangent space $T_{F(p)}(N)$. Choosing $f \in C^\infty(F(p))$, then by definition

$$F_\star E_{ip} f = E_{ip} f \circ F = \frac{\partial}{\partial x_i} f \circ \psi^{-1} \circ \psi \circ F \circ \phi^{-1} \bigg|_{\phi(p)}$$

$$= \sum_{j=1}^n \frac{\partial}{\partial y_j} (f \circ \psi^{-1}) \bigg|_{\psi \circ F(p)} \frac{\partial y_j}{\partial x_i} = \sum_{j=1}^n \frac{\partial y_j}{\partial x_i} \tilde{E}_{jF(p)} f.$$

The Jacobian matrix in local coordinates is $\psi \circ F \circ \phi^{-1}$; the matrix determining the basis transformation of the linear transformation from $F_\star : T_p(M) \mapsto T_{F(p)}(N)$ is the Jacobian matrix.

Example 8.18 (Matrix Group Action) Let $A = \begin{pmatrix} a_{ij} \end{pmatrix} \in \mathbf{GL}(n) : x \mapsto Ax$, then

$$\frac{\partial}{\partial x_i} f \circ (Ax) = \sum_{j=1}^n \frac{\partial f}{\partial y_j}(Ax) \frac{\partial (Ax)_j}{\partial x_i} = \sum_{j=1}^n a_{ij} \frac{\partial f}{\partial y_j}(Ax). \quad (8.31)$$

8.2.2 The Jacobian and Local Diffeomorphic Properties

The Jacobian provides the *necessary condition* (not sufficient) for $F : M \to N$ to be a diffeomorphism from $M \to N$. This is illustrated via the following standard example for mappings between finite dimensional vector spaces; the rank of the Jacobian matrix must equal the dimension of the manifolds being mapped, i.e. $\dim M = \dim N = \text{rank} F$. Let $F : M = \mathbb{R}^m \to N \subset \mathbb{R}^n$ be a linear mapping via matrix multiplication of the vector spaces, F an $n \times m$ matrix: $Fx = y, x \in \mathbb{R}^m, y \in \mathbb{R}^n$. Then, if $\dim N \neq \dim M$ either the forward or reverse mapping cannot be onto all of $\mathbb{R}^n$ (hence not invertible). Examine $\dim M = \dim N = m$, with F an $m \times m$ matrix. Since F is identically the Jacobian, the rank of the Jacobian $< m$ implies that F has a non-trivial null-space, and obviously has no inverse; thus it is not a diffeomorphism of $\mathbb{R}^m \to \mathbb{R}^m$ if $\text{rank} F \neq m$.

The Jacobian of $F : M \to N$ being full rank globally over M does imply that locally F describes a diffeomorphism between M and N, but not globally. Essentially, the global 1-1 nature of the transformation can fail.

Examine the mapping from the real line onto the circle, a subset of $\mathbb{R}^2$, $F : \mathbb{R} \to F(\mathbb{R}) \subset \mathbb{R}^2$ according to $F(t) = (\cos t, \sin t)$. The Jacobian globally has rank 1, i.e. $DF(t) = \begin{pmatrix} -\sin t \\ \cos t \end{pmatrix}$. Locally, this describes a diffeomorphism from small neighborhoods of diameter strictly less than 2π to the image of F, subsets of $\mathbb{R}^2$.

The local result is essentially the *inverse function theorem* for $\mathbb{R}^n$, stated here, with the proof found in Boothby (see p. 42, [148]).

Theorem 8.19 (Local Diffeomorphisms via the Inverse Function Theorem) *Let W be an open subset of $\mathbb{R}^n$, $F : W \to \mathbb{R}^n$ a smooth mapping. If the Jacobian of F at $w \in W$ is nonsingular, then there exists an open neighborhood $O \subset W$ of w such that $V = F(O)$ is open and $F : O \to V$ is a diffeomorphism.*

See proof in Boothby, p. 42 [148].

Corollary 8.20 *Then, let M, N be manifolds both of dimension $\dim M = \dim N = m$ with smooth mapping $F : M \to N$, and an open set $W \subset M$. If $w \in W$ and the Jacobian of F at w has rank $DF(w) = m$, then there exists an open neighborhood $O \subset W$ with $V = F(O)$ open and $F : O \to V$ a diffeomorphism.*

Proof For the manifolds apply the definition of the local coordinates, choose ϕ and ψ to be the local coordinate maps for M, N, then $\hat{F} = \psi \circ F \circ \phi^{-1} : \mathbb{R}^m \to \mathbb{R}^m$. Apply the inverse function theorem to the point $\phi(w) \in \mathbb{R}^m$; then there exists an open neighborhood $O' \subset \mathbb{R}^m$ containing $\phi(w)$ such that $V' = \hat{F}(O')$ is open, and $\hat{F} : O' \to V'$ is a diffeomorphism. Then, $F : O = \phi^{-1}O' \to \psi^{-1}V'$ is a diffeomorphism from an open set $O \subset M$ containing w to an open set $F(W) \subset N$. $\square$

Knowing only properties of the Jacobian implies that we will have to accept the fact that our transformations may not be globally invertible. This motivates our understanding of mappings which are only locally diffeomorphisms, such as are associated with *immersions*. See the example below.

Definition 8.21 *Given manifolds M, N, then the mapping $F : M \to N$ is said to be an* **immersion** *if* $\text{rank} F = \dim M = m$ *everywhere.*

Example 8.22 (Circle, Figure Eight) The *circle* corresponding to $F : \mathbb{R} \to \mathbb{R}^2$, $F(t) = (\cos 2\pi t, \sin 2\pi t)$, is an immersion.

The *figure eight* corresponding to $F : \mathbb{R} \to \mathbb{R}^2$, $F(t) = (2\cos(t - (1/2)\pi), \sin(t - (1/2)\pi))$, is not.

8.3 Matrix Groups are Diffeomorphisms which are a Smooth Manifold

We shall work extensively to require the mappings of manifolds in the study of shape $F : M \subset \mathbb{R}^m \to N \subset \mathbb{R}^n$ to be not just smooth but to be diffeomorphisms. Why should we be so fascinated with diffeomorphisms? Well, one of the more intuitive reasons is that diffeomorphic maps preserve features. For us this will correspond to such intuitive ideas as connected sets stay connected (structures are not broken apart), and local maxima in the images remain local maxima under the transformations (both proved below).

8.3.1 Diffeomorphisms

Almost all of the subsequent geometric work concentrates on submanifolds of $\mathbb{R}^n$, corresponding to points, lines, surfaces, and subvolumes, and actions upon them by diffeomorphisms. The most important property that we shall exploit is that diffeomorphisms carry the submanifolds diffeomorphically maintaining their topological structure.

For this we need to first define more exactly the interpretation of the submanifolds and the open sets. Since we will almost exclusively study manifolds which are subsets of $\mathbb{R}^n$, we shall be thinking of the *subspace toplogy* as the collection of open sets.

Definition 8.23 *We shall say M* **is an** *m***-dimensional submanifold** *of $\mathbb{R}^n$ with subspace topology $\mathcal{T}_M$ if it is a set which is locally Euclidean of dimension m, with the topology $\mathcal{T}_M$ of open sets given by*

$$\mathcal{T}_M = \{O \subset M : O = M \cap U, U \text{ open in } \mathbb{R}^n\}. \tag{8.32}$$

The subspace topology is consistent with our everyday notions. The open sets in the submanifold are just the open sets associated with the extrinsic background space intersected with the submanifold.

Then the most celebrated property that we will require is that diffeomorphic mappings carry submanifolds so as to preserve their topology (open sets remain open sets). This of course corresponds to such intuitively natural properties as connected sets remain connected, i.e. two eyes remain two, and do not become three.

Theorem 8.24 *Diffeomorphisms map connected sets to connected sets.*

Proof Let $F : M \to N$ be a diffeomorphism and let M be a connected set. We need to show that $F(M) \subset N$ is also a connected set. Suppose not, i.e. there exist two sets N_1 and N_2, open in $F(M)$, such that $N_1 \cap N_2 = \emptyset$ and $N_1 \cup N_2 = F(M)$. Therefore, $F^{-1}(N_1 \cap N_2) = F^{-1}(N_1) \cap F^{-1}(N_2) = \emptyset$. Also, as F is a diffeomorphsim $F^{-1}(N_1)$ and $F^{-1}(N_2)$ open sets in M. Since $N_1 \cup N_2 = F(M)$, $F^{-1}(N_1) \cup F^{-1}(N_2) = M$, union being a disjoint union. So we have two disjoint open subsets of M whose union is M, i.e. M is not a connected set. Contradiction. □

More generally, diffeomorphisms map smooth submanifolds diffeomorphically.

Theorem 8.25 *Let F be a diffeomorphism from $X \to Y$ with $M \subset X$ a submanifold with subspace topology.*

Then $F : M \subset X \to F(M) \subset Y$ is a diffeomorphism from $M \to F(M)$ with the subspace topology.

Proof Global 1-1, onto and differentiability properties of $F : M \to F(M)$ follows from the fact that F is a diffeomorphism from $X \to Y$. We really need only show continuity of F, F^{-1} and thus the homeomorphism properties. Let O be open in the subspace topology of $F(M)$, implying $O = V \cap F(M)$, V open in the topology of Y. Then

$$F^{-1}(O) = F^{-1}(V) \cap M \tag{8.33}$$

which is in the subspace topology of M (since F is continuous in X giving $F^{-1}(V)$ is open). So we have shown F is continuous on M. In precisely the same way it follows F^{-1} is continuous on the subspace topology of $F(M)$, so F is a homeomorphism from $M \to F(M)$. □

Clearly definiteness of the Jacobian and full rank condition, although not implying global properties, implies that such features as maxima of functions on M are preserved under the mapping. To obtain a global property, the 1-1 condition needs to be added.

Theorem 8.26 *Let M, N be manifolds both of dimension $\dim M = \dim N = m$. Then a smooth mapping $F : M \to N$ is a diffeomorphism if and only if F is a bijection (1-1 and onto) and $\mathrm{rank} F = m$ everywhere.*

8.3.2 Matrix Group Actions are Diffeomorphisms on the Background Space

Group transformations of background spaces which are manifolds are central to the pattern theory.

The major examples of the matrix groups which shall be used are obtained from subgroups of the generalized linear group $\mathbf{GL}(n)$. Now matrix groups acting as transformations $S : X \to X$ are diffeomorphisms on the backround space X. For this reason we examine the group action as a transformation of the background spaces.

Theorem 8.27 *Let $\mathcal{S} = \mathbf{GL}(n)$ and $\mathcal{S} = \mathbf{R}^n$ the generalized linear group and the translation group, and $\mathbb{R}^n$ the differentiable manifold with the group actions viewed as linear mappings $F : \mathcal{S} \times \mathbb{R}^n \to \mathbb{R}^n$ by $F_A(x) = Ax$ or $F_a(x) = x + a$, multiplication by the $n \times n$ matrix $A \in \mathbf{GL}(n)$ or translation by $a \in \mathbb{R}^n$.*

Then these transformations F are group actions which are smooth (C^∞) maps from $\mathbb{R}^n \to \mathbb{R}^n$, and $F_A, F_a : \mathbb{R}^n \to \mathbb{R}^n$ are diffeomorphisms.

Proof That these are group actions follows from $e = I \in \mathbf{GL}(n)$ and $e = 0 \in \mathbb{R}^n$ with $Ix = x, x + 0 = x$. Associativity of matrix multiplication $(A \circ B) \cdot x = A \cdot (B \cdot x)$ and associativity and commutativity of addition of vectors satisfies the group action.

These are smooth maps since $Ax = \sum_{j=1}^m a_{1j} x_j, \sum_{j=1}^m a_{2j} x_j, \ldots, \sum_{j=1}^m a_{mj} x_j$ viewed as a column vector is just a polynomial in the entries of A. Addition is also obviously a smooth map.

The generalized linear group action F_A or translation F_a are $1 - 1$ and onto. The onto property follows since for all $y \in \mathbb{R}^n$, there exists an $x = F_A^{-1} y$ with $F_A x = y$. The translation group is 1-1 and onto as well.

Clearly F_s is 1-1 since for any x, x', if $F_s(x) = F_s(x')$ then

$$x = F_{s^{-1} \circ s}(x) = F_{s^{-1}}(F_s(x)) = F_{s^{-1}}(F_s(x')) = x'. \tag{8.34}$$

For subgroups of $\mathbf{GL}(n)$, $s = A = \left(a_{ij}\right)$, then $F_A(x) = Ax$ and

$$\left(\frac{\partial F_A(x)}{\partial x_i}\right)_j = \left(\frac{\partial Ax}{\partial x_i}\right)_j = a_{ij}. \tag{8.35}$$

Since A is invertible, rank $F_{(A,b)} s(\cdot) = n$. For $s = (A, b)$ with $F_{(A,b)}(x) = Ax + b$, the Jacobian is the same.

Using Theorem 8.25 then it defines a diffeomorphism on smooth submanifolds. □

One of the most important implications of the matrix group action is that it defines a diffeomorphism, and therefore by Theorem 8.25 it carries smooth submanifolds smoothly maintaining their submanifold structure. Examine such a result for the subgroups of the affine group applied to submanifolds of $\mathbb{R}^n$.

Corollary 8.28 *Let S be a subgroup of the affine group. Let M be a submanifold of $\mathbb{R}^n$. Then the linear mappings $F_s : M \to F_s(M)$ are for all $s \in S$ diffeomorphisms from $M \to F_s(M)$.*

8.3.3 The Matrix Groups are Smooth Manifolds (Lie Groups)

Now the matrix groups are differentiable manifolds, providing the opportunity to perform differentiation of group elements.

Definition 8.29 *Let S be a group which is at the same time a differentiable manifold. For $x, y \in S$, let $x \circ y$ denote their product and x^{-1} its inverse.*

*Then S is a **Lie group** provided that the mapping from $S \times S \to S$ defined by $(x, y) \mapsto x \circ y$ and the mapping $S \to S$ defined by $x \to x^{-1}$ are smooth (C^∞) mappings.*

Example 8.30 (Vector Lie Groups) Let $S = \mathbb{R}^n$ with addition. The mapping $\mathbb{R}^n \times \mathbb{R}^n \to \mathbb{R}^n$ given by $(x, y) \to x \circ y = x + y$, and inverse $\mathbb{R}^n \to \mathbb{R}^n$ defined by $x \to x^{-1} = -x$. To show that these are smooth maps, define the local coordinates by the identity map $\phi_1 : \mathbb{R}^n \to \mathbb{R}^n : \phi_1(x) = x$. Let $M = \mathbb{R}^n \times \mathbb{R}^n$ have the product topology with the local coordinate map $\phi : M \to \mathbb{R}^n$ the identity map componentwise $\phi(x, y) = (x, y)$, $F : M \to \mathbb{R}^n$. Then, $F(x, y) = x + y$ is smooth since $\hat{F} = \phi_1 \circ F \circ \phi^{-1} = x + y$ which is smooth in x, y!

The inverse map $F : \mathbb{R}^n \to \mathbb{R}^n$, $F(x) = -x$ is similarly smooth.

Example 8.31 (Multiplicative and Complex Number Lie Group) Let $S = \mathbb{R}^\times$ the multiplication group of reals (not including 0) is a Lie group, $x \circ y = xy$, $x^{-1} = 1/x$. Identical argument as in example 8.30.

The set of non-zero complex numbers $S = \mathbb{C}^\times$, the multiplication group of complex numbers (not including 0) is a Lie group. $\mathbb{C}^\times$ is a group with respect to matrix multiplication of complex numbers, the inverse being $z^{-1} = 1/z$. Also, $\mathbb{C}^\times$ is a smooth (C^∞) manifold covered by a single coordinate neighborhood $U = \mathbb{C}^\times$ with the coordinate map given by $\phi(z) = (x, y)$ for $z = x + iy$. Using these coordinates the product $w = zz'$, $z = x + iy$, $z' = x' + iy'$, is given by

$$((x, y), (x', y')) \mapsto (xx' - yy', xy' + yx')$$

and the mapping $z \mapsto z^{-1}$ by

$$(x, y) \mapsto \left(\frac{x}{x^2 + y^2}, \frac{-y}{x^2 + y^2} \right).$$

These two maps are smooth; therefore $\mathbb{C}^\times$ is a Lie group.

Our two major sources of Lie groups will correspond to the affine group: that composed of subgroups of the generalized linear group $\mathbf{GL}(n)$ and the translation group $\mathbf{R}^n$.

Theorem 8.32 *Then*

1. *$\mathbf{GL}(n)$ as the matrix product group with group operation $A \circ B = AB$ defined by matrix product for $A, B \in \mathbf{GL}(n)$, and A^{-1} matrix inverse; and*

2. $\mathbf{R}^n$ as the translation group with group operation addition $a \circ b = a + b$ and inverse $a^{-1} = -a$

are both differentiable manifolds which are Lie groups.

Proof To see that $\mathbf{GL}(n)$ is a differentiable manifold identify $A \in \mathbf{GL}(n)$ with $\mathbb{R}^{n^2}$ and we need only prove that it is an open subset of a differentiable manifold. To see that it is open, use the distance $d(A, B) = \sum_{i,j} |A_{i,j} - B_{i,j}|^2$. Clearly if $\det A \neq 0$, then there exists a neighborhood $N_\epsilon(A) \subset \mathbf{GL}(n)$ of A since det is a continuous function.

To see that the matrix product is a smooth map of $\mathbf{GL}(n) \times \mathbf{GL}(n)$ it suffices to see that the components of the product matrix are polynomial evaluations in the elements of matrices under multiplication and polynomials are C^∞-functions. Similarly, matrix inversion is also a polynomial function (see below) and hence a smooth map. □

Example 8.33 (Special Orthogonal Group) The matrix subgroups are Lie groups. It is informative to prove this for the heavily used case of $\mathbf{SO}(3) \subset \mathbf{GL}(3)$.

Theorem 8.34 $\mathbf{SO}(3)$ *is a Lie group.*

Proof It is a topological manifold with the subset topology inherited from the space of 3×3 matrices with non-zero determinants, the generalized linear group $\mathbf{GL}(3)$. It is very often studied as the submanifold of $\mathbf{GL}(3)$, the metric used being the regular matrix 2-norm, $\|A-B\| = \sum_{ij}(a_{ij}-b_{ij})^2$. Also, $\mathbf{SO}(3)$ is locally euclidean, of dimensions 3, with the local coordinate chart given by the mapping

$$\phi^{-1}(x_1, x_2, x_3) = \begin{bmatrix} 1 & 0 & 0 \\ 0 & \cos x_1 & \sin x_1 \\ 0 & -\sin x_1 & \cos x_1 \end{bmatrix} \begin{bmatrix} \cos x_2 & 0 & -\sin x_2 \\ 0 & 1 & 0 \\ \sin x_2 & 0 & \cos x_2 \end{bmatrix}$$

$$\times \begin{bmatrix} \cos x_3 & \sin x_3 & 0 \\ -\sin x_3 & \cos x_3 & 0 \\ 0 & 0 & 1 \end{bmatrix}. \tag{8.36}$$

We need to prove that this map is a homeomorphism on small neighborhoods. Sines and cosines are homeomorphic away from the integer multiple of $\pi/2$ and near the integer multiples of $\pi/2$ sine or cosine is homeomorphic depending upon whether it is the even or odd multiple. Hence on neighborhoods small enough this map is a homeomorphism. Besides, $\mathbf{SO}(3)$ also has a group structure with the group operation being the regular matrix product, as elements of $\mathbb{R}^{3\times 3}$, and the inverse given by matrix inversion.

Now we must show that matrix multiplication and inversion in $\mathbf{SO}(3)$ are C^∞ maps. For $A, B \in \mathbf{SO}(3)$ the product AB has entries which are polynomial in the entries of A and B. The entries of A and B are C^∞ maps from the local coordinates of A and B (by our choice of local charts which are made up of sines and cosines terms). Similarly, the entries of the product AB are C^∞ in terms of its local coordinates. By concatenation, there is C^∞ map from the local coordinates of A and B to the local coordinates of AB. The inverse of $A = (a_{i,j})$ may be written as $A^{-1} = (1/\det(A))(\tilde{a}_{i,j})$, where $(\tilde{a}_{i,j})$ are cofactors of A (hence polynomials in the entries of A). The $\det(A)$ is a polynomial in the entries of A, which does not vanish on $\mathbf{SO}(3)$. It follows that A^{-1} has entries which are rational functions with non-vanishing denominators, hence C^∞. Therefore, $\mathbf{SO}(3)$ is a Lie group.

□

8.4 Active Models and Deformable Templates as Immersions

Groups transforming generators which are manifolds locally are an area of active research in the computer vision and pattern theory literature. For single global groups actions — scale, rotation, translation — they are diffeomorphisms; for products of groups which act locally such as for active models they are immersions.

A great deal of work has been done in the area of segmentation via Active Deformable Models, including active snakes and contours [137, 150, 184–195], active surfaces and deformable models [56, 196–203]. These are global shape methods in that they define global generators which aggregate multiple local features into single objects. In this approach generators $g \in \mathcal{G}$ are defined as submanifolds, curves, surfaces, and subvolumes. The transformations making them active fill out the orbit of all shapes as defined through the vector fields which move the boundary manifolds and their connected interiors.

For using these models the image is assumed to be a collection of generators or submanifolds (object) defined parametrically. The collection of generating submanifolds form a disjoint partition of the image.

8.4.1 Snakes and Active Contours

Here is a transformation defined by $\mathbf{SO}(2)^{[0,L]}$ acting on a straight line generator corresponding to *snakes*. Let $F : (0, L) \subset \mathbb{R} \to \mathbb{R}^2$,

$$F(t) = \int_0^t \begin{pmatrix} \cos\theta(l) & -\sin\theta(l) \\ \sin\theta(l) & \cos\theta(l) \end{pmatrix} \begin{pmatrix} 1 \\ 0 \end{pmatrix} dl = \int_0^t \begin{pmatrix} \cos\theta(l) \\ \sin\theta(l) \end{pmatrix} dl. \tag{8.37}$$

Then F is an immersion in $\mathbb{R}^2$ since $\text{rank} F = 1$. Clearly, this is not necessarily a global diffeomorphism onto its image $F(0, L)$. Allow the curve to turn 2π radians in length $L/2$. For example, choose $\theta(l) = 2 \cdot 2\pi l, l \in [L/4, 3/4L]$.

Adding scale, then we have the following.

Theorem 8.35 *Let the mapping $F : M \subset \mathbb{R}^2 \to N \subset \mathbb{R}^2$ generated from the product group $(\mathbf{US} \times \mathbf{SO}(2))^{[0,T]}$ according to*

$$F(t) = \int_0^t \rho(l) \begin{pmatrix} \cos\theta(l) & -\sin\theta(l) \\ \sin\theta(l) & \cos\theta(l) \end{pmatrix} \begin{pmatrix} 1 \\ 0 \end{pmatrix} dl. \tag{8.38}$$

Then if $\rho \neq 0$, the mapping F is an immersion.

Proof The Jacobian matrix $DF(t) = \rho(t) \begin{pmatrix} \cos\theta(t) & -\sin\theta(t) \\ \sin\theta(t) & \cos\theta(t) \end{pmatrix} \begin{pmatrix} 1 \\ 0 \end{pmatrix}$. ☐

8.4.2 Deforming Closed Contours in the Plane

Here is an interesting example of *deforming closed contours* as mapping of circular templates.

Corollary 8.36 *Let the mapping $F : M = S^1 \subset \mathbb{R}^2 \to N \subset \mathbb{R}^2$ generated from the product group $(\mathbf{US} \times \mathbf{SO}(2))^{[0,T]}$ according to*

$$F(t) = \int_0^t \rho(l) \begin{pmatrix} \cos\theta(l) & -\sin\theta(l) \\ \sin\theta(l) & \cos\theta(l) \end{pmatrix} \begin{pmatrix} \cos\theta(l) \\ -\sin\theta(l) \end{pmatrix} dl, \quad t \in [0, T]. \tag{8.39}$$

Then this is an immersion if it satisfies $\dot{\rho} \neq 0$ from above; the mapping $F : S^1 \to M$ is an immersion to a closed curve in the plane if the processes $\rho(t), \theta(t), t \in [0, 2\pi]$ satisfy the the Fourier transform condition

$$\int_0^{2\pi} e^{-jt}\rho(t)e^{j\theta(t)}\, dt = 0. \tag{8.40}$$

Proof Expanding the Fourier transform in its real and imaginary parts gives

$$\int_0^{2\pi} (\cos t - j \sin t)\rho(t)(\cos\theta(t) + j\sin\theta(t))dt$$

$$= \int_0^{2\pi} (\rho(t)\cos\theta(t)\cos t + \rho(t)\sin\theta(t)\sin t)dt$$

$$+ j \int_0^{2\pi} (\rho(t)\sin\theta(t)\cos t - \rho(t)\cos\theta(t)\sin t)dt. \tag{8.41}$$

The Fourier transform Eqn. 8.40 set to 0 means that the real and imaginary parts are zero giving the closure condition

$$F(2\pi) = \int_0^{2\pi} \rho(t) \begin{pmatrix} \cos\theta(t) & -\sin\theta(t) \\ \sin\theta(t) & \cos\theta(t) \end{pmatrix} \begin{pmatrix} \cos t \\ -\sin t \end{pmatrix} dt = 0. \tag{8.42}$$

□

8.4.3 Normal Deformable Surfaces

Thus far we have examined group action on the tangents. Now examine normal deformations of curves and surfaces as defined by products of translation groups.

Theorem 8.37 *Given is smooth surface submanifold $M \subset \mathbb{R}^3$, with local coordinate patch $x : D \subset \mathbb{R}^2 \to O \subset M$*

$$x(u,v) = p + uE_1 + vE_2 + E_3 f(u,v), \tag{8.43}$$

with orthogonal frame E_1, E_2, E_3 and normal n and shape operator $S_{u,v}$. Let the mapping $F : M \to N$ generated from the translation group defined by

$$F(u,v) = x(u,v) + h(u,v)n(u,v), \tag{8.44}$$

with h a scalar field specifying the normal translation motion. Then $F : M \to N$ is an immersion with Jacobian matrix $DF(u,v) = (\partial F/\partial u), (\partial F/\partial v))$ with columns

$$\frac{\partial F}{\partial u} = \begin{pmatrix} 1 - (f_u h_u/\sqrt{1+f_u^2+f_v^2}) + hS_{11} \\ -(f_v h_u/\sqrt{1+f_u^2+f_v^2}) + hS_{21} \\ +f_u + (h_u/\sqrt{1+f_u^2+f_v^2}) \end{pmatrix}, \frac{\partial F}{\partial v} = \begin{pmatrix} (-f_u h_v/\sqrt{1+f_u^2+f_v^2}) + hS_{12} \\ 1 - (f_v h_v/\sqrt{1+f_u^2+f_v^2}) + hS_{22} \\ +f_v + (h_v/\sqrt{1+f_u^2+f_v^2}) \end{pmatrix}. \tag{8.45}$$

Proof The tangents $(\partial x/\partial u) = E_1 + f_u E_3$, $(\partial x/\partial v) = E_2 + f_v E_3$, and normal $n = (-f_u E_1 - f_v E_2 + E_3)/\sqrt{1 + fu^2 + fv^2}$, then

8 MANIFOLDS, ACTIVE MODELS, AND DEFORMABLE TEMPLATES

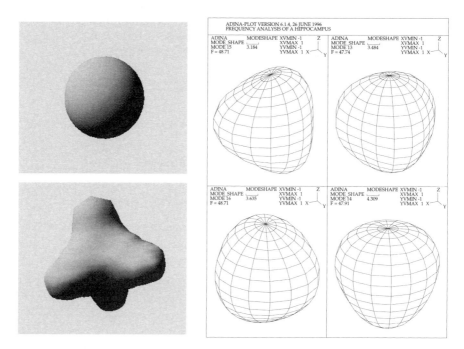

Figure 8.3 Left: shows the spherical template (top panel) and its deformed version (bottom panel) resulting from translation group applied to 4096 generators on the template. Right: Panels show deformations via the first four surface harmonics of the surface of the sphere (see also Plate 11).

$$\frac{\partial F}{\partial u} = \frac{\partial x}{\partial u} + \frac{\partial h}{\partial u}n + h\frac{\partial n}{\partial u} \tag{8.46}$$

$$= E_1 + f_u E_3 + h_u \frac{-f_u E_1 - f_v E_2 + E_3}{\sqrt{1 + f_u^2 + f_v^2}} + h \begin{pmatrix} S_{11} \\ S_{21} \end{pmatrix}_1 E_1 + h \begin{pmatrix} S_{12} \\ S_{22} \end{pmatrix}_2 E_2. \tag{8.47}$$

Similarly as above, $\dfrac{\partial F}{\partial v}$ follows. □

Example 8.38 (Active Deformable Spheres) Joshi and McNally [33, 198] have used active deformable spheres for the study of 3D cell motion.

Generate new shapes via normal deformation with local coordinate representation (see example 7.46)

$$x(u,v) + h(u,v)n(u,v) = (1 + h(u,v))x(u,v), \tag{8.48}$$

with $x(u,v) = (u, v, \sqrt{r^2 - u^2 - v^2})$, $x_u = (1, 0, -(u/\sqrt{r^2 - u^2 - v^2}))$, $x_v = (0, 1, -(v/\sqrt{r^2 - u^2 - v^2}))$, $n(u,v) = (1/r)(u, v, \sqrt{r^2 - u^2 - v^2})$. The Jacobian matrix in local coordinates is in the form

$$\frac{\partial F}{\partial u} = h_u(u,v)x(u,v) + (1 + h(u,v))x_u(u,v),$$

$$\frac{\partial F}{\partial v} = h_v(u,v)x(u,v) + (1 + h(u,v))x_v(u,v). \tag{8.49}$$

8.4 ACTIVE MODELS AND DEFORMABLE TEMPLATES AS IMMERSIONS

Figure 8.3 shows the spherical template containing 4096 nodes. Column 1 shows the sphere (top) and an example of the template shape deformed via the normal vector fields. Column 2 shows examples of the sphere deformed via the first four eigenfunctions of the Laplacian operator.

8.5 Activating Shapes in Deformable Models

Now examine the calculus for activating Active Deformable Models. Define generators $g \in \mathcal{G}$ which are smooth submanifolds; activating them arise by constructing transformations which are vector fields which move the protypical shape to fill out the orbit of all shapes.

8.5.1 Likelihood of Shapes Partitioning Image

For using these models for segmentation, the image is assumed to be a collection of generators or objects defined parametrically. Define the background space $D \subset \mathbb{Z}^d, \mathbb{R}^d$ to be a collection of objects $g_1, g_2, \ldots$, with textured interiors forming a disjoint partition of the image, $D = \cup_j D(g_j)$. The goal is to infer the objects $g_j, j = 1, 2, \ldots$ forming a covering of the complete image. Such a disjoint partition is illustrated by panel 1 of Figure 8.4 showing an electron micrograph of the mitochondria. Model the interiors of the shapes of real-valued images as scalar random fields, either Markov or Gaussian fields defined on the index sets of the integer lattices. Because of splitting, the partitions are conditionally independent given their boundaries. To formulate the inference of the active shapes $g_j, j = 1, 2, \ldots$ in the continuum, associate an energy density with each shape; the energy under any model for shape g takes the form $\int_{D(g)} E(x)dx$, $E(\cdot)$ representing the interior model for shape g as a density in space. Of course, the smooth manifold of the boundary $\partial D(g)$ of each shape intersects obliquely the discrete lattice $\mathbb{Z}^d$, and in calculating the energy associated with each shape the partial voxels are interpolated in the calculation.

Then the conditional probabililty of the partition $D(g_j), j = 1, 2, \ldots$ takes the form

$$p(I|D(g_1), D(g_2), \ldots) \propto \prod_j e^{-\int_{D(g_j)} E_j(x)dx}. \tag{8.50}$$

8.5.2 A General Calculus for Shape Activation

Active Models for segmentation of curves, surfaces, and subvolumes are made active by associating with the bounding manifold of the shapes some velocity of motion which is essentially a vector field acting on the shape. The vector field is constructed so that the shape covers with its interior the region in the image corresponding to the model of the interior of the shape. A principled way to do this is to send the parameters of the active model along a trajectory which follows the gradient of the energy representing the Bayesian likelihood of the image given the parametric representation as depicted in Eqn. 8.50.

Without loss of generality assume that there is one global model defined by $g(\gamma)$ with interior and parameterize its interior directly in the parameter $D(\gamma)$ and complement $D(\gamma)^c = D \setminus D(\gamma)$; the parameterization of g is associated to γ. For group parameters γ represents the local coordinates. To activate the shape so that it evolves to an optimum covering of the background space we now calculate the general form for the variation of the energy densities making up the potential so as

8 MANIFOLDS, ACTIVE MODELS, AND DEFORMABLE TEMPLATES

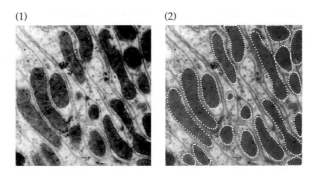

Figure 8.4 Panel 1 depicts an electron micrograph at 30,000 times magnification showing connected subregions of mitochondria. Panel 2 shows the disjoint partition of the connected subregion micrograph images into $\cup_j D(g_j)$ and the bounding contours $\partial D(g_j)$ representing the closed active contour models representing these regions. Data taken from the laboratory of Dr. Jeffrey Saffitz of Washington University.

to evolve the shape towards a minimum. Model the energy density for the interior as $E_1(x;\gamma), x \in D(\gamma)$ and exterior $E_2(x), x \in D(\gamma)^c$ the energy becomes

$$H(\gamma) = \int_{D(\gamma)} E_1(x;\gamma)\,dx + \int_{D \setminus D(\gamma)} E_2(x)\,dx. \tag{8.51}$$

Notice, the interior energy density of the model is generally an explicit function of γ the parameter to be inferred, the exterior is not.

Let $(\partial_1, \partial_2, \ldots, \partial_\gamma, \ldots)$ be a basis of the tangent space of the transformation group. A differential ∂_γ will correspond to the generation of some vector field $V(x), x \in D$ so that a point in the shape $x \in D(\gamma)$ is mapped by the vector field $x \to x + V(x)\epsilon$ under a small perturbation $\gamma \to \gamma + \epsilon$. Thus $V(\cdot)$ is essentially the velocity of the particle being acted upon by the 1-parameter variation of the group transforming the shape under the perturbation of $\gamma \to \gamma + \epsilon$. Now interestingly, the variation of the energy representing the active shapes will have two components arising essentially from the fact that both the region of integration of the active model as well as the energy density contained within the interior change with the group transformation.

For our derivation we will need to interpret the delta-dirac acting through a function following the standard developments in [204].

Lemma 8.39 *With $\Phi(\cdot)$ a real-valued function on $\mathbb{R}^d$ with smooth level set $\{x : \Phi(x) = 0\}$ a smooth submanifold with $\|\nabla \Phi\| = 1$ and surface measure ds, then*

$$\int \phi(x)\delta(\Phi(x))\,dx = \int_{\Phi=0} \phi(s)\,ds. \tag{8.52}$$

Proof Define the curvilinear coordinates $u_1, \ldots, u_{d-1}$ on the smooth manifold $\{(x_1, \ldots, x_d) : \Phi(x) = 0\}$. Then introduce the change of coordinate system $x(u_1, \ldots, u_{d-1}, u_d = \Phi)$; notice $x(u_1, \ldots, u_{d-1}, u_d = 0)$ is a point on the manifold $\Phi = 0$. Locally, $u_d = \Phi$ is the distance to the level set, since Φ is constant along $u_1, \ldots, u_{d-1}$ and $\partial \Phi / \partial u_d = 1$. In these coordinates,

$$dx(u) = |\det Dx(u)| du_1, du_2, \ldots, du_d, \tag{8.53}$$

where $Dx(u)$ is the Jacobian matrix of the transformation $x(u)$ evaluated at $u = (u_1, \ldots, u_d)$. Then the integral becomes

$$\int \phi(x)\delta(\Phi(x))dx = \int \phi(x(u_1,\ldots,u_d))\delta(u_d)|\det Dx(u_1,\ldots,u_{d-1},u_d)|$$
$$\times du_1 \cdots du_{d-1} du_d \qquad (8.54)$$
$$= \int_{\Phi=0} \phi(x(u_1,\ldots,u_{d-1},0))|\det Dx(u_1,\ldots,u_{d-1},0)|$$
$$\times du_1 \cdots du_{d-1} \qquad (8.55)$$
$$= \int_{\Phi=0} \phi(s)\,ds, \qquad (8.56)$$

where the surface measure ds is $du_1, \cdots, du_{d-1}$ weighted by the determinant of the Jacobian Dx evaluated at $(u_1, \ldots, u_{d-1}, 0)$. □

Example 8.40 To illustrate, take the unit circle $\Phi = \sqrt{x^2 + y^2} - 1$, then choose $u_1 \in [0, 2\pi)$, $u_2 = \Phi$, then $x(u_1, u_2) = (u_2 + 1)\cos u_1$, $y(u_1, u_2) = (u_2 + 1)\sin u_1$ and

$$Dx(u_1, u_2) = \begin{pmatrix} -(u_2+1)\sin u_1 & (u_2+1)\cos u_1 \\ \cos u_1 & \sin u_1 \end{pmatrix}. \qquad (8.57)$$

Then $ds = du_1$, and

$$\int \phi(x)\delta(\Phi(x))dx = \int_0^{2\pi} \phi(\cos u_1, \sin u_1)du_1. \qquad (8.58)$$

Now do the sphere, $\Phi = \sqrt{x^2 + y^2 + z^2} - 1$, then choose $u_1 \in [0, 2\pi)$, $u_2 \in [0, \pi]$, $u_3 = \Phi$, with

$$x(u_1, u_2, u_3) = (u_3 + 1)\cos u_1 \sin u_2 \qquad (8.59)$$
$$y(u_1, u_2, u_3) = (u_3 + 1)\sin u_1 \sin u_2 \qquad (8.60)$$
$$z(u_1, u_2, u_3) = (u_3 + 1)\cos u_2. \qquad (8.61)$$

The Jacobian matrix becomes

$$Dx(u_1, u_2, u_3) = \begin{pmatrix} -(u_3+1)\sin u_1 \sin u_2 & (u_3+1)\cos u_1 \sin u_2 & 0 \\ (u_3+1)\cos u_1 \cos u_2 & (u_3+1)\sin u_1 \cos u_2 & -(u_3+1)\sin u_2 \\ \cos u_1 \sin u_2 & \sin u_1 \sin u_2 & \cos u_2 \end{pmatrix}. \qquad (8.62)$$

Then $ds = \sin u_2 du_1 du_2$, and

$$\int \phi(x)\delta(\Phi(x))dx = \int_0^{2\pi}\int_0^{\pi} \phi(x(u_1,u_2,0), y(u_1,u_2,0)) \sin u_2 du_2 du_1. \qquad (8.63)$$

Now we compute the variation of the deformable template with the transport theorem perturbing the shape $g(\gamma)$ parametrically according to $v(\gamma) = dg(\gamma)/d\gamma$.

Theorem 8.41 (Transport Theorem for Deformable Shapes) *Given simply connected regions $g(\gamma)$ with interior $D(\gamma)$ and smooth boundary $\partial D(\gamma)$ and associated smoothly varying unit normal n, vector perturbation $v = \partial g(\gamma)/\partial \gamma$ and associated surface measure ds. Then with interior and exterior models $E_1(\gamma)$, E_2 smooth in space and parameterization with energy*

$H(\gamma) = \int_{D(\gamma)} E_1(x;\gamma)dx + \int_{D(\gamma)^c} E_2(x)dx$ then the differential of the energy is given by the sum of the boundary and interior integrals

$$\frac{\partial H(\gamma)}{\partial \gamma} = \int_{\partial D(\gamma)} (E_1(s;\gamma) - E_2(s))\langle n(s), v(s)\rangle_{\mathbb{R}^d}\, ds + \int_{D(\gamma)} \frac{\partial}{\partial \gamma} E_1(x;\gamma)dx. \quad (8.64)$$

Proof Define the function $\Phi(\cdot)$ with level set $\{x : \Phi(x) = 0\}$ the boundary $\partial D(\gamma)$ with $\Phi > 0$ in D, and the normal to the boundary $\nabla \Phi = n$ ($\|\nabla \Phi\| = 1$). Rewrite the potential using the heaviside function $h(x) = 1$ for $x > 0$, $h(x) = 0$ for $x < 0$ according to

$$H(\gamma) = \int h(\Phi(x))E_1(x;\gamma)dx + \int (1 - h(\Phi(x)))E_2(x)dx. \quad (8.65)$$

The exterior model is not a function of γ, therefore differentiating by parts gives

$$\frac{\partial H(\gamma)}{\partial \gamma} = \int \frac{\partial}{\partial \gamma} h(\Phi(x))(E_1(x) - E_2(x))dx + \int h(\Phi(x))\frac{\partial}{\partial \gamma}E_1(x)dx \quad (8.66)$$

$$= \int \delta(\Phi(x))\langle \nabla\Phi(x), v(x)\rangle_{\mathbb{R}^d}(E_1(x) - E_2(x))dx + \int_{D(\gamma)} \frac{\partial}{\partial \gamma}E_1(x)dx. \quad (8.67)$$

Now use the lemma, $\int \delta(\Phi(x))g(x)dx = \int_{\partial D} g(s)ds$ (since $\|\nabla\Phi\| = 1$) to give

$$\frac{\partial H(\gamma)}{\partial \gamma} = \int_{\partial D(\gamma)} \langle n(s), v(s)\rangle_{\mathbb{R}^d}(E_1(s) - E_2(s))ds + \int_{D(\gamma)} \frac{\partial}{\partial \gamma}E_1(x)dx. \quad (8.68)$$

□

The calculation of the surface measure for integration on ∂D has to be done for each particular case. What is clearly the case is that $\langle n, (\partial g(\gamma)/\partial \gamma)\rangle_{\mathbb{R}^d} ds$ is the local volume element that gets integrated around the boundary parameterization against the difference in energies under the two models.

8.5.3 Active Closed Contours in $\mathbb{R}^2$

Assume that the continuum $D \subset \mathbb{R}^d$ is partitioned into a set of closed connected regions with smooth boundary manifolds simple C^1 closed curves in the plane $g_t(\gamma), t \in [0,1) \in \mathbb{R}^2$. The potential takes the form

$$H(\gamma) = \int_{D(\gamma)} E_1(x;\gamma)dx + \int_{D\setminus D(\gamma)} E_2(x)dx, \quad (8.69)$$

where the potential E_1 is associated with the interior $x \in D(g(\gamma))$ of g, and E_2 with the exterior $x \in D(g(\gamma))^c$. Depicted in Figure 8.5 is an example of a small perturbation of the parameter causing a change in energy which involves the integral along the boundary of the difference of the models. Notice, in this model, that the action of the transformation does not affect the interior energy density.

The one parameter variations of the potential with respect to the parameters γ are given by the following curvilinear integral.

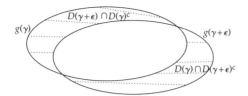

Figure 8.5 Panel shows the computation required for a small perturbation of the active model $g(\gamma) \to g(\gamma + \epsilon)$ depicts the regions over which the energy density in the integral is computed.

Theorem 8.42 *For smooth C^1 curve $g_s(\gamma), s \in [0,1]$ with simply connected interior and smooth potentials E_1, E_2, then with the normal $n = \begin{pmatrix} -\dot{g}_y \\ \dot{g}_x \end{pmatrix}$ and velocity $v = \partial g/\partial \gamma$, then the 1-parameter variation of the potential is the integral*

$$\frac{\partial H(\gamma)}{\partial \gamma} = \int_0^1 (E_1(g_s) - E_2(g_s)) \langle n(s), v(s) \rangle \, ds \,. \tag{8.70}$$

Let us examine the Chan–Vese model [205] in which only mean interior and exterior values between the objects are modelled with the random field having identity covariance.

Corollary 8.43 *Let the potential representation of the interior–exterior modelling of $g_s(\gamma), s \in [0,1]$ taking the form*

$$H(\gamma) = \int_{D(\gamma)} (I(x) - \mu_1)^2 \, dx + \int_{D \setminus D(\gamma)} (I(x) - \mu_2)^2 \, dx. \tag{8.71}$$

The optimizing μ_1 and μ_2 are the mean intensities of the image inside the interior and exterior:

$$\mu_1 = \frac{\int_{D(\gamma)} I(x) dx}{\int_{D(\gamma)} dx}, \quad \mu_2 = \frac{\int_{D \setminus D(\gamma)} I(x) dx}{\int_{D \setminus D(\gamma)} dx}; \tag{8.72}$$

the variation of the potential takes the form

$$\frac{\partial H(\gamma)}{\partial \gamma} = \int_{\partial D(\gamma)} \langle n(s), v(s) \rangle [(I(x) - \mu_1)^2 - (I(x) - \mu_2)^2] ds. \tag{8.73}$$

The functional perturbation in the steepest descent direction $g \to g + \epsilon \psi$ is given by

$$\psi(s) = (\mu_1 - \mu_2)(2I(x) - \mu_1 - \mu_2)n(s). \tag{8.74}$$

Proof From Theorem 8.41, we have $E_1(x; \gamma) = (I(x) - \mu_1)^2$, and $E_2(x; \gamma) = (I(x) - \mu_2)^2$, giving the variation of $H(\gamma)$ as

$$\frac{\partial H(\gamma)}{\partial \gamma} = \int_{\partial D(\gamma)} \langle n(s), v(s) \rangle [(I(x) - \mu_1)^2 - (I(x) - \mu_2)^2] ds$$

$$+ \int_{D(\gamma)} \frac{\partial}{\partial \gamma} (I(x) - \mu_1)^2 dx + \int_{D \setminus D(\gamma)} \frac{\partial}{\partial \gamma} (I(x) - \mu_2)^2 \, dx. \tag{8.75}$$

The second and third terms equal zero since

$$\int_{D(\gamma)} \frac{\partial}{\partial \gamma} (I(x) - \mu_1)^2 dx = -2 \int_{D(\gamma)} (I(x) - \mu_1) \frac{\partial}{\partial \gamma} \mu_1 \, dx$$

$$= -2 \frac{\partial}{\partial \gamma} \mu_1 \left(\int_{D(\gamma)} I(x) dx - \mu_1 \int_{D(\gamma)} dx \right) = 0; \tag{8.76}$$

similarly for the third term.

Using the definition of μ_1 and the Schwartz inequality [206], the steepest descent direction that reduces the energy functional H most rapidly is given by Eqn. 8.74. □

In the literature often an additional boundary integral term is added to the energy functional to improve taking the form $H_b(g(\gamma)) = \int_{\partial D(\gamma)} q\, ds$ where $q(x)$ is a scalar function defined on the image domain D. The boundary integral can be interpreted as the weighted length of the boundary contour, when $q(x) = 1$ it gives the Euclidean length of the contour.

Corollary 8.44 *Adding the boundary integral for smoothness*

$$H_b(g(\gamma)) = \int_{\partial D(\gamma)} q\, ds, \qquad (8.77)$$

the variation of $H_b(g(\gamma))$ with $v = \partial g/\partial \gamma$ is given by

$$\frac{\partial H_b(g(\gamma))}{\partial \gamma} = \int_{\partial D(\gamma)} -(\nabla q \cdot n(s) + q \nabla \cdot n) < n(s),\ v(s) > ds. \qquad (8.78)$$

The steepest descent direction for boundary contour evolution becomes

$$\Psi(s) = (\nabla q \cdot n(s) + q \nabla \cdot n) n(s); \qquad (8.79)$$

combined with the Chan–Vese region functional, the steepest descent direction is given by

$$\Psi(s) = [(\mu_1 - \mu_2)(2I(x) - \mu_1 - \mu_2) + \nabla q \cdot n(s) + q \nabla \cdot n] n(s). \qquad (8.80)$$

See [207] for proof of this; the derivation of the variation in terms of level sets is derived below in section 8.6.

8.5.4 Active Unclosed Snakes and Roads

Linear structures such as roads or membranes with constant narrow width relative to their length have been examined extensively in the community. Model the roads as structures of constant, known width $2w$ pixels parameterized by its midcurve $f_s, s \in [0, L]$ (see Figure 8.6), assumed to be of unit speed $\|(\partial/\partial s) f_s\| = \text{constant}$, with length encoded through the index length $[0, L]$. Perturbing the boundary corresponds to wiggling a structure of constant width (therefore constant area) with position which snakes through $\mathbb{R}^2$.

Perturbing the boundary wiggles the linear structure of constant width and length L, with position snaking through $\mathbb{R}^2$ and parameterized via its midcurve $f_s, s \in [0, L]$. To compute the curvilinear integral around the boundary of the membrane it is divided into two major components f_s^{+w}, f_s^{-w} (see left panel of Figure 8.6) determined by the midcurve f_s and its normal n_s:

$$f_s^{+w} = f_s + w n_s,\quad f_s^{-w} = f_s - w n_s,\quad \text{with } n_s = \begin{pmatrix} -\partial f_{ys}/\partial s \\ \partial f_{xs}/\partial s \end{pmatrix}. \qquad (8.81)$$

Then a simple formula arises for computing the variation of these linear roads with respect to 1-parameter perturbations of the midcurve. Essentially since the Jacobian along either of the long boundaries are equal, then variation amounts to comparing the difference between texture potentials representing the interior and exterior of the roads $\Delta E = E_1 - E_2$ as it runs along each of the boundaries.

8.5 ACTIVATING SHAPES IN DEFORMABLE MODELS

Theorem 8.45 *Given is a simply connected road of fixed width $w \ll L$ (much less than its length) with smooth boundary and interior and exterior potentials satisfying Theorem 8.42 with $\Delta E = E_1 - E_2$. Parameterize the shape via its smooth midcurve in unit speed $f_s = \begin{pmatrix} f_{xs} \\ f_{ys} \end{pmatrix}, s \in [0, L]$ having normal n_s, $\|\dot{f}\| = 1$, and curvature $\|\ddot{f}\| = \kappa$ with $\ddot{f} = \kappa n$. Then the plus and minus boundary normal-velocity terms are given by*

$$\langle n^w, v^w \rangle = (1 - \kappa w)\langle n, v \rangle, \qquad \langle n^{-w}, v^{-w} \rangle = (1 + \kappa w)\langle n, v \rangle. \tag{8.82}$$

The 1-parameter variation of the potential is given by the integral

$$\frac{\partial H(f(\gamma))}{\partial \gamma} = \int_0^L \langle n(s), v(s) \rangle [(1 - w\kappa(s))\Delta E(f_s^{+w})$$

$$- (1 + w\kappa(s))\Delta E(f_s^{-w})]ds + O(w). \tag{8.83}$$

Proof Let the midcurve and normal $f = \begin{pmatrix} f_x \\ f_y \end{pmatrix}$, $n = \begin{pmatrix} -\dot{f}_y \\ \dot{f}_x \end{pmatrix}$, then the plus w boundary and tangent are given by

$$f^w = f + wn = \begin{pmatrix} f_x - w\dot{f}_y \\ f_y + w\dot{f}_x \end{pmatrix}, \qquad \dot{f}^w = \dot{f} + w\dot{n} = \begin{pmatrix} \dot{f}_x - w\ddot{f}_y \\ \dot{f}_y + w\ddot{f}_x \end{pmatrix}. \tag{8.84}$$

Using the fact that $\ddot{f} = \kappa n$, the plus w normal becomes

$$n^w = \begin{pmatrix} -\dot{f}_y - w\ddot{f}_x \\ \dot{f}_x - w\ddot{f}_y \end{pmatrix} = (1 - w\kappa)n. \tag{8.85}$$

Since $(\partial/\partial\gamma)\langle n, n \rangle = 0$ we have

$$\langle n^w, v^w \rangle = \left\langle (1 - w\kappa)n, \frac{\partial}{\partial \gamma} f^w \right\rangle \tag{8.86}$$

$$= \left\langle (1 - w\kappa)n, \frac{\partial}{\partial \gamma} (f + wn) \right\rangle = (1 - w\kappa)\langle n, v \rangle. \tag{8.87}$$

Similarly for the $-w$ boundary. Substituting into the closed curve Theorem 8.42 and integrating along the midcurve give the result. □

Example 8.46 (Piecewise Linear Roads) Let the roads have zero curvature and be built from linear sections as in the left panel of Figure 8.6. Let the midcurve parameterization be

$$f_s = (s - n)\begin{pmatrix} \cos\theta_{n+1} \\ \sin\theta_{n+1} \end{pmatrix} + \sum_{k=1}^n \begin{pmatrix} \cos\theta_k \\ \sin\theta_k \end{pmatrix} + \begin{pmatrix} x_0 \\ y_0 \end{pmatrix} \quad s \in [n, n+1). \tag{8.88}$$

Then the variation $\partial f_s/\partial \theta_m = \begin{pmatrix} -\sin\theta_m \\ \cos\theta_m \end{pmatrix} 1_{>m}(s)$ with the normal $n_s = \begin{pmatrix} -\sin\theta_k \\ \cos\theta_k \end{pmatrix}$ for $s \in [k-1, k)$ giving $\langle n(s; \theta_m), v(s; \theta_m) \rangle = \cos(\theta_m - \theta_s)1_{>m}(s)$; the 1-parameter variation becomes

$$\frac{\partial H(\theta_m)}{\partial \theta_m} = \sum_{k=m+1}^L 2\cos(\theta_k - \theta_m) \int_{k-1}^k [\Delta E(f_s^{+w}) - \Delta E(f_s^{-w})]ds + O(2w). \tag{8.89}$$

Shown in the right panel of Figure 8.6 is a linear membrane depicted using the road model.

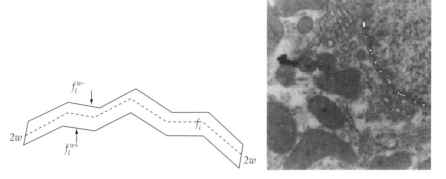

Figure 8.6 Left panel shows the linear membrane model with midline curve f_s shown dashed, and two boundary components defined by the midline curve and normals $f_s^{+w} = f_s + wn_s$ and $f_s^{-w} = f_s - wn_s$. Right panel shows results from the linear active contour model.

8.5.5 Normal Deformation of Circles and Spheres

Corollary 8.47 *For normal deformation of the circle, with $u_s = \sum_{n=0}^{\infty} u_n e^{j2\pi ns}$, then*

$$g_s = \begin{pmatrix} x_0 \\ y_0 \end{pmatrix} + \begin{pmatrix} \cos s \\ \sin s \end{pmatrix} + u_s \begin{pmatrix} \cos s \\ \sin s \end{pmatrix}, \quad s \in [0, 2\pi], \tag{8.90}$$

and the gradient is given by

$$\begin{pmatrix} \partial H(\gamma)/\partial x_0 \\ \partial H(\gamma)/\partial y_0 \end{pmatrix} = \int_0^{2\pi} (E_1(s) - E_2(s)) \begin{pmatrix} -\cos s \\ -\sin s \end{pmatrix} ds;$$

$$\frac{\partial H(\gamma)}{\partial u_n} = \int_0^1 (E_1(s) - E_2(s)) e^{j2\pi ns} ds. \tag{8.91}$$

Proof The normal to the curve is given by $n(s) = \begin{pmatrix} -\partial g_{ys}/\partial s \\ \partial g_{xs}/\partial s \end{pmatrix} = -\begin{bmatrix} \cos s \\ \sin s \end{bmatrix}$, with the velocity $v(s) = \partial g/\partial u_n = -e^{j2\pi ns} n(s)$. Substituting gives the result. □

8.5.6 Active Deformable Spheres

Return to example 8.38 on active spheres under normal deformation with active models. Joshi and McNally [33, 198] have used active deformable spheres to study 3D cell motion during embryogenesis in the slime mold *Dictyostelium discoideum* in which individual cells are labeled with a fluorescent dye and analyzed during an aggregation phase of their life cycle. Joshi uses azimuth-elevation representation and spherical harmonics. Let

$$g(\theta, \psi) = \begin{pmatrix} \cos\theta \sin\psi \\ \sin\theta \sin\psi \\ \cos\psi \end{pmatrix} + u(\theta, \psi) \begin{pmatrix} n_x(\theta, \psi) \\ n_y(\theta, \psi) \\ n_z(\theta, \psi) \end{pmatrix} + \begin{pmatrix} x_0 \\ y_0 \\ z_0 \end{pmatrix}, \quad \psi \in [0, \pi], \; \theta \in [0, 2\pi). \tag{8.92}$$

with $n(\theta, \psi) \in \mathbb{R}^3$ the normal to the sphere at θ, ψ, and $u(\theta, \psi)$ is the scalar field parameterizing the translation vector fields applied to the generator spheres. The scalar field on the sphere is often parametrically defined via a complete orthonormal spherical-harmonic basis analogous to the complex exponentials on the circle (see the subsequent Chapter 9, Section 9.8).

Corollary 8.48 (The Sphere) *Assuming the interior and exterior models E_1, E_2, then the likelihood of a spherical active shapes with simply connected interior $D(g(\gamma))$ has 1-parameter variation of the energy*

$$\frac{\partial H(\gamma)}{\partial \gamma} = \int_{\theta \in [0,2\pi), \psi \in [0,\pi]} (E_1(s) - E_2(s))\langle n(s(\theta, \psi)), v(s(\theta, \psi))\rangle \, ds. \tag{8.93}$$

For scalar fields in the spherical harmonic basis expansion

$$u(\theta, \psi) = \sum_{n=1}^{N} (u_{n0}\phi_{n0}(\theta, \psi) + (u_{nm}^e \phi_{nm}^e(\theta, \psi) + u_{nm}^o \phi_{nm}^o(\theta, \psi)) \tag{8.94}$$

$\phi_{nm}^{e,o}$ *the even (cosine) and odd (sine) spherical-harmonics (see Definition 9.57, Chapter 9), then for $v(\theta, \psi) = \partial g(\theta, \psi)/\partial u_{nm}^{e,o} = \phi_{nm}^{e,o}(\theta, \psi) n(\theta, \psi)$ and*

$$\begin{pmatrix} \partial H(\gamma)/\partial x_0 \\ \partial H(\gamma)/\partial y_0 \\ \partial H(\gamma)/\partial z_0 \end{pmatrix} = \int_{\theta \in [0,2\pi), \psi \in [0,\pi]} (E_1(\theta, \psi) - E_2(\theta, \psi)) n(\theta, \psi) \sin \psi \, d\psi \, d\theta, \tag{8.95}$$

$$\frac{\partial H(\gamma)}{\partial u_{nm}} = \int_{\theta \in [0,2\pi), \psi \in [0,\pi]} (E_1(\theta, \psi) - E_2(\theta, \psi)) \phi_{nm}(\theta, \psi) \sin \psi \, d\psi \, d\theta. \tag{8.96}$$

8.6 Level Set Active Contour Models

Now let us examine level set methods championed in the Osher school as a computational approach for generating solutions to active shape problems. Starting from an initial partition of the domain $x \in D$ as specified by an initial submanifold, active shapes $g(\gamma)$ evolve by generating a sequence of shapes which converge ultimately to a partition at infinite simulation time. In the level set framework [208] for implementation, this is achieved by embedding the explicit representations of the evolutions of the active shapes $g(\gamma)$ in their embedding background spaces. This is obtained by formulating the solutions as the zero level set of a higher dimensional Lipschitz-continuous scalar function $\Phi_t(x), x \in X$; the *zero level set function* at any time instant of simulation time t corresponds to $\{x \mid \Phi_t(x) = 0\}$. Although there are infinitely many choices of the level set function, in practice, the signed distance function is preferred for its stability in numerical computations. The *fast marching method* proposed in [209, 210] provides an efficient algorithm for constructing the signed distance function from a given contour. Alternatively, an explicit representation of the embedded contour can be recovered from the signed distance function by applying any isocontour algorithm.

To relate the evolution of the explicit representation of the bounding manifold of the deformable shape $g(\gamma)$ with the level set formulation we examine the evolution of the level sets of the extrinsic embedding space. For this we shall "interchange between" $\delta(\Phi)$ with the norm of the gradient of the embedding level set function $\|\nabla \Phi\|$. This rescaling corresponds to a natural extension of the evolution of the zero level set to all the other level sets of Φ [211], which does not affect the evolution of the zero level set itself.

In this setting, define the level set function $\Phi(x), x \in X$ with natural boundary conditions along the boundary of the computational domain $\nabla\Phi \cdot n = 0$. To calculate the variational solutions perturbations of $\Phi \to \Phi + \epsilon\Psi$, with Ψ satisfying boundary conditions $\nabla\Psi \cdot n = 0$.

Theorem 8.49 *For the Chan-Vese Mean Model with energy of Theorem 8.43, the steepest-descent direction is given by*

$$\Psi_t(x) = \delta(\Phi_t(x))(\mu_1 - \mu_2)(2I - \mu_1 - \mu_2). \tag{8.97}$$

Defining $\delta(\Phi)$ *by* $|\nabla\Phi|, n = \nabla\Phi/\|\nabla\Phi\|$, *the level set evolution in terms of the extrinsic boundary contour evolution with initial estimate* $\Phi(x)$ *is the directional derivative in the direction of the gradient of the level set*

$$\frac{\partial \Phi_t(x)}{\partial t} = \langle \Psi, \nabla\Phi_t \rangle \quad \text{where } \Psi = (\mu_1 - \mu_2)(2I - \mu_1 - \mu_2)n. \tag{8.98}$$

Proof The proof of Eqn. 8.97 follows techniques similar to that in [205, 211] to derive the variation of $H(\Phi)$. The Frechet derivative of $\partial_\Psi H(\Phi)$ in the direction $\Psi(x)$ is computed as

$$\partial_\Psi H(\Phi) = \int \Psi \delta(\Phi)(I - \mu_1)^2 \, dx - \int \Psi \delta(\Phi)(I - \mu_2)^2 \, dx \tag{8.99}$$

$$- 2\int h(\Phi)(I - \mu_1)\partial(\mu_1, \Psi) dx - 2\int (1 - h(\Phi))(I - \mu_2)\partial(\mu_2, \Psi) \, dx. \tag{8.100}$$

$$\stackrel{(a)}{=} \langle \delta(\Phi)(\mu_1 - \mu_2)(2I - \mu_1 - \mu_2), \Psi \rangle, \tag{8.101}$$

where (a) follows from the fact that the last two terms in Eqn. 8.100 evaluate to zero (similarly to Eqn. 8.76). By the Schwartz inequality the variation in the steepest descent direction Ψ is therefore given by Eqn. 8.97. □

Oftentimes in the literature an additional boundary integral is added to the energy functional to improve the smoothness of the segmented boundary contour.

Corollary 8.50 *Adding the boundary integral* $H_b(\Phi) = \int q(x)\delta(\Phi(x))dx$, *the level set function following the steepest-descent direction*

$$\Psi = \delta(\Phi(x))(\mu_1 - \mu_2)(2I - \mu_1 - \mu_2) + \delta(\Phi)\left(\frac{\nabla q \cdot \nabla\Phi}{\|\nabla\Phi\|} + q\nabla \cdot \left(\frac{\nabla\Phi}{\|\nabla\Phi\|}\right)\right). \tag{8.102}$$

The level set evolution equation becomes

$$\Phi_t = (\mu_1 - \mu_2)(2I(x) - \mu_1 - \mu_2)\|\nabla\Phi\| + \nabla q \cdot \nabla\Phi + q\nabla \cdot \left(\frac{\nabla\Phi}{\|\nabla\Phi\|}\right)\|\nabla\Phi\| \tag{8.103}$$

$$= (\mu_1 - \mu_2)(2I(x) - \mu_1 - \mu_2)\|\nabla\Phi\| + \nabla q \cdot \nabla\Phi + q\kappa\|\nabla\Phi\|, \tag{8.104}$$

where κ *is the curvature of the level sets of* Φ *corresponding to the divergence of the normal.*

8.6 LEVEL SET ACTIVE CONTOUR MODELS

Proof To prove the second boundary part, the Frechet derivative $\partial_\Psi H_b(\Phi)$ in the direction $\Psi(x)$ is computed as

$$\partial_\Psi H_b(\Phi) = \int_\Omega \Psi \delta'(\Phi) \|\nabla \Phi\| q \, dx + \int_\Omega \delta(\Phi) q \frac{\nabla \Phi \cdot \nabla \Psi}{\|\nabla \Phi\|} \, dx,$$

where $\delta'(\cdot)$ denotes the first derivative of the delta function. Applying Green's formula [206] to the second term yields

$$\partial_\Psi H_b(\Phi) = \int_\Omega \Psi \delta'(\Phi) \|\nabla \Phi\| q \, dx + \oint_{\partial \Omega} \Psi \delta(\Phi) q \frac{\nabla \Phi \cdot n}{\|\nabla \Phi\|} \, ds$$

$$- \int_\Omega \Psi \nabla \cdot \left(\delta(\Phi) q \frac{\nabla \Phi}{\|\nabla \Phi\|} \right) d\mathbf{x},$$

where $\nabla \cdot$ is the divergence operator, n is the normal vector to the boundary and ds is a differential element on the boundary. Since

$$\nabla \cdot \left(\delta(\Phi) q \frac{\nabla \Phi}{\|\nabla \Phi\|} \right) = q \delta'(\Phi) \|\nabla \Phi\| + \delta(\Phi) \nabla \cdot \left(q \frac{\nabla \Phi}{\|\nabla \Phi\|} \right),$$

under the boundary conditions $\nabla \Phi \cdot n = 0$ obtains

$$\partial_\Psi H_b(\Phi) = -\int_\Omega \delta(\Phi) \nabla \cdot \left(q \frac{\nabla \Phi}{\|\nabla \Phi\|} \right) \Psi \, d\mathbf{x}$$

$$= \left\langle -\delta(\Phi) \nabla \cdot \left(q \frac{\nabla \Phi}{\|\nabla \Phi\|} \right), \Psi \right\rangle.$$

Figure 8.7 Results from level set evolution showing different iterations for a single face (see also Plate 12).

The steepest descent direction of $H_b(\Phi)$ is given by

$$\Psi = \delta(\Phi)\nabla \cdot \left(q\frac{\nabla\Phi}{\|\nabla\Phi\|}\right) = \delta(\Phi)\left(\frac{\nabla q \cdot \nabla\Phi}{\|\nabla\Phi\|} + q\nabla \cdot \left(\frac{\nabla\Phi}{\|\nabla\Phi\|}\right)\right).$$

□

Example 8.51 (Xiao Han) Here are results from Xiao Han where he studied level set implementations in $\mathbb{R}^2, \mathbb{R}^3$. Here the examples are restricted to the plane. Han's implementations have the smoothing boundary term with $q = 1$, so that $\nabla q = 0$ and the evolution follows the simpler form

$$\Phi_t = (\mu_1 - \mu_2)(2I(x) - \mu_1 - \mu_2)\|\nabla\Phi\| + q\nabla \cdot \left(\frac{\nabla\Phi}{\|\nabla\Phi\|}\right)\|\nabla\Phi\| \tag{8.105}$$

$$= (\mu_1 - \mu_2)(2I(x) - \mu_1 - \mu_2)\|\nabla\Phi\| + q\kappa\|\nabla\Phi\|. \tag{8.106}$$

Shown in Figure 8.7 are results from Han's algorithm showing level set evolutions as a function of iteration number.

8.7 Gaussian Random Field Models for Active Shapes

Model the interiors of the shapes of real-valued images as scalar random fields defined on the index sets of integer lattices. Because of splitting, the partitions are conditionally independent given their boundaries. The Bayesian approach presents a substantial computational hurdle. Since the normalizing partition function must be calculated for each partition $D(g_j) \subset \mathbb{Z}^d, d = 2,3$, $j = 1,\ldots,J$ the number of object regions. The partition function and log-normalizer must be calculated for each shape. For this we use the asymptotic representation of the log-normalizer via the Fourier transform of Theorem 5.17 of Chapter 5. Let I_D be a real-valued Gaussian field on $D \subset \mathbb{Z}^d$, with mean and covariance μ, K_D. Model the inverse covariance via the difference operator which is zero at the boundary of the shape so that $L_D L_D^* = K_D^{-1}$; the probability density requires the computation of the determinant of the covariance corresponding to the log-partition function for Gaussian fields:

$$p(I_D) = (2\pi)^{-|D|/2} \det{}^{-1/2} K_D e^{-1/2\|L_D(I_D-\mu)\|^2}, \tag{8.107}$$

with $\|L_D(X - \mu)\|^2$ the integral-square of the field over the domain D. Notice the role that the partition functions play; segmentation requires computation of $\log K_D$ for the random shape.

The Gaussian Markov random field is used for segmentation of the micrographs.

Theorem 8.52 *Given is the simple parametric shape model with the interior D of shape $g_t, t \in [0,1]$ a random realization of a Gaussian random field with density*

$$p(I_D) = (2\pi)^{-|D|/2} \det{}^{-1/2} K_D e^{-1/2\|L_D(I_D-\mu)\|^2}, \tag{8.108}$$

with operators for the interior and exterior L_1, L_2 solving the difference equation

$$L_1 I_i = \sum_{s \in S} a_s^{(1)} I_{i+s}, \quad L_2 I_i = \sum_{s \in S} a_s^{(2)} I_{i+s}. \tag{8.109}$$

8.7 GAUSSIAN RANDOM FIELD MODELS FOR ACTIVE SHAPES

Then the 1-parameter variation of the potential $H(\gamma)$ is given by Theorem 8.42 with difference in energies for an asymptotic in size shape

$$E_1 - E_2 \simeq \frac{1}{2}\frac{1}{(2\pi)^d}\int_{[-\pi,\pi]^d} \log \frac{\sigma_1(\omega)}{\sigma_2(\omega)} d\omega + \frac{1}{2}\left(\|L_1(I_s - \mu_1)\|^2 - \|L_2(I_s - \mu_2)\|^2\right), \tag{8.110}$$

with $\sigma_1(\omega) = \sum_s a_s^{(1)} e^{j\langle \omega, s\rangle}, \quad \sigma_2(\omega) = \sum_s a_s^{(2)} e^{j\langle \omega, s\rangle}.$ (8.111)

Proof The asymptotic in size partition Theorem 5.17 of Chapter 5, gives the log-determinant covariance by the spectrum:

$$\log \det^{1/2} K_D = \frac{|D|}{2}\frac{1}{(2\pi)^d}\int_{[-\pi,\pi]^d} \log \sigma(\omega) d\omega + O(|\partial D|)$$

with $\sigma(\omega) = \sum_s a_s e^{j\langle \omega, s\rangle}.$ (8.112)

The energy spectrum becomes

$$\int_{D(\gamma)} E(x)dx = \frac{|D(\gamma)|}{2}\frac{1}{(2\pi)^d}\int_{[-\pi,\pi]^d} \log \sigma(\omega) d\omega + \frac{1}{2}\|L(I_{D(\gamma)} - \mu)\|^2. \tag{8.113}$$

$\square$

Example 8.53 (Segmenting Mitochondria) Now examine the multiple shape problem partitioning the background space as multiple object domains. Construct the background space $D \subset \mathbb{Z}^d$ as a disjoint partition of simply connected regions $D(g_j)$ with random image field $I_{D(g_j)}$. Conditioned on the parametric models, $g_1, g_2, \ldots$, the subregion Markov random fields are assumed conditionally independent. Write the potential via energy densities $E_j(x), x \in D(g_j)$ representing the log-normalizer and quadratic data statistic according to

$$p(I|g_1, g_2, \ldots) = \prod_{j=1}^{J}(2\pi)^{-|D(g_j)|/2} \det^{-1/2} K_{D(g_j)} e^{-1/2\|L(I_{D(g_j)}-\mu_j)\|^2} \tag{8.114}$$

$$\propto \prod_{j=1}^{J}(2\pi)^{-|D(g_j)|/2} e^{-\int_{D(g_j)} E_j(x)dx}. \tag{8.115}$$

Clearly, if the regions corresponding to the interiors of an unknown shape which are to be inferred the asymptotic partition approximations must be used. Notice $E_j(x), x \in D(g_j)$ the energy density representing the log-normalizer using the asymptotic representation via the Fourier transform of Theorem 5.17 of Chapter 5.

The Gaussian Markov random field is used for segmentation of the micrographs. Assume the mitochondrial interiors are random realization of the stochastic difference equation for $i = (i_1, i_2) \in D \subset \mathbb{Z}^2$:

$$L(I_i - \mu) = (-\nabla^2 + a)(I_i - \mu) = W_i \text{ with} \tag{8.116}$$

$$\nabla^2 I_{i_1,i_2} = I_{i_1-1,i_2} + I_{i_1+1,i_2} + I_{i_1,i_2-1} + I_{i_1,i_2+1} - 4I_{i_1,i_2}, \tag{8.117}$$

where μ is the mean and a is the constant restoring force, with white Gaussian noise variance σ^2. The Laplacian induces nearest-neighbor dependence between pixels on the discrete lattice of points. The three parameters, a, μ and σ, completely specify the model. The energy density resulting from the spectrum of the 2D Laplacian becomes

$$\log \det{}^{1/2} K_{D_g} = \frac{|D(g_j)|}{2} \frac{1}{4\pi^2} \int_{[-\pi,\pi]^2} \log\left(2 \sum_{k=1}^{2}(1 - \cos \omega_k) + a\right) d\omega$$

$$+ O(|\partial D_g|) \text{ where} \tag{8.118}$$

$$\int_{D(g(\gamma))} E(x) dx = \frac{|D(g(\gamma))|}{2} \left(\frac{1}{4\pi^2} \int_{[-\pi,\pi]^2} \log 2\pi\sigma^2 \left(\sum_{k=1}^{2}(2 - 2\cos \omega_k) + a\right) d\omega\right)$$

$$+ \frac{1}{2\sigma^2} \|LI_{D(g(\gamma))}\|^2.$$

Using the asymptotic maximum-likelihood parameter estimation of Chapter 5, Example 5.26 of Theorem 5.17, parameters are estimated from sets of hand-labelled mitochondria and background cytoplasm region types. The parameters estimated from the micrograph data assuming the Gaussian model with $L = \Delta - a$ having parameters of mean μ, noise power σ^2, and restoring force a are shown in Table 8.1.

Let us use Corollary 8.42 for computing gradients of the shapes for closed curves as given in Example 7.29 then

$$g_t(\gamma) = \begin{pmatrix} x_0 \\ y_0 \end{pmatrix} + \int_0^t \begin{pmatrix} u_{1s} & -u_{2s} \\ u_{2s} & u_{1s} \end{pmatrix} \begin{pmatrix} -\sin s \\ \cos s \end{pmatrix} ds, \quad t \in [0, 2\pi], \tag{8.119}$$

with $\gamma \in \{u_{1s}, u_{2s}, x_0, y_0\}$, then

$$\frac{\partial g_s}{\partial s} = \begin{pmatrix} -u_{1s} \sin s - u_{2s} \cos s \\ u_{1s} \cos s - u_{2s} \sin s \end{pmatrix}, \quad \frac{\partial g_s}{\partial x_0} = \begin{pmatrix} 1 \\ 0 \end{pmatrix}, \quad \frac{\partial g_s}{\partial y_0} = \begin{pmatrix} 0 \\ 1 \end{pmatrix} \tag{8.120}$$

$$\frac{\partial g_t}{\partial u_{1s}} = 1_{\geq t}(s) \begin{pmatrix} -\sin s \\ \cos s \end{pmatrix}, \quad \frac{\partial g_t}{\partial u_{2s}} = 1_{\geq t}(s) \begin{pmatrix} -\cos s \\ -\sin s \end{pmatrix}. \tag{8.121}$$

For computation there are n unique parameter vectors, $\{u_{1k}, u_{2k}\}_{k=1}^{n}$ constant over intervals $l \in ((k-1)/n, k/n)$, corresponding to a polygonal approximation to the circle. The n-similarities become the scales and rotations applied to the chords, and adding global translation gives

$$g_{t=\frac{l}{n}} = \begin{pmatrix} x_0 \\ y_0 \end{pmatrix} + \sum_{k=1}^{l} \begin{pmatrix} u_{1k} & u_{2k} \\ -u_{2k} & u_{1k} \end{pmatrix} \begin{pmatrix} \cos 2\pi k/n - \cos 2\pi(k-1)/n \\ \sin 2\pi k/n - \sin 2\pi(k-1)/n \end{pmatrix}. \tag{8.122}$$

Table 8.1 Maximum-likelihood estimates of the parameters μ, noise variance σ^2, and restoring force as estimated from the mitochondria data using the algorithm from Example 5.26 of Chapter 5.

	μ Mean gray level	σ^2 Variance of White Noise	a restoring Force
Mitochondria	58.7	847.3	0.62
Background	130.3	2809.9	0.15

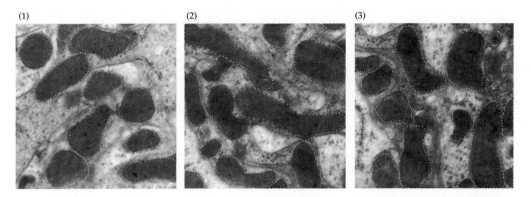

Figure 8.8 Panels 1, 2, and 3 show the segmentation via the closed active contour models. Mitochondria interiors and cytoplasmic exteriors were represented using the Gaussian random field asymptotic partition function. Each circular template shape was manually placed; data were taken from the laboratory of Dr. Jeffrey Saffitz at Washington University.

Figure 8.8 shows examples of active model segmentation of the EM micrographs. The active shape model is deformed to maximize the "probability" or minimize the energy by varying the parameters of the deformable contour to match the interior compartments of mitochondria. Examining the results it appears as if the shape models are flexible enough to accomodate the variation in organelle shapes.

9 SECOND ORDER AND GAUSSIAN FIELDS

ABSTRACT This chapter studies second order and Gaussian fields on the background spaces which are the continuum limits of the finite graphs. For this random processes in Hilbert spaces are examined. Orthogonal expansions such as Karhunen–Loeve are examined, with spectral representations of the processes established. Gaussian processes induced by differential operators representing physical processes in the world are studied.

In several of the examples studied in previous chapters the graphs have all been discrete, at most countable. Oftentimes these graphs are representations of shapes with generators which are sub-manifolds sewn together, with each of the groups acting only locally on submanifolds of the whole. Then, the graph-based regularity on the transformations corresponds to the notion of constraining the local action of these transformations so that the total manifold stays regular or connected. For many of the examples, the number of submanifolds (generators) are chosen in some sense arbitrarily, oftentimes based on computational convenience. However, the continuum limits are in most applications in image processing the regular structures, not discrete sets. This important distinction was not made systematically in the literature until the early 1980s, and is based on the obvious observation that *natural objects live in $\mathbb{R}^d$-spaces*, not on lattices. The lattices often appear for computational convenience. A theory which is consistent only on lattices is not sufficient since they do not allow most of the natural invariances that are so important in the physical world. Of course, the computations will be done discretely in the final stage, thus intimately connecting the continuum representations with the finite graph structures and discrete generators of previous sections. *It is our belief*, however, that whenever the representations support a continuum limit, the discrete representation which is often of implementation convenience must be consistent with that continuum limit. For this reason, we require representations on the continuum, processes built from products of the Lie groups which take their actions on the manifolds and sub-manifolds. These form transformations on the continuum, requiring the study of vector and group valued fields.

As probabilistic structures on the representations will allow us to express the variation of natural patterns, we now move from probabilistic model representations on discrete sets to those on the continuum. Therefore the study of random processes and random fields on both lattices as well as the continuum becomes essential.

9.1 Second Order Processes (SOP) and the Hilbert Space of Random Variables

First for classical measurability and continuity assumptions.

9.1.1 Measurability, Separability, Continuity

All the random processes will be examined in their coordinate representation assuming standard measurability with respect to the underlying probability space. As the random processes begin to be indexed on countably infinite and uncountable sets, properties of measurablity, separability and continuity become important. These are discussed in Appendix A.2 in some detail (see the texts [212–214] for indepth development). Basic definitions are provided below for clarity. Since

9.1 SOP AND THE HILBERT SPACE OF RANDOM VARIABLES

some of the index spaces will be associated with time, we will sometimes index with the 1D time variable $t \in T \subset \mathbb{R}$; for multiple dimensions use $x \in X \subset \mathbb{R}^d$.

As we extend to fields on the continuum technical conditions arise requiring the processes to be well behaved, measurable and separable. For this reason, we define the basic measurability and separability properties which will be assumed throughout. We shall be interested in m-dimensional fields, real and complex $\mathbb{R}^m, \mathbb{C}^m$.

Definition 9.1 *A real-valued* **random process** $Y(x), x \in X \subset \mathbb{R}(\mathbb{Z})$ *and is a* **random field** *if* $X \subset \mathbb{R}^d(\mathbb{Z}^d)$ *on the probability space* $(\Omega, \mathcal{A}, P)$ *is said to be* $\mathcal{A}$-**measurable** *if* $\forall r \in \mathbb{R}, \forall x \in X$ *then*

$$\{\omega : Y(\omega, x) \leq r\} \in \mathcal{A}. \tag{9.1}$$

*An m-***vector-valued random process (field)** *is* $\mathcal{A}$ − *measurable for each of the components* $Y_1, \ldots, Y_m$. *In particular, a* **complex-valued** *process* $Y_R + jY_I$ *is $\mathcal{A}$-measurable with respect to each of the real and imaginary components,* Y_R, Y_I.

In moving from finite graphs to countable and uncountable index sets, we need to reduce questions involving the uncountable sets associated with the continuum and countable subsets. This is the separability program.

Suppose we were to ask about probabilities on such events as

$$\Pr\{\omega : 0 \leq Y(\omega, x) \leq 1, \forall x \in X \subset \mathbb{R}\} = \Pr \bigcap_{x \in X} \{\omega : 0 \leq Y(\omega, x) \leq 1\}. \tag{9.2}$$

In general, this cannot be evaluated as $\bigcap_{x \in X} \{\omega : 0 \leq Y(\omega, x) \leq 1\}$ may not necessarily be in the σ-algebra since X is uncountable. Now suppose the question is changed to what is

$$\Pr\{\omega : 0 \leq Y(\omega, s) \leq 1, s \in S\} = \Pr \bigcap_{s \in S} \{\omega : 0 \leq Y(\omega, s) \leq 1\}, \tag{9.3}$$

with $S \subset X$ a countable dense subset. This quantity can be computed because $\bigcap_{s \in S} \{\omega : 0 \leq Y(\omega, s) \leq 1\} \in \mathcal{A}$ is a countable intersection of measurable sets.

The point is, for separable processes, such questions as given by Eqns. 9.2 and 9.3 are probabilistically equivalent. This relies on separability allowing questions involving non-countable unions or intersections to be reduced to questions involving countable ones; not all processes are separable (see Example A.1 in Appendix A). More precisely, here is the definition of separability.

Definition 9.2 *A set* $E \subset E_1$ *is said to be dense in* E_1 *if for every* $x_1 \in E_1$, *every open subset of* E_1 *containing* x_1 *also contains an element of* E.

A process (field) $Y(x), x \in X \subset \mathbb{R}^1(\mathbb{R}^d)$ *is said to be separable if there exists a countable set* $S \subset X$ *and a set* $\Lambda \subset X$ *with* $P(\Lambda) = 0$ *such that for any closed set* $K \subset \mathbb{R}^1$ *and any open interval I the two sets*

$$\left\{\omega : Y(\omega, x) \in K, x \in I \bigcap X\right\}, \quad \left\{\omega : Y(\omega, x) \in K, x \in I \bigcap S\right\} \tag{9.4}$$

differ by a subset of Λ.

The good news is that with continuity in probability, we can assume we work on realizations of the process with the particular probability model which is separable. As well for this process, any countable dense set in the time or space index is sufficient on which to ask such probability questions (e.g. see Doob [212]).

Throughout the development we assume all our processes are both measurable and separable, hence we drop such qualifications in our specification of the process. Now for various forms of convergence.

Definition 9.3 *Various forms of convergence and continuity of the process will be important.*

1. *Given are a set of P-measurable random variables $Y_n, n = 1, \ldots,$ and Y.* **Convergence in probability (in p.), convergence in quadratic mean (q.m.), and converge almost surely (a.s.)** *for Y_n to Y are defined to be*

$$\forall \epsilon > 0 \quad \lim_n \Pr\{\omega : |Y_n - Y| > \epsilon\} = 0 \ (in\ p.),$$

$$\lim_n E|Y_n - Y|^2 = 0 \ (q.m.),$$

$$\Pr\{\omega : \lim_n |Y_n - Y| = 0\} = 1 \ (a.s.).$$

We write $\lim_n Y_n \stackrel{p}{=} Y$, $\lim_n Y_n \stackrel{q.m.}{=} Y$, $\lim_n Y_n \stackrel{a.s.}{=} Y$, respectively.

2. *For a stochastic process $Y(x), x \in X$ on $(\Omega, \mathcal{A}, P)$,* **continuity at x in probability, quadratic mean, and almost sure** *are defined to be*

$$\forall \epsilon > 0, \quad \lim_{x \to y} \Pr\{\omega : |Y(x) - Y(y)| > \epsilon\} = 0,$$

$$\lim_{x \to y} E|Y(x) - Y(y)|^2 = 0,$$

$$\Pr\{\omega : \lim_{x \to y} |Y(x) - Y(y)| = 0\} = 1.$$

For m-vector valued random processes, change the distance measure $|\cdot|$ to the version $\|\cdot\|$ for vectors in $\mathbb{R}^d$ or $\mathbb{R}^m$, with **continuity in probability, quadratic mean, and almost sure** *remaining otherwise the same.*

Example 9.4 (Measurability) Given a probability space $(\Omega, \mathcal{A}, P)$, $\Omega = [0,1]$, $\mathcal{A}$ is the Borel σ-algebra of $[0,1]$, and $P\{(a,b)\} = b - a$, with $X = \mathbb{N}$ the naturals. Let $Y(\omega, n), n \in \mathbb{N}$ with $Y(\omega, n) = n$th ordinate of the dyadic expansion (binary expansion) of ω restricted to have an infinite number of zeros:

ω	dyadic expansion of ω
0.0	0.00000000...
0.25	0.01000000...
0.5	0.10000000...
0.75	0.11000000...

This stochastic process is $\mathcal{A}$-measurable because for example

$$\{\omega : Y(\omega, 0) = 1\} = [0.5, 1) \in \mathcal{A},$$

$$\{\omega : Y(\omega, 0) = 0\} = [0, 0.5) \in \mathcal{A},$$

$$\{\omega : Y(\omega, 1) = 1\} = \left[[0, 0.5) \bigcap [0.25, 0.5)\right] \bigcup \left[[0.5, 1) \bigcap [0.75, 1)\right] \in \mathcal{A}.$$

Example 9.5 (Separability) The rationals are dense in the reals. In addition, the rationals are countable, making the reals separable. There exists a bijection from the naturals to the rationals, with the rationals generated by the following sequence

$$0, \frac{1}{1}, \frac{1}{2}, \frac{2}{2}, \frac{2}{1}, \frac{1}{3}, \frac{2}{3}, \frac{3}{3}, \frac{3}{2}, \frac{3}{1}, \frac{1}{4}, \frac{2}{4}, \frac{3}{4}, \frac{4}{4}, \frac{4}{3}, \frac{4}{2}, \frac{4}{1}, \frac{1}{5} \ldots \tag{9.5}$$

A one-to-one mapping of the naturals to the rationals is generated by assigning 1 to the first element above, 2 to the next, and so on, skipping any rationals that are repeated.

9.1 SOP AND THE HILBERT SPACE OF RANDOM VARIABLES

Let $Y(\omega,t), t \in [0,1]$ be a real-valued process, separable and continuous in probability. Then, using the separating set $S = \bigcup_{k=1}^{\infty}\{0, 2^{-k}, 2 \cdot 2^{-k}, \ldots, 2^k \cdot 2^{-k}\}$ which is dense in $[0,1]$, the probability $\Pr\{Y(\omega,t) \geq 0, 0 \leq t \leq 1\}$ can be computed:

$$\Pr\{Y(t) \geq 0, 0 \leq t \leq 1\} = \Pr\left\{Y\left(\frac{k}{2^n}\right) \geq 0, \ 0 \leq k \leq 2^n, \ n = 0, 1, \ldots\right\}$$

$$= \Pr\bigcap_{n=0}^{\infty}\left\{\omega : Y\left(\omega, \frac{k}{2^n}\right) \geq 0, \ 0 \leq k \leq 2^n\right\}) = \Pr\bigcap_{n=0}^{\infty} A_n,$$

where $A_n = \{\omega : Y(\omega, k/2^n) \geq 0, \ 0 \leq k \leq 2^n\}$ is a decreasing sequence in n. Therefore, by the continuity of probability,

$$\Pr\{Y(t) \geq 0, 0 \leq t \leq 1\} = \Pr\{\lim_{n \to \infty} A_n\} = \lim_{n \to \infty} \Pr\{A_n\}$$

$$= \lim_{n \to \infty} \Pr\left\{\omega : Y\left(\omega, \frac{k}{2^n}\right) \geq 0, \ 0 \leq k \leq 2^n\right\}.$$

9.1.2 Hilbert space of random variables

There will be several Hilbert spaces which shall be worked with, in particular the Hilbert space, H_Y, generated from linear combinations of random variables generated from the process $Y(\cdot)$. Statements about second order representations require understanding the Hilbert space of mean squared normed random variables and vectors.

Definition 9.6 *Define $L^2(P)$ to be the set of P-measurable random m-vectors* $Y = \begin{pmatrix} Y_1 \\ Y_2 \\ \vdots \end{pmatrix}$

with inner product and finite norm for all $Y, Z \in L^2(P)$,

$$E\langle Y, Z\rangle = EY^*Z, \quad E\|Y\|^2 = EY^*Y < \infty, \quad (9.6)$$

where $(\cdot)^*$ denotes complex transpose for vectors (for scalars use $(\cdot)^*$).

For the scalar case, the norm and inner product reduce to $E|Y|^2, EY^*Z$.

That $L^2(P)$ is a Hilbert space of random variables requires showing completeness, i.e. quadratic mean Cauchy sequences converging in quadratic mean.

Theorem 9.7 *$L^2(P)$ is a Hilbert space with norm and inner product as in Eqn. 9.6.*

Proof See Theorem B.5 in Appendix B.1 for the proof. □

Second order processes and the Hilbert space constructed from their linear combinations are as follows.

Definition 9.8

1. Define $Y(x), x \in X \subset \mathbb{R}^d$) to be a **second order process (field) with respect to** P if for all $x \in X$, $Y(x) \in L^2(P)$.

2. Given a second order process (field) $Y(x), x \in X \subset \mathbb{R}^d$) then Z is said to be a **random variable derived from a linear operation** on $Y(\cdot)$ if Z is a random variable expressible in the form

$$Z(\omega) = \sum_{n=1}^{N} \alpha_n Y(\omega, x_n) \quad \text{or if } Z(\omega) \stackrel{q.m.}{=} \lim_{N \to \infty} \sum_{n=1}^{N} \alpha_n Y(\omega, x_n). \quad (9.7)$$

Theorem 9.9 *For $Y(x), x \in X \subset \mathbb{R}^d$ second order with respect to P, let H_Y denote the* **set of all random variables derived from linear operations** *on $Y(\cdot)$.*

Then H_Y is a Hilbert space. (Furthermore, it is a Hilbert subspace of the Hilbert space $L^2(P)$.)

Proof Since $Y(\cdot)$ is second order then for all $t \in T(x \in X)$, $Y(t)\overline{Y(x)} \in L^2(P)$. Because $L^2(P)$ is a vector space, finite linear combinations are also in $L^2(P)$, so $Z(\omega)$ of the first form Eqn. 9.7 are in $L^2(P)$. Then recall that q.m. convergence of a sequence of random variables corresponds to convergence in the norm of $L^2(P)$, so $Z(\omega)$ of the second form Eqn. 9.7 are also in $L^2(P)$ because it is complete. Thus, $H_Y \subset L^2(P)$. Now let $Z_1(\omega), Z_2(\omega) \in H_Y$ and $a \in \mathbb{C}$. By definition there exist sequences $Z_1^{(N)} = \sum_{n=1}^N \alpha_n Y(\omega, x_n)$ and $Z_2^{(N)} = \sum_{k=1}^N \beta_k Y(\omega, x_k)$ which converge in quadratic mean to Z_1 and Z_2, respectively, (if Z_1 or Z_2 is derived from finite linear combinations then for large n, k take $\alpha_n = \beta_k = 0$ and arbitrary x_k, x_n). Clearly, $aZ_1^{(N)} + Z_2^{(N)}$ is in H_Y for each N, and is a finite linear combination derived from $Y(\cdot)$. By continuity of addition and scalar multiplication $aZ_1^{(N)} + Z_2^{(N)}$ converges to $aZ_1 + Z_2$ in quadratic mean. Therefore $aZ_1 + Z_2 \in H_Y$ by definition and we thus conclude that H_Y is a vector subspace of $L^2(P)$. Furthermore, since Z was arbitrary and $Z_1^{(N)} \to Z_1$ in quadratic mean, H_Y is closed with respect to the norm inherited from $L^2(P)$. It is left to show that H_Y is complete. Let $\{Z_1^{(n)}\}$ be a Cauchy sequence in H_Y. Then $Z_1^{(n)}$ is also Cauchy in $L^2(P)$, and since $L^2(P)$ is complete $Z_1^{(n)}$ converges to a random variable, say Z_1, which must belong to H_Y since H_Y is closed. Thus, H_Y is a complete vector subspace with the norm and inner product inherited from $L^2(P)$. □

Having a complete orthonormal (CON) basis[21] for generating random variables in the Hilbert space is very useful. Therefore notions of basis representation and Gram-Schmidt orthogonalization are important. Now clearly, if $F = \{Z_k\}$ is a countable CON set in H_Y then every $Y \in H_Y$ has the property $Y \stackrel{q.m.}{=} \lim_{n\to\infty} \sum_{k=1}^n \langle Z_k, Y \rangle Z_k$ (q.m. convergence and convergence in H_Y are the same). But can a countable CON basis in H_Y always be found? As proved in the appendix, Theorem B.6, if a second order field $Y(x), x \in X$ is q.m. continuous and $S \subseteq X$ is dense in X then the set of all random variables derived from linear operations on $Y(x), x \in S$ is dense in H_Y. That is, for all $Z(\omega) \in H_Y$, $Z(\omega) \stackrel{q.m.}{=} \lim_{n\to\infty} \tilde{Z}_n(\omega) \stackrel{q.m.}{=} \lim_{n\to\infty} \sum_{k=1}^{N_n} \alpha_{n,k} Y(\omega, \tilde{x}_{n,k})$, where for all $n, \tilde{x}_{n,k} \in S$.

Thus, supposing X is separable and $Y(\cdot)$ is q.m. continuous, then theorem 9.9 and appendix B.6 establish the existence of a countable basis for H_Y, so any CON set in H_Y must be countable. One may be constructed using the Gram–Schmidt procedure on the countable basis set guaranteed to exist by the previous theorem. Thus, given these conditions, we can express any random variable in the Hilbert space H_Y in the form of an orthogonal expansion. Since we are often interested in cases where $X \subset \mathbb{R}^d$ which is separable, these expansions will be useful to us.

Example 9.10 Here is an example illustrating a 2D Hilbert space of random variables H_Y generated by the random process $Y(\cdot)$. Let

$$Y(t) = Y_0(\omega)\cos 2\pi t - Y_1(\omega) \sin 2\pi t, \quad t \in [0,1] \quad (9.8)$$

with $E|Y_0|^2 = E|Y_1|^2 = 1$, and $EY_0 Y_1^* = EY_0^* Y_1 = 0$. The covariance function for this random process is $K(t,s) = EY(t)Y^*(s) = \cos 2\pi(t-s)$. The Hilbert space H_Y generated by the given process is the space of all random variables derived from linear operations on $Y(\cdot)$. This space is 2D; it has the random variables Y_0, Y_1 as a basis. The basis is orthonormal since $E|Y_0|^2 = E|Y_1|^2 = 1$, and $EY_0 Y_1^* = EY_0^* Y_1 = 0$. That H_Y is two dimensional is proved in Exercise F.7.45.

---

[21] An orthonormal sequence in a Hilbert space is said to be a complete orthonormal (CON) basis if the closed subspace it generates is the whole space. The only vector orthogonal to each vector in the basis is the 0 vector.

9.1.3 Covariance and Second Order Properties

Smoothness of the processes is required which can be studied through the second order statistics summarized via the covariance function of the process [9]. We examine the scalar case more intensely.

Definition 9.11 Let $Y = \begin{pmatrix} Y_1 \\ Y_2 \\ \vdots \end{pmatrix}$ on $X \subset \mathbb{R}^d(\mathbb{Z}^d)$ be a second order m-vector process. The associated **covariance function** for fields Y on $X \subset \mathbb{R}^d, \mathbb{Z}^d$ is defined to be

$$K_Y : X \times X \to \mathbb{C}^m \times \mathbb{C}^m \quad \text{where} \quad K_Y(x,y) = EY(x)Y(y)^*.$$

For the scalar case, the K_Y is defined with the hermitian transpose symbol $(\cdot)^*$ meaning complex conjugation for non vectors.

Stationarity of the second order (wide-sense) will be used extensively.

Definition 9.12 A second-order zero-mean m-vector random process (field) $Y = \begin{pmatrix} Y_1 \\ Y_2 \\ \vdots \end{pmatrix}$ on $X \subset \mathbb{R}^d(\mathbb{Z}^d)$ is **wide-sense stationary** if its covariance function is only a function of the time difference:

$$EY(x)Y^*(y) = K(x - y). \tag{9.9}$$

Properties of the covariance determine continuity and differentiability properties.

Theorem 9.13

(i) A scalar second order random field $Y(x), x \in X \subset \mathbb{R}^d$ with $d \geq 1$ is quadratic mean continuous at $x \in X$, if and only if K_Y is continuous at (x,x).

(ii) If $Y(x)$ is quadratic mean continuous then $K_Y(x,y)$ is continuous at every point (x,y) in $X \times X$.

(iii) For $Y(x), x \in X \subset \mathbb{R}^d$ an m-vector field, then (i),(ii) hold.

Proof
Proof (i) if: Assume $K = K_Y$ is continuous at (y,y). Then

$$\lim_{x \to y} E|Y(x) - Y(y)|^2$$
$$= \lim_{x \to y} (EY(x)Y(x)^* - EY(x)Y(y)^* - EY(y)Y(x)^* + EY(y)Y(y)^*)$$
$$= \lim_{x \to y} (K(x,x) - K(x,y) - K(y,x) + K(y,y) - K(y,y))$$
$$= \lim_{x \to y} (K(x,x) - K(y,y)) - \lim_{x \to y} (K(x,y) - K(y,y))$$
$$\quad - \lim_{x \to y} (K(y,x) - K(y,y))$$
$$= 0 \quad \text{by continuity of } K \text{ at } (x,x).$$

(i) *only if:* By the Cauchy-Schwarz inequality $|EYZ^*|^2 \leq E|Y|^2 E|Z|^2$ implying

$$|K(x_1, x_2) - K(y, y)| = |EY(x_1)Y(x_2)^* - EY(y)Y(y)^*|$$
$$= |E(Y(x_1) - Y(y))Y(x_2)^* + EY(y)(Y(x_2) - Y(y))^*|$$
$$\leq \sqrt{E|Y(x_1) - Y(y)|^2 E|Y(x_2)|^2}$$
$$+ \sqrt{E|Y(y)|^2 E|Y(x_2) - Y(y)|^2}.$$

Taking limits as $(x_1, x_2) \to (y, y)$ in $X \times X$, the right side goes to zero by the q.m. continuity of $Y(y)$ at t. Thus, K is continuous at (y, y).

(ii) : Consider $(x_1, x_2) \to (y_1, y_2)$. Then the exact same derivation as in the proof of (i) only if yields

$$|K(x_1, x_2) - K(y_1, y_2)| \leq \sqrt{E|Y(x_1) - Y(y_1)|^2 E|Y(x_2)|^2}$$
$$+ \sqrt{E|Y(y_1)|^2 E|Y(x_2) - Y(y_2)|^2}. \qquad (9.10)$$

Taking limits as $(x_1, x_2) \to (y_1, y_2)$ in $X \times X$, the right side goes to zero by the q.m. continuity of $Y(y)$ at y_1 and at y_2. Thus, K is continuous at (y_1, y_2) for every $(y_1, y_2) \in X$.

(iii) *vector case if:* For the vector case, if K is continuous at (x, x) then

$$\lim_{x \to y} E\|Y(x) - Y(y)\|^2$$
$$= \lim_{x \to y} EY(x)^* Y(x) - EY(x)^* Y(y) - EY(y)^* Y(x) + EY(y)^* Y(y)$$
$$= \lim_{x \to y} \sum_{i=1}^{m} [EY_i(x)Y_i(x)^* - EY_i(x)Y_i(y)^* - EY_i(y)Y_i(x)^* + EY_i(y)Y_i(y)^*]$$
$$= \sum_{i=1}^{m} [K_{ii}(x, x) - K_{ii}(x, y) - K_{ii}(y, x) + K_{ii}(y, y)].$$

Since K_{ii} is continuous at (y,y) for each $i = 1, \ldots, m$, the right-hand limit equals 0 by the same argument as in the scalar case.
(i) only if and (ii) for the vector case are proved identically as above just componentwise. □

Example 9.14 (Covariance and Differentiation) Covariance properties on derivatives in the quadratic mean sense imply continuity almost surely. To illustrate, for example, a scalar SOP $Y(t), t \in [0, 1]$ is q.m. differentiable,

$$\lim_{h \to 0} \frac{Y(t+h) - Y(t)}{h} \stackrel{q.m.}{=} Y'(t) \in L^2(P), \qquad (9.11)$$

if and only if the covariance $K(t, s)$ has a generalized second derivative:

$$\frac{\partial^2 K(t, s)}{\partial t \partial s} = \lim_{h \to 0} \lim_{k \to 0} \frac{K(t+h, s+k) - K(t+h, s) - K(t, s+k) + K(t, s)}{hk}. \qquad (9.12)$$

The if part follows from if $Y(t)$ is q.m. differentiable, then defining $Y_h(t) = (Y(t+h) - Y(t)/h)$ then

$$\lim_{h \to 0, k \to 0} \frac{K(t+h, s+k) - K(t+h, s) - K(t, s+k) + K(t, s)}{hk}$$

$$= \lim_{h \to 0, k \to 0} E Y_h(t) Y_k^*(s) \stackrel{(a)}{=} E|Y'(t)|^2. \tag{9.13}$$

with (a) following from continuity of the inner product. If the second derivative of the covariance exists then $Y(t), t \in [0, 1]$ has $E|Y'(t)|^2 < \infty$. Similarly the reverse direction follows as well. If $E|Y'(t)|^2 < \infty$ then the second derivative of covariance exists.

The only if converse part follows if the generalized derivative exists, then $EY_h(t)Y_k(t)^* \to$ limit c, so that

$$E|Y_h(t) - Y_k(t)|^2 = E|Y_h(t)|^2 - EY_h(t)Y_k(t)^* - EY_k(t)Y_h(t)^* + E|Y_k(t)|^2 \to 0. \tag{9.14}$$

Since this Cauchy condition holds, the limit of $Y_h(t)$ exists as an element in $L_2(P)$, and the proposition is true.

In section 9.10 almost-sure continuity of sample-paths is examined more completely.

9.1.4 Quadratic Mean Continuity and Integration

Computing integrals of the random process are linked to the quadratic mean continuity. Examine the random process case with $X \subset \mathbb{R}^1$ closed bounded. Associated with the index sets $X \subset \mathbb{R}^d$ there will be a Hilbert space of square-integrable functions, $L^2(X)$, norm $\|\cdot\|_2$. Throughout, the random processes will have a basic covariance property, essentially a finite mean-square property, which will imply the existence of linear functionals of the process. Assuming the processes $Y(\cdot)$ are quadratic mean continuous, then the covariance is continuous along the diagonal. Clearly, index set X compact, then $\int_X K(x,x)dx < \infty$, since K must be bounded from its continuity. This is the so-called *trace-class* property (see discussion below, definition 9.30).

That the integral is well defined in the sense that the limit is independent of the choice of sequence of partitions has still to be established. For this, examine the joint measurability with respect to both Ω as well as X.

Definition 9.15 *For processes $Y(\omega, x), x \in X$ on $(\Omega, \mathcal{A}, P)$ for which X is Lebesgue measurable with measure m, then* **the process is said to be $\mathcal{L} \otimes \mathcal{A}$ measurable** *with respect to $m \times P$ on the product topology of $X \times \Omega$ (the open sets generated from the topologies $\mathcal{L}, \mathcal{A}$) if $\forall r \in \mathbb{R}$,*

$$\{(\omega, x) : Y(\omega, x) \leq r\} \in \mathcal{L} \otimes \mathcal{A}. \tag{9.15}$$

Joint measurability allows the understanding of $Y(\omega, x), x \in X, \omega \in \Omega$ in several ways, along with integrals of the type $\langle \phi, Y \rangle_2$ given by $\langle \phi, Y \rangle_2 = \int_X \phi(x) Y(\omega, x) \, dx$. Clearly,

$$E\|Y(\cdot)\|_2^2 = \int_{\Omega \times X} |Y(\omega, x)|^2 \, d(P(\omega) \times m(x)) \tag{9.16}$$

and joint measurability allows for the swapping of the order of integration via Fubini's theorem. From the second order process condition on bounded domains X implies the trace class condition $E\|Y(\omega, \cdot)\|_2^2 < \infty$ which with joint measurability and Fubini's theorem implies $P(\cdot)$ almost surely,

$\int_X Y^2(\cdot, t)\,dt < \infty$ so that $Y \in L^2(X)$ almost surely. Thus $P(\cdot)$ almost surely, $\int_X |Y(\cdot, t)|\,dt < \infty$ giving $Y \in L^1(X)$. For orthogonal expansions, integrals of the random process are computed against functions $\phi \in L^2$ and existence of the integrals as quadratic mean limits will be required. Then $\int_X Y(\cdot, t) e^{j2\pi ft}\,dt$ exists almost surely and integrals $\langle \phi, Y \rangle_2$ well defined. Being able to compute projections onto eigenfunctions of random processes will be used quite often.

Example 9.16 (Integrals in q.m.) Defining the partition $T_n = \{a = t_0^{(n)} < t_1^{(n)} < \cdots < t_n^{(n)}\}$ with the sum defined as

$$I_n(Y, \phi) = \sum_{i=1}^{n} Y(t_i^{(n)}) \phi(t_i^{(n)}) (t_i^{(n)} - t_{i-1}^{(n)}).$$

Clearly, $I_n(Y, \phi) \in H_Y$ and the q.m. limit is as well $I(Y, \phi) \stackrel{q.m.}{=} \lim_{n\to\infty} I_n(Y, \phi) \in H_Y$ since

$$\lim_{n\to\infty} E|I_n(Y, \phi)|^2 \leq \lim_{n\to\infty} \sum_{i=1}^{n} E|Y(t_i^{(n)})|^2 (t_i^{(n)} - t_{i-1}^{(n)}) \sum_{i=1}^{n} |\phi(t_i^{(n)})|^2 (t_i^{(n)} - t_{i-1}^{(n)})$$

$$= \int |\phi(t)|^2\,dt \int K(t, t)\,dt < \infty. \tag{9.17}$$

Example 9.17 Computing the expectation of an integral of the process involves joint measurability.

Brownian motion is an example of an $\mathcal{L} \otimes \mathcal{A}$ measurable. Let

$$W^{(n)}(\omega, t) = W(\omega, k2^{-n}), \qquad k2^{-n} \leq t < (k+1)2^{-n}, \tag{9.18}$$

where $k = 0, 1, 2, \ldots, 2^n - 1$. Then, $W^{(n)}(\omega, t)$ is measurable because

$$\{(\omega, t) : W^{(n)}(\omega, t) \leq x\} = \left\{ \bigcup_{k=0}^{2^n-1} [k2^{-n}, (k+1)2^{-n}) \right\}$$

$$\times \{\omega : W(\omega, k2^{-n}) \leq x, \ k = 1, 2, \ldots, 2^n - 1\}.$$

Since a countable union of measurable sets is measurable, the set $\{(\omega, t) : W^{(n)}(\omega, t) \leq x\} \in \mathcal{L} \otimes \mathcal{A}$ is measurable.

9.2 Orthogonal Process Representations on Bounded Domains

The SOP processes which are studied associated with bounded domains have covariances which are supported on closed and bounded sets and are therefore compact as operators. This implies they have a discrete set of eigenvalues and associated orthogonal expansion. This is important as various orthogonal process representations such as Mercer's representation and Karhunen-Loeve expansions.

9.2.1 Compact Operators and Covariances

For this we follow Reed and Simon's [215] particularly incisive development. Assume operators $A : F \to G$ on Banach spaces [22] F and G with norms $\|\cdot\|_F, \|\cdot\|_G$. In almost all cases these are Hilbert spaces denoted by H_F, H_G, with associated inner products $\langle \cdot, \cdot \rangle_F, \langle \cdot, \cdot \rangle_G$. Then the bounded operators taking their action [23] on these spaces are what shall be studied.

Definition 9.18 *The* **operator norm** $\|A\|$ *is defined to be*

$$\|A\| = \sup_{f \in F} \frac{\|Af\|_G}{\|f\|_F} = \sup_{f \in F : \|f\|_F = 1} \|Af\|_G \, ; \tag{9.19}$$

the **space of bounded norm, linear operators** *are denoted as* $\mathcal{B}(F, G)$; *when* $F = G$ *the bounded operators are denoted as* $\mathcal{B}(F)$.

The operators will be expanded via their eigenfunctions, which for the **self-adjoint** and **normal** operators will be orthogonal. All of the covariances will be self-adjoint, and differential operators will be normal.

Definition 9.19
1. Let $A \in \mathcal{B}(H)$; then $\phi \neq 0 \in H$ is an **eigenfunction** of A if it satisfies $A\phi = \lambda\phi$ for some **eigenvalue** $\lambda \in \mathbb{C}$. The **eigen-spectrum** $\sigma(A)$ are the set of eigenvalues.
2. The **adjoint** operator A^* of $A \in \mathcal{B}(H_2, H_1)$ is the unique operator such that

$$\langle Af, g \rangle_{H_2} = \langle f, A^*g \rangle_{H_1} \quad \forall f \in H_1, \, \forall g \in H_2 \, . \tag{9.20}$$

When the operators A are **self-adjoint** $A = A^*$ mapping from one Hilbert space to itself.

3. A **normal operator** A is one for which $A^*A = AA^*$.

For these cases the eigenfunctions are orthogonal.

Lemma 9.20
(i) *For A self adjoint, the eigenvalues $\{\lambda\}$ are real with eigenfunctions $\{\phi\}$ orthogonal.*
(ii) *For A normal, the eigenfunctions are orthogonal and if A has eigenelements (λ, ϕ) then the operator A^* has eigenelements (λ^*, ϕ).*

Proof *Part (i):* For A self-adjoint, the eigenvalues are real valued:

$$\langle A\phi_1, \phi_1 \rangle = \lambda^* \|\phi_1\|^2 = \langle \phi_1, A\phi_1 \rangle = \lambda \|\phi_1\|^2 \, . \tag{9.21}$$

The eigenfunctions ϕ_1, ϕ_2 corresponding to distinct non-zero eigenvalues $\lambda_1 \neq \lambda_2$, are orthogonal:

$$\langle A\phi_1, \phi_2 \rangle = \langle \lambda_1 \phi_1, \phi_2 \rangle$$
$$= \langle \phi_1, A^* \phi_2 \rangle \stackrel{(a)}{=} \langle \phi_1, A\phi_2 \rangle = \langle \phi_1, \lambda_2 \phi_2 \rangle, \tag{9.22}$$

---
[22] A **Banach space** is a normed vector space that is complete with respect to the norm metric.
[23] The operators $A : f \in F \mapsto Af \in G$ will most often define mappings between Hilbert spaces. For example $L^2([0,1])$, A will have a kernel representation $A = (A(x,y))$ on $[0,1]^2$, with the action defined through the kernel, $Af(x) = \int_0^1 A(x,y)f(y)\,dy$. For l^2, the kernel (A_{ij}) is on $\mathbb{Z}^2$ and $Af_i = \sum_{j \in \mathbb{Z}} A_{ij}f_j$; for $\mathbb{R}^n$ its an $n \times n$ matrix kernel $A = (A_{ij})$ with $Af_i = \sum_{j=1}^n A_{ij}f_j$.

with (a) following from the self-adjoint property. This implies $\langle \phi_1, \phi_2 \rangle = 0$ if $\lambda_1 \neq \lambda_2 \neq 0$. *Part (ii):* If the normal operator $A \in L(H)$ has eigenelement (λ, ϕ) then the operator A^* has the eigenelement (λ^*, ϕ). This follows since $A\phi = \lambda\phi, \lambda \in \mathbb{C}$ for normal operator A, i.e. $\|(A - \lambda\,\mathrm{id})\phi\| = 0$ giving

$$\|(A - \lambda\,\mathrm{id})\phi\|^2 = \langle (A - \lambda\,\mathrm{id})\phi, (A - \lambda\,\mathrm{id})\phi \rangle = \langle (A - \lambda\,\mathrm{id})^*(A - \lambda\,\mathrm{id})\phi, \phi \rangle \tag{9.23}$$

$$= \langle (A - \lambda\,\mathrm{id})(A - \lambda\,\mathrm{id})^*\phi, \phi \rangle = \langle (A - \lambda\,\mathrm{id})^*\phi, (A - \lambda\,\mathrm{id})^*\phi \rangle \tag{9.24}$$

$$= \|(A - \lambda\,\mathrm{id})^*\phi\|^2 = 0, \tag{9.25}$$

which implies $A^*\phi = \lambda^*\phi$. That the eigenfunctions are orthogonal for distinct eigenvalues,

$$\lambda_1^* \langle \phi_1, \phi_2 \rangle = \langle \lambda_1 \phi_1, \phi_2 \rangle = \langle A\phi_1, \phi_2 \rangle \tag{9.26}$$

$$= \langle \phi_1, A^*\phi_2 \rangle = \langle \phi_1, \lambda_2^* \phi_2 \rangle = \lambda_2^* \langle \phi_1, \phi_2 \rangle, \tag{9.27}$$

giving $\lambda_1^* \langle \phi_1, \phi_2 \rangle - \lambda_2^* \langle \phi_1, \phi_2 \rangle = (\lambda_1^* - \lambda_2^*)\langle \phi_1, \phi_2 \rangle = 0$. Since $\lambda_1 \neq \lambda_2$, this implies that $\langle \phi_1, \phi_2 \rangle = 0$, hence proving orthogonality for eigenfunctions ϕ_1, ϕ_2 of a normal operator A. □

Thus, for self-adjoint bounded linear operators, $A = A^* \in \mathcal{B}(H)$, then the eigen-spectrum is real $\sigma(T) \subset \mathbb{R}$ and bounded since

$$\sup_{\lambda \in \sigma(T)} |\lambda| = \sup_{\lambda \in \sigma(T)} \frac{\|A\phi_\lambda\|}{\|\phi_\lambda\|} < \|A\| < \infty. \tag{9.28}$$

The subclass of bounded, linear operators, those which are compact, will play an important role in the process representations. One of the most useful properties of compact operators is that their spectrum is discrete. It is the most important property of compact operators that they can essentially be represented via the superposition of a finite rank operator and a remainder which has arbitrarily small norm.

Definition 9.21 *An operator $A \in \mathcal{B}(F, G)$ is compact if A takes bounded subsets of F into precompact sets in G. Equivalently, A is compact if for every bounded sequence $\{f_n\} \subset F$, $\{Af_n\}$ has a subsequence convergent in G.*

There are two kinds of operators which are compact for which we shall exploit the discreteness of their spectrum for orthogonal expansions. These are (i) finite rank operators which can be identified with matrices, for example, and (ii) continuous covariances which support eigenfunction expansions through Karhunen-Loeve.

Theorem 9.22

1. *Let A be a finite dimensional operator on L^2 with norm $\|\cdot\|_2$ and inner product $\langle \cdot, \cdot \rangle_2$, then $A : L^2 \to L^2$ according to*

$$Af = \sum_{k=1}^{n} \alpha_k \langle \phi_k, f \rangle_2 \phi_k, \tag{9.29}$$

where ϕ_k is a CON basis and $\alpha_k \in \mathbb{C}$. Then A is compact.

2. Let A be a continuous operator with kernel $A(t,s), (t,s) \in [0,1]^2$ mapping square-integrable functions to square-integrable functions, $A : L^2[0,1] \to L^2[0,1]$, according to

$$Af(t) = \int_0^1 A(t,s)f(s)\,ds. \tag{9.30}$$

Then A is compact. Similarly for fields, $A(x,y), (x,y) \in X^2, X \subset \mathbb{R}^d$.

Proof For the proof, see Appendix section C, Theorem C.3. $\square$

Now it follows that such compact operators can be written as the superposition of finite rank approximations. Self-adjoint operators (such as covariances) have finite rank approximations constructed from the span of their eigenvectors.

Theorem 9.23   Riesz-Schauder and Hilbert-Schmidt Theorem
(a) Let A be compact in $\mathcal{B}(H)$. Then $\sigma(A)$ is a discrete set of eigenvalues having no limit points except possibly $\lambda = 0$. Any $\lambda \in \sigma(A)$ is an eigenvalue, with the corresponding space of eigenvectors of at most finite dimension.
(b) Let A be compact in $\mathcal{B}(H)$, and $\{\phi_k\}$ an orthonormal set in H. Then

$$A = \lim_{n \to \infty} \sum_{k=1}^n A\phi_k \langle \phi_k, \cdot \rangle. \tag{9.31}$$

(c) Let A be self adjoint and compact on $\mathcal{B}(H)$. Then the set of eigenvectors $\{\phi_k\}$ satisfying $A\phi_k = \lambda_k \phi_k$ are a C.O.N. basis for H with $\lambda_k \to 0$ as $k \to \infty$, and

$$A = \lim_{n \to \infty} \sum_{k=1}^n \lambda_k \phi_k \langle \phi_k, \cdot \rangle. \tag{9.32}$$

Here the limit is in the operator norm sense $\lim_{n \to \infty} \sup_f \|Af - \sum_{k=1}^n \lambda_k \langle \phi_k, f \rangle \phi_k\| = 0$.

Proof For proof see the Appendix section C.2, Theorem C.4. $\square$

This gives the basic characterization of self adjoint and compact operators which we will use in numerous places for L^2 representations of stochastic processes on compact domains familiar to the communication engineer.

Theorem 9.24 (Mercer's Theorem) *(1909) Consider a symmetric (Hermitian) nonnegative definite operator K with continuous kernel $K(x,y)$ on $X \times X$ for $X \subset \mathbb{R}^d$, $d \geq 1$ with eigenelements $\{\lambda_k, \phi_k\}$ satisfying the integral equation $\lambda \phi(x) = \int_X K(x,y)\phi(y)dy$ then*

$$K(x,y) = \sum_{k=1}^\infty \lambda_k \phi_k(x) \phi_k^*(y), \text{ converging absolutely and uniformly.} \tag{9.33}$$

Similarly for vector fields with eigenelements $\{\lambda_k, \begin{pmatrix} \phi_{1k} \\ \phi_{2k} \\ \vdots \end{pmatrix}\}$ and $m \times m$ matrix covariance $K(x,y), x,y \in X$.

Proof K is continuous and compact by Theorem 9.22. It is also self-adjoint, for by swapping order of integrals (Fubini's theorem) gives

$$\langle f, Kg \rangle = \int f^*(y) Kg(y) \, dy = \int f^*(y) \int K(y,x) g(x) \, dx \, dy \qquad (9.34)$$

$$= \int \int f^*(y) K(y,x) \, dy g(x) \, dx = \int \left(\int K^*(y,x) f(y) \, dy \right)^* g(x) \, dx \qquad (9.35)$$

$$= \int \left(\int K(x,y) f(y) \, dy \right)^* g(x) \, dx = \langle Kf, g \rangle .$$

Thus, by the *Hilbert–Schmidt* theorem, $\{\phi_i\}$ forms a C.O.N. basis of $L^2(X)$. Since $K(\cdot,\cdot) \in L^2(X \times X)$, by Fubini's theorem $K(\cdot,y)$ is in $L^2(X)$ for almost all y. Thus,

$$K(\cdot,y) = \sum_{i=1}^{\infty} \langle \phi_i(\cdot), K(\cdot,y) \rangle \phi_i(\cdot) = \sum_{i=1}^{\infty} \lambda_i \phi_i(\cdot) \phi_i^*(y) \in L^2(X) , \qquad (9.36)$$

Also, by the triangle inequality we have

$$\left\| K(\cdot,y) - \sum_{i=1}^{n} \lambda_i \phi_i(\cdot) \phi_i^*(y) \right\| \leq \|K(\cdot,y)\| + \left\| \sum_{i=1}^{n} \lambda_i \phi_i(\cdot) \phi_i^*(y) \right\| \qquad (9.37)$$

$$= \sum_{i=1}^{n} \lambda_i |\phi_i^*(y)| \leq 2\|K(\cdot,y)\|, \qquad (9.38)$$

and since $4\|K(\cdot,y)\|^2 = 4 \int |K(x,y)|^2 dx \in L^1(X)$ by Fubini's theorem, we have

$$\lim_{n \to \infty} \int \int \left| K(x,y) - \sum_{i=1}^{n} \lambda_i \phi_i(x) \phi_i^*(y) \right|^2 dx \, dy$$

$$= \int \lim_{n \to \infty} \left(\int \left| K(x,y) - \sum_{i=1}^{n} \lambda_i \phi_i(x) \phi_i^*(y) \right|^2 dx \right) dy = 0$$

by the dominated convergence theorem. Hence $K(\cdot,\cdot) = \sum_{i=1}^{\infty} \lambda_i \phi_i(\cdot) \phi_i^*(\cdot)$ in $L^2(X \times X)$. Consider the remainders

$$K_n(x,y) = K(x,y) - \sum_{i=1}^{n} \lambda_i \phi_i(x) \phi_i^*(y)$$

for $n = 1, \ldots$. By 9.22, $Kf(x)$ is a continuous function for all $f \in L^2$. In particular, for all $\lambda_i \neq 0$, the corresponding eigenfunctions, ϕ_i are continuous and thus it follows that the partial sums of the series are continous, and hence so is $K_n(\cdot,\cdot)$. By Fubini's theorem we have $\langle f, K_n f \rangle = \langle f(\cdot) f^*(\cdot), K_n(\cdot,\cdot) \rangle$, where the second inner product is the one associated with the joint product space, and since $K_n(x,y) = \sum_{i=n+1}^{\infty} \lambda_i \phi_i(x) \phi^*(y)$ in $L^2(X \times X)$, it follows from continuity of the inner product that

$$\int \int K_n(x,y) f(y) f^*(x) \, dx \, dy = \sum_{i=n+1}^{\infty} \int \int \lambda_i \phi_i(x) \phi_i^*(y) f(y) f^*(x) \, dx \, dy$$

$$= \sum_{i=n+1}^{\infty} \lambda_i |\langle \phi_i, f \rangle|^2 \geq 0.$$

Thus, $K_n(\cdot,\cdot)$ is also nonnegative definite. This implies that $K_n(x,x) \geq 0$, for otherwise $K_n(s_0,s_0) < 0$ for some (s_0,s_0), and by continuity of $K_n(\cdot,\cdot)$ we must have $K_n(x,y) < 0$ in some neighborhood N of (s_0,s_0). Setting $f(x) = 1$ for all x such that $(s,s_0) \in N$ and setting $f(x) = 0$ elsewhere, we find $\langle K_n f, f \rangle < 0$ which contradicts the non-negative definite property of $K_n(\cdot,\cdot)$. So, by the positivity of $K_n(x,x)$ and by the definition of $K_n(\cdot,\cdot)$, we have $\sum_{i=1}^{\infty} \lambda_i |\phi_i(x)|^2 \leq K(x,x)$. Now, let $M = \max_{x \in X} K(x,x)$. By Cauchy's inequality we have

$$\left| \sum_{i=n}^{m} \lambda_i \phi_i(x) \phi_i^*(y) \right|^2 \leq \sum_{i=n}^{m} \lambda_i |\phi_i(x)|^2 \sum_{i=n}^{m} \lambda_i |\phi_i(y)|^2 \leq M \sum_{i=n}^{m} \lambda_i |\phi_i(x)|^2. \tag{9.39}$$

Therefore, $\sum_{i=1}^{\infty} \lambda_i \phi_i(x) \phi_i^*(y)$ converges uniformly in y for every fixed value of x, and thus the series converges to a continuous function of y, say $R(x,y)$. We have

$$\int R(x,y) f(y)\, dy = \sum_{i=1}^{\infty} \lambda_i \phi_i(x) \int \phi_i^*(y) f(y)\, dy = \int K(x,y) f(y)\, dy,$$

for all continuous functions f, where the first equality follows from uniform convergence and the second by the *Hilbert-Schmidt* theorem. In particular, for $f(y) = (R(x,y) - K(x,y))^*$ we have

$$\int |R(x,y) - K(x,y)|^2\, dy = 0,$$

and thus we conclude that for fixed x, $R(x,y) = K(x,y)$ for almost all y. Thus $K(x,x) = R(x,x) = \sum_{i=1}^{\infty} \lambda_i |\phi_i(x)|^2$, and in fact the series is uniformly convergent by Dini's Theorem[24] since $K(x,x)$ and the partial sums of the series are continuous. Finally, from (9.39) we now have joint uniform convergence in x and y.

Vector case: See homework problem F.7.46 for proof based on introduction of Frobenius norm on matrices $\langle K, K' \rangle = \operatorname{tr} K^* K'$. □

9.2.2 Orthogonal Representations for Random Processes and Fields

Many of the processes examined in the pattern theoretic representations are quadratic mean continuous, and we are now in a position to state the Karhunen-Loeve expansion of second order quadratic mean continuous processes. Not surprisingly, since the covariance kernel is continuous on a compact interval, its discrete spectrum plays the fundamental role in the expansion.

Theorem 9.25 (Karhunen-Loeve Expansion) *Let $Y(\omega,x), x \in X \subset \mathbb{R}^d, d \geq 1$, be a q.m. continuous, second order field with continuous covariance kernel $K(y,x)$. Then $\{\lambda_k, \phi_k\}$ are the eigenelements satisfying*

$$\int_X K(y,x) \phi_k(x)\, dx = \lambda_k \phi_k(y) \quad \text{if and only if } Y(y) \stackrel{q.m.}{=} \lim_{n \to \infty} \sum_{k=1}^{n} Y_k \phi_k(y) \tag{9.40}$$

uniformly in $y \in X$, with $Y_k = \int_X \phi_k^*(y) Y(y)\, dy$ orthogonal random variables according to $EY_k^* Y_{k'} = \lambda_k \delta(k-k')$.

---

[24] **Dini's Theorem** If $\{f_j\}$ is a sequence of non-negative continuous functions on a compact set T such that $f_1(x) \leq f_2(x) \leq \ldots$ for all $x \in X$, and $\lim_{j \to \infty} f_j = f$ is continous then $\{f_j\}$ converges uniformly.

The m-vector case is identical as above with $Y(\omega, x) = \begin{pmatrix} Y_1(\omega, x) \\ Y_2(\omega, x) \\ \vdots \end{pmatrix}, x \in X \subset \mathbb{R}^d$, and the continuous covariance an $m \times m$ matrix $K(x,y) = (K_{ij}(x,y))$, with $m \times 1$ eigenvectors

$$\left\{ \lambda_k, \phi_k(x) = \begin{pmatrix} \phi_{1k}(x) \\ \phi_{2k}(x) \\ \vdots \end{pmatrix} \right\}.$$

Proof If the $\{\phi_k\}$ are eigenfunctions of Eqn. 9.40 then

$$E \left| Y(y) - \sum_{k=1}^{n} Y_k \phi_k(y) \right|^2 = K(y,y) - EY^*(y) \sum_{k=1}^{n} Y_k \phi_k(y) - E \sum_{k=1}^{n} Y_k^* \phi_k^*(y) Y(y)$$

$$+ \sum_{k=1}^{n} \sum_{k'=1}^{n} EY_k^* Y_{k'} \phi_k^*(y) \phi_{k'}(y)$$

$$= K(y,y) - \sum_{k=1}^{n} \lambda_k \phi_k(y) \phi_k^*(y) \qquad (9.41)$$

which goes to zero as $n \to \infty$ uniformly in y by Mercer's theorem.

Conversely, suppose $Y(y)$ has the orthogonal expansion of Eqn. 9.40, then $EY(y)Y^*(x) = \sum_{k=1}^{\infty} \lambda_k \phi_k(y) \phi_k^*(x)$ giving

$$\int_X K(y,x) \phi_m(x) \, dx = \int_X \sum_{k=1}^{\infty} \lambda_k \phi_k(y) \phi_k^*(x) \phi_m(x) \, dx \qquad (9.42)$$

$$\stackrel{(a)}{=} \sum_{k=1}^{\infty} \lambda_k \phi_k(y) \int_X \phi_k^*(x) \phi_m(x) \, dx \stackrel{(b)}{=} \lambda_m \phi_m(y), \qquad (9.43)$$

with (a) following from uniform convergence and (b) from the orthogonality of the functions $\{\phi_k\}$, completing the proof.

The vector case is proved in Homework problem F.7. $\square$

Notice, continuity of the covariance implies boundedness over the compact interval giving that the process is of trace class $\sum_{k=1}^{\infty} \lambda_k = \int_X K(y,y) \, dy < \infty$. Also, in the vector case q.m. continuity and finite trace corresponds to $\int_X \text{tr } K(x,x) \, dx < \infty$.

For the vector case, block orthogonal expansions are useful owing to incremental expansion of the covariance and random processes generating vectors of random variables which are independent across vectors, but correlated within a vector. See Homework problem F.7.

9.2.3 Stationary Periodic Processes and Fields on Bounded Domains

Examples of processes on bounded domains which are used often in the study of patterns are periodic processes, processes on the sphere and others. It is well known from communications that the eigenfunctions for shift are the Fourier series implying covariances on compact domains which are shift invariant have countable eigenfunctions for their Karhunen–Loeve representation which are *the complex exponentials*.

On compact intervals, cyclo-stationarity is associated with periodicity of the process.

9.2 ORTHOGONAL PROCESS REPRESENTATIONS

Definition 9.26 *A second-order zero-mean periodic vector field* $Y(x) = \begin{pmatrix} Y_1(x) \\ Y_2(x) \\ \vdots \end{pmatrix}, x \in X = [0,1]^d$, *which is* **cyclo (wide-sense) stationary** *if it is periodic* $Y(x) = Y(x+m), m \in \mathbb{Z}^d$ *which covariance a function of shift* $EY(y)Y^*(x) = K(y-x, 0)$.

Now examine periodic processes on $[0,1]^d$ which are cyclo-stationary. Stationary on the Torus means the covariance is Toeplitz and a function only of modulo shift $t - s$; this gives the Fourier representation of the periodic covariance process with the eigenfunctions of the covariance, the complex exponentials. This is Egevary's representation for the cyclo-stationary fields. Use $\omega_k = 2\pi k$ to denote the scalar angular radian frequency for the kth harmonic, and $\omega_k = (2\pi k_1, 2\pi k_2, \ldots)$ to denote the vector version. The interpretation will be made obvious by the context.

Corollary 9.27 *For* $Y(\omega, x), x \in X = [0,1]^d$ *a quadratic mean continuous periodic vector field with continuous covariance kernel* $K(x, y) = K(x - y, 0)$, *then the Karhunen-Loeve expansion of Theorem 9.25 holds with the eigenelements*

$$\lambda_k = \int_{[0,1]^d} K(\tau, 0) e^{-j\langle \omega_k, \tau \rangle_{\mathbb{R}^d}} d\tau, \quad \phi_k(\tau) = e^{j\langle \omega_k, \tau \rangle_{\mathbb{R}^d}},$$

$$\omega_k = (2\pi k_1, 2\pi k_2, \ldots), \quad k \in \mathbb{Z}^d \,.. \tag{9.44}$$

Proof Karhunen Loeve holds giving the orthogonal expansion of Eqn. 9.44. The eigenfunctions, ϕ_k, of the covariance are the complex exponentials:

$$\int_{[0,1]^d} K(t,s) e^{j\langle \omega_k, s \rangle_{\mathbb{R}^d}} ds = \int_{[0,1]^d} K(t-s, 0) e^{j\langle \omega_k, s \rangle_{\mathbb{R}^d}} ds$$

$$= \int_{[0,1]^d} K(\tau, 0) e^{j\langle \omega_k, (t-\tau) \rangle_{\mathbb{R}^d}} d\tau \tag{9.45}$$

$$= e^{j\langle \omega_k, t \rangle_{\mathbb{R}^d}} \int_{[0,1]^d} K(\tau, 0) e^{-j\langle \omega_k, \tau \rangle_{\mathbb{R}^d}} d\tau. \tag{9.46}$$

$\square$

Remark 9.2.0 (FFT) Notice, for the discrete periodic process case (period N), then $\phi_k(n) = e^{j(2\pi/N)\langle k,n \rangle_{\mathbb{R}^d}}$, $\lambda_k = \sum_{\tau=1}^{N} K(\tau, 0) e^{-j2\pi \langle k, \tau \rangle_{\mathbb{R}^d}}$ with $Y_k = \langle \phi_k, Y \rangle, k \in \mathbb{Z}_N$ an orthogonal process with spectrum $EY_k Y_k^* = \lambda_k$.

Example 9.28 For representing biological shapes in electron micrographs examine Figure 9.1 (left panel) illustrating mitochondrial connected shapes with closed boundaries curves in $\mathbb{R}^2$. Stationary processes have been used as in [137, 216] with Fourier expansions arising for representing closed contours in [185].

For closed curves of arbitrary lengths introduce scales and rotations of the form

$$\begin{pmatrix} \rho(t) & 0 \\ 0 & \rho(t) \end{pmatrix} \begin{pmatrix} \cos \theta(t) & \sin \theta(t) \\ -n \sin \theta(t) & \cos \theta(t) \end{pmatrix} = \begin{pmatrix} U_1(t) & U_2(t) \\ -U_2(t) & U_1(t) \end{pmatrix}. \tag{9.47}$$

Curves are generated by applying these to the tangent vectors of a template circle giving the closed curves $f(t), t \in [0,1], f(0) = f(1)$ according to

$$f(m) = 2\pi \int_0^m \begin{pmatrix} U_1(t) & U_2(t) \\ -U_2(t) & U_1(t) \end{pmatrix} \begin{pmatrix} -\sin 2\pi t \\ \cos 2\pi t \end{pmatrix} dt + \begin{pmatrix} x_0 \\ y_0 \end{pmatrix}, \quad m \in [0,1]. \tag{9.48}$$

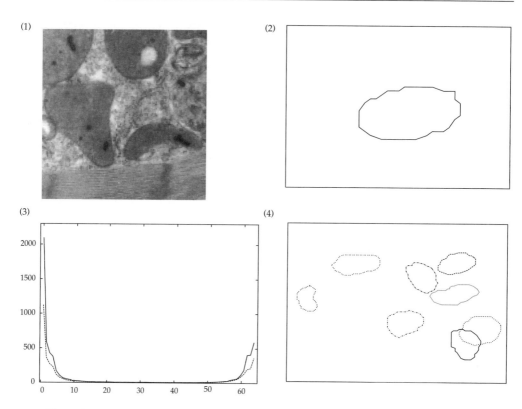

Figure 9.1 Top row: Panel 1 shows an electron-micrograph of rat-heart myocyte containing a linear organelle at 20,000× magnification taken from the laboratory of Jeffrey Saffitz of the department of Pathology at Washington University. Panel 2 shows the average mitochondria generated from 41 micrographs and hand tracing 497 mitochondria. Bottom row: Panel 3 shows the spectrum of the covariance of the stationary Gaussian prior on the deformations of the mitochondria. Panel 4 shows 8 mitochondria generated from the prior distribution placed uniformly in the scene.

Now use Egevary's result Corollary 9.27 in the block form for representing the random 2-vector process on the Torus $\{U(t) = \begin{pmatrix} U_1(t) \\ U_2(t) \end{pmatrix} \in \mathbb{R}^2, t \in T = [0,1]\}$. Model the process $U(\cdot)$ as cyclo-stationary with 2×2 covariance matrices which are Toeplitz $K(t,s) = K(t-s, 0) = EU(t)U(s)^*$.

Periodicity implies the expansion of the random vector fields $U(t), t \in [0,1]$ in the C.O.N. block diagonalizing basis constructed from the complex exponentials $\phi_k^{(1)} = \underbrace{\begin{pmatrix} 1 \\ 0 \end{pmatrix}}_{\mathbf{1}^{(1)}} e^{j2\pi kt}, \phi_k^{(2)} = \underbrace{\begin{pmatrix} 0 \\ 1 \end{pmatrix}}_{\mathbf{1}^{(2)}} e^{j2\pi kt}$ giving the orthogonal expansion

$$\begin{pmatrix} U_1(t) \\ U_2(t) \end{pmatrix} \stackrel{q.m.}{=} \sum_{k \in \mathbb{Z}} \begin{pmatrix} U_{1k} \\ U_{2k} \end{pmatrix} e^{j2\pi kt}, \text{ where } \begin{aligned} U_{1k} &= \int_0^1 U_1(t) e^{-j2\pi kt} \, dt \\ U_{2k} &= \int_0^1 U_2(t) e^{-j2\pi kt} \, dt \end{aligned}.$$
(9.49)

Notice $U_k = \begin{pmatrix} U_{1k} \\ U_{2k} \end{pmatrix} \in \mathbb{C}^2$ are complex orthogonal vector process (Homework problem F.7, Corollary F.2) with spectrum $EU_k U_{k'}^* = \Lambda_k \delta(k-k')$, and

$\Lambda_k = \int_0^1 K(t) e^{-j2\pi k t} \, dt$. Discretizing to n unique scale, rotation groups gives piecewise polygonal curves $f(l/n), l = 0, 1, \ldots, n-1$, with $f(0) = f(1)$ and keeping the scale/rotation parameters over constant sampling intervals

$$f\left(\frac{l}{n}\right) = \begin{pmatrix} x_0 \\ y_0 \end{pmatrix} + \sum_{i=1}^{l} \begin{pmatrix} U_1(i) & U_2(i) \\ -U_2(i) & U_1(i) \end{pmatrix} \begin{pmatrix} \cos(2\pi i/n) - \cos(2\pi(i-1)/n) \\ \sin(2\pi i/n) - \sin(2\pi(i-1)/n) \end{pmatrix}. \quad (9.50)$$

The Toeplitz covariance has n, 2×2 blocks $K(k,j) = EU(k)U(j)^* = K(k-j,0)$; the block diagonalizing transformation is the block DFT matrix. Then $U_k = \sum_{i=1}^n U(i) e^{-j2\pi i k/n}$ is a block-orthogonal process with 2×2 block covariances $\Lambda_k = \begin{pmatrix} \lambda_{11k} & \lambda_{21k}^* \\ \lambda_{21k} & \lambda_{22k} \end{pmatrix} = \sum_{i=1}^n e^{-j2\pi i k/N} K(i)$, $k = 1, \ldots, n$. A method of synthesis is via the set of complex 2×1 Gaussian vectors $\{U_k, k \in \mathbb{Z}_n\}$ of Goodman class [217]. The closure constraint on the simple curves $f(0) = f(1)$ implies the 2D linear manifold constraint associated with the discrete Fourier transform (DFT) is zero:

$$\begin{pmatrix} 0 \\ 0 \end{pmatrix} = \sum_{i=1}^n \begin{pmatrix} U_1(i) & -U_2(i) \\ U_2(i) & U_1(i) \end{pmatrix} \begin{pmatrix} \cos(2\pi i/n) - \cos(2\pi(i-1)/n) \\ \sin(2\pi i/n) - \sin(2\pi(i-1)/n) \end{pmatrix}, \quad (9.51)$$

implying the complex zero equation

$$0 = \sum_{i=1}^n \left(U_1(i) - jU_2(i)\right)\left(e^{j(2\pi i/n)} - e^{j(2\pi(i-1)/n)}\right). \quad (9.52)$$

This implies the highest frequency discrete Fourier transform coefficient is zero since

$$\sum_{i=1}^n \left(U_1(i) - jU_2(i)\right) e^{-j(2\pi i(n-1)/n)} = e^{-j(2\pi/n)} \sum_{i=1}^n \left(U_1(i) - jU_2(i)\right) e^{-j(2\pi i(n-1)/n)}. \quad (9.53)$$

The effect of closure is to reduce the dimension of the $2n$, U-variables to $2n - 2$ dimensions.

Figure 9.1 shows the results of experimental generation of mitochondrial boundary shapes representing the empirical covariance structure of 41 images containing 497 mitochondria which were hand-traced. Each boundary was discretized and sampled to constant arc length. The template was mapped to the hand tracing from which the parameters were fitted and discrete Fourier transformed for all M of the mitochondria. The 2×2 blocks $\Lambda_k, k \in \mathbb{Z}_n$ are calculated according to

$$\Lambda_k = \begin{pmatrix} a & b^* \\ b & d \end{pmatrix} \quad \text{where} \quad \begin{aligned} a &= \tfrac{1}{M} \sum_{m=1}^M |U_{1k}^{(m)} - \bar{U}_{1k}|^2 \\ b &= \tfrac{1}{M} \sum_{m=1}^M (U_{1k}^{(m)*} - \bar{U}_{1k})^* (U_{2k}^m - \bar{U}_{2k}) \\ d &= \tfrac{1}{M} \sum_{m=1}^M |U_{2k}^{(m)} - \bar{U}_{2k}|^2 \end{aligned}, \quad (9.54)$$

where $\bar{U}$ represents the sample means of the complex Fourier transforms $\bar{U}_{1k} = (1/M) \sum_{m=1}^M U_{1k}^{(m)}, \bar{U}_{2k} = (1/M) \sum_{m=1}^M U_{2k}^{(m)}$.

Panel 2 of Figure 9.1 shows the average mitochondria generated by hand tracing the population of mitochondria and estimating the covariance parameters. Panel 3 (bottom row) shows the spectrum of the covariance of the stationary Gaussian process on the mitochondria shapes. Since there are two parameters at each boundary point, the spectrum has two components at each Fourier transform frequency corresponding to the eigenvalues of Λ_k. Panel 3 shows a plot of these two spectral components.

Because the shapes are smooth, the spectra have low frequency content, with the familiar symmetry around the midpoint of the spectrum being due to the fact that the scale/rotation process is real. The mean shape and its covariance represents the typical structure and its range of variation. Panel 4 (bottom row) shows 8 mitochondria generated from the empirical distribution placed uniformly throughout the scene.

9.3 Gaussian Fields on the Continuum

Orthogonal expansions require specification of second order properties, mean and covariance; it is natural to return to Gaussian fields, completely determined by their mean and covariance operators. The familiar view of Gaussian processes is that arbitrary linear combinations of samples of the field are Gaussian. The Gaussian fields studied here are associated with Hilbert spaces $H = L^2, l^2$; therefore it will be powerful to define the Gaussian fields through their properties in inner products with elements of the Hilbert space. We give both definitions.

Definition 9.29 *Then $Y(\omega, x), x \in X$ is an m-dimensional* **Gaussian vector field** *on $X \subset \mathbb{R}^d(\mathbb{Z}^d), m \geq 1, d \geq 1$ with* **mean and covariance** *fields $m_Y, K_Y, m_Y : X \to \mathbb{C}^m, K_Y : X \times X \to \mathbb{C}^m \times \mathbb{C}^m$, if for any n and n-tuple of m-vectors $a_1, \ldots, a_n \in \mathbb{C}^m$, points $x_1, \ldots, x_n \in X$, $Z(\omega) = \sum_{k=1}^n a_k^* Y(\omega, x_k)$ is Gaussian with mean and variance*

$$m_a = \sum_{i=1}^n a_i^* m_Y(x_i), \quad \sigma_a^2 = \sum_{i,j=1}^n a_i^* K_Y(x_i, x_j) a_j. \tag{9.55}$$

The Hilbert space definition is useful.

Definition 9.30 *Define the complex m–vector Hilbert space H on $X \subset \mathbb{R}^d, m \geq 1, d \geq 1$ with*

$$\langle f, g \rangle_H = \int_X \langle f(x), g(x) \rangle_{\mathbb{C}^m} dx = \sum_{i=1}^m \int_X f_i^*(x) g_i(x) dx. \tag{9.56}$$

Then Y is a **Gaussian m-vector field** *on H with* **mean and covariance** *$m_Y \in H, K_Y : H \to H$, if for all $f \in H$, $\langle f, Y \rangle_H$ is normally distributed with mean and variance m_f, σ_f^2 given by*

$$m_f = \langle f, m_Y \rangle_H, \quad \sigma_f^2 = \langle f, K_Y f \rangle_H. \tag{9.57}$$

We shall say Y **is trace class** (**covariance operator** K_Y **is trace class**) if for any C.O.N. basis $\{\phi_k\} \subset H$,

$$\mathrm{tr} K_Y = \sum_{k=1}^\infty \langle \phi_k, K_Y \phi_k \rangle_H < \infty. \tag{9.58}$$

K_Y is a positive definite and self-adjoint operator; this is homework problem F.7.52. The covariance properties required for the orthogonal expansions require the trace class condition. The covariance trace class connects to the notion of SOP's, $E\|Y(\omega)\|^2 < \infty$.

Theorem 9.31 *Given a process on H (zero mean) with covariance trace class, $\mathrm{tr}\, K_Y < \infty$, then*

$$\mathrm{tr}\, K_Y = E\|Y\|^2 < \infty. \tag{9.59}$$

Proof Define the C.O.N. base of H to be $\{\phi_k\}$, then

$$E\|Y\|^2 = E \lim_{n\to\infty} \sum_{k=1}^n |\langle Y, \phi_k\rangle|^2 = \lim_{n\to\infty} \sum_{k=1}^n E|\langle Y, \phi_k\rangle|^2 \text{ (monotone convergence)}$$

$$= \lim_{n\to\infty} \sum_{k=1}^n \langle \phi_k, K_Y \phi_k\rangle = \sum_{k=1}^\infty \langle \phi_k, K_Y \phi_k\rangle = \operatorname{tr} K_Y < \infty.$$

$\square$

Of particular interest is $L^2(X)$, $X \subset \mathbb{R}^d$ compact. Take $X = [0,1]$ for example, then Y on X being trace class implies $E \int_X |Y(\omega, t)|^2 \, dt < \infty$. The connection to q.m. continuity is clear, Y being q.m. continuous implies K is bounded over X (Theorem 9.13), implying $\int_X K(t, t) \, dt < \infty$ which by Fubini's theorem implies $Y(\omega, \cdot)$ is of trace class!

Notice, integrals against functions in the Hilbert space are a direct extension from finite linear combinations. Homework problem F.7 examines the equivalence of the two definitions 9.29 and 9.30.

Example 9.32 (Brownian Motion) The **Wiener process** is such an important example of a random process with so many beautiful properties: let us single it out. Let $Y(t), t \geq 0$ be a real-valued Gaussian process. Then it is a **standard Wiener process** or a **standard Brownian motion** if it has mean and covariance

$$EY(t) = 0 \quad \text{and} \quad EY(t)Y(s) = \min(t, s) . \tag{9.60}$$

Since it is a Gaussian process, finite linear combinations of the form $Z = \sum_{k=1}^N \alpha_k Y(t_k)$ are Gaussian. The nth order density on samples are Gaussian (see Remark 5.1.1 of the Gaussian fields Section 5.1, Chapter 5) giving

$$p(y(t_1), \cdots, y(t_n)) = \det^{-1/2} 2\pi K e^{-1/2 \sum_{ij} y(t_i) y(t_j) (K^{-1})_{ij}} \tag{9.61}$$

with $n \times n$ covariance $K = \bigl(\min(t_i, t_j)\bigr)$.

Observe that this process has independent increments, i.e. $Y(t_1) \sim N(0, t_1)$, $Y(t_2) - Y(t_1) \sim N(0, t_2 - t_1)$, $Y(t_n) - Y(t_{n-1}) \sim N(0, t_n - t_{n-1})$, and if $s < t < u < v$ then

$$E(Y(t) - Y(s))(Y(v) - Y(u)) = 0 . \tag{9.62}$$

There is a much stronger property of Brownian motion which actually provides the basis for its role in diffusions and stochastic integration. The variance of an increment varies as the length of the increment, $E(Y(t) - Y(s))^2 = t - s$, suggesting that $(dY(t))^2 \sim O(dt)$. This is in fact the case in an almost sure sense, and provides the basis for many results on diffusions and stochastic integrals (see Theorem A.2 in the Appendix).

Example 9.33 (Eigen Spectrum of Brownian Motion) Brownian motion with boundary condition $Y(0) = 0$ is a second order process, and is quadratic mean continuous as well since $K(t, t) = \min(t, t) = t$ is continuous. Then $K(t, s) = EY(t)Y(s)$ is continuous on $[0, T]^2$ (see Theorem 9.13), and as an operator is compact and self adjoint:

$$Kf(t) = \int_0^T K(t, s) f(s) \, ds = \int_0^t s f(s) \, ds + t \int_t^T f(s) \, ds, \quad t \in [0, T]. \tag{9.63}$$

The eigenfunctions of the covariance satisfy

$$\lambda \phi(t) = \int_0^T K(t, s) \phi(s) \, ds = \int_0^t s \phi(s) \, ds + t \int_t^T \phi(s) \, ds , \quad t \in [0, T] . \tag{9.64}$$

Differentiation twice gives

$$\int_t^T \phi(s)\,ds = \lambda \dot\phi(t) \quad 0 < t < T$$

$$-\phi(t) = \lambda \ddot\phi(t) \quad 0 < t < T.$$

To expand the Brownian motion around $Y(0) = 0$, then the first boundary condition gives $\phi(0) = 0$, with the second boundary condition $\dot\phi(T) = 0$, then $\phi(t) = A \sin t/\sqrt{\lambda}$, $\cos T/\sqrt{\lambda} = 0$, and the eigenelements become $\lambda_n = T^2/((n + (1/2))^2 \pi^2)$, $\phi_n(t) = \sqrt{2/T} \sin((n + (1/2))\pi t)/T$.

Example 9.34 (White Noise) White-noise is tricky, its not continuous anywhere. Assume $Y(t), t \in [0,1]$ on $(\Omega, \mathcal{A}, P)$ is an i.i.d. collection of normal random variables. Let $Y(t)$ be continuous at some point $t_0 \in [0,1]$. The i.i.d. condition implies

$$\Pr\{\omega : Y(\omega, t) > \epsilon, Y(\omega, s) < -\epsilon\} = \left(\int_\epsilon^\infty \frac{1}{\sqrt{2\pi}} e^{-z^2/2}\,dz \right)^2, \qquad (9.65)$$

with the event $\Pr\{\omega : Y(\omega, t) > \epsilon, Y(\omega, s) < -\epsilon\}$ not depending on s and t, implying as s approaches t the limit is a non-zero constant for any choice of $\epsilon > 0$. Then if $Y(\omega, t_0)$ is continuous at $t_0 \in [0,1]$, it implies

$$\lim_{n \to \infty} A_n = \lim_{n \to \infty} \left\{ \omega : Y(\omega, t_0) > \epsilon, Y\left(\omega, t_0 + \frac{1}{n}\right) < -\epsilon \right\} = \emptyset. \qquad (9.66)$$

Thus,

$$P\left(\lim_{n \to \infty} A_n\right) = P(\emptyset) = 0 \neq \lim_{n \to \infty} P(A_n) = \left(\int_\epsilon^\infty \frac{1}{\sqrt{2\pi}} e^{-z^2/2}\,dz \right)^2. \qquad (9.67)$$

P is not sequentially continuous. Impossible! Apparently there is no such continuity point $t_0 \in [0,1]$.

9.4 Sobolev Spaces, Green's Functions, and Reproducing Kernel Hilbert Spaces

For vector fields differential operators which are *physically based* will be used to model the structure of physical systems. The operators express physical properties which represent the transformations of the dynamical system being modeled.

Herein linear differential operators are studied supporting orthogonal process descriptions. Assume throughout the domain is compact. The Gaussian random fields are induced as solutions of partial differential equations of the type

$$LU(x) = W(x), \quad x \in X \subset \mathbb{R}^d, \qquad (9.68)$$

with W a Gaussian noise process, and L a linear differential operator. Smoothness properties on the solution U are inherited from smoothness properties associated with the differential operators L.

We have already seen from our various decomposition theorems that smoothness on covariance (continuity for example) and existence of smooth eigenelements are fundamental. This smoothness will be determined by the existence of the Green's function and its smoothness. Assume G is the Green's function of L, so that $LG = \delta$. For the engineers amongst us, think of

9.4 REPRODUCING KERNEL HILBERT SPACES

L as representing the transformation of the linear time invariant system on the input W transforming to the output U. Let us assume for the moment that W is well defined, a simple smooth function. In this case G is the "impulse response" of the linear time invariant system, thus linear solutions (initial conditions at minus infinity) take the form

$$U(x) = \int G(x,y) W(y)\, dy . \tag{9.69}$$

Thus, the smoothness of the Green's function G determines the smoothness of U. Even in the "white noise case", if we interpret U weakly, then clearly the covariance of U is determined by the smoothness of the Green's function according to $K_U = \int G(\cdot, y) G(y, \cdot)\, dy$.

9.4.1 Reproducing Kernel Hilbert Spaces

So what is the space of functions associated with solutions of the equations associated with the differential operators? They are a Hilbert space $H \subset L^2$, generally smoother than L^2 functions, which we shall call a reproducing kernel Hilbert space (RKHS) which can be generated by operating a smooth function (the Green's function) on the L^2 functions.

The general definition of an RKHS is defined by the existence of a continuous linear functional which sifts on functions.

Definition 9.35 *A Hilbert space $H(X, \mathbb{R})$ of real-valued scalar functions on $X \subset \mathbb{R}^d$ is said to be a **reproducing kernel Hilbert space (RKHS)** if for each $x \in X$, the linear evaluation functional given by $\delta_x : h \in H \mapsto \delta_x(h) = h(x) \in \mathbb{R}$ is continuous [25] and there exists a unique element $K(x, \cdot) \in H$ such that*

$$\delta_x(h) = \langle K(x, \cdot), h \rangle_H = h(x), \quad \text{for all } h \in H; \tag{9.70}$$

it is reproducing since

$$\langle K(x, \cdot), K(y, \cdot) \rangle_H = K(y, x). \tag{9.71}$$

If $H(X, \mathbb{R}^m)$ is the space of m-vector functions then it is an RKHS if for each $x \in X, \alpha \in \mathbb{R}^m$; then $\delta_x^\alpha(h) = \langle h(x), \alpha \rangle_{\mathbb{R}^m}$ is continuous and there exists a unique element $K(x, \cdot)\alpha \in H$ such that

$$\delta_x^\alpha(h) = \langle K(x, \cdot)\alpha, h \rangle_H = \langle \alpha, h(x) \rangle_{\mathbb{R}^m} \quad \text{for all } h \in H; \tag{9.72}$$

it reproduces since

$$\langle K(x, \cdot)\alpha, K(y, \cdot)\beta \rangle_H = \langle \alpha, K(x,y)\beta \rangle_{\mathbb{R}^m}. \tag{9.73}$$

The kernels for such spaces come in many forms, for us they will often be the convolution of the Green's kernel of a shift-invariant differential operator with itself. In the one dimensional signal analysis case this corresponds to the convolution of the impulse response with itself.

Remark 9.4.1 So how do we see that this Hilbert space $H \subset L^2$ is smoother than L^2 functions. Well, essentially they are functions which can be generated by operating a

---

[25] Continuity of a linear function $F : H \to H$ is equivalent to boundedness in the sense that $\exists M \in \mathbb{R}$ such that $|\delta_x^\alpha(h)| \leq M \|h\|_H$ for all $h \in H$, implying $|F(h - h_n)| \leq M \|h - h_n\|_H \to 0$ as $\|h_n \to h\|_H$. The existence of the kernel follows from the Riesz Fréchet Theorem; if $F : H \to \mathbb{R}$ is a continuous linear functional on a Hilbert space H then there exists a unique $f \in H$ such that $F(h) = \langle f, h \rangle_H$ for all $h \in H$. Since the linear functional δ_x is continuous then there exists a unique element $k_x(\cdot) \in H$ such that $\delta_x(h) = \langle k_x, h \rangle_H = h(x)$ for all $h \in H$. The same argument holds for the vector case.

smooth function on the L^2 functions. When the kernel is sufficiently smooth so that its square root is smooth, then there exists a $G(x, \cdot), x \in X$ which is continuous and positive definite with reproducing kernel $K(\cdot, \cdot) = \int_X G(\cdot, z) G(z, \cdot) \, dz$ such that the Hilbert space H is constructed according to

$$H(X, \mathbb{R}^m) = \left\{ h \in L^2(X, \mathbb{R}^m) : \exists f \in L^2(X, \mathbb{R}^m), h(\cdot) = \int_X G(\cdot, x) f(x) \, dx \right\}. \quad (9.74)$$

The inner product on H is defined as $< h, h' >_H = < f, f' >_2$.

9.4.2 Sobolev Normed Spaces

Sobolev spaces and the Sobolev inclusion theorems will allow us to relate the L^2 existence of the derivatives to the pointwise differentiabilty of continuous and differentiable functions with sup-norm. The connection to RKHS is with sufficient derivative then the space has a smooth kernel.

Definition 9.36 *Define the **Sobolev space** $H_0^s(X, \mathbb{R}^m)$ to be the closure of the compactly supported with zero-boundary continuously differentiable m-vector functions on the $X \subset \mathbb{R}^d$ with $s > 0$ the integer order of the Sobolev norm; $\| \cdot \|_{H_0^s}$ given by*

$$\|h\|_{H_0^s}^2 = \sum_{i=1}^m \sum_{\alpha \leq s} \|D^\alpha h_i\|_2^2, \quad (9.75)$$

with the notation D^α representing the derivative on the scalar function with the index α selecting the derivatives and cross derivatives up to order α.

Now to guarantee the existence of smooth kernels we use the Sobolev embedding theorems allowing us to show that if the order s of differentiability in the Sobolev norm satisfies $s > k + (d/2)$ then functions $h \in H_0^s$ agree almost everywhere with functions in C_0^k, written as $H_0^s \hookrightarrow C_0^k$. This condition we say is the inclusion map being continuous.

Theorem 9.37 *If X is bounded in $\mathbb{R}^d$, $k \geq 0$ and $s > k + (d/2)$ then $H_0^s(X) \hookrightarrow C_0^k(X)$ is continuous and $H_0^s(X) \hookrightarrow C^k(X)$ is compact.*

In particular $X \subset \mathbb{R}^d$ open, $s > (d/2)$, so that $H_0^s(X) \hookrightarrow C_0^0(X)$; then H_0^s is a reproducing kernel Hilbert space with $m \times m$ reproducing kernel K satisfying, for all $h \in H$, $\alpha \in \mathbb{R}^m$,

$$\langle K(x, \cdot) \alpha, h \rangle_H = \langle \alpha, h(x) \rangle_{\mathbb{R}^m}. \quad (9.76)$$

It follows from Theorem 9.37 that if $s > d/2$ then the evaluation functionals on H_0^s are continuous. Indeed, for any $x \in X$, the continuity of the inclusion map implies that there exists a constant M such that for all $h \in H_0^s$,[26]

$$|h(x)| \leq \sup_y |h(y)| \leq M \|h\|_{H_0^s}.$$

---

[26] By the inclusion map of H into B where $H \subset B$, we mean the map id : $H \to M$ with id $h = h$. To say that the inclusion map is continuous, (we write this concisely as $H \hookrightarrow B$) we mean that operator id is continuous in the usual sense (which is equivalent to bounded as defined above, i.e. there exists a constant M such that $\|h\|_B \leq M \|h\|_H$ for all $h \in H$.) To say that the inclusion map is compact, we mean that the operator id is compact. That is, id is compact if id maps bounded sets in H to precompact sets (sets whose closure is compact) in B. id is just the identity; so this means that sets in H which are bounded with respect to the norm on H are precompact with respect to the norm on B. Another equivalent statement is that bounded sequences in H have convergent subsequences with respect to the norm on B.

9.4.3 Relation to Green's Functions

For modeling purposes, it is useful to use the differential operators L to dominate the Sobolev norm of sufficient derivatives, since then, from the above *Sobolev embedding theorem* 9.37, the space is a reproducing kernel Hilbert space with the kernel of smoothness determined by s. This kernel is generated from the Green's operator according to LL^* since for all $h \in H^s$, then if $\|h\|_{H^s}^2 = \|Lh\|_2^2$ implying if there exists a kernel K then

$$\|h\|_{H^s}^2 = \int_X \langle L^*LK(x,y)\alpha, h(y) \rangle_{\mathbb{R}^m} \, dy = \langle K(x, \cdot)\alpha, h(\cdot) \rangle_H = \langle \alpha, h(x) \rangle_{\mathbb{R}^m}. \tag{9.77}$$

Thus we have, if the kernel exists, then $K = GG^*$ generated from the Green's kernels.

To illustrate the condition which satisfies the $s > k + d/2$ condition, with $s = 2, k = 1, d = 1$. Let $L = id - \alpha^2 \nabla^2$ in $\mathbb{R}^1$, then the Green's kernel satisfying $L^*LK = \delta$ is given by

$$K(x,y) = \frac{1}{4\alpha^2} \left(e^{-\frac{1}{\alpha}|x-y|}(|x-y| + \alpha) \right)$$

and $K(x,y)$ is C^1 and an element of the Sobolev space of two derivatives (reproducing kernel space).

To see this, to calculate the Green's kernel $L^*L = (id - \alpha^2 \Delta)^2$, then let $K(x,y) = g(x-y)$ and define the Fourier transform pair $g(x) \leftrightarrow G(f)$. Then $L^*LK = \delta$, $G(f)$ is the inverse spectrum of LL^* given by

$$G(f) = \frac{1}{(\alpha^2 f^2 + 1)^2}.$$

Clearly, the kernel is in the Sobolev space of two derivatives since differentiating twice gives a Fourier transform which is integrable going as f^{-2}.

Since the transform is the square, the kernel is given by the convolution

$$g(x) = \frac{1}{2\alpha} e^{-\frac{|x|}{\alpha}} * \frac{1}{2\alpha} e^{-\frac{|x|}{\alpha}}$$

$$= \frac{1}{4\alpha^2} \int_{-\infty}^{\infty} \left(e^{-\frac{(t-\tau)}{\alpha}} u(t-\tau) + e^{\frac{(t-\tau)}{\alpha}} u(-t+\tau) \right) \left(e^{-\frac{\tau}{\alpha}} u(\tau) + e^{\frac{\tau}{\alpha}} u(-\tau) \right) d\tau$$

$$= \frac{1}{4\alpha^2} \left(e^{-\frac{1}{\alpha}(|x| + \alpha)} \right)$$

To show that $K(x,y)$ is C^1, we observe $g'(x)$ is continuous according to

$$g'(x) = \begin{cases} -\frac{1}{4\alpha^3} e^{-\frac{x}{\alpha}}(x + \alpha) + \frac{1}{4\alpha^2} e^{-\frac{x}{\alpha}} & x \geq 0 \\ \frac{1}{4\alpha^3} e^{\frac{x}{\alpha}}(-x + \alpha) - \frac{1}{4\alpha^2} e^{-\frac{x}{\alpha}} & x < 0 \end{cases}$$

with $g'(0^+) = g'(0^-) = 0$

9.4.4 Gradient and Laplacian Induced Green's Kernels

Here are three examples deriving the Green's operator; in $\mathbb{R}^1, \mathbb{R}^3$ $k = 0$ and the evaluation kernels are continuous only with $s > d/2$ condition; in $\mathbb{R}^2$, then $k = 1$ with $s = 2$; so the evaluation functional has a non-continuous derivative.

Theorem 9.38 Let $L = \text{id} + \alpha(\partial/\partial x)$ in $\mathbb{R}^1$ and let $L = \text{id} - \alpha^2\nabla^2$ in $\mathbb{R}^3$, then the Green's kernel satisfying $L^*LK = \delta$ is given by

$$K(x,y) = \beta e^{-1/\alpha \|x-y\|_{\mathbb{R}^{1,3}}}, \qquad (9.78)$$

with β providing the normalizer given by $\beta = 1/2\alpha$ in $\mathbb{R}^1$, $\beta = 1/(2\alpha)^3\pi$ in $\mathbb{R}^3$, respectively. For $L = \text{id} - \alpha^2\nabla^2$ in $\mathbb{R}^2$,

$$K(x,y) = \frac{1}{4\alpha^2\pi}\int_0^\infty e^{-t-\|x-y\|_{\mathbb{R}^2}^2/4\alpha^2 t}\,dt = \frac{\|x-y\|_{\mathbb{R}^2}}{|\alpha|}B_1\left(\frac{\|x-y\|_{\mathbb{R}^2}}{|\alpha|}\right),$$

where B_1 denotes the modified Bessel function of the second kind of order 1[218].

Proof In 1-dimension $\mathbb{R}^1$, the adjoint L^* is required to compute the Green's kernel; its defined by

$$\langle Lg, h\rangle_2 = \langle g, L^*h\rangle_2 \qquad (9.79)$$

with g, h differentiable functions vanishing at infinity. Solving gives

$$\int \left(\text{id} + \alpha\frac{\partial}{\partial x}\right)g(x)h(x)\,dx = \int g(x)h(x)\,dx - \int g(x)\alpha\frac{\partial}{\partial x}h(x)\,dx \qquad (9.80)$$

$$= \int g(x)\left(\text{id} - \alpha\frac{\partial}{\partial x}\right)h(x)\,dx, \qquad (9.81)$$

implying $L^* = \text{id} - \alpha(\partial/\partial x)$. Then $L^*L = \text{id} - \alpha^2(\partial^2/\partial x^2)$. That the Green's kernel in 1-D is the exponential spline with $\beta = 1/2\alpha$ follows from the second derivative:

$$\frac{\partial}{\partial x}\frac{1}{2\alpha}e^{-\frac{1}{\alpha}|x|} = \frac{1}{2\alpha}\frac{1}{\alpha}e^{\frac{1}{\alpha}x}u(-x) - \frac{1}{2\alpha}\frac{1}{\alpha}e^{-\frac{1}{\alpha}x}u(x), \qquad (9.82)$$

$$\frac{\partial^2}{\partial x^2}\frac{1}{2\alpha}e^{-\frac{1}{\alpha}|x|} = \frac{1}{\alpha^2}K(x,0) - \frac{1}{\alpha^2}\delta(x), \qquad (9.83)$$

with $u(x)$ the unit step function. Then $L^*LK = \delta$ as an operator.

To calculate the Green's kernel in $\mathbb{R}^3$ use Fourier identities. Start with $L^*L = (\text{id} - \alpha^2\nabla^2)^2$, then let $K(x,y) = g(x-y)$, and define the Fourier transform pair $g(x) \leftrightarrow G(f)$, $x = (x_1, x_2, x_3)$ and $f = (f_1, f_2, f_3)$. Since $L^*Lg = \delta$, then $G(f)$ is the inverse spectrum of L^*L given by $G(f) = 1/(c^2\|f\|_{\mathbb{R}^3}^2 + 1)^2$ with $c = 2\pi\alpha$ implying the Green's function is given by the following integral:

$$g(x) = \int_{-\infty}^\infty \int_{-\infty}^\infty \int_{-\infty}^\infty \frac{1}{(c^2\|f\|_{\mathbb{R}^3}^2 + 1)^2}e^{j2\pi\langle x,f\rangle_{\mathbb{R}^3}}\,df. \qquad (9.84)$$

Note that $G(\cdot)$ is rotationally invariant, so $g(\cdot)$ is rotationally invariant, and only needs to be evaluated at $x = (0, 0, z), z > 0$. Making the change to spherical coordinates

$f_1 = r \sin \psi \sin \theta$, $f_2 = r \sin \psi \cos \theta$, $f_3 = r \cos \psi$, then the integral becomes

$$\frac{1}{2\pi} \int_0^\infty \int_0^{2\pi} \int_0^\pi \frac{1}{(c^2 r^2 + 1)^2} e^{j 2\pi r z \cos \psi} r^2 \sin \psi \, d\psi \, d\theta \, dr$$

$$\stackrel{(a)}{=} 2\pi \int_0^\infty \int_{-1}^1 \frac{r^2}{(c^2 r^2 + 1)^2} e^{j 2\pi r z y} \, dy \, dr$$

$$\stackrel{(b)}{=} \frac{1}{z} \int_0^\infty \frac{r}{(c^2 r^2 + 1)^2} \left(\frac{e^{j 2\pi r z} - e^{-j 2\pi r z}}{j} \right) dr$$

$$= \frac{1}{2z} \int_{-\infty}^\infty \frac{r}{(c^2 r^2 + 1)^2} \left(\frac{e^{j 2\pi r z} - e^{-j 2\pi r z}}{j} \right) dr$$

$$= -\frac{1}{z} \int_{-\infty}^\infty \frac{jr}{(c^2 r^2 + 1)^2} e^{j 2\pi r z} \, dr,$$

with (a) since the integrand does not depend on ψ and making the change of variable $y = \cos \theta$, and (b) integrating with respect to y. Now from the tables

$$-\frac{1}{z} \int_{-\infty}^\infty \frac{jr}{(c^2 r^2 + 1)^2} e^{j 2\pi r z} \, dr = \frac{1}{8\pi \alpha^3 z} \int_{-\infty}^\infty \frac{-8\pi jr}{(4\pi^2 r^2 + (1/\alpha)^2)^2} e^{j 2\pi r z} \, dr \quad (9.85)$$

$$= \frac{1}{(2\alpha)^3 \pi z} z e^{-(1/\alpha) \|z\|_{\mathbb{R}^3}}$$

$$= \frac{1}{(2\alpha)^3 \pi} e^{-(1/\alpha) \|z\|_{\mathbb{R}^3}}, \quad (9.86)$$

giving the result $g(x) = 1/(2\alpha)^3 \pi e^{-1/\alpha \|x\|_{\mathbb{R}^3}}$.

For $\mathbb{R}^2$, then examine the Fourier transform of $g(x, 0) \leftrightarrow G(f) = (1 - 4\alpha^2 \pi^2 \|f\|_{\mathbb{R}^2}^2)^{-2}$. We have $g \in L^1$, for

$$\|g\|_{L^1}^2 \leq \frac{1}{4\alpha^2 \pi} \int_0^\infty \int_{\mathbb{R}^2} e^{-t} e^{-\|x\|_{\mathbb{R}^2}^2 / 4\alpha^2 t} \, dx \, dt = \int_0^\infty t e^{-t} \, dt = 1,$$

where the first equality follows from the Gaussian integral identity, $\int_{\mathbb{R}^n} e^{-\pi a \|x\|_{\mathbb{R}^n}^2} dx = a^{-n/2}$. Thus, by Fubini's Theorem we have

$$G(f) = \frac{1}{4\alpha^2 \pi} \int_{\mathbb{R}^2} \left(\int_0^\infty e^{-t - \|x\|_{\mathbb{R}^2}^2 / 4\alpha^2 t} \, dt \right) e^{-2\pi i \langle f, x \rangle_{\mathbb{R}^2}} \, dx$$

$$= \frac{1}{4\alpha^2 \pi} \int_0^\infty e^{-t} \left(\int_{\mathbb{R}^2} e^{-\|x\|_{\mathbb{R}^2}^2 / 4\alpha^2 t} e^{-2\pi i \langle f, x \rangle_{\mathbb{R}^2}} \, dx \right) dt$$

$$= \int_0^\infty t e^{-(1 + 4\alpha^2 \pi^2 \|f\|_{\mathbb{R}^2}^2) t} \, dt$$

$$= (1 + 4\alpha^2 \pi^2 \|f\|_{\mathbb{R}^2}^2)^{-2},$$

where the third equality follows from the Fourier Transform pair of the Gauss kernel, $e^{-\pi a \|x\|_{\mathbb{R}^n}^2} \leftrightarrow a^{-n/2} e^{-\pi \|f\|_{\mathbb{R}^n}^2 / a}$. □

Example 9.39 (Auto-regressive Models) Let $dY(t)/dt = -cY(t) + W(t)$, $Y(0) = 0$ with $W(\cdot)$ white-noise. Solving for the G function gives

$$LG(t, s) = \left(\frac{d}{dt} + c \right) G(t, s) = \delta(t - s) \quad (9.87)$$

implying $G(t,s) = e^{-c(t-s)}u(t-s)$. The adjoint becomes

$$\langle Gf, h \rangle = \int_0^\infty \int_0^\infty G(t,s)f(s)\,ds h(t)\,dt$$

$$= \int_0^\infty \int_0^\infty e^{-c(t-s)}u(t-s)f(s)\,ds h(t)\,dt$$

$$= \int_0^\infty f(s) \int_0^\infty \underbrace{e^{-c(t-s)}u(t-s)}_{G^*(s,t)} h(t)\,dt\,ds, \qquad (9.88)$$

implying $G^*(s,t) = e^{c(s-t)}u(-(s-t))$. The covariance becomes

$$K(t,s) = GG^*(t,s) = \int_0^\infty e^{-c(t-z)}u(t-z)e^{c(z-s)}u(-(z-s))\,dz$$

$$= \int_0^s e^{-c(t-z)}u(t-z)e^{c(z-s)}\,dz$$

$$\stackrel{t \geq s}{=} \int_0^s e^{-c(t-z)}e^{c(z-s)}\,dz = e^{-c(t+s)} \frac{e^{2cz}}{2c}\Big|_0^s = \frac{e^{-c(t-s)} - e^{-c(t+s)}}{2c}$$

$$\stackrel{t \leq s}{=} \int_0^s e^{-c(t-z)}u(t-z)e^{c(z-s)}\,dz$$

$$= \int_0^t e^{-c(t-z)}e^{c(z-s)}\,dz = \frac{e^{+c(t-s)} - e^{-c(t+s)}}{2c}\,.$$

Homework problem F.7.49 shows that the covariance using the sifting property of white-noise gives the same result.

Example 9.40 (Wiener Process) Let $(dB(t))/(dt) = W(t)$ weakly with $B(0) = 0, t \geq 0$. Then the Green's function satisfies

$$\frac{dG(t,s)}{dt} = \delta(t-s) \qquad (9.89)$$

implying $G(t,s) = u(t-s)$ the unit step. Notice, for $s \geq 0$, $u(0-s) = 0$ satisfying the boundary condition $G(0,s) = 0$. G is not self-adjoint since d/dt is not self adjoint:

$$\langle Gf, g \rangle = \int_0^\infty g(t) \int_0^\infty u(t-s)f(s)\,ds\,dt$$

$$= \int_0^\infty f(s) \int_0^\infty \underbrace{u(t-s)g(t)}_{G^*(s,t)}\,dt\,ds = \langle f, G^*g \rangle \qquad (9.90)$$

proving $G \neq G^*$. The covariance satisfies

$$K(s,t) = GG^*(s,t) = \int_0^\infty u(s-\tau)u(t-\tau)\,d\tau \quad t,s > 0$$

$$= \int_0^{\min(t,s)} d\tau = \min(t,s) \quad t,s > 0. \qquad (9.91)$$

To see $K(s,t) = \min(t,s)$ satisfies the equation $LK_UL^* = \delta$ identity as an operator on function $f \in \hat{C}(0,\infty)$ vanishing on the boundary, $f(0) = f(\infty) = 0$, then

$$\min(t,s) = -\frac{1}{2}|t-s| + \frac{1}{2}s + \frac{1}{2}t\,. \qquad (9.92)$$

Now $L = d/dt, L^* = -d/dt$, giving

$$LK_U L^*(t,s) = \frac{d}{dt}\left(-\frac{1}{2}|t-s| + \frac{1}{2}s + \frac{1}{2}t\right)\frac{-d}{ds} = \left(-\frac{1}{2}u(t-s) + \frac{1}{2}\right)\frac{-d}{ds}, \quad (9.93)$$

implying via integration by parts of the sifting property on functions $f \in \hat{C}$: giving

$$LK_U L^* f(t) = \int_0^\infty LK_U L^*(t,s) f(s)\, ds = \int_0^\infty \underbrace{\left(-\frac{1}{2}u(t-s) + \frac{1}{2}\right)\frac{-d}{ds}f(s)ds}_{g(t,s)}$$

$$= -(g(t,\infty)f(\infty) - g(t,0)f(0))$$

$$+ \left(\int_0^\infty \frac{d}{ds}\left(-\frac{1}{2}u(t-s) + \frac{1}{2}\right)f(s)\, ds = f(t). \quad (9.94)$$

9.5 Gaussian Processes Induced via Linear Differential Operators

As in Chapter 5 we are going to induce Gaussian fields using differential operators (rather than difference operators). The background space $X \subset \mathbb{R}^d$ is assumed compact so that the random process will have discrete spectrum. This approach generalizes auto-regressive modeling for causal time-series applications, and allows the parameterization of the covariance of the induced fields to be parametric. This also completes the approach taken in the previous chapters, but extended from discrete graphs to the continuum. If L is a shift invariant operator with cyclic boundary conditions, then we shall be inducing cyclo-stationary Gaussian fields with covariance functions having a spectral decomposition based on the complex exponentials.

In the finite dimensional setting, the differential operators were driven by "white nois", $LU = W$, W = white noise. Then the covariance of the field is directly given by the inverse operators $K_U = (LL^*)^{-1}$. In the infinite dimensional setting, the inverse operators are given by the **Green's operator**, and care must be taken in dealing with white noise as it does not exist in the pointwise function sense. Contrast this to the finite dimensional Problem F.5.15, Chapter 5, where the covariance of white noise is identity. White noise is clearly not a second-order process.

For this treat white noise as a *generalized Gaussian random process* as in [219]. Introduce the family of test functions, $C^\infty(X)$ infinitely differentiable functions for $X \subset \mathbb{R}^d$ bounded domains (vanishing at infinity $\hat{C}^\infty$ for unbounded domains).

Definition 9.41 *Then $W(x), x \in X \subset \mathbb{R}^d$ is a **generalized random process or random field** if it is a continuous linear mapping that associates with each $f \in C^\infty$, the random variable $\langle W, f \rangle \in L^2(P)$.*

*We shall say $W(\cdot)$ is **Gaussian distributed in the generalized sense** with **mean and covariance**, m_W, K_W, if for all smooth test functions $f \in C^\infty$, $\langle W, f \rangle$ is Gaussian distributed with mean $\langle m_W, f \rangle$ and variance $\langle f, K_W f \rangle$.*

Return to white noise. Since it is nowhere continuous (see Example 9.34), the stochastic differential equation cannot be defined pointwise. Rather it is defined in the *weak sense* via its distribution in integration against test functions. The weak interpretation of the equality in Eqn. 9.68 gives for all smooth $f \in C^\infty$,

$$\langle W, f \rangle = \langle LU, f \rangle = \langle U, L^* f \rangle \quad (9.95)$$

which implies

$$\|f\|^2 = E|\langle W, f\rangle|^2 = \langle L^*f, K_U L^*f\rangle \tag{9.96}$$
$$= \langle f, LK_U L^*f\rangle. \tag{9.97}$$

Since this is true for all f this defines the operator $L^* K_U L$ to be the identity operator, $LK_U L^* = \delta$. Thus choose $K_U = GG^*$ the covariance satisfying Eqn. 9.97.

Definition 9.42 *Let L be a differential operator defining the norm $\|f\|_L^2 = \int_X \|Lf(x)\|^2\, dx$ with associated Hilbert space of functions H_L with inner product $\langle f, g\rangle_L = \int_X (Lf(x))^* Lg(x)\, dx$. Suppose H_L supports a sifting function $K(x_i, \cdot) \in H_L$ such that for all $f \in H_L$ then $\langle K(x_i, \cdot), f\rangle_L = f(x_i)$.*

Then K with kernels $K(x, \cdot) \in H_L$ for all $x \in X$ is the Green's operator of $L^* L$ according to for all $f \in H_L$,

$$f(x_i) = \langle K(x_i, \cdot), f\rangle_L = \int_X (LK(x_i, x))^* Lf(x)\, dx \tag{9.98}$$
$$= \int_X L^* L K(x_i, x) f(x)\, dx. \tag{9.99}$$

Theorem 9.43 *Let L be a normal invertible differential operator on $X \subset \mathbb{R}^d$ with continuous eigenelements $\{\lambda_k, \phi_k\}$, $L\phi_k = \lambda_k \phi_k$, admitting Greens operator G. Define $U^{(n)} = \sum_{k=0}^{n} U_k \phi_k$, U_k orthogonal Gaussian variates variance $E|U_k|^2 = 1/|\lambda_k|^2$. Then $U(\cdot) \stackrel{q.m.}{=} \lim_{n\to\infty} U^{(n)}(\cdot)$ is a quadratic mean continuous Gaussian field with trace class covariance $K_U = GG^* = \sum_{k=0}^{\infty} 1/|\lambda_k|^2 \phi_k \langle \phi_k, \cdot\rangle$ satisfying*

$$LU(\cdot) = W(\cdot), \quad W(\cdot) = \text{white noise}, \tag{9.100}$$

if and only if GG^* is continuous along the diagonal.

Proof Assume GG^* is continuous along the diagonal. Then GG^* is trace-class since X is compact so that

$$\operatorname{tr} GG^* = \int_X GG^*(x, x)\, dx < \infty. \tag{9.101}$$

Thus with $G = \sum_{k=0}^{\infty} \frac{1}{\lambda_k} \phi_k \langle \phi_k, \cdot\rangle$ and GG^* finite trace implies trace $GG^* = \sum_{k=0}^{\infty} 1/|\lambda_k|^2 < \infty$.

For every n and $j = 0, 1, \ldots, n$,

$$\lambda_j U_j = \int_X \phi_j^*(x) LU^{(n)}(x)\, dx = \int_X \phi_j^*(x) W^{(n)}(x)\, dx \tag{9.102}$$
$$= W_j. \tag{9.103}$$

Thus, $U_j = W_j/\lambda_j$ are independent Gaussian variates with variance $1/|\lambda_j|^2$ and $U^{(n)}$ is an n-dimensional Gaussian field since inner products with test functions $\langle f, U^{(n)}\rangle = \sum_{k=0}^{n} W_k/\lambda_k \langle f, \phi_k\rangle$ are Gaussian.

Then for Eqn. 9.100 to hold it must be shown that the left-and right-hand sides are equal when integrated against smooth functions. Expanding $W(\cdot) = \sum_{k=0}^{\infty} W_k \phi_k(\cdot)$,

9.5 GAUSSIAN PROCESSES INDUCED VIA LINEAR DIFFERENTIAL OPERATORS

W_k zero-mean orthogonal Gaussian variates variance $E|W_k|^2 = 1$, then

$$E|\int f^*(x)(LU^{(n)}(x) - W(x))\,dx|^2 = E\left|\sum_{k=n+1}^{\infty} W_k \langle f, \phi_k \rangle\right|^2$$

$$\stackrel{(a)}{=} \sum_{k=n+1}^{\infty} |\langle f, \phi_k \rangle|^2 \stackrel{(b)}{\to} 0 \text{ as } n \to \infty. \quad (9.104)$$

where (a) holds since $W(\cdot)$ is white noise and (b) since $f \in C^\infty(X)$ is square integrable on X.

Then the covariance $K_U = GG^*$ directly from the definitions of the Green's operator. □

Theorem 9.44 *Let L be a normal invertible linear differential operator on $D \subset \mathbb{R}^d$ with continuous eigenelements $L\phi_k = \lambda_k \phi_k$, with $W^{(n)}(\cdot) = \sum_{k=0}^n W_k \phi_k(\cdot)$, W_k zero-mean independent Gaussian variates with variances $E|W_k|^2 = |\alpha_k|^2$, with $\sum_{k=0}^\infty |\alpha_k|^2 < \infty$.* [27]

Defining $U^{(n)}(\cdot) = \sum_k U_k \phi_k(\cdot)$ with U_k zero-mean independent Gaussian variates with variance $E|U_k|^2 = |\alpha_k|^2/|\lambda_k|^2$ satisfying $\sum_{k=0}^\infty |\alpha_k|^2/|\lambda_k|^2 < \infty$ then $\lim_{n\to\infty} U^{(n)}(x) \stackrel{q.m.}{=} U(x)$ is a quadratic mean continuous process solving

$$LU(\cdot) = W(\cdot), \quad W(\cdot) \stackrel{q.m.}{=} \lim W^{(n)}(\cdot), \quad (9.105)$$

with covariance operator $K_U = \sum_{k=0}^\infty (|\alpha_k|^2/|\lambda_k|^2)\phi_k \langle \phi_k, \cdot \rangle$.

Proof For every n and $j = 0, 1, \ldots, n$,

$$\lambda_j U_j = \int_X \phi_j^*(x) LU^{(n)}(x)\,dx = \int_X \phi_j^*(x) W^{(n)}(x)\,dx \quad (9.106)$$

$$= W_j. \quad (9.107)$$

Thus, $U_j = W_j/\lambda_j$ are independent Gaussian variates with variance $|\alpha_j|^2/|\lambda_j|^2$ and $U^{(n)}$ is an n-dimensional Gaussian field since inner products with test functions $\langle f, U^{(n)} \rangle = \sum_{k=0}^n W_k/\lambda_k \langle f, \phi_k \rangle$ are Gaussian.

With the trace class assumption $\sum_{k=0}^\infty |\alpha_k|^2/|\lambda_k|^2 < \infty$ holds, then $U^{(n)}$ converges in quadratic mean since $E\|\sum_{k=n+1}^\infty U_k \phi_k\|^2 \to 0$; call the limit U satisfying $LU = W$, examine $LU_n - W$ as $n \to \infty$. Since W is trace class then $\sum_k |\alpha_k|^2 < \infty$ implies

$$E\|LU_n - W\|^2 = \sum_{k=n+1}^\infty |\alpha_k|^2 \stackrel{(a)}{\to} 0 \quad \text{as } n \to \infty, \quad (9.108)$$

with (a) the finite trace condition on W.

That U is a quadratic mean continuous process follows from the continuity of the eigenfunctions with the covariance definition K_U given by for all functions f,

$$\langle f, K_U f \rangle = E|\langle U, \phi \rangle|^2 \quad (9.109)$$

$$= \sum_{k=0}^\infty \frac{|\alpha_k|^2}{|\lambda_k|^2} \langle \phi_k, f \rangle^2 = \left(f, \sum_{k=0}^\infty \frac{|\alpha_k|^2}{|\lambda_k|^2} \phi_k \langle \phi_k, f \rangle \right). \quad (9.110)$$

□

---

[27] Such $W^{(n)}$ (finite n) could be called "bandlimited" since they are in the span of a finite number of "frequencies" ϕ_k.

9.6 Gaussian Fields in the Unit Cube

Now examine 3-valued vector fields on the unit cube $[0,1]^3$, with the covariance and random structure viewed as arising through solutions to the stochastic equation of the type $LU(x) = W(x)$, L a cyclic, shift invariant (constant coefficient) linear differential operator.

Consider the vector valued functions $f : X \to \mathbb{R}^3$ made into a Hilbert space $L^2(X)$ with the inner product $\langle f, g \rangle_{(L^2(X))^3} = \sum_{i=1}^{3} \int_X f_i(x) g_i(x)\, dx$. Various operators arise and have been used by the community, including 3D small deformation elasticity operator, $L = -a\nabla^2 - b\nabla\nabla \cdot + cI$ [220–222], as well as the bi-harmonic (describing small deformations energetics of thin plates) [223,224], Laplacian [225], etc. Since cyclic shift invariant operators (also termed cyclo-stationary) on the unit-cube are assumed, they have eigenfunctions arising from complex exponentials. With $x = (x_1, x_2, x_3) \in X = [0,1]^3$, the complementary variable becomes $\omega_k = (\omega_{k_1}, \omega_{k_2}, \omega_{k_3})$, $\omega_{k_i} = 2\pi k_i$, $i = 1, 2, 3$, and the Fourier basis for periodic functions on $[0,1]^3$ takes the form $e^{j\langle \omega_k, x \rangle}$, $\langle \omega_k, x \rangle = \sum_{i=1}^{3} \omega_{k_i} x_i$.

Since the random U-fields are vector valued, for each ω_k there will be corresponding three orthogonal eigenfunctions. It leads to a natural indexing of the eigenfunctions and eigenvalues according to $\{\phi_k^{(i)}, \lambda_k^{(i)}, i = 1, 2, 3\}$.

The general form for the eigenfunctions and eigenvalues can be derived for general linear differential operators which determines the covariance of the resulting Gaussian process.

Theorem 9.45 *Let L be a non-singular normal shift invariant (constant coefficient) linear differential operator on $X = [0,1]^3$ with eigenelements $\{\lambda_k^{(i)}, \phi_k^{(i)}\}$ satisfying $L\phi_k^{(i)} = \lambda_k^{(i)} \phi_k^{(i)}$ and of the form*

$$L = \begin{pmatrix} A_{11} & A_{12} & A_{13} \\ A_{21} & A_{22} & A_{23} \\ A_{31} & A_{32} & A_{33} \end{pmatrix}, \qquad (9.111)$$

where $A_{jl} = \sum_{m=1}^{n_{jl}} a_{jl}(m) (\partial^{p_{jl}(m)}) \Big/ (\partial x_1^{p_{jl}^{(1)}(m)} \partial x_2^{p_{jl}^{(2)}(m)} \partial x_3^{p_{jl}^{(3)}(m)})$ *and* $p_{jl}(m) = \sum_{i=1}^{3} p_{jl}^{(i)}(m)$.

1. *The eigenelements* $\left\{ \lambda_k^{(i)}, e^{j\langle \omega_k, x \rangle} \begin{pmatrix} c_{k1}^{(i)} \\ c_{k2}^{(i)} \\ c_{k3}^{(i)} \end{pmatrix} \right\}$ *with normalizing constants* $\|c_k\|^2 = 1$ *satisfy*

$$\lambda_k^{(i)} \begin{pmatrix} c_{k1}^{(i)} \\ c_{k2}^{(i)} \\ c_{k3}^{(i)} \end{pmatrix} = \begin{pmatrix} \mathcal{A}_{11}(\omega_k) & \mathcal{A}_{12}(\omega_k) & \mathcal{A}_{13}(\omega_k) \\ \mathcal{A}_{21}(\omega_k) & \mathcal{A}_{22}(\omega_k) & \mathcal{A}_{23}(\omega_k) \\ \mathcal{A}_{31}(\omega_k) & \mathcal{A}_{32}(\omega_k) & \mathcal{A}_{33}(\omega_k) \end{pmatrix} \begin{pmatrix} c_{k1}^{(i)} \\ c_{k2}^{(i)} \\ c_{k3}^{(i)} \end{pmatrix}, \quad i = 1, 2, 3,$$

$$(9.112)$$

$$\mathcal{A}_{jl}(\omega_k) = \sum_{m=1}^{n_{jl}} a_{jl}(m) (j\omega_{k_1})^{p_{jl}^{(1)}(m)} (j\omega_{k_2})^{p_{jl}^{(2)}(m)} (j\omega_{k_3})^{p_{jl}^{(3)}(m)}, \quad j, l = 1, 2, 3.$$

$$(9.113)$$

2. Let $W(x), x \in X$ be a random Gaussian process with covariance operator $K_W = \sum_{k=0}^{\infty} \sum_{i=1}^{3} |\alpha_k^{(i)}|^2 \phi_k^{(i)} \langle \phi_k^{(i)*}, \cdot \rangle$. Then if L is such that $\sum_{k=0}^{\infty} \sum_{i=1}^{3} \left(\frac{|\alpha_k^{(i)}|^2}{|\lambda_k^{(i)}|^2} \right) < \infty$ the solution U of the random equation

$$LU(x) = W(x) \tag{9.114}$$

is a quadratic mean continuous Gaussian process with orthogonal expansion

$$U(x) \stackrel{q.m.}{=} \sum_{k=0}^{\infty} \sum_{i=1}^{3} U_k^{(i)} \phi_k^{(i)}(x). \tag{9.115}$$

$U_k^{(i)}$ zero-mean orthogonal Gaussian variates, variances $\left(\frac{|\alpha_k^{(i)}|^2}{|\lambda_k^{(i)}|^2} \right)$, with $U(\cdot)$ having covariance operator

$$K_U = \sum_{k=0}^{\infty} \sum_{i=1}^{3} \frac{|\alpha_k^{(i)}|^2}{|\lambda_k^{(i)}|^2} \phi_k^{(i)} \langle \phi_k^{(i)}, \cdot \rangle. \tag{9.116}$$

Proof Applying $L\phi_k^{(i)} = \lambda_k^{(i)} \phi_k^{(i)}$ gives Eqn. 9.112. That the quadratic mean limit exists follows from the trace class assumption $\sum_k \sum_{i=1}^{3} \left(\frac{|\alpha_k^{(i)}|^2}{|\lambda_k^{(i)}|^2} \right) < \infty$; the rest follows from Theorem 9.44 and Corollary 9.43 above with U satisfying $LU = W$. □

Remark 9.6.2 (Real expansions) In general the eigenfunctions are complex. For expansions of real-valued fields, $U_k = U_{-k}^*$, so that the quadratic mean expansions reduce to a real expansion in sines and cosines. The reduced set of real expansion functions correspond to $\phi_k + \phi_k^*, -j(\phi_k - \phi_k^*)$, with eigenvalues $L(\phi_k + \phi_k^*) = \lambda_k + \lambda_k^*$, and $L(\phi_k - \phi_k^*) = (\lambda_k - \lambda_k^*)$.

Example 9.46 (Boundary Conditions) For small deformation elasticity Christensen [221, 226] has used

$$L = -a\nabla^2 - b\nabla\nabla + \epsilon \text{ id}, \tag{9.117}$$

where ∇^2 and ∇ are the Laplacian and divergence operators, $\nabla^2 = (\partial^2/\partial x_1^2) + (\partial^2/\partial x_2^2) + (\partial^2/\partial x_3^2)$, $\nabla = (\partial/\partial x_1, \partial/\partial x_2, \partial/\partial x_2)^*$, id is the identity operator, and a, b in terms of the the Lame elasticity constants are $a = \mu_0, b = \lambda_0 + \mu_0$. The particular mixed boundary conditions which have been studied correspond to the Von-Neuman and Dirichlet boundary conditions (first used by Amit et al. [225]) mapping the boundary to the boundary with sliding along the sides of the cube.

Amit, Grenander, Piccioni For the 2D square domain $X = [0,1]^2$, examine the operator, let $u = (u_1, u_2)^*$ and consider the Laplacian operator $Lu = \nabla^2 u = (\partial^2/\partial x_1^2)u + (\partial^2/\partial x_2^2)u$ satisfying the boundary conditions:

$$u_1(0, x_2) = \partial u_1/\partial x_1(0, x_2) = u_1(1, x_2) = \partial u_1/\partial x_1(1, x_2) = 0$$
$$u_2(x_1, 0) = \partial u_2/\partial x_2(x_1, 0) = u_2(x_1, 1) = \partial u_2/\partial x_2(x_1, 1) = 0. \tag{9.118}$$

Classical Fourier analysis suggests eigenfunctions can be expressed as a linear combination of $\sin i\pi x_1 \sin j\pi x_2$, $\sin i\pi x_1 \cos j\pi x_2$, $\cos i\pi x_1 \sin j\pi x_2$ and $\cos i\pi x_1 \cos j\pi x_2$ where $i, j = 0, 1, \ldots$. Inspection of the boundary conditions suggest the two eigenvectors are

$$\phi_{i,j}^{(1)} = (\sin i\pi x_1 \cos j\pi x_2, 0)^*, \quad \phi_{i,j}^{(2)} = (0, \cos i\pi x_1 \sin j\pi x_2)^*. \qquad (9.119)$$

It is easy to show $L\phi_{i,j}^{(k)} = \lambda_{i,j}\phi_{i,j}^{(k)}$ where $\lambda_{i,j} = -\pi^2(i^2 + j^2)$.

Christensen: Laplacian Any integer multiple of the above eigenfunctions is also an eigenfunction. Thus we can have

$$\phi_{i,j}^{(1)} = (i \sin i\pi x_1 \cos j\pi x_2, 0)^*, \quad \phi_{i,j}^{(2)} = (0, i \cos i\pi x_1 \sin j\pi x_2)^*. \qquad (9.120)$$

It is easy to show $L\phi_{i,j}^{(k)} = \lambda_{i,j}\phi_{i,j}^{(k)}$ where $\lambda_{i,j} = -\pi^2(i^2 + j^2)$.

Christensen: Cauchy-Navier Let $\nabla = \left(\dfrac{\partial}{\partial x_1}, \dfrac{\partial}{\partial x_2}\right)^*$. Now consider the operator $L = \nabla\nabla$. So Lu becomes

$$\begin{pmatrix} \dfrac{\partial^2 u_1}{\partial x_1^2} + \dfrac{\partial^2 u_2}{\partial x_1 \partial x_2} \\ \dfrac{\partial^2 u_2}{\partial x_2^2} + \dfrac{\partial^2 u_1}{\partial x_1 \partial x_2} \end{pmatrix}. \qquad (9.121)$$

Using Christensen's idea for the Laplacian, the eigenvectors are

$$\phi_{i,j}^{(1)} = (i \sin i\pi x_1 \cos j\pi x_2, j \cos i\pi x_1 \sin j\pi x_2)^*,$$

$$\phi_{i,j}^{(2)} = (-j \cos i\pi x_1 \sin j\pi x_2, i \cos i\pi x_1 \sin j\pi x_2)^*. \qquad (9.122)$$

The trick is to note that the cross-derivatives involving the other component suggest that the component of the other eigenvector must play a role in satisfying $L\phi_{i,j}^{(k)} = \lambda_{i,j}\phi_{i,j}^{(k)}$ where $\lambda_{i,j} = -\pi^2(i^2 + j^2)$. The same result is achieved for $L = \nabla^2 + b\nabla\nabla$.

Example 9.47 (1D case) Examine the 1D setting $U(t), t \in [0,1]$ satisfying

$$LU(t) = W(t), \quad t \in [0,1]. \qquad (9.123)$$

Choosing L a constant coefficient differential operator with circulant boundary conditions, and $W(\cdot)$ cyclo-stationary Gaussian white noise process with covariance $K_W(t,s) = \sum_k e^{j2\pi k(t-s)}$, gives U cyclo-stationary with spectrum induced by the structure of the differential operator.

Let $L = \sum_{m=0}^{p} a_m (\partial^m/\partial t^m)$ with circulant bondary conditions be invertible with eigenvalues non-zero: $\forall k \in \mathbb{Z}, \sum_{m=0}^{p} a_m (j2\pi k)^m \neq 0$. Then $U(t), t \in [0,1]$ solving Eqn. 9.123 is a Gaussian process with Green's function and covariance

$$G(t,s) = \sum_{k=-\infty}^{\infty} \frac{1}{\sum_{m=0}^{p} a_m (j2\pi k)^m} e^{j2\pi k(t-s)}, \qquad (9.124)$$

$$K_U(t,s) = G^* K_W G(t,s) = \sum_{k=-\infty}^{\infty} \frac{1}{|\sum_{m=0}^{p} a_m (j2\pi k)^m|^2} e^{j2\pi k(t-s)}. \qquad (9.125)$$

9.6 GAUSSIAN FIELDS IN THE UNIT CUBE

Example 9.48 (Laplacian circulant boundary) Let L be the Laplacian operator over $t \in [0,1]$, $L = -(\partial^2/\partial t^2) + c$, with random process $(-\partial^2 U(t)/\partial t^2) + cU(t) = W(t)$.

The eigenfunctions $\phi_k(t) = e^{j2\pi kt}$ have eigenvalues $\lambda_k = (2\pi k)^2 + c$; notice the operator is invertible for $c > 0$. Then $L = L^*$, with Green's function and covariance

$$G(t,s) = \sum_{k=-\infty}^{\infty} \frac{1}{(2\pi k)^2 + c} e^{j2\pi k(t-s)}, \qquad (9.126)$$

$$K_U(t,s) = \int_{[0,1]} G(t,r) G^*(r,s)\, dr = \sum_{k=-\infty}^{\infty} \frac{1}{((2\pi k)^2 + c)^2} e^{j2\pi k(t-s)}. \qquad (9.127)$$

Notice, LG is the identity operator since

$$LG(t,s) = \sum_{k=-\infty}^{\infty} L \frac{1}{\lambda_k} \phi_k(t) \phi_k^*(s) = \sum_{k=-\infty}^{\infty} e^{j2\pi k(t-s)} = \delta(t-s). \qquad (9.128)$$

Example 9.49 (Green's Operator Laplacian, Zero Boundary) Take $t \in [0,1]$ and the Laplacian operator $L = \partial^2/\partial t^2$ defining the S.D.E. $\partial^2 U(t)/\partial t^2 = W(t)$ with boundary conditions $U(0) = U(1) = 0$ and W white noise. The Green's function is given by

$$G(t,s) = \tfrac{1}{2}|t-s| + a + b(t-s), \qquad (9.129)$$

with a, b satisfying the boundary conditions $G(0,s) = G(1,s) = 0, s \in [0,1]$.

This gives $a = s^2 - s, b = s - (1/2)$ which implies the Green's function is

$$G(t,s) = \tfrac{1}{2}|t-s| + t(s - \tfrac{1}{2}) - \tfrac{1}{2}s. \qquad (9.130)$$

Checking at $(0,s)$ or $(1,s)$ shows the boundary conditions are satisfied. To show $LG(t,s) = \delta(t-s)$, taking the first derivative gives

$$\frac{\partial G(t,s)}{\partial t} = \tfrac{1}{2}(u(t-s) - u(-t+s)) + \left(s - \tfrac{1}{2}\right), \qquad (9.131)$$

with the second derivative giving $(\partial^2 G(t,s))/(\partial t^2) = \delta(t-s)$.

The eigenfunctions and eigenvalues are $\phi_k(t) = (1/\sqrt{2}) \sin 2\pi kt$, $\lambda_k = -(2\pi k)^2$. For W a white noise field, then

$$K_U(t,s) = \sum_k \frac{1}{(2\pi k)^4} \sin 2\pi kt \, \sin 2\pi ks \qquad (9.132)$$

and

$$\frac{\partial^2 K_U(t,s)}{\partial t \partial s} = \sum_k \frac{1}{(2\pi k)^2} \cos 2\pi kt \, \cos 2\pi ks. \qquad (9.133)$$

Thus the second mixed partial derivative covariance condition exists on the diagonal, and it follows that U is q.m. differentiable by Theorem 9.13.

9.6.1 Maximum Likelihood Estimation of the Fields: Generalized ARMA Modelling

The operator L plays the role of a *pre-whitening operator*. To be empirical introduce the basic generating operator L_0, and an associated polynomial operator consisting of powers of L_0, generalizing AR modeling:

$$L = p(L_0) = a_d L_0^d + a_{d-1} L_0^{d-1} + \cdots + a_0 \text{id} \tag{9.134}$$

with the unknown parameters $a_d, a_{d-1}, \ldots a_0$ estimated from the data. Assume the set $\{U^{(1)}, \ldots, U^{(N)}\}$ of vector fields are given and are generated: $U_k^{(n)} = \langle \phi_k, U^{(n)} \rangle, n = 1, \ldots, N$. It follows that the eigenvalues are polynomials of the original eigenvalues.

Corollary 9.50 *Let $L = p(L_0) = \sum_{m=0}^{d} a_m L_0^m$ with L_0 a linear shift invariant normal invertible operator on $X = [0,1]^3$ with eigenvalues and eigenfunctions $\{\lambda_k^0, \phi_k\}$.*

Then L has identical eigenfunctions $\{\phi_k\}$ as given by Theorem 9.45, with eigenvalues satisfying

$$\lambda_k = p(\lambda_k^0) = \sum_{m=0}^{d} a_m (\lambda_k^0)^m. \tag{9.135}$$

If in turn $\sum_k \sum_{i=1}^{3} (|\alpha_k^{(i)}|^2 / |\lambda_k|^2) < \infty$, with W a Gaussian process with covariance $K_W = \sum_k \sum_{i=1}^{3} |\alpha_k^{(i)}|^2 \phi_k^{(i)} \langle \phi_k^{(i)}, \cdot \rangle$, then Theorem 9.45 is in force and $U \stackrel{q.m.}{=} \lim_{n \to \infty} \sum_{k=0}^{\infty} \sum_{i=1}^{3} U_k^{(i)} \phi_k^{(i)}$.

The maximum-likelihood estimate of the spectrum is given by $|\hat{\sigma}_k^{(i)}|^2 = |\alpha_k^{(i)}|^2 / |\hat{\lambda}_k|^2$ where $\hat{\lambda}_k = \hat{p}(\lambda_k^0) = \sum_{m=0}^{d} \hat{a}_m (\lambda_k^0)^m$ satisfying the MLE equations for each $j = 0, 1, \ldots, d$:

$$0 = \frac{\partial}{\partial a_j} \log p(U^{(1)}, \ldots, U^{(n)}; a) \tag{9.136}$$

$$= -N \sum_k \frac{(\lambda_k^0)^j}{\hat{p}(\lambda_k^0)} + \sum_k \frac{\hat{p}(\lambda_k^0)(\lambda_k^0)^j}{\alpha_k^{(i)2}} \sum_{n=1}^{N} |U_k^{(n)} - \bar{U}_k|^2 = 0. \tag{9.137}$$

Differential operators which are mixtures $L = \sum_i a_i L^{(i)}$ depend upon the parameters and to solve the maximum likelihood problem iteratively in general requires re-computation of the eigenvectors. However, if the operators commute, this will not be necessary.

Corollary 9.51 *Let $L = \sum_m a_m L^{(m)}$, $L^{(m)}$ a shift invariant normal differential operator of the form given in Theorem 9.45, with the property that each of the operators commute $L^{(m)} L^{(m')} = L^{(m')} L^{(m)}$. Then $L^{(m)}, m = 1, 2, \ldots$ have identical eigenfunctions $L^{(m)} \phi_k = \lambda_k^{(m)} \phi_k$ for all m. The eigenvalues of $L = \sum_m a_m L^{(m)}$ are $\lambda_k = \sum_m a_m \lambda_k^{(m)}$.*

The MLE equations for each $j = 0, 1, \ldots, d$

$$0 = \frac{\partial}{\partial a_j} \log p(U^{(1)}, \ldots, U^{(N)}; a) \tag{9.138}$$

$$= -N \sum_k \frac{\lambda_k^{(j)}}{\left(\sum_m \hat{a}_m \lambda_k^{(m)}\right)} + \sum_k \frac{\left(\sum_m \hat{a}_m \lambda_k^{(m)}\right) \lambda_k^{(j)}}{|\alpha_k^{(i)}|^2} \sum_{n=1}^{N} |U_k^{(n)} - \bar{U}_k|^2. \tag{9.139}$$

Proof The only thing to be proved is that the eigenfunctions are equal for the various operators $L^{(m)}, m = 1, 2, \ldots$. The eigenfunctions take the form $e^{j\langle \omega_k, x \rangle} \left(c_{k1}^{(m)}, c_{k2}^{(m)}, c_{k3}^{(m)}\right)^*$, implying that for the operators to have the same eigenfunctions it must be shown $c_k^{(m)} = c_k^{(m')}$. Since the operators $L^{(m)}, L^{(m')}$ commute, the matrices $\left(\mathcal{A}^{(m)}\right), \left(\mathcal{A}^{(m')}\right)$ from Eqn. 9.113 commute implying they have the same simple eigenvectors associated with their simple eigenvalues, and $c_k^{(m)} = c_k^{(m')}$. □

Example 9.52 (Stationary Navier Elasticity Operator) Examine the elasticity operator $L = -a\nabla^2 - b\nabla \cdot \nabla + c\,\mathrm{id}$ studied extensively [220–222]. Then the operator matrix from Theorem 9.45 has entries $A_{ii} = -a\nabla^2 - b(\partial^2/\partial x_i^2) + c$, $i = 1, 2, 3$, and $A_{il} = -b(\partial^2/\partial x_i \partial x_l), i \neq l$, with

$$\mathcal{A}_{ii}(\omega_k) = -a \sum_{m=1}^{3} (j\omega_{k_m})^2 - b(j\omega_{k_i})^2 + c \quad i = 1, 2, 3, \tag{9.140}$$

$$\mathcal{A}_{il}(\omega_k) = -b(j\omega_{k_i})(j\omega_{k_l}) \quad i \neq l. \tag{9.141}$$

The eigenfunctions and eigenvalues take the form

$$\phi_k^{(1)}(x) = \alpha_1 \left[\omega_{k_1} e^{j\langle \omega_k, x \rangle}, \omega_{k_2} e^{j\langle \omega_k, x \rangle}, \omega_{k_3} e^{j\langle \omega_k, x \rangle}\right]^*,$$

$$\lambda_k^{(1)} = (2a+b)\left(\omega_{k_1}^2 + \omega_{k_2}^2 + \omega_{k_3}^2\right) + c,$$

$$\phi_k^{(2)}(x) = \alpha_2 \left[-\omega_{k_2} e^{j\langle \omega_k, x \rangle}, \omega_{k_1} e^{j\langle \omega_k, x \rangle}, 0\right]^*, \quad \lambda_k^{(2)} = a\left(\omega_{k_1}^2 + \omega_{k_2}^2 + \omega_{k_3}^2\right) + c,$$
$$\tag{9.142}$$

$$\phi_k^{(3)}(x) = \alpha_3 \left[\omega_{k_1} \omega_{k_3} e^{j\langle \omega_k, x \rangle}, \omega_{k_2} \omega_{k_3} e^{j\langle \omega_k, x \rangle}, -\left(\omega_{k_1}^2 + \omega_{k_2}^2\right) e^{j\langle \omega_k, x \rangle}\right]^*,$$

$$\lambda_k^{(3)} = a\left(\omega_{k_1}^2 + \omega_{k_2}^2 + \omega_{k_3}^2\right) + c,$$

with the coefficients α scaling each eigenvector to unit energy

$$\alpha^1 = \sqrt{1/(\omega_{k_1}^2 + \omega_{k_2}^2 + \omega_{k_3}^2)}, \quad \alpha^2 = \sqrt{1/(\omega_{k_1}^2 + \omega_{k_2}^2)},$$

$$\alpha^3 = \sqrt{1/((\omega_{k_1}^2 + \omega_{k_2}^2)(\omega_{k_1}^2 + \omega_{k_2}^2 + \omega_{k_3}^2))}. \tag{9.143}$$

Since the operator is self-adjoint, $\lambda_k = \lambda_{-k}$, the real eigenelements become $\{\phi_k^{(i)} + \phi_{-k}^{(i)}, 2\lambda_k^{(i)}\}$. Now consider the maximum-likelihood estimation of the parameters associated with the linear differential operator $L = -aL^A - bL^B + c\,\mathrm{id}$, with $L^A U = \nabla^2 U$, $L^B U = \nabla \nabla \cdot U$. Examine the method for calculating the eigenvalues. L depends upon parameters and to solve the maximization problem iteratively would require re-computation of the eigenfunctions. This is not necessary due to the fact that

L^A, L^B, id are normal and commute implying they have identical eigenfunctions. Notice the eigenfunctions are independent of a, b, c as predicted by Corollary 9.51.

Denote the eigenvalues of A and B by $\{\lambda_k^{(i)A}\}$ and $\{\lambda_k^{(i)B}\}$, then the eigenvalues just add because of the commutativity, Corollary 9.51,

$$\lambda_k^{(i)} = (a\lambda_k^{(i)A} + b\lambda_k^{(i)B} + c). \tag{9.144}$$

In this case

$$\lambda_k^{(1)A} = -(\omega_{k_1}^2 + \omega_{k_2}^2 + \omega_{k_3}^2), \quad \lambda_k^{(2)A} = -(\omega_{k_1}^2 + \omega_{k_2}^2 + \omega_{k_3}^2),$$
$$\lambda_k^{(3)A} = -(\omega_{k_1}^2 + \omega_{k_2}^2 + \omega_{k_3}^2), \quad \lambda_k^{(1)B} = -(\omega_{k_1}^2 + \omega_{k_2}^2 + \omega_{k_3}^2),$$
$$\lambda_k^{(2)B} = 0, \quad \lambda_k^{(3)B} = 0, \quad \lambda_k^{(1)C} = 1, \quad \lambda_k^{(2)C} = 1, \quad \lambda_k^{(3)C} = 1.$$

9.6.2 Small Deformation Vector Fields Models in the Plane and Cube

Now we examine a problem from Computational Anatomy corresponding to small deformations in which the Gaussian random fields act as deformation of the underlying coordinate system, so that $U : x \mapsto x - U(x)$. Chapter 16 addresses this problem in great detail. Examine the unit cube for the background space $X = [0,1]^3$. Since only finite many anatomies can be observed, the class of covariances may be restricted using symmetry properties associated with the physical deformation of the tissues. For the random models, the covariance and random structure is viewed as arising through the fact that $\{U(x), x \in X\}$ is thought to be the solution of a stochastic PDE of the type,

$$L U(x) = W(x), \tag{9.145}$$

$\{W(x), x \in X\}$ a Gaussian random process with covariance K_W. Assume the noise W is "white" with covariance I, I the identity operator. Herein we focus on the Laplacian and elasticity operator, arising via a continuum mechanics construction, and corresponding to various mixtures of differential operators. As well, various other forms arise, including the bi-harmonic (describing small deformations energetics of thin plates), Laplacian, etc.

To illustrate the use of the differential operator examine how it describes shape change corresponding to the above equation. Say that a planar anatomy is shrinking or expanding, then the group of transformations $\mathcal{G}$ take the form $\mathcal{G} : x = (x_1, x_2) \to x - (U_1(x), U_2(x))$. Introduce the dilatation vector $d = (d_1, d_2) \in \mathbb{R}^2$, then an anatomical shift is generated via the solution of the random equation $LU = W$, with (L^{H_0}, W^{H_0}) associated with normal, and (L^{H_1}, W^{H_1}) associated with disease. Under the two hypotheses π_0, π_1 correspond to the stochastic equations

$$H_0 : (\nabla^2 U_j)(x_1, x_2) = W_j(x_1, x_2), \quad j = 1, 2, \tag{9.146}$$

$$H_1 : (\nabla^2 U_j)(x_1, x_2) = W_j(x_1, x_2) + d_1, \quad j = 1, 2, \tag{9.147}$$

where $W(\cdot, \cdot)$ is noise and assume cyclic boundary conditions throughout. Then under the two hypotheses, the log-priors take the form

$$H_0 : \log \pi \simeq -\sum_{i=1}^{2} \|\nabla^2 U_i\|^2 \quad H_1 : \log \pi \simeq -\sum_{i=1}^{2} \|\nabla^2 U_i - d_i\|^2, \text{ with} \tag{9.148}$$

$\nabla^2 U_i(x_1, x_2) = U_i(x_1+1, x_2) - 2U_i(x_1, x_2) + U_i(x_1-1, x_2) + U_i(x_1, x_2+1) - 2U_i(x_1, x_2) + U_i(x_1, x_2-1)$.

A slight modification of this would be to allow the vectors d_1, d_2 to be space dependent, for example pointing outwards or inwards from a center of the abnormality.

To illustrate such shifts, shown in the left panel of Figure 9.2 are deformations on an MRI section from a human brain for various choices of the dilatation vector. The top left panel (a) shows the original image with the area of dilatation depicted; the top right panel (b) shows a contracting field $d_1 < 0$ and $d_2 < 0$; the bottom left panel (c) shows an expanding field $d_1 > 0$ and $d_2 > 0$; the bottom right panel (d) shows a shearing field $d_1 < 0$ and $d_2 > 0$. The right column shows similar results for the ventricles.

Example 9.53 (**Maximum-Likelihood Estimation of Elasticity Parameterized Gaussian Vector Fields**) To associate the orthogonal expansion of the Gaussian process, Eqn. 17.1, with the stochastic PDE of Eqn. 9.145, choose $\{\phi, \lambda\}$ the eigenelements of the variability operator $L = -a\nabla^2 - b\nabla \cdot \nabla + cI$, according to

$$L\phi_k(x) = \lambda_k \phi_k(x). \tag{9.149}$$

Then with the random variables $\{U_k^{(d)}, k = 1, 2, \dots\}$ orthogonal Gaussian random variables with mean and variance $EU_k^{(d)} = \bar{U}_k^{(d)}$, $E|U_k^{(d)} - \bar{U}_k^{(d)}|^2 = 1/|\lambda_k^{(d)}|^2$ then $U(x) \stackrel{q.m.}{=} \sum_{k=0}^{\infty} \sum_{d=1}^{3} U_k^{(d)} \phi_k^{(d)}(x)$ is a quadratic mean Gaussian process satisfying $LU(x) = E(x)$, with mean and covariance

$$\bar{U}(x) = \sum_{k=0}^{\infty} \sum_{d=1}^{3} \bar{U}_k^{(d)} \phi_k^{(d)}(x), \quad K_U(x,y) = \sum_{k=0}^{\infty} \sum_{d=1}^{3} \frac{1}{|\lambda_k^{(d)}|^2} \phi_k^{(d)}(x) \phi_k^{(d)T}(y). \tag{9.150}$$

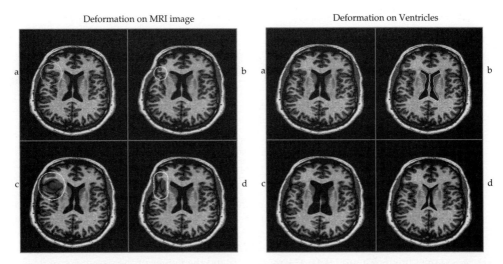

Figure 9.2 Figures show deformations corresponding to the solution of the random equation, $LU = W$. Left figure: Top left panel (a) shows the original image with the area of dilatation depicted; the top right panel (b) shows a contracting field $d_1 < 0$ and $d_2 < 0$; the bottom left panel (c) shows an expanding field $d_1 > 0$ and $d_2 > 0$; the bottom right panel (d) shows a shearing field $d_1 < 0$ and $d_2 > 0$. Right figure shows analogous deformations to the ventricles (see also Plate 13).

Let $u_k^{(n)} = \langle u^{(n)}, \phi_k \rangle$ be the coefficients generated from the nth map, $n = 1, 2, \ldots,$ then the log-likelihood written in the K-dimensional basis becomes

$$\ell(a,b,c,\bar{U};u^{(1)},\ldots,u^{(N)}) = -\frac{1}{2}\sum_{k=0}^{K}\sum_{d=1}^{3}\log\left(\frac{2\pi}{|\lambda_k^{(d)}|^2}\right) - \frac{1}{2}\sum_{k=0}^{K}\sum_{d=1}^{3}|\lambda_k^{(d)}|^2$$

$$\times \left|\frac{1}{N}\sum_{n=1}^{N}u_k^{(d)}(n) - \bar{U}_k^{(d)}\right|^2. \qquad (9.151)$$

The elasticity coefficients and other parameters parameterize the eigenvalues λ_k. The extremum conditions become

$$(\hat{a},\hat{b},\hat{c},\hat{\bar{U}}) \longleftarrow \arg\max_{a,b,c,\bar{U}_k} \ell(a,b,c,\bar{U};u^{(1)},\ldots,u^{(N)}).$$

As the maps of the various anatomies are assumed independent, then the log likelihood function is additive across the anatomical maps $\{u_n, n = 1, 2, \ldots\}$.

Parameters can be estimated from the familiy of maps modeling the random fields using the elasticity operator $L = -a\nabla^2 - b\nabla \cdot \nabla + cI$, with the particular Von-Neuman and Dirichlet mixed boundary conditions chosen mapping the boundary to the boundary [227]:

$$u_1(0,x_2,x_3) = u_1(1,x_2,x_3) = 0,$$

$$\frac{\partial u_1(x_1,0,x_3)}{\partial x_2} = \frac{\partial u_1(x_1,1,x_3)}{\partial x_2} = \frac{\partial u_1(x_1,x_2,0)}{\partial x_3} = \frac{\partial u_1(x_1,x_2,1)}{\partial x_3} = 0.$$

$$u_2(x_1,0,x_3) = u_2(x_1,1,x_3) = 0$$

$$\frac{\partial u_2(0,x_2,x_3)}{\partial x_1} = \frac{\partial u_2(1,x_2,x_3)}{\partial x_1} = \frac{\partial u_2(x_1,x_2,0)}{\partial x_3} = \frac{\partial u_2(x_1,x_2,1)}{\partial x_3} = 0. \qquad (9.152)$$

$$u_3(x_1,x_2,0) = u_3(x_1,x_2,1) = 0$$

$$\frac{\partial u_3(0,x_2,x_3)}{\partial x_1} = \frac{\partial u_3(1,x_2,x_3)}{\partial x_1} = \frac{\partial u_3(x_1,0,x_3)}{\partial x_2} = \frac{\partial u_3(x_1,1,x_3)}{\partial x_2} = 0.$$

The eigenvectors and eigenvalues of the elasticity operator L are mixtures of sines and cosines (see example 9.46, Chapter 9 or [226]). The empirical estimation of the variability was performed on cryosection monkey brain volumes. The three brains used in this experiment were labeled 87A, 90C and 93G. 87A was mapped to both the target brains 90C and 93G. The means $\{\bar{U}\}$ and the values of the parameters a, b were estimated from these empirical maps. The basis coefficients were generated determining the displacement fields so as to correspond to the elasticity coefficients (a, b) estimated from the population. Random brains were generated from the empirical template with these displacement vectors. Figure 9.3 shows three section (left to right) in three randomly generated brains (top to bottom).

9.6 GAUSSIAN FIELDS IN THE UNIT CUBE

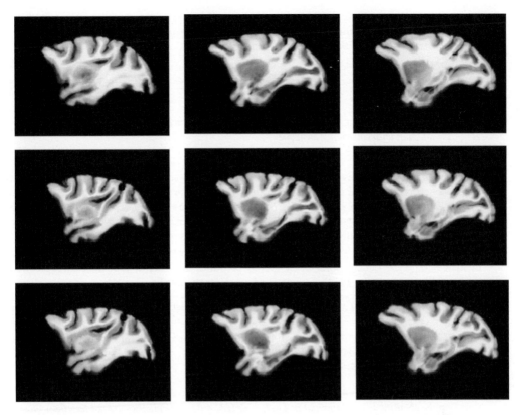

Figure 9.3 Three sections for each of three randomly generated macaque monkey brains.

9.7 Discrete Lattices and Reachability of Cyclo-Stationary Spectra

While the representations are formulated on the continuum, the computations are on the finite lattices associated with the discrete images. Similar results hold as above. To illuminate, restrict for convenience to the class of discrete operators with periodic boundary conditions. The background space becomes a discrete 3-torus, $X = \mathbb{Z}_{N^3}$ and the operators are difference operators which are cyclo-stationary in the sense that addition is done modulo N.

The eigenfunctions and eigenvalues are of the same form as in Theorem 9.45, but with the constants given by the discrete Fourier series of the finite difference coefficients.

Corollary 9.54 *Let the discrete shift invariant (cyclo-stationary) normal invertible operators L be of the form*

$$LU(n) = \begin{pmatrix} LU_1(n) \\ LU_2(n) \\ LU_3(n) \end{pmatrix} = \begin{pmatrix} \sum_{l=1}^{3} \sum_{h \in H} a_{1l}(h) U_l(n+h) \\ \sum_{l=1}^{3} \sum_{h \in H} a_{2l}(h) U_l(n+h) \\ \sum_{l=1}^{3} \sum_{h \in H} a_{3l}(h) U_l(n+h) \end{pmatrix},$$

$$n = (n_1, n_2, n_3) \in \{0, 1, \ldots, N-1\}^3, \tag{9.153}$$

with the operators having finite support, $H \subset \mathbb{Z}_N^3$. The eigenfunctions and eigenvalues $\{\phi_k, \lambda_k\}$ solve the matrix equation of Theorem 9.45, Eqn. 9.112, with the constants A_{jl}, the discrete Fourier transforms of the difference coefficients:

$$\mathcal{A}_{jl}(\omega_k) = \sum_{h \in H} a_{jl}(h) e^{j\langle \omega_k, h \rangle}, \quad j, l = 1, 2, 3, \quad \omega_k = \left(\frac{2\pi k_1}{N}, \frac{2\pi k_2}{N}, \frac{2\pi k_2}{N} \right). \quad (9.154)$$

Then $\{U(n), n \in \mathbb{Z}_{N^3}\}$ satisfies the random difference equation $LU(n) = W(n)$; $U(\cdot)$ is Gaussian with covariance as above.

The question arises, *how general a family of operators can be obtained from such differential or difference operators*, or equivalently, *how general a family of cyclo-stationary spectra?*. Restrict to the class $\mathcal{L}$ of discrete normal operators with periodic boundary conditions on $X = \mathbb{Z}_{N^3}$. Examine operators generated from polynomials of normal operators L_0 as $\mathcal{L}(L_0) \subset \mathcal{L}$. It will be shown that if an operator has lower multiplicities in its eigenspectrum than the generating operator L_0 it cannot be obtained as a polynomial from L_0, otherwise it is reachable.

The operators commute with identical eigenvectors $\phi_k = \phi_{k_1 k_2 k_3}$ and eigenvalues of L_0 given by $\lambda_k, k = (k_1, k_2, k_3) \in \mathbb{Z}_N^3$. Some of them may be multiple, which will be the case typically if the operator has symmetries. Let the corresponding multiplicities be

$$m_k = m_{k_1 k_2 k_3}, \quad k_1, k_2, k_3 = 1, 2, \ldots, N. \quad (9.155)$$

Now, if L_0 has the eigenvalues $(2,2,3,3,4,4,5,5)$ so that the multiplicities are 2, then an operator $L \in \mathcal{L}$ with the eigenvalues $(1,1,5,5,4,4,3,3)$ or $(1,1,1,1,7,7,6,6)$ can be obtained, however one with eigenvalues $(1,2,9,9,8,8,7,7)$ cannot.

Theorem 9.55 *Given $L \in \mathcal{L}$, and generating operator L_0 of $\mathcal{L}(L_0)$. Then an operator $L \in \mathcal{L}$ can be obtained as a polynomial in L_0, $L \in \mathcal{L}(L_0)$, if its multiplicities are not less than those of L_0:*

$$m_k \geq m_k^0, \quad k \in \mathbb{Z}_{N^3}. \quad (9.156)$$

Proof Write the operators L_0 and L in spectral decomposition:

$$L_0 = \sum_k \lambda_k^0 P_k, \quad L = \sum_k \lambda_k P_k, \quad (9.157)$$

where P_k is the projection operator down to the sub-space spanned by ϕ_k:

$$P_k = \phi_k \langle \phi_k, \cdot \rangle : (\ell^2(X))^3 \to \text{span}\{\phi_k\}. \quad (9.158)$$

Notice, operators being normal implies eigenfunctions are orthogonal. For any polynomial $p(\cdot)$ transforming L_0 to L, $L = p(L_0)$, we obtain the transformed eigenvalues $\lambda_k^L = p(\lambda_k^0)$. It remains to show that there is a polynomial that makes the transformed eigenvalues equal to the ones prescribed for L: $p(\lambda_k^0) = \lambda_k, k = 1, \ldots, N^3$. Given two finite sets $\Lambda^0 = \{\lambda_k^0\}, \Lambda = \{\lambda_k\}$ of numbers, the *distinct* eigenvalues of L_0 and L, respectively; therefore find a polynomial of some degree d such that the numbers in the first set are mapped into the numbers in the second. The only case when this is not possible is when $|\Lambda| > |\Lambda^0|$ but this can be ruled out if the multiplicities behave as required in the proposition. □

Essentially, this theorem is a statement of operator symmetry. The new variability operator L should not have less symmetry structure than the original one L_0. This answers definitively how general a class of auto-regressive spectra can be obtained.

Example 9.56 (1-D and 3-D Laplacian examples) Examine two examples, 1D and 3D cases. In 1-D consider the second difference operator;

$$L_0 u(n) = u(n+1) - 2u(n) + u(n-1); \quad n \in \mathbb{Z}_N. \tag{9.159}$$

The eigenvalues are then

$$\lambda_k^0 = 2\left(\cos\frac{2\pi kn}{N} - 1\right); \quad k = 0, 1 \ldots, N-1. \tag{9.160}$$

Say that N is even; the odd case is dealt with similarly. Then the multiplicities $m_k^0 = 2; \forall k$ so that if $L \in \mathcal{L}$ is to be expressed as a polynomial in L_0 it must have double eigenvalues. But the eigenvalues of any symmetric circulant matrix $L = (c_{n-m}; n, m = 1, 2 \ldots, N)$ are proportional to $\lambda_k \propto \sum_n c_n \exp^{j2\pi kn/N}$. Since the matrix is symmetric, $c_{n-m} = c_{n-m}$, the c-sequence is even and the eigenvalues appear in pairs. Hence L can be expressed as a polynomial and to use the class $\mathcal{L}(L_0)$ implies no loss of generality.

The 3-D discrete Laplacian $L_0 = \Delta$ has the eigenvalues

$$\lambda_k^0 = 2\left(\cos\frac{2\pi k_1}{N} + \cos\frac{2\pi k_2}{N} + \cos\frac{2\pi k_3}{N} - 6\right). \tag{9.161}$$

This operator has more symmetry, as not only is there the symmetry $\lambda_{-k_1 k_2 k_3}^0 = \lambda_{k_1 k_2 k_3}^0$ but the symmetries $\lambda_{k_2 k_1 k_3}^0 = \lambda_{k_1 k_2 k_3}^0$ as well, and so on. In this case the restriction to the class $\mathcal{L}(L_0)$ is essential.

9.8 Stationary Processes on the Sphere

Stationarity can be generalized to other background spaces. Examine the 2D manifold the sphere $X = S^2$, assumed centered at the origin defined in the standard azimuth-elevation representation. Here the generalization of translation to the spherical background space is rotation of points with the orthogonal group $\mathbf{SO}(3)$. Random processes which are shift invariant with respect to the orthogonal group on the sphere we shall call *isotropic or stationary on the sphere*. It will follow that all stationary processes on the sphere can be written via the spherical harmonic basis; spherical harmonics play the role of the complex exponentials in that they become the eigenfunction of covariances which are shift invariant on the sphere.

To establish the Karhunen-Loeve representation, define the inner product on the sphere in azimuth-elevation coordinates according to which

$$x(\theta, \psi) = (\cos\theta \sin\psi, \sin\theta \sin\psi, \cos\psi), \quad \theta \in [0, 2\pi), \quad \psi \in [0, \pi]. \tag{9.162}$$

The surface measure in local azimuth-elevation θ, ψ coordinates given by $dx(\theta, \psi) = \sin\theta \, d\theta \, d\psi$:

$$\langle g, h \rangle_{L^2(S^2)} = \int_0^{2\pi} \int_0^{\pi} g(x(\theta, \psi)) h(x(\theta, \psi)) \sin\psi \, d\psi \, d\theta. \tag{9.163}$$

The complete orthonormal base on the sphere is constructed from the even and odd spherical harmonics. First define the normalized Legendre functions $P_n^m, m = 1, \ldots, n$, according to

$$P_n^m(\cos\psi) = \frac{1}{2^n n!} \sin^m\psi \left(\frac{d^{n+m}((x^2-1)^n)}{dx^{n+m}}\right)\bigg|_{x=\cos\psi}, \quad \text{with } P_n(x) = P_n^0(x). \tag{9.164}$$

Then the complex spherical harmonics are defined as direct analog of the cosines and sines.

Definition 9.57 *Define the* **even, odd spherical harmonics** *to be*

for $m = 1, \ldots, n,$

$$\phi_{nm}^e(\theta, \psi) = \sqrt{((2n+1)(n-m)!)/(2\pi(n+m)!)} \; P_n^m(\cos\psi) \cos m\theta \qquad (9.165)$$

$$\phi_{nm}^o(\theta, \psi) = \sqrt{((2n+1)(n-m)!)/(2\pi(n+m)!)} \; P_n^m(\cos\psi) \sin m\theta, \qquad (9.166)$$

with $\phi_{n0}(\theta, \psi) = \sqrt{(2n+1)/4\pi} \; P_n(\cos\psi).$ \hfill (9.167)

Lemma 9.58 *Then the complex spherical harmonics defined as* ϕ_{n0} *with, for* $m = 0, 1, \ldots n, n = 1, 2, \ldots,$

$$\phi_{nm}(\theta, \psi) = \frac{1}{\sqrt{2}} \left(\phi_{nm}^e(\theta, \psi) + j\phi_{nm}^o(\theta, \psi) \right), \qquad (9.168)$$

form a complete orthonormal base.

Proof For completeness see [228]; for orthonormality within an order n, there are $n+1$ even harmonics and n odd harmonics which are orthogonal for $m \neq m'$, $\phi_{nm}, \phi_{nm'}$ since the complex exponentials are orthogonal for different frequencies $m \neq m'$ over multiple cycles,

$$\int_0^{2\pi} \phi_{nm}(\theta, \psi) \phi_{nm'}(\theta, \psi)^* \, d\theta = \delta(m - m').$$

To show orthogonality across orders $n \neq n'$, use the orthogonality of Legendre polynomials:

$$\int_0^{2\pi} \int_0^{\pi} P_n^m(\cos\psi) P_{n'}^{m'}(\cos\psi) \sin\psi \, d\psi \, d\theta = \int_{-1}^{1} P_n^m(x) P_{n'}^{m'}(x) \, dx = 0 \text{ for } n \neq n', \qquad (9.169)$$

which gives $\langle \phi_{n'}, \phi_n \rangle_{L^2(S^2)} = 0$ for $n \neq n'$. □

Figure 9.4 shows spherical harmonics visualized on the sphere. For wide-sense stationarity shift and **distance** between two points on the surface of the unit sphere $x, y \in S^2$ is defined to be the arc with cosine the angle between their vector representation in $\mathbb{R}^3$:

$$d(x, y) = \text{arc} \cos\langle \mathbf{x}, \mathbf{y} \rangle_{\mathbb{R}^3}, \qquad (9.170)$$

where $\langle \mathbf{x}, \mathbf{y} \rangle_{\mathbb{R}^3}$ is the inner product of the two vectors connecting the origin to the points x, y. From this stationarity is defined as follows.

Definition 9.59 *The random field* $\{U(\omega, x), x \in S^2\}$ *is* **isotropic or wide-sense stationary on the sphere** *if the covariance is a function of only the distance* $d(x, y)$ *between the points* x, y *on the sphere:*

$$K_U(x, y) = EU(\omega, x)U(\omega, y)^* = K_U(Ox, Oy), \quad O \in \mathbf{SO}(3). \qquad (9.171)$$

Just as the complex exponentials forming the Fourier series are eigenfunctions of the shift operator on the cube, the spherical harmonics are eigenfunctions of the shift operator on the sphere (see Theorem 9.84 below). Shift is operation by the orthogonal group. It follows that all stationary correlated processes on the sphere can be written via the spherical harmonic orthogonal expansion, the so-called orthogonal *Oboukhov expansion* [9]. For characterizing the variation of 2D manifolds, vector fields on smooth surfaces have been used [33, 229]. Associate with the surfaces real-valued random 3-vector fields $\{U(\omega, x) \in \mathbb{R}^3, x \in S^2\}$, and expand the vector fields in orthogonal blocks using the C.O.N. basis constructed from the spherical harmonics $\phi_{nm}^{(i)} = \phi_{nm}\mathbf{1}^{(i)}$, $i = 1, 2, 3$.

9.8 STATIONARY PROCESSES ON THE SPHERE

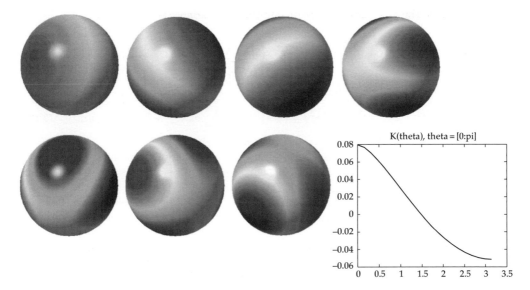

Figure 9.4 Panels 1–7 show the spherical harmonics 2–8 on the unit sphere computed numerically from the Laplacian operator on the sphere. Panel 8 shows the eigenvalues of the shift invariant covariance associated with the Laplacian operator (see also Plate 14).

Theorem 9.60 *Let the random scalar field* $\{U(\omega, x), x \in S^2\}$ *be q.m. continuous on the sphere with continuous covariance kernel* $K(x, y), x, y \in S^2$. *Then* $\{U(x), x \in S^2\}$ *is stationary on the sphere with eigenelements* $\{\sigma_{nm}^2, \phi_{nm}\}$ *the spherical harmonics and covariance kernel*

$$K_U(x,y) = \sum_{n=1}^{\infty} \sigma_n^2 \frac{2n+1}{2\pi} P_n(\cos d(x,y)) \quad \text{if and only if}$$

$$U(x) \stackrel{q.m.}{=} \lim_{N \to \infty} \sum_{n=1}^{N} \sum_{m=0}^{n} U_{nm} \phi_{nm}(x) \qquad (9.172)$$

uniformly $x \in S^2$ *with* $U_{nm} = \int_{S^2} \phi_{nm}^*(x) U(x) \, dx$ *zero-mean orthogonal random variables*

$$E U_{nm} U_{n'm'}^* = \sigma_{nm}^2 = \delta(n-n')\delta(m-m')\sigma_n^2 \sigma_n^2, \text{ with } \sum_n \sigma_n^2 < \infty.$$

Let $\{U(x) = (U_1(x), U_2(x), U_3(x))^*, x \in S^2\}$ be a vector valued random field on the sphere. Expanding the vector field in orthogonal vectors using the C.O.N. basis constructed from the spherical harmonics $\phi_{nm}^{(i)} = \phi_{nm} \mathbf{1}^{(i)}$, $i = 1, 2, 3$, then the q.m. representation

$$U(x) \stackrel{q.m.}{=} \sum_{n=1}^{\infty} \sum_{m=0}^{n} \sum_{i=1}^{3} U_{nm}^{(i)} \phi_{nm}^{(i)}(x) \qquad (9.173)$$

with the orthogonal vectors $\{U_{nm} = (U_{nm}^{(1)}, U_{nm}^{(2)}, U_{nm}^{(3)})^*\}$ *is stationary on the sphere.*

Proof The quadratic mean continuity corresponds to the summability condition $\sum_n \sigma_n^2 < \infty$ with Karhunen–Loeve holding since the Legendre polynomials in the eigenfunctions are continuous on the interval $[-1, 1]$.

The orthogonal expansion implies the covariance is of the form

$$K_U(x,y) = EU(x)U^*(y) = E \lim_{N,N'\to\infty} \sum_{n=1}^{N}\sum_{n'=1}^{N'}\sum_{m=0}^{n} U_{nm}\phi_{nm}(x) \sum_{m'=0}^{n'} U_{n'm'}\phi_{n'm'}(y) \tag{9.174}$$

$$= \sum_{n=1}^{\infty}\sum_{m=0}^{n} \sigma_n^2 \phi_{nm}(x)\phi_{nm}^*(y) \stackrel{(a)}{=} \sum_{n=1}^{\infty} \sigma_n^2 \frac{2n+1}{2\pi} P_n(\cos d(x,y)), \tag{9.175}$$

with (a) given by the Addition Theorem of spherical harmonics.[28] To show stationarity, then for all rotation elements $O \in \mathbf{SO}(3)$, we must show $K_U(Ox, Oy) = K_U(x,y)$ which follows since $d(x,y) = d(Ox, Oy)$.

In part (2), let the random field $\{U(x), x \in S^2\}$ on the sphere be a stochastic process stationary with respect to the group $\mathbf{SO}(3)$ and also *stationarily correlated* so that

$$EU(x)U^*(y) = EU(Ox)U^*(Oy), \quad x,y \in S^2, \quad O \in \mathbf{SO}(3). \tag{9.177}$$

Oboukhov's representation theorem establishes the expansion of the components of the U-field in terms of the spherical harmonics $\phi_{nm}(\cdot)$ as

$$U_i(x) \stackrel{q.m.}{=} \sum_{n=0}^{\infty}\sum_{m=0}^{n} U_{nm}^{(i)} \phi_{nm}(x), \quad x \in S^2, \quad i=1,2,3, \tag{9.178}$$

with covariances $E(U_{nm}^{(i)} U_{n'm'}^{(i)}) = \delta(n-n')\delta(m-m')\sigma_n^{(i)2}$.

For an arbitrary 3-vector $a = (a_1, a_2, a_3)$ consider the scalar stochastic process on the sphere $U^a(x) = \sum_{i=1}^{3} a_i U_i(x)$. Since $U_i(\cdot), i=1,2,3$ are stationarily correlated it follows that $U^a(\cdot)$ is also stationary on the sphere:

$$EU^a(Ox)U^a(Oy) = \sum_{i,j=1}^{3} a_i a_j EU_i(Ox)U_j(Oy) \stackrel{(a)}{=} \sum_{i,j=1}^{3} a_i a_j EU_i(x)U_j(y) \tag{9.179}$$

$$= EU^a(x)U^a(y), \tag{9.180}$$

with (a) following from the stationarity of $U(\cdot)$. We can then apply the Oboukhov orthogonal representation to it giving

$$U^a(x) = \sum_{n=1}^{\infty}\sum_{m=0}^{n} U_{nm}^a \phi_{nm}(x), \quad EU_{nm}^a U_{n'm'}^a = \delta(n-n')\delta(m-m')\sigma_n^{a2}. \tag{9.181}$$

Since the $\{\phi_{nm}\}$ are orthogonal the expansion coefficients are uniquely determined; this gives $U_{nm}^a = \sum_{i=1}^{3} a_i U_{nm}^{(i)}$ and $EU_{nm}^a U_{n'm'}^a = \sum_{i,j=1}^{3} a_i a_j EU_{nm}^{(i)} U_{n'm'}^{(j)}$. Now let one of the components of a be zero, the others 1, for example $a = (1,1,0)$. Then

$$EU_{nm}^a U_{n'm'}^a = EU_{nm}^{(1)} U_{n'm'}^{(1)} + EU_{nm}^{(2)} U_{n'm'}^{(2)} + 2EU_{nm}^{(1)} U_{n'm'}^{(2)}. \tag{9.182}$$

---

[28] The *Addition Theorem* ([228], p. 268) of spherical harmonics corresponds to for any two points $x,y \in S^2$ with $d(x,y) = \arccos\langle \mathbf{x}, \mathbf{y}\rangle$ denoting the solid angle between x,y denoted as elements $\mathbb{R}^3$, then

$$\frac{2n+1}{4\pi} P_n[\cos d(x,y)] = \sum_{m=0}^{n}(\phi_{nm}^e(x)\phi_{nm}^e(y) + \phi_{nm}^o(x)\phi_{nm}^o(y)) = \frac{1}{2}\sum_{m=0}^{n}(\phi_{nm}(x)\phi_{nm}(y)^*). \tag{9.176}$$

But if $(nm) \neq (n'm')$, the left-hand side vanishes as well as the two first terms on the right-hand side. Thus we have shown the orthogonality of $U_{nm}^{(1)}, U_{n'm'}^{(2)}$, and in the same way we show orthogonality for any pair $U_{nm}^{(i_1)}, U_{n'm'}^{(i_2)}; i_1 \neq i_2$.

Introduce the 3×3-matrix valued spectral density

$$\Sigma_{nm} = EU_{nm}U_{nm}^* \tag{9.183}$$

of the uncorrelated random 3-vectors $U_{nm} = (U_{nm}^{(1)}, U_{nm}^{(2)}, U_{nm}^{(3)})^*$. Then we have obtained a spectral representation of a vector valued stationary process on the sphere $EU_{nm}U_{n'm'}^* = \delta(n - n')\delta(m - m')\Sigma_n$. $\square$

9.8.1 Laplacian Operator Induced Gaussian Fields on the Sphere

Examine random fields induced on the surface of the sphere through elastic shell deformations. Spherical harmonics are eigenfunctions of the Laplacian operator on the sphere (see [228], pp. 258–263) with eigenvalue $n(n+1)$:

$$\nabla^2 \phi_{nm} = \frac{1}{\sin\psi} \frac{\partial}{\partial \psi}\left[\sin\psi \frac{\partial \phi_{nm}}{\partial \psi}\right] + \frac{1}{\sin^2 \psi} \frac{\partial^2 \phi_{nm}}{\partial \theta^2} = -n(n+1)\phi_{nm}. \tag{9.184}$$

Corollary 9.61 *Then the random fields $\{U(\omega, x), x \in S^2\}$ solving*

$$LU(x) = W(x), \quad \text{where } L = \nabla^2 = \frac{1}{\sin\psi}\frac{\partial}{\partial \psi}\left[\sin\psi \frac{\partial}{\partial \psi}\right] + \frac{1}{\sin^2\psi}\frac{\partial^2}{\partial \theta^2}; \tag{9.185}$$

$W(\cdot)$ *white Gaussian noise on the sphere are quadratic mean continuous stationary on the sphere*

$$K_U(x, y) = \sum_{n=1}^{\infty} \frac{1}{(n(n+1))^2} \frac{2n+1}{2\pi} P_n(\cos d(x, y)). \tag{9.186}$$

Proof The spherical harmonics of order n, ϕ_{nm} have eigenvalues $\lambda_{nm} = n(n+1)$, implying with $W(\cdot)$ a white Gaussian process on the sphere, then from Theorem 9.43 $U(\cdot)$ is Gaussian with covariance kernel

$$K_U(x, y) = \sum_{n=1}^{\infty} \sum_{m=0}^{n} \underbrace{\frac{1}{(n(n+1))^2}}_{\sigma_{nm}^2} \phi_{nm}(x)\phi_{nm}(y) \tag{9.187}$$

$$\stackrel{(a)}{=} \sum_{n=1}^{\infty} \frac{1}{(n(n+1))^2} \frac{2n+1}{2\pi} P_n(\cos d(x,y)), \tag{9.188}$$

where $d(x, y)$ is the solid angle between the points $x, y \in S^2$ and P_n is the Legendre polynomial (p. 325 [230]) and (a) follows via the addition Theorem Eqn. 9.176 for spherical harmonics [228]. Then the field satisfies the trace class condition. $\square$

Bakircioglu [231] uses the following algorithm for computing the covariance. Given the distance on the sphere of radius R between two points $\rho(x, y)$, then the solid angle is calculated in degrees according to $\theta(x,y) = (\rho(x,y))/(2\pi R) \times 360$ with the Legendre function in the covariance calculation, Eqn. 9.188, given by $P_n(\cos\theta(x,y))$ where $P_n(x)$ is calculated using the recursion [228]

$$P_0(x) = 1, \quad P_1(x) = x, \quad P_2(x) = \frac{1}{2}(3x^2 - 1) \tag{9.189}$$

$$nP_n(x) = x(2n-1)P_{n-1}(x) - (n-1)P_{n-2}(x). \tag{9.190}$$

Shown in the last panel 8 of Figure 9.4 is the eigen spectrum of the shift invariant associated with the Laplacian operator.

Example 9.62 (Active Deformable Surface Spheres for Amoeba) Optical-sectioning microscopy is proving to be a powerful tool for 3-D visualization of living biological specimens for the study of cell motion which plays a critical role during embryogenesis, and is crucial to a wide range of physiological processes, including embryonic development, and wound healing [232]. In computational optical-sectioning microscopy (COSM) conventional flourescence images are acquired in a series of focal planes spanning the specimen. As described in Chapter 2, Example 2.55, the optical characteristics and statistical models for various microscope/CCD/optical systems have been characterized (e.g. see [30,235]) based on Poisson statistical models for the data. To visualize *in vivo* 3D motion of cells in the slime mold *Dictyostelium discoideum*, individual cells are labeled with a fluorescent dye. The amoebae aggregate to form a multicellular mass that undergoes dramatic shape changes.

Single cells under motion are modeled as active deformable spheres by Joshi [33] generated via global translation and normal deformation of the sphere according to

$$g(\theta,\psi) = \begin{pmatrix} x_0 \\ y_0 \\ z_0 \end{pmatrix} + \begin{pmatrix} \cos\theta\sin\psi \\ \sin\theta\sin\psi \\ \cos\psi \end{pmatrix} + U(\theta,\psi) \begin{pmatrix} n_x(\theta,\psi) \\ n_y(\theta,\psi) \\ n_z(\theta,\psi) \end{pmatrix},$$

$$\psi \in [0,\pi], \quad \theta \in [0,2\pi). \tag{9.191}$$

with $n(\theta,\psi) \in \mathbb{R}^3$ the normal to the sphere at θ,ψ, and $U(\theta,\psi)$ is the random scalar field parameterizing the translation groups.

Now choose as the prior that is induced by the energetics of elastic membranes. From Poisson's equation for pressure fields acting on flat membranes, the energy $\int_{S^2} \|\nabla^2 U(x(\theta,\psi))\|^2 \, dx(\theta,\psi)$, with $dx(\theta,\psi) = \sin\theta \, d\theta \, d\psi$, denotes that which induces a stationary random field. The random scalar transformation field $U(\theta,\psi)$ expanded via the real orthogonal spherical harmonic expansion as a stationary process with covariance induced according to Corollary 9.61 by the Laplacian operator for elasticity. Then U is the quadratic mean limit

$$U(\theta,\psi) \stackrel{q.m.}{=} \lim_{N\to\infty} \sum_{n=0}^{N} \sum_{m=-n}^{n} U_{nm}\phi_{nm}(\theta,\psi), \tag{9.192}$$

where the ϕ_{nm} are spherical harmonics (even and odd) on the surface of the sphere and the U_{nm} are real valued, Gaussian random variables with $E\{U_{nm}U_{n'm'}\} = \delta(n-n')\delta(m-m')\lambda_n \propto (1/(n(n+1))^2)$.

The simplest model Joshi has explored assumes the measurements are a simple Poisson counting process $\{M(dx), x \in \Omega\}$ with intensity $\lambda(x) = \lambda_{\text{in}}, x \in D(g(\gamma))$, and

$\lambda(x) = \lambda_{\text{out}}, x \in D(g(\gamma))^c$, the log-posterior with Poisson data term and Gaussian prior, becomes

$$H(\gamma) = -\int_{D(g(\gamma))} \lambda_{\text{in}}(x)\, dx + \int_{D(g(\gamma))} \log \lambda_{\text{in}}(x) M(dx) + \alpha \sum_{mn} (n(n+1))^2 (U_{nm})^2$$

$$- \int_{D(g(\gamma))^c} \lambda_{\text{out}} x\, dx + \int_{D(g(\gamma))^c} \log \lambda_{\text{out}}(x) M(dx).$$

Then with Theorem 8.41 and Corollary 8.48 from Chapter 6, compute the Jacobian $J(\theta, \psi; \gamma) = |\det D_{\theta, \psi, \gamma} g(\gamma)|$, and compute the gradients in parameters:

$$\begin{pmatrix} \partial H(\gamma)/\partial x_0 \\ \partial H(\gamma)/\partial y_0 \\ \frac{\partial H(\gamma)}{\partial z_0} \end{pmatrix} = -\int_\theta \int_\psi (\lambda_{\text{in}}(\theta, \psi) - \lambda_{\text{out}}(\theta, \psi)) n(\theta, \psi) J(\theta, \psi)\, d\psi\, d\theta$$

$$+ \int_\theta \int_\psi \left(\log \frac{\lambda_{\text{in}}(\theta, \psi)}{\lambda_{\text{out}}(\theta, \psi)} \right) n(\theta, \psi) J(\theta, \psi) M(d\theta\, d\psi)$$

$$\frac{\partial H(\gamma)}{\partial u_{nm}} = -\int_\theta \int_\psi (\lambda_{\text{in}}(\theta, \psi) - \lambda_{\text{out}}(\theta, \psi)) \phi_{nm}(\theta, \psi) J(\theta, \psi)\, d\psi\, d\theta$$

$$- \alpha n(n+1) U_{nm}$$

$$+ \int_\theta \int_\psi \left(\log \frac{\lambda_{\text{in}}(\theta, \psi)}{\lambda_{\text{out}}(\theta, \psi)} \right) \phi_{nm}(\theta, \psi) J(\theta, \psi) M(d\theta\, d\psi),$$

where the surface, $g(\theta, \psi)$, is parameterized with the azimuth angle, θ, and the zenith angle, ψ. For optical sectioning microscopy the set of measurements is determined by the physics of data collection. The emission intensity is assumed constant intensity, λ_{in}, inside the shapes, with known fixed background intensity, λ_{out}, and due to imperfect optical focusing the measurements $\{M_j(du), u \in \mathcal{D}\}$ correspond to projections of emissions in the 3D space onto the 2D CCD detector plane $\mathcal{D}$ intensity of the jth measurement, $\mu_j(u)$, determined by the optical point-spread function focused on plane j, denoted as $p_j(u|x)$ focused on position x:

$$\mu_j(du) = \int_{\mathbb{R}^3} p_j(u|x) \lambda(x) dx, \quad \text{where} \tag{9.193}$$

$$\lambda_{\text{in}} = \lambda(x) \quad \forall x \in D(g(\gamma)), \quad \lambda_{\text{out}} = \lambda(x), \quad \forall x \in D(g(\gamma))^c. \tag{9.194}$$

The posterior density includes the normal density on the expansion coefficients in the spherical harmonic basis, giving

$$H(\gamma) = \sum_{j=1}^J \left(\int_\mathcal{D} \int_{D(g(\gamma))} p_j(y|x)(\lambda_{\text{in}} - \lambda_{\text{out}})\, dx\, dy \right) + \sum_{mn} (n(n+1))^2 (U_{nm})^2 \tag{9.195}$$

$$- \sum_{j=1}^J \left(\int_\mathcal{D} \log \left[\int_{D(g(\gamma))} p_j(y|x) \lambda_{\text{in}}\, dx + \int_{\mathcal{D}\setminus D(g(\gamma))} p_j(y|x) \lambda_{\text{out}}\, dx \right] M_j(dy) \right).$$

Now to compute the gradient with respect to parameters of the surface; analogous to the case in $\mathbb{R}^2$ for active contours the variations have the elegant property that volume integrals reduce to surface integrals around the respective shapes. Assume the one-parameter family of surfaces $f(\gamma)$ with $f = 0$ corresponding to the boundary $\partial D(g(\gamma))$ of the interior $D(g(\gamma))$.

Theorem 9.63 *The derivative of the posterior potential $\partial H(g(\gamma))/\partial \gamma$ of Eqn. (9.260) with respect to parameters $\gamma \in \{U_{nm}\}$, the spherical harmonic coefficients, becomes*

$$\frac{\partial H(\gamma)}{\partial \gamma} = -\sum_{j=1}^{J}\left(\int_{\mathcal{D}} \frac{\int_{[0,2\pi)\times[0,\pi]}(\lambda_{\text{in}} - \lambda_{\text{out}})p_j(y|x(\theta,\phi,\gamma))J(\theta,\phi,\gamma)\,ds}{\int_{D(g(\gamma))} p_j(y|x)\lambda_{\text{in}}\,dx + \int_{D\setminus D(g(\gamma))} p_j(y|x)\lambda_{\text{out}}\,dx}\,|M_j(dy)\right)$$

$$+ \int_{[0,2\pi)\times[0,\pi]}(\lambda_{\text{in}} - \lambda_{\text{out}})J(\theta,\phi,\gamma)\,ds + \frac{\partial}{\partial \gamma}\sum_{mn}(n(n+1))^2(U_{nm})^2.$$

(9.196)

The algorithm was implemented by Joshi [33, 198]. The surface of the sphere on which the deformations are defined is discretized on to a lattice to the lattice $L = \{[0,1,\cdots,N-1] \times [0,1,\cdots,N-1]\}$ in the set of discretization points

$$\left\{U(l_1,l_2) = \left[\sin\left(\frac{2\pi l_1}{N}\right)\cos\left(\frac{2\pi l_2}{N}\right), \sin\left(\frac{2\pi l_1}{N}\right)\sin\left(\frac{2\pi l_2}{N}\right), \cos\left(\frac{2\pi l_1}{N}\right)\right],\right.$$

$$\left.(l_1,l_2) \in L\right\}.$$

(9.197)

The integrals and the partial derivatives in Eqn. 9.196 are approximated on this mesh by finite sums and differences on this lattice. Data for the shape reconstruction was a 3D phantom with a "pseudopod" extension. Different contrasts were generated to evaluate the performance of the algorithm with different counting rates. Panel 1 of Figure 9.5 shows the 3D phantom section. Panels 2 and 3 in Figure 9.5 show X–Z sections through the 3-D data set from the optical sectioning microscope for the high

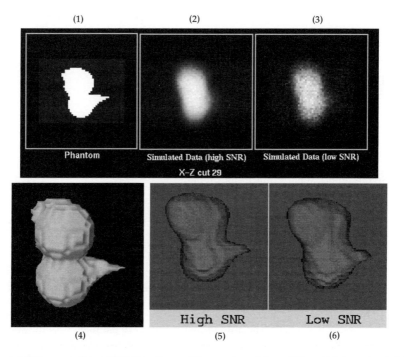

Figure 9.5 Top row shows 2D X-Z sections with pseudopod (panel 1) and high contrast data (panel 2) and low contrast data (panel 3). Bottom row (panel 4) shows 3D surface; (panel 5) shows high contrast, and (panel 6) low contrast surface reconstruction using active sphere.

contrast (panel 2) and low contrast (panel 3) data. Notice the profound blurriness of the data resulting from the optical sectioning point-spread function. Also notice the speckle effect due to the Poisson nature of the CCD noise at low contrast. Notice also that the structure of the phantom is lost due to the point-spread function and the CCD noise. The bottom row of 9.5 show the 3D object surface (panel 4) along with the reconstruction via active spheres of the high contrast (panel 5) and low contrast (panel 6) data.

9.9 Gaussian Random Fields on an Arbitrary Smooth Surface

For the past 15 years investigators have been studying patterns via orthonormal bases on smooth surfaces. Pentland and Sclaroff [236] pioneered such approaches in computer vision via deterministic approaches, and Joshi et al. [229] extended these ideas to the construction of Gaussian random fields on anatomical manifolds. Essentially the geometry of arbitrary surfaces are studied using a generalized notion of complete orthonormal bases on surfaces called, "surface harmonics". These generalize spherical harmonics to smooth 2D submanifold in $\mathbb{R}^3$. From such a complete orthonormal base, smoothing, interpolation, and estimation driven by mean fields and covariance fields can be performed completely analogously to the more regular cases in $\mathbb{R}^3$ and the sphere.

In this section we study this idea for arbitrary smooth surfaces following the work of Qiu and Bitouk by constructing complete orthonormal bases and Gaussian random fields via the Laplace Beltrami operator, the generalized version of the Laplacian operator. Define a smooth surface $\mathcal{M}$ parameterized by $u = (u^1, u^2)$. Any point on $\mathcal{M}$ has coordinates $x = \{x_1(u), x_2(u), x_3(u)\}$. Random fields are constructed as a quadratic mean limit with continuous covariance kernel on the manifold embedded into $\mathbb{R}^3$, which can be represented by a complete set of orthonormal bases of the Laplace-Beltrami operator in the Hilbert space. In the following, we shall construct the Gaussian random fields as a quadratic mean limit using a complete orthonormal bases of Laplace-Beltrami operator $\{\phi_k, k = 1, 2, \ldots\}$. The Gaussian field is defined to be the quadratic mean limit as $Y(u) \stackrel{q.m.}{=} \sum_{k=1}^{\infty} Y_k \phi_k(u)$, with mean and covariance fields

$$\bar{Y} = \sum_k \mu_k \phi_k, \quad K_Y = \sum_k \sigma_k^2 \phi_k \phi_k^*, \tag{9.198}$$

where $Y_k, k = 1, 2, \ldots$ are independent, Gaussian random variables with means and variances, $EY_k = \mu_k, E|Z_k|^2 = \sigma_k^2$.

9.9.1 Laplace-Beltrami Operator with Neumann Boundary Conditions

Qiu and Bitouk solve for the orthonormal eigenelements of the Laplace-Beltrami operator $\lambda, \phi(\cdot)$ under Neumann boundary conditions. The basis solves the eigen function problem:

$$-\Delta \phi(u) = \lambda \phi(u), \text{ subject to } \int \phi(u)^2 \, d\mathcal{M} = 1, \int_{\mathcal{M}} \phi_i(u) \phi_j(u) \, d\mathcal{M} = \delta(i-j) \tag{9.199}$$

with boundary conditions $\langle \nabla \phi(u), n \rangle$ for all $u \in |_{\partial \mathcal{M}} = 0$, where Δ is the Laplace-Beltrami operator, and n is the normal vector on the boundary of $\mathcal{M}$.

As a consequence of the divergence theorem, finding the solution to the above partial differential equation is equivalent to solving the associated variational problem.

Theorem 9.64 *The eigenfunctions of Eqn (9.199) for fixed eigenvalues λ with associated boundary conditions satisfy the variational problem*

$$E(\phi(u)) = \int_{\mathcal{M}} \|\nabla \phi(u)\|^2 \, d\mathcal{M} - \lambda \int_{\mathcal{M}} \phi(u)^2 \, d\mathcal{M}. \tag{9.200}$$

Proof For a fixed eigenvalue λ, take the variation $\phi \to \phi + \epsilon v$ according to

$$\partial_v E(\phi) = 2 \int_{\mathcal{M}} \langle \nabla \phi(u), \nabla v(u) \rangle \, d\mathcal{M} - 2\lambda \int_{\mathcal{M}} \langle \phi(u), v(u) \rangle \, d\mathcal{M} \tag{9.201}$$

$$\stackrel{(a)}{=} -2 \int_{\mathcal{M}} \Delta \phi(u) v(u) \, d\mathcal{M} - 2\lambda \int_{\mathcal{M}} \langle \phi(u), v(u) \rangle \, d\mathcal{M} = 0, \tag{9.202}$$

with (a) following from the divergence theorem and the application of the boundary conditions.

Because of convexity, it is straightforward to see that this optimization problem always has a unique solution. □

To construct the orthonormal bases on smooth surfaces of the Laplace-Beltrami operator Qiu and Bitouk use the finite element method. For this, in Figure 9.6, define three vertices in a triangle T as P_1, P_2, P_3 with coordinates $(x_1, y_1), (x_2, y_2), (x_3, y_3)$, respectively. For this define triangle T_i with area A_i. The algorithm takes the following form.

Algorithm 9.65 (Finite Element Method for Generation of the CON)

Approximation via Shape Functions. Assume the polynomial approximation for function $\phi(x, y)$ on the plane given by

$$\phi(x, y) = a + bx + cy, \tag{9.203}$$

where coefficients a, b, and c are unknown and determined by function values at these three vertices as

$$\begin{bmatrix} a \\ b \\ c \end{bmatrix} = \begin{bmatrix} 1 & x_1 & y_1 \\ 1 & x_2 & y_2 \\ 1 & x_3 & y_3 \end{bmatrix}^{-1} \begin{bmatrix} \phi(x_1, y_1) \\ \phi(x_2, y_2) \\ \phi(x_3, y_3) \end{bmatrix}. \tag{9.204}$$

The determinant of the 3×3 matrix in Eqn. 9.204 is the area of triangle T_i, denoted as A_i. The function value $\phi(x, y)$ at any point x, y within the triangle is approximated by a convex combination of weights $\phi(x, y) = \sum_{i=1}^{3} \alpha_i(x, y) \phi(x_i, y_i)$, determined by the shape functions

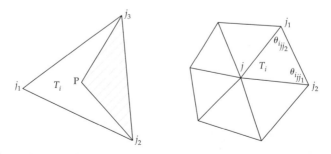

Figure 9.6 Left panel shows triangle element with vertices P_1, P_2, P_3 and arbitrary point within the triangle P. Right panel shows one ring centered at vertex j with triangle T_i having three vertices j, j_1, j_2 and angle θ_{ijj_1} and θ_{ijj_2} opposite to edges jj_1 and jj_2, respectively.

α_i satisfying $\sum_{i=1}^{3} \alpha_i(x,y) = 1$ given by the ratio of areas [237,238] of contributing triangles according to

$$\alpha_1(x,y) = \frac{1}{2A}\left[(x_2y_3 - x_3y_2) + (y_2 - y_3)x + (x_3 - x_2)y\right],$$

$$\alpha_2(x,y) = \frac{1}{2A}\left[(x_3y_1 - x_1y_3) + (y_3 - y_1)x + (x_1 - x_2)y\right], \quad (9.205)$$

$$\alpha_3(x,y) = \frac{1}{2A}\left[(x_1y_2 - x_2y_1) + (y_1 - y_2)x + (x_2 - x_1)y\right].$$

For example in the left panel of Figure 9.6 α_1 is equal to the ratio of the area of the shaded triangle to that of the total area A_i.

Within triangle T_i, the discrete energy $E_{T_i}(\phi(u))$ becomes

$$E_{T_i}(\phi(u)) = \int_{T_i} \left(\|\nabla \phi_i(u)\|^2 - \lambda \phi_i(u)^2\right) d\mathcal{M}$$

$$= \sum_{j=1}^{3}\sum_{k=1}^{3} \int_{T_i} \left[\langle \nabla \alpha_{i_j}, \nabla \alpha_{i_k}\rangle - \lambda \alpha_{i_j}\alpha_{i_k}\right]\phi(v_{i_j})\phi(v_{i_k}) dS. \quad (9.206)$$

1. **Construct Discrete Version of the Energy on each Triangle**. Simplifying each of the terms, first $\int_T \alpha_i \alpha_j dS$ becomes by change of variables [239]

$$\int_T \alpha_i \alpha_j \, dS = 2A \int_0^1 \int_0^{1-\alpha_j} \alpha_i \alpha_j \, d\alpha_i \, d\alpha_j = \begin{cases} A/6 & i=j \\ A/12 & i \neq j \end{cases},$$

and taking in partial derivatives of α_i gives

$$\int_T \langle \nabla \alpha_i, \nabla \alpha_i \rangle dS = \int_T \frac{|P_jP_k|^2}{4A^2} dS = \frac{|P_jP_k|^2}{4A} \quad (9.207)$$

$$= \frac{1}{2}(\cot \angle P_j + \cot \angle P_k), \quad (9.208)$$

and for $i \neq j \neq k$,

$$\int_T \langle \nabla \alpha_j, \nabla \alpha_k \rangle dS = \int_T \frac{\langle \overrightarrow{P_iP_j}, \overrightarrow{P_kP_j}\rangle}{4A^2} dS \quad (9.209)$$

$$= -\frac{1}{2}\cot \angle P_i. \quad (9.210)$$

2. **Matrix form of the Total Energy:** Define N_T the number of triangles on the triangulated mesh, and N_v the number of vertices. with vertex locations $v_{i_1}, v_{i_2}, v_{i_3}$, Define $N_T(j)$ as the set of triangles containing vertex j, and $N_T(i,j)$ as the set of triangles containing edge (i,j). $\theta_{i_{23}}$ denotes the angle opposite to the edge j_2j_3 in triangle T_i (shown in the left panel of Figure 9.6). Let the indices of three vertices in triangle T_i be j_1, j_2, and j_3, respectively. Assume T_i on the

plane. The left panel of Figure 9.6 shows a triangle T_i with three vertices $j_1, j_2,$ and j_3. Define the 3×3 matrices $K_i = \left(K_i(j,k) = \int_{T_i} \alpha_{i_j} \alpha_{i_k} \, dS \right)$, $G_i = \left(G_i(j,k) = \int_{T_i} \langle \nabla \alpha_{i_j}, \nabla \alpha_{i_k} \rangle dS \right)$, along with the vectors

$$\Phi_i = \begin{bmatrix} \phi(v_{i_1}) \\ \phi(v_{i_2}) \\ \phi(v_{i_3}) \end{bmatrix}^T, \quad K_i = \frac{A_i}{12} \begin{bmatrix} 2 & 1 & 1 \\ 1 & 2 & 1 \\ 1 & 1 & 2 \end{bmatrix},$$

$$G_i = \frac{1}{2} \begin{bmatrix} \cot \theta_{i_{12}} + \cot \theta_{i_{13}} & -\cot \theta_{i_{12}} & -\cot \theta_{i_{13}} \\ -\cot \theta_{i_{12}} & \cot \theta_{i_{12}} + \cot \theta_{i_{23}} & -\cot \theta_{i_{23}} \\ -\cot \theta_{i_{13}} & -\cot \theta_{i_{23}} & \cot \theta_{i_{13}} + \cot \theta_{i_{23}} \end{bmatrix}.$$

The total energy is obtained by assembling all triangle elements. The energy within each triangle becomes $E_{T_i}(\phi(u)) = \Phi_i^* G_i \Phi_i - \lambda \Phi_i^* K_i \Phi_i$ giving the total energy

$$E(\Phi) = \sum_{i=1}^{N_T} E_{T_i}(\phi(u)) = \sum_{i=1}^{N_T} \left[\Phi_{T_i}^* G_i \Phi_{T_i} - \lambda \Phi_{T_i}^* K_i \Phi_{T_i} \right]. \quad (9.211)$$

3. **Reorganize in matrix form via vertex contributions.** Define the $N_v \times N_v$ sparse matrices $G = \left(G_{ij} \right), K = \left(K_{ij} \right)$ with entries

$$G_{jj} = \frac{1}{2} \sum_{i \in N_T(j)} \left(\cot \theta_{i_{jj_1}} + \cot \theta_{i_{jj_2}} \right), \quad G_{jk} = -\frac{1}{2} \left(\sum_{i \in N_T(jk)} \cot \theta_{i_{jk}} \right) \quad (9.212)$$

$$K_{jj} = \frac{1}{6} \sum_{i \in N_T(j)} A_i, \quad K_{jk} = \frac{1}{12} \sum_{i \in N_T(jk)} A_i. \quad (9.213)$$

G is a semi-positive definite matrix. As depicted in the right panel of Figure 9.6 the jth rows of matrices G, K correspond to vertex j. K is a positive definite matrix since it is strictly diagonally dominant and diagonal entries are positive. The total energy in matrix form becomes

$$E(\Phi) = \Phi^* G \Phi - \lambda \Phi^* K \Phi. \quad (9.214)$$

4. **Solve variational problem for CON basis of eigen-elements.**
Define Σ_K a diagonal matrix with all positive eigenvalues of matrix K as entries, and V is a normal matrix with eigenfunctions as columns associated with eigenvalues. Then the eigenelements (λ, Φ_λ) subject to normalization minimizing the total energy in Eq. 9.214 is given by

$$\Phi_\lambda = \inf_{\Phi_\lambda : \Phi_\lambda^* K \Phi_\lambda = 1} \Phi_\lambda^* G \Phi_\lambda - \lambda \Phi_\lambda^* K \Phi_\lambda$$

$$= V \Sigma^{-1/2} \Psi_\lambda,$$

where Ψ_λ satisfy

$$\Sigma_K^{-1/2} V^* G V \Sigma_K^{-1/2} \Psi_\lambda = \lambda \Psi_\lambda.$$

$\Sigma_K^{-1/2} V^* G V \Sigma_K^{-1/2}$ is a semi-positive definite matrix.

Example 9.66 (Qiu and Bitouk Surface Harmonics via Laplace Beltrami) Qiu and Bitouk generate these bases on the neocortex of the brain. For this they construct the triangulated mesh for each volume using three steps: (i) segmentation of the MR tissue as white matter, gray matter, and cerebrospinal fluid (CSF) voxels using Bayesian

9.9 GAUSSIAN RANDOM FIELDS

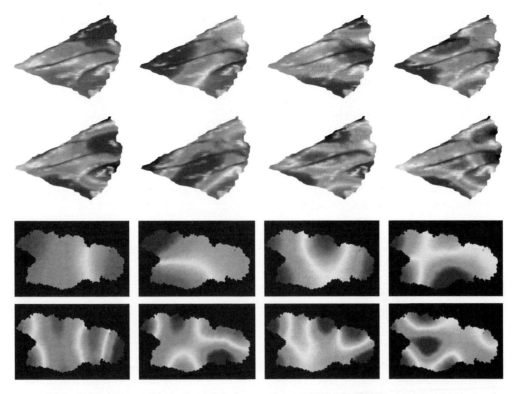

Figure 9.7 Rows 1 and 2 show surface harmonics 1–8 of the Laplace Beltrami operator on the planum temporale. Rows 3 and 4 are identical for the central sulcus. Central sulcus results are visualized via a bijection to the plane. Surface harmonics taken from Qiu and Bitouk (see also Plate 15).

segmentation [53,54], (ii) 3D isocontouring [54], and (iii) dynamic programming delineation of the boundary of the associated gyral and sulcal submanifolds, from which the submanifold of interest is automatically extracted from the surface masked by subvolume ROI [154,183],

Shown in the top two rows of Figure 9.7 are results showing the first eight surface harmonics corresponding to the eigenfunctions of the Laplace Beltrami operator computed for the planum temporale. The bottom two rows show similar results for the central sulcus. The central sulcus is a highly curved sulcus in the brian and is visualized by defining a bijection to the plane. All the surface harmonics are from Qiu and Bitouk.

9.9.2 Smoothing an Arbitrary Function on Manifolds by Orthonormal Bases of the Laplace-Beltrami Operator

The function model on the manifold $\mathcal{M}$ associated with the smoothing problem is

$$Y(u) = \bar{Y}(u) + W(u),$$

where $\tilde{Y}(u)$ is the unknown function to be estimated from the observable Y, and W is zero mean Gaussian noise with variance σ^2. Defining the Green's function

$$G(u_1, u_2) = \sum_{i=1}^{\infty} \frac{1}{\lambda_i} \phi_i(u_1) \phi_i(u_2), \tag{9.215}$$

the estimator of $\tilde{Y}$ is given by

$$\hat{\tilde{Y}} = \arg\min_{\tilde{Y}(\cdot)} \int_{\mathcal{M}} \|\nabla \tilde{Y}\|^2 \, d\mathcal{M} + \gamma \sum_{i=1}^{N_v} (Y(v_i) - \tilde{Y}(v_i))^2, \tag{9.216}$$

$$= \sum_{k=1}^{N_v} \beta_k G(u, u_k), \tag{9.217}$$

$\tilde{Y}(v_i)$ and b_i are the function value and the observational data at vertex i, respectively. Defining the N_v vector β and $N_v \times N_v$ matrix G on the vertices, then

$$\beta = \arg\min_{\beta} \beta^*(G + \gamma G^* G)\beta - \gamma \beta^* G^* Y$$

$$= \left(\frac{1}{\gamma} G + G^* G\right)^{-1} G^* Y.$$

In practice, the constant component of $Y(u)$ is removed by extracting the average value from $Y(u)$. In addition, when λ_i/γ becomes large, the corresponding bases no longer contribute to $\tilde{Y}(u)$. As for a positive function, such as cortical thickness, the logarithm of the function is smoothed.

Example 9.67 (Qiu and Bitouk) Examine the application to the expansion of curvature fields on the neocortex. Notice the first eigenvalue is close to zero, which leads to the trivial solution of the associated orthonormal basis. Shown in Figure 9.8 are the curvature profiles expanded in the complete orthonormal basis of the Laplace Beltrami operator. In practice, the constant component is excluded in the estimate of $\tilde{Y}(u)$ by extracting the average value $\int_X Y \, dx$ from $Y(u)$.

Example 9.68 Examine the Laplacian operator on the discrete triangulated graph representing a surface. Discretizing the Laplacian, assume $\mathcal{N}(x)$ consists of the six closest neighbors to x and define the operator

$$Mf(x) = n(x)[f(x) - \bar{f}(x)], \tag{9.218}$$

where $\bar{f}(x)$ is the average of f over $\mathcal{N}(x) \cap S$, i.e. $\bar{f}(x) = \sum_{y \in \mathcal{N}(x)} f(y)$ with $n(x) = \text{card}[\mathcal{N}(x)]$.

The discrete Laplacian becomes $L = M + aI$, $a > 0$ a strictly positive constant. Assume the neighborhood structure is symmetric so that $y \in \mathcal{N}(x)$ implies $x \in \mathcal{N}(y)$. Then M is self-adjoint:

$$\langle Mf, g \rangle = \sum_x n(x) f(x) g(x) - \sum_x n(x) \bar{f}(x) g(x) \tag{9.219}$$

$$= \sum_x \sum_{y \in \mathcal{N}(x)} f(x) g(x) - \sum_x \sum_{y \in \mathcal{N}(x)} f(y) g(x) \tag{9.220}$$

$$= \sum_x \sum_{y \in \mathcal{N}(x)} f(x) g(x) - \sum_y \sum_{x \in \mathcal{N}(y)} f(y) g(x). \tag{9.221}$$

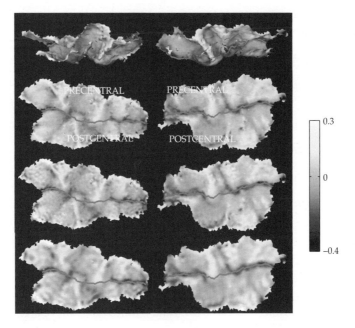

Figure 9.8 Figure shows planum temporal. Shown are the curvature profiles expanded in the complete orthonormal basis of the Laplace Beltrami operator (see also Plate 16).

Rewriting the last term using the symmetric neighborhood relation gives the self adjoint relationship

$$\langle Mf, g \rangle = \sum_x \sum_{y \in \mathcal{N}(x)} f(x)g(x) - \sum_x \sum_{y \in \mathcal{N}(x)} f(x)g(y) \tag{9.222}$$

$$= \langle f, Mg \rangle. \tag{9.223}$$

That the stochastic equation $LU = W$ has a unique solution follows from the fact that M is non-negative definite, implying $M + aI$ is positive definite. For this examine

$$\langle Mf, f \rangle = \sum_x \sum_{y \in \mathcal{N}(x)} f^2(x) - \sum_x \sum_{y \in \mathcal{N}(x)} f(x)f(y) \tag{9.224}$$

$$= \frac{1}{2} \sum_x \sum_{y \in \mathcal{N}(x)} (f(x) - f(y))^2 \geq 0. \tag{9.225}$$

Thus, $A = M + aI$ is a positive-definite self-adjoint operator.

9.10 Sample Path Properties and Continuity

Thus far we have examined continuity in the mean-squared sense. We will be interested in continuity almost surely.

Definition 9.69 *The process* $\{Y(\omega, t), t \in T\}$ *is said to be* **almost surely sample path continuous** *if*

$$\Pr\{\omega : \lim_{h \to 0} \sup_{t \in T} |Y(t+h) - Y(t)| = 0\}$$

$$= 1 - \Pr \bigcup_{t \in T} \{\omega : \lim_{h \to 0} |Y(t+h) - Y(t)| \neq 0\} \quad (9.226)$$

$$= 1. \quad (9.227)$$

Similarly for fields, $Y(\omega, x), x \in X$,

$$\Pr\{\omega : \lim_{h \to 0} \sup_{x \in X} |Y(x+h) - Y(x)| = 0\}$$

$$= 1 - \Pr \bigcup_{x \in X} \{\omega : \lim_{h \to 0} |Y(x+h) - Y(x)| \neq 0\} \quad (9.228)$$

$$= 1. \quad (9.229)$$

For the m-vector case, almost sure continuity is taken componentwise.

The following theorem on almost-sure continuity of sample paths of the random processes and fields are taken from Wong and Hajek [213] and Karatzas and Shreve [214].

Theorem 9.70

1. *(Continuity for Processes [213]) Given random process* $\{Y(t), t \in T \subset \mathbb{R}\}$, *if there exists strictly positive constants* α, β, C *such that*

$$E|Y(t+h) - Y(t)|^\alpha \leq Ch^{1+\beta} \quad (9.230)$$

then

$$\sup_{t \in T} |Y(t+h) - Y(t)| \xrightarrow{a.s} 0. \quad (9.231)$$

2. *(Continuity for Fields [214]) Let* $\{Y(x), x \in X \subset \mathbb{R}^d\}$, *if there exists strictly positive constants* α, β, C *such that*

$$E|Y(x+h) - Y(x)|^\alpha \leq C \|h\|^{d+\beta} \quad (9.232)$$

then

$$\sup_{x \in X} |Y(x+h) - Y(x)| \xrightarrow{a.s} 0. \quad (9.233)$$

Proof The proof of (1) follows Proposition 4.2 of Wong and Hajek [213]; see Appendix A.1. For the proof of the multi-dimensional field case, see Karatzas and Shreve [214]. □

The random fields are modeled as Gaussian random vector fields $\{U(x), x \in [0, 1]^d \subset \mathbb{R}^d\}$. Smoothness on the fields is forced by assuming they are solutions of stochastic partial differential equations of the type $LU = W$, L a normal differential operator. Index the variables in the d-dimensional cube $x \in [0, 1]^d$, where L is a constant coefficient differential operator of the form:

$$L = \sum_{m=1}^{n} a_m \frac{\partial^{p(m)}}{\partial x_1^{p_1(m)}, \ldots, \partial x_d^{p_d(m)}} + \epsilon, \quad \text{with} \quad \sum_{l=1}^{d} p_l(m) = p(m). \quad (9.234)$$

9.10 SAMPLE PATH PROPERTIES AND CONTINUITY

The eigenvalues and eigenfunctions $L\phi_k(x) = \lambda_k \phi_k(x)$ are of the form

$$\phi_k(x) = e^{j2\pi \langle k, x \rangle}, \quad \lambda_k = \sum_{m=1}^{n} a_m \sum_{l=1}^{d} (j2\pi k_l)^{p_l(m)} + \epsilon, \quad k \in \mathbb{Z}^d. \tag{9.235}$$

We will require the operator to be strongly elliptic.

Definition 9.71 *Define the operator L to be* **strongly elliptic** *of order q (Reed and Simon [240]) if for all but a finite k*

$$Re\{\lambda_k\} = Re\left\{\sum_{m=1}^{n} a_m \sum_{l=1}^{d} (j2\pi k_l)^{p_l(m)}\right\} \geq \|k\|^{2q}. \tag{9.236}$$

Example 9.72 A simple example of an elliptic operator is the q power of the Laplacian, $L = (-\nabla^2 + I)^q$ which has such propertes, since with

$$\nabla^2 = \frac{\partial^2}{\partial x_1^2} + \frac{\partial^2}{\partial x_2^2} + \frac{\partial^2}{\partial x_3^2} + \cdots + \frac{\partial^2}{\partial x_d^2}, \tag{9.237}$$

then the eigenvalues are $\lambda_k = \|2\pi k\|^2 + 1$; for L^q, $\lambda_k = (\|2\pi k\|^2 + 1)^q, k \in \mathbb{Z}^d$.

For almost sure-continuity we follow Amit and Piccioni [241] and do the scalar valued case.

Theorem 9.73 (Joshi [242]) *Let $U(x), x = (x_1, x_2, x_3, \ldots, x_d) \in X = [0,1]^d$ be a real-valued Gaussian random field defined in the quadratic mean sense as above to satisfy the partial differential equation*

$$LU(x) = W(x), x \in [0,1]^d, \tag{9.238}$$

where L is strongly elliptic of order q and the noise process $W(x) = \sum_k W_k \phi_k(x)$ is assumed to have a spectral representation with spectral variances $E|W_k|^2 = |\alpha_k|^2$ bounded above by $\|k\|^{2p}$ except on a finite set $F \subset \mathbb{Z}^d$.
(i) If $4q - 2p - 2 > d$ then $U(x)$ has sample paths which are almost surely continuous,
(ii) and if $4q - 2p - 4 > d$ then it is almost surely differentiable; similarly it is n-times differentiable if $4q - 2p - 2(n+1) > d$.

Proof To prove sample path continuity use the Kolmogorov criteria from Theorem 9.70 following Amit and Piccioni [241]; then $U(x), x \in \mathbb{R}^d$ is *a.s.* sample path continuous if there exists constants $\alpha, \beta, c > 0$ such that

$$E|U(x) - U(y)|^\alpha \leq c\|x - y\|^{d+\beta}. \tag{9.239}$$

The proof exploits the following Lemma.

Lemma 9.74 *If $4q - 2p - 2 > d$ then for all $x, y \in [0,1]^d$,*

$$E|U(x) - U(y)|^2 \leq c\|x - y\|^2. \tag{9.240}$$

Proof *Proof of Lemma.* Since U satisfies the stochastic partial differential equation $LU(x) = W(x)$, $x \in [0,1]^d$, use the spectral representation giving

$$E|U(x) - U(y)|^2 = \left| \sum_{k \in \mathbb{Z}^d} \frac{W_k}{\lambda_k} (\phi_k(x) - \phi_k(y)) \right| \tag{9.241}$$

$$\leq \sum_{k \in \mathbb{Z}^d} \frac{|\alpha_k|^2}{|\lambda_k|^2} |e^{j2\pi \langle k,x \rangle} - e^{j2\pi \langle k,y \rangle}|^2 = \sum_{k \in \mathbb{Z}^d} \frac{|\alpha_k|^2}{|\lambda_k|^2} (2 - 2\cos 2\pi \langle k, x-y \rangle)$$

$$= \sum_{k \in \mathbb{Z}^d} \frac{|\alpha_k|^2}{|\lambda_k|^2} (4 \sin^2 \pi \langle k, x-y \rangle) \leq \sum_{k \in \mathbb{Z}^d} \frac{|\alpha_k|^2}{|\lambda_k|^2} c |\langle k, x-y \rangle|^2$$

$$\leq \sum_{k \in \mathbb{Z}^d} \frac{|\alpha_k|^2}{|\lambda_k|^2} c \|x-y\|^2 \|k\|^2$$

$$\overset{(a)}{\leq} c \|x - y\|^2 \left(c_1 + \sum_{k \in \mathbb{Z}^d / F} \frac{\|k\|^{2(p+1)}}{\|k\|^{4q}} \right), \tag{9.242}$$

where (a) uses the fact that L is strongly elliptical of order q so that $|\lambda_k|^2 \geq \|k\|^{4q}$, and $|\alpha_k|^2 \leq \|k\|^{2p}$, and the series converges for $4q - 2p - 2 > d$. □

Proof To establish the Kolmogorof conditions with α and $d + \beta$, since $U(x)$ is a zero mean Gaussian field, then $U(x) - U(y)$ is a zero mean Gaussian random variable with the moment generating function given by $\Psi(s) = e^{\sigma^2 s^2 / 2} = \sum_{l=1}^{\infty} (\sigma^{2l} s^{2l})/(2^l l!)$, where $\sigma^2 = E|U(x) - U(y)|^2$. Then for any positive integer l since $4q - 2 - 2p > d$ we have

$$E|U(x) - U(y)|^{2l} = \frac{2l!}{2^l l!} (E|U(x) - U(y)|^2)^l$$

$$\leq c(\|x - y\|^2)^l = c\|x - y\|^{2l}. \tag{9.243}$$

Then the process has a.s. sample path continuity as on choosing $\alpha = 2l$, $\beta = 2l - d$.

To show it has a.s. continuously differentiablility, examine the components of the derivative of the vector field $U'(x)$ with mean square derivative, $U'(x)$ having spectral representation given by

$$U'(x) = \left(\frac{\partial U(x)}{\partial x_1}, \ldots, \frac{\partial U(x)}{\partial x_d} \right) = \sum_{k \in \mathbb{Z}^d} \frac{j2\pi k W_k(\omega)}{\lambda_k} e^{j2\pi \langle k,x \rangle}. \tag{9.244}$$

Then we need to show that each component is almost surely continuous, and since U' is Gaussian this requires us to show that for each j, $E|\frac{\partial U(x)}{\partial x_j} - \frac{\partial U(y)}{\partial x_j}|^2 \leq c\|x-y\|^2$. Following the same property as above,

$$E \left| \frac{\partial U(x)}{\partial x_j} - \frac{\partial U(y)}{\partial x_j} \right|^2 \leq c \sum_{k \in \mathbb{Z}^d} \frac{|\alpha_k|^2}{|\lambda_k|^2} \|2\pi k\|^2 |e^{j2\pi \langle k,x \rangle} - e^{j2\pi \langle k,y \rangle}|^2$$

$$\leq c\|x - y\|^2 \left(c_1 + \sum_{k \in \mathbb{Z}^d / F} \frac{\|k\|^{2(p+2)}}{\|k\|^{4q}} \right). \tag{9.245}$$

The series converges for $4q - 2p - 4 > d$.

Similarly, for n-derivatives. □

Example 9.75 (Wiener Process) The Kolmogorov-Centsov Theorem 9.70 applies for showing Brownian motion is almost surely sample path continuous. The increments $Y(t+h) - Y(t)$ are zero-mean Gaussian with variance h, implying $|Y(t+h) - Y(t)|^2$ is chi-squared with variance $3h^2$. This implies $E|Y(t+h) - Y(t)|^4 = 3h^2$, and Theorem 9.70 is satisfied for $\alpha = 4, \beta = 1, C = 3$.

Example 9.76 It is helpful to examine the dependence on dimensions for various differential operators For 1D fields $x \in X = [0,1]$, and for the equation $dU/dx = W(x)$, $L = d/dx$, and for W of trace class fall quadratically in spectrum $E|W_k|^2 = 1/k^2$, then $d = 1, q = 1/2, p = -1$ and $4q - 2 - 2p = 2 > d = 1$. Thus U has a.s. sample path continuity.

Example 9.77 Let $U(\omega,t)$ satisfy the random equation $\triangle^p U = W$ with boundary conditions $U(0) = U(1) = 0$, with $W(t) = \sum_k W_k(\omega)\phi_k(t)$, ϕ_k eigen functions of the Laplacian and W_k independent, zero-mean normal variances, α_k^2.

If the Gaussian random process, W, has noise spectrum dominated above by $\alpha_k^2 \leq ck^{2m}$ for all but finitely many k, and $\triangle^p$ a polynomial power of the Laplacian with $2p \geq r + m + 2$, then the U-field has sample paths which are almost surely continuously r-times differentiable. To see this, the covariance of the U-field takes the form

$$K_U(t,s) = \sum_k \frac{\alpha_k^2}{(2\pi k)^{2(2p)}} \sin 2\pi kt \, \sin 2\pi ks, \tag{9.246}$$

implying the covariance of the rth derivative

$$K_{\frac{\partial^r U}{\partial t^r}}(t,s) = \sum_k \frac{\alpha_k^2}{(2\pi k)^{2(2p-r)}} \sin\left(2\pi kt + \frac{\pi r}{2}\right) \sin\left(2\pi ks + \frac{\pi r}{2}\right). \tag{9.247}$$

The second mixed derivative becomes bounded by the inequality for some finite integer l and constant $c_1 < \infty$

$$\frac{\partial^2 K_{\partial^r U/\partial t^r}/\partial t^r (t,s)}{\partial t \partial s} \leq c_1 + c_2 \sum_{k>l} \frac{k^{2m}}{(2\pi k)^{2(2p-r-1)}} \sin\left(2\pi kt + \frac{\pi(r+1)}{2}\right)$$
$$\times \sin\left(2\pi ks + \pi(r+1)/2\right). \tag{9.248}$$

But this exists along the diagonal for $2p - r - m - 1 \geq 1$, implying the process is almost surely continuous.

9.11 Gaussian Random Fields as Prior Distributions in Point Process Image Reconstruction

During the early 1980s it was recognized that image reconstruction as a density estimation problem in functions spaces was fundamentally inconsistent with purely likelihood based methods. An unconstrained maximization of likelihood from finite data fails when maximizing in function spaces. There are simply too many parameters floating around. In emission tomography images which have been produced via purely unconstrained maximum-likelihood methods have been observed by many investigators to have seemingly random artifacts (sharp peaks and valleys distributed throughout the image field) [243–254]. We point out that although the image reconstruction problem has been described via a continuous model, the problem does not go away if the image is constrained to histogram bins. Then the implementation has the undesirable property that as it becomes finer (decreasing pixel size) the problem becomes worse, a situation termed *dimensional instability* by Tapia and Thompson [255]. To remedy the difficulty in emission tomography, the Method of Sieves [9], and penalized likelihood methods [255–258] are used to estimate infinite dimensional functions from finite data sets. Here we examine penalty methods (appealing to Good's work [257,258]) and make explicit their connection to Gaussian random field priors.

9.11.1 The Need for Regularization in Image Reconstruction

What is the role of regularization or smoothing via a Gaussian prior in image reconstruction? Examine the classic density estimation problem from Poisson measurements with mean $\lambda(x), x \in \mathbb{R}^d$. Assume measurements points occur at random places $x_1, x_2, \ldots, x_N$, with intensity $\lambda(\cdot)$. The measurement processes $N(dx)$ is a counting process (see [42], for more details on the development of counting processes) with measurement locations of the counting process $x_i, i = 1, \ldots, N$, with intensity $EN(A) = \int_A \lambda(x)\,dx, A \subset \mathbb{R}^d$, and maximizer given by

$$\hat{\lambda} = \arg\max_{\lambda(\cdot)} -\int \lambda(x)\,dx + \int N(dx)\ln\lambda(x) = \sum_{i=1}^{N} \delta_{x_i}(dx)\,. \qquad (9.249)$$

Eqn. 9.249 is meant in the weak sense, so that the unconstrained MLE is not a function, but rather the unconstrained maximizer is a Dirac measure centered at the points of the N observations. Obviously this is unacceptable because it contradicts one's *a priori* knowledge about almost all imaging scenarios, where it is expected that $\lambda(\cdot)$ is bounded, and at least piecewise continuous. An unconstrained maximization of 9.249 fails to produce meaningful estimates. The fundamental problem is that the likelihood is unbounded above over the set of functions (densities). For finite data sets there are too many parameters floating around. As noted in [259] the function spaces are too large for direct estimation. Notice, placing the problem into histogram bins does not make it go away, only results in a series of estimated pulses with heights proportional to the number of observations in each pixel, $\Lambda(i) = \int_{i\Delta}^{(i+1)\Delta} N(dt)$. As $\Delta \to 0$, then the estimator converges to a set of Dirac delta-functions, centered at the points of the N observations.

We also emphasize that while the formulation in this section for the image reconstruction problem is placed in the the Poisson setting precisely these issues arise in maximum-likelihood reconstruction in function spaces from Gaussian processes. In the Gaussian setting for optical flow, see for example Poggio et al. [260] and Yuille and Grzywach [261]. In Medical imaging Gaussian priors have been used by Levitan and Herman [262], Hart and Liang [263], and Herman and Odhner [264]. Joyce and Root [265] have written an exquisite paper on Precision Gauges, direcly analogous to Grenander's Method of Sieves but restricted to the least-squares setting. Preza et al. [235] employed these Precision Gauges to control the dimensional stability problem in the Gaussian Random field setting for image deconvolution problem.

9.11.2 Smoothness and Gaussian Priors

To constrain the maximization problem, introduce the roughness penalized estimators. Good and Gaskins have argued for penalties based both on first and second derivatives (curvature). Lanterman has examined what he terms generalized Good's roughness [254] introducing powers of the Laplacian to increase the smoothing and connecting the penalty in the least-squares setting to the theory of Butterworth filters. We state that penalty here.

9.11 POINT PROCESS IMAGE RECONSTRUCTION

Definition 9.78 *Define the penalty*

$$\Phi(Y) = \|LY\|_2^2 \text{ with } L = \sum_i a_i \nabla^{2q_i}. \tag{9.250}$$

Given Z a Gaussian random field with mean $EZ = Y$, then the **penalized estimator** $\hat{Y}$ *is defined to be*

$$\hat{Y} = \arg\max_Y 2\langle Z, Y\rangle_2 - \Phi(Y). \tag{9.251}$$

Given N a Poisson random field with mean $\lambda = Y^2$, then the **penalized estimator** $\hat{Y}$ *is given by*

$$\hat{Y} = \arg\max_Y \int N(dx) \ln Y^2(x) - \Phi(Y). \tag{9.252}$$

In optical flow [260] and emission tomography [243] the smoothing is gradient only, $q_1 = 1/2$. Good and Gaskins argued for the gradient and Laplacian penalty, $q_1 = 1/2, q_2 = 1$ (first and second derivatives). In the context of Medical imaging, [266] has principally examined curvature constraints, $q_2 = 1$. Lanterman [254] has examined generalized Good's roughness introducing powers of the Laplacian to increase the smoothing in the resulting penalized methods. The introduction of these generalized smoothing methods is quantified in part by several of the following results characterizing the smoothness introduced by the penalty when it is viewed as specifying the potential of a Gaussian random field.

Theorem 9.79 *Let Y be random solutions of*

$$LY = W \quad \text{with circulant b.c. on } X = [0,1]^d \subset \mathbb{R}^d, \tag{9.253}$$

W circulant white-noise on Y, variance 1.

Then for smoothness operators which are strongly elliptic of order q (the Laplacian to the q power in the penalty), we have the following conditions for almost surely continuous samples paths: (i) for $d = 1$, $X = [0,1] \subset \mathbb{R}$, then $q > 0.75$; (ii) for $X = [0,1]^2 \subset \mathbb{R}^2$, $d = 2$ then $q > 1.0$; (iii) for $d = 3$, $X = [0,1]^3 \subset \mathbb{R}^3$, then $q > 1.25$; (iv) for $d = 4$, $X = [0,1]^4 \subset \mathbb{R}^4$, then $q > 1.5$.

Proof The proof is Theorem 9.73 given by the $4q - 2 > d$ condition.
□

9.11.3 Good's Roughness as a Gaussian Prior

Good was one of the earliest to argue [257, 258] that a principled basis for introducing regularity could be derived via principles of information. Good's roughness measure (norm-squared gradient normalized by variance) is a direct measure of the Fisher information on the location of a known waveform at unknown position. Subsequently, regularization via the norm-square gradient penalty (for constant variance fields) has been used extensively in computer vision [260, 261] and the norm-squared gradient normalized by the variance for Poisson fields [243].

Good and Gaskin originally derived their first-order penalty via Fisher discrimination arguments. For $Z(\cdot)$ a Gaussian random field on $\mathbb{R}^d$, variance 1 with mean field $\mu(\cdot) = Y(\cdot)$ a fixed waveform at unknown position $p \in \mathbb{R}^d$ which is smooth (twice differentiable) with bounded support, then the $d \times d$ Fisher information matrix has entries $F_{ij} = \langle (\partial Y/\partial p_i), (\partial Y/\partial p_j) \rangle_2$ with $\mathrm{tr} F = \|\nabla Y\|_2^2$. This follows from

$$-E \frac{\partial^2}{\partial p_i \partial p_j}\left(-\|Z-\mu\|_2^2\right) = E\left(\left\langle \frac{\partial Y}{\partial p_i}, \frac{\partial Y}{\partial p_i}\right\rangle_2 - 2\left\langle (Z-Y), \frac{\partial^2 Y}{\partial p_i \partial p_j}\right\rangle_2\right) = \left\langle \frac{\partial Y}{\partial p_i}, \frac{\partial Y}{\partial p_j}\right\rangle_2. \quad (9.254)$$

For $N(\cdot)$ a Poisson random field on $\mathbb{R}^d$ with mean field (smooth as above, bounded support) $\lambda(\cdot) = Y^2(\cdot)$ a fixed waveform at unknown position $p \in \mathbb{R}^d$, then the $d \times d$ Fisher information matrix has entries $F_{ij} = \langle (1/\sqrt{\lambda})(\partial \lambda/\partial p_i), (1/\sqrt{\lambda})(\partial \lambda/\partial p_j) \rangle_2$, with $\mathrm{tr} F = \|(\nabla \lambda/\sqrt{\lambda}\|_2^2 = \|\nabla Y\|_2^2$. This follows from

$$-\frac{\partial^2}{\partial p_i \partial p_j}\left(-\int \lambda(x)dx + \int N(dx)\ln \lambda(x)\right) = \int \frac{\partial^2}{\partial p_i \partial p_j}\lambda(x)dx - \int \frac{N(dx)}{\lambda(x)}\frac{\partial^2}{\partial p_i \partial p_j}\lambda(x)$$
$$+ \int \frac{N(dx)}{\lambda(x)^2}\frac{\partial \lambda(x)}{\partial p_i}\frac{\partial \lambda(x)}{\partial p_j}, \quad (9.255)$$

and taking exceptions gives the result.

In 1-D the first order Good-roughness constraint is the norm-square derivative; its analog in 2,3-D involves curvature via the Laplacian. In such cases they give exponential splines, and are equivalently smooth in the sense (as proven below) that when viewed as a Gaussian prior disribution sample-paths are quadratic mean-continuous and not almost-surely continuous.

9.11.4 Exponential Spline Smoothing via Good's Roughness

Adding the potential of the Gaussian prior density to the maximum-likelihood problem gives Good's penalized methods explicitly involving the Green's kernel of the Gaussian prior. The solutions to the penalized estimator for Good's roughness in 1-dimension, and extended to 2,3 dimensions via the Laplacian are as derived by Snyder and miller [243] for emission tomography exponential splines of the data.

Corollary 9.80 *The penalized estimators for the Gaussian and Poisson cases with potential as given in the above Theorem 9.38 with Green's kernel $K(x-y) = \beta \exp^{-1/\alpha \|x-y\|_{\mathbb{R}^d}^2}$ are given by the convolution with the Green's kernel:*

$$\hat{Y}(\cdot) = \arg\max_Y 2\langle Z, Y\rangle_2 - \Phi(Y) = \int K(\cdot - y)Z(y)dy, \quad (9.256)$$

$$\hat{Y}(\cdot) = \arg\max_Y \int N(dx)\ln Y^2(x) - \Phi(Y) = \int \frac{K(\cdot - y)}{Y(y)}N(dy) = \sum_{i=1}^N \frac{K(\cdot - x_i)}{\hat{Y}(x_i)}. \quad (9.257)$$

Proof For the Gaussian case, the the maximization solves the quadratic form. For the nonlinear Poisson case, the penalized likelihood becomes

$$-\|LY(\cdot)\|_2^2 + \int N(dx)\ln Y^2(x) = -\langle L^*LY, Y\rangle_2 + \int N(dx)\ln Y^2(x), \quad (9.258)$$

where L^*L has Green's kernel K. Maximizing gives the solution of 9.257. □

The regularized estimator is a sum of exponential splines with knots at the data points. Figure 9.9 shows results in 1 dimension illustrating the smoothing by the GRF. A Poisson-process

9.11 POINT PROCESS IMAGE RECONSTRUCTION

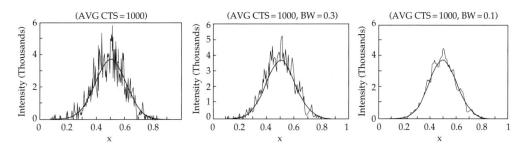

Figure 9.9 Panel 1 shows unconstrained likelihood estimator demonstrating dimensional instability. Panels 2 and 3 show the likelihood estimators with Gaussian prior. Results taken from Snyder and Miller [243].

was generated with a mean in each pixel of $\Lambda(i)$, where $\Lambda(i) = \int_{i\Delta}^{(i+1)\Delta} \lambda(x)dx$, Δ the bin size. Panel 1 shows the likelihood estimates of the smooth profile based on a Poisson simulation containing an average of 1000 counts in the 512 bin simulation. The histogram shown in Figure 9.9 results from a direct maximization of the unconstrained discrete likelihood of 9.249.

The 1D histogram of Figure 9.9 demonstrates the "dimensional instability" that the MLEs exhibit. Notice the occurrence of large variations between adjacent pixel estimates of Λ; this effect gets worse if the pixel size or the number of measurement points are decreased. Plotted in panels 2 and 3 are the results of applying the Gaussian prior. The estimates of Figure 9.9 were obtained by solving the exponential spline estimate of 9.257 iteratively according to

$$\lambda^{\text{new}}(x) = \sqrt{\lambda^{\text{old}}(x)} \sum_{i=1}^{N} \frac{K(x-x_i)}{\sqrt{\lambda^{\text{old}}(x_i)}}, \quad x \in [0,T], \qquad (9.259)$$

where $K(\cdot)$ is given in Eqn. 9.78. For the initial estimate $\sqrt{\lambda^{(0)}}$ the square root of the histogram estimate was used. Panels 2 and 3 show the estimates derived with increasing weight α implying exponential splines of inreasing width. Clearly as the exponential width increases the effect is to smooth the variations between adjacent estimates in the histogram.

Example 9.81 (Image Deconvolution in Space $\mathbb{R}^2, \mathbb{R}^3$) Examine the algorithms for image deconvolution in time-of-flight positron emission tomography as originally developed by Snyder [35, 35, 46, 245] (see Chapter 2, Example 2.55). The point-spread function $p(\cdot)$ reflects both the line-of-flight and the perpendicular to the line of flight ambiguity. For time-of-flight PET [35] they vary as a function of the projection direction. Emissions occur at random places $X_i \in \mathbb{R}^3$ with intensity $\lambda(x), x \in \mathbb{R}^3$ and are observed with random errors, the random errors ϵ_i giving rise to independent measurements $Y_i = X_i + \epsilon_i$. Each measurement has a different line of flight with the value corresponding to its point-spread function resulting in the multiple measurement processes $N_1(dy), N_2(dy), \ldots$ each with mean intensity $EN_j(A) = \int_A \int_{\mathbb{R}^3} p_j(y|x)\lambda(x)dx\,dy$ generalizing Eqn. 2.173, Chapter 2.

Parameterizing in $Y = \sqrt{\lambda}$ and adding the Gaussian prior $\|\nabla^{2q}X\|_2^2$ gives the penalized estimator

$$\hat{Y} = \arg\max_{Y} -\|Y\|_2^2 - \alpha^2 \|\nabla^{2q}Y\|_2^2 + \sum_{j=1}^{J} \int N_j(dy) \ln \int p_j(y|x)Y^2(x)dx. \qquad (9.260)$$

The EM algorithm used for maximizing the posterior [30] generalizes Eqn. 2.174 from Chapter 2:

$$Y^{\text{new}} = \arg\max_{\lambda = Y^2 \geq 0} -\|Y\|_2^2 - \alpha^2 \|\nabla^{2q} Y\|_2^2 + \int_{\mathbb{R}^3} Y^{\text{old2}}(x) \sum_{j=1}^{J}$$

$$\times \int_{y} \left(\frac{p_j(y|x)}{\int_{\mathbb{R}^3} p_j(y|x)(Y^{\text{old}})^2(x)dx} N_j(dy) \right) \ln Y^2(x) dx \,.$$

(9.261)

Shown in Figure 9.10 are results on the Hoffman brain phantom from Lanterman with Gaussian random field Good's smoothing. Column 1 shows the Hoffman brain phantom with the 1000th iteration unconstrained reconstruction shown in the bottom panel below. Columns 2 and 3 row show reconstructions using Good's first-order roughness penalty with $q = 1/2$ (column 2) and the Laplacian-squared for $q = 2$ (column 3). The penalty has $q = 1/2$ (column 1) and $q = 2$ (column 2) as above.

Example 9.82 (**Deblurring and Denoising via Gaussian Random Fields Priors in Gaussian Noise Models**) We have in the previous section studied regularization in point process image reconstruction. Now turn to the huge area where a great deal of work has been done in which the observations Y are a Gaussian random field with mean field, Y. The degraded image Z is observed as a function of the continuous background space and the goal is to reconstruct $Y(x), x \in D$; the optimizer is given by

$$\inf_{Y} \int_{D} (Z(x) - Y(x))^2 dx + \alpha \int_{D} \|\nabla Y(x)\|^2 dx. \qquad (9.262)$$

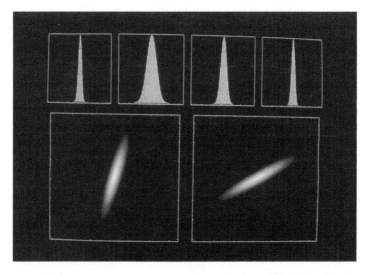

Figure 9.10 Panel 1 shows example point-spread functions from time-of-flight PET. Panel 2 (top row) show the Pie Phantom. Panel 3 (bottom row, right column) shows results of 1,000 EM iterations with no smoothing ($\alpha = 0$) of Eqn. 2.174 from Chapter 2. Results taken from Lanterman.

Then non-linear diffusion has been studied by many (see Guichard and Morel [267]) corresponding to functional gradient descent taking the form

$$\frac{\partial Y}{\partial t} = \mathrm{div}\nabla Y + c(Z - Y). \tag{9.263}$$

Clearly, fixed points correspond to the necessary conditions for a minimizer 9.262.

9.12 Non-Compact Operators and Orthogonal Representations

There are clearly many examples of great interest in which the index spaces are not compact. Then, Riesz Schauder and Hilbert Schmidt do not hold; the spectrum is not discrete. This is well appreciated by communications engineers; the Fourier transform on the entire real line is built from a continuum of complex exponentials. Thus, the projection operations of transforms are constructed using integration; the projections turn out to be measures which are integrated over non discrete index sets.

This takes us towards orthogonal representations over non bounded index sets such as the Cramer representation of stationary processes. This is a familiar setting as Fourier representations which correspond to the spectral decomposition of the shift operator over the real line.

First define *unitary* operators on the Hilbert space, and *projection* operators.

Definition 9.83 *A bounded linear operator $U \in L(H)$ is called a* **unitary operator** *if it satisfies $\forall g, h \in H$,*

$$\langle g, h \rangle_H = \langle Ug, Uh \rangle_H. \tag{9.264}$$

An operator $P \in L(H)$ is called a **projection operator** *if $P^2 = P$, and is* **orthogonal** *if $P = P^*$.*

Remark 9.12.3 Eigenvalues of unitary operators have magnitude 1, and eigenvectors corresponding to distinct non-zero eigenvalues are orthogonal, for if ϕ, ψ are two distinct eigenvectors, then

$$\langle \phi, \psi \rangle_H = \langle U\phi, U\psi \rangle_H = \lambda_1^* \lambda_2 \langle \phi, \psi \rangle_H. \tag{9.265}$$

This implies if $\lambda_1 \neq \lambda_2 \neq 0$ then $\langle \phi, \psi \rangle_H = 0$; if $\phi = \psi$ then $|\lambda_1|^2 = 1$.

Here are several operators which are unitary; shift on the real line and scale. It is their eigenfunctions which form the basis for the orthogonal expansions studied.

Theorem 9.84 *Let $U_s \in L(H)$ be a unitary operator, with eigenfunctions and eigenvalues satisfying the fundamental equation $U_s \phi_f(t) = \lambda_f^{(s)} \phi_f(t)$. Then* **shift** *and* **scale** *are unitary operators with the following eigenelements.*

Shift: *Let $H = L^2(\mathbb{R}, dt)$ be the Hilbert spaces with, for each $s \in \mathbb{R}^d$, U_s being the* **shift operator** *defined for all $g \in H$,*

$$U_{s+s'}g(t) = U_s U_{s'} g(t) = g(s + s' + t). \tag{9.266}$$

Then U_s is a unitary operator on H, with eigenelements

$$\text{for } s \in \mathbb{R}, \quad \lambda_f^{(s)} = e^{j2\pi fs}, \quad \phi_f(t) = e^{j2\pi ft}, \quad f \in \mathbb{R}. \tag{9.267}$$

Scale: Let H be the Hilbert space $L^2((0,\infty),(1/t)dt)$ with inner product $\langle g, h \rangle_{L^2((1/t)dt)} = \int_0^\infty g(t)h(t)(1/t)dt$. Then for each $s \in (0,\infty)$, let U_s be the **scale operator** defined by for all $g \in H$,

$$U_{s \circ s'} g(t) = U_s U_{s'} g(t) = g(s \circ s' t) \tag{9.268}$$

Then U_s is a unitary operator on H with eigenelements

$$\text{for } s \in (0,\infty), \quad \lambda_f^{(s)} = s^{jf}, \quad \phi_f(t) = t^{jf}. \tag{9.269}$$

Proof *Proof of shift:* The unitary property is well known, just following from the substitution of variable and Jacobian of shift being 1:

$$\int g(t+s)h(t+s)dt = \int g(t)h(t)dt. \tag{9.270}$$

For the Torus, the domain is compact, therefore there is a discrete spectrum corresponding to Fourier series; the complex exponentials give the eigenfunctions and eigenvalues.

Proof of scale: The unitary property follows:

$$\langle U_s g, U_s h \rangle = \int_0^\infty h(st)h(st)\frac{1}{t}dt \stackrel{(a)}{=} \int_0^\infty g(r)h(r)\frac{1}{r}dr$$

$$= \langle g, h \rangle, \tag{9.271}$$

with (a) following from the substitutions $r = st$.

The eigenelements $(\lambda_f^{(j)} = s^{jf}, \phi_f(t) = t^{jf}$ satisfying the fundamental equation

$$U_s \phi_f(t) = \phi_f(st) = (st)^{jf} = \lambda_f^{(s)} \phi_f(t). \tag{9.272}$$

The eigenvalues have magnitude 1, and to see orthogonality of the eigenvectors, use the log trick. For $g \in L^2[(0,\infty),\frac{1}{t}dt]$, then $\tilde{g} \in L^2[(-\infty,\infty),dt]$ defined by the relation $\tilde{g} \circ \log = g$ so that

$$\tilde{g}(\log t) = g(t), \quad \log t \in (-\infty, \infty). \tag{9.273}$$

Then, the orthogonality of $\phi_f(t) = t^{jf}, \phi_{f'}(t) = t^{jf'}$ reduces to the orthogonality of $\tilde{\phi}_f$ as complex exponentials, $\tilde{\phi}_f(r) = e^{jrf}$. Orthogonality holds in the sense of $L^2[(-\infty,\infty)]$ functions; the inner product of two basis components acts as a delta dirac in its actions on functions in L^2. To see this, define $g(s) \leftrightarrow G(f)$ to form a scale transform pair. Then

$$\int_{-\infty}^\infty \langle \phi_f, \phi_{f'} \rangle G(f') df' = \int_{-\infty}^\infty \int_0^\infty s^{-j2\pi(f-f')} \frac{1}{s} ds G(f') df' \tag{9.274}$$

$$= \int_0^\infty s^{-j2\pi f} \underbrace{\int_{-\infty}^\infty s^{j2\pi f'} G(f') df'}_{g(s)} \frac{1}{s} ds = G(f). \tag{9.275}$$

□

9.12 NON-COMPACT OPERATORS

Definition 9.85 *The second order random process $\{Z(df), f \in \mathbb{R}\}$ is called an **orthogonal process with spectral density** $S(f), f \in \mathbb{R}$ if for all F_1, F_2 measurable in $\mathbb{R}$,*

1. $F_1 \cap F_2 = \varnothing \implies Z(F_1 \bigcup F_2) = Z(F_1) + Z(F_2);$ (9.276)
2. $EZ(F_1)Z^*(F_2) = \int_{F_1} \int_{F_2} EZ(df)Z^*(df') = \int_{F_1 \cap F_2} S(f)\, df.$ (9.277)

Example 9.86 (Compact Covariances) On compact domains we have already seen orthogonal expansions. **Karhunen-Loeve** is an orthogonal process representation. Let $Y(t), t \in [0,1]$ have covariance $K(t,s), t, s \in [0,1]$ with eigenelements $\{\phi_n, \lambda_n\}$, $K\phi_n = \lambda_n \phi_n$. Then the orthogonal decomposition becomes

$$Y(t) = \sum_n \phi_n(t) Z_n, \quad (9.278)$$

with $Z_n = \langle \phi_n, Y \rangle$ with spectrum $E\{Z_n Z^*_{n'}\} = \lambda_n \delta(n - n')$.

For **cyclo-stationary processes** $\{Y(\omega, t), t \in [0,1]\}$ with covariance $K(t,s) = K(t-s, 0)$, then the eigenfunctions are $\phi_n(t) = e^{j2\pi n t}$, and orthogonal process representation becomes

$$Y(\omega, t) = \sum_n e^{j2\pi n t} Z_n(\omega), \quad (9.279)$$

with $Z_n(\omega)$ an orthogonal process with $EZ_n Z_{n'} = S_n \delta(n' - n)$. To see orthogonality, use the fact that $Z_n = \langle \phi_n, X \rangle$ implying that for all $F_1, F_2 \subset \mathbb{Z}, F_1 \cap F_2 = \varnothing$,

$$Z(F_1 \bigcup F_2) = \sum_{n \in F_1 \bigcup F_2} \langle \phi_n, Y \rangle = Z(F_1) + Z(F_2). \quad (9.280)$$

Orthogonality follows from the eigenfunction property:

$$EZ_n Z_{n'} = \langle \phi_n, K_Y \phi_{n'} \rangle = S_n \delta(n - n'). \quad (9.281)$$

9.12.1 Cramer Decomposition for Stationary Processes

The Cramer decomposition in the non-compact setting now follows. Assume $\{Y(\omega, t), t \in \mathbb{R}\}$ is a second order process which is stationary so that the covariances is only a function of shift $K(t,s) = K(t-s, 0)$. The Fourier representation implies the eigenfunctions of the covariance are the complex exponentials:

$$K(t-s, 0) = \int_{\mathbb{R}} e^{j2\pi f(t-s)} S(f) df. \quad (9.282)$$

As long as the Fourier transforms exists, the complex exponential provides the orthogonal process representation. Here is the Cramer decomposition.

Theorem 9.87 (Cramer) *Assume $\{Y(\omega, t), t \in \mathbb{R}\}$ is a stationary second order process with covariance $K(t,s) = K(t-s, 0)$ whose Fourier transform exists:*

$$K(t, 0) = \int_{\mathbb{R}} e^{j2\pi f t} S(f) df. \quad (9.283)$$

Then the orthogonal representation,

$$Y(\omega, t) = \int_{\mathbb{R}} e^{j2\pi ft} Z(\omega, df), \text{ with orthogonal process } Z(\omega, df)$$
$$= \int_{\mathbb{R}} e^{-j2\pi ft} Y(\omega, t) dt\, df, \qquad (9.284)$$

has spectral density $S(f) = \int_{\mathbb{R}} e^{-j2\pi ft} K(t, 0) dt$.

Proof The eigenvalues and eigenfunctions for the shift operator U_t on $\mathbb{R}$ are $\lambda_f^{(t)} = e^{+j2\pi ft}, \phi_f(t) = e^{j2\pi ft}$. We need to show that the process being shift invariant implies $Z(\omega, df)$ is an orthogonal increments process with variance $EZ(df)Z^*(df) = S(f)df$. The additivity and orthogonal process properties follows from the properties of the complex exponentials as eigenfunction of the covariance: that is for $F_1, F_2 \in \mathbb{R}$,

$$F_1 \cap F_2 = \emptyset, \; Z(F_1 \cup F_2) = \int_{F_1 \cup F_2} Z(\omega, df) = Z(F_1) + Z(F_2) \qquad (9.285)$$

$$EZ(F_1)Z(F_2) = \int_{F_1} \int_{F_2} EZ(df)Z^*(df') = \int_{F_1} \int_{F_2} \langle \phi_f, K_Y \phi_{f'} \rangle df\, df'$$
$$= \int_{F_1} \int_{F_2} S(f') \underbrace{\langle \phi_f, \phi_{f'} \rangle}_{\text{sifting property}} df\, df' = \int_{F_1 \cap F_2} S(f) df. \qquad (9.286)$$

□

9.12.2 Orthogonal Scale Representation

The identical approach holds for scale.

Definition 9.88 *Let $\{Y(\omega, t), t \in (0, \infty)\}$ be a second order process. Then $Y(t)$ is said to be* **second-order stationary to scale** *if*

$$EY(st)Y^*(t) = K(st, t) = K(s, 1), \quad t, s \in (0, \infty). \qquad (9.287)$$

Theorem 9.89 *Let $\{Y(\omega, t), t \in (0, \infty)\}$ be a second order process with covariance which is stationary with respect to scale $K(st, t) = K(s, 1)$ whose transform exists*

$$K(t, 1) = \int_{-\infty}^{\infty} t^{j2\pi f} S(f) df. \qquad (9.288)$$

Then the orthogonal representation,

$$Y(\omega, t) = \int_{-\infty}^{\infty} t^{j2\pi f} Z(\omega, df), \text{ with orthogonal process,}$$
$$Z(\omega, df) = \int_{0}^{\infty} t^{-j2\pi f} Y(\omega, t) \frac{1}{t} dt\, df, \qquad (9.289)$$

has spectral density $S(f) = \int_0^\infty t^{-j2\pi f} K(t, 1) \frac{1}{t} dt$.

Proof The eigenvalues and eigenfunctions for the scale operator U_t are $\lambda_f^{(t)} = t^{j2\pi f}, \phi_f(t) = t^{j2\pi f}$. Now we must show that $Z(\omega, df)$ is an orthogonal increments process with spectral density $S(f)$. The additivity property follows from the integral. For $F_1, F_2 \in \mathbb{R}$,

$$F_1 \cap F_2 = \varnothing, \quad Z(F_1 \cup F_2) = \int_{F_1 \cup F_2} Z(df) = Z(F_1) + Z(F_2) \tag{9.290}$$

$$EZ(F_1)Z(F_2) = \int_{F_1}\int_{F_2} EZ(df)Z^*(df') = \int_{F_1}\int_{F_2} \langle \phi_f, K_Y \phi_{f'}\rangle df\, df'$$

$$\stackrel{(a)}{=} \int_{F_1}\int_{F_2} S(f) \langle \phi_f, \phi_{f'}\rangle df\, df' \stackrel{(b)}{=} \int_{F_1 \cap F_2} S(f) df. \tag{9.291}$$

Property (b) follows from the sifting property; property (a) must be justified that $t^{j2\pi f}$ is an eigenfunction of the scale invariant covariance, $K_Y \phi_f = S(f)\phi_f$. Stationary with respect to scale implies the covariance operator has the property $K_Y(t,s) = K_Y(s^{-1}t, 1)$ which follows directly from the orthogonal representation theorem:

$$K_Y(t,s) = EY(t)Y^*(s) = E\int_{-\infty}^{\infty} t^{j2\pi f} Z(df) \int_{-\infty}^{\infty} s^{-j2\pi f'} Z^*(df')$$

$$= \int_{-\infty}^{\infty} t^{j2\pi f} \int_{-\infty}^{\infty} s^{-j2\pi f'} EZ(df)Z^*(df') = \int_{-\infty}^{\infty} (ts^{-1})^{j2\pi f} S(f)\, df. \tag{9.292}$$

Then,

$$K_Y \phi_f(t) = \int_0^{\infty} K_Y(t,s)\phi_f(s)\frac{1}{s}\, ds \stackrel{(a)}{=} \int_0^{\infty} \int_{-\infty}^{\infty} (ts^{-1})^{j2\pi f'} S(f')\, df'\, s^{j2\pi f} \frac{1}{s}\, ds$$

$$\tag{9.293}$$

$$= \int_{-\infty}^{\infty} t^{j2\pi f'} S(f') \underbrace{\int_0^{\infty} s^{-j2\pi f'} s^{j2\pi f} \frac{1}{s}\, ds}_{\langle \phi_f, \phi_{f'}'\rangle_{L^2(0,\infty)}} df' = \int_{-\infty}^{\infty} t^{j2\pi f'} S(f')\delta(f - f')\, df',$$

$$\tag{9.294}$$

where (a) follows from Eqn. 9.292 proving the eigenfunction property. Thus $\{Z(\omega, \cdot)\}$ is an orthogonal process with spectral density $S(f)$. $\square$

Example 9.90 (Stone's representation.) Stone's representation represents a continuous unitary operator $U_s, s \in T$ on $H(T)$ as

$$U_s = \int_{\mathbb{R}} \lambda_f^{(s)} P(df) \quad \text{with action} \quad U_s g(t) = \int_{\mathbb{R}} \lambda_f^{(s)} P(df) g(\cdot)|_t. \tag{9.295}$$

Here are several familiar examples: the Fourier and scale transforms.
1. **Fourier transform** Let $\{U_s, s \in \mathbb{R}\}$ form a unitary group on $s, s' \in \mathbb{R}$ satisfying $U_{s+s'} = U_s U_{s'}$; then with $H = L^2(\mathbb{R}, \mathbb{Z})$ and inner product

$$\langle g, h\rangle_{L^2((-\infty,\infty), dt)} = \int_{-\infty}^{\infty} g^*(t)h(t)\, dt. \tag{9.296}$$

Then the Fourier transform of the unitary shift operator is

$$g(t+s) = U_t g(s) = \int_{\mathbb{R}} e^{j2\pi ft} \left(\langle e^{j2\pi f \cdot}, g(\cdot)\rangle_{L^2(dt)} e^{j2\pi f \cdot}\, df\right)\bigg|_s \quad \text{for} \quad s, t \in \mathbb{R}.$$

$$\tag{9.297}$$

2. **Scale Transform** Let $H = L^2(\frac{1}{t} dt)$, with inner product

$$\langle g, h \rangle_{L^2(1/t)\, dt)} = \int_0^\infty g^*(t) h(t) \frac{1}{t}\, dt.$$

For the unitary scale operator $U_s, s \in (0, \infty)$, then the scale transform is given by

$$g(ts) = U_t g(s) = \int_{\mathbb{R}} t^{j2\pi f} \left(\langle e^{(j2\pi f) \ln \cdot}, g(\cdot) \rangle_{L^2(1/tdt)} e^{(j2\pi f) \ln s} \, df \right). \quad (9.298)$$

Example 9.91 (Elasticity Operator Induced Stationary Priors on $\mathbb{R}^3$) Assume the transformations are defined to be shift invariant $h : \mathbb{R}^3 \to \mathbb{R}^3$, again parameterized via the vector field relation $h(x) = x - u(x)$. For this the domain is not compact, and we allow for global registration via the rigid translation group as well as the various subgroups of **GL**(3). Assume again the prior is induced via the differential operators of continuum mechanics, as above small deformation linear elasticity with the resulting stochastic PDE again of the type.

The equation in vector form is

$$-b\nabla^2 U - c \nabla \cdot \nabla U + aU = W \quad (9.299)$$

with $A = -b\nabla^2 - c\nabla\nabla + aI$. The term aI represents the restoring elastic forces towards the static equilibrium. The minus signs are used for notational convenience later on.

We shall show that this is satisfied for the random Navier equations driven by a stochastic Gaussian process. The Cramer representation of the vector field from Section 9.12.1, Theorem 9.87, is appropriate since our operator is shift invariant. For this, introduce the orthogonal process representation of the noise process so that

$$W(x) = \int_{\omega \in \mathbb{R}^3} e^{j\langle x, \omega \rangle} dZ_W(\omega), \quad (9.300)$$

with $\langle x, \omega \rangle = \omega_1 x_1 + \omega_2 x_2 + \omega_3 x_3$ and $\{Z_W(\omega) \in \mathbb{C}^3, \omega \in \mathbb{R}^3\}$ a complex orthogonal process $Z_E(\Omega_1 \bigcup \Omega_2) = Z_W(\Omega_1) + Z(\Omega_2)$ for $\Omega_1 \cap \Omega_2 = \phi$, and $E\{Z_W(\Omega_1) Z_E^*(\Omega_2)\} = \int_{\Omega_1 \cap \Omega_2} S_w(\omega) d\omega$. $S_w(\omega)$ is the spectral density, matrix i.e. $E\{dZ_W(\omega) dZ_W^*(\omega)\} = S_w(\omega) d\omega$. For $w(x)$ a white noise process with power spectral density, σ^2, $S_w(\omega) = \sigma^2 I$. More general forms are useful.

The operator A is shift invariant, so let

$$U(x) = \int_{\omega \in \mathbb{R}^3} e^{j\langle x, \omega \rangle} dZ_W(\omega), \quad (9.301)$$

where $S(\omega) d\omega = E\{dZ(\omega) dZ^*(\omega)\}$ with covariance

$$K(x, y) = \int_{\omega \in \mathbb{R}^3} e^{j\langle x-y, \omega \rangle} S(\omega) \, d\omega. \quad (9.302)$$

Applying for example, the differential operator $(\partial/\partial x_1)$ to the U- process corresponds to multiplication by $i\omega_1$ in the frequency domain, and so on. Rewriting in component form gives

$$-b\left(\frac{\partial^2 U_1}{\partial x_1^2} + \frac{\partial^2 U_1}{\partial^2 x_2} + \frac{\partial^2 U_1}{\partial^2 x_3} \right)$$

$$- c \frac{\partial}{\partial x_1} \left(\frac{\partial U_1}{\partial x_1} + \frac{\partial U_2}{\partial x_2} + \frac{\partial U_3}{\partial x_3} \right) + aU_1 = W \quad (9.303)$$

and two more similar equations.

9.12 NON-COMPACT OPERATORS

Solving the stochastic Navier equation gives

$$AU(x) = \int_{\omega \in \mathbb{R}^3} e^{j\langle x, \omega \rangle} M(\omega) \, dZ(\omega), \quad (9.304)$$

with the function $M(\omega)$ a 3×3 matrix. Via straightforward formal manipulation of the stochastic equation $AU = W$ gives $M(\omega) dZ(\omega) = dZ_W(\omega)$ revealing

$$S(\omega) = M^{-1} S_w(\omega) M^{-1}. \quad (9.305)$$

To simplify the analysis, assume $W(x)$ is white, i.e. $S_w(\omega) = \sigma^2 \text{id}$. The actual form for $M(\omega)$ can be written as the sum

$$M(\omega) = M_1(\omega) + M_2(\omega) + M_3(\omega), \quad (9.306)$$

with

$$M_1(\omega) = b(\omega_1^2 + \omega_2^2 + \omega_3^2) \, \text{id}, \quad (9.307)$$
$$M_2(\omega) = c(\omega_i \omega_j; i, j = 1, 2, 3), \quad (9.308)$$
$$M_3(\omega) = a \, \text{id}. \quad (9.309)$$

Note that all M-matrices are non-negative and the third one strictly positive so that $M(\omega)$ is non-singular. In order to show that the u-field is defined we show that its spectral density matrix $\sigma M^{-2}(\omega)$ is integrable w.r.t. Lebesgue measure over $\mathbb{R}^3$. To do this we note that

$$M(\omega) \geq M_1(\omega) + M_3(\omega) \quad (9.310)$$

with the ususal partial order for matrices; just recall that $M_2(\omega)$ is non-negative. Hence

$$M^{-2} \leq \frac{1}{[a + b(\omega_1^2 + \omega_2^2 + \omega_3^2)]^2} \, \text{id}. \quad (9.311)$$

In polar coordinates this function behaves like r^{-4} for large r and is therefore integrable since the volume element is $r^2 \, d\phi \, d\psi$. Thus the prior has a meaningful continuous limit and is therefore acceptable.

10 METRICS SPACES FOR THE MATRIX GROUPS

ABSTRACT In this chapter the metric space structure of shape is developed. We do this by first studying the action of the matrix groups on the coordinate systems of shape. We begin by reviewing the well-known properties of the finite-dimensional matrix groups, including their properties as smooth Riemannian manifolds, allowing us to develop metric distances between the group elements. We explore the construction of the metric structure of these diffeomorphisms and develop many of the properties which hold for the finite dimensional matrix groups and subsequently in the infinite dimensional setting as well.

10.1 Riemannian Manifolds as Metric Spaces

Our principal study in this chapter is shapes studied via their transformation via diffeomorphisms. Both the finite dimensional matrix group and infinite dimensional diffeomorphisms will be examined. Although they cannot be added, they form a group which is as well a Riemannian manifold on which a metric space structure can be defined. The metric distance in the groups of diffeomorphisms is the length of the shortest curve geodesic connecting them. This induces the distance between shapes, becoming the distance between elements of the group which generate them. We begin by constructing metrics spaces and Riemannian manifolds.

10.1.1 Metric Spaces and Smooth Manifolds

Definition 10.1 A **metric space** is a set M with a function $\rho : M \times M \to \mathbb{R}^+$ which is a **metric** on set M if it satisfies the following three properties:

1. $\rho(g,h) \geq 0$ with $\rho(g,h) = 0$ if and only if $g = h$ for each $g, h \in M$;
2. $\rho(g,h) = \rho(h,g)$ for each $g, h \in M$ (symmetry);
3. $\rho(g,h) \leq \rho(g,f) + \rho(f,h)$ for each $g, h, f \in M$ (triangle inequality).

One of the most familiar examples of metric spaces are normed linear vector spaces as discussed in Section 2.4.1, Chapter 2. The associated metric $\rho : \mathcal{M} \times \mathcal{M} \to \mathbb{R}^+$ is given by the norm, $\rho(g,h) = \|g - h\|$. Notice, the norm satisfying triangle inequality implies the associated metric satisfies the triangle inequality; for all $f, g, h \in M$,

$$\rho(g,h) = \|g - h\| = \|g - f + f - h\| \tag{10.1}$$

$$\leq \|g - f\| + \|f - h\| = \rho(g,f) + \rho(f,h). \tag{10.2}$$

The classic examples of normed metric spaces which are familiar are the Euclidean space of n-tuples with sum-square norm, l^2, L^2 with sum-square and integral square norm, and continuous functions with sup-norm. Geometric shape is studied via transformations which are products of mappings which do not in general form a normed vector space; they will be intrinsically curved manifolds which do not support addition. To establish length and metric distance in such a setting Riemannian or geodesic length is required. Distance between elements is given by shortest or geodesic length of the curves connecting elements of the set.

10.1.2 Riemannian Manifold, Geodesic Metric, and Minimum Energy

To measure length in these perhaps curved manifolds, define paths or curves $g : [0,1] \to M$ connecting exists a smooth curve $g_t, t \in [0,1]$ with $g_0 = g, g_1 = h$, with smooth tangent (velocity) field $v_t = dg_t/dt$ well defined for all $t \in [0,1]$.

If the manifold M is as well a *Riemannian manifold*, then the tangent spaces have inner product $\langle \cdot, \cdot \rangle_{T_g}^{1/2}$ and norm $\| \cdot \|_{T_g}$ enabling the measurement of angles between curves (dot products) and lengths of curves (norms). This allows for the definition of the length.

Definition 10.2 *A manifold M is a* **Riemannian manifold** *if there is a symmetric, positive definite, inner product associated with each of the points of the tangent space; for all $g \in M$ there exists $\langle \cdot, \cdot \rangle_{T_g} : T_g(M) \times T_g(M) \to \mathbb{R}$ with associated norm $\| \cdot \|_{T_g} = \langle \cdot, \cdot \rangle_{T_g}^{1/2}$.*

Then curve $g : [0,1] \to M$ has **velocity** *$v_t = dg_t/dt, t \in [0,1]$ with* **length** *defined as*

$$\text{Length}(v) = \int_0^1 \|v_t = \frac{dg_t}{dt}\|_T \, dt. \tag{10.3}$$

Define g to be a **minimal geodesic**[29] *between $g, h \in M$ if*

$$g = \arg\inf_{v: g_0 = g, g_1 = h} \int_0^1 \|v_t = \frac{dg_t}{dt}\|_T \, dt. \tag{10.4}$$

It is convenient to work with curves in arc-length parameterization with constant speed. Since length does not change under reparameterization (see Lemma 7.38 of Chapter 7), geodesics will be in the so-called arc-length parameterization.

Lemma 10.3 *The infimum energy curves satisfying*

$$\inf_{v: g_0 = g, g_1 = h} \left(\int_0^1 \|v_t\|_T^2 \, dt \right) \tag{10.5}$$

are attained with constant speed $\|v_t = dg_t/dt\|_T = \text{constant}, t \in [0,1]$:

Proof The Cauchy–Schwartz inequality gives the ordering on the energy and length of the paths on the manifold with $v_t = dg_t/dt, t \in [0,1]$:

$$\text{Length}(v)^2 = \left(\int_0^1 \|v_t\|_T \, dt \right)^2 \leq \int_0^1 \|v_t\|_T^2 \, dt = E(v) \tag{10.6}$$

with equality only if (non-zero paths) $\|v_t\|_T = \text{constant}$ for all $t \in [0,1]$. Thus, the square of length $\text{Length}(v)$ of any path is a lower bound for the energy of the path, the minimum of energy being equal to square of length of the path if the path is traversed at constant speed $\|\dot{g}_t = v_t\|_T = c$. □

Such shortest length paths provide the metric to make the Riemannian manifold into a metric space. Also, minimum energy curves because of their constant speed property are geodesics, demonstrating the connection of the variational minimizers of quadratic forms and the associated geodesic length metric.

---

[29] There may be many length minimizing curves, in which case a minimal geodesic is one in the set.

Theorem 10.4 Let M be a Riemannian manifold with inner products $\langle \cdot, \cdot \rangle_{T(M)}$ and norm $\| \cdot \|_{T(M)}$ at every point of the tangent space. Define the positive function $\rho: M \times M \to \mathbb{R}^+$ on pairs g, h defined as the length of the minimal geodesic $g_t, t \in [0,1]$ on M connecting g to h:

$$\rho(g,h) = \inf_{v: g_0=g, g_1=h} \left(\int_0^1 \|v_t\|_T \, dt \right). \tag{10.7}$$

Then $\rho(\cdot,\cdot)$ satisfies symmetry and the triangle inequality, and makes M the Riemannian manifold into a metric space.

The infimum of the energy of all paths connecting g to h is the square of the metric (geodesic) distance:

$$E = \inf_{\substack{v \\ g_0=g, g_1=h}} \left(\int_0^1 \|v_t\|_T^2 \, dt \right) = \rho^2(g,h). \tag{10.8}$$

Proof To establish that ρ on $M \times M$ is in fact a metric distance, both symmetry and triangle inequality follow from the geodesic property. For symmetry, let g attain the infimum length connecting $g_0 = g$ to $g_1 = h$. Then generating a path $\tilde{g}_t, t \in [0,1]$, $\tilde{g}_0 = h$ with velocity $\tilde{v}_t = -v_{1-t}$ connects h to g so that $\tilde{g}_1 = g$ and has identical length. If there were an alternative candidate connecting h to g with shorter length then it should be used as the solution to connecting g to h.

The triangle inequality exploits the geodesic property; we must show that for all points $g, h, k \in M$,

$$\rho(g,k) \leq \rho(g,h) + \rho(h,k). \tag{10.9}$$

Define the curves $g^{0,1}, g^{1,2}$ to be the shortest length curves connecting pairs $(g,h), (h,k)$, respectively. Then define the new curve $\hat{g}$ connecting $\hat{g}_0 = g, \hat{g}_1 = k$ according to

$$\hat{g}_t = g^{0,1}_{2t}, \quad t \in \left[0, \tfrac{1}{2}\right], \tag{10.10}$$

$$= g^{1,2}_{2t-1}, \quad t \in \left[\tfrac{1}{2}, 1\right]. \tag{10.11}$$

Then $\hat{g}_0 = g, \hat{g}_1 = k$ and has length

$$\int_0^1 \|\hat{v}_t\|_T \, dt = \int_0^{1/2} 2\|v^{0,1}_{2t}\|_T \, dt + \int_{1/2}^1 2\|v^{1,2}_{2t-1}\|_T \, dt \tag{10.12}$$

$$\stackrel{(a)}{=} \rho(g,h) + \rho(h,k) \stackrel{(b)}{\geq} \rho(g,k), \tag{10.13}$$

with (a) coming from the constant speed Lemma 7.38 and (b) following from the fact that $\rho(g,k)$ attains an infimum over all possible curves.

The second half of the proof requires the constant speed Lemma 10.3 of geodesics demonstrating minimum energy paths are at constant speed:

$$E = \inf_{\substack{\|v\|=\|\dot{g}\|=c: \\ g_0=g, g_1=h}} E(v) = \inf_{\substack{\|v\|=\|\dot{g}\|=c: \\ g_0=g, g_1=h}} \left(\text{Length}(v)\right)^2 = \rho^2(g,h). \tag{10.14}$$

□

10.2 Vector Spaces as Metric Spaces

One of the normed spaces which shall be worked with in describing the patterns of random vectors are l^2, L^2 the Hilbert space of square-summable and integrable sequences and functions. The Hilbert space l^2 is the set of countable vector functions $g(\cdot) = (g_1, g_2, \dots)$ having finite two norm $\|g\|_{l^2} = \left(\sum_i |g_i|^2\right)^{\frac{1}{2}}$ induced by the inner product: $\langle g, h\rangle_{l^2} = \sum_i g_i h_i^\star$. The square-integrable Hilbert space L^2 are the functions $g(\cdot)$ having finite two norm $\|g\|_{L^2} = \left(\int |g_t|^2 dt\right)^{\frac{1}{2}}$ and inner product inducing the norm $\langle g, h\rangle_{L^2} = \int_0^1 g_t h_t^\star \, dt$. Thinking on the continuum puts us on L^2; discretizing to pixels associates us to l^2.

If H is a Hilbert space with the inner product $\langle \cdot, \cdot \rangle_H$ and associated norm $\|\cdot\|_H$, then it is a Riemannian manifold; the tangent spaces are the spaces themselves. To see this construct the tangent space at each point $g \in H$ by generating curves

$$g_t = (h - g)t + g, \quad \text{for all } g, h \in H. \tag{10.15}$$

Then the tangent vectors $v^g \in T_g(H)$ are the entire space H since $\|h - g\|_H \leq \|g\|_H + \|h\|_H < \infty$.

For these structures shortest paths are not curved, but rather are straight lines. The minimal geodesic lengths are then just the normed differences between the elements.

Theorem 10.5 *Given the Hilbert space $(H, \|\cdot\|_H)$ then, the minimal geodesics in H with tangent norm $\|\frac{d}{dt}g_t = v_t\|_H$ minimizing $\int_0^1 \|v_t\|_H^2 \, dt$ connecting $g_0 = g, g_1 = h$ satisfy the Euler equation*

$$\frac{dv_t}{dt} = 0, \quad \text{implying } v_t = h - g. \tag{10.16}$$

The geodesics are straight lines and the metric distance $\rho : H \times H \to \mathbb{R}^+$ is given by

$$\rho(g, h) = \inf_{\substack{v = \frac{d}{dt}g: \, g_0 = g \\ g_1 = h}} \left(\int_0^1 \|v_t\|_H \, dt\right) = \|h - g\|_H. \tag{10.17}$$

Proof To calculate the minimal geodesic with minimizing energy $\int_0^1 \|v_t\|^2 dt$, we use the calculus of variations to generate a smooth perturbation of $g_t \to g_t(\epsilon) = g_t + \epsilon \eta_t$ giving the perturbed vector field $v_t \to v_t + \epsilon (d/dt)\eta_t$ and satisfying the boundary conditions $\eta_0 = \eta_1 = 0$ so that $g_0(\epsilon) = g_0, g_1(\epsilon) = g_1$. Then

$$0 = \frac{d}{d\epsilon}\int_0^1 \|v_t + \epsilon\frac{d}{dt}\eta_t\|_H^2 \, dt|_{\epsilon=0} = 2\int_0^1 \left\langle v_t, \frac{d\eta_t}{dt}\right\rangle_H dt \tag{10.18}$$

$$\stackrel{(a)}{=} \int_0^1 \langle \frac{dv_t}{dt}, \eta_t\rangle_H \, dt = 0, \tag{10.19}$$

where (a) follows via integration by parts and the boundary condition. This gives $v_t =$ constant, and the form for the minimizing geodesics $g_t = (h - g)t + g$ with total length given by the norm. □

This suggests that any positive definite quadratic form can be introduced in the inner product $\langle \cdot, \cdot\rangle_Q$ and associated norm $\|\cdot\|_Q$ giving the familiar representation of the metric.

Corollary 10.6 *Let Q be a positive definite quadratic form defining the inner product in the tangent space $\langle g, h\rangle_Q$, then*

$$\rho(g, h)^2 = \|g - h\|_Q^2. \tag{10.20}$$

If the Hilbert space has closed and bounded background space, then it has a countable orthonormal base diagonalizing Q with eigenelements $\{q_i, \phi_i\}$, $Q\phi_i = q_i \phi_i$ and the metric reduces to an l^2 metric:

$$\rho(g,h)^2 = \sum_i q_i |g_i - h_i|^2, \quad \text{where } g = \sum_i g_i \phi_i, \ h = \sum_i h_i \phi_i. \tag{10.21}$$

10.3 Coordinate Frames on the Matrix Groups and the Exponential Map

We will now study curves, in particular the shortest geodesic curves on the manifolds of the matrix groups. These will provide the metrics and allow us to define the tangent spaces for taking derivatives.

10.3.1 Left and Right Group Action

Notice, from the curves emanating from any matrix group element then the tangent space and coordinate frames that can be generated as in Theorem 8.12 of Chapter 8. This is how the exponential mapping will be used at $t = 0$ to generate the tangent space at the identity for the subgroups of **GL**(n).

For this define the group action as a transformation of the manifold.

Definition 10.7 *Define the* **left and right group actions** *transforming the generalized linear group* $\Phi^l(B, \cdot), \Phi^r(B, \cdot) : \mathbf{GL}(n) \to \mathbf{GL}(n)$ *for* $B \in \mathbf{GL}(n)$:

$$\Phi^l(B, \cdot) : A \in \mathbf{GL}(n) \mapsto \Phi^l(B, A) = B \circ A \tag{10.22}$$

$$\Phi^r(B, \cdot) : A \in \mathbf{GL}(n) \mapsto \Phi^r(B, A) = A \circ B^*, \tag{10.23}$$

where $\circ$ denotes group operation of matrix multiplication. It is usual convention to drop the composition symbol $\circ$ with the understanding that $A \circ B = AB$ is matrix multiplication.

For the affine motions in homogeneous coordinates $\bar{A} \in \mathbf{A}(n)$, the group actions are defined similarly as above $\Phi^l(\bar{A}, \cdot), \Phi^r(\bar{A}, \cdot) : \mathbf{A}(n) \to \mathbf{A}(n)$.

Notice, for the right transformation to be a group action, the transpose (or inverse for that matter) must be applied according to

$$\Phi^r(C, \Phi^r(B, A)) = (A \circ B^*) \circ C^* = A \circ (B^* \circ C^*) \tag{10.24}$$

$$= A \circ (C \circ B)^* = \Phi^r(C \circ B, A). \tag{10.25}$$

The left group action is the usual convention in rigid body mechanics (see Arnold). The right group action will be consistent with our examinations of the matrix groups applied to functions such as the images. In particular we will use the right action to define the coordinate frames associated with curves emanating from points on the manifold.

The curves used to generate the tangent space will be generated from the exponential map.

Definition 10.8 *Define set of* $n \times n$ **real-valued matrices** $\mathcal{M}(n)$. *Then the* **exponential** e^X *of the matrix* $X = (X_{ij}) \in \mathcal{M}(n)$ *is defined to be the matrix given by the following series when the series converges:*

$$e^X = I + X + \frac{1}{2!}X^2 + \frac{1}{3!}X^3 + \cdots. \tag{10.26}$$

10.3 COORDINATE FRAMES ON THE MATRIX GROUPS

The **exponential integral curve** rooted at the identity in the direction of $V = (V_{ij}) \in \mathcal{M}(n)$ is denoted as $e^{tV}, t \in [0,1]$. The **left** and **right** group action is defined as

$$\Phi^l(e^{tV}, A) = e^{tV}A, \quad \Phi^r(e^{tV}, A) = Ae^{-tV}. \tag{10.27}$$

As proved in Boothby pages 147–150 e^X converges for all $X \in \mathcal{M}(n)$ implying that e^{tV}, $t \in [0,1]$ and $V \in \mathcal{M}(n)$ are curves through $\mathbf{GL}(n)$ and can be used to generate the tangent space. This follows from the fact that it converges absolutely (proved in Boothby), implying that $e^{Y+Z} = e^Y e^Z$ if the matrices commute $YZ = ZY$. Clearly, V and $-V$ commute, implying $e^{tV-tV} = e^{tV}e^{-tV} = I$. Thus, e^{tV} has an inverse implying it is an element of $\mathbf{GL}(n)$.

10.3.2 The Coordinate Frames

For the generalized linear group the tangent spaces are generated from the exponential curves. For this we use the right action applied to the manifold of matrix group elements.

Theorem 10.9 *For the generalized linear group $\mathbf{GL}(n)$ the elements of the tangent space $T_A(\mathbf{GL}(n))$ at $A \in \mathbf{GL}(n)$ are constructed from curves generated from the right action of the exponential map. Identifying elements of v_{ij} with coordinate frames $\partial/(\partial x_{ij})$, then the tangent elements are given by*

$$\sum_{ij}(AV)_{ij}\frac{\partial}{\partial x_{ij}} \in T_A(\mathbf{GL}(n)). \tag{10.28}$$

Defining the $n \times n$ matrix $\mathbf{1}_{ij} = (\delta_{ij}(k,l))$ having ij-entry of 1, and 0 otherwise, then the $m \leq n^2$ coordinate frames at the identity $E_{\cdot,I}, i$ of the subgroups are

1. *for $\mathbf{GL}(n)$, there are $m = n^2$ coordinate frames at the identity transformed by A:*

$$E_{ijI} = \mathbf{1}_{ij}, \; i,j = 1,\dots,n; \tag{10.29}$$

2. *for $\mathbf{O}(n)$, there are $m = n(n-1)/2$ coordinate frames,*

$$E_{ijI} = \mathbf{1}_{ij} - \mathbf{1}_{ji}, \quad j = 1+i,\dots,n, \quad i = 1,\dots,n-1; \tag{10.30}$$

3. *for $\mathbf{SL}(n)$, there are $n^2 - 1$ coordinate frames; define the $n-1 \times n-1$ full rank matrix $\left(v_{ij}\right)$, then the coordinate frames become*

$$E_{ijI} = \mathbf{1}_{ij}, \quad 1 \leq i \neq j \leq n; \tag{10.31}$$

$$E_{iiI} = \sum_{j=1}^{n-1} v_{ij}\mathbf{1}_{jj} - \left(\sum_{j=1}^{n-1}v_{ij}\right)\mathbf{1}_{nn}, \quad i=1\dots n-1. \tag{10.32}$$

4. *for $\mathbf{US}(n)$, there is 1 coordinate frame at the identity matrix $E_I = \sum_{i=1}^n \mathbf{1}_{ii} = I$.*

Proof The coordinate frames for the tangent spaces $T_A(\mathbf{GL}(n))$ for the vector fields are transferred to arbitrary points $A \in \mathbf{GL}(n)$ on the entire manifold by examining the curves emanating from A using the property of the derivative of the right group action

$$\frac{d}{dt}\Phi^r(e^{tV}, A)|_{t=0} = \frac{d}{dt}Ae^{-tV}|_{t=0} = -AV. \tag{10.33}$$

Then for all $V = (v_{ij}) \in \mathcal{M}(n)$, identifying elements of v_{ij} with coordinate frames $\partial/\partial x_{ij}$, then the tangent elements are given by

$$\frac{d}{dt}f \circ (Ae^{-tV})|_{t=0} = \sum_{ij} \frac{\partial f}{\partial x_{ij}}(Ae^{-tV})\bigg|_{t=0} \left(\frac{d}{dt}Ae^{-tV}|_{t=0}\right)_{ij}$$

$$= -\sum_{ij}(AV)_{ij} \frac{\partial f}{\partial x_{ij}}(A). \tag{10.34}$$

Proof of $\mathbf{GL}(n)$: Let $E_{ijI} = \mathbf{1}_{ij}$, then $e^{tE_{ijI}} \in \mathbf{GL}(n)$ (Theorem 10.9 above) therefore $\frac{d\Phi_t}{dt}|_{t=0} = E_{ijI} \in T_I(\mathbf{GL}(n))$. Since the $E_{ijI} = \mathbf{1}_{ij}, i = 1, \ldots n, j = 1, \ldots n$ are independent, and n^2 in number, they span the n^2 dimensional tangent space $T_I(\mathbf{GL}(n))$ and are a basis.

Proof of $\mathbf{O}(n)$: Let $\mathbf{O}(n) \subset \mathbf{GL}(n)$ be the orthogonal group. Then $e^{tV}(e^{tV})^* = I$, giving $e^{tV}e^{tV^*} = I$, which implies since V commutes with itself, and $e^{tV}e^{-tV} = e^{t(V-V)} = I$, $V^* = -V$. V is a skew symmetric matrix. Clearly V has dimension $n(n-1)/2$ the number of free parameters in a skew symmetric matrix, implying $\mathbf{O}(n)$ has dimension $n(n-1)/2$.

Proof of $\mathbf{SL}(n)$: For A non-singular, then

$$Ae^X A^{-1} = AA^{-1} + AXA^{-1} + \frac{1}{2}AX^2A^{-1} + \cdots$$

$$= I + AXA^{-1} + \frac{1}{2}AX^2A^{-1} + \cdots \stackrel{(a)}{=} e^{AXA^{-1}}, \tag{10.35}$$

and (a) follows since $AXA^{-1} \in \mathcal{M}(n)$ and Theorem 10.9, e^X converges absolutely. This implies $\det e^X = \det e^{XAX^{-1}}$ giving $\det e^X = e^{\text{tr}X}$. Thus, $X \in \mathbf{SL}(n)$ implies $\text{tr}X = 0$. Thus one parameter subgroups of $\mathbf{SL}(n)$ takes the form e^{tA}, where A is an element of $n^2 - 1$ dimensional manifold subset $\mathcal{M}(n)$ satisfying $\text{tr}A = \sum_{i=1}^n a_{ii} = 0$. □

Corollary 10.10 *The coordinate frames of the tangent spaces at any point $A \in \mathcal{S}$ are determined by the coordinate frames E_{iI} at the identity:*

$$E_{iA} = A \circ E_{iI}. \tag{10.36}$$

The vector fields for $\mathbf{GL}(n)$ and $\mathbf{O}(n)$ take the following form:
1. *for $\mathbf{GL}(n)$ the generalized linear group*

$$E_{ijA} = A \circ \mathbf{1}_{ij} = \begin{pmatrix} 0 & \cdots & 0 & a_{1i} & 0 & \cdots & 0 \\ \vdots & \vdots & \vdots & \vdots & \vdots & \vdots & \vdots \\ 0 & \cdots & 0 & a_{n-1\,i} & 0 & \cdots & 0 \\ 0 & \cdots & 0 & a_{ni} & 0 & \cdots & 0 \\ & & & i\text{th column} & & & \end{pmatrix};$$

10.3 COORDINATE FRAMES ON THE MATRIX GROUPS

2. for **O**(n) the orthogonal group

$$E_{ij}A = A \circ (\mathbf{1}_{ij} - \mathbf{1}_{ji}) = \begin{pmatrix} 0 & \cdots & 0 & a_{1i} & 0 & \cdots & 0 & -a_{1j} & 0 & \cdots \\ \vdots & \vdots & \vdots & \vdots & \vdots & \vdots & \vdots & \vdots & \vdots & \vdots \\ 0 & \cdots & 0 & a_{n-1\,i} & 0 & \cdots & 0 & -a_{n-1\,j} & 0 & \cdots \\ 0 & \cdots & 0 & a_{ni} & 0 & \cdots & 0 & -a_{nj} & 0 & \cdots \\ & & & i\text{th column} & & & & i\text{th column} & & \end{pmatrix};$$

(10.37)

3. for **US**(n) the scale group, $E_{ij}A = A$.

Example 10.11 (Integral Curves for Scale and Rotation) For scaling, $e^{tV} = \begin{pmatrix} e^{tv} & 0 \\ 0 & e^{tv} \end{pmatrix}$ where $V = \begin{pmatrix} v & 0 \\ 0 & v \end{pmatrix}$, then the curve rooted at $A = (a_{ij}) \in \mathbf{GL}(n)$ gives

$$\Phi^r(e^{tV}, A) = \begin{pmatrix} a_{11} & a_{12} \\ a_{21} & a_{22} \end{pmatrix} \begin{pmatrix} e^{-tv} & 0 \\ 0 & e^{-tv} \end{pmatrix}, \quad \text{with } V_{Ae^{-tV}} = -\sum_{ij} a_{ij} e^{-tv} v \frac{\partial}{\partial x_{ij}}.$$

(10.38)

For rotation in **SO**(2) at the identiy $e^{tV} = \begin{pmatrix} \cos t & \sin t \\ -\sin t & \cos t \end{pmatrix}$ where $V = \begin{pmatrix} 0 & 1 \\ -1 & 0 \end{pmatrix}$, then

$$\Phi^r(e^{tV}, A) = \begin{pmatrix} a_{11} & a_{12} \\ a_{21} & a_{22} \end{pmatrix} \begin{pmatrix} \cos t & -\sin t \\ \sin t & \cos t \end{pmatrix}, \quad \text{with } V_{Ae^{-tV}}$$

$$= \sum_{ij} \left(\begin{pmatrix} a_{11} & a_{12} \\ a_{21} & a_{22} \end{pmatrix} \begin{pmatrix} -\sin t & -\cos t \\ \cos t & -\sin t \end{pmatrix} \right)_{ij} \frac{\partial}{\partial x_{ij}}.$$

(10.39)

10.3.3 Local Optimization via Directional Derivatives and the Exponential Map

To generalize our local variational calculus for extreme points on curved matrix groups (viewed as smooth manifolds) directional derivatives at any point of the tangent space must be computed. This will become possible as we characterize the basis of the tangent space on the matrix groups from which the directional derivatives are generated. We have all of the tools to generalize our local variational calculus for optimization of functions on manifolds which are matrix groups. We assume throughout that the extreme points are interior to the manifold. With the ability to generate tangents via the curves, local optimization falls out immediately on the matrix groups as requiring the directional derivatives with the coordinate frames being zero. The tangents to these curves will be elements of the tangent space of the Lie group $\mathcal{S}$.

Theorem 10.12 *Let H be a smooth function on M, and let $\{E_i, i = 1, \ldots, m\}$ be the set of m-coordinate frames from M. The necessary condition for $p \in O \subset M$ to be a local maximizer of H becomes*

$$E_{ip} H = 0, \quad i = 1, \ldots m.$$

(10.40)

Proof To calculate the optimizer condition, let $t_0 = 0$, with curve rooted in $p \in M$ according to $\Phi_t(p) \in M$, $\Phi_0(p) = p$ and $f = H$, then from Theorem 8.12

$$\frac{d}{dt} H \circ \Phi_t = \sum_{i=1}^{m} \dot{x}_i(0) E_{i\Phi_0(p)} H = 0. \tag{10.41}$$

Since this is true for all curves, the necessary condition on the coordinate frames follows. □

10.4 Metric Space Structure for the Linear Matrix Groups

We shall study shapes and structures using the low-dimensional matrix groups and high-dimensional diffeomorphisms. First we study the metric space structure for the matrix groups.

10.4.1 Geodesics in the Matrix Groups

Now examine the geodesics within the subgroups of the matrix groups. Measure the length in the tangent space with the quadratic form defined by the $d \times d$ positive definite matrix M according to

$$\|f\|_M^2 = \langle Mf, f \rangle_{\mathbb{R}^{d \times d}} = \operatorname{tr} M ff^*. \tag{10.42}$$

For calculating the geodesics we take the variation in the group via perturbation of the optimum trajectory. The significant departure from the approach taken for the Hilbert vector space (e.g. see Theorem 10.5), is that the perturbation argument is not done via addition, but rather via group action. Define $g_t, t \in [0,1]$ to be the optimizing geodesic connecting $g_0 = g$ to $g_1 = h$; then each point along the curve is perturbed in the direction of η_t according to $e^{\epsilon \eta_t}$: $g_t \to g_t(\epsilon) = e^{\epsilon \eta_t} g_t = (\mathrm{id} + \epsilon \eta_t) g_t + o(\epsilon)$. The perturbation must satisfy the boundary condition $g_0(\epsilon) = g, g_1(\epsilon) = h$, implying $\eta_0 = \eta_1 = 0$.[30] This is depicted in Figure 10.1.

The following Lemma calculates the perturbation of v_t along the flow g_t. It explicitly involves the Lie bracket which dictates how elements in the tangent space transform along the flow.

Lemma 10.13 *With $\frac{dg_t}{dt} = v_t g_t$, then transporting tangent element w_0 by the group as $w_t = g_t w_0 g_t^{-1}$ gives the Lie bracket*[31]

$$\frac{dw_t}{dt} = (v_t w_t - w_t v_t) = [v_t, w_t]. \tag{10.43}$$

The perturbation of the group element

$$g_t \to g_t(\epsilon) = (\mathrm{id} + \epsilon \eta_t) g_t + o(\epsilon) = g_t + \epsilon \eta_t g_t + o(\epsilon). \tag{10.44}$$

gives the perturbed vector field $v_t \to v_t(\epsilon) = v_t + \epsilon(dv_t(\epsilon)/d\epsilon) + o(\epsilon)$ satisfying the Lie bracket equation

$$\frac{dv_t(\epsilon)}{d\epsilon} = \frac{d\eta_t}{dt} - (v_t \eta_t - \eta_t v_t). \tag{10.45}$$

---

[30] Notice, in the vector space case this is addition of η rather than matrix multiplication; see Theorem 10.5.
[31] The Lie bracket for matrices $A, B \in \mathbb{R}^{d \times d}$ is familiar as $[A, B] = AB - BA$.

10.4 METRIC SPACE STRUCTURE

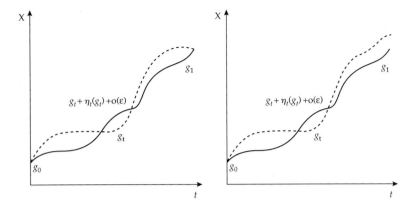

Figure 10.1 Panel 1 shows perturbation of the flow $g_t \to g_t(\epsilon) = g_t + \epsilon \eta_t(g_t) + o(\epsilon)$ in exact matching with $g_1(\epsilon) = g_1$. For the matrix case $g_t \to g_t(\epsilon) = (\text{id} + \epsilon \eta_t) g_t + o(\epsilon) = g_t + \eta_t g_t + o(\epsilon)$. Panel 2 shows the variation of the flow g_t in inexact matching with $g_1(\epsilon)$ free, i.e. $g_1(\epsilon) \neq g_1$.

Proof *Part 1:* Define $dg_t/dt = v_t g_t$, then $dg_t^{-1}/dt = -g_t^{-1} v_t$, implying

$$\frac{dw_t}{dt} = \frac{d}{dt}(g_t w_0 g_t^{-1}) = v_t w_t - w_t v_t = [v_t, w_t]. \tag{10.46}$$

Part 2: The Lemma follows according to

$$\frac{d}{dt}(g_t + \epsilon \eta_t g_t) = \left(v_t + \epsilon \frac{dv_t(\epsilon)}{d\epsilon}\right)(g_t + \epsilon \eta_t g_t) + o(\epsilon)$$

$$\stackrel{(a)}{=} v_t g_t + \epsilon v_t \eta_t g_t + \epsilon \frac{dv_t(\epsilon)}{d\epsilon} g_t + o(\epsilon)$$

$$= \frac{dg_t}{dt} + \epsilon \frac{d\eta_t}{dt} g_t + \epsilon \eta_t \frac{dg_t}{dt} \stackrel{(b)}{=} v_t g_t + \epsilon \frac{d\eta_t}{dt} g_t + \epsilon \eta_t v_t g_t,$$

and equating (a),(b) and applying g_t^{-1} on the right to both terms gives

$$\left.\frac{dv_t(\epsilon)}{d\epsilon}\right|_{\epsilon=0} = \frac{d\eta_t}{dt} + \eta_t v_t - v_t \eta_t.$$

□

With this we can calculate the geodesic equation for the matrix groups.

Theorem 10.14 *The minimal geodesics connecting $g, h \in \mathbf{GL}(d)$ with tangent norm-square $\|v_t\|_M^2 = \langle Mv_t, v_t \rangle_{\mathbb{R}^{d \times d}}$ minimizing $\int_0^1 \|v_t\|_M^2 dt$ satisfying $dg_t/dt = v_t g_t$ with $g_0 = g$, $g_1 = h$ are solutions of the Euler equation given by*

$$\frac{dMv_t}{dt} = Mv_t v_t^* - v_t^* Mv_t. \tag{10.47}$$

Proof The proof on the geodesic follows a path shown to one of the authors by Trouve and Younes. Examine a smooth perturbation of the shortest length curve (extremal energy) $g_t, t \in [0, 1]$ to the perturbed flow $g_t(\epsilon) = (\text{id} + \epsilon \eta_t + o(\epsilon)) g_t$, as depicted in Figure 10.1. The perturbation has the property that $\eta_0 = \eta_1 = 0$ so that the boundary conditions remain satisfied at $g_0(\epsilon) = g, g_1(\epsilon) = h$. Using Lemma 10.13 we can identify how v_t satisfying $(d/dt) g_t = v_t g_t$ changes to $v_t(\epsilon) = v_t + \epsilon (dv_t(\epsilon)/d\epsilon) + o(\epsilon)$ satisfying $(d/dt) g_t(\epsilon) = v_t(\epsilon) g_t(\epsilon)$. This is given according to Lemma 10.13.

To complete the theorem, the variational equation which the minimizing energy geodesic satisfies

$$0 = \frac{d}{d\epsilon} \int_0^1 \left\| v_t + \epsilon \frac{dv_t(\epsilon)}{d\epsilon} \right\|_M^2 dt \bigg|_{\epsilon=0} = 2 \int_0^1 \left\langle v_t, \frac{dv_t(\epsilon)}{d\epsilon} \right\rangle_M dt \bigg|_{\epsilon=0} \quad (10.48)$$

$$= 2 \int_0^1 \left\langle Mv_t, \frac{d\eta_t}{dt} + \eta_t v_t - v_t \eta_t \right\rangle_{\mathbb{R}^{d \times d}} dt \quad (10.49)$$

$$= 2 \int_0^1 \left\langle Mv_t, \frac{d\eta_t}{dt} \right\rangle_{\mathbb{R}^{d \times d}} dt + 2 \int_0^1 \langle Mv_t v_t^* - v_t^* Mv_t, \eta_t \rangle_{\mathbb{R}^{d \times d}} dt \quad (10.50)$$

$$\stackrel{(a)}{=} 2 \int_0^1 \left\langle -\frac{dMv_t}{dt} + Mv_t v_t^* - v_t^* Mv_t, \eta_t \right\rangle_{\mathbb{R}^{d \times d}} dt, \quad (10.51)$$

with (a) using integration by parts and the boundary conditions on the perturbation. Since this is zero over all perturbations, the Euler Eqn. 10.47 on the geodesics. □

10.5 Conservation of Momentum and Geodesic Evolution of the Matrix Groups via the Tangent at the Identity

Along the geodesic there is a conservation law; essentially the momentum Mv_t is constant. This implies that by knowing momentum acting against matrix fields at the identity gives the matrix field and momentum along the geodesic.

Theorem 10.15 *Defining the matrix field transported from the identity as $w_t = g_t w_0 g_t^{-1}$, then along the geodesics satisfying the Euler equation (Theorem 10.14) the momentum is conserved:*

$$\frac{d}{dt} \langle Mv_t, w_t \rangle_{\mathbb{R}^{d \times d}} = 0, \quad t \in [0,1]. \quad (10.52)$$

Along the geodesic, the momentum is determined by the momentum at the identity:

$$Mv_t = g_t^{-1*} Mv_0 g_t^*. \quad (10.53)$$

Proof Use the geodesic equation equalling 0 to evaluate the derivative

$$\frac{d}{dt} \langle Mv_t, w_t \rangle_{\mathbb{R}^{d \times d}} = \left\langle \frac{dMv_t}{dt}, w_t \right\rangle_{\mathbb{R}^{d \times d}} + \left\langle Mv_t, \frac{dw_t}{dt} \right\rangle_{\mathbb{R}^{d \times d}} \quad (10.54)$$

$$\stackrel{(a)}{=} \left\langle \frac{dMv_t}{dt}, w_t \right\rangle_{\mathbb{R}^{d \times d}} + \langle Mv_t, (v_t w_t - w_t v_t) \rangle_{\mathbb{R}^{d \times d}} \quad (10.55)$$

$$\stackrel{(b)}{=} \left\langle \frac{d}{dt} Mv_t + v_t^* Mv_t - Mv_t v_t^*, w_t \right\rangle_{\mathbb{R}^{d \times d}} \stackrel{(c)}{=} 0, \quad (10.56)$$

with (a) from Lemma 10.13, and (b) definition of the transpose, with (c) exactly following from the Euler Eqn. 10.47 equalling 0.

Part 2: Since $w_t = g_t w_0 g_t^{-1}$ this implies $g_t^{-1} w_t g_t = w_0$ with conservation of momentum implying

$$\langle Mv_t, w_t \rangle_{\mathbb{R}^{d \times d}} = \langle Mv_0, w_0 \rangle_{\mathbb{R}^{d \times d}} = \langle Mv_0, g_t^{-1} w_t g_t \rangle_{\mathbb{R}^{d \times d}} \quad (10.57)$$

$$= \langle g_t^{-1*} Mv_0 g_t^*, w_t \rangle_{\mathbb{R}^{d \times d}}, \quad (10.58)$$

which reduces to Eqn. 10.53. □

10.6 Metrics in the Matrix Groups

In particular, various metrics can be calculated in the subgroups.

Theorem 10.16 *Let the norm in the tangent space have $M = \mathrm{id}$.*

1. *For the orthogonal motions $g_t \in \mathbf{SO}(d) \subset \mathbf{GL}(d)$, then the velocity on the geodesic is skew-symmetric, $v_t = -v_t^*$ implying the velocity is constant $\partial v_t / \partial t = 0$. With A the $d \times d$ skew-symmetric matrix satisfying $O' = e^A O \in \mathbf{O}(d)$, then geodesic distance between O, O' is given by*

$$\rho^2(O,O') = 2 \sum_{i=1}^{(d-1)d/2} a_i^2, \quad \text{where } A = \begin{pmatrix} 0 & a_1 & \cdots & a_{d-1} \\ -a_1 & 0 & \cdots & a_{d-2} \\ \vdots & \vdots & \vdots & \vdots \\ -a_{d-1} & \cdots & -a_1 & 0 \end{pmatrix}. \tag{10.59}$$

For $\mathbf{SO}(2), \mathbf{SO}(3)$ the distances become, respectively,

$$\rho^2_{\mathbf{SO}(2)}(O,O') = 2\arccos^2\left(\tfrac{1}{2} \mathrm{tr} O' O^*\right), \tag{10.60}$$

$$\rho^2_{\mathbf{SO}(3)}(O,O') = 2\arccos^2\left(\tfrac{1}{2} \mathrm{tr} O' O^* - \tfrac{1}{2}\right). \tag{10.61}$$

2. *The geodesic distance between diagonal scale matrices $\mathrm{diag}(s_1, s_2, \ldots), \mathrm{diag}(s'_1, s'_2, \ldots)$ is*

$$\rho^2(\mathrm{diag}(s_1,s_2,\ldots),\mathrm{diag}(s'_1,s'_2,\ldots)) = \sum_{i=1}^{d} |\log \tfrac{s_i}{s'_i}|^2. \tag{10.62}$$

3. *The geodesic distance between similitudes $(s, O), (s', O') \in \mathbf{Sim} = (0, \infty) \times \mathbf{SO}(d)$ where $O' = e^A O$ (A of Eqn. 10.59) is given by*

$$\rho^2((s,O),(s',O')) = |\log \tfrac{s}{s'}|^2 + 2 \sum_{i=1}^{(d-1)d/2} a_i^2. \tag{10.63}$$

Proof For the orthogonal group with the flow through $\mathbf{O}(d)$, since $\mathrm{id} = g_t g_t^*$, then

$$0 = \frac{dg_t}{dt} g_t^* + g_t \frac{dg_t^*}{dt} = 0, \quad \text{with } g_0 = O. \tag{10.64}$$

Substituting $dg_t/dt = v_t g_t$, this implies that

$$0 = v_t g_t g_t^* + g_t g_t^* v_t^* = v_t + v_t^*, \tag{10.65}$$

giving that v_t is skew-symmetric. Clearly then the speed along v_t is constant since

$$\frac{dv_t}{dt} = v_t v_t^* - v_t^* v_t = 0. \tag{10.66}$$

Thus the general solution for the geodesics becomes $g_t = e^{At}O, g_0 = O$, with A an $n \times n$ skew-symmetric matrix $A = -A^*$. To calculate the distances determined by the $((d-1)/2)d$ entries of A explicitly,

$$\rho^2(O, O') = \rho^2(O, e^A O) = \int_0^1 \|v_t\|_{\mathbb{R}^{d \times d}}^2 \, dt = \int_0^1 \text{tr} AA^* \, dt \tag{10.67}$$

$$= \text{tr} AA^* = 2 \sum_{i=1}^{(d-1)d/2} a_i^2 = 2\beta^2 \quad \text{where } \beta^2 = \sum_{i=1}^{(d-1)d/2} a_i^2. \tag{10.68}$$

In **SO**(3) with $O' = e^A O$ with $A = \begin{pmatrix} 0 & a_1 & a_2 \\ -a_1 & 0 & a_3 \\ -a_2 & -a_3 & 0 \end{pmatrix}$, define $\beta^2 = a_1^2 + a_2^2 + a_3^2$, then $\rho^2(O, O') = 2\beta^2$. Now to simplify the matrix exponential $e^A = \text{id} + A + \frac{1}{2}A^2 \ldots$, the matrix sum reduces exploiting the fact that $A^3 = -\beta^2 A$ since

$$A^2 = \begin{pmatrix} (-a_1^2 - a_2^2) & -a_2 a_3 & a_1 a_3 \\ -a_2 a_3 & (-a_1^2 - a_3^2) & -a_1 a_2 \\ a_1 a_3 & -a_1 a_2 & (-a_2^2 - a_3^2) \end{pmatrix}. \tag{10.69}$$

Using the alternating sin, cos series gives the Rodriquez formula:

$$e^A = \text{id} + \frac{A}{\beta}\left(\beta - \frac{1}{3!}\beta^3 + \cdots\right) + \frac{A^2}{\beta^2}\left(\frac{\beta^2}{2!} - \frac{1}{4!}\beta^4 + \cdots\right)$$

$$= \text{id} + \frac{A}{\beta} \sin \beta + \frac{A^2}{\beta^2}(1 - \cos \beta). \tag{10.70}$$

Since $\text{tr} A = 0, \text{tr} A^2 = -2\beta^2$, then $\text{tr} e^A = 1 + 2\cos\beta$ giving $\beta = \arccos(\frac{1}{2}\text{tr}\, O'O^* - \frac{1}{2})$ with geodesic distance in the group

$$\rho^2(O, O') = 2\arccos^2\left(\frac{1}{2}\text{tr} O'O^* - \frac{1}{2}\right). \tag{10.71}$$

For **SO**(2), it rotates around only one axis according to $A = \begin{pmatrix} 0 & a \\ -a & 0 \end{pmatrix}$, giving $\beta = a$, with $\text{tr}\, e^A = 2\cos\beta$ giving $\beta = \arccos(\frac{1}{2}\text{tr}\, O'O^*)$.

Scale Subgroup: Let

$$g_t = \text{diag}(e^{t\log \frac{s_1'}{s_1}}, e^{t\log \frac{s_2'}{s_2}}, \ldots)\text{diag}(s_1, s_2, \ldots), \tag{10.72}$$

then $dg_t/dt = \text{diag}(\log(s_1'/s_1), \log(s_2'/s_2), \ldots)g_t$ implying that $v_t = \text{diag}(\log(s_1'/s_1), \log(s_2'/s_2), \ldots) = v_t^*$ and the geodesic condition is satisfied:

$$\frac{dv_t}{dt} = v_t v_t^* - v_t^* v_t = 0. \tag{10.73}$$

The geodesics are exponentials e^{At} with $A = v_t$; the metric distance becomes

$$\rho((s_1, s_2, \ldots), (s_1', s_2', \ldots)) = \inf_{v: g_0 = s, g_1 = s'} \int_0^1 \|v_t\|_{\mathbb{R}^{d \times d}} dt = \sum_{i=1}^d |\log s_i' - \log s_i|. \tag{10.74}$$

10.6 METRICS IN THE MATRIX GROUPS

The Similitudes: Examine geodesics in the similitudes $\mathbf{Sim}(4) = (0, \infty) \times \mathbf{SO}(3)$ with $g_t = (s_t/s)O_t sO$, $t \in [0,1]$, then the derivative becomes

$$dg_t/dt = \left(\frac{ds_t/dt}{s_t}\mathrm{id} + \frac{dO_t}{dt}O_t^{-1}\right)g_t. \tag{10.75}$$

The geodesic distance $\rho((O,s),(O',s'))$ is then

$$\rho^2((O,s),(O',s')) = \inf_{\substack{\dot{g}: g_0 = g \\ g_1 = g'}} \int_0^1 \|v_t\|_T^2 \, dt = \inf_{\substack{\frac{ds_t}{dt}, \frac{dO_t}{dt}: s_0 = s, O_0 = O \\ s_1 = s', O_1 = O'}}$$

$$\int_0^1 \left(\frac{(ds_t/dt)^2}{s_t^2} + 2\mathrm{tr}\frac{(ds_t/dt)}{s_t}\frac{dO_t}{dt}O_t^{-1} + \mathrm{tr}\frac{dO_t}{dt}\frac{dO_t}{dt}^*\right) dt.$$

Then, with $dO_t/dt = AO_t$ where A is skew-symmetric, then clearly $\mathrm{tr}(dO_t/dt)O_t^{-1} = 0$ since $\mathrm{tr}A = 0$. This gives the geodesic

$$\rho^2((O,s),(O',s')) = \inf_{\substack{\dot{s}: s_0 = s \\ s_1 = s'}} \int_0^1 \frac{(ds_t/dt)^2}{s_t^2} dt + \inf_{\substack{\dot{O}: O_0 = O \\ O_1 = O'}} \int_0^1 \mathrm{tr}\frac{dO_t}{dt}\frac{dO_t}{dt}^* dt \tag{10.76}$$

$$= \left|\log\frac{s}{s'}\right|^2 + 2\sum_i a_i^2. \tag{10.77}$$

□

Remark 10.6.0 Clearly logarithm plays a fundamental role since the matrices involve simple multiplication. Look at the scalar case. Defining $u = \log s$, then $\tilde{\rho}: \mathbb{R} \times \mathbb{R} \to \mathbb{R}^+$ is a metric

$$\tilde{\rho}^2(u, u') = \inf_{\substack{u: u_0 = u \\ u_1 = u'}} \int_0^1 |\dot{u}_t|^2 \, dt = |u - u'|^2. \tag{10.78}$$

Clearly $\log : (0, \infty) \to \mathbb{R}$ is a bijection, thus $\rho : (0, \infty) \times (0, \infty) \to \mathbb{R}^+$ defined by $\rho(s, s') = \tilde{\rho}(\log s, \log s') = |\log \frac{s'}{s}|$ is a metric.

10.7 Viewing the Matrix Groups in Extrinsic Euclidean Coordinates

Viewing the groups of transformation as vector subspaces of $\mathbb{R}^{d \times d}$ and $\mathbb{R}^{(d+1) \times (d+1)}$ then the metric distances between elements in the subgroups are just that given by the Euclidean distance.

10.7.1 The Frobenius Metric

Theorem 10.17 *Viewed as vector subspaces of $\mathbb{R}^{d \times d}$ with Frobenius norm $\|A - B\|_{\mathbb{R}^{d \times d}} = \mathrm{tr}(A-B)(A-B)^*$, then the matrix groups of $\mathbf{GL}(d)$ have geodesics $g_t, t \in [0,1], g_0 = g, g_1 = h, g, h \in \mathbf{GL}(d)$ given by*

$$g_t = t(h - g) + g, \quad v_t = h - g. \tag{10.79}$$

The metric ρ is the Frobenius metric, so that for $A, A' \in \mathbf{GL}(d) \subset \mathbb{R}^{d^2}$,

$$\rho^2_{\mathbb{R}^{d \times d}}(A, A') = \|A - A'\|^2_{\mathbb{R}^{d \times d}} = \text{tr}(A - A')(A - A')^*. \tag{10.80}$$

For $\mathbf{SO}(d)$ as a vector subspace of $\mathbb{R}^{d \times d}$, then

$$\rho^2_{\mathbb{R}^{d \times d}}(O, O') = 2d - 2\text{tr}\, OO'^*. \tag{10.81}$$

For $\mathbf{SE}(d)$ in homogeneous coordinates $\bar{g} = \begin{pmatrix} A & a \\ 0 & 1 \end{pmatrix} \in \bar{\mathbf{A}}$, $A \in \mathbf{GL}(d)$, $a \in \mathbb{R}^d$, then

$$\rho^2_{\mathbb{R}^{(d+1)^2}}(\bar{g}, \bar{g}') = \|\bar{g} - \bar{g}'\|^2_{\mathbb{R}^{(d+1)^2}} = \|A - A'\|^2_{\mathbb{R}^{d \times d}} + \|a - a'\|^2_{\mathbb{R}^d}. \tag{10.82}$$

Proof Viewing $\mathbf{GL}(d) \subset \mathbb{R}^{d \times d}$ as a vector subspace, the geodesics are straight lines (Theorem 10.5):

$$g_t = A + t(A' - A), \quad g_0 = A, \quad t \in [0, 1]. \tag{10.83}$$

Notice, the tangent space of $\mathbb{R}^{d \times d}$ itself given by the slope of the line connecting the group elements: $dg_t/dt = A' - A$. Then as it should be

$$\rho^2(A, A') = \int_0^1 \|v_t\|^2_{\mathbb{R}^{d \times d}}\, dt = \|A' - A\|^2_{\mathbb{R}^{d \times d}}. \tag{10.84}$$

For $\mathbf{SO}(d)$, direct calculation shows

$$\rho^2_{\mathbb{R}^{d \times d}}(O, O') = 2d - \text{tr}\, O'O^* - \text{tr}\, OO'^* = 2d - 2\text{tr}\, O'O^*. \tag{10.85}$$

For the homogeneous coordinate representation of shift,

$$\rho^2_{\mathbb{R}^{d \times d}}(\bar{g}, \bar{g}') = \text{tr}(g - g')(g - g')^* = \text{tr}\begin{pmatrix} A - A' & a - a' \\ 0 & 1 \end{pmatrix}\begin{pmatrix} A - A' & a - a' \\ 0 & 1 \end{pmatrix}^* \tag{10.86}$$

$$= \|A - A'\|^2_{\mathbb{R}^{d \times d}} + \|a - a'\|^2_{\mathbb{R}^d}. \tag{10.87}$$

$\square$

10.7.2 Comparing intrinsic and extrinsic metrics in SO(2,3)

To appreciate that the $\mathbf{GL}(d)$ and $\mathbf{SO}(d)$ distances are fundamentally different, one through the Euclidean space associated with $\mathbf{GL}(d)$, and one the geodesic distance through the curved space of $\mathbf{SO}(d)$, examine $\mathbf{SO}(2), \mathbf{SO}(3)$. For $\mathbf{SO}(d)$ the distance based on Frobenius norm for matrices $\|A\|^2_{\mathbb{R}^{d \times d}} = \text{tr}\, AA^*$ becomes $2d - 2\,\text{tr}\, OO'^*$. For $\mathbf{SO}(2)$, identify $O(\theta), O'(\theta') \leftrightarrow \theta, \theta'$. Then $A = (\theta - \theta')\begin{pmatrix} 0 & -1 \\ 1 & 0 \end{pmatrix}$ giving (comparing the distances)

$$\rho^2_{\mathbb{R}^{2 \times 2}}(O, O') = \|O - O'\|^2_{\mathbb{R}^{2 \times 2}} = 4 - 4\cos(\theta - \theta'), \tag{10.88}$$

$$\rho^2_{\mathbf{SO}(2)}(O, O') = \text{tr}\, AA^* = 2(\theta - \theta')^2. \tag{10.89}$$

Notice, for small difference in angles, the two are similar. For **SO**(3), write $O' = e^A O$ where $A = -A^*$ is a skew-symmetric matrix of the form $A = \begin{pmatrix} 0 & a_1 & a_2 \\ -a_1 & 0 & a_3 \\ -a_2 & -a_3 & 0 \end{pmatrix}$. Define the normalized skew-symmetric matrix $Z = (1/\beta)A$ with $\beta^2 = \sum_{i=1}^{3} a_i^2$. Then $O' = e^{\beta Z} O$ in the matrix sum reduces exploiting the fact that $Z^3 = -Z$ from Eqn. 10.69. Using the alternating sin, cos series gives

$$e^{\beta Z} = I + Z \sin\beta + Z^2(1 - \cos\beta). \tag{10.90}$$

This gives a direct formula for obtaining Z from $O \in$ **SO**(3) according to $O = e^{\beta Z}$. Using the facts that $\operatorname{tr} Z = 0$, $\operatorname{tr} Z^2 = -2$ giving $\operatorname{tr} O = 1 + 2\cos\beta$ and $\cos\beta = \frac{1}{2}(\operatorname{tr} O - 1)$. Thus we have obtained the Frobenius norm required which can be compared with the geodesic distance through the group:

$$\rho^2_{\mathbb{R}^{3\times 3}}(O, O') = \rho^2_{\mathbb{R}^{3\times 3}}(O, e^{\beta Z}O) = 6 - 2\operatorname{tr} e^{\beta Z} = 4 - 4\cos\beta; \tag{10.91}$$

$$\rho^2_{\mathbf{SO}(3)}(O, O') = \rho^2_{\mathbf{SO}(3)}(O, e^{\beta Z}O) = 2\beta^2. \tag{10.92}$$

A direct formula for Z may be obtained from $e^{\beta Z} = O'O^*$ with $\beta = \arccos\frac{1}{2}(\operatorname{tr} O - 1)$. Let n be the axis of rotation of O, eigenvector eigenvalue 1 so that $On = n$, then

$$On = \exp(\beta Z)n = n + \sin\beta Zn + (1 - \cos\beta)Z^2 n, \tag{10.93}$$

with $Zn = 0$. As n is the eigenvector of O, let $Z \propto (O - O^*)$, with the proportionality constant given by the normalizing constraint, then

$$Zn \propto On - O^*n = 0.$$

This gives the exact formula for Z,

$$Z = \begin{pmatrix} 0 & -n_z & n_y \\ n_z & 0 & -n_x \\ -n_y & n_x & 0 \end{pmatrix},$$

where $n = [n_x, n_y, n_z]^*$ is the axis of rotation and is given by

$$n_x = \frac{O_{32} - O_{23}}{2\sin\beta}, \quad n_y = \frac{O_{13} - O_{31}}{2\sin\beta}, \quad n_z = \frac{O_{21} - O_{12}}{2\sin\beta}.$$

11 METRICS SPACES FOR THE INFINITE DIMENSIONAL DIFFEOMORPHISMS

ABSTRACT In this chapter the metric space structure of shape is developed by studying the action of the infinite dimensional diffeomorphisms on the coordinate systems of shape. Riemannian manifolds allow us to develop metric distances between the group elements. We examine the natural analog of the finite dimensional matrix groups corresponding to the infinite dimensional diffeomorphisms which are generated as flows of ordinary differential equations. We explore the construction of the metric structure of these diffeomorphisms and develop many of the properties which hold for the finite dimensional matrix groups in this infinite dimensional setting.

11.1 Lagrangian and Eulerian Generation of Diffeomorphisms

Clearly the matrix groups are diffeomorphisms acting on the background space $g: X = \mathbb{R}^d \to X$. They are of finite dimension encoded via the parameters of the matrices. Now examine the construction of the group of infinite dimensional diffeomorphisms acting on bounded $X \subset \mathbb{R}^d$.

Definition 11.1 Let X be the background space, then define $g : X \to X$ is a **diffeomorphism** with inverse g^{-1}, and define the group of transformations $\mathcal{G}, \circ$ as **subgroups of diffeomorphisms** acting on the background space X with the law of composition of functions $g(\cdot) \circ g'(\cdot) = g(g'(\cdot))$.

That this is a group follows from the fact that the identity id $\in \mathcal{G}$, the composition of two diffeomorphisms, is in the group $g \circ g' = g(g') \in \mathcal{G}$ and the inverse is in the group as well $g^{-1} \in \mathcal{G}$.

Unlike the finite dimensional matrix groups, in the infinite dimensional setting it is much less clear how to generate the mappings which are diffeomorphisms. For this, the approach has been to construct the diffeomorphisms as a flow of ordinary differential equations (ODEs) as originally put forward by Christensen et al. [220].

Assume the diffeomorphisms $g \in \mathcal{G}$ evolve in time as a flow $g_t, t \in [0, 1]$ with an associated vector field $v : X \times [0, 1] \to \mathbb{R}^d$ controlling the Lagrangian evolution according to

$$\frac{\partial g_t}{\partial t}(x) = v_t(g_t(x)), \quad g_0(x) = x, \quad x \in X, \quad t \in [0, 1]. \tag{11.1}$$

Figure 11.1 depicts the Lagrangian formulation of the flow of diffeomorphisms.

The forward and inverse maps are linked through the fact that for all $t \in [0, 1], x \in X$, $g_t^{-1}(g_t(x)) = x$. Defining the $d \times d$ Jacobian matrix $Df = \left(\partial f_i / \partial x_j\right)$, then differentiating the identity

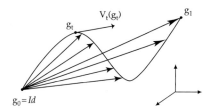

Figure 11.1 Figure shows the Lagrangian description of the flow depicting the ODE $(dg_t/dt) = v_t(g_t), g_0 = \text{id}$.

11.1 LAGRANGIAN AND EULERIAN GENERATION

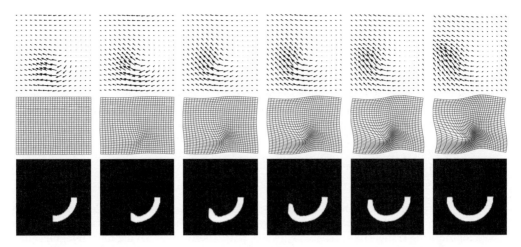

Figure 11.2 Top row: Panels show the vector fields $v_t(\cdot)$, $t_1 = 0, t_2, t_3, t_4, t_5, t_6$ depicting the flow of a sequence of diffeomorphisms satisfying $\partial g_t/\partial t = v_t(g_t)$, $g_0 = \text{id}$. Middle row: Panels depict the flow of diffeomorphisms g_t^{-1} associated with the vector fields in the top row. Bottom row: Panels depict the flow of particles which carry black and white labels corresponding to the circular movement of the patch under the flow of diffeomorphisms $I \circ g_t^{-1}$.

gives the inverse evolution according to

$$0 = \frac{\partial}{\partial t} g_t^{-1}(g_t(x)) = \frac{\partial g_t^{-1}}{\partial t}(g_t(x)) + Dg_t^{-1}(g_t(x)) \frac{\partial g_t}{\partial t}(x) \quad (11.2)$$

$$= \frac{\partial g_t^{-1}}{\partial t}(g_t(x)) + Dg_t^{-1}(g_t(x)) v_t(g_t(x)). \quad (11.3)$$

Substituting $y = g_t(x)$ gives the transport equation, or the evolution of the inverse:

$$\frac{\partial g_t^{-1}}{\partial t}(y) = -Dg_t^{-1}(y) v_t(y), \quad g_0^{-1}(y) = y, \quad y \in X. \quad (11.4)$$

Example 11.2 (Rotation of Grid [268]) This example taken from Beg [268] examines the generation of a flow via the numerical integration of a specified series of vector fields. Depicted in Figure 11.2 are a series of vector fields $v_t(\cdot)$, t is used to construct the flow of the grid. Shown in the middle row are the integration of the flow generating g_t^{-1}; the diffeomorphisms were applied to a discrete lattice and depicted by drawing the movement of the coordinate lines. Shown in the bottom row is the transport of material according to $I \circ g_t^{-1}$ which forms the basis for our modeling of the image orbits in the next chapter.

11.1.1 On Conditions for Generating Flows of Diffeomorphisms

In the matrix groups, the diffeomorphisms are identified with the finite number of parameters specifying the linear action of the group. For the infinite dimensional case generated via the flow of Eqn. 11.1, the vector fields play this role. Naturally, since the vector fields $v_t, t \in [0, 1]$ are not a finite set of parameters, these mappings are infinite dimensional mappings.

To ensure that the ODEs generate diffeomorphisms the vector fields are constrained to be sufficiently smooth so that $v_t \in V, t \in [0,1]$, where V is a Sobolev space (Hilbert space with finite norm on derivatives). The starting point for generating smooth flows is the compactly supported (0-boundary) continuously p-differentiable vector fields $C_0^p(X, \mathbb{R}^d)$ with associated sup-norm $\|\cdot\|_{p,\infty}$. Clearly over finite times, then for $v_t, t \in [0,1]$ sufficiently smooth, Eqn. 11.1 is integrable and generates diffeomorphisms (see Boothby [148]). Dupuis and Trouve have shown that the minimal amount of smoothness required [269, 270] is that for $X \subset \mathbb{R}^d$ bounded the fields be at least continuously 1-times differentiable $C_0^1(X, \mathbb{R}^d)$ with sup-norm

$$\|v\|_{1,\infty} = \sup_{x \in X} \sum_{i=1}^d |v_i(x)| + \sum_{i=1}^d \sum_{j=1}^d \left|\frac{\partial v_i}{\partial x_j}(x)\right|. \tag{11.5}$$

The pointwise continuity and differentiability properties are controlled by controlling the integral square of the derivatives by forcing the indexed family of vector fields $v_t, t \in [0,1]$ to be in a reproducing kernel Hilbert space $(V, \|\cdot\|_V)$, with s-derivatives having finite integral square and zero boundary. Via the Sobolev embedding theorems this then relates to the sup-norm pointwise smoothness. The norm dominates for each of the components $v_{it}, i = 1, \ldots, d$ and dominates the s−derivatives according to

$$\|v\|_V^2 \geq \sum_{i=1}^d \int_X \sum_{\alpha \leq s} |D^\alpha v_i(x)|^2 \, dx. \tag{11.6}$$

The smoothness conditions are dimension dependent: for $X \subset \mathbb{R}^d$, the number of derivatives s required in the Sobolev space norm for the sup-norm $p=1$ condition is $s > d/2 + 1$.[32]

Theorem 11.3 *[Dupuis et al. [269] and Trouve [270]] Consider the Hilbert space of vector fields $(V, \|\cdot\|_V)$ on X compact with zero boundary conditions.*
Then if there exists some constant c, such that for some $p \geq 1$ and all $v \in V$,

$$\|v\|_V \geq c\|v\|_{p,\infty}, \tag{11.7}$$

with the vector fields satisfying $\int_0^1 \|v_t\|_V dt < \infty$, then the ordinary differential equation Eqn. 11.1 can be integrated over $[0,1]$, and $g_1(\cdot) : X \to X$ is a diffeomorphism of X.
The set of all such diffeomorphism solutions obtained from $\mathcal{V} = \{v : \int_0^1 \|v_t\|_V dt < \infty\}$ form the group:

$$\mathcal{G}(\mathcal{V}) = \left\{g_1 : \frac{\partial g_t}{\partial t} = v_t(g_t), t \in [0,1], v \in \mathcal{V}\right\}. \tag{11.8}$$

Proof The diffeomorphic properties rely most heavily on the most basic property of solutions of the ODE Eqn. 11.1. □

Lemma 11.4 *For v sufficiently smooth, for any $t \in [0,1]$, and any $x \in X$, then Eqn. 11.1 admits a unique solution for all $t \in [0,1]$.*

The smoothness required is proved in [269, 270]; see Lemma 2.2 of [269].
That the solutions are 1–1 mapping from X onto itself, first since $v(x) = 0, x \in \partial X$ implies $g_t(x) = x, x \in \partial X$. Uniqueness implies points inside the boundary cannot cross the boundary, thus $g_1(\cdot)$ is a mapping into X. Uniqueness also implies the mapping is 1–1, since if two points cross, $g_1(x) = g_1(x') = y$, then following the negative vector field from y generates two different solutions. The mapping is onto, since given any point y, then by flowing the negative velocity

---

[32] The dominating condition is background space dependent so that for $X \subset \mathbb{R}^d$, the number of derivatives s required in the Sobolev space norm to dominate the sup-norm in C_0^p is $s > (d/2) + p$.

$d\tilde{g}_t/dt = -v_{1-t}(\tilde{g}_t), \tilde{g}_0 = \text{id}$ field from $g_1(y)$ maps back to y (see part of proof below Eqns. 11.10, 11.11). The differentiability property follows from the sufficient differentiability in space of $v_t(\cdot)$ for all $t \in [0, 1]$.

The group properties follow from manipulations on the paths. Denote the solution corresponding to velocity v as g^v. Clearly $v = 0 \in \mathcal{V}$, with the identity element, $\text{id} = g_1(\cdot)$. That $g_1^{v'} \circ g_1^v \in \mathcal{G}$ follows by defining $\tilde{v}$ according to

$$\tilde{v}_t = 2v_{2t}, \quad t \in [0, \tfrac{1}{2}), \qquad \tilde{v}_t = 2v'_{2t-1}, \quad t \in [\tfrac{1}{2}, 1). \tag{11.9}$$

Then $\tilde{v} \in \mathcal{V}$ is finite, and $\tilde{g} \in \mathcal{G}$ and $\tilde{g} = g' \circ g$. Constructing g^{-1} via the negative of the vector field gives the inverse in the group $\mathcal{G}$. To see this, examine the semi-group property of the flows. Define $\phi_{s,t}$ to satisfy $\partial \phi_{s',t}/\partial t = v_t(\phi_{s',t})$, $\phi_{s',s'}(x) = x$. Then $\phi_{s',t} = \phi_{st} \circ \phi_{s',s}$, since $y_t = \phi_{s',t}$ and $\tilde{y}_t = \phi_{s,t} \circ \phi_{s',s}$ satisfy the same differential equation both with initial condition $y_s = \tilde{y}_{s=\phi_{s',s}}$, and the ODE has a unique solution. For g_t^v with $\partial g_t^v/\partial t = v_t(g_t^v)$, let $w_t = -v_{1-t}$ with g_t^w satisfying $\partial g_t^w/\partial t = -v_{1-t}$, then

$$g_{1-t}^w(x) = x - \int_0^{1-t} v_{1-s} \circ g_s^w(y)\, ds = x + \int_1^t v_s \circ g_{1-s}^w\, ds. \tag{11.10}$$

For $t = 0$ then this is precisely the inverse, $g_1^w = \phi_{1,0}$ implying

$$g_1^w \circ g_1^v = \phi_{1,0} \circ \phi_{0,1} = \phi_{0,0} = \text{id}. \tag{11.11}$$

Thus the inverse $\left(g_1^v\right)^{-1} = g_1^w$ is an element of the set. □

11.1.2 Modeling via Differential Operators and the Reproducing Kernel Hilbert Space

As we have seen in Chapter 9, Section 9.4 on Green's kernels and Sobolev Hilbert spaces the pointwise continuity of the vector fields is obtained via control in the square norm on the differentiability properties. The indexed family of vector fields $v_t, t \in [0,1]$ in $X \subset \mathbb{R}^d$ are constrained to be in a reproducing kernel Hilbert space [271] V with s-derivatives having a finite integral square; via the Sobolev embedding theorems this then relates to the sup-norm pointwise continuity and differentiability. For any $f \in V, \alpha \in \mathbb{R}^m$ the kernel K sifts according to

$$\langle K(\cdot, x)\alpha, f \rangle_V = \langle \alpha, f(x) \rangle_{\mathbb{R}^d}. \tag{11.12}$$

The smoothness of $f \in V$ depends on the smoothness of the kernel function K; choosing a proper kernel such that V is embedded in $C^1(X)$ assures the necessary smoothness for the generation of diffeomorphisms. To see the explicit dependence of the smoothness of functions $f \in V$ on the kernel, see Section 9.4.2 with Eqn. 9.74.

Most often the construction of the reproducing Hilbert space (with associated kernel) designs the norm $\|\cdot\|_V$ by starting with a one-to-one $d \times d$ matrix differential operator $L = (L_{ij})$ with null boundary conditions operating on $\mathbb{R}^m$ valued vector fields of the form $(Lv)_j = \sum_{i=1}^d L_{ij}v_i$, $j = 1, \ldots, d$. The inner product in V is defined by the operator and it's adjoint,

$$\langle f, h \rangle_V = \langle Lf, Lh \rangle_2 = \langle Af, h \rangle, \quad \text{with } A = L^*L, \tag{11.13}$$

where L^* is the adjoint operator and $\langle \cdot, \cdot \rangle_2$ is the L^2 vector inner product.[33] The differentiability in L is selected to dominate the sup norm $\|\cdot\|_{p,\infty}$ for at least $p=1$ derivative of Theorem 11.3 for each of the components $v_{it}, i = 1, \ldots, d$.

From a modeling point of view, the differential operators are usually chosen via analogs based on physical models; Christensen et al. [220] originally selected particular forms for the norm based on arguments from physical modeling analogies in fluid mechanics. Powers of the Laplacian for the classic thin-plate splines [223,224,272] have been used, and the Cauchy operator for 3D elasticity $L = -\alpha + \beta \nabla \nabla \cdot + \gamma \nabla^2$ [220, 273–275], the differential operators with sufficient derivatives and proper boundary conditions ensure the existence of solutions of the transport equation in the space of diffeomorphic flows [269,276,277].

11.2 The Metric on the Space of Diffeomorphisms

Examine the construction of metric distances on the group $\mathcal{G}(\mathcal{V})$ of infinite dimensional diffeomorphisms on the closed and bounded background space $X \subset \mathbb{R}^d$. Again the Riemannian length is constructed defining the curves $g_t, t \in [0,1]$ in the Riemannian manifold of diffeomorphisms connecting group elements $g_0 = g, g_1 = h \in \mathcal{G}(\mathcal{V})$. The curves are generated via the flow Eqns. 11.1, 11.4. Each point along any curve is a diffeomorphism in $\mathcal{G}(\mathcal{V})$.

The natural extension from the finite dimensional curves is to consider paths $g : [0,1] \to \mathcal{G}(\mathcal{V})$ connecting elements $g, h \in \mathcal{G}(\mathcal{V})$ with the tangent element of the evolution $\frac{\partial g_t}{\partial t}(g_t^{-1}) = v_t \in V, t \in [0,1]$, with norm $\|\cdot\|_V$ satisfying Theorem 11.3. In contrast to the finite dimensional setting here the tangents to the curve v_t are vector fields, the tangent space a Hilbert space with finite norm $\|v_t\|_V < \infty$. Clearly, unlike the finite dimensional setting of the matrix groups, these are curves through the infinite dimensional space of diffeomorphisms $g_t \in \mathcal{G}(\mathcal{V}), t \in [0,1]$. The length and energy of these infinite dimensional curves through the group of diffeomorphisms are defined via their vector fields:

$$\text{Length}(v) = \int_0^1 \left\| \frac{\partial g_t}{\partial t}(g_t^{-1}) \right\|_V dt = \int_0^1 \|v_t\|_V \, dt. \tag{11.14}$$

With this norm in the tangent space now defined, the Euler equation for infimum length paths as proved in [278] generalize the finite matrix group Euler equation of Theorem 10.14.

First we calculate the analog of the Lie bracket for the matrix case of Lemma 10.13, Eqn. 10.45, in this infinite dimensional setting for how the vector field changes with a perturbation of the group flow. We will also require the perturbation of the group flow with a perturbation of the vector field.

Lemma 11.5 *The perturbation of the group element $g_t \to g_t(\epsilon) = g_t + \epsilon \eta_t(g_t) + o(\epsilon)$ gives the perturbation of the vector field $v_t \to v_t(\epsilon) = v_t + \epsilon(\partial v_t(\epsilon)/\partial \epsilon) + o(\epsilon)$ satisfying the function space equation (analagous to the finite dimensional Lie bracket of Lemma 10.13):*

$$\psi_t = \partial_\eta v_t = \frac{\partial v_t(\epsilon)}{\partial \epsilon} = \frac{\partial \eta_t}{\partial t} - (Dv_t \, \eta_t - D\eta_t \, v_t). \tag{11.15}$$

---

[33] Strictly speaking, although $Lf \in L^2$, in general $Af \notin L^2$ so that $\langle Af, g \rangle$ is not the L^2 inner product; rather Af is a 1-form in the dual.

11.2 THE METRIC ON THE SPACE OF DIFFEOMORPHISMS

The perturbation of the vector field $v_t \to v_t + \epsilon \psi_t + o(\epsilon)$ generates a perturbation of the group element $g_t \to g_t(\epsilon) = (\mathrm{id} + \epsilon \eta_t)(g_t) + o(\epsilon) = g_t + \epsilon \eta_t(g_t) + o(\epsilon)$ where

$$\partial_\psi g_t = \frac{\partial}{\partial \epsilon} g_t(\epsilon) \bigg|_{\epsilon=0} = \eta_t(g_t) = Dg_t \int_0^t (Dg_u)^{-1} \psi_u(g_u) \, du \qquad (11.16)$$

$$= \int_0^t Dg_{u,t}(g_u) \psi_u(g_u) \, du. \qquad (11.17)$$

Proof We have

$$\frac{\partial}{\partial t}(g_t + \epsilon \eta_t(g_t)) = \left(v_t + \epsilon \frac{\partial v_t(\epsilon)}{\partial \epsilon}\right)(g_t + \epsilon \eta_t(g_t)) + o(\epsilon)$$

$$\stackrel{(a)}{=} v_t(g_t) + \epsilon Dv_t(g_t) \eta_t(g_t) + \epsilon \frac{\partial v_t(\epsilon)}{\partial \epsilon}(g_t) + o(\epsilon)$$

$$= \frac{\partial g_t}{\partial t} + \epsilon \frac{\partial \eta_t}{\partial t}(g_t) + \epsilon D\eta_t(g_t) \frac{\partial g_t}{\partial t}$$

$$\stackrel{(b)}{=} v_t(g_t) + \epsilon \frac{\partial \eta_t}{\partial t}(g_t) + \epsilon D\eta_t(g_t) v_t(g_t),$$

and equating (a),(b) and applying g_t^{-1} on the right to both terms gives Eqn. 11.15

$$\frac{\partial v_t(\epsilon)}{\partial \epsilon} = \frac{\partial \eta_t}{\partial t} + D\eta_t \, v_t - Dv_t \, \eta_t.$$

From the definition of the perturbation $\partial_\psi g_t = \eta_t(g_t)$. To prove $\eta_t(g_t)$ equals the right-hand sides of Eqn. (11.16), clearly for $t = 0$ we have $\eta_0(g_0) = 0$, and they satisfy the same differential equation.
Differentiating the left hand side $\eta_t(g_t)$ gives

$$\frac{\partial \eta_t(g_t)}{\partial t} = \frac{\partial \eta_t}{\partial t}(g_t) + D\eta_t(g_t) v_t(g_t) \stackrel{(a)}{=} Dv_t(g_t) \eta_t(g_t) + \psi_t(g_t) \qquad (11.18)$$

where (a) follows from Lemma 11.5, Eqn. 11.15. Differentiating the right hand side of Eqn. 11.16 gives the identical equation

$$\frac{\partial}{\partial t} Dg_t \int_0^t (Dg_u)^{-1} \psi_u(g_u) du = \frac{\partial Dg_t}{\partial t} \int_0^t (Dg_u)^{-1} \psi_u(g_u) \, du + Dg_t (Dg_t)^{-1} \psi_t(g_t)$$

$$= Dv_t(g_t) Dg_t \int_0^t (Dg_u)^{-1} \psi_u(g_u) \, du + \psi_t(g_t). \qquad (11.19)$$

The equality of Eqn. 11.17 follows from the fact that

$$(D(g_t \circ g_u^{-1}))(g_u) = Dg_t(g_u^{-1} \circ g_u) Dg_u^{-1}(g_u) = Dg_t (Dg_u)^{-1}. \qquad (11.20)$$

$\square$

Theorem 11.6 *Assume the inner product $\|v\|_V^2 = \langle Av, v \rangle_2$ satisfies the $p = 1$ condition of Theorem 11.3.[34] Given $g, h \in \mathcal{G}(V)$, then the geodesics minimizing*

$$\int_0^1 \|v_t\|_V^2 dt = \int_0^1 \langle Av_t, v_t \rangle_2 \, dt,$$

---

[34] There exists some constant c such that for some $p \geq 1$ and all $v \in V$, then $\langle Av, v \rangle_2 \geq c \|v\|_{p,\infty}$.

connecting $g, h \in \mathcal{G}(V)$ are solutions of the Euler equation

$$\frac{\partial Av_t}{\partial t} + (Dv_t)^* Av_t + \mathrm{div}(Av_t \otimes v_t) = 0, \tag{11.21}$$

with divergence operator $\mathrm{div}\, v = \sum_{i=1}^{d}(\partial v_i/\partial x_i)$ and $\mathrm{div}(Av \otimes v) = (DAv)v + (\mathrm{div}\, v)Av$.

Proof The proof on the geodesic follows a similar path as for the matrix groups. Examine a smooth perturbation of the shortest length curve (extremal energy) $g_t, t \in [0,1]$ to the perturbed flow $g_t(\epsilon) = (\mathrm{id} + \epsilon \eta_t + o(\epsilon))g_t$ as depicted in the left panel of Figure 10.1. So that the boundary conditions are satisfied at $g_0(\epsilon) = g, g_1(\epsilon) = h$ the perturbation must satisfy $\eta_0 = \eta_1 = 0$, and $\eta(x) = 0, x \in \partial X$. From Lemma 11.5 we require how the vector field changes with the perturbation in the flow. The variational equation which the minimizing energy geodesic satisfies is

$$0 = \frac{\partial}{\partial \epsilon}\int_0^1 \left\| v_t + \epsilon \frac{\partial v_t(\epsilon)}{\partial \epsilon}\right\|_V^2 dt\big|_{\epsilon=0} = 2\int_0^1 \left\langle v_t, \frac{\partial v_t(\epsilon)}{\partial \epsilon}\right\rangle_V dt. \tag{11.22}$$

$$\stackrel{(a)}{=} 2\int_0^1 \left\langle Av_t, \frac{\partial \eta_t}{\partial t} + D\eta_t\, v_t - Dv_t\, \eta_t\right\rangle_2 dt \tag{11.23}$$

$$\stackrel{(b)}{=} 2\int_0^1 \left\langle -\frac{\partial Av_t}{\partial t} - (Dv_t)^* Av_t, \eta_t\right\rangle_2 dt + 2\int_0^1 \langle Av_t, D\eta_t\, v_t\rangle_2\, dt, \tag{11.24}$$

with (a) from Eqn. 11.15 and (b) following from integration by parts and the boundary condition $\eta_0 = \eta_1 = 0$. From Stoke's theorem, since v and η vanish on ∂X,

$$\langle Av_t, D\eta_t\, v_t\rangle_2 = -\langle \mathrm{div}(Av_t \otimes v_t), \eta_t\rangle_2$$

giving

$$0 = \int_0^1 \left\langle -\frac{\partial Av_t}{\partial t} - (Dv_t)^* Av_t - \mathrm{div}(Av_t \otimes v_t), \eta_t\right\rangle_2 dt. \tag{11.25}$$

Since this is zero over all perturbations, it gives the Euler Eqn. 11.21. □

Remark 11.2.0 The approach to deriving Equation 11.21 was first derived by Arnold for the incompressible divergence free flow [equation 1 of [279]], and then by Mumford [280] and Miller et al. [278] and in this setting. The approach shown here which parallels the finite Lie group approach has emerged more recently through the continued work of Trouve et al. [281]. In 1-dimension with $A = \mathrm{id}$ the equation reduces to Burger's equation as discussed by Mumford [280].

11.3 Momentum Conservation for Geodesics

The Euler equation examines the shortest path geodesic corresponding to a perturbation on the flow of diffeomorphisms. Now following [281] examine the variational problem in the tangent space of the group minimizing with respect to the vector fields generating the shortest path flows. As we have already seen for the matrix groups corresponding to finite dimensional mechanics, the geodesic is completely determined by the initial momentum (Theorem 10.15, Chapter 10). For the deformable setting the generalized momentum is naturally defined as Av the function of the vector field which when acting against the vector field gives energy in the metric $\|v\|_V^2 = \langle Av, v\rangle_2$.

11.3 MOMENTUM CONSERVATION FOR GEODESICS

There are however substantive difficulties with the momentum. In general it is not smooth enough to be in L^2, so strictly speaking it cannot be defined via the L^2 inner product $\langle Av, v \rangle_2$. However, it acts as a 1-form on smooth enough objects such as the elements of V, and in particular Av acting on v is well defined.

For the metric mapping of the start and end point, there is a conservation law which requires the momentum to be conserved along the path of the geodesic flow. This of course implies that the momentum at the identiy Av_0 on the geodesic determines the entire flow $g_t, t \in [0,1]$, where $\dot{g}_t = v_t(g_t), g_0 = \text{id}$. More generally for growth, the momentum is constantly being transformed along the shortest path.

Lemma 11.7 *Define the vector field transported from the identity as $w_t = Dg_t(g_t^{-1})w_0(g_t^{-1})$, then*

$$\frac{\partial w_t}{\partial t} = (Dv_t)w_t - (Dw_t)v_t. \tag{11.26}$$

Proof Define the notation $w_t = (Dg_t w_0)(g_t^{-1})$ then

$$\frac{\partial w_t}{\partial t} = \frac{\partial}{\partial t}(Dg_t w_0(g_t^{-1})) = \frac{\partial Dg_t w_0}{\partial t}(g_t^{-1}) + D(Dg_t w_0)(g_t^{-1})\frac{\partial g_t^{-1}}{\partial t} \tag{11.27}$$

$$= (D(v_t(g_t))w_0)(g_t^{-1}) + D(Dg_t w_0)(g_t^{-1})(-Dg_t^{-1}v_t) \tag{11.28}$$

$$= (Dv_t(g_t)Dg_t w_0)(g_t^{-1}) - D((Dg_t w_0)(g_t^{-1}))v_t$$

$$= (Dv_t)w_t - (Dw_t)v_t. \tag{11.29}$$

□

Theorem 11.8 *Defining the vector field transported from the identity as $w_t = Dg_t(g_t^{-1})w_0(g_t^{-1})$, then along the geodesics satisfying the Euler equation (Theorem 11.6, Eqn. 11.21) the momentum is conserved:*

$$\frac{\partial}{\partial t}\langle Av_t, w_t \rangle_2 = 0, \quad t \in [0,1]. \tag{11.30}$$

Along the geodesic, the momentum is determined by the momentum at the identity:

$$Av_t = (Dg_t^{-1})^* Av_0(g_t^{-1})|Dg_t^{-1}|. \tag{11.31}$$

Proof Use the Euler equation to evaluate the derivative

$$\frac{\partial}{\partial t}\langle Av_t, w_t \rangle_2 = \left\langle \frac{\partial}{\partial t}Av_t, w_t \right\rangle_2 + \left\langle Av_t, \frac{\partial}{\partial t}w_t \right\rangle_2 \tag{11.32}$$

$$\stackrel{(a)}{=} \left\langle \frac{\partial}{\partial t}Av_t, w_t \right\rangle_2 + \langle Av_t, (Dv_t w_t - Dw_t v_t)\rangle_2 \tag{11.33}$$

$$\stackrel{(b)}{=} \left\langle \frac{\partial}{\partial t}Av_t + (Dv_t)^* Av_t + \text{div}(Av_t \otimes v_t), w_t \right\rangle_2 = 0 \tag{11.34}$$

with (a) from Lemma 11.5 and (b) exactly Eqn. 11.25 which by integration of parts gives the Euler equation equalling 0.

Part 2: Since $w_t = Dg_t(g_t^{-1})w_0(g_t^{-1})$ this implies $(Dg_t)^{-1}w_t(g_t) = w_0$ with conservation of momentum implying

$$\langle Av_t, w_t \rangle_2 = \langle Av_0, w_0 \rangle_2 = \langle Av_0, (Dg_t)^{-1}w_t(g_t) \rangle_2 \tag{11.35}$$

$$= \langle (Dg_t)^{-1*}Av_0, w_t(g_t) \rangle_2 \tag{11.36}$$

$$= \langle (Dg_t)^{-1*}(g_t^{-1})Av_0(g_t^{-1})|Dg_t^{-1}|, w_t \rangle_2. \tag{11.37}$$

This reduces to Eqn. 11.31 using the inverse function theorem $Dg_t^{-1} = (Dg_t)^{-1}(g_t^{-1})$.

$\square$

11.4 Conservation of Momentum for Diffeomorphism Splines Specified on Sparse Landmark Points

To illustrate the conservation of momentum and the difficulty of viewing the momentum as a function, examine the fundamental question of how to choose the initial momentum to generate mappings of correspondences between one set of points and another. Assume we are given observations of coordinatized objects through finite sets of point locations or features, $x_1, x_2, \ldots, x_N, x'_1, x'_2, \ldots, x'_N$; term these *objects* N-shapes. We return to this again in Section 12.6, Chapter 12 when we study landmarked metric spaces in greater detail. Assume the diffeomorphism is known at subsets of points $g_0(x_n) = x_n, g_1(x_n) = x'_n$, then Joshi argued [242, 282], that if a vector field $v_t \in V, t \in [0, 1]$ is an optimum trajectory satisfying a particular set of constraints then to be of minimum length it must have the property that while the vector fields are constrained along the particle paths $v_t(g_t(x_n))$ it must otherwise be of minimum norm-square $\|v_t\|_V^2$. Thus the diffeomorphism landmark matching problem is essentially the creation of diffeomorphism splines, analogous to the standard spline formulations [283]. The solution to the spline problem has as the minimum energy solution a linear combination of reproducing kernels associated with $(V, \|\cdot\|_V)$ with the linear weights chosen to satisfy constraints.

Joshi [282] first examined the diffeomorphic correspondence of landmarked shapes in Euclidean subsets of $\mathbb{R}^d, d = 2, 3$. The momentum at the identity is transported according to the conservation law above.

Theorem 11.9 (Joshi Splines) *Along the geodesics connecting the N-shapes $x_n = g_0(x_n), x'_n = g_1(x_n), n = 1, \ldots, N$, the vector fields are splines of the form*

$$v_t(\cdot) = \sum_{n=1}^{N} K(g_t(x_n), \cdot)\beta_{nt}, \tag{11.38}$$

where K is the Green's kernel associated with A determining the norm, with the momentum satisfying

$$Av_t(\cdot) = \sum_{n=1}^{N} \delta(g_t(x_n) - \cdot)\beta_{nt}. \tag{11.39}$$

The optimizing flow minimizing the inexact matching problem

$$\int_0^1 \|v_t\|_V^2 \, dt + \sum_{n=1}^{N} \|x'_n - g_1(x_n)\|_{\mathbb{R}^d}^2, \tag{11.40}$$

11.4 CONSERVATION OF MOMENTUM FOR DIFFEOMORPHISM

has momentum of the geodesic at $t = 0$ given by

$$Av_0(\cdot) = \sum_n \delta(x_n - \cdot)Dg_1(x_n)^*(x'_n - g_1(x_n)) ; \qquad (11.41)$$

the momentum transported along the geodesic $Av_t = (Dg_t^{-1})^*Av_0(g_t^{-1})|Dg_t^{-1}|$ is given by

$$Av_t(\cdot) = \sum_n \delta(g_t(x_n) - \cdot)(Dg_{t,1}(g_t(x_n)))^*(x'_n - g_1(x_n)). \qquad (11.42)$$

Proof The proof exploits the fact that if $v_t \in V$ is of minimum norm for each $t \in [0,1]$, then this implies $\int_0^1 \|v_t\|_V^2 dt$ is minimum. Let $v_t, t \in [0,1]$ be a candidate vector field connecting x_n, x'_n of the form $g_0(x_n) = x_n, g_1(x_n) = x'_n, n = 1,\ldots,N$, then v_t is optimal of minimum norm-square if it is of the form

$$v_t(x) = \sum_{n=1}^N K(g_t(x_n), x)\beta_{nt}, \quad \text{implying } Av_t(x) = \sum_{n=1}^N \delta(g_t(x_n) - x)\beta_{nt}. \qquad (11.43)$$

Choose another candidate solution $\tilde{v}_t = v_t + h_t, t \in [0,1]$ with $h_t(g_t(x_n)) = 0$, $t \in [0,1], n = 1, 2, \ldots$ leaving the second constraint term unchanged on the constrained paths. Then

$$\|\tilde{v}_t\|_V^2 = \|v_t + h_t\|_V^2 = \|v_t\|_V^2 + \|h_t\|_V^2 + 2\langle v_t, h_t\rangle_V \qquad (11.44)$$

$$= \|v_t\|_V^2 + \|h_t\|_V^2 + 2\left\langle \sum_{n=1}^N K(g_t(x_n), \cdot)\beta_{nt}, h_t\right\rangle_V \qquad (11.45)$$

$$\stackrel{(a)}{=} \|v_t\|_V^2 + \|h_t\|_V^2 + 2\left\langle \sum_{n=1}^N \beta_{nt}, h_t(g_t(x_n))\right\rangle_{\mathbb{R}^d}$$

$$= \|v_t\|_V^2 + \|h_t\|_V^2 \geq \|v_t\|_V^2, \qquad (11.46)$$

where (a) follows from the reproducing property, and h being zero on the paths of the particles.

To compute the variation, of the inexact matching, the first term contributes Av_t, and the gradient of the second term in the cost Eqn. 11.40 requires the perturbation $\eta_t(g_t) = \int_0^t Dg_{u,t}(g_u)\psi_u(g_u)du$. Then for $t = 1$ gives the variation of the cost term becomes

$$-\sum_{n=1}^N \left\langle x'_n - g_1(x_n), \frac{\partial g_1^{v+\epsilon\psi}(x_n)}{\partial \epsilon}\right\rangle_{\mathbb{R}^d}$$

$$\stackrel{(a)}{=} -\sum_{n=1}^N \left\langle x'_n - g_1(x_n), \int_0^1 Dg_{u,1}(g_u(x_n))\psi_u(g_u(x_n))\,du\right\rangle_{\mathbb{R}^d}$$

$$= -\int_0^1 \sum_{n=1}^N \langle (Dg_{u,1}(g_u(x_n)))^*(x'_n - g_1(x_n)), \psi_u(g_u(x_n))\rangle_{\mathbb{R}^d}\,du$$

$$\stackrel{(b)}{=} -\int_0^1 \sum_{n=1}^N \langle K(g_u(x_n), \cdot)(Dg_{u,1}(g_u(x_n)))^*(x'_n - g_1(x_n)), \psi_u\rangle_V\,du$$

where (a) is the substitution from Eqn. 11.17 of Lemma 11.5 and (b) is the sifting property of the Green's operator, which completes the proof of Eqn. 11.42.

To see the momentum is transported, start with $Av_0(x) = \sum_{n=1}^{N} \delta(x_n - x)$ $Dg_{0,1}(x_n)^*(x'_n - g_1(x_n))$, giving

$$(Dg_t^{-1}(\cdot))^* Av_0(g_t^{-1}(\cdot))|Dg_t^{-1}(\cdot)|$$

$$= \sum_{n=1}^{N} \delta(x_n - g_t^{-1}(\cdot))(Dg_t^{-1}(g_t(x_n)))^* Dg_{0,1}(x_n)^*(x'_n - g_1(x_n))|Dg_t^{-1}(\cdot)|$$

$$= \sum_{n=1}^{N} \delta(x_n - g_t^{-1}(\cdot))(Dg_{t,1}(g_t(x_n)))^*(x'_n - g_1(x_n))|Dg_t^{-1}(\cdot)|. \quad (11.47)$$

Now substituting $y = g_t^{-1}(x)$ with $dx = |Dg_t(y)|dy$ gives

$$(Dg_t^{-1}(\cdot))^* Av_0(g_t^{-1}(\cdot))|Dg_t^{-1}(\cdot)|$$

$$= \sum_{n=1}^{N} \delta(g_t(x_n) - \cdot)(Dg_{t,1}(g_t(x_n)))^* |Dg_t^{-1}(g_t(\cdot))||Dg_t(\cdot)|(x'_n - g_1(x_n))$$

$$\stackrel{(a)}{=} \sum_{n=1}^{N} \delta(g_t(x_n) - \cdot)(Dg_{t,1}(g_t(x_n)))^*(x'_n - g_1(x_n)) = Av_t(\cdot), \quad (11.48)$$

where (a) uses $(Dg_t(\cdot))^{-1} = Dg_t^{-1}(g_t(\cdot))$. □

Example 11.10 (Beg Example) To compute the landmark matching vector fields for numerical stability Faisal Beg computed the gradient directly on the vector fields rather than the momentum. Initialize $v^{old} = 0$, choose constant ϵ, then for all $t \in [0, 1]$,

Step1: $\frac{\partial}{\partial t} g_t^{new} = v_t^{old}(g_t^{new}), \quad \frac{\partial}{\partial t} g_t^{-1 new}$

$\qquad = -Dg_t^{-1 new} v_t^{old}, \quad g_{t,1}^{new} = g_1^{new}(g_t^{-1 new}),$

Step2: Compute $v_t^{new} = v_t^{old} - \epsilon \nabla_v E_t^{old}$, set $v^{old} \leftarrow v^{new}$, (11.49)

$\qquad$ return to Step1 where

$$\nabla_v E_t^{old}(x) = v_t^{old}(x) - \sum_{n=1}^{N} K(g_t^{new}(x_n), x)(Dg_{t,1}^{new}(g_t^{new}(x_n)))^*$$

$$\times (x'_n - g_1^{new}(x_n)), \quad x \in X \quad (11.50)$$

Figure 11.3 shows examples of the landmark solution from the Faisal Beg algorithm a finite grid approximation to the gradient Algorithm 11.10. The tangent space metric for the large deformation is defined through the Laplacian $A = (\text{diag}(-\nabla^2 + c \text{ id}))^2$ with norm-square $\|v\|_V^2 = \langle Av, v \rangle_2$ and circulant boundary conditions. The columns shows the grid deformations for the compression (column 1), swivel (column 2), and circular rotation (column 3). The top and bottom rows show two different values of landmark placement noise $\sigma = 0.3$ (top), $\sigma = 1.0$ (bottom); circles $x_n = \circ$ depict the template landmarks, stars $x'_n = *$ depict the target landmarks. The line emanating from the template landmark is its trajectory to reach the corresponding target landmark. Column 1 shows a scale out of 13 pixels which is matched using four landmarks as shown to a larger ball of radius of 28 pixels. Column 2 shows

11.4 CONSERVATION OF MOMENTUM FOR DIFFEOMORPHISM

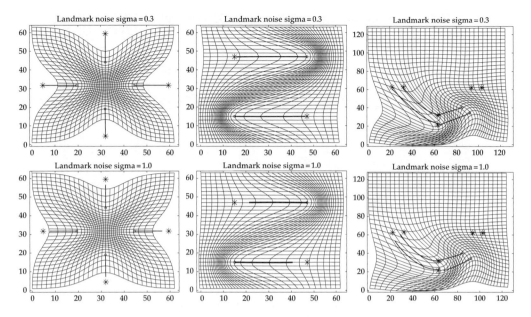

Figure 11.3 Panels show the diffeomorphisms applied to the regular coordinates generated from the landmark metric matching; $x_n = \circ$, $x'_n = *$. Figure shows the scale (column 1), S-curve (column 2), and circular rotation (column 3) for $\sigma = 0.3$ (top row) and $\sigma = 1.0$ (bottom row). The line emanating from the landmarks are the trajectory onto the corresponding landmark.

the S-curve with the 4 landmarks forcing the deformation in opposite directions. Column 3 shows a quarter C parametrized by 6 landmarks and transformed to a similarly parameterized half-C. As the landmark placement noise gets smaller, the matching tends to be more exact.

11.4.1 An ODE for Diffeomorphic Landmark Mapping

The landmark problem is very special; its solution is reduced to a series of N ordinary differential equations [281, 284].

Corollary 11.11 *The landmarks are transported along the geodesic with the vector fields given by*

$$v_t(\cdot) = \sum_{n=1}^{N} K(g_t(x_n), \cdot)\beta_{nt}, \tag{11.51}$$

where K is the Green's kernel associated with A; defining $\beta_{nt} = (Dg_{t,1}(g_t(x_n)))^* (x'_n - g_1(x_n))$, then

$$\frac{d}{dt}\beta_{nt} + Dv_t(g_t(x_n))^*\beta_{nt} = 0 \tag{11.52}$$

$$\frac{d}{dt}g_t(x_n) - v_t(g_t(x_n)) = 0. \tag{11.53}$$

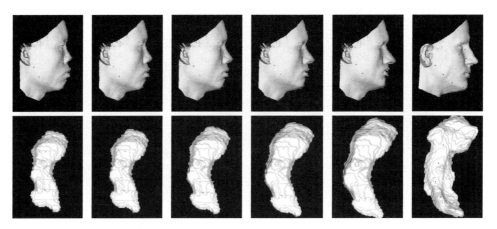

Figure 11.4 Results of landmark mapping of surface geometries. Top row shows the face mapping; bottom row shows the hippocampus mapping. Column 1 shows the starting template; column 6 shows the target surfaces. Columns 2–5 show examples along the geodesic generated by solving the diffeomorphic mapping of the faces. Results taken from Vaillant [284].

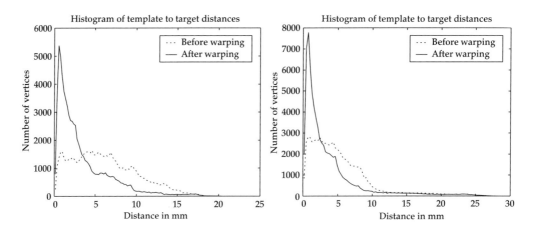

Figure 11.5 Histograms of accuracy showing distance between template and target surface geometries after landmark mapping. Results taken from Vaillant, PhD.

Proof Since $(Dg_{t,1}(g_t(x_n)))^* = (Dg_t^{-1}(g_t(x_n)))^* Dg_{0,1}(x_n)^*$ we have $\beta_{nt} = (Dg_t^{-1}(g_t(x_n)))^* \beta_{n0}$. Computing the derivative gives

$$\frac{d}{dt} \beta_{nt} = \frac{d}{dt} (Dg_t^{-1}(g_t(x_n)))^* \beta_{n0}. \tag{11.54}$$

Now to compute the derivative,

$$\frac{d}{dt} Dg_t^{-1}(g_t(x_n)) = \frac{d}{dt} (Dg_t(x_n))^{-1} = -(Dg_t(x_n))^{-1} \frac{d}{dt} Dg_t(x_n)(Dg_t(x_n))^{-1} \tag{11.55}$$

$$= -(Dg_t(x_n))^{-1} Dv_t(g_t(x_n)) Dg_t(x_n)(Dg_t(x_n))^{-1} \tag{11.56}$$

$$= -Dg_t^{-1}(g_t(x_n)) Dv_t(g_t(x_n)). \tag{11.57}$$

Substituting in Eqn. 11.54 gives

$$\frac{d}{dt}\beta_{nt} = \frac{d}{dt}(Dg_t^{-1}(g_t(x_n)))^*\beta_{n0} = -Dv_t(g_t(x_n))^*(Dg_t^{-1}(g_t(x_n)))^*\beta_{n0}. \quad (11.58)$$

□

Example 11.12 (Landmark Mapping of Faces from Vaillant [284]) Vaillant has implemented the ODE of Corollary 11.11. Shown in Figure 11.4 are examples from Vaillant [284] demonstrating geodesic landmark mapping of faces. Column 1 shows the template starting face. Columns 2–5 show examples along the geodesic match to the target face. Column 6 shows the three target faces.

Shown in Figure 11.5 are histograms demonstrating the distances between triangles on the surface geometry before and after landmark mapping. Notice that after landmark mapping the manifolds are largely within 1–2 mm of each other.

12 METRICS ON PHOTOMETRIC AND GEOMETRIC DEFORMABLE TEMPLATES

ABSTRACT In this chapter the metric structure developed for geometric transformations is extended to define a metric structure on the images which they act. Defined as an orbit under the finite and infinite dimensional groups we establish the necessary properties for measuring distance in the image orbit. We extend the results from not only geometric transformation of the images but also to photometric variation in the images.

12.1 Metrics on Dense Deformable Templates: Geometric Groups Acting on Images

Geometric shape is studied via transformations which are products of subgroups of diffeomorphisms $\mathcal{G}$ acting on the background spaces. To establish metrics on the space of images $I, I' \in \mathcal{I}$, we associate group elements $g, h \in \mathcal{G}$ to them and then define metric distances in group $\mathcal{G}$. This induces the metric on the orbit $\mathcal{I}$. Such metrics defined on the matrix groups and diffeomorphisms require properties of geodesic length on Riemannian manifolds.

Now examine two kinds of transformations, the first accomodating geometric or background shape change, the second involving the photometric or intensity value changes. The actions of the diffeomorphic transformations (invertible differentiable mappings with differentiable inverse) are introduced on the background space $\mathcal{G} : X \to X$. The group of diffeomorphisms divides the ensemble of images into geometrically distinct orbits.

The second kind of transformation is associated with photometric or intensity transformations, acting on the image values themselves. Begin by defining the basic geometric group acting on the images. Throughout the next several chapters, the basic model studied represents the source of imagery $\mathcal{I}$ as an orbit under groups of transformations of exemplar templates: $\mathcal{G} : I_\alpha \to \mathcal{I} = \mathcal{G}I_\alpha$. The model for the source is that the images $I \in \mathcal{I}$ are orbits under actions of transformation $g \in \mathcal{G}$, and depending on the specific problem the transformations can be translations, rotations, rigid motions of translations and rotations, scale, and Cartesian products of them.

12.1.1 Group Actions on the Images

The transformation groups of diffeomorphisms $(\mathcal{G}, \circ)$ generate the orbit of images through the group action defined as follows. The diffeomorphisms will be assumed to act on the infinite background space $X = \mathbb{R}^d$ for the matrix groups, and a bounded subset $X \subset \mathbb{R}^d$ for the infinite dimensional diffeomorphisms.

Definition 12.1 *Define the* **set of images** *$\mathcal{I}$ on X the background space $I : X \to V$ with V the value space.*

The group defines a group action on the set of images as follows.

Theorem 12.2 *Define Φ acting on the the images $\Phi : \mathcal{G} \times \mathcal{I} \to \mathcal{I}$ according to*

$$\Phi(g, \cdot) : I \mapsto g \cdot I = I \circ g^{-1}. \tag{12.1}$$

Then with the group operation composition of functions $g \circ g'(\cdot) = g(g'(\cdot))$ is a group action dividing $\mathcal{I}$ into a set of disjoint orbits.

Proof Clearly since $\Phi(\mathrm{id}, I) = I$ we must show the product property of the group action, $\Phi(g, \Phi(h, I)) = \Phi(g \circ h, I)$ which follows according to

$$\Phi(g, \Phi(h, I)) = \Phi(g, (I \circ h^{-1})) = (I \circ h^{-1}) \circ g^{-1} \tag{12.2}$$

$$= I \circ (g \circ h)^{-1} = \Phi(g \circ h, I). \tag{12.3}$$

Then since it is a group action it defines an equivalence relation $I_1 \sim I_2$ if $\exists g \in \mathcal{G}$ such that $I_1 = gI_2$, dividing the $\mathcal{I}$ into disjoint orbits. $\square$

We can now define the source of images as a deformable template.

Definition 12.3 *The* **deformable template** $\mathcal{I}_\alpha$ *corresponds to the orbit under the group $\mathcal{G}$ of one selected and fixed image, term it the template $I_\alpha \in \mathcal{I}$, such that*

$$\mathcal{I}_\alpha = \mathcal{G} I_\alpha = \{I \in \mathcal{I} : I = I_\alpha \circ g^{-1}, g \in \mathcal{G}\}. \tag{12.4}$$

Corresponding to multiple orbits, not necessarily connected, the full model of the source of images becomes $\mathcal{I} = \cup_\alpha \mathcal{I}_\alpha$.

12.1.2 Invariant Metric Distances

Examine metrics on the orbits of the group actions. For the case where the images are sufficiently complex (devoid of symmetries) and in the same orbit with group actions which are of sufficiently low dimension, then there is a unique identification between images, and group elements, $I = I_\alpha \circ g \leftrightarrow g$. This corresponds to there being exactly one $g \in \mathcal{G}$ such that $gI = I'$, or alternatively $gI = I$ implies $g = \mathrm{id}$. In the unique identifiability case, the metric between images is determined by the metric in the group directly. Thus a natural construction of the metric on images $\tilde\rho : \mathcal{I} \times \mathcal{I} \to \mathbb{R}^+$ via the metric on group elements $\rho : \mathcal{G} \times \mathcal{G}$ analogous to the Euclidean metric in $\mathbb{R}^n$ is to define an origin and compare the distance between group elements required to carry the images back to the origin. The template I_α plays the role of the origin, and the group elements play the role of the vectors defining the mapping of points in $\mathbb{R}^n$ back to the origin. Then the metric takes the form

$$\tilde\rho(I, I') = \inf_{g,h \in \mathcal{G} : gI = hI' = I_\alpha} \rho(g, h) = \rho(g^\star, h^\star), \quad \text{where } g^\star I = h^\star I' = I_\alpha. \tag{12.5}$$

More generally, for high dimensional diffeomorphisms there will not be a unique identification of the image I to a single group element which generates it from another candidate in the orbit. Since there will be in general many maps between I, I' (since the stabilizer in the group can always be used), the image distance will be a set distance. For this define the stabilizer of some particular image or template; the stabilizer leaves invariant the template. The natural construction of the metric simply includes the stabilizer.

Definition 12.4 *Let $\mathcal{I}$ be the orbit under the action of the group $\mathcal{G}$ with group action $g \cdot I = I \circ g^{-1}$. Define $\mathcal{S} \subset \mathcal{G}$ to be the* **stabilizing subgroup** *of the reference object or template $I_\alpha \in \mathcal{I}$:*

$$\mathcal{S} = \{s \in \mathcal{G} : sI_\alpha = I_\alpha\}. \tag{12.6}$$

The natural construction of the metric includes the stabilizer giving a set distance since

$$\inf_{\substack{g,h \in \mathcal{G} \\ gI=hI'=I_\alpha}} \rho(g,h) \geq \inf_{\substack{g' \in [g]_\mathcal{S}, h \in \mathcal{G}: \\ gI=hI'=I_\alpha}} \rho(g',h) \geq \inf_{\substack{g' \in [g]_\mathcal{S}, h' \in [h]_\mathcal{S}: \\ gI=hI'=I_\alpha}} \rho(g',h'). \tag{12.7}$$

For the resulting function to be a metric it must extend to the orbits under the stabilizer as a set distance. The condition for the metric to extend to the orbits is that it is invariant to the stabilizer (see Theorem below).

Theorem 12.5 *Let $\mathcal{I}$ be the orbit under the action of the group $\mathcal{G}$ with group action $g \cdot I = I \circ g^{-1}$.*

Let $\rho(\cdot, \cdot)$ be a distance on $\mathcal{G}$ which is invariant by the left action of the stabilizer $\mathcal{S}$ so that for all $s \in \mathcal{S}, g, g' \in \mathcal{G}$, $\rho(sg, sg') = \rho(g, g')$. The function $\tilde{\rho} : \mathcal{I} \times \mathcal{I} \to \mathbb{R}^+$ defined by

$$\tilde{\rho}(I, I') = \inf_{\substack{g' \in [g]_\mathcal{S}, h' \in [h]_\mathcal{S}: \\ gI = hI' = I_\alpha}} \rho(g', h'), \tag{12.8}$$

is a metric distance satisfying symmetry and the triangle inequality. If it attains the infimum for all I, I' such that $\tilde{\rho}(I, I') = 0$ implies $I = I'$, then $\tilde{\rho}$ is a distance.

Proof For the case where the stabilizer is trivially the identity, then the infimum reduces to the distance in the group which is a metric distance by definition of the theorem. That $\tilde{\rho}$ is symmetric in I, I' follows from the property that ρ is a metric on the group $\mathcal{G}$. To prove the triangular inequality let us show that for all I, I', I'',

$$\tilde{\rho}(I, I'') \leq \tilde{\rho}(I, I') + \tilde{\rho}(I', I''). \tag{12.9}$$

Define $I_\alpha = gI = hI' = kI''$ then for all $g, h, k \in \mathcal{G}, s \in \mathcal{S}$, then

$$\rho(g, sk) \overset{(a)}{\leq} \rho(g, h) + \rho(h, sk) \tag{12.10}$$

$$\overset{(b)}{=} \rho(g, h) + \rho(s^{-1}h, k), \tag{12.11}$$

with (a) following from the triangle inequality property of $\rho(\cdot, \cdot)$ and (b) from its left invariance. Now choose $g^\star, h^\star, h^{\star\star}, k^\star$ to satisfy the infimums:

$$g^\star, h^\star = \arg\inf_{g' \in [g]_\mathcal{S}, h' \in [h]_\mathcal{S}} \rho(g', h') \tag{12.12}$$

$$(h^{\star\star}, k^\star) = \arg\inf_{h' \in [h]_\mathcal{S}, k' \in [k]_\mathcal{S}} \rho(h', k'). \tag{12.13}$$

There exists an $s^\star \in \mathcal{S}$ such that $s^{\star-1}h^\star = h^{\star\star}$ giving

$$\tilde{\rho}(I, I') + \tilde{\rho}(I', I'') = \rho(g^\star, h^\star) + \rho(h^{\star\star}, k^\star)$$

$$= \rho(g^\star, h^\star) + \rho(s^{\star-1}h^\star, k^\star) \tag{12.14}$$

$$\overset{(a)}{\geq} \rho(g^\star, s^\star k^\star) \tag{12.15}$$

$$\overset{(b)}{\geq} \tilde{\rho}(I, I''), \tag{12.16}$$

with (a) following from Eqn. 12.11, and (b) following from the infimum property of the distance. □

For image matching we shall extensively use a metric which satisfies a stronger condition in which the metric is left invariant to the entire group (not just the stabilizer). Then, any point in the metric space can act as the template. This is analogous to the Euclidean setting in which any point in $\mathbb{R}^n$ can play the role of 0; this is because the metric is shift invariant by a rigid motion.

Corollary 12.6 *Let $\mathcal{I}$ be the orbit under the action of the group $\mathcal{G}$ with group action $g \cdot I = I \circ g^{-1}$. If the more restrictive condition holds that the distance is left invariant to the entire group $\mathcal{G}$, then all templates are equivalent, and*

$$\tilde{\rho}(I, I') = \inf_{g \in \mathcal{G} : gI' = I} \rho(\mathrm{id}, g). \tag{12.17}$$

Proof Under the more general condition of left invariance to the group, then let $gI = I_\alpha$ for some $g \in \mathcal{G}$, then for all $g' \in [g]_S$, $\inf_{h' \in [h]_S} \rho(g', h') = \inf_{h' \in [h]_S} \rho(\text{id}, g^{-1}h')$ which implies

$$\inf_{g' \in [g]_S, h' \in [h]_S} \rho(g', h') = \inf_{g \in \mathcal{G}, h' \in [h]_S} \rho(\text{id}, g^{-1}h'). \tag{12.18}$$

This gives

$$\tilde{\rho}(I, I') = \inf_{\substack{g' \in [g]_S, h' \in [h]_S: \\ gI = hI' = I_\alpha}} \rho(g', h') = \inf_{\substack{g \in \mathcal{G}, h' \in [h]_S: \\ gI = hI' = I_\alpha}} \rho(\text{id}, g^{-1}h') \tag{12.19}$$

$$= \inf_{g \in \mathcal{G}: gI' = I} \rho(\text{id}, g). \tag{12.20}$$

$\square$

12.2 The Diffeomorphism Metric for the Image Orbit

The natural extension for calculating metric length between the images is to consider the shortest paths $g_t \in \mathcal{G}$, $t \in [0, 1]$, connecting the group elements which match one image to another via the template with the tangent element of the evolution $(\partial g_t / \partial t)(g_t^{-1})$. Then we have the following invariance properties for inducing metrics on the images in the orbit.

Theorem 12.7 *Given the group of diffeomorphism $\mathcal{G}(\mathcal{V})$ with norm $\|\cdot\|_V$ and vector fields having finite energy $\mathcal{V} = \{v : \int_0^1 \|v_t\|_V^2 \, dt < \infty\}$. Defining $\rho : \mathcal{G} \times \mathcal{G} \to \mathbb{R}^+$ to be*

$$\rho^2(g, h) = \inf_{\substack{v: g_0 = g \\ \dot{g} = v(g)g_1 = h}} \int_0^1 \left(\left\| \frac{\partial g_t}{\partial t} \left(g_t^{-1} \right) \right\|_V^2 = \|v_t\|_V^2 \right) dt, \tag{12.21}$$

then $\rho(\cdot, \cdot)$ is left invariant to $\mathcal{G}(\mathcal{V})$; for all $h \in \mathcal{G}$, $\rho(h, hg) = \rho(\text{id}, g)$. Defining $\mathcal{I}$ to be the orbit with respect to the group with $\tilde{\rho} : \mathcal{I} \times \mathcal{I} \to \mathbb{R}^+$ given by

$$\tilde{\rho}(I, I') = \inf_{\substack{g, h \in \mathcal{G}: \\ gI = hI' = I_\alpha}} \rho(g, h) = \inf_{g \in \mathcal{G}: gI' = I} \rho(\text{id}, g), \tag{12.22}$$

then $\tilde{\rho}$ is symmetric and satisfies the triangle inequality. If it attains the infimum for all I, I' such that $\tilde{\rho}(I, I') = 0$ implies $I = I'$, then $\tilde{\rho}$ is a distance on $\mathcal{I}$.

Proof Eqn. 12.22 follows from the substitution $x = g_t^{-1}(y)$ in $(\partial g_t / \partial t)(x) = v_t(g_t(x))$. The essential property which ρ posesses is that it is not a function of where in the group the elements are, so translation by another group element leaves the distance unchanged; it is left invariant. If g satisfies $(\partial g_t / \partial t) = v_t(g_t), g_0 = \text{id}, g_1 = g$, then left action by $h \in \mathcal{G}$ simply shifts the solution. That is $g_0 \circ h = h, g_1 \circ h = g(h)$, and has identical velocity:

$$\frac{\partial g_t \circ h}{\partial t}(x) = \frac{\partial g_t}{\partial t}(h(x))$$

$$= v_t(g_t(h(x))), \quad g_0 = h, \quad g_1 = g(h).$$

Clearly the distance $\rho(\text{id}, g) = \rho(h, hg)$ are the same distance giving $\sqrt{\int_0^1 \|v_t\|_V^2 \, dt}$. $\square$

The left invariance property of the metric implies all elements are equally good templates. Any element in the orbit can be taken as the template.

12.3 Normal Momentum Motion for Geodesic Connection Via Inexact Matching

In Chapter 11, we have examined the shortest path geodesic flows of diffeomorphisms corresponding to the Euler equation. As well in Section 11.3 we examined the variational problem in the tangent space of the group minimizing with respect to the vector fields generating the shortest path flows.

As we have seen, along the geodesic, the momentum is determined by the momentum at the identity:

$$Av_t = (Dg_t^{-1})^* Av_0(g_t^{-1})|Dg_t^{-1}|. \tag{12.23}$$

The natural question becomes, where does the initial momentum Av_0 specifying the geodesic connection of one shape to another come from? Naturally, it arises by solving the variational problem of *shooting* one object onto another, or inexact matching via geodesic connection. In this setting we do not require an exact boundary condition for g_1 implying that the the perturbation in η satisfies only the $t = 0$ boundary conditions and at the boundary of the background space $x \in \partial X$. The inexact correspondence associated with the boundary conditions are depicted in panel 2 of Figure 10.1 of Chapter 10. Notice how at the endpoint only approximate correspondence is enforced. For this we define an endpoint correspondence condition corresponding to the squared-error mismatch between the mapped exemplar and the image according to $C = \|I' - I \circ g_1^{-1}\|_2^2$. The inexact matching corresponds to minimizing this distance function.

To calculate the optimizing initial momentum for shooting one object to another we work in direct perturbation of the vector field $v \to v + \epsilon \psi$ with a zero boundary $\psi(\partial X) = 0$, with the total boundary conditions taking the form

$$\eta_0(\cdot) = 0, \quad \eta_1(\partial X) = 0, \quad \psi(\partial X) = 0. \tag{12.24}$$

Now we calculate the momentum at the origin solving the inexact matching problem following [285]. We shall require the perturbation of the inverse map with the perturbation of the flow.

Lemma 12.8 *The perturbation via η of the group element $g_t \to g_t(\epsilon) = (\mathrm{id} + \epsilon \eta_t) \circ g_t + o(\epsilon) = g_t + \epsilon \eta_t(g_t) + o(\epsilon)$ generates the variation of the inverse map g^{-1} according to*

$$\partial_\eta g_t^{-1} = \frac{\partial}{\partial \epsilon}(g_t + \epsilon \eta_t(g_t))^{-1}\Big|_{\epsilon=0} = -Dg_t^{-1}\eta_t. \tag{12.25}$$

The perturbation of the vector field $v_t \to v_t + \epsilon \psi_t$ generates a perturbation of the inverse group element $g_t^{-1} \to g_t^{-1}(\epsilon) = g_t^{-1} + \epsilon \partial_\psi g_t^{-1} + o(\epsilon)$ where

$$\partial_\psi g_t^{-1} = \frac{\partial}{\partial \epsilon}(g_t^{v+\epsilon\psi-1})\Big|_{\epsilon=0} = -\int_0^t (Dg_u(g_t^{-1}))^{-1} \psi_u(g_{t,u}) du. \tag{12.26}$$

12.3 NORMAL MOMENTUM MOTION FOR GEODESIC CONNECTION

Proof

Part 1: The Lemma follows from

$$\text{id} = (g_t + \epsilon \eta_t(g_t) + o(\epsilon))((g_t + \epsilon \eta_t(g_t) + o(\epsilon))^{-1}) \quad (12.27)$$

$$= (g_t + \epsilon \eta_t(g_t) + o(\epsilon))(g_t^{-1} + \epsilon \partial_\eta g_t^{-1} + o(\epsilon)) \quad (12.28)$$

$$= \text{id} + \epsilon Dg_t(g_t^{-1}) \partial_\eta g_t^{-1} + \epsilon \eta_t + o(\epsilon), \quad (12.29)$$

yielding Eqn. 12.25.

Part 2: To prove the inverse perturbation, use the identity $g_t^{v-1}(g_t^v) = \text{id}$ giving

$$\text{id} = \left(g_t^{v+\epsilon\psi}\right)^{-1}(g_t^{v+\epsilon\psi}) = g_t^{-1}(g_t + \epsilon \eta_t(g_t) + o(\epsilon)) + \epsilon \partial_\psi g_t^{-1}(g_t) + o(\epsilon) \quad (12.30)$$

$$= \text{id} + \epsilon Dg_t^{-1}(g_t)\eta_t(g_t) + \epsilon \partial_\psi g_t^{-1}(g_t) + o(\epsilon). \quad (12.31)$$

Substituting for $\eta_t(g_t)$ from Eqn. 11.17 gives

$$\partial_\psi g_t^{-1}(g_t) = -Dg_t^{-1}(g_t)\eta_t(g_t) = -Dg_t^{-1}(g_t)Dg_t \int_0^t (Dg_u)^{-1} \psi_u(g_u) du \quad (12.32)$$

$$\stackrel{(a)}{=} \int_0^t (Dg_u)^{-1} \psi_u(g_u) du, \quad (12.33)$$

(a) from the inverse theorem $Dg_t^{-1}(g_t) = (Dg_t(g_t^{-1} \circ g_t))^{-1} = (Dg_t)^{-1}$. Substituting $g_t^{-1}(y) = x$ gives

$$\partial_\psi g_t^{-1} = -\int_0^t (Dg_u(g_t^{-1}))^{-1} \psi_u(g_u \circ g_t^{-1}) du. \quad (12.34)$$

□

Theorem 12.9 (Geodesic Connection for Inexact Dense Matching) *Given distance $C = \|I' - I \circ g_1^{-1}\|_2^2$, with the template I smooth so that $\nabla(I \circ g) = (Dg)^* \nabla I(g)$, and $g_{t,u} = g_u \circ g_t^{-1}$, with $\text{div}(Av \otimes v) = (DAv)v + (\text{div } v)Av$. Assume the observed time-point data I' with the matching functional energy given by*

$$E(v_g) = \int_0^1 \|v_t\|_V^2 dt + \|I' - I(g_1^{-1})\|_2^2. \quad (12.35)$$

1. *The minimizer with respect to perturbations $g_t \to g(\epsilon) = g_t + \epsilon \eta_t(g_t) + o(\epsilon)$ satisfies the Euler equations*

$$\frac{\partial Av_t}{\partial t} + (Dv_t)^* Av_t + \text{div}(Av_t \otimes v_t) = 0, \quad (12.36)$$

$$Av_1 + (I' - I(g_1^{-1}))\nabla(I \circ g_1^{-1}) = 0. \quad (12.37)$$

2. *The momentum of the variational minimizer of the geodesic at the identity is*

$$Av_0 = -(I'(g_1) - I)|Dg_1|\nabla I; \quad (12.38)$$

along the geodesic the momentum satisfies $Av_t = (Dg_t^{-1})^* Av_0(g_t^{-1})|Dg_t^{-1}|$ given by

$$Av_t = -\nabla(I \circ g_t^{-1})|Dg_{t,1}|(I'(g_{t,1}) - I(g_t^{-1})). \tag{12.39}$$

Proof

Part 1: To compute the variation via perturbations $g \to g + \epsilon \eta(g)$ the variation of the second inexact matching term is given by the variation of the inverse of Eqn. 12.25 of Lemma 12.8 according to

$$-2\langle (I' - I(g_1^{-1}))\nabla I(g_1^{-1}), \partial_\eta g_1^{-1}\rangle_2 = 2\langle (I' - I(g_1^{-1}))\nabla I(g_1^{-1}), Dg_1^{-1}\eta_1\rangle_2 \tag{12.40}$$

$$= 2\langle (I' - I(g_1^{-1}))\nabla(I(g_1^{-1})), \eta_1\rangle_2. \tag{12.41}$$

Computing the entire variation gives

$$2\int_0^1 \langle Av_t, \partial_\eta v_t\rangle_2\, dt + \langle (I' - I(g_1^{-1}))\nabla(I \circ g_1^{-1}), \eta_1\rangle_2$$

$$\stackrel{(a)}{=} 2\int_0^1 \left\langle -\frac{\partial Av_t}{\partial t} - (Dv_t)^* Av_t - \text{div}(Av_t \otimes v_t), \eta_t \right\rangle_2 dt$$

$$+ 2\langle Av_1, \eta_1\rangle_2 + \langle (I' - I(g_1^{-1}))\nabla(I \circ g_1^{-1}), \eta_1\rangle_2. \tag{12.42}$$

The first term in Eqn. 12.42 is the Euler equation of Theorem 11.6, giving the variation over the interior $t \in [0,1)$ giving the Euler Eqn. 12.36. The second term is the free boundary term at $t = 1$ from the integration by parts from the perturbation η_1 from Eqn. 11.24 of Theorem 11.6 of Chapter 11. The boundary term is Eqn. 12.37 completing Part 1 of the proof.

Part 2: The variation with respect to perturbations of $v \to v + \epsilon \psi$ of the first term $\|v_t\|_V^2$ gives $2Av_t$. The second term requires the vector field perturbation of the inverse $\partial_\psi g_1^{-1}$ given by Lemma 12.8, Eqn. 12.26 according to

$$\partial_\psi g_1^{-1} = -\int_0^1 (Dg_t(g_1^{-1}))^{-1} \psi_t(g_t \circ g_1^{-1})dt$$

$$= -\int_0^1 Dg_1^{-1} D(g_t \circ g_1^{-1})^{-1} \psi_t(g_t \circ g_1^{-1})dt. \tag{12.43}$$

Now using the substitution $g_{1,t} = g_t \circ g_1^{-1}$, the variation of the endpoint term becomes

$$-2\langle (I' - I(g_1^{-1}))\nabla I(g_1^{-1}), \partial_\psi g_1^{-1}\rangle_2$$

$$= 2\langle (I' - I(g_1^{-1}))\nabla I(g_1^{-1}), \int_0^1 Dg_1^{-1} D(g_t \circ g_1^{-1})^{-1} \psi_t(g_{1,t})\rangle_2 dt$$

$$= 2\langle (I' - I(g_1^{-1}))Dg_1^{-1*}\nabla I(g_1^{-1}), \int_0^1 D(g_t \circ g_1^{-1})^{-1} \psi_t(g_{1,t})\rangle_2 dt$$

$$= 2\langle (I' - I(g_1^{-1}))\nabla(I \circ g_1^{-1}), \int_0^1 D(g_t \circ g_1^{-1})^{-1} \psi_t(g_{1,t})\rangle_2 dt$$

$$\stackrel{(a)}{=} 2\int_0^1 \langle (I'(g_{t,1}) - I(g_t^{-1}))\nabla(I \circ g_t^{-1}), \psi_t|D(g_{t,1})|\rangle_2 dt$$

$$= 2\int_0^1 \langle \nabla(I \circ g_t^{-1})|Dg_{t,1}|(I'(g_{t,1}) - I(g_t^{-1})), \psi_t\rangle_2 dt \tag{12.44}$$

where (a) follows from the change of variable $y = g_{1,t}(x)$ in the inner product.

12.3 NORMAL MOMENTUM MOTION FOR GEODESIC CONNECTION

Combining this with the first variation term $2Av_t$ gives the momentum on the geodesic Eqn. 12.39. □

Example 12.10 (Geodesic Shooting(Beg,Trouve,Vaillant,Younes)) Shown in Figure 12.1 are results from using the initial momentum at the identity Av_0 to generate objects via shooting. Row 1 shows the starting objects before shooting. Row 2 shows the density of momentum at the identity $Av_0 = (I'(g_1) - I)|Dg_1|\|\nabla I\|$ from Eqn. 12.38 as a function of position in the grid matching each image to it's target. Row 3 shows the image transported under the diffeomorphism $I(g_1^{-1})$ generated via shooting the initial momentum according to $\dot{g}_t = v_t(g_t), g_0 = \text{id}$. The density was computed using Faisal Beg's algorithm for solving the Euler equation for geodesic correspondence of the shapes to derive Av_0 (see Algorithm 16.3 in Chapter 16). Notice the density of the momentum is concentrated on the boundaries of the shape.

Figure 12.2 shows three objects studied to demonstrate the momentum at the identity generated via the Beg algorithm for the smooth Gaussian bump for shift, circles for scale, and two mitochondria with both forward and inverse mapping. Shown are comparisons between the momentum at the identity Av_0 and the gradient of the image ∇I. Column 1 shows the results

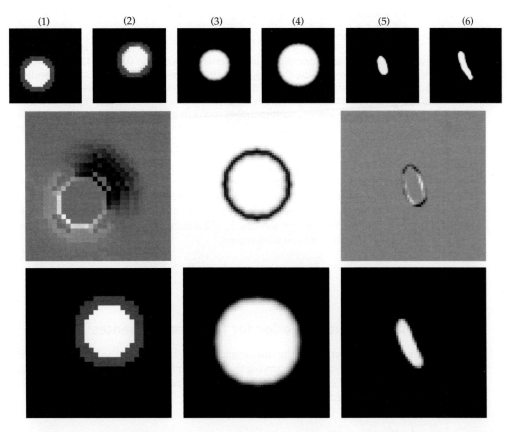

Figure 12.1 Row 1 shows objects for translation via diagonal translation (panels 1,2), scale (panels 3,4), and mitochondria (panels 5,6). Row 2 shows the density of momentum at the identity $Av_0 = (I'(g_1)-I)|Dg_1|\|\nabla I\|$ selected to match object 1 to object 2 constructed via the Faisal Beg large deformation diffeomorphic metric mapping algorithm (see Chapter 16) to satisfy Eqn. 12.38. Row 3 shows the image transported $I \circ g_1^{-1}$ by integrating the vector fields from the momentum at the identity along the flow $\dot{g}_t = v_t(g_t), g_0 = \text{id}$.

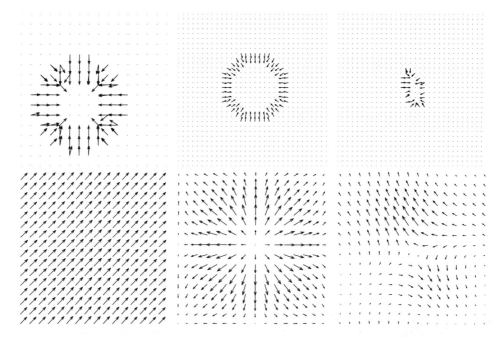

Figure 12.2 Experiments comparing momentum for translation (column 1), scale (column 2), and mitochondria (column 3). Row 1 shows comparison between momentum at the identity Av_0 generated by the Beg large deformation diffeomorphic metric mapping algorithm (see Chapter 16) and the gradient of the image ∇I; arrows depict the direction vector of each. Row 2 shows the vector fields v_0 at the identity corresponding to the momentum Av_0.

for diagonal translation, column 2 shows scale, and column 3 shows the mitochondria mapping. Shown in row 1 at every point in the grid are the arrows depicting the direction vector of the momentum Av_0 of each point in the grid given by the geodesic solution. Superimposed is a second arrow showing the normal level set gradient of the image. Notice how they superimpose almost exactly at every point; for the geodesic the momentum motion is normal to the level sets of the image as given by the conservation law. Row 2 shows the vector fields given by the Green's kernel of the metric operating on the momentum.

12.4 Normal Momentum Motion for Temporal Sequences

Let us now address another one of the fundamental models in which the hidden diffeomorphism sequence $g_t, t \in [0,1]$ corresponds to carrying the exemplar through a time sequence of observables $I'_t, t \in [0,1]$. Here the diffeomorphism is selected to minimize the matching function indexed throughout time according to

$$C = \int_0^1 \|I'_t - I \circ g_t^{-1}\|_2^2 \, dt. \tag{12.45}$$

Assume throughout that $(V, \|\cdot\|_V)$ is a reproducing kernel Hilbert space with operator K.

Theorem 12.11 (Normal Momentum Motion for Growth Connection of Dense Imagery)
Given is the growth sequence $I'_t, t \in [0,1]$ and distance $C = \int_0^1 \|I'_t - I(g_t^{-1})\|_2^2 \, dt$, with the

12.4 NORMAL MOMENTUM MOTION FOR TEMPORAL SEQUENCES

template I continuously differentiable with $\nabla(I \circ g) = (Dg)^* \nabla I(g)$, and $g_{t,u} = g_u \circ g_t^{-1}$ with growth sequence energy

$$\int_0^1 \|v_t\|_V^2 \, dt + \int_0^1 \|I'_t - I(g_t^{-1})\|_2^2 \, dt. \tag{12.46}$$

1. The Euler-equation associated with the minimizer with respect to variations $g_t \to g_t(\epsilon) = g_t + \epsilon \eta_t(g_t) + o(\epsilon)$ satisfies for all $t \in [0,1]$,

$$\frac{\partial Av_t}{\partial t} + (Dv_t)^* Av_t + \mathrm{div}(Av_t \otimes v_t) = -(I'_t - I(g_t^{-1}))\nabla(I \circ g_t^{-1}). \tag{12.47}$$

2. The momentum of the variational minimizer of the geodesic at the identity is

$$Av_0 = -\nabla I \int_0^1 (I'_u(g_u) - I)|Dg_u|du = 0; \tag{12.48}$$

along the geodesic the momentum satisfies

$$Av_t = -\nabla(I \circ g_t^{-1}) \int_t^1 (I'_u(g_{t,u}) - I(g_t^{-1}))|Dg_{t,u}|du = 0. \tag{12.49}$$

Proof

Part 1: The optimizer follows via perturbations $g_t \to g_t + \epsilon \eta_t(g_t)$. The perturbation of the first term $\int \|v_t\|_V^2 dt$ gives the Euler equation. The second term for inexact matching follows using the Lemma 12.8, Eqn. 12.25 of Chapter 12:

$$\delta_\eta C = -2 \int_0^1 \langle (I'_t - I(g_t^{-1}))\nabla I(g_t^{-1}), \partial_{\eta(g)} g_t^{-1} \rangle_2 dt$$

$$= 2 \int_0^1 \langle (I'_t - I_t(g_t^{-1}))\nabla I(g_t^{-1}), Dg_t^{-1} \eta_t \rangle_2 dt$$

$$\stackrel{(a)}{=} 2 \int_0^1 \langle (I'_t - I(g_t^{-1}))\nabla(I \circ g_t^{-1}), \eta_t \rangle_2 dt, \tag{12.50}$$

with (a) writing $\nabla(I \circ g_t^{-1}) = (Dg_t^{-1})^* \nabla I(g_t^{-1})$. Collecting the two components of the gradient, the first corresponding to the Euler equation and the second Eqn. 12.50 gives the proof of Eqn. 12.47.

Part 2: The L^2-norm variation via perturbations of $v \to v + \epsilon \psi$ of the first term $\|v_t\|_V^2$ gives $2Av_t$. The second term requires the vector field perturbation of the inverse $\partial_\psi g_u^{-1}$; from Lemma 12.8, Eqn. 12.26 gives

$$\delta_\psi C = -2 \int_0^1 \langle (I'_u - I(g_u^{-1}))\nabla I(g_u^{-1}), \partial_\psi g_u^{-1} \rangle_2 du \tag{12.51}$$

$$\stackrel{(a)}{=} 2 \int_0^1 \langle (I'_u - I(g_u^{-1}))\nabla(I(g_u^{-1})), \int_0^u (Dg_{u,t})^{-1} \psi_t(g_{u,t}) dt \rangle_2 du$$

$$\stackrel{(b)}{=} 2 \int_0^1 \int_0^u \langle (I'_u(g_{t,u}) - I(g_t^{-1}))\nabla(I \circ g_t^{-1}), \psi_t |Dg_{t,u}| \rangle_2 \, dt \, du$$

$$\stackrel{(c)}{=} 2 \int_0^1 \int_t^1 \langle |Dg_{t,u}|\nabla(I \circ g_t^{-1})(I'_u(g_{t,u}) - I(g_t^{-1})), \psi_t \rangle_2 \, du \, dt, \tag{12.52}$$

with (a) following from $\nabla(I \circ g_u^{-1}) = (Dg_u^{-1})^* \nabla I(g_u^{-1})$, (b) follows by making the substitution $z = g_t(g_u^{-1}(y))$ so that $y = g_u(g_t^{-1}(z))$ giving $dy = |Dg_u(g_t^{-1}(z))|dz$, $g_t^{-1}(z) = g_u^{-1}(y)$, $D(g_t(g_u^{-1}(y))) = Dz = \text{id}$, (c) interchanging integrals in u and t and setting $g_u(g_t^{-1}) = g_{t,u}$ to give the gradient. $\square$

12.5 Metric Distances Between Orbits Defined Through Invariance of the Metric

Fundamental to understanding are invariances. The representation of objects as orbits with respect to particular groups representing the invariances is therefore fundamental.

The construction of metric distances between such orbits becomes essential. For this define $\mathcal{O} \subset \mathcal{G}$ to be the subgroup defining the orbits. Then defining the notion of distances between orbits under $\mathcal{O}$ of group elements $g, g' \in \mathcal{G}$ is natural. The distance should be the set distance defined by the minimum between elements in the orbits. Then a sufficient condition for the distance to extend to the orbits is left invariance.

Theorem 12.12 *Let $\mathcal{O} \subset \mathcal{G}$ be the group defining the orbits, $[g]_{\mathcal{O}}$ in $\mathcal{G}/\mathcal{O}$, with ρ a distance on $\mathcal{G}$ which is invariant by the left action of $\mathcal{O} \subset \mathcal{G}$ so that for all $o \in \mathcal{O}, g, g' \in \mathcal{G}$,*

$$\rho(og, og') = \rho(g, g'). \tag{12.53}$$

Then the function $\tilde{\rho} : \mathcal{G}/\mathcal{O} \times \mathcal{G}/\mathcal{O} \to \mathbb{R}^+$ defined by

$$\tilde{\rho}([g]_{\mathcal{O}}, [h]_{\mathcal{O}}) = \inf_{g' \in [g]_{\mathcal{O}}, h' \in [h]_{\mathcal{O}}} \rho(g', h'), \tag{12.54}$$

is a metric distance on the orbits satisfying symmetry and the triangle inequality.

Proof For the case where $\mathcal{O}$ is trivially the identity, then the infimum reduces to the distance in the group which is a metric distance by definition of the theorem. More generally, that $\tilde{\rho}$ satisfies $\tilde{\rho}([g]_{\mathcal{O}}, [g']_{\mathcal{O}}) = 0$ implies $[g]_{\mathcal{O}} = [g']_{\mathcal{O}}$ and is symmetric in $[g]_{\mathcal{O}}, [g']_{\mathcal{O}}$ follows from the property that ρ is a metric on the group and attains its infimum in the group. To prove the triangular inequality let's show that for all $[g]_{\mathcal{O}}, [h]_{\mathcal{O}}, [k]_{\mathcal{O}}$,

$$\tilde{\rho}([g]_{\mathcal{O}}, [k]_{\mathcal{O}}) \leq \tilde{\rho}([g]_{\mathcal{O}}, [h]_{\mathcal{O}}) + \tilde{\rho}([h]_{\mathcal{O}}, [k]_{\mathcal{O}}). \tag{12.55}$$

For all $g, h, k \in \mathcal{G}, o \in \mathcal{O}$, then

$$\rho(g, ok) \stackrel{(a)}{\leq} \rho(g, h) + \rho(h, ok) \tag{12.56}$$

$$\stackrel{(b)}{=} \rho(g, h) + \rho(o^{-1}h, k), \tag{12.57}$$

with (a) following from the triangle inequality property of $\rho(\cdot, \cdot)$ and (b) from its left invariance. Now choose $g^\star, o^\star, h^\star, k^\star$ to satisfy the infimums:

$$g^\star, h^\star = \arg\inf_{g' \in [g]_{\mathcal{O}}, h' \in [h]_{\mathcal{O}}} \rho(g', h') \tag{12.58}$$

$$(o^\star, k^\star) = \arg\inf_{o \in \mathcal{O}, k' \in [k]_{\mathcal{O}}} \rho(o^{-1}h^\star, k'). \tag{12.59}$$

This gives

$$\tilde{\rho}([g]_{\mathcal{O}},[h]_{\mathcal{O}}) + \tilde{\rho}([h]_{\mathcal{O}},[k]_{\mathcal{O}}) = \rho(g^\star, h^\star) + \rho(o^{\star-1}h^\star, k^\star) \tag{12.60}$$
$$\geq \rho(g^\star, o^\star k^\star) \tag{12.61}$$
$$\geq \tilde{\rho}([g]_{\mathcal{O}},[k]_{\mathcal{O}}). \tag{12.62}$$

□

12.6 Finite Dimensional Landmarked Shape Spaces

In the *deformable template* models, variation in the image space $\mathcal{I}$ are accomodated by introducing groups of transformations carrying individual elements from one to another. Thus far we have examined dense images $\mathcal{I}$ with elements $I \in \mathcal{I}$ which are defined as complete functions on the background space $I : X \subset \mathbb{R}^d \to$ value space. Examine the special class of sparse images corresponding to finite dimensional landmark shapes in which the images are a collection of points or landmarks. Because such points give direct information about the objects shape they will be termed N-shapes. The study of such class of images have been extensively pioneered by Bookstein [223, 224, 272]. Fix an integer $N > 0$ and denote by $\mathcal{I}_N$ the set of all collections of landmarks parametrizing the N-shapes $I_N \in \mathcal{I}_N$ through the points

$$I_N = (x_1, \ldots, x_N), \qquad I'_N = (x'_1, \ldots, x'_N). \tag{12.63}$$

Then the space of N-shapes $\mathcal{I}_N = \mathbb{R}^{dN}$ has metric $\rho : \mathcal{I}_N \times \mathcal{I}_N \to \mathbb{R}^+$ defined by the lengths the paths the points travel to correspond. Length is measured through the geodesic curves $g_t \in \mathbb{R}^{dN}, t \in [0,1]$, with $Q(g_t)$ a positive definite symmetric matrix defining the quadratic form in the tangent space $(T_{g_t}(\mathbb{R}^{dN}), \|\cdot\|_{\mathbb{R}^{dN}})$.

Definition 12.13 *Define the* **manifold of N-shapes** $\mathcal{I}_N \subset \mathbb{R}^{dN}$ *with* **metric distance** $\rho : \mathcal{I}_N \times \mathcal{I}_N \to \mathbb{R}^+$ *between shapes defined via the $Nd \times Nd$ positive definite matrices $Q(g_t) = (Q(g_t)_{ij})$ with $d \times d$ blocks $Q(g_t)_{ij}$ giving the metric*

$$\rho^2(I_N, I'_N) = \inf_{\substack{\dot{g}_t(x_n): g_0(x_n) = x_n \\ n=1,\ldots,N g_1(x_n) = x'_n}} \int_0^1 \left(\left\| \frac{\partial g_t}{\partial t}(g_t^{-1}) \right\|_Q^2 = \sum_{ij=1}^N \dot{g}_t(x_i)^* Q(g_t)_{ij} \dot{g}_t(x_j) \right) dt. \tag{12.64}$$

12.6.1 The Euclidean Metric

The distance metric between N-shapes is determined by the quadratic form giving length to the tangent element as the particles flow from one shape to another. Choosing the quadratic form to correspond to a constant independent of the paths gives the straight lines and Euclidean distance $\|\cdot\|_{\mathbb{R}^d}$. Here is a relatively simple example from landmark matching illustrating left invariance of the Euclidean metric.

Theorem 12.14 *Given N-shapes $I_N = x_1, \ldots, x_N$, $I'_N = x'_1, \ldots x'_N$, with curves $g_t \in \mathbb{R}^{Nd}$ choose the norm-square in the tangent space to be constant $Q(g_t) = Q$, then the metric*

$\rho : \mathcal{I}_N \times \mathcal{I}_N \to \mathbb{R}^+$ measures straight line paths

$$\rho^2(I_N, I'_N) = \|I_N - I'_N\|_Q^2 = \sum_{ij=1}^N (x_i - x'_i)^* Q_{ij}(x_j - x'_j). \tag{12.65}$$

Defining the quadratic form to be identity $Q(g_t) = \mathrm{id}$ gives the Euclidean metric $\rho : \mathcal{I}_N \times \mathcal{I}_N \to \mathbb{R}^+$ according to $\rho(I_N, I'_N) = \sum_{n=1}^N \|x'_n - x_n\|_{\mathbb{R}^d}^2$.

Then, in $\mathbb{R}^3$ for example, ρ is left invariant under the rigid motions $\mathbf{SE}(3)$ so that

$$\rho(I_N, I'_N) = \rho(OI_N + a, OI'_N + a); \tag{12.66}$$

this defines a metric distance $\tilde{\rho}$ between orbits $[I_N]_{\mathbf{SE(3)}}, [I'_N]_{\mathbf{SE(3)}}$ according to

$$\tilde{\rho}([I_N]_{\mathbf{SE(3)}}, [I'_N]_{\mathbf{SE(3)}}) = \inf_{(O,a)\in\mathbf{SE(3)},(O',a')\in\mathbf{SE(3)}} \rho(OI_N + a, O'I'_N + a') \tag{12.67}$$

$$= \inf_{(O,a)\in\mathbf{SE(3)}} \rho(I_N, OI'_N + a). \tag{12.68}$$

With centered points $\tilde{x}_n = x_n - \bar{x}, \tilde{x}'_n = x'_n - \bar{x}'$, and means $\bar{x} = 1/N \sum_{n=1}^N x_n, \bar{x}' = 1/N \sum_{n=1}^N x'_n$, then

$$\tilde{\rho}^2([I_N]_{\mathbf{SE(3)}}, [I'_N]_{\mathbf{SE(3)}}) = \sum_{n=1}^N \|\tilde{x}_n\|_{\mathbb{R}^3}^2 + \sum_{n=1}^N \|\tilde{x}'_n\|_{\mathbb{R}^3}^2$$

$$- 2 \max_{O\in\mathbf{SO(3)}} \mathrm{tr}\left(\sum_{n=1}^N \tilde{x}'_n \tilde{x}_n^* O^*\right). \tag{12.69}$$

Proof This is a flat vector space, with straight line curves the geodesics as in Theorem 10.5 of Chapter 10. The left invariance condition is satisfied since

$$\|I_N - I'_N\|_{\mathbb{R}^{3N}}^2 = \sum_{n=1}^N \|x_n - x'_n\|_{\mathbb{R}^3}^2$$

$$= \sum_{n=1}^N \|Ox_n + a - Ox'_n - a\|_{\mathbb{R}^3}^2 = \rho(OI_N + a, OI'_N + a). \tag{12.70}$$

The joint minimization gives

$$(O, a)_{MS} = \underset{O,O'\in SO(3), a,a'\in\mathbb{R}^3}{\arg\min} \rho(OI_N + a, O'I'_N + a')$$

$$= \underset{O\in SO(3), a\in\mathbb{R}^3}{\arg\min} \rho(I_N, OI'_N + a) \tag{12.71}$$

$$= \underset{O\in SO(3), a\in\mathbb{R}^3}{\arg\min} \sum_{n=1}^N \|Ox'_n - x_n + a\|_{\mathbb{R}^3}^2. \tag{12.72}$$

Minimizing with respect to translation gives the MMSEs

$$a_{MS} = \arg\min_{a \in \mathbb{R}^3} \sum_{n=1}^{N} \|x_n - (Ox'_n + a)\|_{\mathbb{R}^3}^2 \qquad (12.73)$$

$$= \arg\min_{a \in \mathbb{R}^3} \sum_{n=1}^{N} \|x_n - Ox'_n\|^2 - 2\sum_{n=1}^{N} \langle a, x_n - Ox'_n \rangle_{\mathbb{R}^3} + N\|a\|^2 = \bar{x} - O\bar{x}'. \qquad (12.74)$$

Substituting a_{MS} and the centered points $\tilde{x}_n, \tilde{x}'_n$ gives

$$O_{MS} = \arg\min_{O \in SO(3)} \sum_{n=1}^{N} \|\tilde{x}_n - O\tilde{x}'_n\|_{\mathbb{R}^3}^2$$

$$= \arg\min_{O \in SO(3)} \sum_{n=1}^{N} \|\tilde{x}_n\|_{\mathbb{R}^3}^2 - 2\text{tr} \sum_{n=1}^{N} \tilde{x}_n \tilde{x}'^*_n O^* + \sum_{n=1}^{N} \tilde{x}'^*_n O^* O \tilde{x}'_n \qquad (12.75)$$

$$= \arg\max_{O \in SO(3)} \text{tr } AO^*, \quad \text{where } A = \sum_{n=1}^{N} \tilde{x}_n \tilde{x}'^*_n. \qquad (12.76)$$

$\square$

12.6.2 Kendall's Similitude Invariant Distance

Now expand the orbit invariance to the similitudes following Younes developments in [286]. Kendall [287] defines distance between sets of N-shapes which is invariant to uniform scale, rotation, and translation. Define the affine similitudes, the subgroup of the affine group to be matrices $A = sO, s \in (0, \infty), O \in \mathbf{SO}(d)$ with the action on the shapes to be scale-rotation and translation of each point of the shape $(AI_N + a)^* = (Ax_1 + a, \ldots, Ax_N + a)$. Then for all (A, a) in the similitudes the left invariant distance satisfies $\tilde{\rho}(AI_N + a, AI'_N + a) = \tilde{\rho}(I_N, I'_N)$.

Theorem 12.15 (Kendall [287]) *Define the mean shape $\bar{g}_t = 1/N \sum_{n=1}^{N} g_t(x_n)$ and variance*

$$\sigma^2(g_t) = \frac{1}{N} \sum_{n=1}^{N} \|g_t(x_n) - \bar{g}_t\|_{\mathbb{R}^d}^2, \qquad (12.77)$$

with the $Nd \times Nd$ diagonal matrix with varying weights $Q(g_t) = (1/\sigma^2(g_t))\,\text{id}$, then the metric becomes

$$\rho^2(I_N, I'_N) = \inf_{\substack{\dot{g}: g_0(x_n) = x_n \\ g_1(x_n) = x'_n}} \sum_{n=1}^{N} \int_0^1 \frac{1}{\sigma^2(g_t)} \left\|\frac{dg_t(x_n)}{dt}\right\|_{\mathbb{R}^d}^2 dt \qquad (12.78)$$

$$= \rho_0^2((\sigma(g_0), \bar{g}_0), (\sigma(g_1), \bar{g}_1)) + \left(\arccos \sum_{n=1}^{N} \left\langle \frac{g_0(x_n) - \bar{g}_0}{\sigma(g_0)}, \frac{g_1(x_n) - \bar{g}_1}{\sigma(g_1)} \right\rangle_{\mathbb{R}^d}\right)^2 \qquad (12.79)$$

with ρ_0 a metric on $\mathbb{R}^{Nd} \times \mathbb{R}^{Nd}$.

Proof Defining the normalized landmarks $\gamma_t(x_n) = ((g_t(x_n) - \bar{g}_t)/\sigma(g_t))$, then $g_t(x_n) = \sigma(g_t)\gamma_t(x_n) + \bar{g}_t$,

$$\frac{dg_t(x_n)}{dt} = \frac{d\sigma(g_t)}{dt}\gamma_t(x_n) + \sigma(g_t)\frac{d\gamma_t(x_n)}{dt} + \frac{d\bar{g}}{dt}. \qquad (12.80)$$

This implies $\sum_{n=1}^{N}\int_0^1 \frac{1}{\sigma^2(g_t)}\left\|\frac{dg_t(x_n)}{dt}\right\|_{\mathbb{R}^d} dt$ is given by

$$\int_0^1 \frac{1}{\sigma^2(g_t)}\left(\frac{d\sigma(g_t)}{dt}\right)^2 \sum_{n=1}^{N}\|\gamma_t(x_n)\|_{\mathbb{R}^d}^2 dt + N\int_0^1 \frac{1}{\sigma^2(g_t)}\left\|\frac{d\bar{g}}{dt}\right\|_{\mathbb{R}^d}^2 dt + \int_0^1 \sum_{n=1}^{N}\left\|\frac{d\gamma_t(x_n)}{dt}\right\|_{\mathbb{R}^d}^2 dt$$

$$+ 2\int_0^1 \frac{1}{\sigma(g_t)}\frac{d\sigma(g_t)}{dt}\sum_{n=1}^{N}\left\langle\gamma_t(x_n), \frac{d\gamma_t(x_n)}{dt}\right\rangle_{\mathbb{R}^d} dt$$

$$+ 2\int_0^1 \frac{1}{\sigma^2(g_t)}\left\langle\frac{d\bar{g}_t}{dt}, \sum_{n=1}^{N}\gamma_t(x_n)\right\rangle_{\mathbb{R}^d} dt + 2\int_0^1 \frac{1}{\sigma(g_t)}\left\langle\bar{g}_t, \sum_{n=1}^{N}\frac{d\gamma_t(x_n)}{dt}\right\rangle_{\mathbb{R}^d} dt$$

$$\stackrel{(a)}{=} \int_0^1 \frac{1}{\sigma^2(g_t)}\left(\frac{d\sigma(g_t)}{dt}\right)^2 dt + N\int_0^1 \frac{1}{\sigma^2(g_t)}\left\|\frac{d\bar{g}}{dt}\right\|_{\mathbb{R}^d}^2 dt$$

$$+ \int_0^1 \sum_{n=1}^{N}\left\|\frac{d\gamma_t(x_n)}{dt}\right\|_{\mathbb{R}^d}^2 dt, \qquad (12.81)$$

with (a) following from $\|\gamma_t\|_{\mathbb{R}^d}^2 = 1$ implying the 4th term $(d\|\gamma_t\|_{\mathbb{R}^d}^2/dt) = 0$, the 5th term has mean zero, and the final term is the derivative of the mean which is zero. Denote by $\rho_0((\sigma(g_0), \bar{g}_0), (\sigma(g_1), \bar{g}_1))^2$ the minimum of the first two terms, then the minimum of the last term can be explicitly computed (because $(\gamma(x_1), \ldots, \gamma(x_N))$ belongs to a sphere of dimension $N - 2$ and is given by the length of the great circle in the sphere which connects the extremities of the path, namely $\left(\arccos \sum_{n=1}^{N}\langle\gamma_0(x_n), \gamma_1(x_n)\rangle_{\mathbb{R}^d}\right)^2$. This gives Eqn. 12.79. □

Corollary 12.16 *Kendall's similitude metric is left-invariant* $\rho(AI_N + a, AI'_N + a) = \rho(I_N, I'_N)$, *implying the metric between orbits becomes*

$$\tilde{\rho}(I_N, I'_N) = \min_{A\ similitude, a\in\mathbb{R}^d} \rho(AI_N + a, I'_N). \qquad (12.82)$$

Proof Kendall's distance $\tilde{\rho}$ in equation (12.82) requires computing the minimum of $\rho(sOI_N + a, I'_N)$, for $s > 0$, $O \in \mathbf{SO}(d)$ and $a \in \mathbb{R}^d$. Since the action of s and a does not affect $\gamma(x_n)$, one can select them in order to cancel the distance ρ_0 without changing the second term implying that $\tilde{\rho}(I_N, I'_N)$ is the minimum of arccos $\sum_{n=1}^{N}\langle\gamma_0(x_n), \gamma_1(x_n)\rangle_{\mathbb{R}^d}$ for all $O \in \mathbf{SO}(d)$. When $d = 2$, there is an explicit solution

$$\tilde{\rho}(I_N, I'_N) = \arccos\left|\sum_{n=1}^{N}\langle\gamma_0(x_n), \gamma_1(x_n)\rangle_{\mathbb{R}^d}\right|. \qquad (12.83)$$

□

12.7 The Diffeomorphism Metric and Diffeomorphism Splines on Landmark Shapes

Now change the metric on landmarked shapes to the diffeormorphism metric. This returns us to the landmark splines of Section 11.4 examining the metric on the geodesics in which the diffeomorphisms are assumed known only on the subsets of the points defined by the N-shape I_N, I'_N. The group action for landmarked shapes is the most straightforward.

Definition 12.17 *Define the* **group action for N-shapes** *as*

$$g : I_N = (x_1, x_2, \ldots, x_N) \mapsto g \cdot I_N = (g(x_1), g(x_2), \ldots, g(x_N)). \tag{12.84}$$

The metric for diffeomorphic correspondence defined for dense correspondence reduces to terms only involving flows of curves of the N-correspondence points.

Theorem 12.18 (Diffeomorphism Metric for N-Shapes) *Assume $(V, \|\cdot\|_V)$ is a reproducing kernel Hilbert space with Green's operator $K : L^2(X, \mathbb{R}^d) \to V$ which defines the $Nd \times Nd$ positive definite symmetric matrix*

$$Q(g_t) = \Big(K(g_t(x_i), g_t(x_j))\Big)^{-1}. \tag{12.85}$$

The metric distance between N-shapes of Eqn. 12.64 subject to the constraints $g_0(x_n) = x_n, g_1(x_n) = x'_n, n = 1, 2, \ldots$ is given by

$$\rho^2(I_N, I'_N) = \inf_{\substack{v_{nt}: \dot{g}_t(x_n) = v_{nt} \\ g_0(x_n) = x_n, \; g_1(x_n) = x'_n}} \int_0^1 \sum_{i,j=1}^N v_{it}^* Q(g_t)_{ij} v_{jt} \, dt. \tag{12.86}$$

Proof From Theorem 11.9 and Corollary 11.11 of Chapter 11, the optimizing spline vector field is given by superposition of reproducing kernels

$$v_t(x) = \sum_{n=1}^N K(g_t(x_n), x) \beta_{nt}, \quad x \in X. \tag{12.87}$$

The norm-square becomes

$$\|v_t\|_V^2 = \left\langle \sum_{n=1}^N K(g_t(x_n), \cdot) \beta_n, v_t(\cdot) \right\rangle_V = \sum_{n=1}^N \beta_{nt}^* v_t(g_t(x_n)) \tag{12.88}$$

$$= \sum_{i,j=1}^N v_t(g_t(x_i))^* Q(g_t)_{ij} v_t(g_t(x_j)). \tag{12.89}$$

□

12.7.1 Small Deformation Splines

For small deformations, there is a lovely approximation of the diffeomorphism metric problem which has been pioneered by Bookstein [223, 224, 272]. This approximates the metric flow through the tangent flow of the landmark trajectories at the origin. In general this is not a metric; it is not symmetric nor does it satisfy the triangle inequality since it places a special role on the origin.

Corollary 12.19 *Let the mapping $v_0 : x \mapsto x + v(x)$ with vector fields $(V, \|\cdot\|_V)$ be as in Theorem 12.18 the $Nd \times Nd$ positive definite symmetric matrix $Q(g_0) = Q(I_N) = \left(K(x_i, x_j)\right)^{-1}$, then the Bookstein measure $m : \mathcal{I}_N \times \mathcal{I}_N \to \mathbb{R}^+$ satisfies*

$$m^2(I_N, I'_N) = \inf_{\substack{v_0 : x'_n = x_n + v_0(x_n) \\ n=1,2,\dots}} \|v_0\|_V^2 = \sum_{ij=1}^N (x_i - x'_i)^* Q(I_N)_{ij} (x_j - x'_j). \qquad (12.90)$$

Adding the affine motions $x \to x' = Ax + b + v_0(x)$ then

$$m^2(I_N, I'_N) = \inf_{\substack{v_0, A, b: \\ Ax_n + b + v_0(x_n) = x'_n}} \|v_0\|_V^2 = \inf_{A,b} \sum_{ij=1}^N (Ax_i + b - x'_i)^* Q(I_N)_{ij} (Ax_j + b - x'_j). \qquad (12.91)$$

For inexact matching minimizing

$$\|v_0\|_V^2 + \frac{1}{\sigma^2} \sum_{n=1}^N \|Ax_n + b + v_0(x_n) - y_n\|_{\mathbb{R}^d}^2, \qquad (12.92)$$

with $Nd \times Nd$ quadratic form matrix $M = \left(K(x_i, x_j) + \sigma^2 \delta(i-j)\right)^{-1}$ the optimizers satisfy

$$v_0(x) = -\sum_{n=1}^N K(x_n, x) \left(\sum_{m=1}^N M_{nm}(Ax_m + b - y_m)\right), \quad x \in X, \qquad (12.93)$$

with $A, b = \min_{A,b} \sum_{nm=1}^N (Ax_n + b - y_n)^* M_{nm}(Ax_m + b - y_m).$ $\qquad (12.94)$

Proof The proof is a direct result of the diffeomorphism spline proof Theorem 12.18, only applied for the single time $t = 0$. The optimizer minimizing $\|v_0\|_V^2$ superposes Green's kernels $v_0(x) = \sum_n K(x_n, \cdot) \beta_n$, giving

$$\min_{v_0, A, b} \|v_0\|_V^2 + \sum_{n=1}^N \frac{\|Ax_n + b + v_0(x_n) - y_n\|_{\mathbb{R}^d}^2}{\sigma^2}$$

$$= \sum_{ij=1}^N B_i^\star K(x_i, x_j) \beta_j + \sum_{n=1}^N \frac{\|Ax_n + b + v_0(x_n) - y_n\|_{\mathbb{R}^d}^2}{\sigma^2}.$$

Differentiating with respect to β_k gives

$$0 = 2 \sum_{n=1}^N K(x_k, x_n) \beta_n + \frac{2}{\sigma^2} \sum_{n=1}^N K(x_k, x_n)(Ax_n + b + v_0(x_n) - y_n), \qquad (12.95)$$

which implies

$$\sum_{n=1}^{N} K(x_k, x_n)\beta_n + \frac{1}{\sigma^2} \sum_{n=1}^{N} K(x_k, x_n) \sum_{j=1}^{N} K(x_n, x_j)\beta_j$$

$$= -\frac{1}{\sigma^2} \sum_{n=1}^{N} K(x_k, x_n)(Ax_n + b - y_n). \qquad (12.96)$$

Rewriting in matrix notation gives $K = \left(K(x_i, x_j)\right), M = (K + \sigma^2 I)^{-1}$ and

$$KM^{-1}\beta = -K(Ax + b - y), \qquad (12.97)$$

implying $\beta = -M(Ax + b - y)$. Solving for the minimum in A, b gives

$$\|v_0\|_V^2 + \frac{\|Ax + b + v_0 - y\|^2}{\sigma^2} \stackrel{(a)}{=} \beta^* K\beta + \frac{\|K\beta - M^{-1}\beta\|^2}{\sigma^2} = \beta^* M^{-1}\beta \qquad (12.98)$$

$$\stackrel{(b)}{=} (Ax + b - y)^* MM^{-1}M(Ax + b - y). \qquad (12.99)$$

$\square$

Notice, the minimum norm is determined by the residual error after fitting the landmarks with an affine motion.

Example 12.20 (Flows and Small Deformations) Joshi [242] first implemented the original algorithm for large deformation landmark mapping exploiting the fact that the vector fields are in the span of the Green's kernels, comparing the large deformation diffeomorphism metric mapping with the small deformation approximation. The metric was chosen to be $\|v\|_V^2 = \langle(\text{diag}(-\nabla^2 + c\,\text{id}))^2 v, v\rangle_2$ for inducing the Green's kernel $K(x,y) = \text{diag}\left(\kappa e^{-\sqrt{\frac{1}{c}}\|x-y\|_{\mathbb{R}^d}}\right)$, with κ chosen to make the Green's kernel integrate to 1 (see Section 9.4.4, Chapter 9 for calculated examples). Joshi's algorithm [242] implements via successive differences the spline equations reducing the problem to a finite dimensional problem by defining the flows on the finite grid of fixed times of size δ, $t_k = k\delta, k = 0, 1, \ldots, K = 1/\delta$. Figure 12.3 shows results illustrating the flow of diffeomorphisms generated in the image plane $g_t \in \mathbb{R}^{Nd}$, $N = 6$ landmarks, $d = 2$, for the sparse correspondence of points. Column 1 shows the grid test pattern for the correspondence of $A \to B, C \to D$ while fixing the corners of the grid (panel 1) along with the paths of the flow $g_t(x_i)$ projected onto the plane (panel 4). Notice the required twist of the grid. Column 2 compares the large deformation flow (top panel) with the small deformation solution (bottom panel) approximating the diffeomorphism with the vector space solution. Panel 2 shows the metric mapping g_1 applied to the grid; panel 5 shows the small deformation applied to the grid. Column 3 shows the determinant of the Jacobian map for the large deformation flow (panel 3) and small deformation (panel 6). The determinant of the Jacobian of the small deformation is shown in black (negative Jacobian) to white color (positive Jacobian). The variances used were constant $\sigma^2 = 0.01$.

Notice the geometric catastrophe which occurs for the curved transformation which is required. The resulting transformation of the small deformation approximation results in the grid crossing over. The determinant is negative in the region where the grid lines in the landmark deformation cross. The small deformations have no flexibility to create a curved trajectory by integrating the flow.

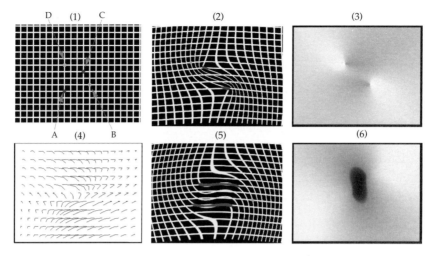

Figure 12.3 Top row: Panel 1 shows the grid test pattern matching $A \to B$, $C \to D$ with the corners fixed. Panel 2 shows the movement of the grid under the diffeomorphism g_1; panel 3 shows the determinant of the Jacobian $|Dg_1|$. Bottom row: Panel 4 shows the trajectories of the particles $g_t(x_i)$, $i = 1, 2, t \in [0, 1]$ traced out by the landmark points A,C and the four corners of the image projected into the plane. Panel 5 shows the small deformation mapping applied to the grid; panel 6 shows the determinant of the Jacobian $|Dg_1|$. Black to white color scale means large negative to large positive Jacobian. The variances were $\sigma^2 = 0.01$; mappings from Joshi [282] (see also Plate 17).

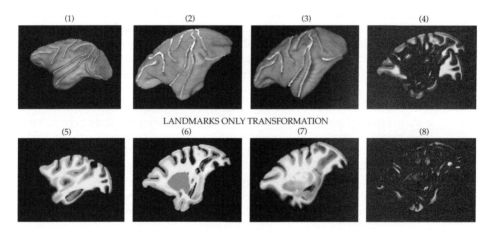

Figure 12.4 Figure shows results of large deformation diffeomorphic landmark matching. Row 1: Panels 1 and 2 show brains 87A and target 90C, panel 3 shows 87A matched to 90C transformed via landmarks. Row 2: Panels 5, 6, 7, show sections through 87A, 90C, and 87A matched to 90C, respectively. Panel 4 shows the difference image between 87A and 90C; panel 8 shows difference image after landmark transformation. Mapping results taken from Johis [282]; data taken from David Van Essen of Washington University (see also Plate 18).

Example 12.21 (Whole Brain Mapping via Landmarks) Examine the whole macaque cryosection brains shown in Figure 12.4 in which the gyri and associated sulci have been labeled. The sulci and gyri can be defined precisely in terms of the geometrical properties of extremal curves on the cortical surface using the notions of ridge curves and crest lines (extremal points of curvature) as in Khaneja [153]. Joshi used the fundus curves on the sulci and gyri to constrain the transformations across brains. Figure 12.4 shows entire macaque cryosection brains ($500^2 \times 200$ voxels). Panels 1 and

2 show brains 87A and 90C in which the nine major principle sulcus curves were identified in the brains and represented as piecewise linear curves with 16 nodes. The gyri and associated sulci have been labeled following the nomenclature used in Felleman and Van Essen [155]. The deformation field mapping the template to the target was constrained so that the corresponding points along the extremal curves were mapped onto each other. Panel 3 of Figure 12.4 shows the landmark matching of 87A matched to 90C using only the linear landmarks to define the transformation. Panels 5–7 show corresponding sections through 87A (panel 5), 90C (panel 6), and the transformed 87A → 90C (panel 7). Panel 4 shows the difference image between 87A and 90C before transformation; panel 8 shows the same after landmark transformation. Notice that there is a large difference in the shape and positions of the major subvolumes (the thalamus and the cortical folds) between the undeformed template and the target. Notice the correspondence in the alignment of the major subvolumes (panel 8) after the deformation.

12.8 The Deformable Template: Orbits of Photometric and Geometric Variation

Thus far the metric depends only on the geometric transformation in the background space. Variability in objects is not solely associated with geometric transformation, but as well as other factors such as lighting conditions, object surface properties and features, texture variations and the operational state of the object. Extend the construction of the metric to be image dependent following [286] by defining the group action to operate on the geometry and image pair.

Now examine models of variation which explicitly construct metric spaces which model photometric variation with geometric variation. The basic model of Figure 12.5 represents the source of imagery $\mathcal{I}$ as the orbit of the product of all photometric variations with geometric variations. The space of photometric intensities are modeled as a Hilbert space $\mathcal{H}$. The deformable template becomes the orbits of all photometric intensities under geometric motions, $\mathcal{A} = \mathcal{H} \times \mathcal{G}$, with elements the pairs $(I, g) \in \mathcal{A} = \mathcal{H} \times \mathcal{G}$. Notice, in this setting there are no fixed exemplars. The resulting images which form the mean fields become $I \circ g^{-1} \in \mathcal{I}$, thus the image formation process before the noise channel is a mapping $(I, g) \in \mathcal{A} \mapsto I \circ g^{-1}$.

12.8.1 Metric Spaces for Photometric Variability

Now examine the metric space of photometric variability. Figure 12.6 depicts the introduction of the metric dependence on both geometry and photometric evolution. Each curve depicts a particular object under geometric deformation. The metric distance on the product space of photometric and

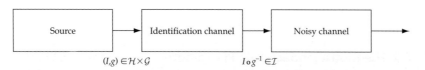

Figure 12.5 The source of images are the pair of photometric intensities and geometric transformation $(I, g) \in \mathcal{A} = \mathcal{H} \times \mathcal{G}$. The identification generate the images $I \circ g^{-1}, g \in \mathcal{G}$.

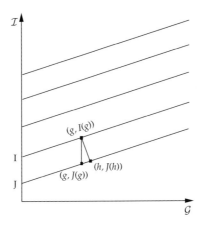

Figure 12.6 Shown is a comparison of objects *I* and *J* by looking for the smallest distance within the set (id, *I*) ∘ $\mathcal{G}$ and (id, *J*) ∘ $\mathcal{G}$.

geometry evolution must take into account both the distance between the geometric change as well as the photometric change, depicted as labelled pairs acting on the right $(h, J(h))$ and $(g, I(g))$. The invariance condition corresponds to parallel translation by the group acting as the right. If $\mathcal{G}$ is parallel translation, then the metric on the elements extends to a metric on the orbits.

Theorem 12.22 *Define the product set* $\mathcal{A} = \mathcal{G} \times \mathcal{I}$, *and define on it the (essentially a right action) action* $\mathcal{G} : \mathcal{A} \to \mathcal{A}$ *according to* $(g, I)h = (gh, I(h))$. *Let* ρ *be a distance on* $\mathcal{A}$ *which is invariant by the action of* $\mathcal{G}$; *for all* $h \in \mathcal{G}, (g, I), (g', I') \in \mathcal{A}$,

$$\rho((g, I)h, (g', I')h) = \rho((g, I), (g', I')). \tag{12.100}$$

Then the function $\tilde{\rho}$ *defined on* $\mathcal{I} \times \mathcal{I}$ *by*

$$\tilde{\rho}(I, I') = \inf_{g, h \in \mathcal{G}} \rho((g, I(g)), (h, I'(h))) \tag{12.101}$$

$$= \inf_{g \in \mathcal{G}} \rho((\mathrm{id}, I), (g, I'(g))), \tag{12.102}$$

is symmetric and satisfies the triangle inequality. If it attains the infimum for all I, I' *such that* $\tilde{\rho}(I, I') = 0$ *implies* $I = I'$, *then* $\tilde{\rho}$ *is a distance.*

Proof Since ρ is invariant, Eqn. 12.102 follows from

$$\inf_{g, h \in \mathcal{G}} \rho((g, I(g)), (h, I'(h))) = \inf_{g, h \in \mathcal{G}} \rho((\mathrm{id}, I), (hg^{-1}, I'(hg^{-1}))) \tag{12.103}$$

$$= \inf_{g \in \mathcal{G}} \rho((\mathrm{id}, I), (g, I'(g))). \tag{12.104}$$

The triangle inequality follows from the same proof as for Theorem 12.5 since ρ is invariant to the entire group. □

12.8.2 The Metrics Induced via Photometric and Geometric Flow

The natural extension for adding the calculation of photometric evolution is to consider shortest paths $a_t = (g_t, I_t) \in \mathcal{A} = \mathcal{G}(V) \times \mathcal{I}, t \in [0, 1]$, connecting elements $(g, I), (h, I') \in \mathcal{A}$ with the tangent element of the evolution

$$\frac{\partial a_t}{\partial t}(g_t^{-1}) = \left(\frac{\partial g_t}{\partial t}(g_t^{-1}), \frac{\partial I_t}{\partial t}(g_t^{-1})\right) \tag{12.105}$$

We shall study exclusively the superposition norm-square in the tangent space.

Definition 12.23 *Define the* **superposition norm-square** *in the tangent space of photometric and geometric variation, as*

$$\left\|\frac{\partial a_t}{\partial t}(g_t^{-1})\right\|^2_{T(\mathcal{A})} = \left\|\frac{\partial g_t}{\partial t}(g_t^{-1})\right\|^2_V + \left\|\frac{\partial I_t}{\partial t}(g_t^{-1})\right\|^2_2. \tag{12.106}$$

Defining the norm in the tangent space of $\mathcal{A}$ of photometric and geometric variation as the superposition of the two norms, then we have the following equivalence of the differential length associated with the tangent space norm. Examine the connection of (g, I) to (h, I').

Lemma 12.24 *Defining the geometric evolution $v_t = \partial g_t/\partial t(g_t^{-1})$, the image evolution $J_t = I_t(g_t^{-1})$, then the following energies are equal:*

$$E = \int_0^1 \left(\left\|\frac{\partial g_t}{\partial t}(g_t^{-1})\right\|^2_V + \left\|\frac{\partial I_t}{\partial t}(g_t^{-1})\right\|^2_2\right) dt, \quad \text{with b.c.} \begin{array}{l} g_0 = g, g_1 = h \\ I_0 = I, I_1 = I' \end{array} \tag{12.107}$$

$$= \int_0^1 \left(\|v_t\|^2_V + \left\|\frac{\partial J_t}{\partial t} + \nabla J_t^* v_t\right\|^2_2\right) dt \quad \text{with b.c. } J_0 = I(g^{-1}), J_1 = I'(h^{-1}). \tag{12.108}$$

Proof With the definition $J_t = I_t(g_t^{-1})$, then the boundary conditions imply with $g_0 = g, g_1 = h, J_0 = I(g^{-1}), J_1 = I'(h^{-1})$, then we have

$$J_0 = I_0(g_0^{-1}) = I(g^{-1}), \quad J_1 = I_1(g_1^{-1}) = I'(h^{-1}). \tag{12.109}$$

Defining $I_t = J_t(g_t)$ then

$$\frac{\partial I_t}{\partial t} = \frac{\partial}{\partial t} J_t(g_t) = \frac{\partial J_t}{\partial t}(g_t) + \nabla J_t^*(g_t) \frac{\partial g_t}{\partial t}. \tag{12.110}$$

Substituting g_t^{-1} gives

$$\frac{\partial I_t}{\partial t}(g_t^{-1}) = \frac{\partial J_t}{\partial t} + \nabla J_t^* v_t, \tag{12.111}$$

implying the norms in the tangent space are identical:

$$\left\|\frac{\partial a_t}{\partial t}(g_t^{-1})\right\|^2_{T(\mathcal{A})} = \|v_t\|^2_V + \left\|\frac{\partial J_t}{\partial t} + \nabla J_t^* v_t\right\|^2_2. \tag{12.112}$$

$\square$

Then, beautifully enough, the metric measuring photometric transport involves the material or total derivative of J_t and is left invariant.

Theorem 12.25 *Defining the geometric and photometric evolutions, in $\mathcal{A} = \mathcal{G}(\mathcal{V}) \times \mathcal{I}$ to be $v_t = \partial g_t / \partial t (g_t^{-1})$, $J_t = I_t(g_t^{-1})$, then the function*

$$\rho^2((g, I), (h, I'))$$

$$= \inf_{\substack{g(\cdot), I(\cdot): g_0 = g, g_1 = h \\ I_0 = I, I_1 = I'}} \int_0^1 \left(\left\| \frac{\partial g_t}{\partial t}(g_t^{-1}) \right\|_V^2 + \left\| \frac{\partial I_t}{\partial t}(g_t^{-1}) \right\|_2^2 \right) dt \quad (12.113)$$

$$= \inf_{\substack{v, J: \\ J_0 = I(g^{-1}), J_1 = I'(h^{-1})}} \int_0^1 \left(\|v_t\|_V^2 + \left\| \frac{\partial J_t}{\partial t} + \nabla J_t^* v_t \right\|_2^2 \right) dt, \quad (12.114)$$

and ρ is invariant so that $\rho((h, I(h)), (gh, I'(gh))) = \rho((\mathrm{id}, I), (g, I'(g)))$. The induced metric $\tilde{\rho} : \mathcal{I} \times \mathcal{I} \to \mathbb{R}^+$ defined as

$$\tilde{\rho}(I, I') = \inf_{g \in \mathcal{G}} \rho((\mathrm{id}, I), (g, I'(g))) \quad (12.115)$$

is a metric distance on $\mathcal{I}$.

Proof Eqn. 12.113 equalling Eqn. 12.114 follows is from Lemma 12.24. To prove invariance, define $\tilde{g}_t = g_t(h)$, $\tilde{v}$ and we first show $\tilde{g}$ satisfies the same O.D.E. as g just shifted by h in initial condition with identical velocity fields $v = \tilde{v}$:

$$\frac{\partial \tilde{g}_t}{\partial t}(y) = \tilde{v}_t(\tilde{g}_t(y)) \quad (12.116)$$

$$= \frac{\partial g_t}{\partial t}(h(y)) = v_t(g_t(h(y))) \quad (12.117)$$

$$= v_t(\tilde{g}_t(y)), \text{ implying } v(\cdot) = \tilde{v}(\cdot). \quad (12.118)$$

Similarly defining $\tilde{I}_t = I_t(h) = J_t(g_t(h))$, then it follows that $\tilde{I}$ and I satisfy the same differential equation:

$$\frac{\partial \tilde{I}_t}{\partial t}(y) = \frac{\partial J_t}{\partial t}(\tilde{g}_t(y)) + \nabla J_t^*(\tilde{g}_t(y)) \frac{\partial \tilde{g}_t}{\partial t}(y), \quad (12.119)$$

$$\frac{\partial \tilde{I}_t}{\partial t}(\tilde{g}_t^{-1}(x)) \stackrel{(a)}{=} \frac{\partial J_t}{\partial t}(x) + \nabla J_t^*(x) \frac{\partial \tilde{g}_t}{\partial t}(\tilde{g}_t^{-1}(x)), \quad (12.120)$$

$$= \frac{\partial J_t}{\partial t}(x) + \nabla J_t^*(x) \tilde{v}_t(x) \quad (12.121)$$

$$= \frac{\partial J_t}{\partial t}(x) + \nabla J_t^*(x) \tilde{v}_t(x) \stackrel{(b)}{=} \frac{\partial I_t}{\partial t}(g_t^{-1}(x)), \quad (12.122)$$

with (a) following from the substitution $\tilde{g}_t^{-1}(x)$ for y, and (b) from $v = \tilde{v}$. Thus the energies of both solutions (g, I), $(\tilde{g}, \tilde{I})$ are identical:

$$\int_0^1 \left(\|v_t\|_V^2 + \left\| \frac{\partial I_t}{\partial t}(g_t^{-1}) \right\|_2^2 \right) dt = \int_0^1 \left(\|\tilde{v}_t\|_V^2 + \left\| \frac{\partial \tilde{I}_t}{\partial t}(\tilde{g}_t^{-1}) \right\|_2^2 \right) dt, \quad (12.123)$$

completing the second part of the proof.

From Theorem 12.22, Eqn. 12.115 satisfying triangle inequality and therefore giving a metric requires the left invariance of ρ which has just been shown. □

12.9 The Euler Equations for Photometric and Geometric Variation

The Euler equation for the photometric and geometric matching now follows from Younes.

Theorem 12.26 *The geodesics minimizing distance in Eqn. 12.114 of Theorem 12.25 with $\|v\|_V^2 = \langle Av, v \rangle_2$ satisfies the Euler equation with respect to variations in the flow gives*

$$\frac{\partial Av_t}{\partial t} + (Dv_t)^* Av_t + \text{div}(Av_t \otimes v_t) = -\left(\frac{\partial J_t}{\partial t} + \nabla J_t^* v_t\right) \nabla \left(\frac{\partial J_t}{\partial t} + \nabla J_t^* v_t\right), \quad (12.124)$$

where $\text{div}(Av \otimes v) = (DAv)v + (\text{div } v)Av$. *The variation with respect to the tangent elements*

$$Av_t + \left(\frac{\partial J_t}{\partial t} + \nabla J_t^* v_t\right) \nabla J_t = 0, \quad (12.125)$$

$$\frac{\partial}{\partial t}\left(\frac{\partial J_t}{\partial t} + \nabla J_t^* v_t\right) + \text{div}\left(\frac{\partial J_t}{\partial t} v_t + (\nabla J_t^* v_t)v_t\right) = 0. \quad (12.126)$$

Proof First we will derive the variation with respect to (v, J) via perturbation $v \to v + \epsilon \psi$, $J \to J + \epsilon \phi$. To calculate the perturbation of the energy $E(v, J) = \int_0^1 (\|v_t\|_V^2 + \|\frac{\partial J_t}{\partial t} + \nabla J_t^* v_t\|_2^2) dt$ of Eqn. 12.114 minimizing jointly in v and J. The variation in v takes the form

$$\partial_\psi E(v, J) = 2 \int_0^1 \left(\langle Av_t, \psi_t \rangle + \left\langle \frac{\partial J_t}{\partial t} + \nabla J_t^* v_t, \nabla J_t^* \psi_t \right\rangle\right) dt, \quad (12.127)$$

giving the first variation of Eqn. 12.125.

The variation in J is obtained by computing the differential in J with v fixed:

$$\partial_\phi E(v, J) = 2 \int_0^1 \left\langle \frac{\partial J_t}{\partial t} + \nabla J_t^* v, \frac{\partial \phi_t}{\partial t} + v_t^* \nabla \phi_t \right\rangle_2 dt$$

$$= -2 \int_0^1 \left\langle \frac{\partial}{\partial t}\left(\frac{\partial J_t}{\partial t} + \nabla J_t^* v_t\right) + \text{div}\left(\frac{\partial J_t}{\partial t} v_t + (\nabla J_t^* v_t)v_t\right), \phi_t \right\rangle_2 dt$$

which gives the variation of Eqn. 12.126.

To compute the Euler equation 12.124 with respect to variations of the flow, first define

$$Z_t = \frac{\partial J_t}{\partial t} + \nabla J_t^* v_t. \quad (12.128)$$

Then we shall require the following identities. □

Lemma 12.27

$$-\nabla(\langle \nabla J, v \rangle)Z = (DAv)^* v + (Dv)^* Av + \nabla Z \langle \nabla J, v \rangle \quad (12.129)$$

$$(DAv)v = (DAv)^* v + \nabla Z \langle \nabla J, v \rangle - \langle \nabla Z, v \rangle \nabla J. \quad (12.130)$$

Proof The first identity follows from

$$\nabla \langle Z \nabla J, v \rangle = (\nabla Z)\langle \nabla J, v \rangle + Z \nabla \langle \nabla J, v \rangle \quad (12.131)$$

implying

$$-\nabla(\langle \nabla J, v \rangle)Z = -\nabla \langle Z \nabla J, v \rangle + (\nabla Z)\langle \nabla J, v \rangle \quad (12.132)$$

$$\stackrel{(a)}{=} (DAv)^* v + (Dv)^* Av + \nabla Z \langle \nabla J, v \rangle. \quad (12.133)$$

where (a) uses Eqn. 12.125. To prove the second identity Eqn. 12.130 use the two equalities

$$(DAv)v = -\nabla J \langle \nabla Z, v \rangle - ZH_J v \tag{12.134}$$

$$(DAv)^*v = -\nabla Z \langle \nabla J, v \rangle - ZH_J^* v, \tag{12.135}$$

where H_J is the Hessian of J. Now since $H_J = H_J^*$, and equating 12.134, 12.135 gives the second identity. □

Now to finish the proof of the Theorem, to prove the rate of change of the momentum with time use Eqn. 12.125 and the Lemma 12.27 according to

$$0 = \frac{\partial}{\partial t}(Av + Z\nabla J) = \frac{\partial Av}{\partial t} + \frac{\partial Z}{\partial t}\nabla J + \nabla \frac{\partial J}{\partial t} Z \tag{12.136}$$

$$\stackrel{(a)}{=} \frac{\partial Av}{\partial t} - (Z\mathrm{div}v + \langle \nabla Z, v \rangle)\nabla J - \nabla(\langle \nabla J, v \rangle)Z + Z\nabla Z \tag{12.137}$$

$$= \frac{\partial Av}{\partial t} + Av\mathrm{div}v - \langle \nabla Z, v \rangle \nabla J - \nabla(\langle \nabla J, v \rangle)Z + Z\nabla Z \tag{12.138}$$

$$\stackrel{(b)}{=} \frac{\partial Av}{\partial t} + Av\mathrm{div}v + (DAv)^*v + (Dv)^*Av + \nabla Z \langle \nabla J, v \rangle$$

$$- \langle \nabla Z, v \rangle \nabla J + Z\nabla Z$$

$$\stackrel{(c)}{=} \frac{\partial Av}{\partial t} + Av\mathrm{div}v + (Dv)^*Av + (DAv)v + Z\nabla Z, \tag{12.139}$$

where (a) follows from Eqn. 12.126 giving $\frac{\partial Z}{\partial t} = -\mathrm{div}(Zv) = -(Z\mathrm{div}v + \langle \nabla Z, v \rangle)$ and from the definition of Z according to $\nabla \frac{\partial J}{\partial t} = -\nabla(\langle \nabla J, v \rangle) + \nabla Z$; the equalities (b,c) follow from Eqns. 12.129, 12.130 of Lemma 12.27, respectively.

The first part of the integral penalizes large variations of the homeomorphism, and the second one penalizes variations in the material which are not due to the deformation. Both are expressed as norms which depend only on the current object $J_t(\cdot)$, a consequence of the left-invariance requirement. The energy does not track the accumulated stress from time $t = 0$. For this reason, the matching procedure has been classifed as *viscous matching* ([220]), in opposition to the methods involving *elastic matching* [288,273–296]). From the point of view of elasticity, it should be harder to make a small deformation of an object J if it is considered to already be a deformation of another object, than to operate the same deformation, but considering that J itself is at rest. Technically, this means that the norms should depend in the elastic case, both on the deformed object, and on the stress associated with the deformation, which implies that the left-invariance assumption has to be relaxed. Most applications of elasticity in image analysis essentially assume that small deformation approximations are valid, which means that they do not require the optimization of the deformation path. Standard comparison of objects using elastic matching consider that one object is at equilibrium and the other one is deformed, thus the special role of the initial template state (see e.g, the Bookstein approximate metric 12.7.1). The result depends on the choice on which object is at equilibrium. This is valid when comparing changes around an equilibrium condition, for example facial deformation of expressions [199], or for growth [297].

However, as first established in [286], without the left invariance the measure of mismatch given by the minimum norm-square will not be a metric distance.

Example 12.28 (Computing Infinite Dimensional Geometric and Photometric Variation (Younes)) This section follows closely the implementations of Younes [286]. Let the image mapping be defined on $X = [0,1]^2 \subset \mathbb{R}^2$, with values in $\mathbb{R}$, and let the geometric group $\mathcal{G}$ be the group of infinite dimensional diffeomorphisms of the previous section. Define the finite norm condition for the zero boundary via the Laplacian-squared $A = L^*L$ with $L = \nabla^2 = (\partial/\partial x_1^2) + (\partial/\partial x_2^2)$ giving

12.9 THE EULER EQUATIONS FOR PHOTOMETRIC AND GEOMETRIC VARIATION

Hilbert space for the tangent component of the geometric flow having norm-square $\|v_t\|_v^2 = \sum_{i=1}^2 \|\nabla^2 v_{it}\|_2^2$ with the metric energy

$$\int_0^1 \sum_{i=1}^2 \|\nabla^2 v_{it}\|_2^2 \, dt + \alpha \int_0^1 \left\| \frac{\partial J_t}{\partial t} + \nabla J_t^* v_t \right\|_2^2 dt, \qquad (12.140)$$

$\alpha > 0$ a positive parameter. Comparison between two images I_0, I_1 is performed by minimizing the functional over all paths which satisfy $v_t(y) = 0, t \in [0,1], y \in \partial X$, and $J_0 = I_0, J_1 = I_1$.

The minimization of Eqn. 12.140 is performed after time and space discretization, the derivatives being estimated by finite differences. If N^d is the dimension of the space grid and T the number of time steps, there are, because of the boundary conditions, a little less than $3N^d T$ variables to estimate (v has d coordinates and J has one).

For fixed v (respectively fixed J), the energy is quadratic in J (respectively v). For this reason, Younes uses a two-step relaxation procedure, which alternates a conjugate gradient descent for J with a gradient descent for v. A complication comes from the fact that the discretization of the intensity conservation term $\partial J_t / \partial t + \nabla J_t^* v_t$ has to be done with care. A direct linear discretization with finite difference leads to very unstable numerical procedures, and one has to use a non-linear, but more stable approximation for the last term, of the kind

$$\sum_{k=1}^d (J_t(s+e_k) - J_t(s))v_t^+(k,s)/h - (J_t(s) - J_t(s-e_k))v_t^-(k,s)/h$$

h being the space step, (e_k) the canonical basis of $\mathbb{R}^d$ and v^+ (respectively v^-) the positive (respectively negative) part of v. We also use a hierarchical, multigrid, procedure: the deformations are first estimated on a coarse grid (small N) and then iteratively refined until the finer grid is reached. The choice of the number of time steps, T, is quite interesting. For $T = 2$ ($t = 0$ or 1), the minimization reduces to the regularized intensity-conservation cost function which is of standard use for optical flow estimation.

Shown in Figures 12.7, 12.8 are results demonstrating the transformation process $J_t(\cdot)$ as functions of time $t \in [0,1]$ between two images I_0 and I_1. Each row shows the process of either creating pixel intensity or geometric change when a totally black image is matched to an image containing a white disc in its center. Depending on the choice of the parameter α, the process will allow for the creation of a large quantity of pixel intensity, yielding a transformation which looks like fading, with almost no

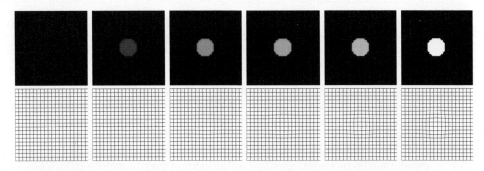

Figure 12.7 Shows two rows from Younes [286] exclusively depicting photometric variation. Top row shows photometric change during matching. Bottom row shows grid undergoing no deformation.

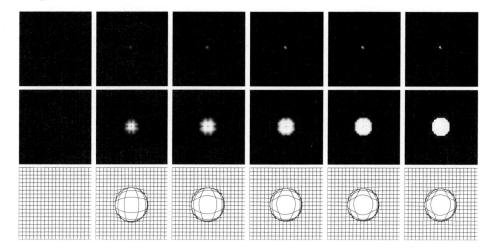

Figure 12.8 Shows all grid deformation from Younes [286]. Top row shows photometric change required to generate matching. Middle row shows photometric change put through geometric variation. Bottom row shows deformation to the grid.

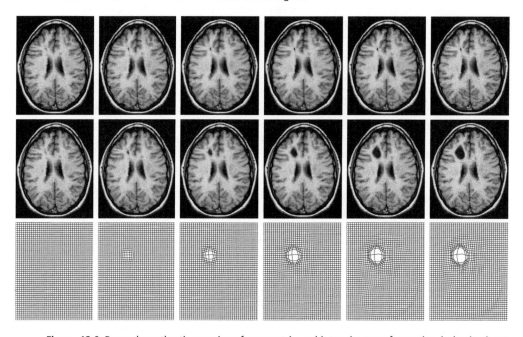

Figure 12.9 Rows show the time series of geometric and intensity transformation in brain tissue used for tumor growth. Row 1 shows the insertion of photometric change depicted by the small black dot. Row 2 shows the tumor developing under mostly geometry deformation. Row 3 shows the geometric deformation of the grid.

deformation at all (α small), or will prefer introducing a small white zone in the center of the disc, and deform it into a disc, yielding a process which resembles an explosion (α large).

The experiments illustrate how luminance can be created during the optimal transformation process. Depending on the value of the parameter α (which penalizes non-conservation of luminance) the results are quite different: for small α, the disk is

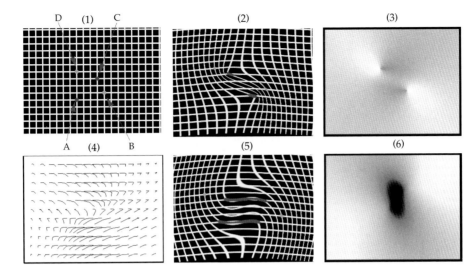

Plate 17 See page 364.

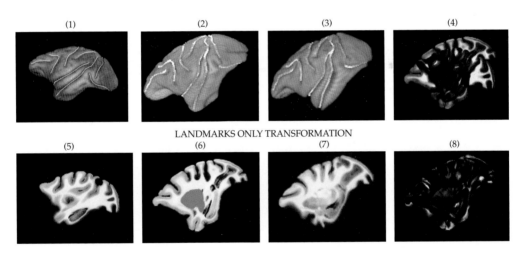

LANDMARKS ONLY TRANSFORMATION

Plate 18 See page 364.

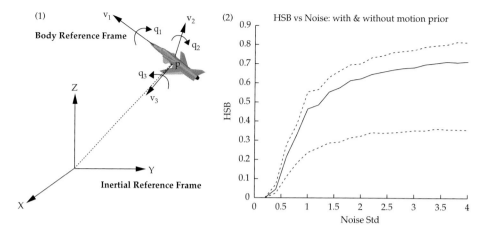

Plate 19 See page 387.

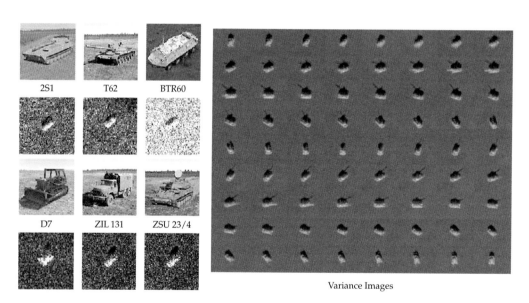

Plate 20 See page 389.

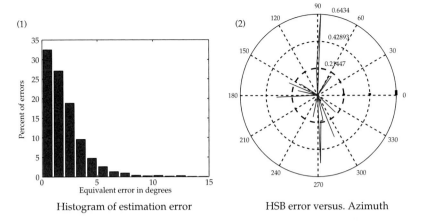

Plate 21 See page 391.

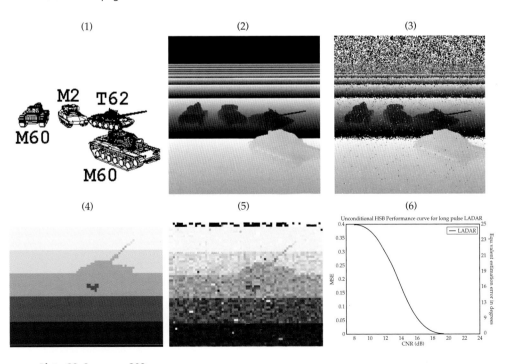

Plate 22 See page 392.

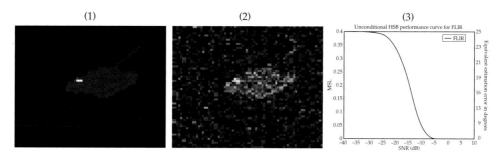

Plate 23 See page 395.

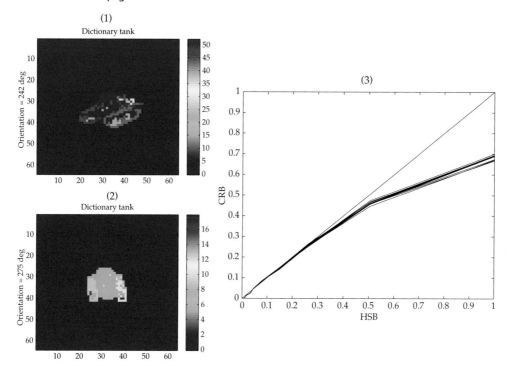

Plate 24 See page 401.

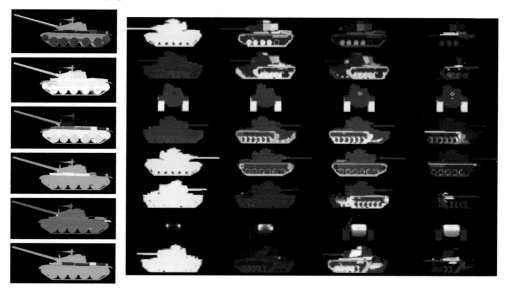

Plate 25 See page 421.

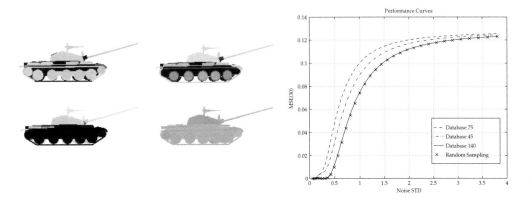

Plate 26 See page 428.

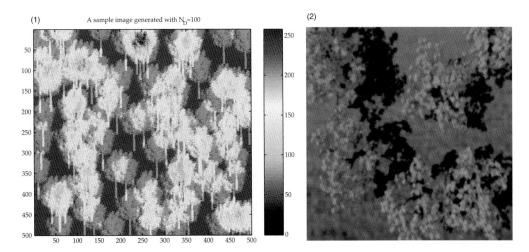

Plate 27 See page 432.

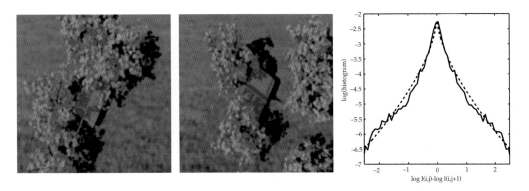

Plate 28 See page 437.

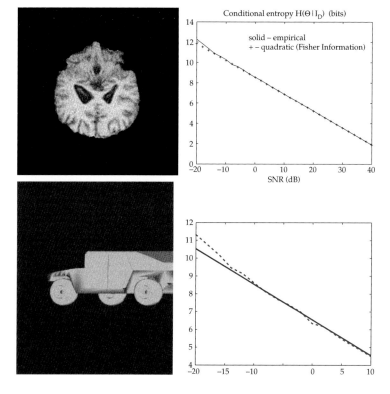

Plate 29 See page 453.

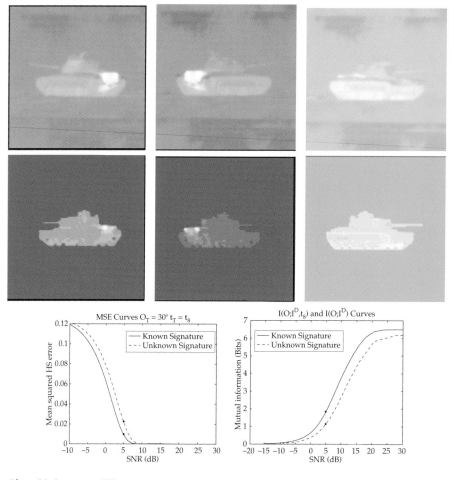

Plate 30 See page 455.

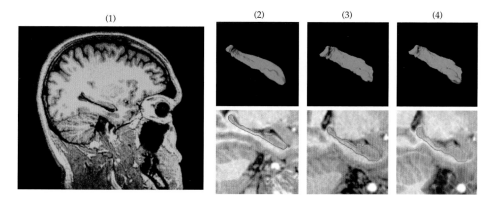

Plate 31 See page 482.

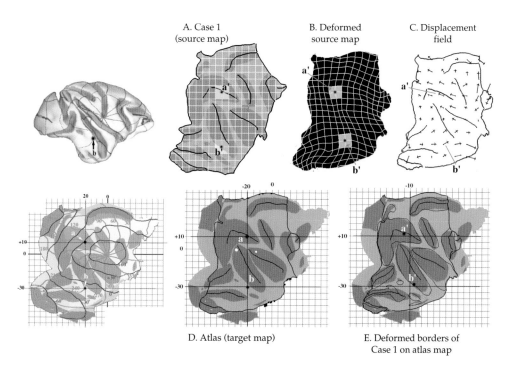

Plate 32 See page 484.

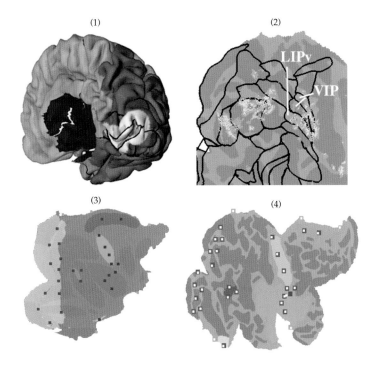

Plate 33 See page 485.

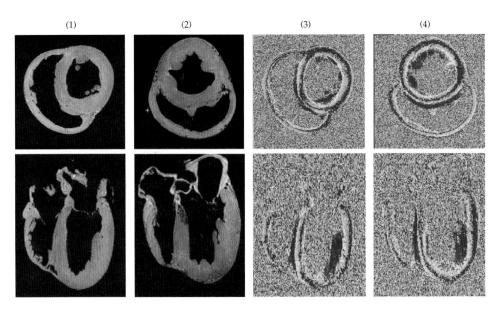

Plate 34 See page 485.

12.10 METRICS BETWEEN ORBITS OF THE SPECIAL EUCLIDEAN GROUP

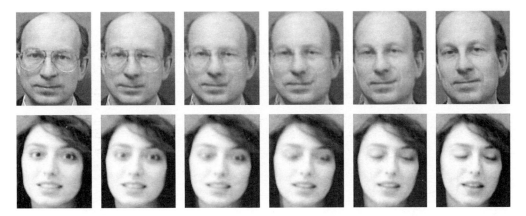

Figure 12.10 Top row shows photometric and high-dimensional geometric motion for the glasses experiment. Bottom rows show the eye opening video. Results taken from Younes showing image flow $J_{t_k}(x)$.

appearing without any deformation, simply the gray levels vary. For large α, a white spot first appears at the center of the disk and the deformation generates the expansion.

Shown in Figure 12.7 is the time series associated with the geometric transformation and intensity transformation for the matching of the totally black image to the white disk centered. The first and last picture in each row are the data provided to the algorithm, and the other ones are synthesized intermediate stages based on photometric and geometry transformation.

To illustrate such methods on anatomical tissue, shown in Figure 12.9 is the result of matching sections of a brain with a tumorous shape. The top row shows the insertion of photometric change depicted by the small black dot. The middle row shows the tumor developing under mostly geometry deformation. The botom row shows the geometric deformation of the grid. Notice in the rightmost panel the centered black spot.

Example 12.29 (Glasses and Eyelids (Younes)) Shown in Figure 12.10 are results from the glasses experiment and eye opening experiments from Younes. The top row shows the glasses experiment results. Shown in the bottom row of Figure 12.10 are results from Younes for the woman closing her eye.

12.10 Metrics between Orbits of the Special Euclidean Group

Return to Theorem 12.12 examining understanding through the invariances of the special Euclidean group. In the context of the infinite dimensional setting we construct a distance between the orbits, the natural definition becoming

$$\tilde{\rho}([I]_{\mathbf{SE}(d)}, [I']_{\mathbf{SE}(d)}) = \inf_{g,h \in \mathbf{SE}(d)} \rho(gI, hI'). \tag{12.141}$$

This clearly places the requirement on the distance function that it be invariant to the group $\mathbf{SE}(d)$ defining the orbit.

Theorem 12.30 *Let*

$$\rho^2(I, I') = \inf_{\substack{g(\cdot), J(\cdot): g_0 = \mathrm{id}, g_1 = g \\ J_0 = I, J_1 = I'}} \int_0^1 \left(\|v_t\|_V^2 + \left\| \frac{\partial J_t}{\partial t} + \nabla J_t^* v_t \right\|_2^2 \right) dt, \quad (12.142)$$

with $A = L^*L$ and $\tilde\rho$ defined as

$$\tilde\rho([I]_{\mathrm{SE}(d)}, [I']_{\mathrm{SE}(d)}) = \inf_{g, h \in \mathrm{SE}(d)} \rho((gI), (hI')) \quad (12.143)$$

If for all $g \in \mathrm{SE}(d)$,

$$\int_0^1 \int_X \|Lg^{-1}v_t(g(y))\|_{\mathbb{R}^d}^2 \, dy\, dt = \int_0^1 \int_X \|Lv_t(y)\|_{\mathbb{R}^d}^2 \, dy\, dt, \quad (12.144)$$

then $\tilde\rho : \mathcal{G}/\mathrm{SE}(d) \times \mathcal{G}/\mathrm{SE}(d) \to \mathbb{R}^+$ is a distance on the orbits.

Proof The proof requires showing the invariance $\tilde\rho(gI, gI') = \tilde\rho(I, I')$ under the assumption Eqn. 12.144 on the $\|Lv\|$ energy. Define $J'_t(y) = J_t(gy)$ giving energy

$$\int_0^1 \int_X \left(\|Lv_t(y)\|_{\mathbb{R}^d}^2 + \left| \frac{\partial J_t}{\partial t}(y) + \nabla J_t^*(y) v_t(y) \right|^2 \right) dy\, dt \quad (12.145)$$

$$= \int_0^1 \int_X \left(\|Lv_t(y)\|_{\mathbb{R}^d}^2 + \left| \frac{\partial J'_t}{\partial t}(y) + \nabla J'^*_t(y) g^{-1} v_t(gy) \right|^2 \right) dy\, dt \quad (12.146)$$

$$\stackrel{(a)}{=} \int_0^1 \int_X \left(\|Lgv'_t(g^{-1}y)\|_{\mathbb{R}^d}^2 + \left| \frac{\partial J'_t}{\partial t}(y) + \nabla J'^*_t(y) v'_t(y) \right|^2 \right) dy\, dt, \quad (12.147)$$

with (a) following from the substitution $v'_t(y) = g^{-1} v_t(gy)$. If the first term in the integral is unchanged, then the cost is unchanged in the distance giving the required left invariance condition. □

The examples which have been used extensively including the Laplacian and its powers, biharmonic, satisfy the invariance condition.

12.11 The Matrix Groups (Euclidean and Affine Motions)

This section follows Bitouk [298]. Given images I_0 and I_1 defined on $\mathbb{R}^d$ ($d = 2, 3$), assume there are rigid objects in the background under rotational motion. Introduce the computational time variable generated by the ODE

$$\frac{dO_t}{dt} = \Omega_t O_t, \qquad O_0 = \mathrm{id}, \quad (12.148)$$

where Ω_t is a skew-symmetric matrix representing the roll of the object. In 2D, for example, $\Omega_t = \begin{pmatrix} 0 & \theta_t \\ -\theta_t & 0 \end{pmatrix}$. The following theorem gives the metric between images under rotational motions, assuming rotation is a pure nuisance variable and assigning no metric length for rotation.

12.11 THE MATRIX GROUPS

Theorem 12.31 *Defining the group action to be the rotational motion $g_t : x \mapsto O_t x$, O satisfying Eqn. 12.148 above, then the distance between images I and I' is given by*

$$\rho^2(I, I') = \rho^2((id, I), (O, I'(O))) \tag{12.149}$$

$$= \inf_{\substack{O \in \mathbf{SO}(d)}} \inf_{\substack{O,I: O_0 = id, O_1 = O \\ I_0 = I, I_1 = I'(O)}} \int_0^1 \left\| \frac{\partial I_t}{\partial t}(O_t^{-1}) \right\|_2^2 dt \tag{12.150}$$

$$= \inf_{\substack{\Omega, J: J_0 = I, J_1 = I' \\ O_0 = id, O_1 = O}} \int_0^1 \left\| \frac{\partial J_t}{\partial t} + \langle \nabla J_t, \Omega_t \cdot \rangle_{\mathbb{R}^d} \right\|_2^2 dt = \inf_{O \in \mathbf{SO}(d)} \|I - I'(O)\|_2^2. \tag{12.151}$$

Proof Computing the change of variable for $J_t = I_t(g_t^{-1})$ makes Eqns. 12.150, 12.151 equivalent:

$$\frac{\partial I_t}{\partial t}(x) = \frac{\partial J_t}{\partial t}(O_t x) + \langle \nabla J_t(O_t x), \Omega_t O_t x \rangle_{\mathbb{R}^d} \tag{12.152}$$

$$\frac{\partial I_t}{\partial t}(O_t^{-1} x) = \frac{\partial J_t}{\partial t}(x) + \langle \nabla J_t(x), \Omega_t x \rangle_{\mathbb{R}^d}. \tag{12.153}$$

To find the minimum simply note that the expression to be minimized in Eqn. 12.150 is

$$\int_0^1 \left\| \frac{\partial I_t}{\partial t}(O_t^{-1}) \right\|_2^2 dt = \int_0^1 \left\| \frac{\partial I_t}{\partial t} \right\|_2^2 dt, \tag{12.154}$$

since the determinant of the Jacobian of the change of variables $x = O_t^{-1} y$ is 1. Thus I is the linear interpolation of (since geodesics are straight lines, Theorem 10.5) its boundary conditions

$$I_t(\cdot) = tI_1(\cdot) + (1-t)I_0(\cdot) = tI'(O_1 \cdot) + (1-t)I(\cdot); \tag{12.155}$$

$$J_t(\cdot) = tI_1(O_t^{-1} \cdot) + (1-t)I_0(O_t^{-1}) = tI'(O_1 O_t^{-1} \cdot) + (1-t)I(O_t^{-1}). \tag{12.156}$$

Substituting into the cost of Eqn. 12.154 gives Eqn. 12.152 ☐

A simple algorithm involving Eqn. 12.151 can be devised: alternate minimization in J and in Ω. This will converge to a local minimum of the energy.

Algorithm 12.32

1. *Fixing Ω, O, then minimizing Eqn. 12.151 is linear giving*

$$J_t(\cdot) = tI'(O_1 O_t^{-1} \cdot) + (1-t)I(O_t^{-1} \cdot). \tag{12.157}$$

2. Given $J_t(\cdot), t \in [0,1]$, minimize Eqn. 12.151:

$$\arg\inf_{\Omega} \int_0^1 \left\| \frac{\partial J_t}{\partial t}(\cdot) + \langle \nabla J_t(\cdot), \Omega_t \cdot \rangle_{\mathbb{R}^d} \right\|_2^2 dt. \tag{12.158}$$

Example 12.33 (Adding cost for rotational motion) To add distance within the orthogonal group $\mathcal{G} = SO(d)$ then add the geodesic distance between group elements giving

$$\rho^2((\mathrm{id}, I), (O, I'(O))) = \inf \left(\int_0^1 \left\| \frac{\partial}{\partial t} O_t \right\|_{\mathbb{R}^{d^2}}^2 dt + \int_0^1 \left\| \frac{\partial I_t}{\partial t}(O_t^{-1}(\cdot)) \right\|^2 dt \right) \tag{12.159}$$

$$= \inf \left(\int_0^1 \mathrm{tr}\, \dot{O}_t \dot{O}_t^* \, dt + \int_0^1 \left\| \frac{\partial I_t}{\partial t}(O_t^*) \right\|_2^2 dt \right) \tag{12.160}$$

$$= 2\beta^2 + \|I(\cdot) - I'(O_1 \cdot)\|_2^2, \tag{12.161}$$

where $e^{\beta Z}\mathrm{id} = O$, Z-antisymmetric.

Adding the translation group is straightforward. Let $g_t(x) = O_t x + a(t)$, let $\beta(t) = da(t)/dt$ and minimize

$$\int_0^1 \left\| \frac{\partial J_t}{\partial t} + \langle \nabla J_t, \Omega_t \cdot + \beta(t) \rangle \right\|_2^2 dt$$

The problem still is quadratic in $(\Omega_t, \beta(t))$. For the explicit expression of J, simply replace O_t^{-1} by g_t^{-1} and O, O_t^{-1} by $g_1(g_t^{-1})$.

12.11.1 Computing the Affine Motions

To incorporate scaling, define the similitude $S_t = \rho_t O_t \in \mathbf{Sim}(d)$ by the ODE

$$\frac{dS_t}{dt} = (\lambda_t \mathrm{id} + \Omega_t) S_t, \quad \text{where } \lambda_t = \frac{\dot{\rho}_t}{\rho_t},$$

and Ω_t is as above skew-symmetric. Since the Jacobian is $(\rho_t)^d$,

$$\frac{d}{dt} \log \det S_t = d\frac{d}{dt} \log \rho_t = d\lambda_t. \tag{12.162}$$

Thus with $S_0 = \mathrm{id}$ then

$$\det S_t = e^{d \int_0^t \lambda_u \, du},$$

implying the energy becomes

$$\int_0^1 \left\| \frac{\partial I_t}{\partial t}(S_t^{-1}) \right\|_2^2 dt = \int_0^1 \det S_t^{-1} \left\| \frac{\partial I_t}{\partial t} \right\|_2^2 dt \tag{12.163}$$

$$= \int_0^1 e^{-d \int_0^t \lambda_s \, ds} \left\| \frac{\partial I_t}{\partial t} \right\|_2^2 dt. \tag{12.164}$$

This still can be explicitly minimized, introducing the time change

$$\alpha_t = \frac{\int_0^t \det S_u^{-1}\, du}{\int_0^1 \det S_u^{-1}\, du} \qquad (12.165)$$

yielding $\quad I_t(x) = \alpha_t I_0(x) + (1-\alpha_t) I_1(x).$ (12.166)

In all cases, the energy at the minimum is

$$\int_0^1 |\frac{d\alpha}{dt}|^2\, dt + \|I - I'(O_1)\|_2^2$$

implying that the problem is equivalent to the usual static problem of minimizing $\|I - I'(O_1)\|^2$.

13 ESTIMATION BOUNDS FOR AUTOMATED OBJECT RECOGNITION

ABSTRACT Thus far we have only studied representations of the source. Now we add the channel, pushing us into the frameworks of estimate then examine estimation bounds for understanding rigid object recognition involving the low-dimensional matrix groups. Minimum-mean-squared error bounds are derived for recognition and identification.

13.1 The Communications Model for Image Transmission

In automated object recognition objects are observed at arbitrary locations and orientations via remote sensors. Depending on the relative orientation, distance and position—henceforth called pose—between the object and sensor the observed imagery may vary dramatically. Building object recognition systems which are invariant to pose is a fundamental challenge to such systems. Throughout this chapter the goal is to analyze object recognition from a principled point of view in which estimators are studied which satisfy optimality criteria with bounds derived, which describe optimality independent of the particular algorithm used for analysis.

The basic model for the source is that the images $I \in \mathcal{I} = I_\alpha \circ \mathcal{G}$ are orbits under actions of transformation $g \in \mathcal{G}$ of an exemplar template, I_α the transformations studied throughout this chapter are the matrix groups and affine motions. Assume the objects are complex enough (modulo the symmetries) that they uniquely determine the group elements, so that there is a bijection between $I_\alpha(g) = I_\alpha \circ g \leftrightarrow g$. Then there is a natural identification of the image $I_\alpha(g) \in \mathcal{I}$ with the group element $g \in \mathcal{G}$ which generates it. Thus, the prior density in the Bayes model is induced by the density on the groups of transformations, $\pi(g), g \in \mathcal{G}$. Randomness on the transformation induces randomness on the underlying source of images $\mathcal{I}$.

The ideal images are observed through various nonideal sensors as depicted in Figure 13.1. The observable data $I^D \in \mathcal{I}^D$ may contain multiple components corresponding to several sensors $I^D = \left(I^{D_1}, I^{D_2}, \dots \right)$. The observation process is characterized via the *likelihood (conditional density) function* for the sensors, $p(\cdot|\cdot): \mathcal{I} \times \mathcal{I}^D \to \mathbb{R}^+$, summarizing completely the transition law mapping the input ideal image I to the output I^D; the likelihood of I^D given I.

The importance of the conceptual separation of the source of possible images $\mathcal{I}$ with prior density and the channel with transition law is that there is only one true underlying scene, irrespective of the number of sensor measurements forming $I^D = (I^{D_1}, I^{D_2}, \dots)$. Only one inference problem is solved, with the multiple observations due to multiple sensors viewed as providing additional information in the posterior distribution. *Sensor fusion occurs automatically in this framework.* Certainly, in calculating information gain and channel capacity this view determines these estimation bounds.

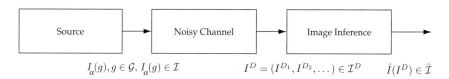

Figure 13.1 The source of images: An orbit under matrix groups of transformations. The observable I^D is from multiple sensors with $\hat{I}$ the estimator generated by the inference machine.

13.1.1 The Source Model: Objects Under Matrix Group Actions

Begin examining the matrix groups of low-dimension such as the generalized linear group and its subgroups. In this chapter the subgroups of the affine group are examined for pose. Defining $g = (A, a)$, the image is identified with transformation on the background space of the template:

$$I_\alpha(g) = I_\alpha(Ax + a), \; g = (A, a) \in \mathcal{G} \subset \mathbf{GL}(d) \otimes \mathbb{R}^d. \tag{13.1}$$

The orbit $\mathcal{I} = \{I_\alpha(g) = I_\alpha \circ g, g \in \mathcal{G}\}$ is the space of all such imagery with prior $\pi(\cdot)$.

13.1.2 The Sensing Models: Projective Transformations in Noise

The estimators are determined by the statistical models (likelihood functions) representing the imaging modalities. Many common imagers are maps $T : \mathcal{I} \to \mathcal{I}^\mathcal{D}$ from d-dimensional scenes, $d = 2, 3, 4$, of objects occupying subvolumes in $\mathbb{R}^d$ to real- and complex-valued measurements in multiple dimensions. Generally, accurate analytical expressions for T may not be available in all situations. In such cases, high quality simulators are used to generate samples from the transformations. For example, for infra-red imaging the PRISM [299] simulator is often used, or for high resolution radar XPATCH is used [300, 301]. Begin with the most ubiquitous mappings through projections from three-dimensions to two.

For many sensors, imaging is essentially a projective mechanism operating by accumulating responses from the scene elements which project to the same measurement position in the image. Such mechanisms are modelled as maps $T : \mathcal{I} \to \mathcal{I}^\mathcal{D}$, the observations in most cases are $\mathbb{R}^d, \mathbb{C}^d$ valued for some fixed number d.

Definition 13.1 *The **generalized projective sensor transformation** $T : I \in \mathcal{I} \to TI \in \mathcal{I}^\mathcal{D}$ is a mapping from $\mathbb{R}^3$ to $\mathbb{R}^2$, taking images to the function defined on the detector plane, $T : I(x), x \in X \subset \mathbb{R}^3 \mapsto TI(y), y \in Y \subset \mathbb{R}^2$.*

The projective transformation of standard optical sensors provides 2D real-valued images of objects sampled in the projective coordinates $I^D(y), y \in Y$. The transformation from the 3D image world to the 2D imaging plane for each of the projective and perspective transformations takes the following form. Aligning the $x_1 \times x_2$ axes with the imaging plane, the orthogonal projections are mappings

$$(x_1, x_2, x_3) \in X \subset \mathbb{R}^3 \mapsto (x_1, x_2) \in Y \subset \mathbb{R}^2 \; \text{(projection)} \tag{13.2}$$

$$(x_1, x_2, x_3) \in X \subset \mathbb{R}^3 \mapsto (\frac{x_1}{x_3}, \frac{x_2}{x_3}) \in Y \subset \mathbb{R}^2 \; \text{(perspective)}. \tag{13.3}$$

Figure 13.2 depicts the perspective projections' 3D objects onto the 2D projective plane.

13.1.3 The Likelihood and Posterior

For video imaging in the high count limit, the additive noise model defines I^D as a Gaussian random field with mean field the projection $I^D = TI_\alpha(g) + W$, W a Gaussian random field. The

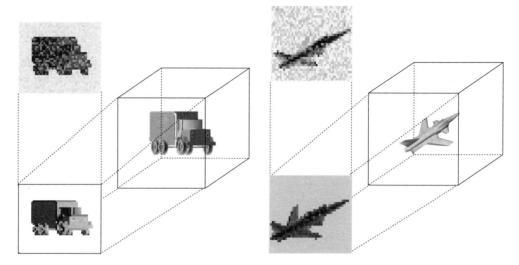

Figure 13.2 Panels show the projective transformations in noise. Cubes depict objects in 3D $I_\alpha \circ g$; columns show projection mean $TI_\alpha(g)$ (bottom panel) and projection in Gaussian noise $I^D = TI_\alpha(g) + W$.

conditional likelihood density of the data I^D conditioned on mean-field $TI(g)$ is given by

$$p(I^D | I_\alpha(g)) = \frac{1}{(2\pi\sigma^2)^{N/2}} e^{-1/2\sigma^2 \|I^D - TI_\alpha(g)\|^2}, \tag{13.4}$$

where $\|I^D - TI_\alpha(g)\|^2 = \sum_{y \in Y} |I^D(y) - TI_\alpha(g)(y)|^2$. $\tag{13.5}$

Note that throughout this chapter the imagery will be assumed to correspond to a discrete lattice of dimension $N = |Y|$, and $\|\cdot\|$ will correspond to a componentwise indexing over the lattice. In subsequent chapters we will extend to the continuum model.

Shown in Figure 13.2 is an example of the Gaussian additive noise projection model. Cubes depict objects in 3D, $I_\alpha \circ g$ for the truck and airplane. Columns show the projective mean $TI_\alpha(g)$ (bottom panel) and projection in Gaussian noise $I^D = TI_\alpha(g) + W$. The rightmost panels in each column show the 3D object.

Figure 13.3 illustrates how the Gaussian log-likelihood encodes the information concerning the objects parameters such as pose $g \in \mathcal{G}$. The top row panel 1 shows the true rendered object. Panels 2–4 show different choices of the object at different poses and different object types. Panel 5 shows the observed projective imagery in noise, $I^D = TI_\alpha(g) + W$. The bottom row panels 6–8 show the difference between the optical data synthesized according to the Gaussian model from the target at its estimated pose $I^D - TI_\alpha(g)$, subtracted from the true measured data.

With the prior density[35] $\pi(g), g \in \mathcal{G}$, the posterior is taken directly on the transformation in the form

$$p(g|I^D) = \frac{\pi(g)p(I^D|I_\alpha(g))}{Z(I^D)}. \tag{13.6}$$

---

[35] The density is the derivative of the probability measure with respect to the base measure, which is Lebesgue on $\mathbb{R}^d$ and the Haar measure more generally on the other subgroups of **GL**(d), such as **SO**(d) (see Boothby [148] for a discussion of Haar measures).

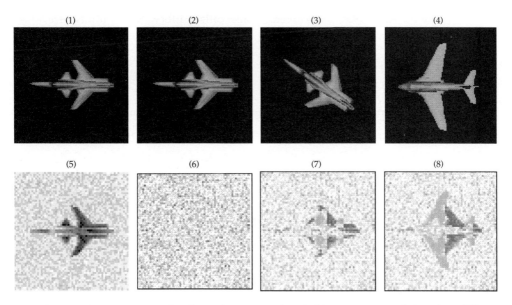

Figure 13.3 Top row: Panel 1 shows the true rendered object $TI_\alpha(g)$. Panels 2–4 show different choices of the object at different poses and different object types. Bottom row: Panel 5 shows the observed Gaussian random field $I^D = TI_\alpha(g) + W$ for the true object. Panels 6–8 show the difference between the optical data synthesized according to the Gaussian model from the target at its estimated pose, subtracted from the true measured data. Panel 6 shows the correct object at correct pose. Panels 7,8 show mismatched pose and object type, respectively.

13.2 Conditional Mean Minimum Risk Estimation

13.2.1 Metrics (Risk) on the Matrix Groups

Using classical Bayesian techniques, optimum estimators can be generated of the object pose, and in addition to the estimator present lower bounds on their estimation accuracy. The minimum risk estimator requires the introduction of the risk function $R : \mathcal{G} \times \mathcal{G} \to \mathbb{R}^+$, then given g^* any estimator a mapping from $I^D \to \mathcal{G}$ the risk $\mathcal{R}$ becomes

$$\mathcal{R} = E\{R(g, g^*(I^D))\}. \tag{13.7}$$

To define a minimum error estimator on pose estimation subgroups of the affine motion are studied utilizing the Euclidean metric from the general linear group. We call this the Frobenius or Hilbert Schmidt metric (distance). Estimators which minimize the square of this distance we shall call minimum-mean-squared error (MMSE) estimators (sometimes Hilbert Schmidt estimator). The estimator is similar to the conventional least-square estimator but extended to account for the geometry of the subgroups of the special Euclidean group.

Examine the affine group $\mathbf{A}(d) = \mathbf{GL}(d) \otimes \mathbb{R}^d$ and its subgroups. It is convenient to identify these group elements with their $(d+1) \times (d+1)$ homogeneous coordinate representation so that group composition is purely matrix multiplication within their homogeneous coordinate representation.

Definition 13.2 *Define elements of $g \in \mathbf{A}(d) = \mathbf{GL}(d) \otimes \mathbb{R}^d$ represented via their $(n+1) \times (n+1)$ homogeneous coordinate representation $\bar{\mathbf{A}}$ according to*

$$g = \begin{pmatrix} A & a \\ 0 & 1 \end{pmatrix} \in \bar{\mathbf{A}}, \quad A \in \mathbf{GL}(d), \quad a \in \mathbb{R}^d. \tag{13.8}$$

This is a metric space with **Frobenius (also called Hilbert–Schmidt) distance** between two elements defined by

$$\rho^2_{\mathbb{R}^{(d+1)^2}}(g,g') = \|g - g'\|^2_{\mathbb{R}^{(d+1)^2}} = \operatorname{tr}(g - g')^*(g - g') \tag{13.9}$$

$$= \operatorname{tr}(A - A')^*(A - A') + \|a - a'\|^2_{\mathbb{R}^d}. \tag{13.10}$$

The **group action of the homogeneous representation on the background space** $X \subset \mathbb{R}^d$ of $g : x \mapsto g\begin{pmatrix} x \\ 1 \end{pmatrix}$ with $g \in \bar{\mathbf{A}}$ operating through matrix multiplication.

13.2.2 Conditional Mean Minimum Risk Estimators

Definition 13.3 *For the posterior density $p(\cdot|I^D)$ on $\mathcal{G} \subseteq \mathbf{GL}(d) \otimes \mathbb{R}^d$, the **MMSE** is defined to be*

$$\hat{g} = \arg\min_{g \in \mathcal{G}} E\left\{\rho^2_{\mathbb{R}^{(d+1)^2}}(g,g')\right\} \tag{13.11}$$

*When $\rho^2_{\mathbb{R}^{(d+1)^2}}(g,g') = \operatorname{tr}(g-g')(g-g')^*$ it is called the **HS estimator**.*

There may be multiple solutions to this equation, and hence the estimator may be set valued (see [302], e.g.).

The Orthogonal Group: The MMSE minimum risk estimator $\hat{O} : I^D \mapsto \mathbf{SO}(d)$ on the orthogonal group has the norm squared risk metric $\|\cdot\|^2_{\mathbb{R}^{d^2}}$ which is continuous in its parameters and since $\mathbf{SO}(d)$ is compact, the minimizer lies in $\mathbf{SO}(d)$. Shown in Figure 13.4 are HS plots of the error metric for $\mathbf{SO}(2)$ (panel 1), $\mathbf{SO}(3)$ (panel 2).

Theorem 13.4 (Srivastava [303]) *The MMSE estimator restricted to $\mathcal{G} = \mathbf{SO}(d)$ defined by*

$$O_{MS} = \arg\min_{O \in \mathbf{SO}(d)} E\left\{\|O - O'\|^2_{\mathbb{R}^{d^2}}\right\}, \tag{13.12}$$

with property that for any other estimator $\hat{O} : \mathcal{I}^D \to \mathbf{SO}(d)$,

$$E\left\{\|\hat{O}(I^D) - O'\|^2_{\mathbb{R}^{d^2}}\right\} \geq E\left\{\|O_{MS}(I^D) - O'\|^2_{\mathbb{R}^{d^2}}\right\}. \tag{13.13}$$

The MMSE estimator $O_{MS}(I^D)$ given data realization $I^D \in \mathcal{I}^D$ is given by,

$$O_{MS}(I^D) = \arg\max_{O \in \mathbf{SO}(d)} \operatorname{tr}(O \cdot A^*) \text{ where } A = E\{O'|I^D\}, \tag{13.14}$$

$$= \begin{cases} UV^\star, & \text{if } determinant(A) \geq 0 \\ ULV^\star, \quad L = \begin{pmatrix} 1 & 0 & \cdots & 0 \\ 0 & 1 & \cdots & 0 \\ \vdots & & & \\ 0 & 0 & \cdots & -1 \end{pmatrix}, & \text{if } determinant(A) < 0, \end{cases} \tag{13.15}$$

and $A = U\Sigma V^\star$ is the singular value decomposition of A. The mean-squared error (MSE) is given by

$$\operatorname{MSE} = E\{\varrho(I^D)\}, \quad \text{where } \varrho(I^D) = 2(d - \operatorname{tr} A^* O_{MS}(I^D)). \tag{13.16}$$

13.2 CONDITIONAL MEAN MINIMUM RISK ESTIMATION

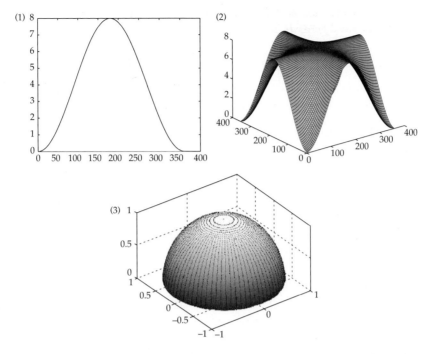

Figure 13.4 Panel 1 shows the Hilbert Schmidt distance for **SO**(2); panel 2 shows the same for **SO**(3). Panel 3 shows tiling of **SO**(3) for discrete integration. Each point on S_+^2 is where a rotation vector passes.

In the case of **SO**(2), *for* $\det(A) \neq 0$ *the MMSE reduces to*

$$O_{MS} = \frac{1}{\sqrt{\det(A)}} A. \tag{13.17}$$

Proof The inequality comes from the definition of the Hilbert–Schmidt estimator.

$$E\|\hat{O} - O'\|_{\mathbb{R}^{d^2}}^2 = E\left\{E\left\{\|\hat{O} - O'\|_{\mathbb{R}^{d^2}}^2 | I^D\right\}\right\} \geq E\left\{E\left\{\|O_{MS} - O'\|_{\mathbb{R}^{d^2}}^2 | I^D\right\}\right\}$$
$$= E\|O_{MS} - O'\|_{\mathbb{R}^{d^2}}^2. \tag{13.18}$$

Calculating O_{MS},

$$O_{MS}(I^D) = \arg\min_{O \in \mathbf{SO}(d)} E\left\{\|O - O'\|_{\mathbb{R}^{d^2}}^2 | I^D\right\} = \arg\min_{O \in \mathbf{SO}(d)} (2d - 2E\{\operatorname{tr} O^* O' | I^D\})$$
$$= \arg\max_{O \in \mathbf{SO}(d)} \operatorname{tr}(O \cdot A^\star),$$

where $A = E\{O'|I^D\}$. The maximizer is given by the singular value decomposition (see [304]). Substituting into the definition of the MSE, Eqn. 13.16 follows. □

The Special Euclidean Group: Examine the special Euclidean group **SE**(2, 3) of dimensions 2 (ground plane) and 3 (air frames). For a fixed ground-plane, objects have positions in $\mathbb{R}^2$ and orientations in **SO**(2), identified with the pair (O, a) taking values in **SE**(2), the special Euclidean group of dimension 3.

Theorem 13.5 *Given is the posterior density $p(\cdot|I^D)$ continuously and compactly supported prior $\pi(\cdot)$ on $\mathcal{C} \subset \mathbf{SE}(d)$. The MMSE estimator in the special Euclidean group is given by*

$$g_{MS} = \begin{pmatrix} O_{MS} & a_{MS} \\ 0 & 1 \end{pmatrix}, \quad \text{where } a_{MS}(I^D) = E\{a|I^D\}, \tag{13.19}$$

$$O_{MS}(I^D) = \arg\max_{O \in \mathbf{SO}(d)} \text{tr}\,(O \cdot E\{O'|I^D\}^\star). \tag{13.20}$$

The mean-squared error is given by

$$MSE = E\{\varrho(I^D)\} \quad \text{where} \tag{13.21}$$

$$\varrho(I^D) = 2d - 2\text{tr}\,A^*O_{MS}(I^D) + E\{\|a_{MS}(I^D) - a\|^2_{\mathbb{R}^d}|I^D\}. \tag{13.22}$$

For any other estimator $\hat{O}, \hat{a} : \mathcal{I}^D \to \mathbf{SE}(d)$, the mean-squared error is larger,

$$E\left\{\|\hat{g} - g\|^2_{\mathbb{R}^{(d+1)^2}}|I^D\right\} \geq E\left\{\|g_{MS} - g\|^2_{\mathbb{R}^{(d+1)^2}}|I^D\right\}. \tag{13.23}$$

Proof Substituting for the MSE gives

$$\|\hat{g} - g\|^2_{\mathbb{R}^{(d+1)^2}} = \|\hat{O} - O\|^2_{\mathbb{R}^{d^2}} + \|\hat{a} - a\|^2_{\mathbb{R}^d}. \tag{13.24}$$

Since the terms separate in their functional dependence on (O, a), the result follows.
□

Note that $a_{MS}(I^D)$ is the usual conditional-mean (or least-squares) estimate of the object position under the marginal posterior density.

13.2.3 Computation of the HSE for SE(2,3)

For most of the situations the integrands are too complicated to be integrated analytically, so the trapezoidal method is used dividing the domains of integration into a uniform grid and evaluating the integrals by finite summations. On **SE**(2) this involves straightforward discretization of the torus. For **SE**(3) the angle–axis representation of the orthogonal group is used and described subsequently (see Section 13.2.4). This reduces the density to a discrete mass function $\tilde{p}(\cdot, \cdot)$ on the discrete set of rotations and translation with the estimators computed as

$$a_{MS}(I^D) = \sum_{j=1}^{M} \sum_{i=1}^{N} a_i \tilde{p}(O_j, a_i | I^D), \tag{13.25}$$

$$O_{MS}(I^D) = \arg\max_{O \in \mathbf{SO}(d)} \text{tr}(O \cdot A^\dagger), \quad \text{where } A = \sum_{j=1}^{M} \sum_{i=1}^{N} O_j \tilde{p}(O_j, a_i | I^D). \tag{13.26}$$

For **SO**(2), $\theta_j = 2\pi j/M$ and so $O_j = \begin{pmatrix} \cos(\theta_j) & -\sin(\theta_j) \\ \sin(\theta_j) & \cos(\theta_j) \end{pmatrix}$, and the average matrix is of the form $A = \begin{pmatrix} \alpha & -\beta \\ \beta & \alpha \end{pmatrix}$ with its singular values are both equal to $\sqrt{\det(A)}$, thus using singular value decomposition to determine O_{MS} gives $UV^T = (1/\sqrt{\det(A)})A$ when $\det(A) \neq 0$. For **SO**(3), Eqn. 13.15 of Theorem 13.4 holds. The algorithm for generating numerical estimates is given as follows.

Algorithm 13.6

1. *Generate elements* $U = (O, a) \in \mathbf{SE}(2,3)$ *uniformly*[36] *and apply each element to the images so that the image is rotated by O and translated by a.*
2. *Compute the posterior density (to within the normalizing constant)* $p(O_i, a_j | I^D)$ *for each* $i = 1, 2, \ldots, M$ *and* $j = 1, 2, \ldots, N$ *to determine*

$$\tilde{p}(O_i, a_j | I^D) = \frac{p(O_i, a_j | I^D)}{\sum_{j=1}^{M} \sum_{i=1}^{N} p(O_i, a_j | I^D)}.$$

3. *Compute* $\hat{a}$ *and the average matrix A as described in Eqns. 13.25 and 13.26, respectively.*

13.2.4 Discrete integration on SO(3)

To compute the average associated with numerical integration of the expectation on $\mathbf{SO}(2)$, the torus $[0, 2\pi]$ is discretized in the usual way. Numerical integration of the expectation required on $\mathbf{SO}(3)$ from Eqn. 13.26 is generated by identifying each element of $\mathbf{SO}(3)$ with an axis of rotation (a unit vector in $\mathbb{R}^3$ or S^2, two degrees of freedom) and an angle (the unit circle S^1, one degree of freedom; see Example 10.7.2, Chapter 11). To make the identification unique restrict to only a clockwise rotation and the axis of rotation only on upper hemi-sphere (S^2_+). The uniform samples for the rotation angle are $\theta_i = 2\pi i / N_2$, $i = 1, 2, \ldots N_2$. To generate equi-spaced points on the upper hemi-sphere S^2_+ let $x_j = 2\pi j / N_1$, $j = 1, 2, \ldots, N_1$, and define $\phi_j = \sin^{-1}(x_j)$. Then there are $N_1 \times N_2$ axes identified with the upper hemi-sphere according to

$$\begin{pmatrix} n_1 \\ n_2 \\ n_3 \end{pmatrix}_{ij} = \begin{pmatrix} \cos \theta_i \cos \phi_j \\ \cos \theta_i \sin \phi_j \\ \sin \theta_i \end{pmatrix}, \qquad (13.27)$$

with an associated set of rotation angles $\psi_k = 2\pi k / N_3$, $k = 1, 2, \ldots, N_3$ around each of the axes. The $N_1 \times N_2 \times N_3$ rotation matrices O_{ijk} are given by the formula

$$O_{ijk} = \begin{pmatrix} \cos(\psi_k) + n_1^2(1 - \cos(\psi_k)) & n_1 n_2 (1 - \cos(\psi_k)) - n_3 \sin(\psi_k) & n_1 n_3 (1 - \cos(\psi_k)) + n_2 \sin(\psi_k) \\ n_2 n_1 (1 - \cos(\psi_k)) + n_3 \sin(\psi_k) & \cos(\psi_k) + n_2^2(1 - \cos(\psi_k)) & n_2 n_3 (1 - \cos(\psi_k)) - n_1 \sin(\psi_k) \\ n_1 n_3 (1 - \cos(\psi_k)) - n_2 \sin(\psi_k) & n_2 n_3 (1 - \cos(\psi_k)) + n_1 \sin(\psi_k) & \cos(\psi_k) + n_3^2(1 - \cos(\psi_k)) \end{pmatrix}.$$

(13.28)

Panel 3 of Figure 13.4 illustrates the tiling of $\mathbf{SO}(3)$ (taken from [305]).

13.3 MMSE Estimators for Projective Imagery Models

13.3.1 3D to 2D Projections in Gaussian Noise

Examine results on minimum-risk bounds for imaging models Gaussian video imagery using perspective projection. The projective transformation studied $T : \mathcal{I} \to \mathcal{I}^D$ is that observed for

---

[36] The translation group is not bounded, thus translations are generated either uniformly over a fixed region, or with respect to a compactly supported prior.

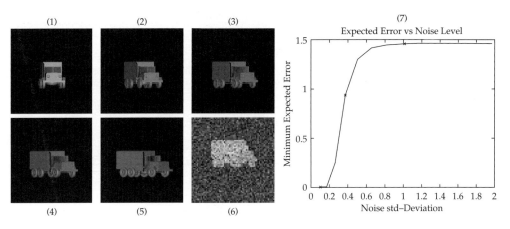

Figure 13.5 Top row: Panels 1–5 show CAD models within one error ball of panel 3 depicted by the middle X at 1.0 unit HS MSE. Bottom row: Panel 6 shows $I^D = TI_\alpha(g) + W$ for noise level $\sigma = 0.4$ corresponding to the middle X in panel 7 for HS performance of MSE=1. Panel 7 shows the mean-squared error bound as measured by the HSB on estimating the orientation using VIDEO projective imagery as a function of noise level.

real-valued video imagery in the Gaussian noise limit. Thus the additive noise model is used, defining I^D as a Gaussian random field with mean field the projection $I^D = TI_\alpha(g) + W$, W a Gaussian random field with conditional density

$$p(I^D|I_\alpha(g)) = \frac{1}{(2\pi\sigma^2)^{N/2}} e^{-\frac{1}{2\sigma^2}\|I^D - TI_\alpha(g)\|^2}. \tag{13.29}$$

Example 13.7 (MMSE for Estimators on SO(2) (Srivastava [303])) Examine experimental results on minimum-risk bounds. Begin with ground vehicles parameterized to axis fixed rotations in **SO**(2) in the high count limit. The imaging model defines I^D as a Gaussian random field with mean field the projection $TI(g)$ with density as given in Eqn. 13.4. Assume a uniform prior probability for orientation of objects $\pi(\cdot) = constant$. Figure 13.5 shows results from computing the mean-squared error performance associated with orientation for the truck in VIDEO imagery. The top row panels 1–5 show renderings $TI(g)$ of a set of CAD models in the database equally spaced around the true object (panel 3). The CAD models shown are within MSE = 1 unit distance from the true CAD model (panel 3) in the database. Shown in the bottom row panel 6 of Figure 13.5 is the truck CAD model at a noise level $\sigma = 1.0$. Panel 7 shows the HSB mean-squared error bound curve for orientation as a function of noise level. The middle cross of panel 7 corresponds to the noise level in the truck image shown in panel 6 at an MSE = 1 unit of error. The MSE was computed over $[0, \pi/2]$ to avoid symmetries; MSE=0 implies perfect placement of the truck, with maximal MSE=1.45 implying maximally unreliable placement of the truck.

Example 13.8 (Adding Dynamical Equations of Motion for SE(2,3)(Srivastava [306])) Examine results on MSE bounds from Srivastava incorporating dynamics. Pose estimation for continuous relative motion between the objects and the camera sensors requires extension to the special Euclidean group **SE**(2, 3). Examine the parametric representations of object motion for airframes composed of motions through **SE**(3) as described in Srivastava [306]. The prior probability on motions are induced using Newtonian dynamics. Assuming standard rigid body motion with respect to

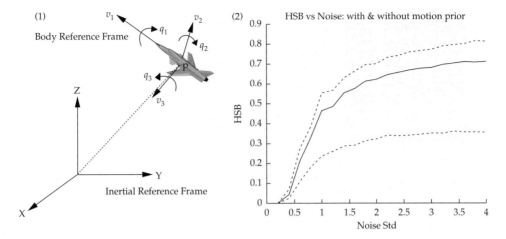

Figure 13.6 Panel 1 shows the CAD airframe with the inertial coordinates with linear and angular velocities. Panel 2 shows MSE as measured by the HSB versus noise in the case of three different motion priors: (i) no prior (broken line), (ii) motion prior, and (iii) strong prior (.-.-)(see also Plate 19).

the inertial frame (see panel 1 of Figure 13.6), $v(t) \in \mathbb{R}^3$ are the linear velocities, $q(t) = \begin{pmatrix} q_1(t) \\ q_2(t) \\ q_3(t) \end{pmatrix} \in \mathbb{R}^3$ are the rotational velocities, $x(t) \in \mathbb{R}^3$ are the inertial positions and $O(t) \in \mathbf{SO}(3)$ are the object orientations with respect to the inertial reference frame. The standard Newtonian laws are used to construct the non-linear differential equations between the linear and angular velocities and the force/torque functions. The displacements relate to the velocities according to the equations, $\dot{O}(t) = O(t)Q(t)$, $\dot{x}(t) = O(t)v(t)$, where Q is the 3×3 skew-symmetric matrix of elements q_1, q_2, q_3,
$$Q(t) = \begin{pmatrix} 0 & q_1(t) & q_2(t) \\ -q_1(t) & 0 & q_3(t) \\ -q_2(t) & -q_3(t) & 0 \end{pmatrix}.$$

Use standard difference equations assuming the angular and linear velocities are piecewise constant over the $\delta = 1$ time increments; identifying discrete time points $t_n = n\delta$, then $v(n) = x(n+1) - x(n)$, $O(n+1) = O(n)e^{Q(n)}$ with $O(0) = $ id the 3×3 identity matrix. Tracking in the plane implies that the rotations are axis-fixed around the z-axis with $q_2 = q_3 = 0$ giving

$$O(n+1) = O(n) \begin{pmatrix} \cos q_1(n) & \sin q_1(n) & 0 \\ -\sin q_1(n) & -\cos q_1(n) & 0 \\ 0 & 0 & 1 \end{pmatrix}$$

$$= \begin{pmatrix} \cos \sum_{i=1}^{n} q_1(i) & \sin \sum_{i=1}^{n} q_1(i) & 0 \\ -\sin \sum_{i=1}^{n} q_1(i) & \cos \sum_{i=1}^{n} q_1(i) & 0 \\ 0 & 0 & 1 \end{pmatrix}. \quad (13.30)$$

Model the differential equations on forces and torques assuming identity moments of inertia. Then
$$v(n+1) + (Q(n) - \mathrm{id})v(n) = f(n),$$
$$q(n+1) + (Q(n) - \mathrm{id})q(n) = t(n). \quad (13.31)$$

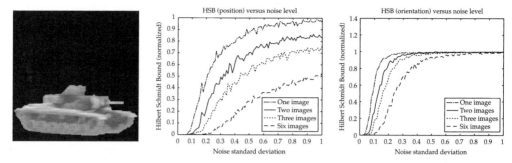

Figure 13.7 Panel 1 shows VIDEO tanks as the camera closes; panel 2 shows the HSB bound for mean-squared error performance of position; panel 3 shows the orientation bound. The different curves correspond to successive estimates of position and orientation as a function of scale as the camera closes. The different lines show the performance curves as a function of number of images used for the estimator. Results from Srivastava [307].

Choosing the forces and torques as a Gaussian processes with covariance Σ_1, Σ_2, then the linear and angular velocities form a Markov process, with joint density

$$p\left(\begin{pmatrix}v(1)\\q(1)\end{pmatrix}, \begin{pmatrix}v(2)\\q(2)\end{pmatrix}, \dots\right)$$

$$= \prod_{i=1}^{n} \frac{1}{\sqrt{(2\pi)^3 \det \Sigma_1}} e^{-1/2(v(i)-(\text{id}-Q(i-1))v(i-1))^*}$$

$$\Sigma_1^{-1}(v(i)-(\text{id}-Q(i-1))v(i-1))$$

$$\times \frac{1}{\sqrt{(2\pi)^3 \det \Sigma_2}} e^{-1/2(q(i)-(\text{id}-Q(i-1))q(i-1))^*}$$

$$(\Sigma_2)^{-1}(q(i)-(\text{id}-Q(i-1))q(i-1)). \tag{13.32}$$

Shown in panel 2 of Figure 13.6 are estimates of the orientation bound for a vehicle moving in the ground plane imaged via the video camera. The dashed line shows the mean-squared error performance as measured by the HSB of airplane orientation in **SO**(2) as a function of the noise level in the imagery. Panel 2 shows three curves demonstrating mean-squared error as it varies with the noise standard deviation for three different prior conditions. The broken line represents a uniform prior on the axis fixed rotations associated with **SO**(2). The solid line represents a moderate dynamics prior (moderate weight relative to conditional probability data term). The dashed-dot (.) curve represents a strong prior (large weight relative to the data likelihood term). Notice the decrease in estimator MSE bound.

Figure 13.7 illustrates results on MSE bounds from Srivastava incorporating translation and rotation bounds of the special Euclidean group of motions for camera motion [307]. Panel 1 shows the object as the camera converges on it. Panels 2 and 3 show the MSE bounds on position and orientation as a function of the noise level for different object distances from the sensor. Due to the perspective projection, increasing the object distance from the sensor results in a smaller number of pixels on the object in the observed image and a pose estimator bound which increases with distance to the camera. Each curve represents a different object distance to the camera. Assuming conditional independence between the collections of observed images $I^D = I^{D_1}, \dots, I^{D_n}$, the likelihood is the product of the likelihoods of the individual images, $p(I^D|O, a) = \prod_{i=1}^{n} p(I^{D_i}|O, a)$.

13.3 MMSE ESTIMATORS

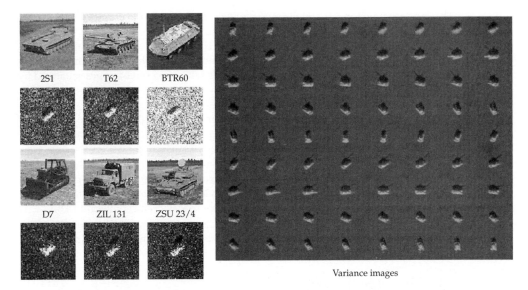

Variance images

Figure 13.8 Left Half: Publicly available SAR data from the MSTAR program showing vehicles (rows 1,3) and SAR datasets (row 2,4). Right Half: Shows estimated variance for each pixel for the 72 azimuth angles spaced from 5° to 360° of a T72. Variances were estimated from training data in the MSTAR database (see also Plate 20).

13.3.2 3D to 2D Synthetic Aperture Radar Imaging

SAR imaging provides information corresponding to projective transformations from $\mathbb{R}^3$ to $\mathbb{R}^2$, $T : I(x), x \in \mathbb{R}^3 \mapsto TI(y), y \in \mathbb{R}^2$, TI the reflectivity of the objects. SAR images are formed by coherently combining range profiles arising from high-resolution radar imagers corresponding to projections onto the slant plane of objects being imaged. X-band range profiles can decorrelate as the object orientation changes by a fraction of a degree. SAR signatures of objects are similarly variable, in addition depending on the imaging windowing used. Target speckle results from the coherent combination of returns from scatterers as those scatterers move through relative distances on the order of a wavelength of the transmitted signal. Shown in Figure 13.8 are examples of vehicles and SAR datasets found in the training datasets of the MSTAR database.

A signal model for SAR imagery must incorporate the sources of such extreme variability and must be robust with respect to the assumptions of the model if it is to be valid under a wide variety of imaging scenarios. O'Sullivan, et al. [308] assume that the scene region corresponding to any given pixel is large enough to contain multiple scattering centers and invoke central limit theorem arguments to suggest a complex conditionally Gaussian model for the voxel values with nonoverlapping regions being conditional independent. Because the returns are complex-valued with the phase modelled as uniformly distributed across $[0, 2\pi)$, the pixel mean is modeled as identically zero and the variance parameter is a function of the target contents and its pose relative to the radar platform. The conditional density on the observed SAR data I^D given $I_\alpha(g)$ is given by

$$p(I^D|I_\alpha(g)) = \left(\prod_{i=1}^{N_S} \frac{1}{\sqrt{\pi \sigma_\alpha^2(g, y_i)}} \right) e^{-\|\frac{I^D}{\sigma_\alpha(g)}\|^2}, \qquad (13.33)$$

where there are $N_S = |Y|$ pixels in the image plane.

The variance function $\sigma_\alpha^2(g)$ is the model which can be estimated empirically from the training data. Shown in the right panel of Figure 13.8 are a collection of variance images $\sigma_{T-72}(g)$ of a T-72 tank estimated from data in the MSTAR collection which were properly registered and correspond to 5° incremental models of the continuum of objects. The image shows estimates for 72 azimuth windows concatenated row-wise with successive images separated by 5° of azimuth.

Example 13.9 (Radar: Synthetic Aperture Radar (O'Sullivan)) Examine results on MSE bounds from O'Sullivan's group studying SAR imaging in the rotation group of axis-fixed rotations $O \in \mathbf{SO}(2)$. A conditionally Gaussian model for the data was assumed with template variances $\sigma_\alpha^2(O)$ at 5° increments. The MSTAR data set was separated into training and testing data sets (using the recommended separation on the MSTAR CD). The variance images in the model were estimated using the training data, examples of which are shown in the bottom row (left half) of Figure 13.8.

Bayesian classification across the vehicle classes in an image I^D can be classified using a Bayesian approach treating pose as a nuisance parameter. The vehicle class is selected as

$$\hat{\alpha} = \arg\max_{\alpha \in \text{model classes}} p(\alpha|I^D) \tag{13.34}$$

An estimate of the pose can be chosen to minimize the expected squared error in terms of the appropriate distance metric. For $O \in \mathbf{SO}(d)$, this becomes

$$O_{MS} = \arg\min_{O \in \mathbf{SO}(\mathbf{d})} E\left\{\|O - O'\|_{\mathbb{R}^{d^2}}^2\right\}. \tag{13.35}$$

For each of the SAR images in the testing data, the orientation of the object was estimated using the conditionally Gaussian model of Eqn. 13.33. Histograms of the Hilbert–Schmidt norm squared values for the BMP and the T-72 for all of the test images are in panel 1 of Figure 13.9. The range of values for the Hilbert–Schmidt norm squared is compressed in order to show more detail. There were 8 out of the 698 BMP images and 32 out of the 463 T-72 images with norm squares outside the interval shown. Six out of eight for the BMP and 29 out of 32 for the T-72 correspond to 180° errors indicating the fundamental resolvability issue with object ambiguity for such orientations. For reference, the Hilbert-Schmidt norm of .1 corresponds to an error of 12°.

The MSTAR dataset was partitioned into two sets of images: a training set for parameter estimation and a testing set for evaluating orientation estimation algorithms. Columns 1,2 of Figure 13.8 show sample images from the training and testing sets for the T-72 tank. Figure13.9 illustrates the results of azimuth estimation and vehicle classification applied to the testing set. Panel 1 shows a relative histogram of the errors, in degrees, between the actual azimuth angles and the estimated azimuth determined according to (13.35). Panel 2 shows a polar plot of the average squared Hilbert–Schmidt error in the azimuth estimate and the average probability of misclassification as a function of true vehicle azimuth, respectively. The plot of estimation error shows spikes at 90° and 270° (vehicle broadside facing left and right, respectively) indicating that these two orientations are ambiguous. The plot of classification error shows that significantly more classification errors are made when the target is facing nominally away from the radar platform.

The table in Figure 13.10 shows the performance for Bayes ID for the 10 class recognition problem. Shown is the confusion matrix depicted correct ID along the diagonal. Row headings name true vehicle classes and the value in any column is the number of images classified as the vehicle named in the column heading.

13.3 MMSE ESTIMATORS

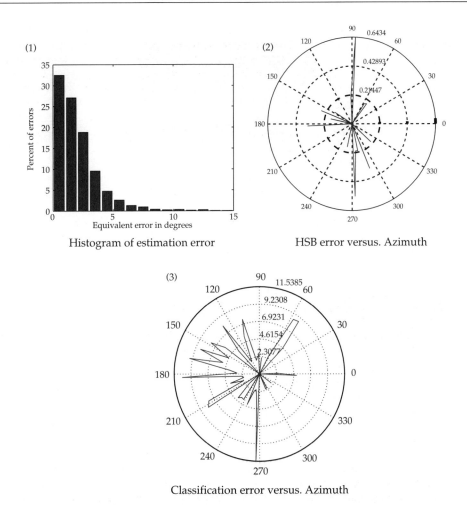

Figure 13.9 Panel 1 shows relative histogram of azimuth estimation errors converted from squared Hilbert–Schmidt norm to an equivalent error in units of degrees. Panel 2 shows average pose estimation error as a function of true vehicle azimuth. Panel 3 shows average classification error rate as a function of true vehicle azimuth. Results from the PhD thesis work of Devore and O'Sullivan [309] (see also Plate 21).

	2S1	BMP 2	BRDM2	BTR60	BTR70	D7	T62	T-72	ZIL131	ZSU234	
2S1	262	0	0	0	0	0	4	8	0	0	95.62%
BMP 2	0	581	0	0	0	0	0	6	0	0	98.98%
BRDM2	5	3	227	1	0	14	3	5	4	1	86.31%
BTR60	1	0	0	193	0	0	0	0	0	1	98.97%
BTR70	4	5	0	0	184	0	0	3	0	0	93.88%
D7	2	0	0	0	0	271	1	0	0	0	98.91%
T62	1	0	0	0	0	0	259	11	2	0	94.87%
T-72	0	0	0	0	0	0	0	582	0	0	100%
ZIL131	0	0	0	0	0	0	2	0	272	0	99.27%
ZSU234	0	0	0	0	0	2	0	1	0	271	98.91%

Figure 13.10 Confusion matrix for the 10-class recognition problem. Row headings name true vehicle classes and the value in any column is the number of images classified as the vehicle named in the column heading. Results from O'Sullivan et al. [308]

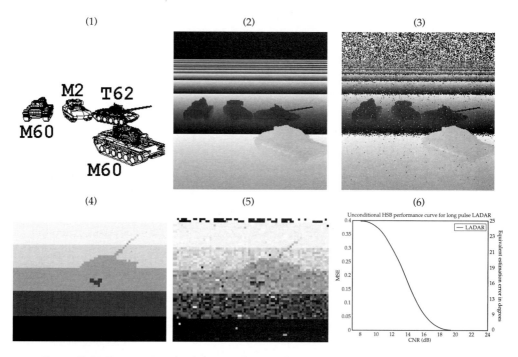

Figure 13.11 Top row: Panel 1 shows wireframes of targets viewed under perspective projection and obscuration. Panel 2 shows noise-free LADAR range image with range ambiguity. Panel 3 shows sample LADAR range image with range ambiguity, anomalous pixels, and range-dependent measurement errors. Bottom row: Panels show results from the LADAR experiments. Panel 4 and 5 show the mean LADAR signal and with noise. Panel 6 shows the HSB mean-squared error performance for pose as a function of CNR (see also Plate 22).

13.3.3 3D to 2D LADAR Imaging

Laser radars (LADARs) sense fixed radiant intensity and range information, LADAR imaging provides vector range and reflectance information corresponding to projective transformations from $\mathbb{R}^3$ to $\mathbb{R}^2$, $(T_1, T_2): I(x), x \in \mathbb{R}^3 \mapsto (T_1 I(y), T_2 I(y)), y \in \mathbb{R}^2(\mathbb{Z}^2)$, $T_1 I$ the range of a particular part of the object, and $T_2 I$ the intensity of reflection.

The laser transmits a scanning sequence of pulses; upon reflection from the object and background and roundtrip propagation through the atmosphere, each pulse is mixed on a photodetector with a strong local oscillator (LO) beam with frequency offset from that of the transmitter by an intermediate frequency (IF). The IF filter's output consists of a signal return corrupted by additive shot noise, which is a Gaussian random processes. The IF filter is followed by a square-law envelope detector and a peak detector, which together produce two random variables, two measurements, for each pixel: the peak intensity and the maximum-likelihood (ML) range-to-reflector estimate. We shall focus on the range-to-reflector field.

Figure 13.11 illustrates a representation of a configuration of ground-based targets being sensed by the LADAR sensor. The LADAR observes the scene through the effects of perspective projection, in which a point (x, y, z) in 3D space is projected onto the 2D detector according to $(x, y, z) \mapsto (x/z, y/z)$. Panel 1 of Figure 13.11 shows a scene of CAD models. Panel 2 presents the mean (noiseless) range image with ambiguity. Panel 3 shows a sample image corrupted with anomalous pixels and range-dependent measurement error.

Let I^D be the laser radar range image with mean field $T_L I_\alpha(g)$, then as modeled in [310,311] the conditional density becomes

$$p(I^D|I_\alpha(g)) = \prod_{i=1}^{N_L} \left(\frac{1 - \Pr_A(y_i)}{\sqrt{2\pi \sigma_L^2(y_i)}} e^{(-(I^D(y_i) - T_L I_\alpha(g)(y_i))^2 / 2\sigma_L^2(y_i))} + \frac{\Pr_A(y_i)}{\Delta R} \right), \quad (13.36)$$

where $N_L = |Y|$ the dimension of the discrete imaging lattice and $R_{\min} \leq I^D, I^D \leq R_{\max}$. Here, $\Pr_A(i)$ is the single-pixel anomaly probability—the probability that speckle and shot-noise effects combine to yield a range measurement that is more than one range resolution cell from the true range, assumed to be the same for all pixels. The first term in the product represents the local Gaussian range behavior for a pixel given that it is not anomalous. The second weights the anomolous behavior with uniform probability over the entire range-uncertainty interval $R_{\min} \leq I^D, I^D \leq R_{\max}$ modeled to include the true range I^D and to have an extent, $\Delta R = R_{\max} - R_{\min}$, that is much larger than the local accuracy σ_L. As established in [312] for a LADAR the range resolution R_{res} is roughly $cT/2$ for a pulse duration of T seconds (c = speed of light) and the number of range resolution bins $N = \Delta R/R_{\text{res}}$, with the local range accuracy and anomaly probability obeying $\sigma_L(y_i) = R_{\text{res}}/\sqrt{\text{CNR}(y_i)}$ and $\Pr_A(y_i) \approx (1/\text{CNR}(y_i))(\ln(N) - 1/N + 0.577)$, respectively.[37]

Example 13.10 (LADAR) Examine experimental results on mean-squared performance in LADAR. Shown in Figure 13.11 are results from the LADAR simulation. The template is the tank $I_\alpha = \text{TANK}$, placed at arbitrary orientations. The imaging model is T_L the projective transformation, with the observed image data I^D with mean $T_L I_\alpha(g)$, $g \in \mathbf{SO}(2)$ a random rotation uniformly distributed. The conditional density $p(I^D|I_\alpha(g))$ is given by Eqn. 13.36. Figure 13.11 shows results from computing the pose bound computed over a range of CNRs. Panel 4 shows the mean, panel 5 shows the LADAR signal I^D with mean $TI_\alpha(g)$ and conditional density obeying Eqn. 13.36. Panel 6 shows HSB mean-squared error performance for pose as a function of CNR.

13.3.4 3D to 2D Poisson Projection Model

CCD detectors in forward looking infrared imaging (FLIR) correspond to projective transformations from $\mathbb{R}^3$ to $\mathbb{R}^2$, $T : I(x), x \in \mathbb{R}^3 \mapsto TI(y), y \in \mathbb{R}^2(\mathbb{Z}^2)$ corresponding to discrete lattices with image noise growing as signal mean. Thus the Poisson noise models of Snyder [31, 32] become appropriate assuming the measured 2D image field I^D is Poisson distributed with mean intensity $TI_\alpha(g)$ the projection on the detector plane [32,317]. Defining $\mathbf{1}$ the all $1's$ vector over the imaging lattice $y \in Y$, then the density takes the form [42]

$$p(I^D|I_\alpha(g)) = \frac{e^{-\langle \mathbf{1}, TI_\alpha(g) \rangle + \langle \log TI_\alpha(g), I^D \rangle}}{\prod_{i=1}^N I^D(y_i)!}, \quad (13.37)$$

---

[37] Following Shapiro [313–315] the CNR for LADAR is the ratio of the average radar-return power and average local-oscillator shot noise power given by the resolved speckle-reflector radar equation [316] according to $CNR(i) = ve^{-2\alpha I(i)}$ where α is the atmospheric extinction coefficient and v is a constant derived from the properties of the laser radar. Notice that the probability of anomaly $Pr_A(i)$ and the local range accuracy $\sigma(i)$ increase with hypothesized distance to the target.

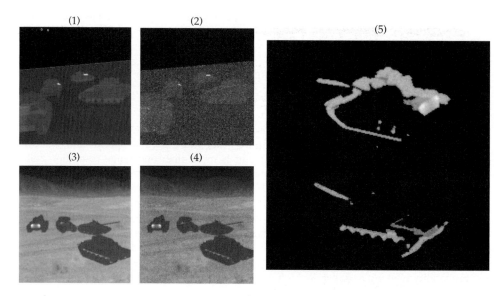

Figure 13.12 Left Half: Panel 1 shows M2 tanks through perspective projection; panel 2 shows the scene in Poisson noise. Panel 3 shows scene with Gaussian blur; panel 4 shows Poisson noise. Right Half: Panel 5 shows the log-likelihood of the data on the pixel array for the two matches $\log p(I^D(y)|I(g))$. Brightness means higher relative log-probability.

$$\text{where } \langle \mathbf{1}, TI_\alpha(g) \rangle = \sum_{y_i \in Y} TI_\alpha(y_i), \quad \langle \log TI_\alpha(g), I^D \rangle = \sum_{y_i \in Y} \log TI_\alpha(g)(y_i) I^D(y_i). \tag{13.38}$$

Note that the discrete lattice has dimension $N = |Y|$. Shown in Figure 13.12 are examples of imagery corresponding to projective imagery via CCD detection. Panel 1 shows M2 tanks through perspective projection; panel 2 shows the random field corresponding to Poisson noise.

Generally the imaging device contains optics in the focal plane resulting in point-spread $p(\cdot)$ (see [31, 32], e.g.). Then the mean field in the above models gets adjusted to $\int_Y p(y - z) TI(z) dz$. Panels 3,4 of Figure 13.12 depict such point spread measurements. Panel 3 shows perspective projection with Gaussian blur in the optics. Panel 4 shows the random field corresponding to Poisson noise.

The right half of Figure 13.12 illustrates how the conditional density $p(I^D|I(g))$ encodes the information about the hypothesized objects. Panel 5 shows the difference in log-likelihood probability density (plotted in gray scale). Bright pixels correspond to high evidence of object classification. The relative probability corresponds to an exponentiation of the integral of the brightness pictures. The top choice M60 is far more favorable because of the bright log-probability values.

Example 13.11 (FLIR) Examine results on MSE bounds for CCD array corresponding to FLIR imagers. Second generation FLIR imagers sense the temperature profile of the object via CCD detection. The infrared (IR) passive sensor performs direct detection of the thermal radiation generated by the spatially-resolved object, with the passive-channel photodetector followed by a low pass filter integrating the resulting photocurrent over the pixel dwell time T_d. The integrated intensity measured in each pixel consists of the radiation power and shot-noise-plus-thermal-noise [318]. Represent the first two moments of the Poisson process corresponding to the mean following the variance. Assuming the constant offset due to dark current has been subtracted

13.3 MMSE ESTIMATORS

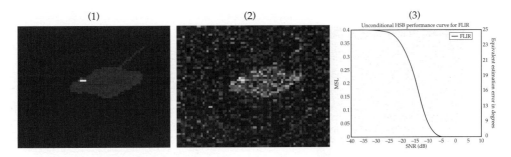

Figure 13.13 FLIR: Panels show results from the FLIR experiments. Panel 1 shows the mean, and panel 2 shows the signal in noise. Panel 3 shows the HSB mean-squared error performance as a function of SNR (see also Plate 23).

from the photocurrent, then I^D is Gaussian with mean field $T_F I_\alpha(g)$ according to

$$p(I^D | I_\alpha(g)) = \left(\prod_{i=1}^{N_F} \frac{1}{\sqrt{2\pi \sigma_F^2(y_i)}} \right) e^{-\|I^D - T_F I_\alpha(g)\|^2 / 2\sigma_F^2}, \qquad (13.39)$$

where the dimension of the discrete CCD lattice is taken as $N_F = |Y|$.[38]

Shown in Figure 13.13 are results from the FLIR experiments. The template is the tank I_α = TANK, placed at arbitrary orientations. Then T_F is the perspective transformation of the FLIR imaging optics, with the observed image data I^D with mean $T_F I_\alpha(g)$, $g \in \mathbf{SO}(2)$ a random rotation uniformly distributed. The conditional density $p(I^D | I(g))$ is given by Eqn. 13.39. Panel 1 shows the mean, and panel 2 shows the signal in noise. Panel 3 shows the HSB mean-squared error performance as a function of SNR (see text for definition).

13.3.5 3D to 1D Projections

There are numerous examples of projective transformations from $\mathbb{R}^3$ to $\mathbb{R}^1$, $T : I(x), x \in \mathbb{R}^3 \mapsto TI(y), y \in \mathbb{R}^1(\mathbb{Z}^1)$. One example is high range-resolution radar (HRR) illuminating the object by a sequence of radar pulses with the received echoes collected at the receiver. The radar cross section (RCS) (effective echoing area of the object) results in reradiation towards the radar receiver. The return signal is filtered through a bandpass filter tuned at the center frequency f_c and then sampled. The resulting range profiles are the discrete Fourier transforms of the filtered return

---

[38] The SNR for FLIR is defined as in [318] and is determined by the noise-equivalent differential temperature NEΔT which for $hv \gg kT_s$ (see [318]) is given approximately by SNR $= (\Delta T / \mathrm{NE}\Delta T)^2$ where

$$\mathrm{NE}\Delta T \approx \frac{kT_s^2}{i_{F,s}^D} \sqrt{\frac{2Bp}{\eta h v}} (i_{F,s}^D + P_d + P_{\mathrm{therm}}), \qquad (13.40)$$

k is Boltzmann's constant, T_s is the absolute temperature of the source (object or background), v is the passive channel's center frequency, $P_d = I_d h v / \eta q$ is the dark current equivalent power for mean dark current I_d, q is the electron charge, and P_{therm} is the thermal-noise equivalent power, $P_{\mathrm{therm}} = 2kT_L hv/\eta q^2 R_L$, with T_L being the absolute temperature and R_L the resistance of the photodetector's load, P_b is the nominal radiation power in the field of view with $\Delta P = P_t - P_b$ the signal power present in a object pixel with differential temperature $\Delta T = T_t - T_b$, and NEΔT is the temperature difference which produces unity SNR.

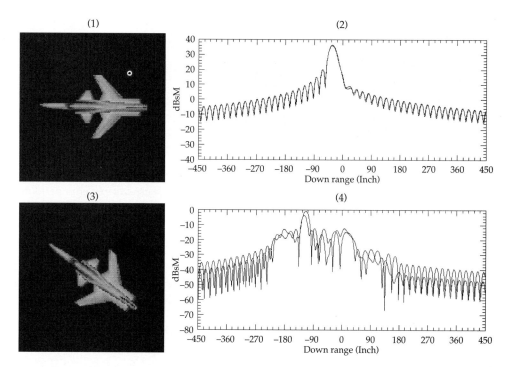

Figure 13.14 Azimuth-elevation power spectrum of the X-29 at several poses. Panels 1,3 show renderings of the object; panels 2,4 show high-resolution radar profiles generated from XPATCH for the X-29 targets at their respective orientations.

signal [319]. The observed range profile $I^D(t)$ is modelled as a complex Gaussian random process with conditional mean and covariance $T_R I_\alpha(g)$, $K_R(g)$, respectively.

Shown in Figure 13.14 are results of azimuth-elevation power spectra of the X-29 object at several pose positions generated via the XPATCH simulator (see [300, 301]). Panels 1,3 show renderings of the object; panels 2,4 show high-resolution radar profiles generated from XPATCH for the X-29 targets at their respective orientations.

The wide-sense-stationary uncorrelated-scatter (WSSUS) model for diffuse radar objects [320] assumes that returns from different delays are statistically uncorrelated. In additive complex white Gaussian noise, the magnitude of the complex envelope for the observed signal is Rice distributed. Let I^D be the random range profile power with mean target RCS profile $I_\alpha(g)$ produced by the HRR, then the likelihood function takes the form

$$p(I^D|I_\alpha(g))$$
$$= \prod_{i=1}^{N_R} J_0 \left(\frac{2\sqrt{I^D(y_i) T_R I_\alpha(g)(y_i)(1-\beta)}}{T_R I_\alpha(g)(y_i)\beta + \sigma_R^2} \right) \frac{e^{((-I^D(y_i) + T_R I_\alpha(g)(y_i)(1-\beta))/(T_R I_\alpha(g)(y_i)\beta + \sigma_R^2))}}{T_R I_\alpha(g)(y_i)\beta + \sigma_R^2}, \quad (13.41)$$

where J_0 is the zeroth-order modified Bessel function of the first kind.[39]

---

[39] The SNR defined for HRR is the ratio of average object cross-section to noise-equivalent cross-section given by

$$\text{SNR} = \frac{1}{N_L} \sum_{l=1}^{N_L} \frac{|m(g_{\text{ref}}, y_l)|^2 + \sigma_R^2(g_{\text{ref}}, y_l)}{\sigma_R^2}, \quad (13.42)$$

where 89° is the reference angle chosen so that the mean reference object cross-sectional area $m(g_{\text{ref}})$ is maximum.

13.3 MMSE ESTIMATORS

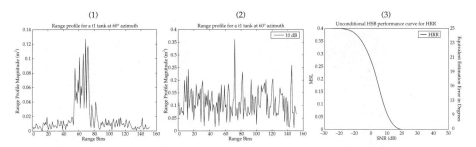

Figure 13.15 Panels 1 and 2 show the HRR range profile mean-fields and samples with noise for a T1 tank at 60°. Panel 3 shows the HSB mean-squared error performance as a function of SNR (see text for definition).

Example 13.12 (HRR) Examine experimental results on the MSE performance as measured by the HS bounds for HRR. Shown in Figure 13.15 are results from the HRR study. The imaging model is T_R the ranging transformation of the HRR device with the observed image data I^D with mean $T_R I_\alpha(g)$, $g \in \mathbf{SO}(2)$ a random rotation uniformly distributed. The conditional density $p(I^D|I_\alpha(g))$ is given by Eqn. 13.41. Panels 1 and 2 show the HRR range profile mean-fields and samples with noise. Panel 3 shows the HSB mean-squared error performance as a function of SNR (see text for definition).

13.3.6 3D(2D) to 3D(2D) Medical Imaging Registration

Many modern high resolution imagers provide fundamental 3D information of the underlying scenes. MR imaging technology is one such example, producing highly detailed representations of the 3D geometry, providing transformations from $\mathbb{R}^3(\mathbb{R}^2)$ to $\mathbb{R}^3(\mathbb{R}^2)$, $T: I(x), x \in \mathbb{R}^d \mapsto TI(y), y \in \mathbb{R}^d(\mathbb{Z}^d)$, $d = 2, 3$. TI represents various physical properties of the tissue including the proton density, T1, or T2 time constants of spin–spin decay and spin-lattic decay.

Image registration associated with estimating parameters in the special Euclidean group is an area of tremendous research in the Medical Imaging context. Model the imagery $I^D \in \mathcal{I}^D$ as a conditionally Gaussian field with mean the MRI brain volumes of the underlying true brain I_α. Then I^D is a conditionally Gaussian random field conditioned on $I_\alpha(g)$, $g \in \mathbf{SE}(2, 3)$, with the orbit of images becoming $\mathcal{I} = \{I_\alpha(g), g \in \mathbf{SE}(2)\}$. The conditional density takes the form

$$p(I^D|I_\alpha(g)) = \frac{1}{(2\pi\sigma^2)^{\frac{N}{2}}} e^{-\|I^D - I_\alpha(g)\|^2/2\sigma^2}, \tag{13.43}$$

where the number of voxels N of the template and the measured data are the same.

Example 13.13 (2D to 2D MRI Registration (Costello [305])) Shown in Figure 13.16 are results from registration in the special Euclidean group from 2D MRI imagery. The top row shows 2D imagery from T2 weighted MRI data collected by Dr. Scott Nadel at Duke University. Panel 1 shows the 2D template image section I_α; panel 2 shows the same section rotated and translated with additive noise $I^D = I_\alpha(g) + W$. The imagery is modelled I^D as a conditionally Gaussian process with mean the brain section in panel 1 randomly rotated and translated in noise element of the orbit of images $I^D \in \mathcal{I}^D = \{I_\alpha(g) + W\}$. Shown in the bottom row panel 3, 4 are results of computing the mean-squared error performance in estimating the unknown rotations and

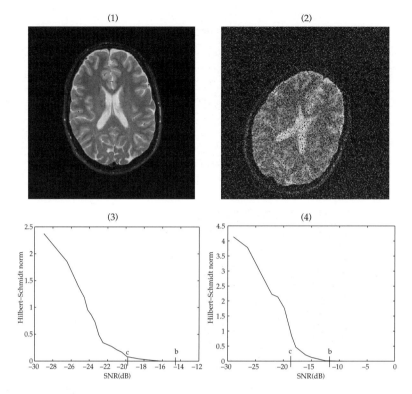

Figure 13.16 Top row: Panels 1, 2 show the template section I_α MRI image, and the image after rotation and translation with noise $I^D = I_\alpha \circ g + W$. Bottom row: Panels 3, 4 show the HSB for rotation, and rotation plus translation **SE**(2) error.

translations as a function of noise level. Panels 3, 4 show the HSB for rotation (panel 3), and rotation plus translation (panel 4) **SE**(2) error as a function of noise SNR.

Figure 13.17 examines the sensitivity of the performance of the HS minimum-mean-squared error estimator on the accuracy of the assumed template model. Panel 1 shows the MRI section taken as the template. Panels 2, 3 show the mean fields used to generate the orbit of actual imagery $I_D \in \mathcal{I}^D$. Panel 2 duplicates an inhomogeneity artifact; panel 3 presents a tumor artifact. Panel 4 shows the degradation in performance of the HSB estimator resulting from the model mismatch attributable to the inaccurate template choices. The solid line shows the MSE performance as measured by the HSB for the true template (panel 1) matched to the orbit of imagery generated from the template $\mathcal{I}^D = \{I_{\text{PANEL1}}(g) + W\}$. The dot–dashed line shows the HSB for the orbit generated from the tumor $\mathcal{I}^D = \{I_{\text{tumor}}(g) + W\}$ but matched with $I_\alpha = \text{PANEL1}$. The dotted line shows the HSB for the orbit generated from the inhomogeneity $\mathcal{I}^D = \{I_{\text{inhomo}}(g) + W\}$ but matched with $I_\alpha = \text{PANEL1}$.

13.4 Parameter Estimation and Fisher Information

We should expect that in the asymptotic consistency setting the matrix of second derivatives of the likelihood function determines the variability of estimation. As well, it will turn out that it will influence the exponential rate of performance for hypothesis testing, linking parameter estimation and identification and recognition.

13.4 PARAMETER ESTIMATION AND FISHER INFORMATION

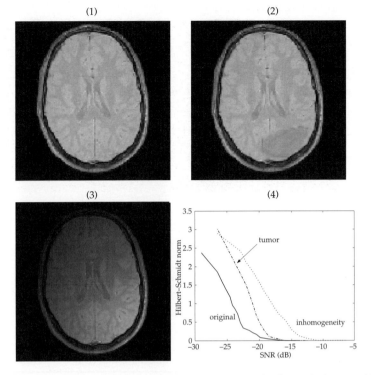

Figure 13.17 Panel 1 shows the T2 weighted MR image; panel 2 shows the image with a simulated tumor. Panel 3 shows the image with simulated inhomogeneity. Panel 4 shows the HSB bounds for rotation error for the template in panel 1 matched to the tumor and inhomogeneity data.

To calculate the Fisher information matrix the first and second derivatives of the log-density must be calculated as given by the following theorem. Throughout we assume the conditional densities are sufficiently smooth so that the derivatives for the Fisher information are well defined.

Theorem 13.14 *Given is the orbit of ideal images $I(g) \in \mathcal{I}$, $g \in \mathcal{G}$ a finite-dimensional group with local parameterization $g : \theta \in \mathbb{R}^d \to g \in \mathcal{G}$ of dimension d observed through generalized projection T with data I^D a conditional random field with mean field $TI(g)$.*

For I^D a conditional Gaussian random field with conditional density

$$p(I^D | I_\alpha(g)) = \frac{1}{(2\pi\sigma^2)^{N/2}} e^{-(1/2\sigma^2)\|I^D - TI_\alpha(g)\|^2}, \tag{13.44}$$

*then the $d \times d$ **Fisher information matrix** and its **sensitivity integrals** ($F_{kj}, k, j = 1, \ldots, d$) defining the $d \times d$ Fisher information matrix*

$$F = [F_{kj}], \quad F_{kj} = \frac{1}{\sigma^2} \left\langle \frac{\partial}{\partial \theta_k} TI_\alpha(g), \frac{\partial}{\partial \theta_j} TI_\alpha(g) \right\rangle. \tag{13.45}$$

For I^D a Poisson field with conditional density

$$p(I^D | I_\alpha(g)) = \frac{e^{-\langle \mathbf{1}, TI_\alpha(g) \rangle + \langle \log TI_\alpha(g), I^D \rangle}}{\prod_{i=1}^N I^D(y_i)!}, \tag{13.46}$$

the Fisher information is given by

$$F_{kl} = \left\langle \frac{\partial/\partial\theta_k TI_\alpha(g)}{\sqrt{TI_\alpha(g)}}, \frac{\partial/\partial\theta_j TI_\alpha(g)}{\sqrt{TI_\alpha(g)}} \right\rangle. \qquad (13.47)$$

Note the similarity between the two cases. In the Poisson setting the variance in each pixel is the normalizer rather than a single fixed constant.

Proof To calculate the Fisher information matrix (Eqn. 2.113 of definition 2.40) the first and second derivatives of the log-likelihood must be calculated:

$$\frac{\partial}{\partial\theta_k} \log p(I^D|I_\alpha(g)) = \frac{1}{\sigma^2}\left\langle I^D - TI_\alpha(g), \frac{\partial}{\partial\theta_k} TI_\alpha(g) \right\rangle \qquad (13.48)$$

$$\frac{\partial^2}{\partial\theta_k \partial\theta_j} \log p(I^D|I_\alpha(g)) = \frac{1}{\sigma^2}\left\langle I^D - TI_\alpha(g), \frac{\partial^2}{\partial\theta_k \partial\theta_j} TI_\alpha(g) \right\rangle$$

$$- \frac{1}{\sigma^2}\left\langle \frac{\partial}{\partial\theta_j} TI_\alpha(g), \frac{\partial}{\partial\theta_k} TI_\alpha(g) \right\rangle. \qquad (13.49)$$

Taking the negative expectation obtains the Fisher information.

In the Poisson case, define $\mathbf{1}(\cdot)$ to be all 1's function over the detector array. The log-probability becomes

$$\log p(I^D|I_\alpha(g)) = -\langle \mathbf{1}, TI_\alpha(g) \rangle + \langle I^D, \log TI_\alpha(g) \rangle + \text{constant}. \qquad (13.50)$$

Then the terms determining the Fisher information become

$$\frac{\partial}{\partial\theta_k} \log p(I^D|I_\alpha(g)) = -\left\langle \mathbf{1}, \frac{\partial}{\partial\theta_k} TI_\alpha(g) \right\rangle + \left\langle I^D, \frac{(\partial/\partial\theta_k) TI_\alpha(g)}{TI_\alpha(g)} \right\rangle, \qquad (13.51)$$

$$\frac{\partial^2}{\partial\theta_k \partial\theta_j} \log p(I^D|I_\alpha(g)) = -\left\langle \mathbf{1}, \frac{\partial^2}{\partial\theta_k \partial\theta_j} TI_\alpha(g) \right\rangle + \left\langle I^D, (\partial^2/\partial\theta_k \partial\theta_j) TI_\alpha(g)/TI_\alpha(g) \right.$$

$$\left. -(\partial/\partial\theta_k) TI_\alpha(g)(\partial/\partial\theta_j) TI_\alpha(g)/(TI_\alpha(g))^2 \right\rangle. \qquad (13.52)$$

Taking the negative expectation gives

$$F_{kl} = \left\langle \mathbf{1}, \frac{\partial^2}{\partial\theta_k \partial\theta_j} TI_\alpha(g) \right\rangle - \left\langle \mathbf{1}, \frac{\partial^2}{\partial\theta_k \partial\theta_j} TI_\alpha(g) \right\rangle + \left\langle \mathbf{1}, \frac{(\partial/\partial\theta_k) TI_\alpha(g)(\partial/\partial\theta_j) TI_\alpha(g)}{TI_\alpha(g)} \right\rangle \qquad (13.53)$$

$$= \left\langle \frac{(\partial/\partial\theta_k) TI_\alpha(g)}{\sqrt{TI_\alpha(g)}}, \frac{(\partial/\partial\theta_j) TI_\alpha(g)}{\sqrt{TI_\alpha(g)}} \right\rangle. \qquad (13.54)$$

$\square$

Example 13.15 (Fisher Information Across Orientation) The mean squared error is bounded below by the inverse Fisher information (see Theorem 2.42, Chapter 2). For pose, choose the local coordinates for the axis fixed rotation as θ, then the mean-squared error is lower bounded according to

$$E\{(\hat{\theta}(I^D) - \theta)^2\} \geq F^{-1} = -\left(E\left\{ \frac{\partial^2 \log p(I^D|\theta)}{\partial\theta^2} \right\} \right)^{-1} \qquad (13.55)$$

$$= \frac{\sigma^2}{\|(\partial/\partial\theta) TI_\alpha(\theta)\|^2}. \qquad (13.56)$$

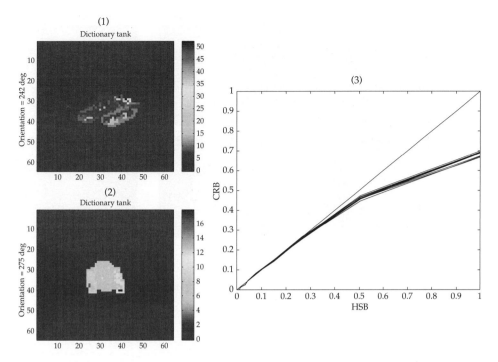

Figure 13.18 Panels 1,2 show tanks at various brightnesses and orientations. Panel 3 shows $1/2HSB(\theta, \sigma)$ for a set of θ (x-axis) versus $\sigma^2/\|\partial_\theta TI(\theta)\|^2$ (y-axis) (see also plate 24).

To compare the mean-squared error of the angle estimator to the HSB, note that

$$\|O(\theta) - \hat{O}(\theta)\|^2_{\mathbb{R}^{n \times n}} = \text{tr}(O - \hat{O}(\theta))(O - \hat{O}(\theta))^*$$
$$= 4 - 4\cos(\theta - \hat{\theta}) \qquad (13.57)$$
$$= 2(\theta - \hat{\theta})^2 + o((\theta - \hat{\theta})^2). \qquad (13.58)$$

Thus, for $\sigma^2 \to 0$, then $\frac{1}{2}HSB \sim E|\theta - \hat{\theta}|^2$.

Figure 13.18 examines use of the inverse Fisher information bound to predict the mean-squared error performance as measured by the Hilbert–Schmidt estimator. Panels 1,2 show examples of vehicles observed via a FLIR imager sensing the temperature profile on the tanks. Panel 3 shows a plot of $\frac{1}{2}HSB$ plotted versus the inverse Fisher information $F^{-1} = (\sigma^2/\|(\partial/\partial_\theta)TI(\theta)\|^2)$. The superimposed lines are the performance plots for many different orientations of the tank. Asymptotically in low noise, the inverse Fisher information and mean-squared error measured by 1/2HSB are identical. They diverge at high error rates.

Example 13.16 (Fisher Information as a Function of Scale) Shown in Figure 13.19 are results of such an analysis as a function of scale and number of pixels on target. Rows 1, 2 show images with 1000 and 250 pixels on target as a function of noise standard deviation. The bottom row shows MSE performance for pose as measured by HSB/2 versus inverse Fisher information $\sigma^2/\|TI(\theta)\|$ as a function of noise σ for 100, 500, 1000 pixels on target. The solid line shows the $\frac{1}{2}HSB$ curve; the dashed line shows the MSE as measured by $E(\theta - \hat{\theta})^2$. Note the divergence at high noise because the Taylor series approximation diverges for the fourth order term in $4 - 4\cos\theta$. Note the near linear relation.

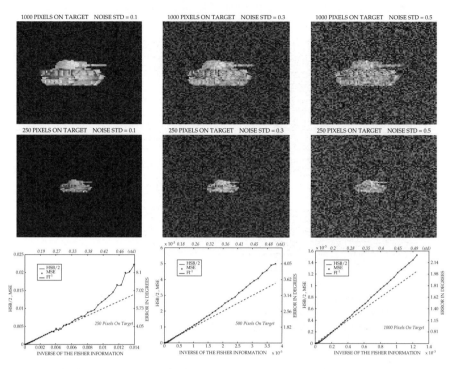

Figure 13.19 Top and middle rows show images from 1000 and 250 pixels on target as a function of $\sigma = 0.1, 0.3, 0.5$ (columns 1, 2, 3), respectively. Bottom row shows pose performance as measured by HSB/2 versus noise $\sigma^2/\|TI_\alpha(\theta)\|$ showing the three curves for 100, 500, 1000 pixels on target. The solid line shows the $\frac{1}{2}$HSB; the dashed line shows the MSE as given by $E|\theta - \hat{\theta}|^2$. Note the near linear relation.

Example 13.17 (Vehicle and Brain registration (Bitouk [321])) Shown in Figure 13.20 are results exploring the relationship between the variance of the estimator and the asymptotic prediction of the inverse Fisher information. Theorem 2.42 predicts that the MSE is lower bounded by the inverse Fisher information. Model the data I^D as conditional Gaussian random field with mean field $I_\alpha(g)$. Columns 1 and 2 show imagery of the vehicle and brain at varying signal to noise (SNR) ratios SNR $= -10, 0, 10$ dB, respectively. Columns 3, 4 explore the dependence of the pose variability in the posterior distribution as a function of the Fisher information. Top panels shows the MSE $E(\theta - \hat{\theta})^2$ as a function of the inverse Fisher Information. Over a broad range of SNRs there is agreement between the standard deviation of object pose with the Fisher information. Bottom row shows the square root of the the inverse Fisher Information plotted versus half the Hilbert–Schmidt error. At low noise these should be equal. Similar results are shown in columns 2,4 of Figure 13.20 from Bitouk [321] for MRI brain registration.

13.5 Bayesian Fusion of Information

The importance of the conceptual separation of the source of possible images $\mathcal{I}$ with prior density $\pi(I_\alpha(g)), I_\alpha(g) \in \mathcal{I}$ and the channel with transition law $p(I^D|I_\alpha(g))$ is that there is only one true underlying scene, irrespective of the number of sensor measurements forming $I^D = (I^{D_1}, I^{D_2}, \dots)$.

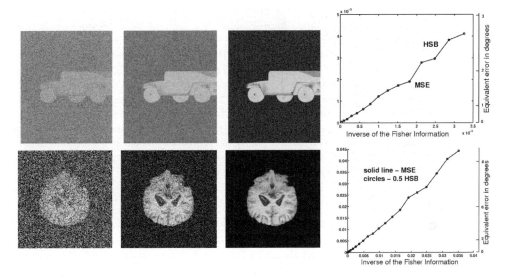

Figure 13.20 Columns 1, 2 and 3 show imagery of the vehicle (top row) and brain (bottom row) as a function of increasing SNR SNR = −10, 0, 10 dB. Column 4: panels show $\frac{1}{2}$ HSB versus inverse Fisher information F^{-1}; dots show the MSE.

For $I^{D_1}, I^{D_2}, \ldots$ conditionally independent given $I_\alpha(g) = (I_1(g), I_2(g) \ldots,)$, the conditional density takes the form

$$p(I^D | I_\alpha(g)) = \prod_{i=1}^{n} p(I^{D_i} | I_\alpha(g)). \tag{13.59}$$

Only one inference problem is solved, with the multiple observations due to multiple sensors viewed as providing additional information in the posterior distribution. *Sensor fusion occurs automatically in this framework.* Certainly, in calculating information gain this view determines these ultimate bounds.

This allows us to prove, for example, that Fisher information is monotonic in a number of sensors. Therefore, mean-squared error bounds have to improve with consistent combination of sensing information.

Theorem 13.18 *Let $I^{D_1}, I^{D_2}, \ldots$ be conditionally independent random fields conditioned on the mean field $I_\alpha(g)$. Then the Fisher information is monotonic in the sensors given by the superposition of each of the Fisher informations:*

$$F(I^{D_1}, \ldots, I^{D_n}; I_\alpha(g)) = \sum_{i=1}^{n} F(I^{D_i}; I_\alpha(g)). \tag{13.60}$$

The covariance lower bound is monotonically decreasing in number of sensors (see Theorem 2.42):

$$\mathrm{Cov}(\hat{I}_\alpha(g); (I^{D_1}, \ldots, I^{D_n})) \geq \left(\sum_{i=1}^{n} F(I^{D_i}; I_\alpha(g)) \right)^{-1}. \tag{13.61}$$

Proof Proof follows directly from Bayes factorization of the conditional density Eqn. 13.59. □

Example 13.19 (Fusion: LADAR, FLIR, HRR) Examine the scenario studied by Shapiro et al. [322, 323] for the multiple sensor suite of MIT Lincoln Laboratory's Infrared Airborne Radar (IRAR) Program [324]. The sensors were selected a range of 2.0 km with a long-wave infrared laser radar (LADAR), a forward looking infrared (FLIR) and a high-resolution radar (HRR). The LADAR and FLIR imaging subsystems share common optics and are pixel-registered sensing fixed radiant intensities, and ranges; the HRR measures 1D cross sectional reflectance profiles.

The Bayesian formulation readily incorporates additional sensors into the inference. Model the sensors as conditionally independent observations of the scene, the product of each of the individual conditional probability terms forms the joint density for all available observations. The sensors are assumed conditionally independent so that the joint density on the sensor data is the product

$$p(I^{D_F}, I^{D_L}, I^{D_R} | I_\alpha(g)) = p(I^{D_F} | I_\alpha(g)) p(I^{D_L} | I_\alpha(g)) p(I^{D_R} | I_\alpha(g)). \quad (13.62)$$

Sensors have different signature representations which must be reflected in the conditional probability functions. The parameters selected for the sensors were selected to model the MIT Lincoln Laboratory Infrared Airborne Radar (IRAR) Program and the URISD dataset parameters.[40]

Figure 13.21 depicts the performance of the multiple sensors FLIR, LADAR, HRR operating together at various carrier-to-noise ratios (CNR) and signal-to-noise ratios (SNR). For each of the sensors, the MSE increases as CNR or SNR decreases. Sensors optimally fused outperform either sensor taken individually. The performance gain that accrues from sensor fusion can be visualized as the decrease of SNRs or CNR required to achieve the chosen performance level. Figure 13.21 depicts this. Panels 1–3 show (CNR, SNR) requirements needed to realize a fixed MSE value of 0.05 (pose estimation error of 9°). The horizontal straight line in panel 1 is the HRR SNR required to achieve the selected pose accuracy from the HRR sensor alone; the vertical straight line is the LADAR CNR required for the active LADAR channel alone to achieve the selected pose accuracy. The solid fusion curve shows the (CNR, SNR) required to realize the pose performance when the HRR and LADAR outputs are optimally

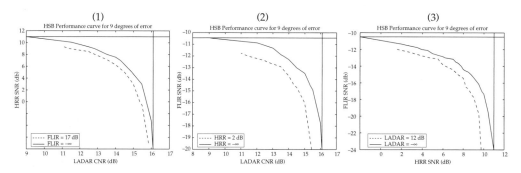

Figure 13.21 HRR,FLIR,LADAR FUSION: Panels show the performance-bound curves for 9° root-mean-square pose estimation error for LADAR, HRR (panel 1), LADAR,FLIR (panel 2), FLIR,HRR (panel 3).

---

[40] The FLIR and LADAR parameters selected are receiver aperture dimension of 13 cm, receiver field-of-view $\phi_R = 2.4$ mrad, detector quantum efficiency 0.25 and atmospheric extinction coefficient 0.5 dB/km. The FLIR noise-equivalent temperature NE$\Delta T = 0.1$ K. The LADAR parameters are average transmitter power of 2 W, pulse repetition frequency of 20 kHz, pulse duration of 200 nsec, peak transmitter power of 500 W, photon energy 1.87×10^{-20} J, IF filter bandwidth 20 MHz, range resolution 30 m, number of range bins 256. The high resolution radar has parameters center frequency 1.5 GHz, bandwidth 739.5 MHz, frequency sample spacing 4.921 MHz, number of frequency samples 151, azimuth angle spacing 0.15°, number of azimuth samples, 2401.

combined. The dashed fusion curve shows the (CNR, SNR) values required to realize the same pose performance with the FLIR operating at SNR = −17 dB.

Panels 2, 3 are similar for the LADAR, FLIR combination and FLIR, HRR, respectively. The dashed curves correspond to adding the third sensor which shifts the entire performance.

13.6 Asymptotic Consistency of Inference and Symmetry Groups

13.6.1 Consistency

We now examine asymptotic consistency and the role of the Fisher information matrix. We examine the similar setting as in Chapter 2, 2.60 in which consistency is obtained through increasing sample size. Since we study the i.i.d. Gaussian setting here, we explicitly parameterize sample size through variance, i.e. as $n \to \infty$, $\sigma^2(n) \to 0$. Throughout the projective imaging model is assumed in Gaussian noise, with I^D a Gaussian random field with mean $TI_\alpha(g)$, and independent variances σ^2. The conditional data density becomes

$$p(I^D | I_\alpha(g), \alpha) = \frac{1}{(2\pi\sigma^2)^{N/2}} e^{-1/2\sigma^2 \|I^D - TI_\alpha(g)\|^2}. \tag{13.63}$$

For the asymptotic study of parameter estimation and identification we shall work in the local coordinates of the group since the density will concentrate on a particular local coordinate chart. Define the coordinate frames to be $\partial/\partial g_i, i = 1, \ldots, d$ the dimension of the transformation group $\mathcal{G}$.

Consistency implies that when the signal-to-noise ratio increases, the estimates restrict to a family of parameters representing the true underlying objects. The sensor map T is, in general, a many-to-one transformation from object occurrences $I_\alpha(g)$ to the observed image I^D. Multiple object occurrences can map into the same observed image because (i) the object may have inherent symmetries so that at different pose it leads to the same image at the sensor, and (ii) the features distinguishing different poses may be lost in the projective transformation constituting T. Define the set of equivalent configurations through the sensor mapping as follows.

Definition 13.20 *Define the* **set of equivalences** *through the sensor map associated with object α_0 at group parameter g_0 as*

$$M(g_0, \alpha_0) = \{(g, \alpha) : TI_\alpha(g) = TI_{\alpha_0}(g_0)\} \subset \mathcal{G} \times \mathcal{A}. \tag{13.64}$$

An assumption of consistency *corresponds to* $|M(g_0, \alpha_0)| = 1$.

Any two pairs (g_1, α_1) and (g_2, α_2) are considered equivalent if $TI_1(g_1) = TI_2(g_2)$, partitioning the set $\mathcal{G} \times \mathcal{A}$ into equivalence classes. The *multiplicity* of the equivalence set $|M(g_0, \alpha_0)|$ is always at least one but can have any finite cardinality as well as that of the continuum. Given an observed image, the estimation is restricted to the space of equivalence classes $\cup_{s,\alpha} M(g, \alpha)$. In the case of multiple sensors, each possibly having a different sensor map $T_1, T_2, \ldots, T_k$, let $T = (T_1, T_2, \ldots, T_k)$ and the above definition holds.

If the object $\alpha^\star$ with the parameter $g^\star$ is observed, then for $\sigma^2 \to 0$ consistency would require that the inference leads to the set $M(g^\star, \alpha^\star)$. This is made precise by the following proposition.

Theorem 13.21 *For any fixed $(\alpha^\star, g^\star) \in \mathcal{A} \times \mathcal{G}$, the support of the likelihood function $p(I^D | I_\alpha(g), \alpha)$ as a function of g and α contracts as $\sigma^2 \downarrow 0$ to the set $M(g^\star, \alpha^\star)$.*

Proof Consider the set

$$\mathcal{G}_\alpha(\epsilon) = \{g : \|I^D - TI_\alpha(g)\|^2 < \epsilon\} \subset \mathcal{G}$$

for an arbitrary positive ϵ. Let $f_\alpha : \mathcal{G} \mapsto \mathbb{R}$ be a bounded and continuous test function which is zero inside $\mathcal{G}_\alpha(\epsilon)$, and evaluate

$$\int_{\mathcal{G}} f_\alpha(g) p(I^D | I_\alpha(g), \alpha) \pi_\alpha(g) \, ds = \int_{\mathcal{G}_\alpha(\epsilon)^c} f_\alpha(g) \frac{1}{(2\pi\sigma^2)^{N/2}} e^{-1/2\sigma^2 \|I^D - TI_\alpha(g)\|^2} \pi_\alpha(g) \, ds. \tag{13.65}$$

This quantity goes to zero for $\sigma^2 \downarrow 0$, for all $\epsilon > 0$ fixed. The limiting set $\lim_{\epsilon \downarrow 0} \mathcal{G}_\alpha(\epsilon) = \{g : \|I^D - TI_\alpha(g)\|^2 = 0\}$, and $\bigcup_\alpha (\{g : I^D = I_\alpha(g)\}, \alpha) = M(g^\star, \alpha^\star)$. □

One consequence of this result is that the MLE converges to the equivalence-class associated with the true parameter $g_\alpha^\star, \alpha^\star$. Accordingly the limiting probability concentrates on the proper equivalence class and it follows that representation can be estimated consistently only if each pair (g, α) has multiplicity one; otherwise the results are accurate only up to the set of equivalent configurations.

13.6.2 Symmetry Groups and Sensor Symmetry

Object symmetries are a basic issue in the studies of patterns. Most man-made objects have well-defined symmetries. The standard notion of symmetry of a object α is studied through the subgroup $\mathcal{G}_{sym} \subset \mathcal{G}$ of the original group $\mathcal{G}$ the parameter space, with the property that the symmetry subgroup is the set such that the imagery looks identical through the transformation:

$$\mathcal{G}_{sym} = \{g \in \mathcal{G} : I_\alpha(g) = I\} \subset \mathcal{G}. \tag{13.66}$$

The inference set is then reduced to the quotient space $\mathcal{G}/\mathcal{G}_{sym}$. This corresponds to the symmetries in underlying object shapes (in R^3) but, in practice, observations are further degraded by the sensors which capture the object features only partially. These become statistical issues of identifiability, requiring the concepts of symmetry to be extended to the power and limitations of the sensor. The concept of symmetry is extended to include identifiability through the sensor as follows.

Definition 13.22 *The* **symmetry set relative to sensor mapping** *T for object α is defined to be*

$$\mathcal{G}_T(\alpha) = \{g : TI_\alpha(g) = TI_\alpha\}. \tag{13.67}$$

For multiple objects indexed over $\mathcal{A}$, the notion of a **symmetry set of object** *α_0 at group parameter g_0, $\mathcal{G}_T(g_0, \alpha_0) \subset \mathcal{G} \times \mathcal{A}$, defined as*

$$\mathcal{G}_T(g_0, \alpha_0) = \{(g, \alpha) : TI_\alpha(g) = TI_{\alpha_0}(g_0)\}. \tag{13.68}$$

This divides the set $\mathcal{G} \times \mathcal{A}$ into equivalence classes.

To illustrate this concept of sensor induced symmetry let us consider various examples assuming the video sensor with $\mathcal{G} = \mathbf{SO}(3)$. If $\alpha = $ missile of ideal cylindrical form with no other distinguishing features then $S[\text{missile}] = \mathbf{SO}(2)$. Viewed via a video sensor, $S_T[\text{missile}] = \mathbf{SO}(2)$, which is to be a subgroup of S leading to the inference space, $\mathbf{SO}(3)/\mathbf{SO}(2) = S^2$. If the missile also has four tail fins separated by the angle 90° the relative symmetry set instead includes the discrete 4-torus.

13.6 ASYMPTOTIC CONSISTENCY

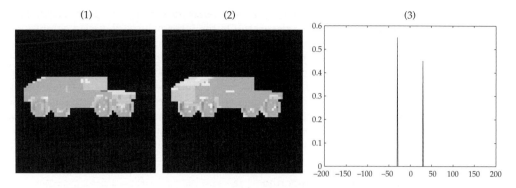

Figure 13.22 Panels 1, 2 show two different object orientations depicting near symmetry visualized through the perspective projection of video imagery $TI(g_1) \approx TI(g_2)$. Panel 3 shows the posterior density of the Gaussian projection model with a uniform prior on the orientation. Note how they are virtually identical.

Assuming object signature $TI_\alpha(g)$ of the object α at pose g viewed through a remote sensor with projective mapping T. Define the quantity

$$\Gamma(i;j) = \inf_{g_i, g_j} \|TI_{\alpha_i}(g_i) - TI_{\alpha_j}(g_j)\|^2, \tag{13.69}$$

quantifying the fundamental separation in objects α_1 and α_2 through the sensor map T. As we shall see in the subsequent sections, asymptotic in small noise the probability of misidentification of objects is exponential with decay rate proportional to $\Gamma(\alpha_i, \alpha_j)$ and inversely proportion to the variance.

This is illustrated by a figure from Srivastava [302] which depicted the posterior density for multiple views. Figure 13.22 depicts via panels 1,2 two orientations of a vehicle which is viewed through an orthographic projection of the video camera. In this low-noise situation, the approximate equivalence of object profiles is depicted in panels 1,2 at the two orientations $-30°$ and $30°$ ($0°$ being exact broadside) and results in the concentration of the posterior measure at these two points. Panel 3 shows that the posterior distribution is concentrated at these two near symmetry points.

13.7 Hypothesis Testing and Asymptotic Error-Exponents

Probability of detection, false alarm, and Bayes risk are the measures which can be calculated for detection/ID. In a Bayesian approach, object detection and identification are instances of discrete hypothesis testing, given $\mathcal{H}$ the family of hypotheses, perform least risk estimation. Define the risk functions $R : \mathcal{H} \times \mathcal{H} \to \mathbb{R}^+$. The detection/recognition performance is specified through the variations of error probabilities, such as false alarm, missed detection, and misclassification, as a function of system parameters. The risk is given by

$$\mathcal{R} = E_{p(H, I^D)} R(H, I^D). \tag{13.70}$$

In a Bayesian approach, object detection and recognition are performed via hypothesis-testing while the unknowns, such as object pose, location, thermal profile, etc., are treated as nuisance parameters. Recognition implies integrating out or estimating these nuisance parameters. In most practical cases, the complicated integrands make the analytical evaluations difficult. One possibility is to evaluate MLEs of the nuisance parameters and perform generalized likelihood-ratio tests, thus avoiding nuisance integrals. On the other hand, a Bayesian solution requires evaluation of the nuisance integral either exactly or through numerical approximation. For approximating, some commonly used techniques are (i) quadrature evaluation (e.g. Newton-Cotes' method), (ii) Markov chain Monte-Carlo (e.g. Metropolis-Hasting's method), and (iii) asymptotical analysis for approximations of the nuisance integral and use them to derive the probabilities of success/failure. Using asymptotics and Laplace's method [4, 61] we will approximate these integrals to study the limiting distributions of the likelihood-ratios and the test performance, similar to techniques used in [60] for approximating integrals to evaluate Bayes' methods.

For binary ID and detection in additive noise there are two object types associated with the hypotheses H_0, H_1:

$$\text{Detection}: \quad I^D = \begin{cases} TI_\alpha(g_1^*) + W & : H_1 \\ W & : H_0 \end{cases} \tag{13.71}$$

$$\text{Identification}: \quad I^D = \begin{cases} TI_1(g_1^*) + W & : H_1 \\ TI_0(g_0^*) + W & : H_0 \end{cases}. \tag{13.72}$$

with W additive Gaussian noise and g_0^*, g_1^* the true parameters for hypotheized objects type H_0, H_1. For an observation $I^D \in \mathcal{I}^D$, a Gaussian random field the Bayesian solution is given by the hypothesis testing problem

$$\frac{p(H_1|I^D)}{p(H_0|I^D)} \overset{H_1}{\underset{H_0}{\gtrless}} 1 \text{ or equivalently,} \quad L(I^D) = \log \frac{p(I^D|H_1)}{p(I^D|H_0)} \overset{H_1}{\underset{H_0}{\gtrless}} \log \frac{\pi(H_0)}{\pi(H_1)} = \nu. \tag{13.73}$$

The two error types to be characterized in the binary hypothesis problem are defined as follows.

Definition 13.23 *Define the* **false alarm rate** *to be given by the probability that H_1 is selected when H_0 is true; define the* **missed detection rate** *to be given by the probability that H_0 is selected when H_1 is true.*

Calculating the error bounds for $m > 2$ object types is extended straightforwardly. Let H_i be the hypothesis associated with the object type i, with data I^D a conditionally Gaussian random field with mean field $TI_i(g_i)$. The Bayes estimate is given by the hypothesis with maximum a-posterior probability,

$$\arg\max_{H_i} p(H_i|I^D) = \arg\max_{H_i} \log p(I^D|H_i)\pi(H_i). \tag{13.74}$$

13.7.1 Analytical Representations of the Error Probabilities and the Bayesian Information Criterion

The evaluation of these log-likelihoods ratios of Eqn. 13.73 involves integrating over the nuisance parameters space. In most practical cases, the complicated integrand makes analytical evaluation difficult. Assume identifiability through the sensor map assuming the size of the equivalence set to be $|M(g,\alpha)| = 1$. We now demonstrate the fundamental interplay of parameter estimation and identification and detection showing that the effective signal to noise ratio and the Hessian associated with the empirical Fisher information play the important role of determining the exponential error rates for probability of detection/ID. Detectability is determined by the effective signal-to-noise ratio of object projection, and identifiability is determined by the cross-correlation of these projections across different objects.

Consider the multihypothesis testing problem of determining model $m = 1, 2, \ldots$ from random observations I^D. This requires evaluation of the likehood ratio tests on the model given the observations. For the deformable template setting, the density involves the random nuisance parameters of pose $g(\theta) \in \mathcal{G}$, $\theta \in \mathbb{R}^d$ the local coordinates of the pose, with some prior density $\pi_m(g(\theta)), \theta \in \Theta \subseteq \mathbb{R}^d$ of dimension d, requiring the calculation of the so-called *nuisance integral* for the conditional density for every m:

$$\hat{m} = \arg\max_m \pi(m) p(I_1^D, \ldots, I_n^D | m) \tag{13.75}$$

$$= \arg\max_m \int_{\Theta_m} p(I^D | g(\theta), m) \pi_m(g(\theta)) \, dg(\theta). \tag{13.76}$$

We follow the the approach taken in Srivastava [325] in which the integrand is approximated via Laplace's method of Chapter 2, Theorem 2.60.

We shall assume the smoothness conditions 2.59 guaranteeing asymptotic consistency, essentially that there is only one maximizer of the likelihood. Then as $n \to \infty$ with

$$p(I_1^D, \ldots, I_n^D | m) \sim p\left(I_1^D, \ldots, I_n^D | \bar{\theta}, m\right) \left(\frac{2\pi}{n}\right)^{d_m/2} \det{}^{-1/2} F_m(n, \bar{\theta}), \tag{13.77}$$

where $\bar{\theta} = \arg\max_{\theta \in \Theta_m} p(I_1^D, \ldots, I_n^D | \theta, m)$ is the MLE with respect to $\theta \in \Theta_m$ and F_m is the Hessian empirical Fisher information.

Theorem 2.60 can be exploited for computing the nuisance integral involved in the calculation of the asymptotics on likelihood ratio testing. We shall be examining the conditional Gaussian model, $I^D = TI_\alpha(g) + W$, W white noise variance σ^2. To denote the asymptotic convergence, we make the variance a function of sample size n, so that as sample size $n \to \infty$, variance $\sigma^2(n) \to 0$.

First exploit a basic Lemma demonstrating the link of performance in hypothesis testing to the Hessian.

Lemma 13.24 *Given is the Gaussian random process $I^D = TI_\alpha(g) + W$, W white Gaussian noise variance $\sigma^2(n)$ with $T : I_\alpha(g) \mapsto TI_\alpha(g)$ satisfying the smoothness properties and identifiability properties 1,2,3 of Theorem 2.60. Denoting the $d \times d$ Fisher information matrix*

$$F_\alpha(g, \sigma^2(n)) = \left(\frac{1}{\sigma^2(n)} \left\langle \frac{\partial}{\partial \theta_i} TI_\alpha(g), \frac{\partial}{\partial \theta_j} TI_\alpha(g) \right\rangle\right), \tag{13.78}$$

and the true parameters under H_0, H_1 as $I_0(g_0^*), I_1(g_1^*)$, respectively, then as the noise variance $\sigma^2(n)$ goes to zero then the likelihood ratio converges in probability:

$$L(I^D) \sim -\frac{\|I^D - TI_1(g_1^*)\|^2}{2\sigma^2(n)} + \frac{\|I^D - TI_0(g_0^*)\|^2}{2\sigma^2(n)} + \log \frac{\det^{1/2} F_0(g_0^*, \sigma^2(n))}{\det^{1/2} F_1(g_1^*, \sigma^2(n))}. \tag{13.79}$$

Proof With $\sigma^2(n)$ going to 0, Theorem 2.60 implies asymptotic as $\sigma^2(n) \to 0$ the log-likelihood ratio goes as

$$L(I^D) = \log \frac{p(I^D|H_1)}{p(I^D|H_0)} \tag{13.80}$$

$$\sim -\frac{\|I^D - TI_1(\bar{g}_1)\|^2}{2\sigma^2(n)} + \frac{\|I^D - TI_0(\bar{g}_0)\|^2}{2\sigma^2(n)}$$

$$+ \frac{1}{2} \log \frac{\det\left((\partial^2/\partial\theta_i\partial\theta_j)\|I^D - I(\bar{g}_0)\|^2/2\sigma^2(n)\right)}{\det\left((\partial^2/\partial\theta_i\partial\theta_j)\|I^D - I(\bar{g}_1)\|^2/2\sigma^2(n)\right)}. \tag{13.81}$$

Theorem 2.60 gives with the maximum likelihood estimates $\bar{g}$ converging in probability to g^* then for $TI : \mathcal{G} \to \mathbb{R}$ a three times differentiable function that $TI_i(\bar{g}_i)$ converges in probability $TI_i(g_i^*)$, thus

$$\log \frac{\det\left((\partial^2/\partial\theta_i\partial\theta_j)\|I^D - I(\bar{g}_0)\|^2/2\sigma^2(n)\right)}{\det\left((\partial^2/\partial\theta_i\partial\theta_j)\|I^D - I(\bar{g}_1)\|^2/2\sigma^2(n)\right)} = \log \frac{\det F_0(g_0^*, \sigma^2(n))}{\det F_1(g_1^*, \sigma^2(n))} + O_p(\sigma(n)). \tag{13.82}$$

□

Theorem 13.25 (Exponential Error Exponents (Srivastava [325])) *Assume I^D is a conditional Gaussian random field mean field $TI_\alpha(g)$ with consistency $|M(g,\alpha)| = 1$ and identifiability and smoothness properties of Theorem 2.60. Under such conditions, and with the noise variance going to zero $\sigma^2(n) \to 0$, then the probability of misclassification selecting H_1 when H_0 is true assuming $\|TI_1(g_1^*) - TI_0(g_0^*)\| > 0$ is given by*

$$Pr\left\{L(I^D) > v|H_0\right\} \sim \sqrt{\frac{2}{\pi}} \frac{\sigma(n)}{\|TI_1(g_1^*) - TI_0(g_0^*)\|} e^{-((\|TI_1(g_1^*) - TI_0(g_0^*)\|^2)/8\sigma(n)^2)}. \tag{13.83}$$

Proof Assume H_0 so that $I^D = TI_0(g_0^*) + W$, then from the previous Lemma 13.24 and substituting $I^D = TI_0(g_0^*) + W$ gives

$$L(I^D) \sim -\frac{\|I^D - TI_1(g_1^*)\|^2}{2\sigma(n)^2} + \frac{\|I^D - TI_0(g_0^*)\|^2}{2\sigma(n)^2} + \frac{1}{2}\log\frac{\det F_0(g_0^*, \sigma(n)^2)}{\det F_1(g_1^*, \sigma(n)^2)}$$

$$\sim -\frac{\|TI_1(g_1^*) - TI_0(g_0^*)\|^2}{2\sigma(n)^2} + 2\frac{\langle W, TI_1(g_1^*) - TI_0(g_0^*)\rangle}{2\sigma(n)^2}$$

$$+ \frac{1}{2}\log\frac{\det F_0(g_0^*, \sigma(n)^2)}{\det F_1(g_1^*, \sigma(n)^2)} + O_p(\sigma(n)). \tag{13.84}$$

The probability of error goes as

$$Pr\left\{L(I^D) \geq v|H_0\right\} = Pr\left\{\left\langle \frac{W}{\sigma(n)^2}, TI_1(g_1^*) - TI_0(g_0^*)\right\rangle \geq v \right.$$

$$\left. + \frac{\|TI_1(g_1^*) - TI_0(g_0^*)\|^2}{2\sigma(n)^2} - \frac{1}{2}\log\frac{\det F_0(g_0^*, \sigma(n)^2)}{\det F_1(g_1^*, \sigma(n)^2)}\right\} + O(\sigma(n)). \tag{13.85}$$

13.7 HYPOTHESIS TESTING

Choose the test statistic to be

$$t(I^D) = \left\langle \frac{W}{\sigma(n)}, \frac{TI_1(g_1^*) - TI_0(g_0^*)}{\|TI_0(g_0^*) - TI_1(g_1^*)\|} \right\rangle, \qquad (13.86)$$

then $t(I^D)$ is standard-normal, mean 0 variance 1. Defining

$$\kappa = \frac{\sigma(n)}{\|TI_0(g_0^*) - TI_1(g_1^*)\|} \left(\nu - \frac{1}{2} \log \frac{\det F_0(g_0^*, \sigma(n)^2)}{\det F_1(g_1^*, \sigma(n)^2)} \right) + \frac{\|TI_1(g_1^*) - TI_0(g_0^*)\|}{2\sigma(n)}, \qquad (13.87)$$

then asymptotically as $\sigma(n)^2 \to 0$ the probability of selecting H_1 when H_0 is true reduces to the probability of a standard-normal random-variable being greater than the threshold value as $\sigma(n) \to 0$ then $\kappa \to (\|TI_1(g_1^*) - TI_0(g_0^*)\|)/2\sigma(n)$ which converges to ∞ giving

$$\Pr\{L(I^D) > \nu | H_0\} = \Pr\{t(I^D) > \kappa\} + O(\sigma(n)) \qquad (13.88)$$

$$= \frac{1}{\sqrt{2\pi}} \int_\kappa^\infty e^{-t^2/2} \, dt + O(\sigma(n)) \sim \frac{1}{\sqrt{2\pi}\kappa} e^{-\kappa^2/2}. \qquad (13.89)$$

□

Remark 13.7.1 (Multiple-Sensor Fusion) For the multi-sensor setting the Fisher information is strengthened as well as the exponential error bound. Defining the multiple observations as

$$I^{D_1} = T_1 I_0(g_0) + W_1, \qquad I^{D_2} = T_2 I_0(g_0) + W_2, \ldots, \qquad (13.90)$$

where $W_1, W_2, \ldots$ are independent and Gaussian with variances $\sigma(n)_1^2, \sigma(n)_2^2, \ldots$. For the asymptotic situation $\min \sigma(n)_d \to 0$, Theorem 13.25 applies as well and the error exponent in Eqn. 13.83 becomes $\sum_d (\|I^{D_d} - T_d I_0(\tilde{g}_0)\|^2 / 2\sigma(n)_d^2)$.

Remark 13.7.2 When $\|TI_0(g_0) - TI_1(g_1)\| = 0$ for $g_0 \neq g_1$ then $M(g_0, \alpha_0) > 1$, and the estimator does not distinguish between the two objects. Hence, the assumption that $\|TI_0(g_0) - TI_1(g_1)\| \neq 0$.

Assuming that the ratio of priors is one, the error rate (of type I error probability) is completely determined by the factor κ (Eqn. 13.87) which taking the general form for positive constants a, b,

$$\kappa = \frac{a}{\sigma(n)} \|TI_1(g_1) - TI_0(g_0)\|^2 - b \sigma(n) \log \frac{\det(\mathrm{CRB}_1)}{\det(\mathrm{CRB}_0)}. \qquad (13.91)$$

The first term quantifies the separation between the true object and the closest occurrence of the incorrect hypothesis. CRB_0 and CRB_1 are the lower-bounds for estimating g under the two hypotheses, respectively. This provides a fundamental connection between the accuracy in nuisance parameter estimation and the associated hypothesis-selection. Depending upon the value of θ, these two terms influence the probability of type I error.

According to Theorem 13.25, the error probability is dependent upon the objects and their images through the correlation coefficients and the projective signal energies. A similar result is derived in [326], where it is concluded that for object-recognition the asymptotic error probability depends on a parameter which characterizes the separation between the most similar but incorrect object and the true object.

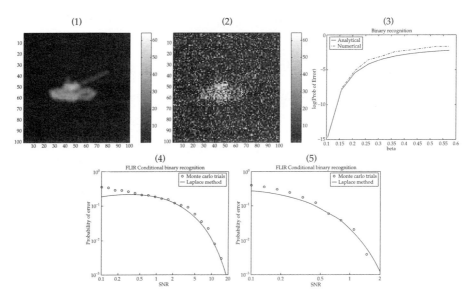

Figure 13.23 Top row: Panels 1,2 show VIDEO images of a tank at noise levels $\sigma/\|TI_\alpha(g)\| = 0.01, 0.1$; panel 3 shows curves denoting the log-probability of misidentifying the tank as a truck for a fixed object pose. Solid curves show asymptotic analytical estimates (solid line); dashed curves show the likelihood ratio test calculation computed via numerical integration. Bottom row: Panels show probability of false alarm comparing Monte-Carlo simulation and asymptotic approximation from Shapiro and Yen [327]. Panel 4 shows FLIR identification between two targets comparing Monte-Carlo (circles) and asymptotic approximation (solid). Panel 5 shows similar results for M-ary recognition.

Example 13.26 (Likelihood Ratio for Tank/Truck Discrimination) Consider a TANK versus TRUCK recognition through the video imager. Shown in the top row panels 1,2 of Figure 13.23 are sample VIDEO images of the tank at two noise levels corresponding to $(\sigma/\|TI_\alpha(g)\|) = 0.01, 0.1$; shown in panel 3 are the plots for the log-probability of misidentifying the TANK video image as the TRUCK. The numerical calculation is based on trapezoidal integration on $\mathbf{SO}(2)$ and random sampling on $\mathcal{I}^\mathcal{D}$. The dashed line plots the result of the numerical integration.

Srivastava has looked at the asymptotic expression for error rate for ID of models from VIDEO imagery. The solid line of Figure 13.23 (panel 3) shows plots of the analytical asymptotic expression for the probability of misidentification. Note how for the entire range of noisy tank images for SNR from $0.01 - 0.2$ the analytical error exponent expression fits the numerical integration of the probability of misidentification error.

13.7.2 m-ary Multiple Hypotheses

Calculating the error bounds for $m > 2$ object types is extended straightforwardly. Let H_i be the hypothesis associated with the object type i, with data I^D a conditionally Gaussian random field with mean field $TI_i(g_i)$. The Bayes estimate is given by the hypothesis with maximum a-posteriori probability,

$$\arg\max_{H_i} p(H_i|I^D) = \arg\max_{H_i} \log p(I^D|H_i)\pi(H_i). \tag{13.92}$$

The probability of error, conditioned on H_0 the true hypothesis, is given by $Pr\{\hat{i} \neq 0|H_0\}$. Defining

$$L_i(I^D) = \log \frac{p(I^D|H_i)}{p(I^D|H_0)} \quad \text{and} \quad \log \frac{\pi(H_0)}{\pi(H_i)} = v_i, \tag{13.93}$$

then the probability of error becomes

$$Pr\{\hat{i} \neq 0|H_0\} = 1 - Pr(\bigcap_{i \neq 0}\{I^D : L_i(I^D) < v_i\}). \tag{13.94}$$

Following Srivastava, the error exponents for exponential probability of decay for the m-ary case are calculated as follows.

Theorem 13.27 *(Srivastava) Given I^D a Gaussian random field with mean $TI_0(g_0^*)$ of object H_0, and define the test statistics and correlations according to*

$$t_i(I^D) = \frac{\langle W/\sigma(n), TI_i(g_i^*) - TI_0(g_0^*)\rangle}{\|TI_i(g_i^*) - TI_0(g_0^*)\|}, \tag{13.95}$$

$$K = \left(K_{ij} = \frac{\langle TI_i(g_i^*) - TI_0(g_0^*), TI_j(g_j^*) - TI_0(g_0^*)\rangle}{\|TI_i(g_i^*) - TI_0(g_0^*)\| \|TI_j(g_j^*) - TI_0(g_0^*)\|}\right). \tag{13.96}$$

Then with the $m \times m$ covariance $K = K_{ij}$ with thresholds given by

$$\kappa_i = \frac{\sigma(n)}{\|TI_i(g_i^*) - TI_0(g_0^*)\|}\left(\log(v_i) - \frac{1}{2}\log\frac{\det F_0(g_0^*, \sigma(n)^2)}{\det F_i(g_i^*, \sigma(n)^2)}\right)$$
$$+ \frac{\|TI_i(g_i^*) - TI_0(g_0^*)\|}{2\sigma(n)}, \tag{13.97}$$

the probability of correct classification as $\sigma(n) \to 0$ is given by

$$\frac{1}{(2\pi)^{m/2} \det^{-1/2} K} \int_{t_1 < \kappa_1, \dots, t_m < \kappa_m} e^{-1/2 \sum_{ij=1}^{m} t_i t_j (K^{-1})_{ij}} dt_1, \dots, dt_m. \tag{13.98}$$

Proof The probability of correct selection in each binary test is evaluated through the set $\{I^D : L_i(I^D) < v_i\}$ can be reduced to evaluating $Pr\{t_i(I^D) < \kappa_i\}$ where the test statistic $t_i(I^D)$ of Eqn. 13.95 is standard normal with thresholds κ_i as given by Eqn. 13.97. Then t_i are a zero-mean Gaussian random vector with the $m \times m$ element covariance matrix K. □

Example 13.28 (Asymptotic Error Exponents: FLIR, LADAR) Shown in the bottom row of Figure 13.23 are similar results from Shapiro and Yen [327] for the FLIR and LADAR sensors. Panel 4 shows the false alarm rates for two different orientations of the tank in noise. The dashed line plots the analytical expression for κ. The solid line plots the numerical integration of the posterior false alarm error probability of Eqn. 13.73. Panels 4 and 5 choose two different orientations of the object.

14 ESTIMATION ON METRIC SPACES WITH PHOTOMETRIC VARIATION

ABSTRACT Model uncertainty comes in many forms. In this chapter we shall examine extensively the variable photometric model in which the underlying image field $I(\cdot)$ is modelled as an element of a Hilbert space $I(\cdot) \in \mathcal{H}(\phi)$ constructed via basis expansion $\{\phi_i\}$. Inference involves both the pose and identity of the objects as well as the photometric intensity itself. This corresponds to making the template random, expanding the deformable template to include both the photometric variations and geometric variations.

14.1 The Deformable Template: Orbits of Signature and Geometric Variation

Now examine models of variation which explicitly show metric spaces on photometric variation with geometry variation. The basic model of Figure 14.1 represents the source of imagery $\mathcal{I}$ as the orbit of the product of all photometric variations with geometric variations. The space of photometric intensities are modeled as a Hilbert space $\mathcal{H}$. The deformable template becomes the orbit of all photometric intensities under geometric motions, $\mathcal{A} = \mathcal{H} \times \mathcal{G}$, with elements the pairs $(I,g) \in \mathcal{A} = \mathcal{H} \times \mathcal{G}$. Notice, in this setting there are no fixed exemplars. The resulting images which form the mean fields become $I(g) = I \circ g \in \mathcal{I}$. The full model on the source requires the construction of the prior model on elements $(I,g), I \in \mathcal{H}, g \in \mathcal{G}$. For this we explicitly model the metric on images via empirical methods for estimating the statistical variation of the imagery. We examine photometric intensities generated in the Hilbert space $\mathcal{H}(\phi)$ spanned via the basis ϕ, and the special Euclidean group $g \in \mathbf{SE}(d)$. The noise model associated with the estimation problems takes the form of the observation I^D as a conditionally Gaussian random field (conditioned on (I,g)) with mean field $I(g), I \in \mathcal{H}(\phi), g \in \mathcal{G}$, with additive white noise W.

14.1.1 The Robust Deformable Templates

We now examine the construction of *robust deformable templates*, which allow the accomodation of both variations of the pose of objects and photometric variations, associated with occlusion, variability of lighting, and other natural effects. Throughout we shall examine the special-Euclidean group $\mathcal{G} = \mathbf{SE}(d) = \mathbf{SO}(d) \otimes \mathbb{R}^d$ as the basic geometric variablility. In the robust deformable template, variability in observed images is not solely associated with geometric transformations

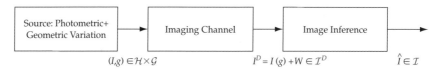

Figure 14.1 The source of images: an orbit under groups of transformations. The source are images and group elements $(I, g), I \in \mathcal{H}, g \in \mathcal{G}$, $\mathcal{H}$ the Hilbert space of photometric intensities, and $\mathcal{G}$ the group of geometric transformations. The observations are $I^D \in \mathcal{I}^D$.

14.1 THE DEFORMABLE TEMPLATE

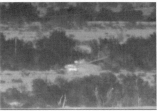

Figure 14.2 Shows photometric variability of FLIR imagery.

defining the pose of targets, but as well other factors such as lighting variations, object surface properties and texture variations corresponding to the presence of clutter.

The variability of target type and pose is accommodated using the group of rigid motions representing geometric variations acting on the background space X according to $g \in \mathcal{G} : x \in X \mapsto g(x) \in X$, with the group action on the image space $\mathcal{I}$ defined by

$$g \in \mathcal{G} : I \in \mathcal{I} \to I \circ g^{-1} \in \mathcal{I}. \qquad (14.1)$$

Photometric variability is accomodated by extending the rigid templates to the robust deformable templates, which explicitly include photometric variations. The challenge is illustrated in Figure 14.2. In natural scenes it is impractical to explicitly represent every object and all of the photometric variation in the scene by a fixed deterministic 3D model. Unquestionably, the main challange to rigid model-based methods in natural imagery is to acquire pose and identification of targets while, at the same time, "seeing through" the confounding nuisance variables of photometric variation. The panels show electro-optic imagery and FLIR imagery of natural and cluttered scenes depicting the variability of object appearance 14.2.

The group of photometric variations is modeled as a Hilbert space $\mathcal{H} \subseteq L^2$. Since not all of the elements of L^2 correspond to observable images, the space of photometric variations $\mathcal{H}$, constructed via basis expansion $\{\phi_n\}$, is bounded to be a very small subset of L^2. Its action on the space of images $\mathcal{I}$ is additive

$$I' = \sum_n I_n \phi_n \in \mathcal{H} : I \in \mathcal{I} \to I + \sum_n I_n \phi_n \in \mathcal{I}. \qquad (14.2)$$

14.1.2 The Metric Space of the Robust Deformable Template

To make the robust deformable template into a metric space, define the metric distance on elements of $\mathcal{I}$ taking into account the action of $\mathcal{G}$ and the change in photometric value.

Definition 14.1 *The group $\mathcal{A}$ of* **geometric and photometric variation** *is a product of groups with elements $\{(g, I) : g \in G, I \in H\}$ with the law of composition*

$$(g', I') \circ (g, I) = (g' \circ g, I + I' \circ g). \qquad (14.3)$$

The **robust deformable template** $\mathcal{A}.I_{\text{temp}}$ *is the orbit under the action of $\mathcal{A}$:*

$$\mathcal{A}.I_{\text{temp}} = \left\{ I \in \mathcal{I} : I = \left(I_{\text{temp}} + \sum_n I_n \phi_n \right) \circ g^{-1} \right\} \text{ with}$$

$$(g, I).I_{\text{temp}} = (I_{\text{temp}} + I) \circ g^{-1}. \qquad (14.4)$$

Defining the energy of a path in $\mathcal{A}$ associated with the family of functional norms $\mathcal{N}_G, \mathcal{N}_I$ on the tangent elements $(\partial g_t/\partial t \circ g_t^{-1}, \partial I_t/\partial t \circ g_t^{-1})$ to be

$$E(\frac{\partial g_\cdot}{\partial t}, I_\cdot) = \int_0^1 \mathcal{N}_G \left(\frac{\partial g_t}{\partial t} \circ g_t^{-1} \right)^2 dt + \int_0^1 \mathcal{N}_I \left(\frac{\partial I_t}{\partial t} \circ g_t^{-1} \right)^2 dt, \qquad (14.5)$$

then the **robust deformable template becomes a metric space** *with distance defined through the infimum over all paths enerrgies $g_t, t \in [0,1]$, connecting elements in $\mathcal{A}$.*

Notice, with the law of composition 14.3, then 14.4 is a group action. In equation 14.5, there are two terms in the metric, which takes I to I'. The first integral penalizes transformations in the background space associated with the group $\mathcal{G}$, and the second one penalized photometric deformations. For the group of rigid motions, $\mathcal{G} = \mathbf{SE}(d) = SO(d) \otimes \mathbb{R}^d$, it is natural to set the energy term due to the rigid motions of the objects, to zero.

14.2 Empirical Covariance of Photometric Variability via Principle Components

We now explore the construction of the metrics corresponding to the functional norm $\mathcal{N}_I$ on photometric variation. We do this by designing the norm to match the empirical statistics of realizations of the patterns being studied. We shall be interested in generating the eigenfunction on submanifolds $X \subset \mathbb{R}^3$ corresponding to surfaces as well as planes and volumes. As long as we have a well defined inner product over the manifold (Riemannian manifold) then the methods extend from the regular intervals, arrays, and cubes to smooth surface submanifolds. Shown in Figure 14.3 are examples of several triangulated graphs $X = S(\triangle) \subset \mathbb{R}^3$ which represent background spaces upon which principal components will be performed.

We begin with the standard definition of the photometric space as a Hilbert space with the metric defined through a positive definite quadratic form.

Definition 14.2 *Let the images I be defined on the closed and bounded background space $X \subset \mathbb{R}^d$. Then the* **Hilbert space of photometric intensities** *$\mathcal{H}(\phi)$ has norm $\|f\|_\mathcal{H}$ defined by a quadratic form Q, with elements constructed through linear combinations of the orthonormal basis ϕ:*

$$\mathcal{H}(\phi) = \{I : I(\cdot) = \sum_n I_n \phi_n(\cdot) \in \mathcal{H}(\phi), \|I\|_\mathcal{H} < \infty\}, \qquad (14.6)$$

where $\|I\|_\mathcal{H}^2 = \sum_n q_n |\langle I, \phi_n \rangle_2|^2.$ $\qquad (14.7)$

Figure 14.3 Examples of faceted models used for studying photometric signature variation; shown depicted are triangulated graphs $X = S(\triangle) \subset \mathbb{R}^3$ which represent the surfaces on which we will want to understand photometric variation along with regular subvolumes.

Identifying images with their l^2 representation $I \in \mathcal{H}(\phi) \leftrightarrow (I_1, I_2, \dots) \in l^2$ the norm reduces to

$$\|I\|_{\mathcal{H}}^2 = \sum_n q_n |I_n|^2, \qquad I_n = \langle I, \phi_n \rangle_2. \tag{14.8}$$

14.2.1 Signatures as a Gaussian Random Field Constructed from Principle Components

Many approaches in pattern representation use principal components and clustering via empirical covariance estimation. The basic idea is to model the space of imagery $I \in \mathcal{H}(\phi)$ as realizations of a Gaussian random field with covariance constructed to fit the empirical variation. The complete orthonormal basis for the Hilbert space representation is generated from the orthonormal eigenfunctions of the empirical covariance; the metric is defined by the eigenvalues of the covariance.

As we cross back and forth between deterministic and probabilistic interpretation it is natural to define the quadratic form through a covariance operator which expresses the statistical occurrence of events. In that case, the operator representing the quadratic form above Q with eigenvalues q_n will correspond to the inverse of the covariance operator K with eigenvalues $\lambda_n = 1/q_n$.

Theorem 14.3 *Let $I(\cdot)$ on X be a zero mean real-valued Gaussian random field in the Hilbert space $\mathcal{H}(X)$ with covariance $K(x, y)$ on closed and bounded background space X. The set of orthonormal functions $\{\phi_n(\cdot), n = 1, \dots, N\}$ minimizing the mean squared error*

$$\arg\min_{\phi_n \in \mathcal{H}} E \left\{ \int_X \left| I(x) - \sum_{n=1}^N I_n \phi_n(x) \right|^2 dx \right\}, \tag{14.9}$$

where $I_n = \int_X I(x) \phi_n(x) \, dx$, satisfy the integral equation

$$\lambda_n \phi_n(x) = \int_X K(x, y) \phi_n(y) \, dy \quad \text{where } \lambda_1 \geq \lambda_2 \geq \dots \lambda_N \geq \lambda_{N+1} \geq \dots . \tag{14.10}$$

Proof The proof follows the standard proof for Karhunen–Loeve expansion [328]. As K is a self adjoint positive operator on X there a exists a set of complete orthonormal functions $\{\psi_n(x), x \in X, n = 1, 2, \dots\}$ such that

$$K(x, y) = \sum_{n=1}^{\infty} \lambda_n \psi_n(x) \psi_n(y), \tag{14.11}$$

with $\lambda_n \geq \lambda_{n+1}, \lambda_n > 0$. Now use the completeness of $\{\psi_n\}$ and orthonormality to express each eigenfunction according to

$$\phi_n(x) = \sum_{j=1}^{\infty} \alpha_j^{(n)} \psi_j(x), \quad \sum_{j=1}^{\infty} |\alpha_j^{(n)}|^2 = 1, \quad n = 1, \dots, N. \tag{14.12}$$

Using this representation, rewrite the minimization as

$$\arg\min_{\phi_n \in \mathcal{H}} E\{\int_X |I(x) - \sum_{n=1}^N I_n \phi_n(x)|^2 \, dx\} \quad (14.13)$$

$$= \arg\max_{\phi_n \in \mathcal{H}} \sum_{i=1}^N \int_X \int_X \phi_i(x) \sum_{n=1}^\infty \lambda_i \psi_n(x) \psi_n(y) \phi_i(y) \, dx \, dy, \quad (14.14)$$

$$= \arg\max_{\alpha_j^{(n)}} \sum_{j=1}^\infty \sum_{n=1}^N \lambda_j (\alpha_j^{(n)})^2 \text{ subject to } \sum_{j=1}^\infty |\alpha_j^{(n)}|^2 = 1, \quad n = 1, \ldots, N.$$

$$(14.15)$$

This is maximized if $\alpha_j^{(n)} = \delta(n-j)$ giving the result $\{\phi_n(\cdot) = \psi_n(\cdot), n = 1, \ldots, N\}$. □

14.2.2 Algorithm for Empirical Construction of Bases

Now examine empirical methods for generating the eigenfunctions. The eigenfunctions are defined on the surface background space $X = S(\Delta)$. Given a database of observed image signatures, $I^{(1)}, I^{(2)}, \ldots$, then the empirical covariance K is an $L \times L$ matrix with elements $K(x_l, y_l), x_l, y_l \in X$. The eigen basis is generated directly from the N-element database of the signatures via Singular Value Decomposition (SVD) as follows.

Algorithm 14.4 *The scalar field is expanded according to $I(\cdot) = \bar{I}(\cdot) + \sum_{n=1}^N I_n \phi_n(\cdot)$, where $I_n = \sum_l (I(x_l) - \bar{I}(x_l)) \phi_n(x_l) \gamma(x_l)$ with discrete pixel size $\gamma(x_l)$. The ϕ_n basis are calculated according to the following:*

1. Discretize continuum $\lambda_n \phi_n(x) = \int_X K(x,y) \phi_n(y) \gamma(dy)$ into the matrix equation:

$$\lambda_n \phi_n(x_k) = \sum_l K(x_k, y_l) \phi_n(y_l) \gamma(y_l). \quad (14.16)$$

2. Defining the sample mean $\bar{I}(x_l) = 1/J \sum_{j=1}^J I^{(j)}(x_l)$, construct the covariance factoring it into its square root $K = QQ^*$ including the measure given by the diagonal matrix $\Gamma = \text{diag}[\gamma(y_1), \ldots, \gamma(y_L)]$:

$$\lambda_n \sqrt{\Gamma} \phi_n = \sqrt{\Gamma} QQ^* \sqrt{\Gamma} \sqrt{\Gamma} \phi_n, \quad \text{where } Q_{lj} = \frac{1}{\sqrt{J}} (I^{(j)}(x_l) - \bar{I}(x_l)), l = 1, \ldots.$$

$$(14.17)$$

3. Via SVD factor $\sqrt{\Gamma} Q$ into orthogonal matrix U, V and diagonal Λ so $(\sqrt{\Gamma} Q) = U \sqrt{\Lambda} V^*$. The columns of U are eigenvectors of $(\sqrt{\Gamma} Q)(\sqrt{\Gamma} Q)^*$; the eigenvalues are the elements of the diagonal matrix Λ:

$$\phi_n = \sqrt{\Gamma}^{-1} U_n, \quad U = [U_1 \cdots U_L], \lambda_n = \Lambda_{nn}, \quad n = 1, \ldots, N. \quad (14.18)$$

Example 14.5 (Eigenfunctions for Illumination Variations in the Image Plane)
The modeling of objects signatures approached from the point of view of empirical statistics has been done by many groups. Examine the work in the image plane on human faces of Belhumeur, Kriegman, Mumford, Yuille and coworkers [329–331]. Here the images are defined on the unit-square background space $X = [0,1]^2 \subset \mathbb{R}^2$

14.2 EMPIRICAL COVARIANCE OF PHOTOMETRIC VARIABILITY

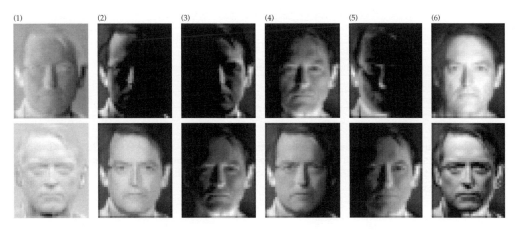

Figure 14.4 Column 1: Panels show eigen-images 2 and 3. Columns 2–6 show images from the illumination variability $\mathcal{I} = \sum_n I_n \phi_n(\cdot)$ corresponding to one (top row) and two (bottom row) light sources from [329].

Shown in Figure 14.4 are results from eigencorrelation experiments of Belhumeur and Kriegman. Column 1 shows eigenfunctions 2 and 3 of the empirical autocorrelation matrix. Shown in columns 2,3,4,5, and 6 are random samples from the span of the eigenfunctions. Each row shows sample images from the illumination variability space corresponding to one (top row) and two (bottom row) light sources. Figures depict elements of the orbit $\mathcal{I} = \sum_n I_n \phi_n(\cdot)$.

Example 14.6 (Eigenfunctions on 2D Surface Manifolds) Examine empirical eigenfunction models associated with 2D surface submanifolds of $\mathbb{R}^3$ as studied by Cooper, Lanterman and Joshi [332–335]. Surface objects are represented as triangulated graphs on the background space $X = S(\triangle) = \{\triangle(x_l), l = 1, \ldots, L\}$, with each triangle having surface measures $\gamma(x_l)$ filling the diagonal matrix Γ. Shown in Figure 14.3 are examples of such triangulated graphs upon which the eignfunctions are constructed. Model the photometric intensity I as a scalar-valued Gaussian random field with mean corresponding to the template at fixed coordinates under the identity transformation. For generating the data base, OpenGL or ray tracing can be used to render 3D objects under variable lighting lumination conditions and material reflective properties. The photometric variations are due to variability in the direction, color, and intensity of light, as well as variability in the ambient and diffuse reflective properties of the object material. Databases can be generated by varying the direction of the illumination and diffuse material reflectivity. A database described in Figure 14.5 for a teacup and a teapot were generated with 2592 lighting directions and diffuse reflectivity 1 for the teacup and 3 for the teapot.

Shown in the top row of Figure 14.6 are the principal components of the teacup data base. The bottom row shows the first eigen signatures for the teapot. Shown in the rightmost panel of Figure 14.6 is the efficiency of the representation as depicted by the eigenvalues of the empirical covariances. The circles (○) show the eigenvalues of the static database, the dashed lines (−−) show the dynamic data base, and the solid lines (−) for the composite data base.

Now examine empirical eigenfunction models 2D submanifolds of $\mathbb{R}^3$ for FLIR imagery [332–335]. On surfaces, the objects are represented as triangulated graphs $S(\triangle) = \triangle(x_l), l = 1, \ldots, L$ the number of triangle, with each triangle having surface measure $\gamma(l)$. The eigenfunctions are indexed over the triangulated graph representing the CAD model; the background space becomes $S(\triangle)$. Shown in Figure 14.3 are

Database Summaries - Video				
Database	Light Directions	Diffuse Reflectivities	Object Vertices
Teacup	2592	1	3749
Teapot	2592	3	1480

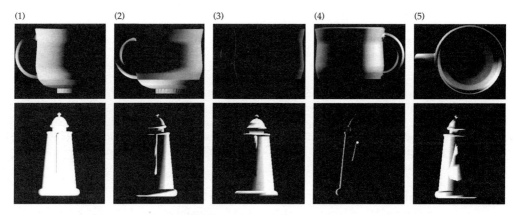

Figure 14.5 Summaries of the contents of the empirical databases used for the KL expansion of the signature random field in the video context. The number of signatures used in the PCA is 2592 for the teacup, and 7776 for the teapot. Right panel shows the normalized power spectrum of the eigenbasis.

Figure 14.6 Top row shows visualizations of the first five eigenfunctions of the teacup database. Bottom row panels show the first five eigenfunctions of the teapot database.

examples of CAD models which have been represented as triangulated graphs showing faceted geometries of the CAD models. For infrared sensors operating in the 8–12 micron band, object material properties, object operational states, and meteorological conditions contribute to photometric variability. Model the template signature I as a scalar-valued Gaussian random field of radiant intensity with mean corresponding to the template at fixed coordinates under the identity transformation. For FLIR, the PRISM [299] simulation package is used to build the databases isolating the primary modes of meteorological and operational photometric variation. The meteorological in weather and atmospheric conditions were isolated in the "static" database. Operational variation resulting from changes in the operational state of the object including land speed, motor speed, and gun state are isolated in the "dynamic" database. The static database was generated from 3, 12-h simulation sessions of weather changes resulting in 343 radiance profiles. Cooper generates the dynamic database from 18-h of simulation sessions consisting of dynamic data resulting in 186 radiance profiles. The composite database was generated using 72, 12-h simulations consisting of 7392 radiance profiles (Figure 14.7).

Figure 14.8 shows the first several eigensignatures $\phi_{1,2,3,4}(\cdot)$. Panels 1 and 2 show the static data base signatures. The first eigen signature of the static database demonstrates uniform heating of the tank as the sun moves across the sky during the twelve hour simulations. Panels 3 and 4 show the eigen signatures for the dynamic

14.2 EMPIRICAL COVARIANCE OF PHOTOMETRIC VARIABILITY

Database Summaries				
Database	Simulations	Obs. Interval	Sets of Weather Data	Radiance Profiles
Static	3	12 h/Each	3	343
Dynamic	6	3 h/Each	1	186
Composite	72	10–12 h/Each	3	7392

Figure 14.7 Summaries of the contents of the empirical databases used. The static database isolates meteorological variability while the dynamic database isolates operational variability. The composite database is composed of the two modes of variability. The Right panel shows the normalized power spectrum of the eigenbasis.

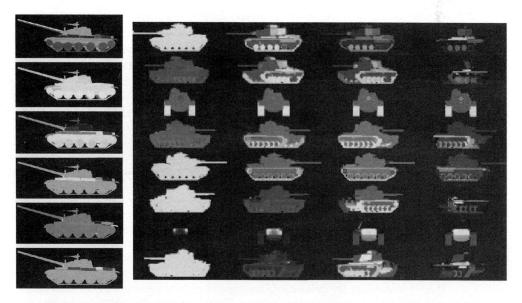

Figure 14.8 Top row shows visualizations of the three eigenfunctions of the static database of the tank for FLIR photometric variation. Left column: Top 3 panels show first 3 static database eigenfunctions; bottom three panels show the dynamic database eigenfunctions. Taken from Cooper [332–334]. Right column: Figure shows *Eigentanks* from Lanterman. The rows show varying orientation; the columns show increasing detail as generated by increasing numbers of eigenfunctions. Shown are estimated eigensignatures for an M60 at different hypothesized orientations and with different numbers of eigentanks. From left to right, columns display signatures with the first 1, 3, 9, and 50 eigentanks used in the expansion (see also Plate 25).

database. Notice the localized variations in the regions of the tank containing and adjacent to the motor.

Shown in the right panel of Figure 14.8 are eight poses (rows) of a model tank constructed by synthesizing the tank via different numbers of principal components (columns). This illustrates that the EigenSurface representations are 2D manifolds associated with the entire surface of the object.

14.3 Estimation of Parameters on the Conditionally Gaussian Random Field Models

Clearly, in this setting the addition of the photometric randomness associated with the parameterization of $I = \sum_i I_i \phi \in \mathcal{H}(\phi)$ increases the difficulty and dimension of estimating the geometric parameters $g \in \mathcal{G}$. Our approach in subsequent sections will be to integrate them out as nuisance variates as well as simultaneously estimate them via MAP and MMSE estimation. The basic technique will be to expand the posterior distribution around the MAP/MMSE estimators.

The orbit is the set of all imagery generated from the template and a linear combination of basis elements spanning the photometric intensities. The imagers are mappings $T: I \in \mathcal{I}(\phi) \mapsto TI \in \mathcal{I}^\mathcal{D}$, then $TI(g)$ is the mean-field conditioned on the geometric transformation $g \in \mathcal{G}$ and signature I. It is natural to model the data I^D as a conditionally Gaussian random field with mean field TI in additive noise, $I^D = TI + W$. Then we can speak of the conditional mean which is the MMSE, or the MAP estimator. Corresponding to the norm square $\| \cdot \|_2^2$ is the additive white noise model, $I^D = TI(g) + W$, with W white noise variance 1. When working on the continuum, then the process I^D will not be of trace class and strictly speaking $\|I^D\|^2$ will not be well-defined in the L^2 norm. This is properly handled by constraining the noise process to have a covariance with the spectrum which has a finite integral. In this case the matching norm-square cost changes to $\| \cdot \|_\mathcal{H}^2$ reflecting the well-defined true covariance. Alternatively, for all of the applications, the noise process is associated with a finite number of detectors (voxels in the image), thereby constraining the image to be well defined. Thus, although the L^2 matching term in the continuum has no rigorous probabilistic interpretation we will continue to use it because it is so convenient and connects us to so many classic formulations in least-squares estimation. We, of course, keep in mind the modeling issues between the two alternatives to make it more precise.

Throughout, the signature and pose are assumed conditionally independent. Take the prior on signature to be a conditional Gaussian random field conditioned on $g \in \mathcal{G}$ with Gaussian densities on the d-dimensional cylinder expansions on $I_1, \ldots, I_d$. The object photometric intensities are modeled as a zero-mean Gaussian random field with eigenfunctions ϕ_i, so that $I_i, i = 1, \ldots, n$ are independent normal with zero mean and variance the eigenvalue of the ith eigensignature ϕ_i, $K\phi_i = \lambda_i \phi_i$.

Clearly, expanding the deformable template to include signature parameters increases the invariance and decreases the discriminability. There is a cost. This is the same cost as for the nuisance parameters in hypothesis testing studied in the asymptotic setting via Theorem 2.60 in Chapter 2. Interestingly, for the Gaussian projective imagery setting, we can calculate an exact version of the asymptotic Theorem 2.60 calculating the complexity (or cost) of introducing the nuisance variables of photometric variation in order to obtain robustness. This requires the projective Gaussian image model, with I^D a conditionally Gaussian field with the mean field $TI(g)$ twice differentiable (Fisher information well defined).

Theorem 14.7 *Given data $I^D = TI(g) + W$ a conditionally Gaussian field with mean field $TI(g)$ twice differentiable, W white noise variance σ^2, and $I(g) = \sum_{i=1}^d I_i \phi_i(g)$ a Gaussian field, with I_i independent Gaussian, mean zero and variance $\lambda_i, i = 1, \ldots, d$.*

Defining the Fisher information (including the prior) and the MAP estimator according to

$$F(g) = \left(\frac{1}{\sigma^2} \langle T\phi_i(g), T\phi_j(g) \rangle + \frac{1}{\lambda_i} \delta(i-j) \right) \qquad (14.19)$$

14.3 ESTIMATION OF PARAMETERS

$$\begin{pmatrix} \bar{I}_1 \\ \vdots \\ \bar{I}_d \end{pmatrix} = \underset{(I_1,\ldots,I_d)\in\mathbb{R}^d}{\arg\max} \; \log p(I_1,\ldots,I_d|I^D,g) = F(g)^{-1} \begin{pmatrix} \frac{\langle I^D, T\phi_1(g)\rangle}{\sigma^2} \\ \vdots \\ \frac{\langle I^D, T\phi_d(g)\rangle}{\sigma^2} \end{pmatrix}, \quad (14.20)$$

then

$$p(I^D|g) = p(I^D, \bar{I}_1, \ldots, \bar{I}_d | g) \det^{-1/2} 2\pi F(g). \quad (14.21)$$

Proof Defining

$$H(I_1,\ldots,I_d;g) = \frac{\|I^D - TI(g)\|^2}{2\sigma^2} + \sum_{i=1}^d \frac{|I_i|^2}{2\lambda_i}, \quad (14.22)$$

then rewriting the conditional density gives

$$p(I^D|g) = \int_{\mathbb{R}^d} p(I^D, I_1, \ldots, I_d | g) dI_1 \ldots dI_d \quad (14.23)$$

$$= \int_{\mathbb{R}^d} p(I^D|I_1,\ldots,I_d, g) \pi(I_1,\ldots,I_d|g) dI_1 \ldots dI_d \quad (14.24)$$

$$= \int_{\mathbb{R}^d} \frac{1}{(2\pi\sigma^2)^{N/2}} e^{-(\|I^D - TI(g)\|^2)/(2\sigma^2)} \frac{1}{(2\pi)^{d/2} \prod_{i=1}^d \sqrt{\lambda_i}}$$

$$\times e^{-(\sum_{i=1}^d |I_i|^2)/(2\lambda_i)} dI_1 \ldots dI_d \quad (14.25)$$

$$= \int_{\mathbb{R}^d} \frac{1}{(2\pi\sigma^2)^{N/2}} \frac{1}{(2\pi)^{d/2} \prod_{i=1}^d \sqrt{\lambda_i}} e^{-H(I_1,\ldots,I_d;g)} dI_1 \ldots dI_d. \quad (14.26)$$

Expanding $H(I_1,\ldots,I_d;g)$ in a Taylor series around the MAP estimator $\bar{I}$ using the fact that the gradient term is zero at the MAP estimator gives

$$H(I_1,\ldots,I_d;g) = H(\bar{I}_1,\ldots,\bar{I}_d;g) + \frac{1}{2} \sum_{i,j=1}^d (I_i - \bar{I}_i)(I_j - \bar{I}_j) F(g)_{ij}. \quad (14.27)$$

Substituting in Eqn. 14.26 gives

$$p(I^D|g) = p(I^D|\bar{I}_1,\ldots,\bar{I}_d, g) \pi(\bar{I}_1,\ldots,\bar{I}_d|g) \int_{\mathbb{R}^d} e^{-1/2 \sum_{i,j=1}^d (I_i - \bar{I}_i)(I_j - \bar{I}_j) F(g)_{ij}} dI_1 \ldots dI_d$$

$$\stackrel{(a)}{=} p(I^D, \bar{I}_1, \ldots, \bar{I}_d | g)(2\pi)^{d/2} \det^{-1/2} F(g), \quad (14.28)$$

where (a) follows from the fact that the function $(2\pi)^{-d/2} \det^{1/2} Q e^{-1/2 x^* Q x}$ integrates over $\mathbb{R}^d$ to 1 where Q is a positive definite $d \times d$ matrix. □

Corollary 14.8 *If for every $g \in \mathcal{G}$, $\{T\phi_i(g)\}$ are orthogonal, then*

$$p(I^D|g) = p(I^D, \bar{I}_1, \ldots, \bar{I}_d | g)(2\pi\sigma^2)^{d/2} \left(\prod_{i=1}^d \frac{\lambda_i}{\lambda_i \|T\phi_i(g)\|^2 + \sigma^2} \right)^{1/2}. \quad (14.29)$$

As $\sigma \to 0$, the complexity penalty goes as $d/2 \log \sigma^2 + O(1)$, as in Eqn. 2.212 in Theorem 2.63.

Example 14.9 (For the Gaussian case, MAP=MMSE) Examine the simple case in which $TI = I$ and there is no projection, and the goal is to directly estimate the photometric parameters alone without any transformation. This is the clearest version of the above problem. Let $I \in \mathcal{H}(\phi)$ be the mean field of the observation I^D, W white noise variance σ^2, then the MMSE $I_{\text{MMSE}} : I^D \in \mathcal{I}^D \to \mathcal{H}$, and the MAP estimator $I_{\text{MAP}} : I^D \in \mathcal{I}^D \to \mathcal{H}$ are identical:

$$I_{\text{MAP}} = I_{\text{MMSE}} = \sum_i \bar{I}_i \phi_i, \quad \bar{I}_i = \frac{\lambda_i}{\lambda_i + \sigma^2} \langle I^D, \phi_i \rangle. \tag{14.30}$$

To see this, solving for the MMSE in l^2 since with I^D Gaussian, then $I_n^D = \langle I^D, \phi_n \rangle$ is Gaussian distributed, zero-mean and variance $\lambda_n + \sigma^2$. Then from Theorem 2.36, Chapter 2 the MMSE $\bar{I}_j = \sum_i \alpha_i^{(j)} I_i^D$ is a linear functional of the data I_n^D and for each j must have the property that the error is orthogonal to the data; for all n,

$$0 = E(\bar{I}_j - I_j)I_n^D = E\left(\sum_i \alpha_i^{(j)} I_i^D - I_j\right) I_n^D \tag{14.31}$$

$$= \sum_i \alpha_i^{(j)} E I_i^D I_n^D - E I_j I_n^D = \sum_i \alpha_i^{(j)}(\sigma^2 + \lambda_i)\delta(i-j) - \lambda_j \delta(j-n) \tag{14.32}$$

implying $\alpha_i^{(j)} = 0$ for $i \neq j$, and $\bar{I}_j = \alpha_j I_j^D$, with $\alpha_j = (\lambda_j/(\lambda_j + \sigma^2))$. The MAP estimator satisfies for each coefficient $\bar{I}_j$, $\bar{I}_j = \arg\min_{I_j : I = \sum_i I_i \phi_i} (\|I^D - \sum_i I_i \phi_i\|^2/2\sigma^2) + (\|I_i\|^2/2\lambda_i^2)$ giving the fixed point condition

$$\bar{I}_j \left(\frac{1}{\sigma^2} + \frac{1}{\lambda_j}\right) = \frac{\langle I^D, \phi_j \rangle}{\sigma^2}. \tag{14.33}$$

14.4 Estimation of Pose by Integrating Out EigenSignatures

Theorem 14.7 Eqn. 14.21 and Corollary 14.8 of Eqn. 14.29 provides the direct mechanism for solving the Bayes integral problem when estimating the pose parameter $g \in \mathcal{G}$. For each pose the Bayes factor is computed and weights the likelihood term evaluated at the MAP estimator. Several investigators have examined this approach in object recognition.

The MAP estimator of pose accomodating photometric variation is given by the following.

Theorem 14.10 Given I^D is a conditionally Gaussian field, mean $TI(g)$ with $I(g) = \sum_{i=1}^d I_i \phi_i(g)$, with priors I_i independent Gaussian variables variance λ_i and $\pi(g), g \in \mathcal{G}$ with conditional mean where $\bar{I}(g) = \sum_{i=1}^d \bar{I}_i \phi_i(g)$, then

$$\hat{g} = \arg\max_{g \in \mathcal{G}} \log p(g|I^D) \tag{14.34}$$

$$= \arg\max_{g \in \mathcal{G}} \log \pi(g) - \frac{\|I^D - T\bar{I}(g)\|^2}{2\sigma^2} - \frac{\sum_{i=1}^d |\bar{I}_i|^2}{2\lambda_i} - \frac{1}{2}\log\det 2\pi F(g). \tag{14.35}$$

The joint estimation of photometric and geometric parameters has been approached by various groups. Examine in detail Cooper's approach [332–334] of representing the photometric variation on the 3D surface parameterization of the CAD models themselves. By simulating a large number of signatures, taken by varying environmental conditions populations of signatures are generated. For simulating temperature radiance profiles, Cooper employs the PRISM software originally developed by the Keweenaw Research Center at Michigan Technological University.[41] For photometric variation associated with Teapots, he uses the openGL software to examine variations of lighting illuminations. To generate the eigentank models, a large database of radiance maps are synthesized under a wide range of conditions, both meteorological (solar irradiance, wind speed, relative humidity, etc.) and operational (vehicle speed, engine speed, gun fired or not, etc.).

Figure 14.8 shows examples of multiple eigensignatures at four different orientations (four columns). Panel 1 of Figure 14.9 depicts an example of a FLIR image of an M60 tank oriented at 270°.[42] The exhaust is the dominant feature, which is typical of operational M60 imagery. Figure 14.9 shows temperature signature profiles resulting from MAP estimates of the expansion coefficients of the M60 principal components model at different hypothesized orientations. The MAP estimates $\bar{I}_1, \ldots, \bar{I}_d = \arg\max_{(I_1,\ldots,I_d) \in \mathbb{R}^d} \log p(I_1, \ldots, I_d | I^D, g)$, computed by solving the least-squares solution for the image signature Eqn. 14.20 for every pose $g \in \mathcal{G}$.[43] For each target

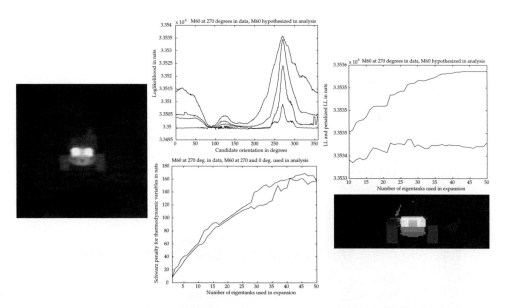

Figure 14.9 Left column: Panel 1 shows a FLIR image of an M60 facing away from the detector provided by NVESD. Middle column: Panel 2 shows the log-likelihood varying with respect to orientation for the NVESD image with an M60 at 270°. The correct target type, an M60, is assumed. Lines correspond to log-likelihoods computed using 1, 3, 9, and 50 eigentanks. Panel 3 shows the log-determinant of the Fisher information penalty, with respect to the number of eigentanks for the M60 oriented at 270° (top line), and 0° (bottom line). Right column: Panel 4 shows the log-likelihood (top line) and penalized log-likelihood (bottom line) with respect to the number of eigentanks used for a tank hypothesized at the correct orientation. Panel 5 shows the MAP estimator of the M60 data (panel 1).

[41] PRISM is currently sold and maintained by ThermoAnalytics, Inc., P.O. Box 66, Calumet, MI 49913, website: www.thermoanalytics.com.
[42] 0° represents the tank facing left, and increasing angle measurements run counter-clockwise looking down on the tank, so 270° is facing directly away from the detector.
[43] In fact Lanterman works with an approximation to the Gaussian density for the projective image model, capturing the first and second moments of the Poisson model by adjusting the projective image to have local variance given by the number of counts in the CCD cell. This corresponds to potential $\|I^D - TI(g)/\sqrt{I^D}\|^2$ where

pose, these equations allow for the computation of the optimum MAP signature $\bar{I}_1, \ldots, \bar{I}_d$. The different rows correspond to hypothesized orientations at 45° increments. The columns correspond to different numbers of EigenTanks ϕ_i: $d = 1, 3, 9, 50$. The 7th row (second to last) corresponds to the 270° correct orientation. Figure 14.8 illustrates various poses (rows) of the model tank constructed by synthesizing the tank via different numbers of principal components (columns). The first eigentank incorporates the exhaust ports, while the remainder brightens the engine region, as well as the treads. Note that in the third row from the top, in which the tank is facing the detector at 90°, the eigentanks cannot describe the data. In fitting the bright part of the data, they can heat up the barrel, but not the remaining part of the tank face. Panel 2 of Figure 14.9 illustrates the log-likelihood surface $\|(I^D - TI(g))/\sqrt{I^D}\|^2$ with respect to orientation of the M60. The lines in Figure. 14.9 correspond to using 1, 3, 9, and 50 eigentanks in the expansion. Each surface peaks sharply around the 270° true orientation.

Panel 3 shows the penalty term varying with the number of eigentanks for different hypothesized orientations. The top line shows the penalty for the correct hypothesized orientation; the bottom line, for a tank facing towards the left. The penalties for these two cases are relatively close. Panel 4 shows the log-likelihood (top lines) and penalized log-likelihood (bottom lines), varying with the number of eigentanks, for tanks oriented at the correct orientation (facing away from the detector, left panel). For the correct orientation, the penalty chooses 28 eigenvalues.

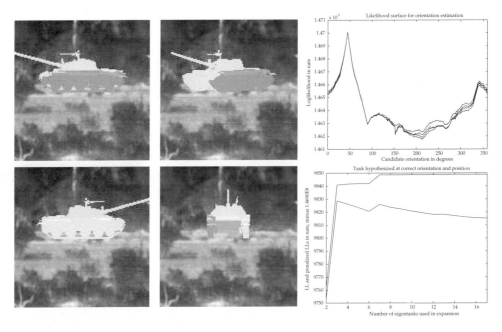

Figure 14.10 Left panels show synthetic T62 superimposed over a real infrared background (courtesy Night Vision and electronic Sensors Directorate). Right column shows log-likelihood varying with respect to orientation for data generated with a true orientation of 45 degrees. From bottom to top, lines correspond to likelihoods computed using 1, 3, 5 and 17 eigentanks. Bottom panel shows the log-likelihood (top line) and penalized log-likelihood (bottom line) with respect to the number of eigentanks employed.

the inner product over the image space does not include voxels with zero counts, $I^D(y_i) = 0$. This modifies Eqn. 14.20 little simply providing a normalization to the inner products arising in the Fisher information:

$$F(I^D; g) = \left(\left\langle \frac{T\phi_i(g)}{\sqrt{I^D}}, \frac{T\phi_j(g)}{\sqrt{I^D}} \right\rangle + \text{diag}\left(\frac{1}{\lambda_1}, \ldots, \frac{1}{\lambda_d}\right) \right). \quad (14.36)$$

Example 14.11 (FLIR with Complexity Term (Lanterman [336])) Figure 14.10 shows 140 × 140 images of a synthetic T62 superimposed on infrared backgrounds. The radiant intensities of the facets were generated from an eigentank model, with the first three coefficients of the T62s eigenexpansion set to −2000, 900, and −700 and the remaining coefficients set to zero. In all of the experiments the background adaptively estimated via its local average over the background. The second frame (top row) shows a synthesized data set with the tank at 45°. Panel 5 of Figure 14.10 shows the log-likelihoods computed using the first 1, 3, 5, and 17 terms in the thermodynamic eigenexpansion. They peak sharply at the correct orientation of 45°. In addition, using more terms in the expansion generates higher log-likelihoods, since the larger number of parameters gives the model greater flexibility in trying to fit the tank to the background.

Panel 6 of Figure 14.10 illustrates the effect of subtracting the penalty term from the log-likelihood. The top line is the log-likelihood associated with a given number of terms; notice that it increases monotonically with the number of terms. The bottom line represents the result of subtracting the logarithm of the inverse Fisher information penalty term. The penalized log-likelihood increases rapidly at first, reaching a peak at three terms, and then slowly decreases.

14.4.1 Bayes Integration

Cooper [335] studies the estimation of pose through lighting and photometric variation. The inference of $I(g)$ reduces to the inference of pose and temperature profile. The orbit is the set of all imagery generated from the template and a linear combination of basis elements $\{\phi_i\}$ spanning the photometric intensities. The basis is constructed as eigenfunctions of the empirical covariance generated from many realizations of actual signatures from varying illuminations of the objects (see Figure 14.8). The openGL simulation software is used for generating the database of illuminations. The top row shows a variety of signatures corresponding to varying illumination. Under the signal model $I = \sum_{i=1}^{d} I_i \phi_i$, and $I(O) = \sum_{i=1}^{d} I_i \phi_i(O)$ and the posterior density on the pose parameter becomes

$$p_d(O, I_1, \ldots, I_d | I^D) = \frac{1}{Z(I^D)} p(I^D | I(O), I_1, \ldots, I_d) \pi(O) \pi(I_1, \ldots, I_d) \tag{14.37}$$

$$= \frac{1}{(2\pi\sigma^2)^{d/2}} e^{-(1/2\sigma^2)\|I^D - TI(O)\|^2} \prod_{i=1}^{d} \frac{1}{\sqrt{2\pi\lambda_i}} e^{-(I_i^2)/(2\lambda_i)}. \tag{14.38}$$

Example 14.12 (Teapot Photometric Variability) Figure 14.11 illustrates experiments on the varying signatures of the teapot. Panels 1–4 of Figure 14.11 show a variety of signatures corresponding to varying illumination. Panel 5 shows MSE curves for each of the teapot signatures demonstrating variability in performance due to variability in the underlying object signature. Panel 5 plots the MSE performance of pose estimators with correct and randomly sampled signature information and this is compared in the video imaging context. The underlying signature of the teapot is $I = 1044$. The y-axis represents the conditional MSE given the underlying object orientation of 30°. The curve labeled "$I = t_{1044}$" is performance with complete signature information, and the curves labeled "Random" correspond to performance using the random sampling approach, without assumption of the object signature. The difference between these two curves represents the cost of knowledge of the object signature, in terms of estimator performance.

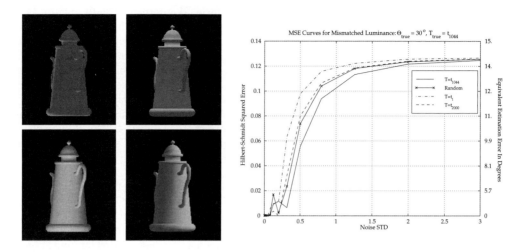

Figure 14.11 Panels 1–4 show four signatures for the teapot $I = 1044, 336, 2000, 1$. Panel 5 shows performance degradation due to lighting signature mismatch. The true signature was $I = 1044$, other signatures studied are $I = 1, 2000$, along with Bayes integration over all signatures denoted with X's and labelled "Random".

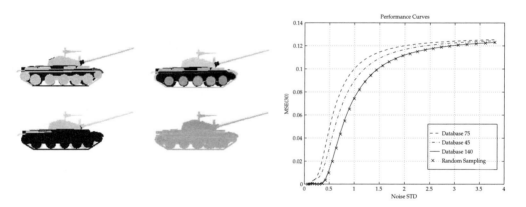

Figure 14.12 Panels 1–4 show T62 tanks with four temperature signatures $I = 8, 45, 75, 140$; panel 5 shows performance degradation due to signature mismatch. The true signature was $I = 140$, other signatures studied are $I = 45, 75$ along with Bayes integration over all temperature signatures depicted via X's and denoted "Random" (see also Plate 26).

Example 14.13 (Cooper FLIR Thermal Variation) Cooper [333, 335] studies photometric variation in infrared imagery. The first several eigensignatures ϕ_1, ϕ_2 are shown in columns 2 and 3 of Figure 14.8. Figure 14.12 shows examples of FLIR signatures and MSE performance for the ATR. Panels 1–4 of Figure 14.12 show various signatures from the Prism simulator projected through the perspective projection of the CCD imager. The rightmost panel of Figure 14.12 shows the loss in estimator performance due to the temperature FLIR photometric variation. Shown are MSE curves for $SO(2)$ estimation as a function of temperature signature, for the correct (140) and mismatched data bases.

14.5 Multiple Modality Signature Registration

Oftentime multiple imaging modalities are used to image the same brain. Examine the setting where the transformations to be estimated in the registration are Euclidean motions between the coordinate systems. In this setting the template and target have different signatures, but often have common assumed homogeneous structuring elements. For example in brain imaging, the brain is made up of bone, CSF, gray matter, white matter all of which images in a relatively homogeneous manner in each of the modalities, however has different signatures for the different modalities CT, MRI, cryosection, etc. For MR, the important parameter is the number of hydrogen protons present while for CT it is linear attenuation coefficient.

Generally the imagery is a union of homogeneous regions all of which appear identically in a particular imaging modality. This is depicted in panel 1 of Figure 14.13 depicting a 2D cryosection of a brain depicting various subregions. The right panel 2 shows the disjoint partition into a series of regions WM=white matter, GM=gray matter, CSF=cerebrospinal fluid, and bone. Denote these disjoint regions via the simple functions $\phi_1, \ldots, \phi_M$, M the number of disjoint compartments covering the background space.

Model the orbit of M-modality imagery as a vector $I = (I^{(1)}, \ldots, I^{(M)})$, each component constructed in span of a disjoint partition of the background coordinates $\phi_1, \phi_2, \ldots$ simple functions defining the coordinate locations

$$I^{(m)}(\cdot) = \sum_i I_i^{(m)} \phi_i(\cdot), \quad I_i^{(m)} \in \mathbb{R}. \tag{14.39}$$

The orbit $\mathcal{I}$ is the vector version of Eqn. 14.5 of Definition 14.1.

Definition 14.14 *Define the **robust multi-modality deformable template** $\mathcal{I}$ as the orbit of all signatures $I = (I^{(1)}, \ldots, I^{(M)}) \in \mathcal{H}(\phi)$ in the span of simple functions defining the homogeneous regions $\{\phi_i\}$ under the group operation $g \in \mathcal{G}$:*

$$\mathcal{I}(\phi) = \{I(g) = (I^{(1)}(g), \ldots, I^{(M)}(g)), \quad I^{(m)} = \sum_i I_i^{(m)} \phi_i, \, g \in \mathcal{G}\}. \tag{14.40}$$

The cross modality registration can be straightforwardly formulated.

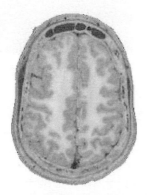
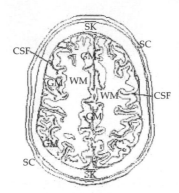

Tissue	T1	T2	PD	CT
ϕ_1 = White	223	140	81	88
ϕ_2 = Gray	134	167	115	88
ϕ_3 = CSF	72	96	202	2
ϕ_4 = Skull	41	36	4	255

Figure 14.13 Panel 1 shows cryosection imagery depicting clearly the regions WM=white matter, GM=gray matter, CSF=cerebrospinal fluid, and bone. Panel 2 depicts the regions of the compartments, a disjoint partition of the full brain. Data taken from the Visible Human Project of the National Library of Medicine. Panel 3 of tissue voxel intensity values for MRI T1, MRI T2, MRI Proton Density (PD), and CT imaging modalities. Values correspond to four disjoint compartments White Matter, Gray Matter, CSF, and skull.

Algorithm 14.15 (MAP Estimation Algorithm) *Given the disjoint partition into homogeneous regions $\{\phi_i\}$ of the multi-modality M-vector $I = I^{(1)}, \ldots, I^{(M)}$, I^{D_m} has mean field $\sum_i I_i^{(m)} \phi_i(g)$ where the compartment ϕ_i has photometric intensity $I_i^{(m)}$ for m-modality.*

The MAP estimator of the registration parameter $g \in \mathcal{G}$ relating the compartments between modalities

$$\hat{g}_m = \arg\min_{g \in \mathcal{G}} \left\| I^{D_m} - \sum_i I_i^{(m)} \phi_i(g) \right\|^2 . \tag{14.41}$$

The $I_i^{(m)}$ are the decoder values transcribing the colors represented by one imaging modality into another. This is depicted in Table 14.13 showing compartment values, White Matter, Gray Matter, CSF, and skull for four imaging modalities T1,T2, proton density, and CT.

Example 14.16 (Cross Modality Registration) Examine experimental results incorporating unknown signatures associated with medical imaging modalities. In this setting the template and target imagery will have different signatures. The deformable template for rigid registration accomodates the geometric properties which are invariant, but must provide enough parameters to accomodate the different photometric intensities.

Shown in Figure 14.14 are results from multiple imaging modalities showing cross-modality registration. Panels 1 and 2 shows T2 and proton density MRI images. Panels 3 and 4 shows multimodality registration bounds for mutual information and the Hilbert–Schmidt estimator. Panel 5 shows the comparison between mutual information and the HSE for T1 and T2 weighted MR images. Panel 6 shows the T2 and proton density registration.

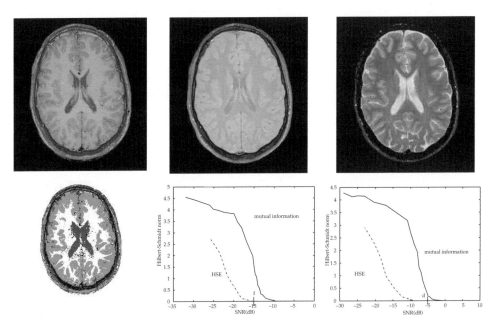

Figure 14.14 Column 1 shows examples of segmentations of the T1 Weighted MR Image. Columns 2 and 3 show T2, and proton density images (top row) with multimodality registration bounds for mutual information and the Hilbert–Schmidt estimator. Panel 5 shows the comparison between mutual information and the HSE for T1 and T2 weighted MR images. Panel 6 shows the T2 and proton density registration. Data generated at the Kennedy Krieger institute.

Panel 4 shows an example segmentation of the T1 Weighted MR Image. The segmentation is produced from the MR image, by Bayes estimation of the means and variances and solution of the likelihood ratio test. Three thresholds are used for mapping the tissues to WM, GM, and CSF; (White matter is white, Gray matter is gray, and CSF is black).

14.6 Models for Clutter: The Transported Generator Model

Now examine the modelling of natural scenes in the world constructed by transporting generators throughout the scene to generate natural patterns. Begin by constructing general models of the world of images motivated by the approach formulated by David Mumford based on the tenet that the world is made up of atoms, or in our terminology generators; the images observed are generators transported throughout the scene. Think in terms of a city scape, for example, so that there are small areas where something "happens", say a city block. Such structures are naturally associated with point processes. Thinking in terms of stochastic fields, such processes must be generated from an ergodic and stationary point process N, with dependence between the random variables $N(E_k)$ where the E_ks are compact Borel sets in the plane. Call the set of random locations W; to each random point W_n generated by N associate a random variable describing a simple geometric object, for example a city block. This will induce a random image.

We are essentially constructing the *transported generator model*, a stochastic model corresponding to marked-filter point processes which transports filters, henceforth called generators throughout the world to construct complex scenes of objects. Model the random field $\{I(x); x \in D \subset \mathbb{R}^d\}$ generated by marking a spatially inhomogeneous Poisson counting process N with spatial intensity $\lambda(x), x \in D$ so that for subsets $A \subset D$, then $N_A, A \subset D$ is Poisson distributed with parameter $\int_A \lambda(x)\,dx$,

$$\Pr\{N_A = n\} = e^{-\int_A \lambda(x)\,dx} \frac{(\int_A \lambda(x)\,dx)^n}{n!}. \tag{14.42}$$

For what follows, we shall exploit the important fact that on condition there being n events in D from the spatial Poisson process intensity λ, the joint density takes the very familiar form.

Lemma 14.17 *Given Poisson occurrences with intensity $\lambda(\cdot)$, then*

$$p(X_1, \ldots, X_n, N_D = n) = \prod_{i=1}^n \lambda(X_i) e^{-\int_D \lambda(x)\,dx}. \tag{14.43}$$

Proof

$$\Pr(X_1 \in \Delta X_1, \ldots, X_n \in \Delta X_n, N_D = n) = \Pr(N_D = n | X_1 \in \Delta X_1, \ldots, X_n \in \Delta X_n)$$

$$\times \prod_{i=1}^n \Pr(N_{\Delta X_i}) \tag{14.44}$$

$$= e^{-\int_{D \setminus \cup_{i=1}^n \Delta X_i} \lambda(x)\,dx} \prod_{i=1}^n \int_{\Delta X_i} \lambda(x)\,dx\, e^{-\int_{\Delta X_i} \lambda(x)\,dx}. \tag{14.45}$$

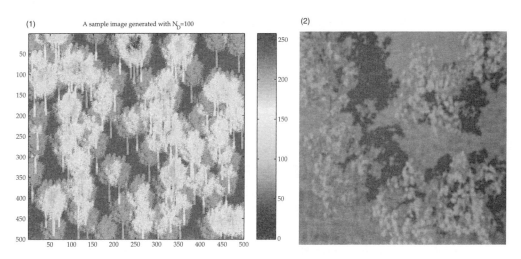

Figure 14.15 Panel 1 shows a scene constructed from CAD models of trees placed via the transported generator clutter model. Panel 2 shows the same for 3D CAD models which were ray-traced; taken from Bitouk (see also Plate 27).

The joint occurrence density is given by the limit

$$p(X_1,\ldots,X_n,N_D = n) = \lim_{|\Delta X_i|\to 0} \frac{1}{\prod \Delta X_i}\Pr(X_1 \in \Delta X_1,\ldots,X_n \in \Delta X_n, N_D = n).$$
(14.46)

□

Then the random image model is defined as follows.

Definition 14.18 *Define $I(x), x \in D$ to be from the random image model termed the* **transported generator model** *if $I(x), x \in D$ is assumed to be of the superposition form*

$$I(x) = \begin{cases} 0 & N_D = 0 \\ \sum_{n=1}^{N_D} h(x, W_n; U_n) & N_D \geq 1 \end{cases},$$
(14.47)

with **generators** $h(\cdot, W_n; U_n)$, *the $W_n, n = 1, 2, \ldots$ occurrence positions from a Poisson process with intensity $\lambda(x), x \in D$ and $U_n, n = 1, 2, \ldots$ the random marks.*

A process in the form 14.47 is referred to as a filtered point process in Snyder [42] with filter functions $h(\cdot)$. Shown in Figure 14.15 are examples of a synthesis via the point process model using tree generators. randomly placed. Panel 1 shows the tree scene with mean $EN_D = 100$.

14.6.1 Characteristic Functions and Cumulants

We work with the characteristic functional of the filtered process integrated against Rieman–Stieltjes functions.

Definition 14.19 *The* **characteristic functional** *for the random process $\{I(x), x \in D\}$ integrated against sufficiently smooth real-valued test functions $v(x), x \in D$ is defined as*

$$M_I(jv(\cdot)) = E[e^{j\int_D I(x)dv(x)}].$$
(14.48)

The **cumulant generating function** $\Gamma_{I(x_i)}(jv)$ is the log-characteristic functional

$$\Gamma_{I(x_i)}(jv) = \log E[e^{jvI(x_i)}]. \tag{14.49}$$

The cumulants $\kappa_k, k = 1, \ldots$ define the cumulant generating function according to

$$\Gamma_{I(x_i)}(jv) = \log E[e^{jvI(x_i)}] = \sum_{k=1}^{\infty} \frac{(jv)^k}{k} \kappa_k ; \tag{14.50}$$

to get the cumulants differentiating implies $(\partial^k \Gamma_{I(x)}(jv))/(\partial (jv)^k)|_{jv=0} = \kappa_k$.

The characteristic functional of the transported generator model has a particularly lovely form.

Theorem 14.20 *Let $\{I(x), x \in D\}$ be from the transported generator model $I(x) = \sum_{n=1}^{N_D} h(x, W_n; U_n)$ corresponding to generators at Poisson points with intensity $\lambda(\cdot)$ and marks U_n independent and identically distributed. Then the characteristic functional and the joint characteristic function on any n-cylinders $I(x_i), i = 1, \ldots, n$ is given by*

$$M_I(jv(\cdot)) = e^{\int_D \lambda(y) E[e^{j \int_D h(x,y;U) dv(x)} - 1] dy}, \tag{14.51}$$

where the expectation is with respect to the distribution of the mark random variable process U. The joint characteristic function on n-cylinders becomes

$$M_I(jv(\cdot))|_{v(\cdot) = \sum_{i=1}^n v_i \delta(\cdot - x_i)} = E e^{\sum_{i=1}^n jv_i I(x_i)} = e^{\int_D \lambda(y) E[e^{j \sum_{i=1}^n h(x_i,y;U)} - 1] dy}. \tag{14.52}$$

Proof Apply the definition

$$M_I(jv(\cdot)) = E[e^{j \int_D I(x) dv(x)}] = Pr(N_D = 0) + \sum_{k=1}^{\infty} Pr(N_D = k)$$

$$\times E[e^{j \sum_{n=1}^k \int_D h(x, W_n; U_n) dv(x)} | N_D = k].$$

Then clearly the summation is unaffected by random re-orderings of the occurrence positions, there being $n!$ realizations of Poisson processes with the same positions for the n-points. Thus the probability of n-events in D at particular positions $x_1, x_2, \ldots x_n$ (see p. 220 of [42]) is given by $\prod_{i=1}^n (\lambda(x_i)/\int_D \lambda(x) dx)$. Evaluating the expectation exploiting the independence of the marks gives

$$M_I(jv(\cdot)) = Pr(N_D = 0) + \sum_{k=1}^{\infty} e^{-\int_D \lambda(x) dx} \frac{(\int_D \lambda(x) dx)^k}{k!} \left(\frac{\int_D \lambda(y) E[e^{j \int_D h(x,y;U) dv(x)}] }{\int_D \lambda(y) dy} \right)^k \tag{14.53}$$

$$= e^{\int_D \lambda(y) E[e^{j \int_D h(x,y;U) dv(x)} - 1] dy}, \tag{14.54}$$

completing the proof. The joint characteristic function on the cylinders is determined by choosing the function f to sift at the n-cylinder locations throughout the process, $v(\cdot) = \sum_{i=1}^n v_i \delta(\cdot - x_i)$. □

These transported generator processes are fundamentally different from Gaussian processes. Examine the cumulant generating function $\Gamma_{I(x)}(jv)$ for the random variable $I(x)$ is given through the series expansion in terms of the cumulants or semi-invariants defined from the log-moment generating function

$$\Gamma_{I(x)}(jv) = \log M_{I(x)}(jv) = \sum_{k=1}^{\infty} \frac{(jv)^k}{k} \kappa_k. \tag{14.55}$$

Corollary 14.21 *Given the transported generator model $I(x) = \sum_{n=1}^{N_D} h(x, W_n; U_n)$ with generators occuring with spatial intensity $\lambda(x), x \in D$ with the mark process U independent, then*

$$\Gamma_{I(x)}(jv) = \int_D \lambda(y) E[e^{jvh(x,y;U)} - 1] \, dy, \tag{14.56}$$

with cumulants $\kappa_k = \int_D \lambda(y) E[h^k(x, y; U)] \, dy. \tag{14.57}$

Proof The form for the cumulant function is generated from the log-characteristic functional for the particular sifting function centered at $x \in D, f(\cdot) = v\delta(\cdot - x)$:

$$\Gamma_{I(x)}(jv) = \log E[e^{jvI(x)}] = \log e^{\int_D \lambda(y) E[e^{jvh(x,y;U)} - 1] dy}$$

$$= \int_D \lambda(y) E[e^{jvh(x,y;U)} - 1] \, dy. \tag{14.58}$$

To get the cumulants use the fact $\Gamma_{I(x)}(jv) = \sum_{k=1}^{\infty} ((jv)^k/k)\kappa_k$ which implies

$$\frac{\partial^k \Gamma_{I(x)}(jv)}{\partial (jv)^k}\bigg|_{jv=0} = \kappa_k. \tag{14.59}$$

$\square$

Irrespective of their covariance properties, such a transport model has **kurtosis** which is lower bounded by 3 the kurtosis of a Gaussian field.

Definition 14.22 *The* **kurtosis** *of the random variable X is given by the fourth moment:*

$$k_X = \frac{E(X - EX)^4}{(E(X - EX)^2)^2}. \tag{14.60}$$

The transported generator model converges in the limit to a Gaussian field with such kurtosis as the intensity of the generators goes to infinity. First define the moments and variances of the random variable $I(x)$ to be

$$m_k = EI(x)^k, \quad \mu_k = E(I(x) - EI(x))^k. \tag{14.61}$$

Then, notice from the series expansion in the cumulant generating function the moments arise:

$$\kappa_n = \frac{\partial^n \Gamma_{I(x)}(jv)}{\partial (jv)^n}\bigg|_{jv=0} \quad \text{implies } m_k = EI(x)^k \tag{14.62}$$

$$m_1 = \kappa_1, \; m_2 = \kappa_2 + \kappa_1^2, \; m_3 = \kappa_3 + 3\kappa_1\kappa_2 + \kappa_1^3, \; m_4 = \kappa_4 + 3\kappa_2^2 + 4\kappa_1\kappa_3 + 6\kappa_1^2\kappa_2 + \kappa_1^4. \tag{14.63}$$

We can now evaluate the kurtosis of the transported generator clutter model for various cases. First for the general case.

14.6 MODELS FOR CLUTTER

Corollary 14.23 *Given the transported generator model $I(x) = \sum_{n=1}^{N_D} h(x, W_n; U_n)$ with generators occuring with spatial intensity $\lambda(x), x \in D$ with the mark process U independent, the kurtosis is given by*

$$k_{I(x)} = \frac{E[I(x) - EI(x)]^4}{\left(E[I(x) - EI(x)]^2\right)^2} = 3 + \frac{\kappa_4}{\kappa_2^2} = 3 + \frac{\int_D \lambda(y) E[h^4(x, y; U)] dy}{\left(\int_D \lambda(y) E[h^2(x, y; U)] dy\right)^2}. \quad (14.64)$$

For the homogeneous transported generator model with $h(x, y; U) = Ag(x - y)$ with

$$I(x) = \sum_{n=1}^{N_D} U_n g(x - W_n), \quad (14.65)$$

and constant spatial intensity λ with U_n independent finite second moments, the kurtosis k_I is bounded below by 3 with excess according to

$$k_{I(x)} = \frac{E[I(x) - EI(x)]^4}{\left(E[I(x) - EI(x)]^2\right)^2} = 3 + \frac{E[A^4] \int_D g^4(x - y) dy}{\left(E[A^2] \lambda \int_D g^2(x - y) dy\right)^2}, \quad (14.66)$$

$$\text{EXCESS} = k_I - 3 \geq 0. \quad (14.67)$$

Proof The numerator becomes

$$E[I(x) - EI(x)]^4 = EI^4(x) - 4EI^3(x)EI(x) + 6EI^2(x)E^2I(x) - 3E^4I(x)$$

$$= m_4 - 4m_3 m_1 + 6m_2 m_1^2 - 3m_1^4$$

$$= \kappa_4 + 3\kappa_2^2 + 4\kappa_1 \kappa_3 + 6\kappa_1^2 \kappa_2 + \kappa_1^4 - 4(\kappa_3 + 3\kappa_1 \kappa_2 + \kappa_1^3)\kappa_1$$

$$\quad + 6(\kappa_2 + \kappa_1^2)\kappa_1^2 - 3\kappa_1^4$$

$$= \kappa_4 + 3\kappa_2^2. \quad (14.68)$$

This completes the proof using the fact that

$$E[I(x) - EI(x)]^2 = m_2 - m_1^2 = \kappa_2. \quad (14.69)$$

For the homogeneous case, use the formula for the kurtosis $3 + (\kappa_4/\kappa_2^2)$. and the covariance. □

Examine the special case of the homogeneous transported generator model with the marking process determining the amplitudes only of the filter functions which are assumed shift invariant, so that

$$h(x, y; U) = Ug(x - y),$$

assumed that the spatial process is homogeneous $\lambda(\cdot) = \lambda$. This analysis suggests that the transported clutter model differs fundamentally from a Gaussian process. The kurtosis of the Gaussian random fields are 3; only as $\lambda \to \infty$ does the process have kurtosis of a Gaussian, $k_I = 3$. This suggests in the limit of densely transported clutter, the process is Gaussian.

Figure 14.16 examines experiments illustrating the class of images produced by the transported generator model. Assume that the counting process is homogeneous with spatial intensity $\lambda(\cdot) = \lambda$, then Eqn. 14.43 of Lemma 14.17 dictates that the events are i.i.d. uniformly distributed over the image plane. Panel 1 of Fig. 14.16 shows the various trees used to study the kurtosis. Panel 2 depicts the effect of changing the tree filtering function on the average kurtosis values. We use the set of 20 trees as shown in the right panel of Fig. 14.16 for the average kurtosis evaluation. The trees in the plot are indexed from left to right and top to bottom in a lexicographic order. Panel 2

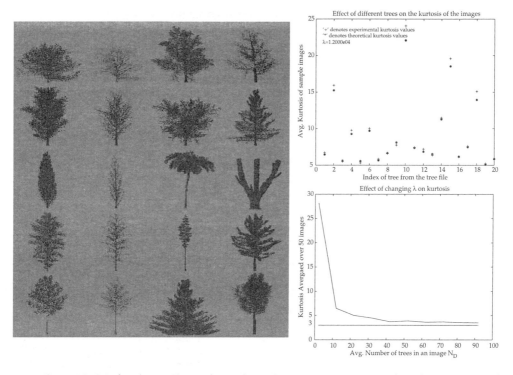

Figure 14.16 Left column: Figure shows the various tree generators. Right column: Top panel 2 shows kurtosis measured for different trees. Bottom panel shows average kurtosis plot as a function of density parameter λ.

shows the values of the kurtosis computed from the corollary are compared between the analytical formula and the empirical calculation (+ versus *). Panel 3 shows the kurtosis as lambda density increases for one of the tree generators.

Example 14.24 (Covariance and Method of Moments Estimation) The covariance from the joint characteristic functional of Theorem 14.20 becomes

$$K_I(x_1, x_2) = \int_D \lambda(y) E[h(x_1, y; U_1) h(x_2, y; U_2)] \, dy. \tag{14.70}$$

For the case $h(x, y; U) = Ug(x - y)$, U's independent, then it is particularly simple as

$$K(x_1, x_2) = \int_D \lambda(x) E h(x_1, y; U_1) h(x_2, y; U_2) \, dy \tag{14.71}$$

$$= \lambda E(U^2) \int_D g(x_1 - y) g(y - x_2) \, dy, \tag{14.72}$$

$$= \lambda E(U^2) \int_D g(x - y) g(y) dy = K(x, 0), \quad \text{where } x = x_1 - x_2. \tag{14.73}$$

The method of moments may be used for estimating the parameters of the model, exploit the Toeplitz structure using properties of the Fourier transformation. Defining $w = (\omega_1, \ldots \omega_d)$, where $d = 2, 3$ the dimension of the image data, define the Fourier transform pairs as

$$G(\omega) = \int_D g(x) e^{-j<\omega, x>} \, dx, \qquad \mathcal{K}(\omega) = \int_D K(x, 0) e^{-j<\omega, x>} \, dx.$$

14.6 MODELS FOR CLUTTER

Exploiting the real, even symmetry of $g(x), x \in D$ according to

$$\mathcal{K}(\omega) = \lambda E(U^2)|G(\omega)|^2 = \lambda E(U^2)G(\omega)^2, \tag{14.74}$$

then the method of moments estimators take the form

$$\lambda E(U^2) = \mathcal{K}(0), \qquad g(x) = \int \sqrt{\frac{\mathcal{K}(\omega)}{\mathcal{K}(0)}} e^{j<\omega,x>} dx. \tag{14.75}$$

Example 14.25 (Ray Traced Imagery) Bitouk has examined a large data set of ray-traced target images which represent natural clutter. Ray-tracing appears to be an especially attractive rendering technique, since it closely resembles the formation process of natural images. The generated dataset consists of more than 10,000 images. The ray-traced images in the dataset contain targets in a randomly synthesized terrain taking into account occlusion, shadows and lighting variation as well as other effects usually encountered in natural scenes. The terrain was generated by random placement of trees on grass backgrounds according to the specified distribution densities. The trees were synthesized using a random growing process controlled by physical parameters [337]. Shown in Figure 14.17 are example images from the synthetic dataset.

Ray-tracing is a very realistic method for generating the low order statistics of natural images. To illustrate that, Bitouk computed single-pixel derivative statistics, recently introduced by Huang and Mumford (see [338] for details). Working with the log intensities, the marginal distribution of the horizontal derivatives were computed $\delta \ln I = \ln I(x_i, x_j) - \ln I(x_i, x_{j+1})$. Figure 14.17 shows the logarithm of the histogram of $\delta \ln I$. The histogram has a sharp peak at 0 and heavy tails. In [338], the density function $f(x)$ of D is modeled as a generalized Laplace distribution

$$f(x) = \frac{1}{Z} \cdot e^{-|x/s|^\alpha}, \tag{14.76}$$

whose parameters α and s are related to the variance and kurtosis of D by

$$\sigma^2 = \frac{\Gamma(3/\alpha)}{\Gamma(1/\alpha)} s^2, \qquad k = \frac{\Gamma(1/\alpha)\Gamma(5/\alpha)}{\Gamma^2(3/\alpha)}. \tag{14.77}$$

The dashed curve in panel 3 (Figure 14.17) presents the model Eqn. 14.76 with the parameters $(\alpha, s) = (0.637, 0.254)$ calculated from the variance and kurtosis using Eqn. 14.77.

Figure 14.17 Panels 1 and 2 show examples from Bitouk of synthesized target chips generated via the ray tracing algorithm. Panel 3 shows the derivative statistic $\delta \ln I$ for synthetic clutter images; solid curve is the observed, dashed curve is the best fit with the generalized Laplace distribution $(\alpha, s) = (0.637, 0.254)$ (see also Plate 28).

14.7 Robust Deformable Templates for Natural Clutter

Let us return to the clutter model of Section 14.6. One of the central problems in Automated Target Recognition is to accommodate the infinite variety of clutter in real military environments. In model-based approaches, identification/classification rates are largely determined by the accuracy of models used to represent real-world scenes. In heavily cluttered environments, it is impractical to explicitly represent each of the objects in the scene by a deterministic 3D model whereas applying low-dimensional statistical description of clutter may improve ATR performance in clutter. *Clearly we must robustify the deformable template approach to accomodate photometric variations which are not explicitly modellable, and perhaps may not be associated with the geometric motions of the objects of interest.* This is an area of increasing interest, and most certainly will emerge as one of the central themes of a good deal of research in the coming years. Unquestionably, the main challenge to model-based methods is to acquire pose and identification of targets while, at the same time, "seeing through" the confounding clutter.

Examine the construction of *robust deformable templates*, following the work of Bitouk which allows the accomodation of both the variations of the pose of objects as well as the photometric variations associated with natural clutter. Construct the metric via the empirical statistics of clutter. The group of photometric variations is modeled as a Hilbert space $\mathcal{H} \subseteq L^2$. Since not all of the elements of L^2 correspond to observable images, the space of photometric variations $\mathcal{H}$ is constructed via basis expansions $\{\phi_n\}$. For the clutter problem, the statistics in the background are modeled explicitly. Shown in Figure 14.18 are examples of imagery from the orbit under geometric transformation and clutter variation.

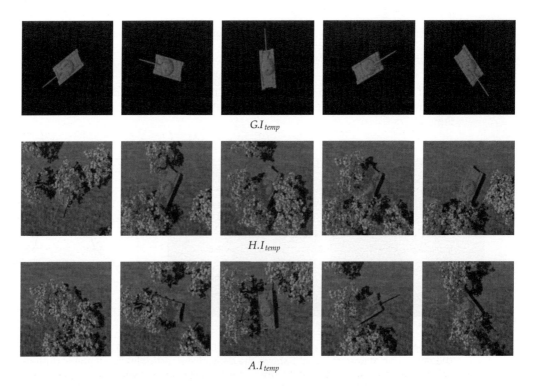

$G.I_{temp}$

$H.I_{temp}$

$A.I_{temp}$

Figure 14.18 Orbits under actions of the groups $\mathcal{G}$, $\mathcal{H}$ and $\mathcal{A} = \mathcal{G} \otimes \mathcal{H}$.

14.7.1 The Euclidean Metric

Let us begin by examining the most straightforward formulation of the metric corresponding to the L^2 norm, so that $\mathcal{N}((\partial I_t/\partial t) \circ g_t^{-1})^2 = \|(\partial I_t/\partial t) \circ g_t^{-1}\|_2^2$. Examine mappings defined by the special-Euclidean group of rigid motions $\mathcal{G} = \mathbf{SE}(d)$ with action $g : x \mapsto Ox + b$ where $O \in \mathbf{SO}(d), b \in \mathbb{R}^d$. Then it is natural to impose no energy on the geometric motion. Defining the energy of the rigid motion transformation to be zero, then the energy of the path (g_t, I_t) can be defined by $\int_0^1 \left\|(\partial I_t/\partial t) \circ g_t^{-1}\right\|^2 dt$. A distance constructed from this energy is clearly left invariant as described in Theorem 12.25 in Chapter 12 under the action of the rigid motion group.

Theorem 14.26 *For the rigid motions $g_t \in \mathbf{SE}(d)$, defining*

$$\rho^2((g,I),(h,I')) = \inf_{\substack{g_t, I_t \,:\, g_0 = g, g_1 = h \\ I_0 = I, I_1 = I'}} \int_0^1 \left\|\frac{dI_t}{dt} \circ g_t^{-1}\right\|_2^2 dt, \qquad (14.78)$$

then the resulting distance between images $I, I' \in \mathcal{I}$ is invariant:

$$\tilde{\rho}^2(I,I') = \inf_{g,h \in \mathbf{SE}(d)} \rho((g,I(g)),(h,I'(h))) = \inf_{g \in \mathbf{SE}(d)} \rho((\mathrm{id},I),(g,I'(g))) \qquad (14.79)$$

and reduces to the squared error minimization

$$\tilde{\rho}^2(I,I') = \inf_{g \in \mathbf{SE}(d)} \|I - I'(g)\|_2^2. \qquad (14.80)$$

Proof Clearly $\|(dI_t/dt) \circ g_t^{-1}\|_2^2 = \|dI_t/dt\|_2^2$ since g_t are rigid motions, Jacobian determinant 1, implying I_t is a linear interpolation of its boundary conditions (Theorem 10.5 of Chapter 10):

$$I_t = (1-t)I(g) + tI'(h) \text{ with } \frac{dI_t}{dt} = I'(h) - I(g). \qquad (14.81)$$

Substituting this back into the definition of $d(I,I')$ completes the proof Eqn. 14.91 establishing the connection of the variational problem for $d(I,I')$ to the maximum likelihood estimation:

$$\tilde{\rho}^2(I,I') = \inf_{g,h \in \mathbf{SE}(d)} \|I(g) - I'(h)\|_2^2 = \inf_{g \in \mathbf{SE}(d)} \|I - I'(g)\|_2^2, \qquad (14.82)$$

with the last equality following from the change of variables in the norm. $\square$

Discriminability clearly involves the shortest metric distance between the orbits under the nuisance parameter between object classes. Figure 14.19 illustrates this for the Euclidean metric calculations for several CAD models in trees in which they were at known, fixed pose. Shown in the table are the metric distances within class and across class models in clutter using the Euclidean norm $\|\cdot\|^2$. The diagonal entries show the within class distance in clutter; smaller numbers mean increased similarity.

14.7.2 Metric Space Norms for Clutter

For the clutter problem, it is clear that the squared error metric cannot adequately represent the orbit of phtometric variations. The textures of clutter do not appear as an independent increment process

Figure 14.19 Figure shows multiple CAD models and metric distances between within class and across class models in clutter. The diagonal entries show the within class distance in clutter; smaller numbers mean increased similarity providing a method for clutter independent identification. These distances are computed by solving the minimum over all rotations.

which would correspond to the L^2 norm of $\|\cdot\|_2$. For this a complementary approach can be taken from the photometric modeling of faces to model the photometric variations of the background space (rather than the objects themselves) for generating metric distances upon which to compute the representations. Our approach here is to craft the Hilbert-space norm $\mathcal{N}((\partial I_t/\partial t) \circ g_t^{-1})^2 = \|((\partial I_t/\partial t) \circ g_t^{-1}\|_{\mathcal{H}}^2$ so as to reflect the statistics of the clutter. Stationary Gaussian Markov Random Field models capture important statistical properties of natural clutter. In these models, the clutter field $W(x), x \in X \subset \mathbb{R}^d$ is characterized by a Toeplitz covariance $K(x, 0), x \in X \subset \mathbb{R}^d$, given by $K(x, y) = K(x - y, 0)$.

To build the Hilbert space of photometric variability with the norm, induced by the empirical covariance kernel, assume the covariance kernel $K(x, y)$ to be Hilbert-Schmidt $\int\int K^2(x, y) dx dy < \infty$; then the Hilbert–Schmidt theorem implies that there exsits a complete orthonormal sequence of eigenfunctions $\{\phi_n(x)\}$ with corresponding eigenvalues $\{\lambda_n\}$ such that

$$\int K(x, y)\phi_n(y)\, dy = \lambda_n \phi_n(x). \tag{14.83}$$

Definition 14.27 *Given a positive-definite covariance kernel $K(x, y)$, define a Hilbert space $H \subseteq L^2$ with the norm $\|\cdot\|_{\mathcal{H}}$, induced by the kernel $K(x, y)$, according to*

$$\mathcal{H} = \left\{ f \in L^2 : \|f\|_{\mathcal{H}}^2 = \sum_n \frac{|\langle f, \phi_n \rangle_{L^2}|^2}{\lambda_n} < \infty \right\}. \tag{14.84}$$

14.7 ROBUST DEFORMABLE TEMPLATES

Consider an element g in the special-Euclidean group of rigid motions $\mathcal{G} = \mathbf{SE}(d)$ with action $g : x \mapsto Ox + b$ where $O \in SO(d), b \in \mathbb{R}^d$. The kernel $K(g(x), g(y))$ is also Hilbert–Schmidt and induces a corresponding Hilbert space $g.\mathcal{H}$. The following lemma establishes the connection between the norms $\|\cdot\|_{\mathcal{H}}$ and $\|\cdot\|_{g.\mathcal{H}}$.

Lemma 14.28 *For any $f \in g.\mathcal{H}$,*

$$\|f\|_{g.\mathcal{H}} = \left\|f \circ g^{-1}\right\|_{\mathcal{H}}. \tag{14.85}$$

Proof Let $\{\tilde{\phi}_n\}$ be a complete orthonormal sequence of eigenfunctions of $K(g(x), g(y))$ with corresponding eigenvalues $\{\tilde{\lambda}_n\}$

$$\int K(g(x), g(y))\tilde{\phi}_n(y)\,dy = \tilde{\lambda}_n \tilde{\phi}_n(x). \tag{14.86}$$

A simple substitution of variables $x \to g^{-1}(x), y \to g^{-1}(y)$ in the integral yields

$$\int K(x,y)\tilde{\phi}_n(g^{-1}(y))\,dy = \tilde{\lambda}_n \tilde{\phi}_n(g^{-1}(x)), \tag{14.87}$$

implying $\tilde{\phi}_n = \phi_n \circ g, \tilde{\lambda}_n = \lambda_n$. Then,

$$\|f\|_{g.\mathcal{H}}^2 = \sum_n \frac{|\langle f, \phi_n \circ g \rangle_2|^2}{\lambda_n} = \sum_n \frac{|\langle f \circ g^{-1}, \phi_n \rangle_2|^2}{\lambda_n} = \|f \circ g^{-1}\|_{\mathcal{H}}^2. \tag{14.88}$$

$\square$

For the rigid motions we set the energy for geometric motion in definition 14.1 to zero, then the energy of the path (g_t, I_t) can be defined by the norm associated with the empirical covariance measuring image motion:

$$E = \int_0^1 \left\|\frac{\partial I_t}{\partial t} \circ g_t^{-1}\right\|_{g_t^{-1}.\mathcal{H}}^2 dt. \tag{14.89}$$

Such a definition of energy allows for the connection of the metric to classical maximum-likelihood estimation. For an arbitrary covariance K defining the quadratic form in the metric $\|\cdot\|_{\mathcal{H}}$, the measure of similarity $d(I, I')$ will not be a metric distance on $\mathcal{I}$, since it will not be symmetric and would not satisfy the triangular inequality. We now establish the connection between rotation and shift invariance properties of $d(I, I')$ and the covariance K. This in turn makes the distance into a metric. In fact the covariance must be rotationally invariant.

Theorem 14.29 *For the rigid motions $g_t \in \mathbf{SE}(d)$, defining the function d between images $I, I' \in \mathcal{I}$*

$$d(I, I')^2 = \inf_{\substack{g_t, I_t : g_0 = \mathrm{id}, \\ I_0 = I, I_1 = I' \circ g_1}} \int_0^1 \left\|\frac{dI_t}{dt} \circ g_t^{-1}\right\|_{g_t^{-1}.\mathcal{H}}^2 dt \tag{14.90}$$

$$= \inf_{g_1} \|I - I' \circ g_1\|_{\mathcal{H}}^2. \tag{14.91}$$

Suppose K is Toeplitz and for all $O \in SO(d)$, $K(Ox, 0) = K(x, 0)$, then for any $h \in \mathbf{SE}(d)$ the function $d(I, I')$ is invariant under its action:

$$d^2(I \circ h, I' \circ h) = d^2(I, I'), \tag{14.92}$$

and the function $d : \mathcal{I} \times \mathcal{I} \to \mathbb{R}^+$ defined by Eqn. 14.90 is a metric distance on $\mathcal{I}$.

Proof Calculating the energy gives

$$d^2(I, I') = \inf_{\substack{g_t, I_t : g_0 = \mathrm{id}, \\ I_0 = I, I_1 = I' \circ g_1}} \int_0^1 \left\| \frac{dI_t}{dt} \circ g_t^{-1} \right\|_{g_t^{-1}.\mathcal{H}}^2 dt$$

$$\stackrel{(a)}{=} \inf_{\substack{g_t, I_t : g_0 = \mathrm{id}, \\ I_0 = I, I_1 = I' \circ g_1}} \int_0^1 \left\| \frac{dI_t}{dt} \right\|_{\mathcal{H}}^2 dt, \tag{14.93}$$

with (a) following from Lemma 14.28. Thus I_t is a linear interpolation of its boundary conditions

$$I_t = (1-t)I + t(I' \circ g_1) \quad \text{with} \quad \frac{dI_t}{dt} = I' \circ g_1 - I. \tag{14.94}$$

Substituting this back into the definition of $d(I, I')$ completes the proof of Eqn. 14.91 establishing the connection of the variational problem for $d(I, I')$ to the maximum likelihood estimation.

For an arbitrary covariance $d(I, I')$ is not a metric distance on $\mathcal{I}$; however, from Theorem 12.5 of Chapter 12 once it is left invariant to the group then it is a metric. To prove left invariance,

$$d^2(I \circ h, I' \circ h) = \inf_g \| I \circ h - I' \circ (h \circ g) \|_{\mathcal{H}}^2 \tag{14.95}$$

$$= \inf_g \sum_n \frac{|\langle I \circ h - I' \circ h \circ g, \psi_n \rangle_2|^2}{\lambda_n} \tag{14.96}$$

$$\stackrel{(a)}{=} \inf_g \sum_n \frac{|\langle I - I' \circ h \circ g \circ h^{-1}, \psi_n \circ h^{-1} \rangle_2|^2}{\lambda_n} \tag{14.97}$$

$$= \inf_g \| I - I' \circ h \circ g \circ h^{-1} \|_{h^{-1}.\mathcal{H}}^2 \tag{14.98}$$

with (a) following from the substitution of variables $x = h^{-1}(y)$. Since the Toeplitz covariance $K(x, 0)$ is invariant with respect to rotations, then $h^{-1}.\mathcal{H} = \mathcal{H}$ and the distance reduces to $d^2(I, I')$.

The proof of symmetry and triangular inequality follows from the invariance of the metric. $\square$

14.7.3 Computational Scheme

Corollary 14.30 *For the rigid motions $g_t \in \mathbf{SE}(d)$ with $J_t = I_t \circ g_t^{-1}$, then defining Ω_t to be the skew-symmetric matrix satisfying $dO_t/dt = \Omega_t O_t$, $\beta_t = db_t/dt - \Omega_t b_t$, then*

$$d(I, I') = \inf_{\substack{\Omega(t), \beta(t) : \Omega_0 = 0, \beta_0 = 0 \\ J_t : J_0 = I, J_1 = I'}} \int_0^1 \left\| \frac{\partial J_t}{\partial t} + \langle \nabla J_t, \Omega_t x + \beta_t \rangle_{\mathbb{R}^d} \right\|_{g_t^{-1}.\mathcal{H}}^2 dt,.$$

$$\tag{14.99}$$

Proof Computing the derivative of $I_t = J_t \circ g_t$ yields

$$\frac{dI_t}{dt} = \frac{\partial J_t}{\partial t} \circ g_t + \langle \nabla J_t \circ g_t, \Omega_t(O_t x + b_t) + \beta_t \rangle_{\mathbb{R}^d}, \qquad (14.100)$$

which substituted gives Eqn. 14.99. □

This property of the metric distance $d(I, I')$ allow an efficient alternating minimization algorithm for the optimization problem 14.29. In fact, Eq. 14.94 implies that flow $J_t = I_t \circ g_t^{-1}$ is also given as a linear combination of its boundary conditions

$$J_t = (1-t)\left(I \circ g_t^{-1}\right) + t\left(I' \circ \left(g_1 \circ g_t^{-1}\right)\right). \qquad (14.101)$$

On the other hand, with J_t fixed, velocities Ω_t and β_t are given as the solution of a simple quadratic optimization problem

$$(\Omega_t, \beta_t) = \arg\inf_{\Omega_t, \beta_t} \int_0^1 \left\| \frac{\partial J_t}{\partial t} + \langle \nabla J_t, \Omega_t x + \beta_t \rangle_{\mathbb{R}^d} \right\|_{\mathcal{H}}^2 dt. \qquad (14.102)$$

Thus, both the minima in J_t and Ω_t, β_t with, respectively, Ω_t, β_t and J_t fixed, are given by closed-form expressions. This suggests the following alternating minimization algorithm for computing $d(I, I')$.

Algorithm 14.31 (Bitouk) *Starting with $g_t^{(0)} = \text{id}$ and $J_t^{(0)} = (1-t)I + tI'$, we apply the following iterative scheme*

1. $\left(\Omega_t^{(n+1)}, \beta_t^{(n+1)}\right) = \arg\inf_{\Omega_t, \beta_t} \int_0^1 \left\| \frac{\partial J_t^{(n)}}{\partial t} + \langle \nabla J_t^{(n)}, \Omega_t x + \beta_t \rangle_{\mathbb{R}^d} \right\|_{\mathcal{H}}^2 dt.$

2. $g_t^{(n+1)} = \left(O_t^{(n+1)}, b_t^{(n+1)}\right),$
   where $O_t^{(n+1)} = \int_0^t \exp\left(\Omega_\sigma^{(n+1)}\right) d\sigma,$
   and $b_t^{(n+1)} = \int_0^t \left[O_t^{(n+1)} - O_\sigma^{(n+1)}\right] \beta_\sigma^{(n+1)} d\sigma.$

3. $J_t^{(n+1)} = (1-t)I \circ \left(g_t^{(n+1)}\right)^{-1} + t\left(I' \circ g_1^{(n+1)}\right) \circ \left(g_t^{(n+1)}\right)^{-1}.$

The iterative process is expected to converge to the values of J_t, Ω_t and β_t which attain the infimum for $d(I, I')$. Shown in Figure 14.20 are examples from the tanks in dense foliage corresponding to obscuring clutter. The panels show image flow along the geodesic generated by the algorithm $J(x, t_k)$ at $t_k = k/5, k = 0, 1, \ldots, 5$.

(1) (2) (3) (4) (5)

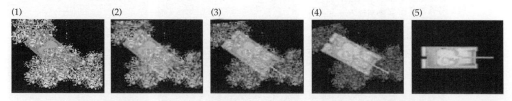

Figure 14.20 Panel 1 shows the tanks in dense foliage corresponding to obscuring clutter. Panels show image flow along the geodesic under the Euclidean norm $\|\cdot\|^2$ for $J_{t_k}(x)$ at $t_k = k/5, k = 0, 1, \ldots, 5$. Results taken from Bitouk.

14.7.4 Empirical Construction of the Metric from Rendered Images

Although it has been pointed out by many authors that natural images are strongly non-Gaussian [338,339], second-order statistics capture important information about spatial correlation between pixels. To verify that, we sampled images from distributions specified by learned Toeplitz covariances.

Given training images, defined on the $N \times N$ lattice of the image plane, $I(x_i, x_j)$, assume periodic boundary conditions $I(x_i, x_j) = I(x_{i+N}, x_j) = I(x_i, x_{j+N})$. The Toeplitz covariance represented by the empirically estimated mean $\mu = 1/N^2 \sum_{i=1}^{N} \sum_{j=1}^{N} I(x_i, x_j)$ and the $N \times N$ covariance matrix

$$K(i,j) = \frac{1}{N^2} \sum_{k=0}^{N-1} \sum_{l=0}^{N-1} \left(I(x_k, x_l) - \mu\right) \left(I(x_{i+k \mod N}, x_{j+l \mod N}) - \mu\right). \tag{14.103}$$

The power spectral density S is given by the discrete Fourier transform of the covariance

$$S(u,v) = \sum_{k=0}^{N-1} \sum_{l=0}^{N-1} K(k,l) e^{(-j2\pi/N)(uk+vl)}. \tag{14.104}$$

The eigenfunctions of the cyclo-stationary covariance $K(i,j)$ are complex exponentials $\phi_{u,v}(k,l) = e^{-j(2\pi/N)(uk+vl)}$, and the eigenvalues $\lambda_{u,v}$ are given by the power spectral density, $\lambda_{u,v} = S(u,v)$.

Images can be synthesized according to

$$I_{synth}(x_i, x_j) = \sum_{u=0}^{N-1} \sum_{v=0}^{N-1} \sqrt{\lambda_{u,v}} X_{u,v} \phi_{u,v}^*(i,j) \tag{14.105}$$

Figure 14.21 Top row: observed images. Bottom row: images synthesized using the second-order model. Images taken from Bitouk.

as second-order random fields specified by Toeplitz covariance $K(i,j)$, where, by the Kahunen–Loeve theorem, $X_{u,v}$ is an independent increment process, $E(X_{u,v}X_{u',v'}) = \delta(u-u')\delta(v-v')$ Figure 14.21 shows pairs of the observed images and the corresponding images synthesized using the GMRF model.

14.8 Target detection/identification in EO imagery

Bitouk has examined the performance of optimum detectors in natural clutter. For the clutter experiments conducted by Bitouk, a dataset of ray-traced target chips were generated to represent natural clutter. The generated dataset consists of more than 10,000 images. The ray-traced images in the dataset contain targets in a randomly synthesized terrain taking into account occlusion, shadows and lighting variation as well as other effects usually encountered in natural scenes. Terrain was generated by random placement of trees on grass backgrounds according to the specified distribution densities. Examples of such imagery are shown in Figure 14.17.

In a statistical framework, target detection is formulated as a hypothesis testing problem and tackled using classic detection theory [328]. Assume a binary scenario with two hypotheses H_0 and H_1, where under the hypothesis H_0, the observed image contains no targets, $I^D = W$, and, under H_1, the target is present, $I^D = I_{\text{temp}}(g) + W$. Note that the pose g is unknown and plays the role of a nuisance parameter. Generalized likelihood ratio test yields the following decision rule written via the generalized likelihood ratio or in terms of metric distances between the observed image I^D and the template:

$$\left\|I^D\right\|_{\mathcal{H}}^2 \underset{H_0}{\overset{H_1}{\gtrless}} \inf_{g \in G} \left\|I^D - I_{\text{temp}} \circ g\right\|_{\mathcal{H}}^2 + \nu \tag{14.106}$$

$$d^2(I^D, 0) \underset{H_0}{\overset{H_1}{\gtrless}} d^2(I^D, I_{\text{temp}}) + \nu. \tag{14.107}$$

In this equation the threshold ν is changed to generate different points along the receiver operating curve (ROC). The probability of correct detection P_D and the probability of a false alarm are given by $P_D = Pr(H_1|H_1)$ and $P_F = Pr(H_1|H_0)$.

The top row of Figure 14.22 displays ROC curves for different clutter densities corresponding to the covariance norm $\|.\|_{\mathcal{H}}$ where the covariance matrix K was estimated from the set of training images of clutter and usual Euclidean norm $\|.\|$.

Let H_0 and H_1 be the hypotheses corresponding to the presence of the class 0 or class 1 target in the observed image, respectively. In our numerical experiment class 0 corresponds to a Jeep vehicle, and class 1 corresponds to a T72 tank. The generalized likelihood ratio test or the GLRT decision rule involving the comparison of metric distances from the observed image I^D to the corresponding templates is given by

$$\inf_{g \in G}\left\|I^D - I_{\text{temp}}^{(0)} \circ g\right\|_{\mathcal{H}}^2 \underset{H_0}{\overset{H_1}{\gtrless}} \inf_{g \in G}\left\|I^D - I_{\text{temp}}^{(1)} \circ g\right\|_{\mathcal{H}}^2 + \nu \tag{14.108}$$

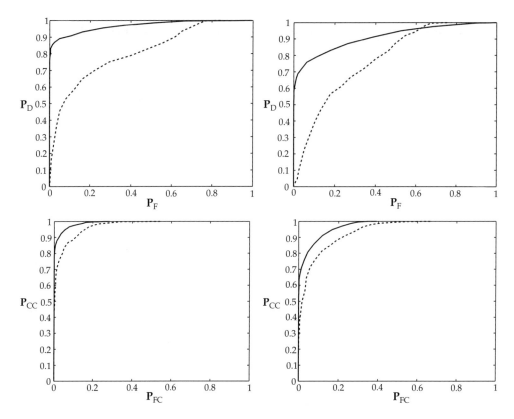

Figure 14.22 Top row: ROC curves for detection of a T72 tank for different densities of clutter. Solid line—covariance norm; dashed line—Euclidean norm. Bottom row: ROC curves for classification of a T72 tank versus a Jeep for different densities of clutter. Solid line—covariance norm; dashed line—Euclidean norm. Results from Bitouk.

$$d^2(I^D, I^{(0)}_{\text{temp}}) \underset{H_0}{\overset{H_1}{\underset{<}{>}}} d^2(I^D, I^{(1)}_{\text{temp}}) + \nu. \tag{14.109}$$

The probability of correct classification P_{CC} and the probability of false classification P_{FC} are given by $P_{CC} = \Pr(H_1|H_1)$ and $P_{FC} = \Pr(H_1|H_0)$. The bottom row of Figure 14.22 shows the ROC curves for target classification using the covariance norm $\|.\|_\mathcal{H}$ where K was estimated from the set of training images of clutter and Euclidean norm $\|.\|$.

15 INFORMATION BOUNDS FOR AUTOMATED OBJECT RECOGNITION

ABSTRACT Examine information bounds for understanding rigid object recognition involving the low-dimensional matrix groups. Minimum-mean-squared error bonds and mutual information bounds are derived for recognition and identification.

15.1 Mutual Information for Sensor Systems

15.1.1 Quantifying Multiple-Sensor Information Gain Via Mutual Information

Examine information bounds for understanding rigid object recognition involving the low-dimensional matrix groups. Minimum-mean-squared error bonds and mutual information bounds may be derived for recognition and identification. The source–channel view provides the absolutely optimal basis for which the source characterization and channel characterization must be combined for optimal information use. This source/channel decomposition allows for the consideration of the information contained in the prior distribution expressing prior knowledge about configurations of patterns before measurement, and the statistical likelihood function associated with the observed image data expressing the way the sensor collects information from the world. This source–channel view of pattern transformation provides the framework for the optimal basis for characterizing the information flow through remote sensors. The posterior density summarizes all of the information that the sensing channel data provides about the source of imagery. The information gain by the observer through the sensor is measured via the mutual information measure, the difference in entropies between the prior distribution before data observations, and after measurements.

Definition 15.1 *The* **mutual information** *between $I \in \mathcal{I}$ and $I^D \in \mathcal{I}^D$ is given for continuous parameter spaces according to*

$$\Delta h(I; I_1^D, \ldots, I_n^D) = h(I) - h(I | I_1^D, \ldots, I_n^D)$$

$$= E_{p(I, I_1^D, \ldots, I_n^D)} \log \frac{p(I | I_1^D, \ldots, I_n^D)}{p(I)}. \quad (15.1)$$

Entropies and mutual information for discrete distributions are denoted in capitals:

$$\Delta H(I; I_1^D, \ldots, I_n^D) = H(I) - H(I | I_1^D, \ldots, I_n^D) \quad (15.2)$$

$$= E_{P(I, I_1^D, \ldots, I_n^D)} \log \frac{P(I | I_1^D, \ldots, I_n^D)}{P(I)}. \quad (15.3)$$

We notice that the inequalities for conditional entropies implies information gain is monotonic in sensor measurements.

Theorem 15.2 *Let $I_1^D, I_2^D, \ldots,$ be independent observations of the image $I \in \mathcal{I}$, then the information gain about I is monotonic in sensors:*

$$\Delta h(I; I_1^D, \ldots, I_n^D) \geq \Delta h(I; I_1^D), \quad n \geq 1, \quad (15.4)$$

$$\Delta H(I; I_1^D, \ldots, I_n^D) \geq \Delta H(I; I_1^D), \quad n \geq 1. \quad (15.5)$$

Proof The monotonicity of the information is implied by the entropy inequality

$$H(I|I_1^D) - H(I|I_1^D, \ldots, I_n^D) = E\{\log \frac{P(I|I_1, \ldots, I_n^D)}{P(I|I_1)} \tag{15.6}$$

$$\geq 0, \tag{15.7}$$

implying the change in information is ordered, $\Delta H(I; I_1^D, \ldots, I_n^D) \geq H(I; I_1^D)$. Similarly for the mutual information involving the densities. □

Operating in the Bayes setting implies that information increases performance. We have already seen this in terms of the Fisher information which is monotonic in the number of observations. The mutual information between two sets of random variables directly measures the dependence between them [7]; we use it to quantify the dependence between the remote observation of the scene and the object parameters of pose, position, and class. Mutual information is a fundamental performance metric providing the bound on communication rate for errorless communication of the parameters describing the scene via the remote-sensing channel. View the observation image data I^D as a random vector statistically processed to infer the random parameters of object pose and signature.

Example 15.3 (Pose estimation Cooper [340]) Take as the random element being transmitted elements of the orthogonal group $O \in \mathbf{SO}(2)$ and identify with the underlying imagery with the rotation group generating it $I(\cdot) = I_\alpha(O \cdot) \leftrightarrow O \in \mathbf{SO}(2)$. The prior density $p(I_\alpha(O)), I \in \mathcal{I}$ is induced via the uniform prior $p(O), O \in \mathbf{SO}(2)$. The likelihood $p(I^D|I_\alpha(O))$ is determined by the various sensors: VIDEO, LADAR, and RADAR, and the posterior summarizes all of the information that the multisensor data provides about the source. The mutual information gain about the orthogonal motion is given by

$$\Delta h(O; I_1^D, \ldots, I_n^D) = h(O) - h(O|I_1^D, \ldots, I_n^D). \tag{15.8}$$

For the imaging model Cooper takes the *perspective projection* according to the mapping $(x_1, x_2, x_3) \mapsto (x_1/x_3, x_2/x_3)$. Then $TI_\alpha(O)$ is the mean-field conditioned on the geometric transformation O and signature I_α. The measurements I^D are modeled as a Gaussian random field with mean the projective transformation of the scene $TI_\alpha(O)$ in noise, variance σ^2 in each pixel $i = 1, 2, \ldots d$. Take the prior $p(\cdot)$ to be uniform on the circle, the posterior density becomes proportional to

$$p(O|I^D) = \frac{1}{(2\pi\sigma^2)^{d/2}} e^{-1/2\sigma^2 \|I^D - TI_\alpha(O)\|^2}, \quad O \in \mathbf{SO}(2). \tag{15.9}$$

Sensor fusion is readily accommodated via the use of the joint likelihood. Figure 15.1 shows entropies and mutual informations for the FLIR and VIDEO sensors. To calculate the entropies Cooper discretizes the integrals over $\mathbf{SO}(2)$ numerically evaluating the discrete set $\mathcal{O} = \{o_m : m = 1, \ldots, M\} \subset \mathbf{SO}(2)$, placing a uniform prior on $\mathcal{O}$. For this reason denote the entropies $H(\cdot)$ with capital symbols. Panel 1 shows the conditional entropies of the object pose for the FLIR sensor $H(O|I_{FLIR}^D)$ (solid) and video sensor $H(O|I_{VIDEO}^D)$ (dashed). The conditional entropy for the fusion case $H(O|I_{FLIR}^D, I_{VIDEO}^D)$ is shown via the x's. Panel 2 shows the mutual information. Note that posterior entropy is decreased by increasing the signal and fusing multiple sensors, implying the mutual information is increased. Panel 3 examines the dependence of the mutual information gain on object geometry ID, $\Delta H(O; I^D)$ is plotted for three CAD models, T62 tank (dot-dashed), HMMV (solid), and truck (dashed). Note the relative variations of the curves for the three objects.

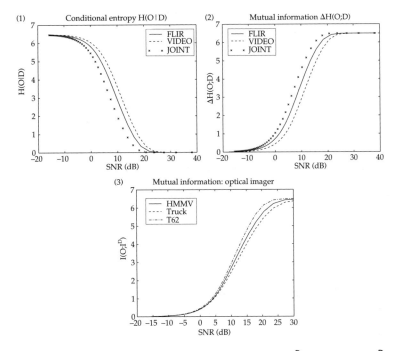

Figure 15.1 Panel 1 shows the conditional entropies $H(O|I_{FLIR}^D)$ (—), $H(O|I_{VIDEO}^D)$ (- - -), $H(O|I_{FLIR}^D, I_{VIDEO}^D)$ (x) are plotted versus increasing SNR. Panel 2 the mutual informations $\Delta H(O; I_{FLIR}^D)$ (—), $\Delta H(O; I_{VIDEO}^D)$ (- - -), $\Delta H(O; I_{FLIR}^D, I_{VIDEO}^D)$ (x) are plotted versus increasing SNR. Panel 3 shows Bits expressed in log base 2.

15.1.2 Quantifying Information Loss with Model Uncertainty

Model uncertainty comes in many forms. We shall examine extensively the variable signature model in which the underlying image field $I(\cdot)$ is modelled as an element of a Hilbert space in the orbit defined by a basis expansion. Inference and information gain involves both the pose and identity of the objects as well as the signature of an infinite dimensional parameter. Model the template I_{temp} as a conditionally scalar-valued Gaussian random field. The orbit is the set of all imagery generated from the template and a linear combination of basis elements $\phi_1, \ldots, \phi_N$ spanning the temperature signatures:

$$\mathcal{I} = \{I(g), \quad I(\cdot) = \sum_{n=1}^{N} I_n \phi_n(\cdot), \ \phi_1, \ldots, \phi_N, \ g \in \mathcal{H}\}. \tag{15.10}$$

The imagers are modeled using *perspective projection*, according to the mapping $(x_1, x_2, x_3) \mapsto (x_1/x_3, x_2/x_3)$. Then $TI(g)$ is the mean-field conditioned on the geometric transformation g and signature I. The measurements I^D are modeled as a Gaussian random field with mean the projective transformation of the scene $TI(g)$ in noise, variance σ^2. The signature and pose are assumed conditionally independent implying that the posterior density is the product of the data-likelihood and the joint prior. Take the prior $p(\cdot)$ to be uniform on the circle with the object signatures modeled as a Gaussian random field; $I_n, n = 1, \ldots, N$ are independent normal with zero mean and variance

λ_n the eigenvalue of the *n*th eigensignature ϕ_n. The posterior density on $I = I_\alpha g$ becomes

$$P(I(g), I | I^D) = \frac{1}{Z(I^D)} p(I^D | I(g), I) p(I(g)) p(I), \tag{15.11}$$

$$= \frac{1}{(2\pi\sigma^2)^{d/2}} e^{-1/2\sigma^2 \|I^D - TI(g)\|^2} \prod_{n=1}^{N} \frac{1}{\sqrt{2\pi\lambda_n}} e^{-(I_n^2/2\lambda_n)}. \tag{15.12}$$

Note, that we are assuming the template signature has a finite expansion. [44] Several forms of the mutual information are used to quantify performance and model uncertainty, these being the mutual information gain given the model parameters. The forms of mutual information which play the key role are $\Delta h(I(g); I^D, I)$ and $\Delta h(I_\alpha(g); I^D)$, these two distinguishing between the situation where the model signature is explicitly known and performance is the average over all conditional models versus a high dimensional nuisance variable which must be integrated out:

$$\Delta h(I(g); I^D, I) = -E\{\log p(I(g))\} + E\{\log p(I(g) | I^D, I)\}, \tag{15.13}$$

$$\Delta h(I(g); I^D) = -E\{\log p(I(g))\} + E\{\log p(I(g) | I^D)\}. \tag{15.14}$$

Naturally, model uncertainty results in loss of information in the image parameters. This is expressed via the Kullback–Liebler under the models.

Corollary 15.4 *The information loss is nonnegative given by the Kullback–Liebler distance:*

$$\Delta h(I(g); I^D, I) - \Delta h(I(g); I^D) = D(p(I(g) | I^D, I) \| p((I(g)) | I^D)) \tag{15.15}$$

$$\geq 0. \tag{15.16}$$

Proof Conditioning decreases entropy, Theorem 15.2. □

Example 15.5 (Pose estimation for signature variability (Cooper[335])) Cooper [335] has studied estimation of pose through lighting and signature variation. Define the image signature indexed over the surface. Inference of $I(s)$ reduces to the inference of pose and temperature profile. Model the signature I as a scalar-valued Gaussian random field with mean corresponding to the template at fixed inertial coordinates under the identity transformation. The orbit is the set of all imagery generated from the template and a linear combination of basis elements $\phi_1, \ldots, \phi_N$ spanning the signatures.

The *N*-dimensional basis is constructed as eigenfunctions of the empirical covariance generated from many realizations of actual signatures from varying illuminations of the objects (see Figure 14.8). The openGL simulation software is used for generating the database of illuminations. The top row shows a variety of signatures corresponding to varying illumination. The first several eigensignatures $\phi_{1,2,3}(\cdot)$ are used in generating the simulations (see Figures 14.6, 14.6, Section 14.2.1 , e.g.).

The posterior density on the pose parameter becomes

$$p(O, I | I^D) = \frac{1}{Z(I^D)} p(I^D | O, I) p(O) p(I), \tag{15.17}$$

$$= \frac{1}{(2\pi\sigma^2)^{d/2}} e^{-1/2\sigma^2 \|I^D - TI(O)\|^2} \prod_{n=1}^{N} \frac{1}{\sqrt{2\pi\lambda_n}} e^{-I_n^2/2\lambda_n}, \tag{15.18}$$

---

[44] Otherwise, we must introduce the finite cylinder expansion $\pi_N = \prod_{n=1}^{N} (1/\sqrt{2\pi\lambda_n}) e^{-(I_n^2/2\lambda_n)}$ and $p_N = p(I^D | I_1, \ldots, I_N)$ and examine the estimation problem in the limit $\lim_{N \to \infty} p_N \pi_N$.

where $I = \sum_{n=1}^{N} I_n \phi_n$. Cooper [333, 335] has studied signature variation in infrared imagery (see below).

Examine the forms of mutual information which quantify model uncertainty, the first $\Delta h(O; I^D, I)$ measuring the *average* information gain conditioned on knowledge of the model signatures, and the second $\Delta h(O; I^D)$ measuring the *average* information gain unconditional:

$$\Delta h(O; I^D, I) = h(O) - h(O|I^D, I), \tag{15.19}$$

$$= -E\{\log p(O)\} + E\{\log p(O|I^D, I)\}. \tag{15.20}$$

$$\Delta h(O; I^D) = h(O) - h(O|I^D), \tag{15.21}$$

$$= -E\{\log p(O)\} + E\{\log p(O|I^D)\}. \tag{15.22}$$

For the second conditional entropy, the conditional density $p(O|I^D)$ is calculated using Bayes rule over all conditioning signature fields:

$$p(O|I^D) = E\{p(O|I^D, I)|I^D\}. \tag{15.23}$$

Cooper performs this calculation using Monte Carlo random sampling in the orbit by exploiting the fact that the signatures $I \in \mathcal{I}$ are modelled as Gaussian in the principle component expansion. An estimate of $p(O|I^D)$ is computed by generating $n = 1, \ldots, N$ signature samples $I^n \in \mathcal{I}$ which are Gaussian distributed $\hat{p}_n$ and then computing the empirical average exploiting the property

$$\lim_{N \to \infty} \frac{1}{N} \sum_{n=1}^{N} p(O|I^D, I^n) = p(O|I^D). \tag{15.24}$$

Panel 1 of Figure 15.2 shows $\Delta H(O; I^D)$ and $\Delta H(O; I^D, I_\alpha)$, the average information gain in object pose given the FLIR imagery and the correct object signature. Conditioning on the signature reduces the entropy, implying that these forms of mutual information are ordered:

$$\Delta H(O; I^D, I) = H(O) - H(O|I^D, I)$$
$$\geq H(O) - H(O|I^D) = \Delta H(O; I^D).$$

The solid curve, $\Delta H(O; I^D, I)$, quantifies the average information gain due to noisy FLIR observations over the space of possible object signatures when correct signature information is available *a priori*. The dashed curve, $\Delta H(O; I^D)$ quantifies the average information gain without *a priori* signature information. The difference between the two, shown in panel 2 represents the average cost of the unknown signature in terms of information regarding the object pose.

The bottom row of Figure 15.2 shows the information loss associated with incomplete model specification. The dashed lines show pose information loss assuming improper template signatures. The x's show the performance loss of the marginalized model. Note that it loses the least information since it hedges its bets by computing the Bayes marginal over all possible template signatures. The information loss is measured in each case via the Kullback–Liebler number between the models $D(p||q)$:

$$\Delta H(O; I^D, I) - \Delta H(O; I^D) = H(O|I^D) - H(O|I^D, I) \tag{15.25}$$

$$= E\left\{\log \frac{p(O|I^D, I)}{p(O|I^D)}\right\}. \tag{15.26}$$

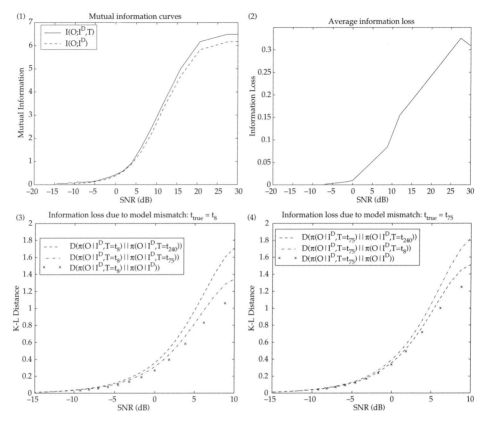

Figure 15.2 Top row: Panel 1 shows the mutual information for the FLIR sensor $\Delta H(O; I^D, I)$ (—) and $\Delta H(O; I^D)$ (- - - -), showing the information loss due to the absence of knowledge of the object signature. Panel 2 shows the average information loss due to signature variation, as measured by the Kullback–Leibler distance, $D(p(O|I^D, I) \| p(O|I^D))$, versus increasing signal to noise ratio. Bottom row: Panels 3, 4 measure information loss between the marginalized model and the true signature model $I = 8$ (panel 3) and $I = 75$ (panel 4). Each panel shows the information loss taken by choosing models with the incorrect signature as indicated by the dashed lines. The marginal model information loss is shown by the x's. In both panels, information loss is minimized by the Bayes signature model without *a priori* assumption of the object signature. Bits expressed in log base 2.

The Bayes signature model minimizes the information loss. Modeling the object signature via marginalizing the joint posterior density $p(O, I|I^D)$ gives the minimal information loss.

15.1.3 Asymptotic Approximation of Information Measures

Example 15.6 (Gaussian Entropy and Fisher Information) We now examine asymptotic approximations to the posterior distribution following [60, 325] and calculate asymptotic forms of the posterior entropy assuming the SNR $\to \infty$. Let θ_T denote the true value of Θ. The Fisher information is computed as

$$F(\theta_T) = \frac{\left\| \frac{d}{d\theta} TI(\theta_T) \right\|^2}{2\sigma^2}. \tag{15.27}$$

15.1 MUTUAL INFORMATION FOR SENSOR SYSTEMS

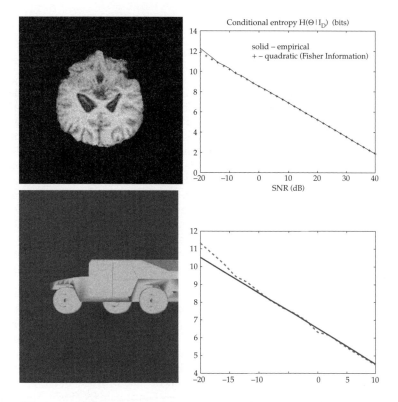

Figure 15.3 Top row shows CAD models. Bottom row shows empirical entropies for pose versus analytic entropy predicted by the variance of the Fisher information (see also Plate 29).

Figure 15.4 shows the empirical entropy (solid curves) and quadratically approximated (dashed curves).

Example 15.7 (Bitouk) Shown in Figure 15.3 are results on approximation of the entropy via the Fisher information. The top row shows the CAD models. The bottom row shows the empirical entropies for pose versus analytic entropy predicted by the variance of the Fisher information.

Example 15.8 (Relationship to the Hilbert Schmidt bound) The second estimate of the Fisher information is based on the HS pose estimator [341, 342]. This estimate is a function of the mean squared error

$$\mathrm{MSE}(O(\theta_T)) = \int_{\mathcal{I}^D} \|\hat{O}_{\mathrm{HS}}(I^D) - O(\theta_T)\|^2 p(I^D|\theta_T) \, d(I^D), \tag{15.28}$$

evaluated via Monte Carlo random sampling. We relate the HS squared error on $\mathbf{SO}(2)$ to squared error on $[0, 2\pi)$ by Taylor's series expansion:

$$\|O(\theta) - O(\theta_T)\|^2 = 4 - 4\cos(\theta - \theta_T) = 4 - 4\left(1 - \frac{1}{2}(\theta - \theta_T)^2 + \cdots\right) \tag{15.29}$$

$$\simeq 2(\theta - \theta_T)^2, \tag{15.30}$$

implying

$$F(\theta_T) \sim \left(\frac{1}{2}\mathrm{MSE}(O(\theta_T))\right)^{-1}. \qquad (15.31)$$

Figure 15.4 shows the Gaussian entropy calculated using the Hilbert–Schmidt bound as an estimate of the Fisher information. The column shows the T62 tank at SNR = 5 dB (top) and 30 dB (bottom). For these experiments the discrete set of possible orientations is $\{\theta_m = \pi m/1800 : m = 0, \ldots, 900\}$ to obtain more accurate results at high SNR. The results appear in column 2 of Figure 15.4 for $\theta_T = 30°$. The numerical results are the solid curves. The dot–dashed curves are the quadratic approximation based on the empirical variance, and the dashed curves are the quadratic approximations based on the HS squared error. Note the range of SNR values for which the quadratic approximations closely fit the numerical results. The accuracy with which the quadratic form approximates the numerical results suggests that for object recognition estimation scenarios with low dimensional parameter spaces, the information measures can be modeled asymptotically for general ranges of SNR. In the simulations of Figure 15.4, $H(\Theta|i^D)$ achieves its maximum of $\log_2(900) = 9.81378$ bits.

Example 15.9 (Commanche FLIR) Now examine the estimation of the signature and pose simultaneously using maximum *a posteriori* (MAP) estimation. The approach is to maximize the posterior probability over the *joint* parameter space, $\mathbf{SO}(2) \times \mathbb{R}^{20}$. The maximization is accomplished via steepest descent using numerical approximations to the derivative of the joint posterior. The observations are generated by PRISM.

These results are extended to real FLIR imagery in Figure 15.5. Shown in Figure 15.5 are MAP estimators for the Commanche data courtesy of Dr. J. Ratches of the Night Vision Laboratory Electronic Sensors Directorate (NVESD). Shown in the top row are FLIR observations of a tank model from the Commanche data set courtesy of Dr. J. Ratches U.S. Army NVESD. The bottom row shows renderings of the MAP estimates of the orientation and signature from the corresponding observation of the top row generated by estimating the principle modes of temperature variation of the PRISM data and estimating rotations, translations, and temperature variation to match the Commanche data set. The parameters of this representation are the principle

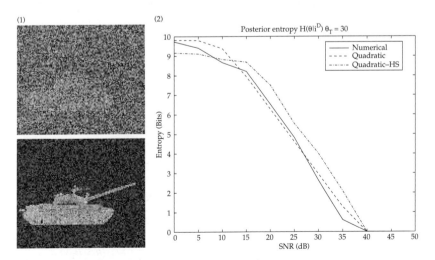

Figure 15.4 Column 1: Video imagery of the T62 tank at SNR = 5 dB (top) and 30 dB (bottom). Column 2: Numerical (—) and analytic (- - -) approximations to $H(\Theta|i^D)$ for $\theta_T = 30°$.

15.1 MUTUAL INFORMATION FOR SENSOR SYSTEMS

eigenfunctions and eigenvalues which represent all of the orientations and translation images from the Commanche data set. The bottom row shows the estimates of pose and temperature. In each case, the algorithm was initialized with orientation broadside, gun facing to the right. The initial object signature was the mean signature from the database. The signature was represented using the $N = 20$ dimensional subspace of signature principal components. In all three cases the correct object orientation was estimated.

Shown in the bottom row of Figure 15.5 are the results of studying the bounds for the Commanche database. The SNR model for FLIR from chapter 13, Eqn. 13.40 based on the mean object and background pixel intensity is used to estimate the operating SNR of 5.07 dB of the FLIR sensor for the NVESD data. The plots in the bottom row of Figure 15.5 show the mean squared error loss (left panel) and

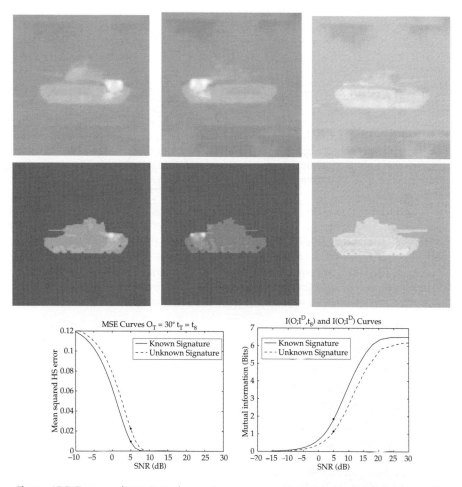

Figure 15.5 Top row shows FLIR observations courtesy of Dr. J. Ratches U.S. Army NVESD. Second row shows renderings of the MAP estimates of orientation and signature from the corresponding observation of the top row. Third row: Panel 7 shows the mean squared error for orientation estimation via FLIR with and without signature information. Panel 8 shows the mutual information curves for FLIR with and without signature information. In both cases, the solid lines correspond to performance with signature information, and the dashed curves without signature information. The operating SNR of the NVESD Commanche FLIR data is indicated on the curves by the solid dots. Bits expressed in log base 2. (see also Plate 30).

information loss (right panel) due to object signature variation, labeling the points on the performance curves corresponding to the operating SNR of the FLIR sensor. The mean-squared-error curves are computed using the Hilbert–Schmidt estimator conditioning on the object orientation $\theta_T = 30°$. Note that the information loss due to signature uncertainty is 0.86 bits with associated mean squared error performance loss by a multiplicative factor of $2.173 = 0.0213/0.0098$. This is consistent with classic MSE bounds which relate mutual information with estimation accuracy. For example, from Shao et al. [343] we have

$$\mathcal{E}\{(\theta - \hat{\theta}(I^D))^2\} \geq \frac{1}{2\pi e} \exp(2 \max_{p(\theta)}[H(\theta) - I(\theta; I^D)]) = \frac{1}{2\pi e} \exp(2 \max_{p(s)} H(\theta|I^D)), \quad (15.32)$$

which would predict a loss by the factor of $2^{0.86} = 1.815$ which is quite similar to the 2.173 loss actually observed.

15.2 Rate-Distortion Theory

Information theory answers directly the question of how complex an object database (source codebook) is and therefore how much computation does the object recognition system (decoder) have to perform to achieve optimal performance. With rate distortion, we can answer the following question precisely: Given limited computational and storage resources for object recognition, *which orientations or codebook elements* are most informative so as to to minimize probability of error in the repesentation?

Consider the problem of determining an estimate $\hat{X}$ of X; clearly any estimation approach which requires searching for the most likely over the parameter spaces $\mathcal{X}$ for the continuum of scale, rotation, and translation groups is too large. Hence we need to define a database (codebook) the set $\hat{\mathcal{X}} \subset \mathcal{X}$ for doing the estimation. If we choose $|\hat{\mathcal{X}}| << \mathcal{X}$, what is the penalty? In general, what is the optimum choice for the database. How should $\hat{\mathcal{X}}$ be chosen? Is the choice sensitive at all to the geometry or reflectivity of the object whose parameters we are trying to estimate? Clearly if the object has symmetries with respect to the sensor, one need not store extra copies of the same object.

15.2.1 The Rate-Distortion Problem

We now examine the problem of constructing codebooks for a random source $X \in \mathcal{X}$ with density $p(x), x \in \mathcal{X}$. One approach would be to brute force generate a codeword for every possible value of $X \in \mathcal{X}$. Clearly, if $\mathcal{X}$ is the continuum this is impossible. Even for discrete sources, the goal may be to define approximating sets which for a minimum complexity (logarithmic size) have optimal approximation properties, i.e. they minimize distortion over all approximations of the same complexity. This places us firmly in the rate-distortion theory in which complexity or rate of codebooks is traded off against approximation accuracy or distortion.

Definition 15.10 *Define $\hat{\mathcal{X}}$ to be the **reproduction alphabet** for $\mathcal{X}$. Define the **penalty** or distortion function*

$$d : \mathcal{X} \times \hat{\mathcal{X}} \to \mathbb{R}^+, \quad (15.33)$$

where $\mathbb{R}^+$ denotes the set of nonnegative reals. *The distortion measure d quantifies how bad it is to assume that $\hat{X}$ is the real parameter when the real parameter is X.*

For each realization $x \in \mathcal{X}$, the goal is to choose that $\hat{x} \in \hat{\mathcal{X}}$ that minimizes d. This is not directly possible since X is a random variable with conditional probability $p(x), x \in \mathcal{X}$. A reasonable thing to do is to minimize the expected distortion. This is the standard rate-distortion theory problem (see, e.g. Chapter 9 of [344], Chapter 13 of [7], or [345]) where $\mathcal{X}$ corresponds to the source alphabet, $\hat{\mathcal{X}}$ the reproduction alphabet and $d(x, \hat{x})$ is the fidelity criterion. Rate-distortion theory dictates that for any specified average cost D there exists a nonnegative number $R(D)$ such that if $R > R(D)$ then for n sufficiently large there exist encoding and decoding functions

$$f: \mathcal{X}^n \to \{1, 2, \ldots, 2^{nR}\}, \tag{15.34}$$

$$g: \{1, 2, \ldots, 2^{nR}\} \to \hat{\mathcal{X}}^n, \tag{15.35}$$

such that

$$E[d_n(X^n, g(f(X^n)))] \leq D, \tag{15.36}$$

where $X^n = X_1, \ldots, X_n$ is a sequence of n-observations from a stationary, ergodic process, and d_n is the average distortion defined via $d(\cdot)$ by

$$d_n(X^n, \hat{X}^n) = \frac{1}{n} \sum_{i=1}^{n} d(X_i, \hat{X}_i). \tag{15.37}$$

Conversely, if $R < R(D)$ then such functions do not necessarily exist.

Rate-distortion theory dictates that in order to estimate a sequence of parameters with average distortion D from a sequence of observations $X_1, \ldots, X_n$, we need at least $2^{nR(D)}$ elements in the n-fold cartesian product $\hat{\mathcal{X}} \times \cdots \times \hat{\mathcal{X}}$.

The function $R(D)$ is the *rate-distortion function*. When $X_1, \ldots, X_n$ is a sequence of independent and identically distributed random variables, then

$$R(D) = \min_{q(\cdot|x): Ed(X,\hat{X}) \leq D} \Delta H(X; \hat{X}) = \min_{q(\cdot|x): Ed(X,\hat{X}) \leq D} \sum_x \sum_{\hat{x}} p(x) q(\hat{x}|x) \log \left(\frac{q(\hat{x}|x)}{q(\hat{x})} \right), \tag{15.38}$$

where $\Delta H(X; \hat{X})$ is the mutual information between X and $\hat{X}$ and the minimum is over all conditional probabilities $q(\hat{x}|x), \hat{x} \in \hat{\mathcal{X}}, x \in \mathcal{X}, \sum_{\hat{x}} q(\hat{x}|x) = 1$ satisfying $Ed(X, \hat{X}) \leq D$. In this case, Eqn. 15.38 is the first order rate-distortion function which is an upper bound to the general rate distortion bound. Throughout we compute the first order $R(D)$ function which will be *the* rate distortion function only in the independent and identically distributed case.

15.3 The Blahut Algorithm

In only a small number of cases can the rate-distortion curve be calculated analytically. Using the alternating minimization of the Blahut algorithm [7, 346–347] the RD function can be calculated. Assume that the observable space is a finite set $\mathcal{X}$ and density $p(x)$ on $\mathcal{X}$. The transition probability from input to reproduction alphabet $q(\hat{x}|x)$ is the conditional probability distribution on the approximating set $\hat{x} \in \mathcal{X}$ with input $x \in \mathcal{X}$. The goal is to find the transition density which minimizes the mutual information between the input and output sets:

$$q^\star(\cdot|x) = \arg \min_{q(\cdot|x): \bar{D} \leq D} \Delta H(X; \hat{X}) = \arg \min_{q(\cdot|x): \bar{D} \leq D} \sum_x \sum_{\hat{x}} p(x) q(\hat{x}|x) \log \left(\frac{q(\hat{x}|x)}{q(\hat{x})} \right), \tag{15.39}$$

where

$$\overline{D} = \sum_x \sum_{\hat{x}} d(\hat{x}, x) q(\hat{x}|x) p(x). \tag{15.40}$$

The problem is solved by the following alternating minimization procedure (Blahut).

Theorem 15.11 *Choose $q^0(\hat{x}) > 0, \hat{x} \in \hat{\mathcal{X}}$, then define the sequence of random variables $\hat{X}^{(1)}, \hat{X}^{(2)}, \ldots$ with densities $q^{(1)}(\cdot), q^{(2)}(\cdot), \ldots$ on $\hat{\mathcal{X}}$ according to*

$$q^{(n)}(\hat{x}|x) = \frac{q^{(n-1)}(\hat{x})e^{-\lambda d(\hat{x},x)}}{\sum_{\hat{x}} q^{(n-1)}(\hat{x})e^{-\lambda d(\hat{x},x)}}, \tag{15.41}$$

$$q^{(n)}(\hat{x}) = \sum_x p(x) q^{(n)}(\hat{x}|x) = q^{(n-1)}(\hat{x}) \sum_x \frac{p(x)e^{-\lambda d(\hat{x},x)}}{\sum_{\hat{x}} q^{(n-1)}(\hat{x})e^{-\lambda d(\hat{x},x)}}, \tag{15.42}$$

Then, $\hat{X}^{(1)}, \hat{X}^{(2)}, \ldots$ has the property that the mutual information is monotonically nonincreasing,

$$\Delta H(X; \hat{X}^{(1)}) \geq \Delta H(X; \hat{X}^{(2)}) \ldots, \tag{15.43}$$

and the sequence of distortions satisfy the constraint $\sum_{\hat{x}} q^{(n)}(\hat{x}, x) d(\hat{x}, x) p(x) \leq D$ for all n. Fixed points satisfy the necessary minimizer conditions for the rate-distortion problem.

Proof Rewrite the mutual information according to

$$\Delta H(X; \hat{X}) = \min \sum_x \sum_{\hat{x}} q(\hat{x}, x) \log\left(\frac{q(\hat{x}, x)}{q(\hat{x}) p(x)}\right). \tag{15.44}$$

Expand the minimization by fixing $q^{(n-1)}(\hat{x})$, and minimize over $q(\hat{x}, x)$ subject to the constraint giving

$$q^{(n)}(\hat{x}, x) = \operatorname*{arg\,min}_{q(\hat{x},x):\overline{D} \leq D} \sum_x \sum_{\hat{x}} q(\hat{x}, x) \log\left(\frac{q(\hat{x}, x)}{q^{(n-1)}(\hat{x}) p(x)}\right) \tag{15.45}$$

$$= q^{(n-1)}(\hat{x}) p(x) e^{\lambda_0 - \lambda d(\hat{x}, x)}, \tag{15.46}$$

with λ_0 satisfying the integrability to 1 constraint, and λ determining the expected distortion constraint. Dividing by $p(x) > 0$ gives the conditional density

$$q^{(n)}(\hat{x}|x) = \frac{q^{(n-1)}(\hat{x})e^{-\lambda d(\hat{x},x)}}{\sum_{\hat{x} \in \hat{\mathcal{X}}} q^{(n-1)}(\hat{x})e^{-\lambda d(\hat{x},x)}}. \tag{15.47}$$

Now fix $q^{(n)}(\hat{x}, x)$ and minimize with respect to $q(\hat{x})$:

$$q^{(n)}(\hat{x}) = \operatorname*{arg\,min}_{q(\hat{x}):\overline{D} \leq D} \sum_x \sum_{\hat{x}} q^{(n)}(\hat{x}, x) \log\left(\frac{q^{(n)}(\hat{x}, x)}{q(\hat{x}) p(x)}\right) \tag{15.48}$$

$$= \operatorname*{arg\,min}_{q(\hat{x}):\overline{D} \leq D} - \sum_{\hat{x}} \left(\sum_x p(x) q^{(n)}(\hat{x}|x)\right) \log q(\hat{x}) \tag{15.49}$$

$$\stackrel{(a)}{=} \left(\sum_x p(x) q^{(n)}(\hat{x}|x)\right), \tag{15.50}$$

with (a) following from the log x inequality implying the lower bound on cross entropy between densities p, f, $E_p\{\log p/f\} \geq 0$ with equality if and only if $f = p, p$ almost everywhere. Thus we have the following set of monotonic inequalities:

$$\Delta H\left(X; X^{(n-1)}\right) = \sum_{\hat{x}} q^{(n-1)}(\hat{x}, x) \log\left(\frac{q^{(n-1)}(\hat{x}, x)}{q^{(n-1)}(\hat{x})p(x)}\right) \tag{15.51}$$

$$\geq \sum_{\hat{x}} q^{(n)}(\hat{x}, x) \log\left(\frac{q^{(n)}(\hat{x}, x)}{q^{(n-1)}(\hat{x})p(x)}\right) \tag{15.52}$$

$$\geq \sum_{\hat{x}} q^{(n)}(\hat{x}, x) \log\left(\frac{q^{(n)}(\hat{x}, x)}{q^{(n)}(\hat{x})p(x)}\right) = \Delta H(X; X^{(n)}). \tag{15.53}$$

That fixed points $q^*(\hat{x}), q^*(\hat{x}, x)$ satisfy the necessary maximizer condition, maximizing Eqn. 15.39 with respect to $q(\hat{x}, x)$ gives

$$q^*(\hat{x}, x) = \frac{p(x) q^*(\hat{x}) e^{-\lambda d(\hat{x}, x)}}{\sum_{\hat{x}} q^*(\hat{x}) e^{-\lambda d(\hat{x}, x)}}. \tag{15.54}$$

Marginalizing gives the fixed point condition for $q^*(\hat{x}) > 0$:

$$1 = \sum_x \frac{p(x) e^{-\lambda d(\hat{x}, x)}}{\sum_{\hat{x}} q^*(\hat{x}) e^{-\lambda d(\hat{x}, x)}} \ ; \tag{15.55}$$

this is satisfied at the fixed point of Eqns. 15.41 and 15.42. □

As has been shown in [346] when these equations converge to a limit point dependent on λ, this limit is the RD function at the point where the slope is equal to λ. By choosing a number of $\lambda \in [0, \infty)$ values, the RD function can be swept out.

15.4 The Remote Rate Distortion Problem

In pattern theory we rarely observe directly the random variables without noise. In fact, rather than observing the underlying source X with realizations $x \in \mathcal{X}$, the observables are almost always noisy versions of the original random variables such as those arising in remote sensing. Let $X \in \mathcal{X}$ be a random variable with probability $p(\cdot)$, and $Y \in \mathcal{Y}$ be such a random observation containing information about X with transition density $p(y|x), y \in \mathcal{Y}, x \in \mathcal{X}$ assumed known. Upon observation of Y, we would like to choose that $\hat{x} \in \hat{\mathcal{X}}$ that minimizes d. This is not possible since given that $Y = y$, X is a random variable with conditional probability $p(x|y), x \in \mathcal{X}$. A reasonable thing to do is to minimize the expected distortion

$$d(y, \hat{x}) = E_{p(x|y)} d(X, \hat{x}) = \sum_x d(x, \hat{x}) p(x|y). \tag{15.56}$$

For the pattern rate distortion problem, assume the underlying space of random patterns $I \in \mathcal{I}$ with prior density $p(\cdot)$ on the pattern space $\mathcal{I}$. Identify the patterns with the transformation which generates it, and assume the images are in the orbit of the special Euclidean group. Writing the group elements in homogeneous coordinates, then the random transformation $X \in \mathcal{X} \subset \mathbf{SE}(n)$ is an $(n+1) \times (n+1)$ matrix. We make observations $I^D \in \mathcal{I}^D$, where I^D contains information about

I with transition density $p(\cdot|I)$ on $\mathcal{I}^\mathcal{D}$. The goal is to decode the original element in the pattern class $I \in \mathcal{I}$, i.e. choose the group element. One approach would be to generate the MAP estimator

$$\arg\max_{I \in \mathcal{I}} p(I|I^D). \tag{15.57}$$

This requires searching for the maximum over the entire space $\mathcal{I}$, which is generally impossible. Because of the identification to the group element, approximating $\mathcal{I}$ with $\hat{\mathcal{I}}$ is constructing the approximating subset $\hat{\mathcal{X}}$ of the special Euclidean group $\mathbf{SE}(n)$. We are now in the remote rate-distortion setting described in Section 15.2.

We seek to characterize the approximating subset of the special Euclidean group $\hat{\mathcal{X}} \subset \mathbf{SE}(n)$ which sits on the rate-distortion curve. Rate-distortion theory dictates that if $R > R(D)$ then for n sufficiently large there exists an approximating set $\hat{\mathcal{X}}^n$ of $\mathcal{X}^n$ such that the average distortion

$$\frac{1}{n}\sum_{i=1}^{n} d(I_i^D, \hat{I}(X_i)). \tag{15.58}$$

Conversely, if $R < R(D)$ then such functions do not necessarily exist.

Rate-distortion theory dictates that in order to approximate a sequence of images $I(X_1), \ldots, I(X_n)$ with average distortion

$$D = \frac{1}{n}\sum_{i=1}^{n} d(I_i^D, \hat{I}(X_i)). \tag{15.59}$$

from a sequence of observations $I_1^D, \ldots, I_n^D$, we need at least $2^{nR(D)}$ elements in the n-fold cartesian product $\hat{\mathcal{I}}(\mathcal{X}) \times \cdots \times \hat{\mathcal{I}}(\mathcal{X})$.

When $I_1^D, \ldots, I_n^D$ is a sequence of independent and identically distributed random variables, then

$$R(D) = \min_{p_{\hat{I}|I^D}(\cdot|\cdot):\mathrm{E}d(I^D,\hat{I})\leq D} \Delta H(I^D;\hat{I}), \tag{15.60}$$

where $\Delta H(I^D;\hat{I}(X))$ is the mutual information between I^D and $\hat{I}(X)$ and the minimum is over all conditional probabilities $p_{\hat{I}|I^D}(\cdot|\cdot)$, satisfying $\mathrm{E}d(I^D,\hat{I}) \leq D$.

15.4.1 Blahut Algorithm extended

Using the Blahut algorithm, Section 15.3, the RD function is calculated. Assume that the observable space $\mathcal{I}^\mathcal{D}$ is a finite set quantized to gray level images with input probability $p(I^D)$ on $\mathcal{I}^\mathcal{D}$. The reproduction alphabet reproduces the images with density $q(\hat{I}), \hat{I} \in \hat{\mathcal{I}}$. The transition probability from input to reproduction alphabet $q(\cdot|I^D)$ is the conditional probability distribution on the approximating set $\hat{I} \in \mathcal{I}$. The goal is to find the transition densities which minimize the mutual information between the input and output sets:

$$q^\star(\cdot|I^D) = \arg\min_{q(\cdot|I^D):\bar{D}\leq D} \Delta H(I^D;\hat{I}) = \arg\min_{q(\cdot|I^D):\bar{D}\leq D} \sum_{I^D}\sum_{\hat{I}} p(I^D)q(\hat{I}|I^D)\log\left(\frac{q(\hat{I}|I^D)}{q(\hat{I})}\right), \tag{15.61}$$

15.4 THE REMOTE RATE DISTORTION PROBLEM

where

$$\bar{D} = \sum_{I^D} \sum_{\hat{I}} d(\hat{I}, I^D) q(\hat{I}|I^D) p(I^D), \tag{15.62}$$

with $d(\hat{I}, I^D) = E_{p(I|I^D)} d(I, \hat{I}) = \sum_{I} d(I, \hat{I}) p(I|I^D). \tag{15.63}$

The problem is solved by an alternating minimization procedure (Blahut). Choose $q^{\text{init}}(\hat{I}) > 0$, $\hat{I} \in \hat{\mathcal{I}}$, then fixed points of the following iteration satisfy the necessary mutual information rate-distortion conditions:

$$q^{\text{new}}(\hat{I}|I^D) = \frac{q^{\text{old}}(\hat{I}) e^{-\lambda d(\hat{I}, I^D)}}{\sum_{\hat{I}} q^{\text{old}}(\hat{I}) e^{-\lambda d(\hat{I}, I^D)}}, \tag{15.64}$$

$$q^{\text{new}}(\hat{I}) = \sum_{I^D} p(I^D) q^{\text{new}}(\hat{I}|I^D) = q^{\text{old}}(\hat{I}) \sum_{I^D} \frac{p(I^D) e^{-\lambda d(\hat{I}, I^D)}}{\sum_{\hat{I}} q^{\text{old}}(\hat{I}) e^{-\lambda d(\hat{I}, I^D)}}. \tag{15.65}$$

The point on the RD curve is given by

$$\bar{D} = \sum_{\hat{I}} \sum_{I^D} p(I^D) \frac{q(\hat{I}) e^{-\lambda d(\hat{I}, I^D)}}{\sum_{\hat{I}} q(\hat{I}) e^{-\lambda d(\hat{I}, I^D)}} d(\hat{I}, I^D), \tag{15.66}$$

$$R(\bar{D}) = s\bar{D} - \sum_{I^D} p(I^D) \log \left(\sum_{\hat{I}} q(\hat{I}) e^{-\lambda d(\hat{I}, I^D)} \right)$$

$$- \sum_{\hat{I}} q(\hat{I}) \log \left(\sum_{I^D} p(I^D) \frac{e^{-\lambda d(\hat{I}, I^D)}}{\sum_{\hat{I}'} q(\hat{I}') e^{-\lambda d(\hat{I}', I^D)}} \right). \tag{15.67}$$

By choosing a number of $s \in [0, \infty)$ values, the RD function can be swept out.

Example 15.12 (Code Book Design for CUP CAD Models) Examine the ground based imaging problem in which the object is at a fixed position in space with unknown axis-fixed pose (orientation) studied extensively by Shusterman, et al. [348]. Assume the space of imagery $I \in \mathcal{I}$ is the orbit under the axis fixed orthogonal group with no translation: $X(\theta) = \begin{pmatrix} \cos\theta & \sin\theta & 0 \\ -\sin\theta & \cos\theta & 0 \\ 0 & 0 & 1 \end{pmatrix}$. The prior probability on rotations is taken as uniform $p(x(\theta)) = 1/2\pi$ inducing a uniform prior probability $p(I(\theta)), I(\theta) \in \mathcal{I}$. The observations $I^D \in \mathcal{I}^D$ are conditionally Gaussian with mean field $TI(\theta)$, and likelihood

$$p(I^D|I(\theta)) = \frac{1}{(2\pi\sigma^2)^{d/2}} e^{-1/2\sigma^2 \|I^D - TI(\theta)\|^2}; \tag{15.68}$$

the detector array will be $d = 32 \times 32$ throughout the simulations.

Throughout the gray level images are discretized with the Gaussian error function used to represent the posterior density $p(I^D|I)$.

Two distortion measures are examined. The first is the Froebenius or Hilbert–Schmidt distortion given by

$$d_1(X(\theta), \hat{X}(\hat{\theta})) \doteq \|X - \hat{X}\|_{HS}^2 = \text{trace}(X - \hat{X})(X - \hat{X})^t$$

$$= 2(n - \text{trace} X\hat{X}^t) \tag{15.69}$$

$$= 4 - 4\cos(\theta - \hat{\theta}). \tag{15.70}$$

The second is directly in the angles themselves according to

$$d_2(\theta,\hat{\theta}) = (\theta - \hat{\theta})^2. \tag{15.71}$$

Note that for small errors d_1 is approximately $2d_2$:

$$d_2(\theta,\hat{\theta}) = 4\left(1 - \left(1 - \frac{(\theta-\hat{\theta})^2}{2!} + \frac{(\theta-\hat{\theta})^4}{4!} - \cdots\right)\right) \tag{15.72}$$

$$= 2d_2(\theta,\hat{\theta}) + o((\theta-\hat{\theta})^2). \tag{15.73}$$

Remark 15.4.1 As shown in [341], in **SO**(3), the rotations correspond to three Euler angles ϕ, θ, ψ giving trace $X\hat{X}^t = (1 + \cos\theta)(\cos\phi\cos\psi - \sin\phi\sin\psi) + \cos\theta = (1 + \cos\theta)\cos(\phi + \psi) + \cos\theta$, which depends only upon the angles θ and $\gamma = \phi + \psi$. This gives the Hilbert–Schmidt distortion

$$d_{HS}(X,\hat{X}) = 6 - 2\cos\theta - 2(1 + \cos\theta)\cos\gamma.$$

Throughout the simulations the orientation space is tiled in fractions of degrees, thus $\mathcal{X}$ is a discrete subset of rotations. The measurement space $\mathcal{Y} \doteq R^{32 \times 32}$ is the continuum corresponding to the conditional Gaussian random fields on the imaging lattice; the measurement density is a Gaussian mixture across the discrete set of angles:

$$p(y) = \sum_{\theta \in \mathcal{X}} p(y|\theta) p(\theta).$$

The iteration of Eqns. 15.64 and 15.65 may be rewritten in the more convenient form

$$c(\hat{\theta}) = \sum_y p(y) \frac{e^{-\lambda d(\hat{\theta},y)}}{\sum_{\hat{\theta}} q^{\text{old}}(\hat{\theta})e^{-\lambda d(\hat{\theta},y)}}, \tag{15.74}$$

$$q^{\text{new}}(\hat{\theta}) = q^{\text{old}}(\hat{\theta})c(\hat{\theta}). \tag{15.75}$$

An estimate of the sum in 15.74 is done by the Monte–Carlo by simulating N image samples $y_1(\theta), \ldots, y_N(\theta)$ for each possible value of θ:

$$\widehat{c}(\hat{\theta}) = \sum_\theta p(\theta) \frac{1}{N} \sum_{i=1}^N \frac{e^{-\lambda d(\hat{\theta},y_i(\theta))}}{\sum_{\hat{\theta}} q^{\text{old}}(\hat{\theta})e^{-\lambda d(\hat{\theta},y_i(\theta))}} = \sum_\theta p(\theta)\widehat{c}(\hat{\theta}|\theta). \tag{15.76}$$

Unlike the original Blahut algorithm, here only estimates $\widehat{c}$ of the coefficients are generated via empirical averages over the N samples.

Shown in the accompanying figures are results for various CAD models from [348]. Imagery at SNR $= \infty$ dB (zero-noise), SNR $= 0$ dB, SNR $= -6$ dB, *and* SNR $= -12$ dB was simulated for

15.4 THE REMOTE RATE DISTORTION PROBLEM

three objects; the "JEEP", the "T62," and the "CUP." Figures 15.6 and 15.7 show the 3 CAD models at SNR = ∞ and SNR = −6 dB, respectively.

For the calculation of the RD curves, the resolution of the orientation grid was taken at 1° for a total of 360 total poses in the database. For the first two CAD models, the JEEP and the T62, all orientations are different and identifiable through the sensor so that the RD curve does not exhibit any "gaps" due to ambiguity as would be expected for a symmetry. This is depicted in Figure 15.8. As we shall see below for the CUP with a handle feature there is a gap in the rate-distortion function since there are a set of orientations which are not identifiable due to obscuration of the sensor. A remarkable thing shown in these plots is the fact that even for −6 dB SNR, the corresponding RD curve coincides with the the curve drawn for the zero noise case. For increased noise levels, the performance decreases rapidly as is illustrated by the −12 dB curve. Only at the highest noise level is there a difference between curves of these two different CAD models. Such noise immunity demonstrates the power of the global modelling approach of deformable templates in which an entire array of pixels enter into the inference of a low-dimensional parametric representation. This is consistent with that shown in [341] for orientation estimator bounds demonstrating that such parametric models behave as if they are asymptotically high in signal to noise ratio over a broad range of noise levels.

For the "JEEP" and "T62" examples the difference between various orientations is visually apparent. The third CAD model "CUP" has a different behavior due to its symmetry properties. Only the cup handle breaks the symmetry. Since it is a "fragile" feature consisting of a small number of pixels at low SNR the handle is masked. The RD curves for the third CAD model are drawn in Figure 15.9. Note that even without noise the distortion never goes to zero because there is a range of about 85° over which the handle is hidden behind the cup, making all orientations in that range indistinguishable. Assuming that the CUP model is positioned with zero degrees with the handle hidden from the sensor, and assuming that all orientations inside the identifiable range

Figure 15.6 CAD models JEEP, T62, and CUP—4 orientations of CAD models without noise through camera projection (image size 32 × 32, 8 bit/pixel).

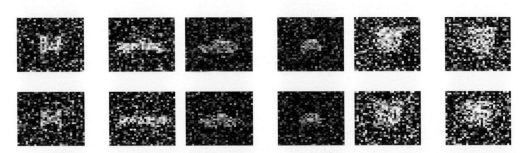

Figure 15.7 Orientations of each model at SNR = −6 dB.

are decoded without error, the minimum average distortion is calculated by

$$D_{\min}^{\text{MSE}} = \frac{1}{2\pi} \int_{-85\pi/360}^{85\pi/360} \theta^2 \, d\theta \approx 0.0433,$$

$$D_{\min}^{\text{HSE}} = \frac{1}{2\pi} \int_{-85\pi/360}^{85\pi/360} 4(1 - \cos \theta) \, d\theta \approx 0.0211$$

(after normalization $D_{\min}^{\text{MSE}} \approx 0.0132$). RD curves in Figure 15.9 come close to these numbers.

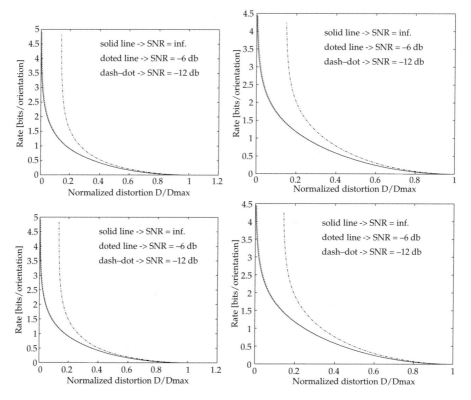

Figure 15.8 Left column shows rate-distortion functions for MSE distortion for objects "JEEP"(top row) and T62"(bottom row). Right column shows the same for HS distortion.

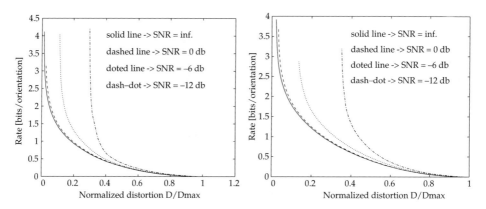

Figure 15.9 Rate-Distortion function for the model "CUP". Left panel MSE; right panel HSE.

15.4 THE REMOTE RATE DISTORTION PROBLEM

This figure shows the sensitivity of the RD function to the noise level. Unlike the previous cases, even at 0 dB SNR noticeable degradation of performance occurs, which is entirely due to the fact that the handle feature is fragile; 0-dB SNR makes it disappear.

We have calculated rate-distortion functions of the first order for orientation alphabets under assumptions that object orientations are distributed uniformly over the circle, they are statistically independent at each sample, and the noise is zero-mean Gaussian and white. For orientations which are temporally dependent on past orientations such as those modelled in [303] for aircraft dynamics, the RD function may be calculated by building the input alphabet from vectors of input samples. The size of the input and output alphabets will be geometric in vector length, substantially increasing the computation. The RD function of the first order is an upper bound for RD functions of the higher orders and is achieved if the input process samples are statistically independent [344].

15.5 Output Symbol Distribution

So far, only the optimal size distortion tradeoff of the output codebook has been calculated, not the code itself. Such performance is achieved for long sequences of data; for short sequences or scalar codebooks the calculated performance becomes a lower bound. In the remaining figures we examine and interpret the fixed point $q^\star(\hat{\theta}), \hat{\theta} \in \hat{\Theta}$ of the iteration Eqn. 15.75 of the Blahut algorithm. Although the distribution density of output symbols is a function with support on the interval $[0, 2\pi]$ and can be plotted using Cartesian coordinates, we prefer to use polar coordinates to give insight into the dependency of the output distribution density on the object geometry. The results are closed curves corresponding to probability densities with area 1 and radii which represent the distribution mass at each angle. For illustration, all plots are scaled independently, so no absolute quantitative information can be deduced across plots. All sub-plots are indexed by their SNR; upper left panel refers to an SNR $= \infty$ dB, upper right SNR $= 0$ dB, lower left SNR $= -6$ dB, and lower right SNR $= -12$ dB.

Shown in Figure 15.10 is the output distribution $q^\star$ for the Hilbert–Schmidt distortion error for the "JEEP" at the rates of 0.5 (left), 1.0 (middle), 2.0 (right) bits/orientation. At the rate 0.5 (columns 1,2) and 1.0 bit/orientation (columns 3,4), the distribution densities at most noise levels are circles, meaning that each one of the orientations should appear in the output alphabet with the same frequency. The situation changes in sub-plot d, where the distribution density is spread along orientations of 90° and 270°.

If one decides to build a scalar quantizer at a particular rate based on the density in sub-plots INF dB, 0 dB then one can choose any two opposite orientations as codebook words (examine

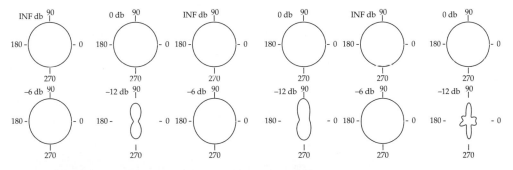

Figure 15.10 Output distribution densities for HSE distortion for object "JEEP" at rate 0.5 (columns 1,2), 1.0 (columns 3,4), 2.0 (columns 5,6) bits.

upper left, right panels). However if they are chosen near 90° and 270° the codebook will be close to optimal for higher noise levels too. The property determining the code orientation is the norm-squared measure on object as demonstrated in [341] establishing $\|T(\theta)\|$ the norm-squared of the mean projective transformation of the object at orientation θ as one of the crucial metric for Bayes probability of error. A simple observation of the object images shows that the highest values of the $\|TI(\theta)\|^2$ measure are near 90° and 270° (see "JEEP" column 2 of Figure 15.6).

Increasing the bit rate further to 2.0 bit/orientation (columns 5,6) demonstrate allocation of codewords at a set of discrete orientations around the circle. At SNR=−12 dB, (lower right panel) provides evidence that the distance between two highest norm-squared valued orientations is also important. Figure 15.11 shows plots of the output symbols distribution density for the "T62" model for the same rates and the Hibert–Schmidt distortion at 0.5 (left), 1.0 (middle), and 2.0 (right) bits/symbol. The results are similar as above. Note that in this example, the highest norm-squared values are at orientations 0° and 180° (see "T62" columns 3 in Figure 15.6).

Both the "JEEP" and "T62" CAD models produce different images for every orientation. If the noise level is not very high the output symbols distribution density is a circle, with each output orientation having the same frequency of appearance in the output vector. The last example, the CAD model "CUP", has a high symmetry with a number of orientations which are indistinguishable. Thus, even without noise, the distribution density of the output is not a circle. Its shape for various rates is shown in Figure 15.12. Note that the orientation of zero degrees is selected when the cup handle is hidden by the cup. Therefore, for all orientations in the range of approximately ±42.5° are mapped to zero degrees.

These figures show that the distribution density changes much faster with the noise and that for a high noise level the best we can do is to decide whether the "CUP" handle is to the left or to

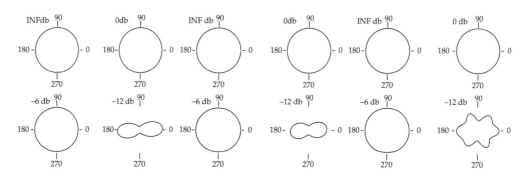

Figure 15.11 Output distribution density for object "T62" at rates 0.5 (columns 1,2), 1.0 (columns 3,4), 2.0 (columns 5,6) bits/orientation.

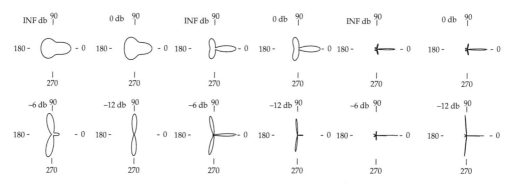

Figure 15.12 Output distribution densities for the HSE distortion measure for the CAD object "CUP" at rates 0.5 (columns 1,2), 1.0 (columns 3,4), 2.0 (columns 5,6) bits/orientation.

the right. For illustration the concentrated nature of the density has been scaled although the total density is 1.

Interestingly, if the model has a high degree of symmetry, then the output symbols density should also be symmetric. Note that sub-plots (c) in Figure 15.12 illustrate slight asymmetries. The light in all our experiments was placed at a position of $90°$; the right part of the cup in Figure 15.6 is brighter and could give rise to the small asymmetry.

16 COMPUTATIONAL ANATOMY: SHAPE, GROWTH AND ATROPHY COMPARISON VIA DIFFEOMORPHISMS

ABSTRACT Computational Anatomy (CA) is the study of human anatomy $I \in \mathcal{I} = I_\alpha \circ \mathcal{G}$, an orbit under groups of diffeomorphisms of anatomical exemplars $I_\alpha \in \mathcal{I}$. The observable images $I^D \in \mathcal{I}^D$ are the output of medical imaging devices. There are three components CA examines: (i) constructions of the anatomical submanifolds under study in anatomy, (ii) estimation of the underlying diffeomorphisms $g \in \mathcal{G}$ defining the shape or geometry of the anatomical manifolds, and (iii) generation of probability laws of anatomical variation $P(\cdot)$ on the images $\mathcal{I}$ for inference and disease testing within the anatomical models. This chapter focuses on the metric comparison of anatomical manifolds.

16.1 Computational Anatomy

Revolutionary advances in the development of digital imaging modalities combined with advances in digital computation are enabling researchers to advance the precise study of the awesome biological variability of human anatomy. This is emerging as the exciting new field of computational anatomy (CA) [227,349]. CA has three principal aspects: (*a*) automated construction of anatomical manifolds, points, curves, surfaces, and subvolumes; (*b*) comparison of these manifolds; and (*c*) the statistical codification of the variability of anatomy via probability measures allowing for inference and hypothesis testing of disease states. This chapter will focus on several of these aspects. Although the study of structural variability of such manifolds can certainly be traced back to the beginnings of modern science, in his influential treatise "On Growth and Form" in 1917, D'Arcy Thompson had the clearest vision of what lay ahead.

In a very large part of morphology, our essential task lies in the comparison of related forms rather than in the precise definition of each; and the *deformation* of a complicated figure may be a phenomenon easy of comprehension, though the figure itself may have to be left unanalyzed and undefined. This process of comparison, of recognizing in one form a definite permutation or *deformation* of another, apart altogether from a precise and adequate understanding of the original "type" or standard of comparison, lies within the immediate province of mathematics and finds its solution in the elementary use of a certain method of the mathematician. This method is the Method of Coordinates, on which is based the Theory of Transformations.

The study of shape and structure has certainly progressed a long way. Thompson's vision is precisely the mathematical structure we term anatomy in CA [227,349], a deformable template in which the space of anatomical imagery is an orbit under groups of transformations. The transformations have already been studied in the previous Chapters 10, 11, and 12 in which the metric space structure of two types of transformations have been examined. The first is of the geometric type studied through diffeomorphic mappings of the coordinate systems of the anatomies; the second is of the photometric type accomodating the appearance or creation of new structures. The metric is calculated via equations of motion describing the geodesic connection between the elements.

Now towards the first of the three principal aspects, the automated construction of anatomical manifolds of points, curves, surfaces, and subvolumes, the general area of the mathematical codification of biological and anatomical structure has expanded rapidly over the past several decades. The automated construction of anatomical manifolds is receiving tremendous focus by many of the groups throughout the world supporting neuromorphometric analyses which are becoming available with large numbers of anatomical samples. Deformable and active models are being used for generating 1D manifold curves in 2 and 3 dimensions [154,164,170,188,202,350–354]. The differential and curvature characteristics of curves and surfaces have been examined as well with active and deformable surface models for the neocortex and cardiac systems [173, 174, 192, 193, 202, 229, 355–358]. Local coordinatized representations for cortical manifolds have included both spherical and planar representations for local coordinates in studying the brain [166, 167, 173–176, 178, 179, 182, 231, 355, 359–366]. A great deal of work has also been focused on methods of segmentation of anatomical volumes into 3D submanifolds. Automatic methods for maximum-likelihood and Bayesian segmentation are being developed across the whole brain as well as focusing on particular gyri [51, 53, 54, 183, 356, 367–378].

Towards comparison of anatomical manifolds, because of the fundamental structural variability of biological shape there has been great emphasis by groups on the study of anatomical shape via vector and metric comparison. The earliest vector mapping of biological coordinates via landmarks and dense imagery was pioneered in the early 1980s and is continued today by Bookstein [223, 224, 272, 379–381], and Bajcsy, Gee and co-workers [274, 288, 290, 291, 382–384]. Comparison via vector maps based on dense imagery is being carried on by many of the aforementioned groups. Most of the efforts define a preferred origin modeling deformations "elastically" based on small-deformation vector maps on the dense volumes; vector mappings restricted to the cortical manifold are being computed as well [294, 295, 350, 385–400].

Our own efforts in comparison of anatomical manifolds have been largely focussed on via large deformation diffeomorphic flows, originally proposed in [220, 401]. Unlike the vector maps, these are not additive but provide guaranteed bijections between anatomical configurations. This flow framework has since 1997 been developed into a complete metric space formulation [269, 276–278, 282, 286, 349, 402, 403], providing the mapping methodology with a least-energy (shortest path) and symmetry property.

Concerning inference on statistical representations of shape, studies in CA of growth, atrophy, and disease have literally exploded over the past five years since the advent of automated methods for manifold construction. Applications to growth and statistical atlas building are evidenced by [275, 284, 297, 381, 404–407]. Applications in CA to normal and abnormal aging in the dementia of the Alzheimer's type have been examined in both cortex and deep brain structures by many groups [408–417]. Also investigators have been looking at the neuropsychiatric illness of schizophrenia [418–433]. Researchers in CA continue to develop new methods of analysis in many of the neuropsychiatric illnesses including but not exclusive to ADHD [434], autism [435, 436], major depressive disorder [437], psychosis [438], alcohol effects [439], neurologic disorders Huntington's disease [440], and multiple-sclerosis [441–443], as well as comparison across illnesses [444].

16.1.1 Diffeomorphic Study of Anatomical Submanifolds

Because of the sheer complexity of human neuroanatomy, in particular the human brain, the study of brain geometry has emerged as the study of the submanifolds of anatomical significance including its landmarks, curves, surfaces, and subvolumes all taken together forming the complete volume. The complete methodology combines mappings that carry all of the submanifolds of points, curves, surfaces, and subvolumes together. This is precisely why the fundamental transformation groups are subgroups of diffeomorphisms, as they carry these submanifolds consistently. The transformations are constrained to be 1–1 and onto, and differentiable with differentiable inverse, so that connected sets remain connected, submanifolds such as surfaces are mapped as surfaces, and the global relationships between structures are maintained. Shown in Figure 16.1 is the methodology the CA community is using for studying coordinatized manifolds on the brain via diffeomorphisms and bijections to local coordinates. Depicted in the top row are diffeomorphic mappings which are used for studying the 3D structures of the brain. These mappings are defined on subvolume manifolds $X \subset \mathbb{R}^3$. Shown in the bottom row are diffeomorphic mapping between manifolds in $X \subset \mathbb{R}^2$. These especially arise in the study of cortical surfaces such as the neocortex

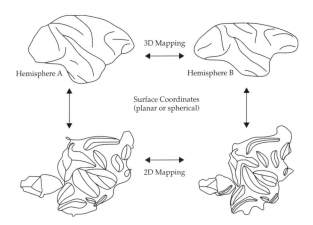

Figure 16.1 Figure shows the diffeomorphisms between brain volumes and submanifolds within the brain; figure taken from Van Essen et al. [167].

and the hippocampus and other subvolumes in which the surface of the subvolume becomes a natural index system. The diffeomorphisms between $\mathbb{R}^3$ provide smooth bijections to the submanifolds thereby allowing for the study of correspondences between the different structures.

16.2 The Anatomical Source Model of CA

In CA many models are being examined by investigators throughout the world, in which data enter through dense imagery as well as landmarked points, and shape themselves. In this chapter we shall examine several of these. In almost all cases the study of anatomy will be accomplished by identifying the anatomical picture with an *unknown, hidden diffeomorphism* on the background space of coordinates. As these diffeomorphisms can almost never be directly observed, only the observable imagery, they must be estimated; this is the quintessential hidden structure problem.

There are essentially two models being examined, the **static metric mapping model**, and the **dynamic metric growth model**. For the static metric mapping model, the anatomical source is a deformable template $\mathcal{I}_\alpha$ corresponding to the orbit under the group $\mathcal{G}$ of one selected and fixed image $I_\alpha \in \mathcal{I}$. A single observation from one time instant is observed, and the goal is to compare an exemplar with it via diffeomorphic transformation. The diffeomorphic transformations on the coordinate systems are defined through infinite dimensional diffeomorphisms generated via the Lagrangian and Eulerian formulation of the flows on $X \subset \mathbb{R}^d$ the closed and bounded background space:

$$\frac{\partial g_t}{\partial t}(x) = v_t(g_t(x)), \quad \frac{\partial g_t^{-1}}{\partial t}(x) = -Dg_t^{-1}(x)v_t(x), \quad x \in X \subset \mathbb{R}^d, \tag{16.1}$$

with boundary conditions $g_0 = g_0^{-1} = \text{id}$. The space of allowable vector fields $\mathcal{V}$ is restricted by working with the V norm-square

$$\int_0^1 \|v_t\|_V^2 \, dt = \int_0^1 \langle Av_t, v_t \rangle_2 \, dt. \tag{16.2}$$

16.2 THE ANATOMICAL SOURCE MODEL OF CA

Assume that $(V, \|\cdot\|_V)$ is a reproducing kernel Hilbert space (see Chapter 9, Section 9.4) with operator $K : L^2(X, \mathbb{R}^d) \to V$, satisfying for all $f, h \in L^2(X, \mathbb{R}^d)$, $\alpha, \beta \in \mathbb{R}^m$,

$$\langle K(x, \cdot)\alpha, f\rangle_V = \alpha^* f(x) = \sum_i \alpha_i f_i(x), \quad \langle K(x_j, \cdot)\alpha, K(x_i, \cdot)\beta\rangle_V = \alpha^* K(x_i, x_j)\beta. \quad (16.3)$$

The orbit of imagery becomes

$$\mathcal{I}_\alpha = \left\{ g_1 \cdot I_\alpha : \frac{\partial g_t}{\partial t} = v_t(g_t),\ t \in [0,1],\ g_0 = \mathrm{id},\ v \in V \right\}. \quad (16.4)$$

The complete anatomy is the union over multiple orbits generated from different templates I_α, $\mathcal{I} = \cup_\alpha \mathcal{I}_\alpha$. Throughout this chapter it is assumed that the exemplar or initial starting template is known and fixed.

Figure 16.2 depicts the basic static metric correspondence model studied throughout this chapter. The goal is to infer the geometric image evolution connecting the observation I^D to the exemplar $I_0 = I_\alpha \in \mathcal{I}$.

In the dynamic growth model, an observation time sequence is given and the goal is to infer the time flow of geometric change which carries the known exemplar through the growth period $t \in [0,1]$. Here the orbits are flows of imagery, with growth corresponding to actual real-time, and comparison is performed by comparing the observables $I_t, t \in [0,1]$ to its start point I_0. Here $I_0 = I_\alpha$ plays the role of the so-called template or exemplar. In static matching time is simulation time only, with the single target observable at time $t = 1$. Throughout we take as the dynamic growth model orbits of the exemplars under space–time flows

$$\mathcal{I}_\alpha[0,1] = \left\{ g_t \cdot I_\alpha, t \in [0,1] : \frac{\partial g_t}{\partial t} = v_t(g_t), g_0 = \mathrm{id} \right\}. \quad (16.5)$$

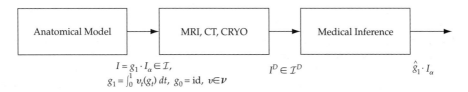

Figure 16.2 The source for the **anatomical model** is the set of all images $\mathcal{I}$ of the exemplars $I_\alpha(g_1^{-1})$. The observed data I^D are observations seen via the sensors (MRI, CT, CRYOSECTION pictures) and are conditional Gaussian random fields with mean fields $g_1 \cdot I_\alpha$. For the **static metric mapping model** there is only one observation I^D the mean field of the exemplar $I_\alpha(g_1^{-1})$.

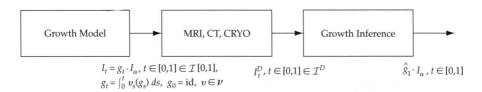

Figure 16.3 The source for the **dynamic growth model** is the set of all growth flows $\mathcal{I}[0,1]$ of the exemplars $g_t \cdot I_\alpha, t \in [0,1]$. The observed data $I_t^D, t \in [0,1]$ are flows of observations seen via the sensors (MRI, CT, CRYOSECTION pictures) and are conditional Gaussian random fields with mean fields $g_t \cdot I_\alpha, t \in [0,1]$.

Each element $g_t \cdot I_\alpha, t \in [0,1] \in \mathcal{I}_\alpha[0,1]$ in the orbit is modeled as corresponding to a different growth trajectory of the exemplar. Figure 16.3 depicts the growth model studied throughout this chapter.

16.2.1 Group Actions for the Anatomical Source Model

Specifying the source model amounts to specifying the group actions on the elements of the orbit. We shall look at several group actions in this chapter defined on the sparse landmark points, the dense scalar valued imagery, as well as normalized vectors from tensor measurements involving diffusion tensor magnetic resonance imaging (DTI). The measurement at each voxel in the DTI image volume is a symmetric second order tensor; the principal direction of the diffusion tensor corresponds to the fiber orientation in heart tissue and neuron orientation in the central nervous system. Cao and Younes define the images to be functions $I: X \to S^2 \subset \mathbb{R}^3$ that associate to each point a vector, the principal direction of the diffusion tensor (outside the object the vector is zero).

The group actions studied in this chapter are as follows.

Lemma 16.1 1. *For the sparse landmarks points, take the orbit of imagery as N-shapes*

$$I_N = \begin{pmatrix} x_1 \\ \vdots \\ x_N \end{pmatrix} \in \mathcal{I}_N = \mathbb{R}^{dN} \text{ with the group action defined as, for any } I_N \in \mathcal{I}_N$$

and $g \in \mathcal{G}$,

$$g \cdot I_N = \begin{pmatrix} g(x_1) \\ \vdots \\ g(x_N) \end{pmatrix}. \tag{16.6}$$

2. *Take the orbit the dense imagery scalar-valued functions $I: X \to \mathbb{R}$ in orbit $I \in \mathcal{I}$ with group action, for any $I \in \mathcal{I}$ and $g \in \mathcal{G}$,*

$$g \cdot I = I \circ g^{-1}. \tag{16.7}$$

3. *Let the images be unit normal vectors $I: X \to \mathbb{R}^3$ that associate to each point a vector, the principal direction of the diffusion tensor (outside the object the vector is zero). Define the group actions according to, for any $I \in \mathcal{I}$ and $g \in \mathcal{G}$,*

$$g \cdot I = \begin{cases} \dfrac{D_{g^{-1}}g I \circ g^{-1} \|I \circ g^{-1}\|}{\|D_{g^{-1}}g I \circ g^{-1}\|} & \text{when } I \circ g^{-1} \neq 0 \\ 0 & \text{when } I \circ g^{-1} = 0 \end{cases}, \tag{16.8}$$

with $D_f g$ denoting the Jacobian matrix of g evaluated at f.

These are group actions.

Proof We only prove the normalized vector version, the others have been shown elsewhere. Clearly, $\text{id} \cdot I = I$. For g and h in $\mathcal{G}$, $g \cdot (h \cdot I) = (g \circ h) \cdot I$ if $I \circ g^{-1} = 0$.

If $I \circ g^{-1} \neq 0$,

$$g \cdot (h \cdot I) = \frac{D_{g^{-1}}g(h \cdot I \circ g^{-1})\|h \cdot I \circ g^{-1}\|}{\|D_{g^{-1}}g(h \cdot I \circ g^{-1})\|}$$

$$= \frac{D_{g^{-1}}g(D_{h^{-1} \circ g^{-1}}hI \circ h^{-1} \circ g^{-1}\|I \circ h^{-1} \circ g^{-1}\|/\|D_{h^{-1} \circ g^{-1}}hI \circ h^{-1} \circ g^{-1}\|)\|I \circ h^{-1} \circ g^{-1}\|}{\left\|D_{g^{-1}}g(D_{h^{-1} \circ g^{-1}}hI \circ h^{-1} \circ g^{-1}\|I \circ h^{-1} \circ g^{-1}\|/\|D_{h^{-1} \circ g^{-1}}hI \circ h^{-1} \circ g^{-1}\|)\right\|}$$

$$= \frac{D_{g^{-1}}g D_{h^{-1} \circ g^{-1}}hI \circ h^{-1} \circ g^{-1}\|I \circ h^{-1} \circ g^{-1}\|}{\left\|D_{g^{-1}}g D_{h^{-1} \circ g^{-1}}hI \circ h^{-1} \circ g^{-1}\right\|}$$

$$= \frac{D_{h^{-1} \circ g^{-1}}(g \circ h) I \circ h^{-1} \circ g^{-1}\|I \circ h^{-1} \circ g^{-1}\|}{\|D_{h^{-1} \circ g^{-1}}(g \circ h) I \circ h^{-1} \circ g^{-1}\|}$$

$$= (g \circ h) \cdot I. \qquad \square$$

16.2.2 The Data Channel Model

In computational anatomy observations are made of the underlying coordinates. Generally the data is observed in multiple forms, including both measurements of material properties of tissue such as scalar and vector valued imagery derived from MRI, CT, nuclear emission, and optical and acoustic scanners, as well as measurement made in the form of geometric properties of objects associated with submanifolds, curves, surfaces, and subvolumes. The general model we pursue was shown in Figure 16.2.

Take the observable $I^D : X \subset \mathbb{R}^d \to \mathbb{R}^m$, in general a vector valued function. Define a distance measure between the underlying source or anatomical structure giving rise to the measurements, and the observable measurement $C : \mathcal{I} \times \mathcal{I}^D \to \mathbb{R}^+$. We will generally study through the Gaussian random field model, in which the distance function arises as the potential associated with such a model. For this, model the observables I^D as a Gaussian random field with mean field $I \in \mathcal{I}$ generated via unknown diffeomorphism. Then the distance takes the form

$$C = \frac{1}{2\sigma^2} \|I^D - g \cdot I_\alpha\|_2^2, \qquad (16.9)$$

where $\|\cdot\|_2$ is the L^2 vector norm $\|f\|_2 = \sum_{i=1}^m \int_X |f_i(x)|^2 \, dx$.

For the growth problem, the goal is to extract the hidden diffeomorphism sequence $g_t, t \in [0,1]$ carrying the exemplar onto the static target observation sequence $I_t^D, t \in [0,1]$. The diffeomorphism is selected to minimize a matching distance function of time

$$C = \frac{1}{2\sigma^2} \int_0^1 \|I_t^D - g_t \cdot I_\alpha\|_2^2 \, dt. \qquad (16.10)$$

For the norm squared cost then the stochastic model corresponds to the additive white noise model, $I^D = g \cdot I_\alpha + W$, with W white noise variance 1. Strictly speaking then $\|I^D - g \cdot I_\alpha\|_2^2$ is not well defined on the continuum, since I^D is not of trace class. This is properly handled by constraining the noise process to have covariance with spectrum which has finite integral. In this case the matching norm-square cost changes to $\|I^D\|_H^2 = \langle I^D, I^D \rangle_\Sigma$ with Σ covariance trace class reflecting the finite bandwith of the noise or the tapering of the spectrum.

Alternatively, for all of the applications, the noise process is associated with a finite number of detectors (voxels in the image). Assume that the measured data $I_n^D, n = 1, 2, \ldots$ is a Gaussian random field with each component representing a noisy measurement of a scalar (e.g. gray level) integrated over a subset of the background space X forming the voxel measurements; the nth voxel measures the integrated response from the voxel $\Delta x_n \subset X$. Assume the high count limit and independence between distinct voxels, in which case the measurement is modeled as a Gaussian random vector of fixed isotropic variance σ^2, with the nth component I_n^D having mean $\int_{\Delta x_n} g_1 \cdot I_\alpha(x)\, dx$.

We use the L^2 formulation (with covariance identity) in the continuum connecting us to so many classic formulations in least-squares estimation. For a strict interpretation we must keep in mind the piecewise continuous model just articulated or the trace class covariance.

The unifying inference problem solved throughout the chapter is to estimate properties of the underlying anatomical structure by minimization of the data cost with the running cost associated with geodesic correspondence. Given observables I^D with mean field $g \cdot I_\alpha \in \mathcal{I}$ and matching cost C, the goal is to find the diffeomorphism g which minimizes C. As with many minimum norm variational problems, there are an infinite number of possible paths which minimize the distance function; we choose the ones which minimize the metric length and energy in the subgroup of diffeomorphisms solving the infimum

$$\int_0^1 \|v_t\|_V^2 dt + C(I^D, g \cdot I_\alpha) . \tag{16.11}$$

16.3 Normal Momentum Motion for Large Deformation Metric Mapping (LDDMM) for Growth and Atrophy

In Chapter 11, we examined the shortest path geodesic flows of diffeomorphisms corresponding to the Euler equation. As well in Section 11.3 we examined the variational problem in the tangent space of the group minimizing with respect to the vector fields generating the shortest path flows. As we shall see, for metric mapping of images in which the data enters only through the two static observations at the start and endpoints of the flow, the geodesic is completely determined by the initial momentum, where in the deformable setting the generalized momentum is naturally defined as Av the function of the vector field which is integrated against the vector field to given energy in the metric $\|v\|_V^2 = \langle Av, v \rangle_2$.

For the metric mapping of the start and end point, there is a conservation law which requires the momentum to be conserved along the path of the geodesic flow. This of course implies that the momentum at the identity Av_0 on the geodesic determines the entire flow $g_t, t \in [0, 1]$, where $\dot{g}_t = v_t(g_t), g_0 = \mathrm{id}$. As proved in Theorem 11.8 defining the vector field transported from the identity as $w_t = Dg_t(g_t^{-1})w_0(g_t^{-1})$, then along the geodesics satisfying the Euler equation (Theorem 11.6, Eqn. 11.21, Chapter 11) the momentum is conserved according to Theorem 11.8, Eqn. 11.31:

$$\frac{\partial}{\partial t}\langle Av_t, w_t \rangle_2 = 0, \quad t \in [0, 1] . \tag{16.12}$$

Along the geodesic, the momentum is determined by the momentum at the identity:

$$Av_t = (Dg_t^{-1})^* Av_0(g_t^{-1}) |Dg_t^{-1}| . \tag{16.13}$$

The natural question becomes, where does the initial momentum Av_0 arise to specify the geodesic connection of one shape to another? Naturally, it arises by solving the variational problem of *shooting* one object onto another. In this setting we do not require an exact boundary condition for g_1 implying that the perturbation in η satisfies only the $t = 0$ boundary conditions and at the boundary of the background space $x \in \partial X$. In previous Chapter 10 we described only the Euler equations associated with variation of the diffeomorphisms. Examine the dense matching problem calculating the variation of the energy with respect to perturbation of the vector field directly with the perturbation $v \to v + \epsilon \psi$ zero on the boundary $\psi(\partial X) = 0$. The perturbations respecting the boundary conditions take the form

$$\eta_0(\cdot) = 0, \quad \eta_1(\partial X) = 0, \quad \psi(\partial X) = 0. \tag{16.14}$$

The inexact correspondence associated with the boundary conditions are depicted in panel 2 of Figure 10.1 of Chapter 10. Notice how at the endpoint only approximate correspondence is enforced. Now we follow Beg's work [285] and calculate the momentum at the origin which solves the minimum matching condition.

From Theorems 12.9 and 12.11 of Chapter 12 we have the Euler equations and the normal momentum motion. Beg [285] computes not on the momentum Av_t for shooting one object on to another but on the more stable object the vector fields. This introduces the smoothing of the Green's operator in the solution. The variation equation essentially corresponds to the Sobolev derivative on the vector field. We term the algorithm of Beg the LDDMM algorithm.

Theorem 16.2 (Large Deformation Diffeomorphic Metric Mapping: Dense Imagery [285]) *Given template I continuously differentiable with $\nabla(I \circ g) = (Dg)^* \nabla I(g)$, and $g_{t,u} = g_u \circ g_t^{-1}$.*

1. *Given static image data I^D with cost $C = (1/2\sigma^2)\|I^D - I(g_1^{-1})\|_2^2$, and flow minimizing energy*

$$\int_0^1 \|v_t\|_V^2 \, dt + \frac{1}{2\sigma^2} \|I^D - I(g_1^{-1})\|_2^2. \tag{16.15}$$

Then the V-norm variational minimizer has vector fields satisfying

$$v_t + K\left(\nabla(I \circ g_t^{-1}) \frac{1}{2\sigma^2}(I^D(g_{t,1}) - I(g_t^{-1}))|Dg_{t,1}|\right) = 0, \tag{16.16}$$

where $Kf(\cdot) = \int K(\cdot, y) f(y) \, dy$.

2. *For the growth problem, given image data $I_t^D, t \in [0,1]$ with cost $C = (1/2\sigma^2)(\int_0^1 \|I_t^D - I(g_t^{-1})\|_2^2) \, dt$, and flow minimizing energy*

$$\int_0^1 \|v_t\|_V^2 \, dt + \frac{1}{2\sigma^2} \int_0^1 \|I_t^D - I(g_t^{-1})\|_2^2 \, dt. \tag{16.17}$$

The V-norm minimizer with respect to variations of the vector field satisfies

$$v_t + K\left(\nabla(I \circ g_t^{-1}) \frac{1}{2\sigma^2} \int_t^1 (I_u^D(g_{t,u}) - I(g_t^{-1}))|Dg_{t,u}| \, du\right) = 0. \tag{16.18}$$

Proof It is only necessary to show the growth version, the other version is similar. To calculate the V-norm variation, then the first term is $2v_t$ (rather than $2Av_t$), and Eqn. 12.52 of Theorem 12.11, Chapter 12 for the second term V-norm variation becomes

$$-\frac{2}{2\sigma^2} \int_0^1 \langle (I_u^D - I(g_u^{-1}))\nabla I(g_u^{-1}), \partial_\psi g_u^{-1} \rangle_2 \, du$$

$$= \frac{2}{2\sigma^2} \int_0^1 \left\langle K\left(\nabla(I \circ g_t^{-1}) \int_t^1 |Dg_{t,u}|(I_u^D(g_{t,u}) - I(g_t^{-1})) \, du\right), \psi_t \right\rangle_V dt. \quad \square$$

Beg's algorithm [285] for the static case with observation I^D at one time instant, with $C = \|I^D - I(g_1^{-1})\|_2^2$ computes as a fixed point a gradient solution of the inexact metric matching equation. We term this the LDDMM algorithm; when applied to dense images the dense image LDDMM algorithm. For numerical issues, Beg computes the optimizing vector field rather than the momentum.

Algorithm 16.3 (Beg Algorithm [285])

1. **Static Matching:** *Fixed points of the following algorithm satisfy Eqn. 16.16. Initialize* $v^{\text{old}} = 0$, *choose constant* ϵ, *then for all* $t \in [0,1]$,

   Step 1 : $\frac{\partial}{\partial t} g_t^{\text{new}} = v_t^{\text{old}}(g_t^{\text{new}})$, $\quad \frac{\partial}{\partial t} g_t^{-1\text{new}} = -Dg_t^{-1\text{new}} v_t^{\text{old}}$, $\quad g_{t,1}^{\text{new}} = g_1^{\text{new}}(g_t^{-1\text{new}})$,

   Step 2 : Compute $v_t^{\text{new}} = v_t^{\text{old}} - \epsilon \nabla_v E_t^{\text{old}}$, set $v^{\text{old}} \leftarrow v^{\text{new}}$, return to Step 1 where

   $$\nabla_v E_t^{\text{old}} = v_t^{\text{old}} + \frac{1}{2\sigma^2} K\left(\nabla(I \circ g_t^{-1\text{new}}) |Dg_{t,1}^{\text{new}}| (I^D(g_{t,1}^{\text{new}}) - I(g_t^{-1\text{new}})) \right), \quad t \in [0,1]. \tag{16.19}$$

2. **Growth:** *The growth model algorithm for dense imagery is same as above with gradient*

   $$\nabla_v E_t^{\text{old}} = v_t^{\text{old}} + \frac{1}{\sigma^2} K\left(\nabla(I \circ g_t^{-1\text{new}}) \left(\int_t^1 |Dg_{t,u}^{\text{new}}| (I_u^D(g_{t,u}^{\text{new}}) - I(g_t^{-1\text{new}})) \, du \right) \right). \tag{16.20}$$

Beg's gradient descent algorithm discretizes Eqns. 16.19. At the fixed point, the discrete version of the Euler–Lagrange equation is satisfied.

Examine the static case with observation I^D at one time instant, with $C = \|I^D - I(g_1^{-1})\|_2^2$. The first three rows of Figure 16.4 show the mapping of multiple simple shapes involving translation, scale and rotation. The operator was chosen to be $A = (-0.01\nabla^2 + \text{id})^2$ for the experiments. The flow was discretized into 20 timesteps each of size $\delta t = 0.1$. The gradient algorithm was run until convergence. Shown in each column are different points along the geodesic $I(g_t^{-1}), t_0, t_6, t_{12}, t_{19}$, *target*. Plotted below each row is the metric distance. Beg's algorithm was implemented with circulant boundary conditions. Shown in the last two rows are the results of the Heart and Macaque cortex experiments (Figure 16.5 and 16.6).

For large deformations such as the circular flow the vector fields are not static over time on the geodesic. Clearly the momentum evolved under the map according to $Av_t = (Dg_t^{-1})^* Av_0(g_t^{-1})|Dg_t^{-1}|$. Therefore the vector fields will have non-overlapping support. This is depicted in Figure 11.2 of Chapter 11. This flow was generated using the LDDMM algorithm of Beg to match the $\frac{1}{2}C$ (panel 1, row 3) to the C (panel 6, row 3).

Example 16.4 (Hippocampus Mapping) Shown in Figure 16.7 are results from the 2-dimensional(2D) hippocampus mapping experiment. Row 1 shows the images being transported $I_0 \circ g_t^{-1}$ for mapping the young control to the Schizophrenia subject; row 2 shows the Alzheimer's subject being mapped. The last panel of each row shows the vector fields for the Schizophrenic (columns 1–3) and Alzheimer's (columns 4–6). Data taken from the laboratory of Dr. John Csernansky; mapping results from Beg [285].

Shown in Figure 16.8 are examples of 3D maps of the hippocampus generated by Faisal Beg via the diffeomorphic flow algorithm generating $g_1(\cdot)$ solving the equation $g_1(x) = \int_0^1 v_t(g_t(x)) \, dt + x$. The time-interval along the flow was discretized into 20 timesteps. Also shown in the figure are the mappings g_1^{-1}, g_1 and the template, target images deformed via these mappings. The mappings are smooth in space.

Example 16.5 (Mitochondria) Shown in Figure 16.9 are examples of metric computations in 2D on high resolution electron-micrographs. Column 1 shows the mitochondria and the shapes for metric comparison. Columns 2–5 of Figure 16.9 show

16.3 LDDMM FOR GROWTH AND ATROPHY

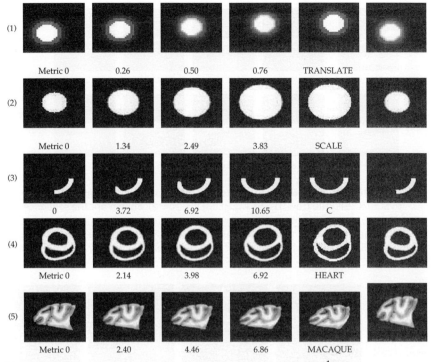

Figure 16.4 Rows 1,2, and 3 show the sequence of metric maps $I(g_t^{-1})$ for t_0, t_6, t_{12}, t_{19}, target for the TRANSLATE, SCALE, and C experiments with metric $\int_0^t \|v_s\|_V \, ds$. Column 6 shows the target through the forward map $I^D(g_1)$. Rows 4 and 5 show the HEART and MACAQUE. Mappings taken from Beg [285]; heart data taken from Dr. Raimond Winslow of Johns Hopkins University and macaque taken from Dr. David Van Essen of Washington University.

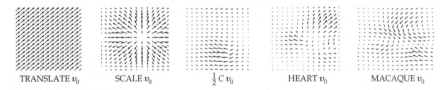

Figure 16.5 Columns 1,2, and 3 show the vector fields v_0 (row 1), v_{10} (row 2), and v_{19} (row 3) for TRANSLATION, SCALE, and C experiments. Columns 4 and 5 show the vector fields for the HEART experiment and MACAQUE for v_0, v_{10}, v_{19}. The mapping results are from Beg [285]; the heart data is courtesy of Dr. Raimond Winslow, The Johns Hopkins University. The macaque data is taken from David Van Essen.

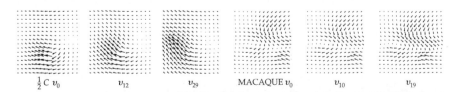

Figure 16.6 Figure shows vector fields for the $\frac{1}{2}C$ (panels 1,2, and 3) and macaque (panels 4,5, and 6) experiments.

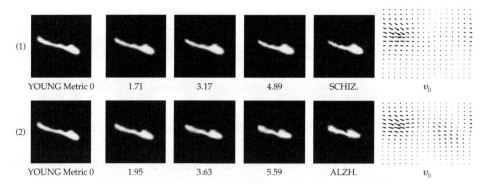

Figure 16.7 Rows 1 and 2 show results from the hippocampus mapping experiment for the YOUNG to SCHIZOPHRENIC (row 1) and ALZHEIMER'S (row 2). Shown are the sequence of geodesic mappings $I(g_t^{-1})$ connecting the YOUNG to the targets for $t_0, t_3, t_6, t_9, t_{12}, t_{15}, t_{19}$. Plotted below each is the metric distance. Column 6 shows the vector fields v_0 at the identity. Mapping results from Beg [285]. Data taken from the laboratory of Dr. John Csernansky of Washington University.

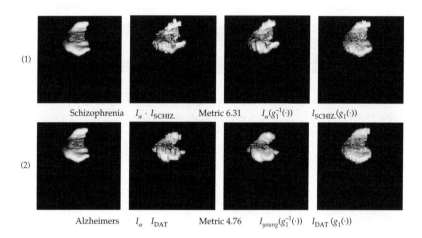

Figure 16.8 Figure shows 3D hippocampus mapping results from Schizophrenia (row 1) and Alzheimer's (row 2). Top row shows template I_α, panel 2 shows the Schizophrenic hippocampus hand labeled I_{SCHIZ}, panel 3 shows $I_\alpha(g_1^{-1}(\cdot))$, and panel 4 shows $I_{\text{SCHIZ}}(g_1(\cdot))$. Row 2 shows similar results for the Alzheimer's. Data taken from the laboratory of Dr. John Csernansky of Washington University.

the shapes ordered in geodesic distances from the templates (column 1). Shown below each panel is the metric distance.

16.4 Christensen Non-Geodesic Mapping Algorithm

What if the flow through the space of flows is not the shortest one? Christensen and Rabbitt were the first to introduce flows for large deformation matching of images [220–222, 273, 445]. Christensen implemented a computationally efficient algorithm for image matching which generates a flow

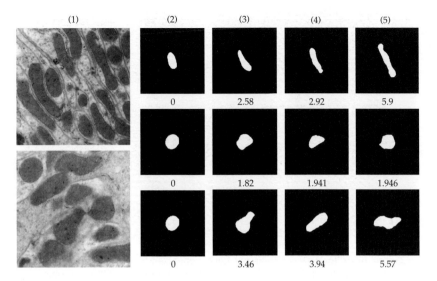

Figure 16.9 Column 1 shows electron microscopy images. Columns 2–5 show computed geodesic distances between the template shapes (column 2) and other mitochondrial shapes (columns 3, 4, and 5).

through the space of diffeomorphisms. The algorithm mapping anatomic configurations separate the joint optimization in space–time into a sequence of indexed in time optimizations solving for the locally optimal at each time transformation and then forward integrating the solution. This is only a locally-in-time optimal method reducing the dimension of the optimization. It does not revisit the entire flow attempting to find the flow of shortest metric length. Assume the deformation fields are generated from vector fields which are assumed piecewise constant within quantized time increments of size δ, $t_k = k\delta, k = 0, \ldots, K = 1/\delta$ giving $v_t = v_{t_k}$, for $t \in [t_{k-1}, t_k)$.

Algorithm 16.6 (Christensen [220]) *The sequence of local-in-time transformations $g_{t_k}^{-1}(y), k = 1, 2, \ldots,$ is given by $g_0^{-1}(y) = y$ with*

$$g_{t_k}^{-1} = \int_{t_{k-1}}^{t_k} -Dg_\sigma^{-1} v_{t_k}\, d\sigma + g_{t_{k-1}}^{-1} \qquad (16.21)$$

and $\quad v_{t_k} = \arg\inf_{v_{t_k}} (t_k - t_{k-1})\|v_{t_k}\|_V^2 + \frac{1}{2\sigma^2}\|I^D - I(g_{t_k}^{-1})\|_2^2\,. \qquad (16.22)$

For each $k = 1, \ldots, K$, solve via Algorithm 16.3 for v_{t_k} in Eqn. 16.22 for local in time updating of Eqn. 16.21, initializing with $v = 0$.

After stopping, define the final iterate as $v_{t_k}^\star(y)$ with $g_{t_k}^{-1}$ satisfying Eqn. 16.21.

The Christensen algorithm generates a particular flow through the space of diffeomorphisms which connects the images. Beg compared various solutions generated by the Christensen Algorithm 16.6 to those generated by the geodesic Algorithm 16.3. Shown in Figure 16.10 are results from the mappings. The top row shows vector fields generated from the metric correspondence algorithm; bottom row shows vector fields from the Christensen algorithm. Column 1 shows translation, column 2 shows the heart, and column 3 shows the macaque sections. Shown in each panel are the superposition of twenty vector fields along the flow. Shown in Table 16.4.1 are the relative distances between Beg's geodesic algorithm and the Christensen algorithm. The final image matching errors are virtually identical. The length of the flow in general is not

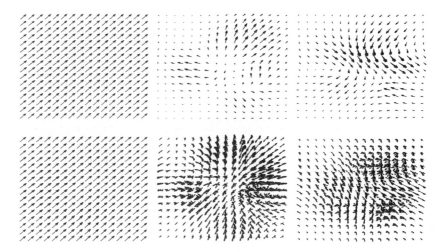

Figure 16.10 Top row shows vector fields generated from the metric correspondence algorithm; bottom row shows vector fields from the Christensen algorithm. Column 1 shows translation, column 2 shows the heart, and column 3 shows the macaque sections. Shown in each panel are the superposition of 20 vector fields along the flow.

Table 16.1 Matching distances and geodesic distance for Christensen algorithm and LDDMM

Experiment	Image Error $C(I^D, I \circ g^{-1})$		Manifold Distance $\int \|v_t\|_V \, dt$	
	"Geodesic"	"GEC"	"Geodesic"	"GEC"
"Parallel Translation"	5.648%	5.65%	0.7223	0.722757
"Heart Mapping"	11.558%	11.799%	4.631	6.7145
"Macaque Cortex Mapping"	9.55%	9.53%	4.636	5.45

similar. The right columns show the distance between the geodesic algorithm and the Christensen algorithm as a percentage for the parallel translation, heart, and macaque experiments. Notice for the linear translation there is no difference. For the curved deformation there is significant difference.

16.5 Extrinsic Mapping of Surface and Volume Submanifolds

Almost all of the geometric work in CA concentrates on smooth submanifolds of $\mathbb{R}^n$, as curved and surface submanifolds abound in human anatomy. Diffeomorphisms restricted to smooth submanifolds carry the submanifolds smoothly, see Theorems 8.24 and 8.25, Chapter 8 on diffeomorphic mappings of submanifolds. If g is a diffeomorphism from $X \to X$ with $M \subset X$ a submanifold with subspace topology, then $g : M \subset X \to g(M) \subset X$ is a diffeomorphism from $M \to g(M)$ with the subspace topology.

Examine the space of 2D smooth surface manifolds M such as the cortical and hippocampul surface. Diffeomorphisms on X carry submanifolds $M \subset X$ smoothly, the smooth tangent structures, curvature, etc. The geometry of the quadratic charts transform in the standard way with the Jacobian of the mapping transforming the tangent space, and curvature transformed by the Hessian of the transformation [229].

16.5 EXTRINSIC MAPPING OF SURFACE AND VOLUME SUBMANIFOLDS

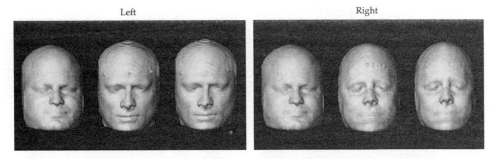

Figure 16.11 Figure shows 3D MRI–MPRAGE head matching. Left three panels show template and target, and the 3D transformation of the template matched to the target. Right three panels show the same for the second individual. Figure reproduced from Christensen et al., Volumetric Transformation of Brain Anatomy, 1997

16.5.1 Diffeomorphic Mapping of the Face

The large deformation Christensen Algorithm 16.6 has been applied for generating whole head maps imaged via MRI–MPRAGE sequences. Figure 16.11 shows results of such experiments for two different subjects. Shown are surface renderings of the surface of the mapped volumes forming the MRI brain volumes. Panels 1 and 2 show the template and target surface renderings. The whole brain volume is being studied via the mapping, but because of the use of the surface rendering only the observable external boundary of the volume is seen manifesting as the human face. Panel 3 shows the rendering of the volume after transformation of the template matched to the target. Panels 4, 5, and 6 show the same for a second individual. Only the faces are depicted although the entire head has been transformed.

16.5.2 Diffeomorphic Mapping of Brain Submanifolds

Restricting the diffeomorphism on $\mathbb{R}^3$ to the submanifolds of the brain provides smooth mappings of the brain substructure. The Csernansky group has been mapping the hippocampus in populations of schizophrenic and Alzheimer's patients [162, 408, 422, 423, 446]. Shown in Figure 16.12 are results depicting the hippocampus mappings of Washington University. Column 1 shows the MRI and associated template hippocampus. Column 2 shows the template surface (above) and embedded in the volume (below). Column 3 shows the target surface (above) and the target surface embedded in the target volume (below). The target surface was generated by hand contouring. Column 4 shows the results of mapping the surface from the template into the target.

16.5.3 Extrinsic Mapping of Subvolumes for Automated Segmentation

The smooth maps carry the segmentation labels of the volumes consistently. Shown in Figure 16.13 are examples illustrating mapping of subvolumes for automated segmentation from Christensen from the occipital lobe in the macaque emphasizing the importance of fusing landmarks and dense image information for estimating the diffeomorphisms. The macaque cryosections were generated

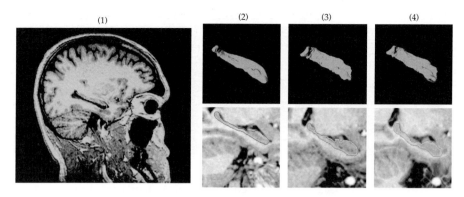

Figure 16.12 Column 1: Panel shows the template hippocampus (green) embedded in the MRI volume. Columns 2,3,4: Panels show the template surface (top), the target surface (middle), and the mapped template surface (bottom). Top row shows the hippocampus surface through the mapping; bottom row shows the surface embedded in the mapped volumes. Mapping results from Christensen et al. [220] and Haller et al. [162]. (see Plate 31)

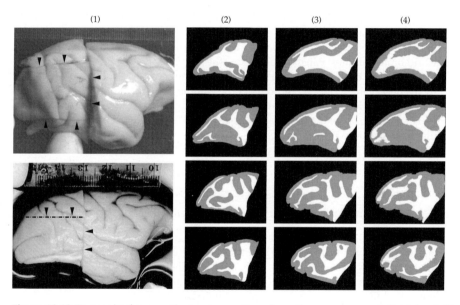

Figure 16.13 Parasagittal cryosection cortex sections from the macaque occipital lobe. Left column: Panels show photographs of the right hemisphere of macaque 92K (top) and 92L (bottom). Arrows show the cuts that were made to remove part of the visual cortex three. Right columns show automated segmentations of sections 23, 37, 52, 61 showing 92K, 92L, and 92K → 92L. Columns 2 and 3 show hand segmentations; column 4 shows automated segmentation from mapping 92K → 92L. Mapping results taken from Christensen et al. [220]; data collected in the laboratory of David Van Essen of Washington University.

in David Van Essen's laboratory. Both occipital lobes were hand traced into complete gray and white matter segmentations.

Figure 16.13 shows automated segmentations of tissue volumes generated by Christensen applying the diffeomorphisms to hand segmentations of the template exemplars. Panels 1 and 2 of Figure 16.13 show the macaque cryosection volumes generated in David Van Essen's laboratory.

16.5 EXTRINSIC MAPPING OF SURFACE AND VOLUME SUBMANIFOLDS

Figure 16.13 show the automated 3D segmentation of 92K, 92L cortex volumes. Top row panels 1 and 2 show photographs of the right hemisphere of macaque monkey 92K template and 92L target. Arrows show the cuts that were made to remove part of the visual cortex. Bottom rows show section under the mapping. Columns 1 and 2 show sections of 92K and 92L; column 3 shows $92K \to 92L$. Columns 4 and 5 show the same sections of the gray and white matter hand segmentations of 92K and 92L. Column 6 shows the corresponding three sections automatically segmented by transforming the 92K hand segmentation to 92L with the large deformation diffeomorphic transformation resulting from the sulcus landmarks and image matching diffeomorphism.

16.5.4 Metric Mapping of Cortical Atlases

The mammalian cerebral cortex has the form of a layered, highly convoluted thin shell of gray matter surrounding white matter. The cortex contains a complex mosaic of anatomically and functionally distinct areas which play a tremendously important role in understanding brain functions [155]. As championed by Van Essen [175], to aid in the understanding of the geometry and the functional topography of the cortex the convoluted cerebral cortex may be mapped to planar coordinates for visualization of the deep and buried gyri. To understand individual variation in the cortical topography the CA community has been using the large deformation metric landmark mapping tools on spherical representations and planar representations of the cortex to establish correspondences between the flat maps of various individual cortical surfaces.

David Van Essen has been studying cortical atlases via both spherical and planar bijections. The left column of Figure 16.14 shows an atlas of the macaque visual cortex constructed by David Van Essen. The top panel shows the macaque surface in 3D and the bottom panel shows the flat atlas. The right five panels of Figure 16.14 show results from the metric transformation of the planar representations of the macaque from Van Essen. A shows the flat map of Case 1. Shading indicates cortical geography; black lines indicate the landmark contours used to constrain the deformation (along sulcal fundi and along the map perimeter); and white lines indicate grid lines at intervals of 5 map-mm. Panel B shows the pattern of grid lines after deformation of Case 1 to bring it into register with the target atlas map in panel D. Gray boxes highlight the locations of deformed marker points a' and b'. Panel C shows the vector field for selected grid points (at intervals of 10 map-mm). Arrow bases indicate grid positions in the source map, and arrow tips indicate the location of grid points in the deformed source map. For example, displacements are largely to the right in the lower part of the map and largely to the left in the upper left of the map. Panel D shows the map of geography (shading) and target registration contours (black lines) on the atlas map. Panel E shows contours from the deformed source map (black lines) and deformed reference marks a', b' are overlaid on the target atlas map.

> **Example 16.7 (Human and Macaque Cortical Mapping)** Figure 16.15 shows an interspecies comparison via cortical surface deformation between the macaque and visible human (see [362]). Panel 1 shows the 3D Visible Human Male. Panels 2 and 3 show landmarks on macaque flat maps and human flat maps to bring them into correspondence. Panel 4 shows the boundaries of deformed macaque visual areas (black lines) superimposed on the fMRI activation pattern from an attentional task from the study of Corbetta et al. [447] after deformation to the Visible Man atlas by Drury et al. [448]).

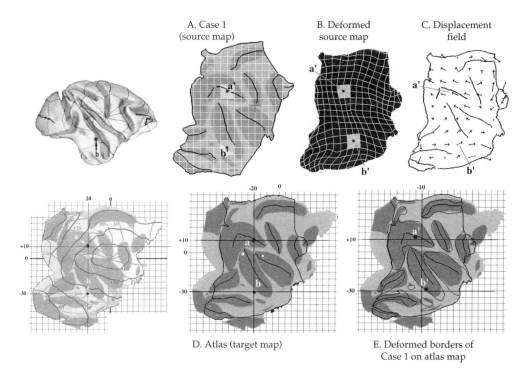

Figure 16.14 Van Essen atlas of macaque visual cortex. Left column shows the macaque atlas (top row) and flattened version (bottom row) generated by David Van Essen. Right column shows flattened surface-based mapping from an individual flat map via landmark matching in the plane to the macaque atlas. **A** shows flat map of Case 1, **B** pattern of grid lines after deformation of Case 1, **C** shows the displacement vector field, **D** shows the map of geography (shading) and target registration contours (black lines) on the atlas map, and **E** shows the deformed borders of Case 1 on the atlas map. Results taken from Van Essen et al. [362] (See Plate 32).

16.6 Heart Mapping and Diffusion Tensor Magnetic Resonance Imaging

Now we examine extending the mapping work to vector and tensor valued imagery. Vector and tensor valued imagery arise in many different imaging modalities including diffusion tensor imaging. Diffusion tensor magnetic resonance imaging (DT-MRI) quantifies the anisotropic diffusion of water molecules in biological tissues, making it possible to non-invasively infer the architecture of the underlying structures. The measurement at each voxel in a DT-MRI image volume is a symmetric second order tensor. The principal direction of the diffusion tensor corresponds to the fiber orientation of the heart. We now examine the work of Cao and Younes, working with Raimond Winslow of the Johns Hopkins University, who have been mapping DTI volumes of the heart based on the principal eigenvectors from DT-MRIs [449]. Figure 16.16 illustrates the use of scalar and vector DTI for studying geometry of heart tissue. Columns 1 and 2 illustrate normal anatomical variation as depicted in conventional MRI scalar imagery. Columns 3 and 4 show the use of DTI for studying tissue geometry. The orientation of the DTI tensor measurements in these sections are color coded by depicting the direction of the principal eigenvectors in the diffusion tensor.

16.6 HEART MAPPING AND DT-MRI

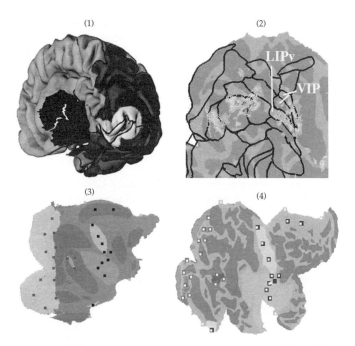

Figure 16.15 Panel 1 shows 3D Visible Human Male. Panel 2 shows the boundaries of deformed macaque visual areas (black lines) superimposed on the fMRI activation pattern from an attentional task from the study of Corbetta et al. [447] after deformation to the Visible Man atlas by Drury et al. [448] and mapping results from [362]. Panels 3 and 4 show landmarks on macaque flat map and human flat map, respectively, used for performing the mappings shown in panel 2 (see Plate 33).

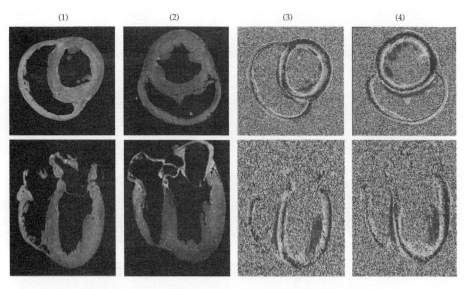

Figure 16.16 Columns 1 and 2 show hearts in grayscale MRI in normal (column 1) and failing (column 2) hearts; columns 3 and 4 show DTI of those sections depicting orientation of the principal eigenvectors by color. The top row shows coronal sections, the bottom row shows axial sections (see Plate 34).

To model the orbit of imagery, define the images to be functions $I: X \to \mathbb{R}^3$ that associate to each point a vector, the principal direction of the diffusion tensor (outside the object the vector is zero). Cao and Younes [449] define the action $\mathcal{G}$ on the set $\mathcal{I}$ of all principal direction images as follows.

With $D_f g$ denoting the Jacobian matrix of g evaluated at f, the group action used for LDDMM is for any $I \in \mathcal{I}$ and $g \in \mathcal{G}$,

$$g \cdot I = \begin{cases} \dfrac{D_{g^{-1}} g I \circ g^{-1} \| I \circ g^{-1} \|}{\| D_{g^{-1}} g I \circ g^{-1} \|} & \text{when } I \circ g^{-1} \neq 0 \\ 0 & \text{when } I \circ g^{-1} = 0, \end{cases} \quad (16.23)$$

with $D_f g$ denoting the Jacobian matrix of g evaluated at f. Given two elements I and data I^D linked via $\dot{g}_t = v_t(g_t), t \in [0,1]$, the LDDMM algorithm minimizes the matching function

$$\int_0^1 \|v_t\|_V^2 \, dt + \alpha \|I^D - I \circ g_1^{-1}\|^2 + 4(\|I\| \circ g_1^{-1} - \|I^D\|)^2. \quad (16.24)$$

Cao computes the optimizer performing steepest descent by computing the gradient with the step size ϵ computed via golden section search for the line search. The vector fields generating the maps are integrated using second-order semi-Lagrangian schemes for the transport equation.

Example 16.8 (Scalar MRI Mapping of Hearts) Beg and Helm have now mapped the geometries of several normal and failing hearts based on only the scalar MRI imagery [450] using the last term (not the vector term, $\alpha = 0$) in the

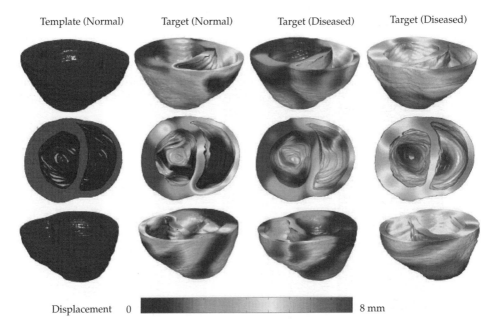

Figure 16.17 Figure shows the template heart and the target heart images after transformation into the coordinate system of the template. The determinant of the Jacobian of the transformation is superimposed as a colormap on the surface rendering. Blue colors indicate regions where the determinant is less than unity whereas red regions are where the determinant is greater than unity. Data courtesy of Dr. Raimond Winslow; mappings courtesy of Faisal Beg and Pat Helm (see Plate 35).

energy minimization of Eqn. 16.24. As depicted in Figure 16.17, the heart geometry displays significant anatomical variability. Using the diffeomorphic metric mapping algorithm Beg registered the hearts to a common template coordinate system, accomodating the large deformations associated with heart disease. The heart datasets were both rigidly registered and diffeomorphically registered based on input landmarks providing an initial condition for the diffeomorphism metric mapping algorithm.

Errors in the mapping were quantified by defining a normalized mismatch defined as a percentage of the error of the template heart segmentation with respect to the target heart segmentation before the mapping given by

$$\epsilon = \frac{\|I_0(g^{-1}) - I_1\|_2^2}{\|I_0 - I_1\|_2^2} . \tag{16.25}$$

Error rates on the order of 11.5% were found in the three mappings. Shown in Figure 16.17 are three views of the diffeomorphic transformations of the hearts into the template coordinates. The operator in the metric was chosen to be $A = (-0.01\nabla^2 + \text{id})^2$, and the flow $t \in [0, 1]$ was discretized into 20 timesteps each of size $\delta t = 0.05$. Row 1 shows the normal and rows 2 and 3 show the diseased hearts after transformation. Notice the similarity of the hearts to the template geometry. Overlaid on the transformed geometry is the determinant of the Jacobian of the transformation at that point. Blue colors indicate regions where the determinant is less than unity indicating regions of contraction, whereas red regions are where determinant is greater than unity, indicating expansion.

Example 16.9 (DTI Heart Mapping (Cao)) Cao and Younes have been mapping heart geometries based on minimizing the cost of Eqn. 16.24. A rigid motion was first calculated to provide global registration generated by aligning the centroids and the principal axes of the two objects. Figure 16.18 shows sections along the geodesic as the algorithm aligns the vector images corresponding to the largest eigen value of the DT data.

Shown in Figure 16.19 are the 3D LDDMM-DT showing the geometries of the two hearts superimposed after rigid motion and LDDMM. The red color in the bottom right panel demonstrates almost total alignment as measured via the dot product of the corresponding principal eigenvectors.

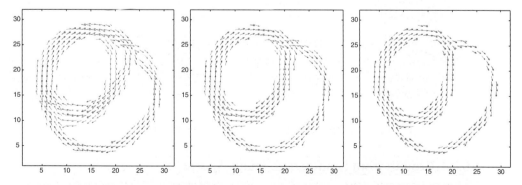

Figure 16.18 Panels show points along the geodesic for the LDDMM of the DT vector data, with the red vectors showing the template, and the blue vectors showing the target. Taken from [449].

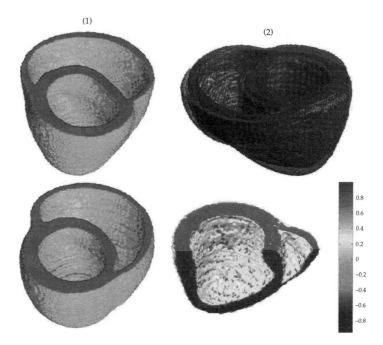

Figure 16.19 Column 1 shows the geometries of the two hearts. Column 2, top panel shows the two geometries superimposed after rigid motion; blue is template, red is target. Column 2, bottom panel shows the LDDMM solution color representing the dot product of the corresponding vectors after alignment. Red color means total alignment; taken from [449] (see Plate 36).

16.7 Vector Fields for Growth

In the dynamic growth model, an observation time sequence is given and the goal is to infer the time flow of geometric change which carries the known exemplar through the growth period $t \in [0, 1]$. Here the orbits are flows of imagery, with growth corresponding to actual real-time, and comparison is performed by comparing the observables $I_t, t \in [0, 1]$ to its start point I_0. Here $I_0 = I_\alpha$ plays the role of the so-called template or exemplar. In static matching time is simulation time only, with the single target observable at time $t = 1$. Throughout we take as the dynamic growth model orbits of the exemplars under space–time flows

$$\mathcal{I}_\alpha[0,1] = \left\{ I_\alpha(g_t^{-1}), t \in [0,1] : \frac{\partial g_t}{\partial t} = v_t(g_t), g_0 = \text{id} \right\}. \tag{16.26}$$

Each element $I_\alpha(g_t^{-1}), t \in [0, 1] \in \mathcal{I}_\alpha[0,1]$ in the orbit is modeled as corresponding to a different growth trajectory of the exemplar. Figure 16.3 depicts the basic static inexact matching model studied throughout this chapter.

16.7.1 Growth from Landmarked Shape Spaces

Now examine the setting where the observables are time sequences of landmarks. Throughout assume $(V, \|\cdot\|_V)$ is a reproducing kernel Hilbert space with operator K.

Theorem 16.10 (Growth via Metric Mapping: Sparse Landmarked Shapes) *Given the sparse landmarks $y_{nt}, n = 1, 2, \ldots,$ with data matching term $C = \frac{1}{\sigma^2} \sum_{n=1}^{N} \int_0^1 \|y_{tn} - g_t(x_n)\|_{\mathbb{R}^d}^2 \, dt$, then the optimizing growth flow minimizing the energy*

$$\int_0^1 \|v_t\|_V^2 \, dt + \frac{1}{\sigma^2} \int_0^1 \sum_{n=1}^{N} \|y_{tn} - g_t(x_n)\|_{\mathbb{R}^d}^2 \, dt, \tag{16.27}$$

has momentum satisfying

$$Av_t(x) = \frac{1}{\sigma^2} \sum_{n=1}^{N} \delta(g_t(x_n) - x) \int_t^1 (Dg_{t,u}(g_t(x_n)))^* (y_{uk} - g_u(x_n)) \, du = 0, \quad x \in X. \tag{16.28}$$

Proof The V-gradient of the second term in the cost requires the differential $\partial_\psi g_t(x_k)$ given by Eqn. 11.20 giving the variation of the second term according to

$$-\frac{2}{\sigma^2} \int_0^1 \sum_{n=1}^{N} \langle y_{un} - g_u(x_n), \partial_\psi g_u(x_n) \rangle_{\mathbb{R}^d} \, du$$

$$\stackrel{(a)}{=} -\frac{2}{\sigma^2} \int_0^1 \int_0^u \sum_{n=1}^{N} \langle y_{un} - g_u(x_n), Dg_{t,u}(g_t(x_n)) \psi_t(g_t(x_n)) \rangle_{\mathbb{R}^d} \, dt \, du$$

$$\stackrel{(b)}{=} -\frac{2}{\sigma^2} \int_0^1 \int_t^1 \sum_{n=1}^{N} \langle (Dg_{t,u}(g_t(x_n)))^*(y_{un} - g_u(x_n)), \psi_t(g_t(x_n)) \rangle_{\mathbb{R}^d} \, du \, dt,$$

$$\stackrel{(c)}{=} -\frac{2}{\sigma^2} \int_0^1 \left\langle \sum_{n=1}^{N} K(g_t(x_n), \cdot) \int_t^1 (Dg_{t,u}(g_t(x_n)))^* \left(y_{un} - g_u(x_n)\right) du, \psi_t(\cdot) \right\rangle_V dt,$$

where (a) follows from Eqn. 11.17 of Lemma 11.5, and rearranging the integral and inner-product, (b) by interchanging the order of integration in t and u, and (c) from the sifting property of the Green's kernel. □

Algorithm 16.11 (Gradient for Growth of Landmarked Shape) *The growth model gradient algorithm has gradient*

$$\nabla_v E_t^{old}(x) = v_t^{old}(x) - \frac{1}{\sigma^2} \sum_{n=1}^{N} K(g_t^{new}(x_n), x)$$

$$\int_t^1 (Dg_{t,u}^{new}(g_t^{new}(x_n)))^* (y_{un} - g_u^{new}(x_n)) \, du, \quad x \in X. \tag{16.29}$$

Example 16.12 (Growth in Mouse) Jiangyiang Zhang, Susumu Mori, and Peter Van Zijl have been characterizing brain development with DTI via 3D, digitized images of ex-vivo mouse brain samples with high resolution (100 μm per pixel) and high tissue contrast[451, 452]. These technique have paved the way for quantitative characterization of mouse brain anatomy during development. Brain development involves complex sequences of tissue growth and movements in three dimensions. During postnatal development, the mouse brain undergoes significant volume and shape changes. Figure 16.20 shows example of the evolving mouse brain structures from the date of birth (P0) to postnatal day 45 (P45) in MR T_2 and diffusion tensor imagery. Volumetric measurements show that cortical and cerebellar volumes increase approximately four and six fold, respectively, from P0 to P10. After P10, volumetric change stabilizes,

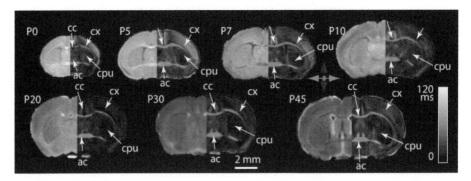

Figure 16.20 MR images of postnatal mouse brains. Coronal MR T_2 and DT images were shown at the level of anterior commissure. Images have been aligned to ensure proper orientation and position. Intensity in MR T_2 images is related to tissue properties, such as the content of myelin protein. MR diffusion tensor images reveal local tissue orientation. Both types of images were utilized in our mouse brain developmental study. The color scheme for diffusion tensor images was illustrated by color arrows in the figure, with red for local tissue whose orientation is anterior-to-posterior (perpendicular to the current plane), green for horizontal orientation, and blue for vertical orientation. The scalar bar represents 2 mm. Structure abbreviations are—ac: anterior commissure; cc: corpus callosum; CPu: caudate putamen; CX: cortex (see Plate 37).

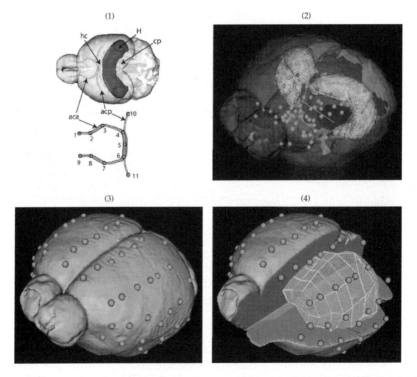

Figure 16.21 Panel 1 shows white matter tracts, anterior part of anterior commisure (aca), posterior (acp), cerebral peduncle (cb), hippocampus (H), and hippocampal commisure (hc). Panels 2,3, and 4 show visualization of landmarks on white matter tracts (green dots) and landmarks inside hippocampus (yellow structures) (blue dots), outer brain surface (orange) with boundary of white matter and cortical region shown in green, and landmarks depicted as intersections of white lines on the green structure (see Plate 38).

while changes in shape can still be appreciated from P10 to P30 [453]. These results suggest that mouse brain development is a dynamic process with complex changes in both volume and shape being involved. In order to capture detailed morphological changes during development, it is necessary to employ a large number of landmarks along structures of interest. As depicted in Figure 16.21, landmarks are placed along several brain compartments, including white matter tracts hippocampus, cortex and boundaries of white matter and cortex. Shown in panel 1 are 3D visualization of landmarks on white matter tracts (green dots) and landmarks inside hippocampus (yellow structures) (blue dots). Shown in panel 2 are 3D visualization of landmarks (orange dots) on outer brain surface. Panel 3 shows a schematic drawing showing several white matter tracts, anterior part of anterior commisure (aca), posterior (acp), cerebral peduncle (cb), hippocampus (H), and hippocampal commisure (hc). Panel 4 shows the visualization of landmarks on outer brain surface and on boundary of white matter and cortical region (the green structure). Landmarks on the green structure are not shown as orange dots as landmarks on the outer brain surface, but as intersections of white lines on the green structure.

Figure 16.22 demonstrates the landmark based LDDMM of a P7 mouse brain to a P30 mouse brain based on 270 landmarks that covered the entire forebrain region. The accuracy of the mappings was examined by plotting manually segmented

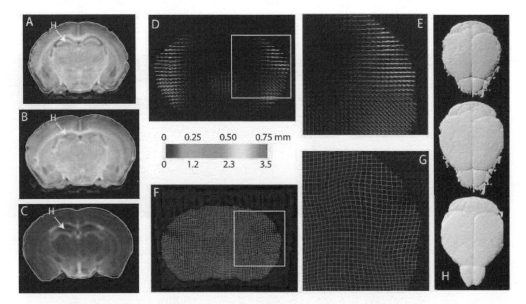

Figure 16.22 LDDMM of P7 mouse brain images to P30 images. Coronal MR images of P7 (A) and P30 (C) were shown with brain boundaries marked by orange curves. Deformed P7 (B) was overlaid with the brain boundary of P30. White arrows in A, B, and C are pointing to hippocampus (H). The transformation was visualized as color-coded vector plot (D) and Jacobian plot (F), with enlarged local areas shown in E and G, respectively. For vector plots, color of vectors represents the magnitude of local displacement. For Jacobian plot, color on deformed grids represents changes in local volume. Surface of P7, deformed P7, and P30 mouse brains were visualized in H, from top to bottom. Vector maps show how transformations deform a local voxel from its original location by taking the original location as the start of the vector and the destination as the end; length and color of vectors reflect the magnitude of tissue displacement. The Jacobian map shows local volume changes; a value greater than unity corresponds to volume increase, a value less than unity corresponds to volume loss, and unity for no volume change (see Plate 39).

tissue boundaries of P30 mouse brains over the deformed P7 mouse brain images. Figure 16.22B shows the deformed P7 mouse brain images together with the outer boundary of the P30 mouse brain in orange. The P30 brain boundary agrees well with the deformed P7 mouse brain images. The shape of inner structures such as the hippocampus (H) also agrees with corresponding structures in the P30 brain images. The lateral cortical regions show significant growth from P7 to P30, with large tissue displacement coupled with near three-fold volume increase. The average volume increase for the entire forebrain in this period is two-fold.

Using LDDMM, various aspects of mouse brain growth may be characterized, including structural evolution and morphological variations. Figure 16.23 shows the growth of the hippocampal surface measured using LDDMM methods. Displacements of the surface of the mouse hippocampus are computed and visualized. Although this result contains not only the effects of growth, but the effects of landmark placement variations as well, it suggests that the hippocampal growth is not spatially uniform. The ventral part of the hippocampus experienced greater tissue displacement during development than the dorsal part.

Example 16.13 (Huntington's Disease) Dr. Elizabeth Aylward of the University of Washington is studying the patterns of neuro-degeneration in Huntington's disease following a time course of atrophy. Over time individuals affected display a marked decrease in the ability to walk, think, talk, and reason. The caudate and putamen are known to be affected by the neuro-degenerative processes. While it is well known that

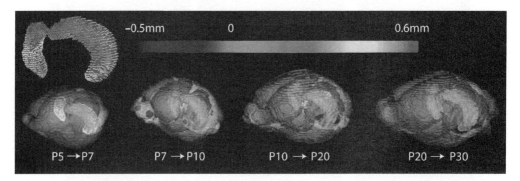

Figure 16.23 Outer-shape changes of mouse hippocampus measured using LDDMM. The displacement due to growth of hippocampus was color coded and the direction of growth was visualized by 3D glyphs (see plate 40).

Figure 16.24 Series of volumes depicting aging in Huntington's disease patient over time as manifest in the Caudate. Data taken from the laboratory of Dr. Elizabeth Aylward of the University of Washington (see Plate 41).

there is a marked decrease in volume of these affected organs, less is known about the pattern of neuro-degeneration, in particular whether particular regions are affected more preferentially than others.

The caudate volumes of pre-symptomatic individuals — those with the mutated-gene who have not yet manifested the clinical symptoms of the disease — are being mapped from MRIs taken subsequent years apart. Preliminary observations suggest that the process of neuro-degeneration in the disease, although affecting the entire caudate, is marked in the medial region with considerable atrophy taking place in the head and the tail regions. Recent research from other groups has shown that regions in the posterior cortex are significantly atrophied. Taken with the neuro-degenerative pattern observed in the caudate, this suggests that the axonal tracts connecting the posterior regions of the caudate and the cortex are perhaps disrupted as well. Shown in Figure 16.24 are results from the mapping of an individual patient with Huntington's disease for over four years.

17 COMPUTATIONAL ANATOMY: HYPOTHESIS TESTING ON DISEASE

ABSTRACT Computational Anatomy (CA) is the study of human anatomy $I \in \mathcal{I} = I_\alpha \circ \mathcal{G}$, an orbit under groups of diffeomorphisms of anatomical exemplars $I_\alpha \in \mathcal{I}$. The observable images $I^D \in \mathcal{I}^D$ are the output of Medical imaging devices. This chapter focuses on the third of the three components of CA: (iii) generation of probability laws of anatomical variation $P(\cdot)$ on the images $\mathcal{I}$ within the anatomical orbits and inference of disease models. The basic random objects studied are the diffeomorphic transformations encoding the anatomical objects in the orbit; Gaussian random fields are constructed based on empirical observations of the transformations. Hypothesis testing on various neuromorphometric changes are studied.

17.1 Statistics Analysis for Shape Spaces

The most common statistical methods applied for the study of shape spaces have been those based on second-order representations and Gaussian random fields. The basis representations are often constructed using principle components analysis (PCA) (examined in the context of Karhunen–Loeve decompositions 14.2.1 of Chapter 14), which compute variance maximizing bases of the linear vector (Hilbert) space modeling the random elements. These approaches fundamentally model the measurements as random variables which are representing to second order (mean and covariance), thus the fundamental link to Gaussian random field models. The basis vectors spanning the Hilbert space are oriented along directions of maximal variance, and ordered such that the first vector spans the 1D (ID) subspace of largest variation, and the second vector captures the next largest variation, etc. It is therefore effective for analyzing variation in the data and for dimensionality reduction. For example, by projecting a data point onto a subspace spanned by the first D basis vectors, we obtain an optimal (to second order) D dimensional representation of the data. If a large percentage of the variation is captured in this D dimensional subspace, then the approximation can be expected to be quite good. If in addition, D is small then the subspace may offer considerable dimensionality reduction of the sample space.

While such second-order representations are commonly used in the analysis of shape, they are only valid for linear spaces. However, they have already found wide applicability, in active shape models (ASM)[199] for estimating shape variability from training sets. Of course we should expect that for modeling ensembles which only exhibit small deformations linear subspace approximations will be adequate. However, for the study of large deformation diffeomorphisms of anatomical models such a methodology cannot be extended. The shape spaces are not a nonlinear metric space. Simply stated, transformations can be composed but not added; vector space addition of the random variables will not generate elements in the anatomical model space. There is no rigorous way, for example, to add two landmark configurations and be guaranteed that the resulting configuration is a meaningful combination of the originals.

Fortunately, interpreting the shape space from the Lie group point of view, its tangent space at the identity provides the proper bridge to the linear approximation methods. For this we formulate the random objects being modeled to second-order as elements associated with the tangent space at the identity of the transformation. We have already seen that the idea for comparing shapes is to model comparison between elements in the orbit $\mathcal{I}_\alpha$ via the diffeomorphic transformations in $\mathcal{G}$. The optimal diffeomorphism which matches two arbitrary elements in the orbit is chosen from all curves $g_t, t \in [0, 1]$ in $\mathcal{G}$ connecting the two elements via the group action. It is chosen as the curve which minimizes an energy with respect to a measure of infinitesimal variations in $\mathcal{G}$. These energy minimizing paths (geodesics) induce a metric on $\mathcal{I}_\alpha$ [278]. Now we focus on the implications of this fundamental property in providing the powerful capability to represent the entire flow of a geodesic in the orbit by a template configuration and a momentum configuration at a single instant in time. Viewing the space of momenta as the random objects provides us the connection to the Gaussian random process models on vector spaces. For this the anatomical template formulation and elements of the orbit are represented by a template configuration and the initial momenta. The template remains fixed and the initial momenta are statistically characterized.

We have already remarked that this approach is founded in the Lie group point of view. The tangent space V at the identity id $\in \mathcal{G}$ is considered the "Lie algebra" of the group. The idea will be to use its vector field elements and members of its dual space V^* the space of momenta to model deformations, given that geodesics can be generated from elements of V or V^*. The power of the approach comes from the dimensionality reduction of geodesic flow to a single representative element, and the fact that the representative space is linear. The linear statistical setting provides a natural mechanism for coping with the non-linear nature of the diffeomorphic shape space. Statistics on manifolds, in particular shape manifolds have been studied in, for example, [454,455]. The representation of shape via their Lie algebra has been applied in the statistical learning setting in [456] and [457]. However, we now examine the diffeomorphic setting. We remark that the approach to estimating the mean we have presented is analogous to the Lie group approaches of computing, the so-called *intrinsic mean* of a probability measure on a Riemannian manifold [455] which has been applied to shape in, for example, [456] and [457]. These other approaches differ in that their Lie algebra representations do not necessarily generate diffeomorphisms of the template.

17.2 Gaussian Random Fields

For constructing probability measures of anatomical variation, we shall characterize the maps as Gaussian fields indexed over the manifolds on which the vector fields are defined.

Specify the vector fields as a d-dimensional Gaussian random field $\{U(x), x \in X \subset \mathbb{R}^d\}$ completely specified by its covariance matrix field which is a mapping $K : X \times X \to \mathbb{R}^d \times \mathbb{R}^d$ such that $K(x,y) = K^*(y,x)$ and for any integer n and any n-tuple of d-vectors $w_1, \ldots, w_n \in \mathbb{R}^d$ and points $x_1, \ldots, x_n \in \mathcal{X}$, $\sum_{i,j=1}^{n} w_i^* K(x_i, x_j) w_j \geq 0$.

Define $\{U(x), x \in X\}$ to be a Gaussian random field on the Hilbert space with mean field $\bar{U}$ and covariance field K_U if for all $f \in H$, $\langle f, U \rangle$ is Gaussian distributed with mean $\langle f, \bar{U} \rangle$, and variance $\langle f, K_U f \rangle$. Construct the $\{U(x), x \in X\}$ as a quadratic mean limit using a complete $\mathbb{R}^d$-valued orthonormal basis $\{\psi_k, k = 0, 1, \ldots\}$, and the U-field given according to

$$U(x) \stackrel{q.m.}{=} \sum_{k=0}^{\infty} U_k \psi_k(x), \qquad (17.1)$$

where U_k are independent Gaussian random variables with fixed means $\bar{U}_k$ and variances σ_k^2. The mean and covariance operator of the field becomes

$$\bar{U} = \sum_{k=0}^{\infty} \bar{U}_k \psi_k, \qquad K_U = \sum_{k=0}^{\infty} \sigma_k^2 \psi_k \langle \psi_k, \cdot \rangle. \qquad (17.2)$$

Eqn. 17.1 is meant as the quadratic mean limit minimally requiring the process to have covariance K_U with finite trace so that $\sum_k \sigma_k^2 < \infty$, $\sum_k |\bar{U}_k|^2 < \infty$.

Since the random U-fields are $\mathbb{R}^d$ valued, for each k there will correspond d-orthogonal eigenfunctions, leading to a natural indexing of the eigenfunctions and variances according to $\{\phi_k^{(i)}, \sigma_k^{(i)}, i = 1, \ldots, d\}$.

17.2.1 Empirical Estimation of Random Variables

Given the set of displacement fields $\{u^{(1)}(\cdot), u^{(2)}(\cdot), \ldots u^{(n)}(\cdot)\}$, then *maximum-likelihood estimation* can be used directly to estimate from the anatomical maps the mean and covariance operators of the prior distribution. Let $u_k^{(i)} = <u^{(i)}, \phi_k>$ be the coefficients generated from the ith map, $i = 1, \ldots n$, then the log-likelihood written in the K-dimensional basis gives

$$\ell(\bar{U}, \sigma; u^{(1)}, \ldots, u^{(n)}) = -\frac{n}{2} \sum_{k=0}^{K} \log\left(2\pi \sigma_k^2\right) - \frac{1}{2} \sum_{k=0}^{K} \frac{\sum_{i=1}^{n} |u_k^{(i)} - \bar{U}_k|^2}{\sigma_k^2}. \tag{17.3}$$

The maximum-likelihood estimates for the mean and variances of the orthogonal Gaussian random variables $U_k \sim N(\bar{U}_k, \sigma_k^2)$ given the set of anatomical maps with mean-field and covariance $(\bar{U}, K_U)$ given by Eqn. 17.2 become

$$\hat{\bar{U}}_k = \frac{1}{n} \sum_{i=1}^{n} \langle \psi_k, U^{(i)} \rangle, \quad \hat{\sigma}_k^2 = \frac{1}{n} \sum_{i=1}^{n} |\langle \psi_k, U^{(i)} \rangle - \hat{\bar{U}}_k|^2. \tag{17.4}$$

17.3 Shape Representation of the Anatomical Orbit Under Large Deformation Diffeomorphisms

Now we examine the representation of typicality and deviation from typicality in the orbit of shapes and images. We study throughout the metric connection model in which single instants are observed, from the orbit of imagery

$$\mathcal{I}_\alpha = \left\{ g \cdot I_\alpha = I_\alpha(g_1^{-1}) : \frac{\partial g_t}{\partial t} = v_t(g_t), t \in [0,1], g_0 = \mathrm{id}, v \in \mathcal{V} \right\}. \tag{17.5}$$

It will be helpful to index the flow explicitly by the vector which generates it, so that g_t^v satisfies $\dot{g}_t^v = v_t(g_t^v)$. According to Theorem 11.8 of Chapter 11 we know from the fundamental conservation law $(\partial/\partial t)\langle Av_t, w_t \rangle_2 = 0, t \in [0,1]$, thus the vector field and momentum acting against vector fields at the identity completely determine the geodesic carrying one object to another:

$$Av_t = (Dg_t^{v-1})^* Av_0(g_t^{v-1})|Dg_t^{v-1}|, \quad t \in [0,1]. \tag{17.6}$$

Understanding variation in the anatomical orbit is reduced to understanding the vector field and momentum at the identity of $t = 0$ of the flow. Thus we model the empirical space of mappings by reducing them to their encoding via the vector field and momentum at $t = 0$. Identifying the image in the anatomical orbit I with the vector V_0, which generates it accordingly, motivates our model on V_0 for anatomical ensembles as Gaussian vector fields with mean and covariances. Then we shall model the family of vector fields as a Gaussian random field with mean and covariance $\bar{V}, K_V$. The statistical law P on $\mathcal{I}$ is induced by the law on V_0. Specializing the study of statistics via the initial momentum is a a natural setting for linear statistical analysis.

Construction 17.1 (The Large Deformation Statistical Model) *Given is the anatomical orbit*

$$\mathcal{I} = \{I = I_\alpha \circ g_1^{-1} : \dot{g}_t = v_t(g_t), g_0 = \mathrm{id}, v \in V\}. \tag{17.7}$$

Associating to each I the V_0 such that

$$I(V_0) = I_\alpha \circ g_1^{-1} \quad \text{where } AV_t$$

$$= (Dg_t^{V-1})^* AV_0(g_t^{V-1})|Dg_t^{V-1}|, \quad \dot{g}_t^V = V_t(g_t^V), \quad g_0^V = \mathrm{id}. \tag{17.8}$$

Then V_0 is a Gaussian random field with mean and covariance $\bar{V}, K_V$. The statistical law P on $\mathcal{I}$ is induced by the law of V_0.

17.3.1 Principal Component Selection of the Basis from Empirical Observations

We have already explored pattern representation via principle components for empirical covariance estimation. Model the space of empirical vector fields as realizations of a Gaussian random field with covariance constructed to fit the empirical variation. The complete orthonormal basis for the Hilbert space representation is generated from the orthonormal eigenfunctions of the empirical covariance.

Let $V(\cdot)$ on X be a zero mean real-valued Gaussian random field with covariance $K(x,y)$ on closed and bounded background space X. The set of orthonormal functions $\{\phi_k(\cdot), k = 1, \ldots, n\}$ minimizing the mean squared error

$$\underset{\phi_k \in \mathcal{H}}{\arg\min} E \left\{ \int_X \left| V(x) - \sum_{k=1}^n V_k \phi_k(x) \right|^2 dx \right\}, \tag{17.9}$$

where $V_k = \int_X V(x)\phi_k(x)\,dx$, satisfies the integral equation

$$\lambda_k \phi_k(x) = \int_X K(x,y)\phi_k(y)\,dy. \tag{17.10}$$

The eigenvalues estimate the variance along the axis of the corresponding eigenvector. So, if D is chosen such that

$$\sum_{k=1}^D \lambda_k \geq \frac{\alpha}{100} \sum_k \lambda_k,$$

then the subspace spanned by $\phi_1, \ldots, \phi_D$ retains $\alpha\%$ of the variation in the training set, and we can hope to model the class of shapes described by the training set by D parameters $\gamma_1, \ldots, \gamma_D$.

Equipped with the evolution equations for the diffeomorphism from the initial vector fields it is straight-forward to state PCA of shapes via the initial vector field and momentum at $t = 0$.

Algorithm 17.2 (Principle Components) *Consider a set of $n+1$ images $I_0, I_1, I_2, \ldots, I_n \in \mathcal{I}$ from the statistical orbit model.*

1. Set initial template to $I_\alpha = I_0$.
2. Determine the vector fields $v_t^{(i)}, t \in [0,1], i = 1, \ldots, n$ satisfying $\dot{g}_t^{(i)} = v_t^{(i)}(g_t^{(i)})$ with minimum $\|I_\alpha \circ g_1^{(i)} - I_i\|_2^2$.
3. Compute the mean vector fields

$$\bar{v}_0 = \frac{1}{n} \sum_{i=1}^n v_0^{(i)}. \qquad (17.11)$$

4. Procrustees [458] Mean Generation: Regenerate template $I_\alpha = I_\alpha \circ \bar{g}_1^{-1}$ where $\bar{g}_t, t \in [0,1], \bar{g}_0 = \text{id}$ is generated from the mean vector field:

$$\frac{\partial \bar{g}_t}{\partial t} = \bar{v}_t(\bar{g}_t), \qquad A\bar{v}_t = (D\bar{g}_t^{-1})^* A\bar{v}_0(\bar{g}_t^{-1})|D\bar{g}_t^{-1}|. \qquad (17.12)$$

5. If the template changes return to step 2. Otherwise continue.
6. Since the space of initial momenta is linear, apply PCA to the resulting $v_0^{(i)}, i = 1, \ldots, n$.

17.4 The Momentum of Landmarked Shape Spaces

For specificity, return to landmarked shape spaces $I_N = (x_1, \ldots, x_N) \in \mathcal{I}_N \subset \mathbb{R}^{dN}$. One of the most important applications of the initial momentum point of view is in effective linearization of the space of shapes. Because of the finite N-dimensional nature of landmarked imagery, the momentum is particularly simple and PCA boils down to the familiar SVD factorization of finite matrices generated from the momenta.

17.4.1 Geodesic evolution equations for landmarks

Given a template configuration $I_{N-\alpha}(x_1, \ldots, x_N)$ and the initial velocity v_0 of a geodesic in the orbit, we have from Chapter 11, Theorem 11.9, the momentum satisfying $Av_0(x) = \sum_{n=1}^N \delta(x_n - x)\beta_{nt}$. The elements $\beta_{n0}, n = 1, \ldots, N$ represent the initial momenta. The evolution equations describing the transport of the template along the geodesic are given as well by Theorem 11.9, Corollary 11.11 according to

$$v_t(g_t(x_i)) = \sum_{j=1}^N K(g_t(x_j), g_t(x_i))\beta_{jt} \qquad (17.13)$$

$$-\dot{\beta}_{it} = (Dv_t(g_{it}))^* \beta_{it}. \qquad (17.14)$$

We recognize that Eqns. 17.13 and 17.14 are an initial value ODE system. So, given initial values $g_0(x_n), \beta_{n0}, n = 1, \ldots, N$, we can solve this system to give the unique solution $(g_t(x_n), \beta_{nt}, n = 1, \ldots, N$ for all $t \in [0, 1]$, and hence $v_t(x)$ over all $x \in \Omega$.

Algorithm 17.3 (Principle Components of Landmark Momentum) *The configuration space of landmarks, $\mathcal{I}_N \subset \mathbb{R}^{Nd}$ with initial template configuration $I_{N-\alpha} = (x_1, x_2, \ldots, x_N)$ and a set N vector momenta, $\beta^{(i)} = \beta_{10}^{(i)}, \ldots, \beta_{N0}^{(i)}, i = 1, \ldots, n$, the PCA landmark momentum construction is as follows:*

1. *Estimate the mean and covariance:*

$$\bar{\beta}_0 = \frac{1}{n} \sum_{i=1}^{n} \beta_0^{(i)}$$

$$K_{jk} = \frac{1}{n} \sum_{i=1}^{n} (\beta_{j0}^{(i)} - \bar{\beta}_{j0})(\beta_{k0}^{(i)} - \bar{\beta}_{k0}).$$

The mean is typically estimated in an iterative procedure which couples an alignment procedure, called Procrustes alignment, with the computation above to give what is called the Procrustes mean.

2. *The desired basis is given by the eigenvectors, $\phi_i, i = 1, \ldots, Nd$, of the covariance matrix K_{ij}, where the order is determined by the decreasing order of the corresponding eigenvalues, $\lambda_1 \geq \lambda_2 \geq \cdots \geq \lambda_{Nd}$.*

3. *A particular landmark configuration or shape $I_N = (y_1, \ldots, y_N)$ is approximated by finding the initial momentum $\beta_0 = (\beta_{10}, \ldots, \beta_{N0})$ at $t = 0$ which maps the template $I_{N-\alpha}$ to the shape and then computing its orthogonal projection onto the subspace spanned by the eigenfunctions and the approximation:*

$$\hat{\beta}_0 = \sum_{j=1}^{D} \gamma_j \phi_j \quad \text{where } \gamma_j = \sum_{i=1}^{NK} \langle (\beta_0 - \bar{\beta}), \phi_i \rangle. \tag{17.15}$$

17.4.2 Small Deformation PCA Versus Large Deformation PCA

Standard small deformation linear vector space PCA approaches fail in the large deformation setting. Figures 17.1 and 17.2 show an example from Vaillant [284] illustrating the application of landmark based LD-DMM and the PCA analysis. Shown in Figure 17.1 is an example of 2D face characature Vaillant has studied. Here simple curves in the plane are generated to represent the 2D geometry (2D); shown in the panels are examples of the simple points on the curves being moved under the geodesic diffeomorphic metric landmark matching. For this 2D example in the plane the landmarks are annotated features in photographs of the face (data is from the AAM database [459]). The left panel shows the template landmark configuration (straight lines connecting landmarks) overlayed on the corresponding image of the face. The far right panel shows the overlay of the target landmark configuration on its corresponding image, and the center panels show snapshots from the time sequence of deformation of the template configuration. We see a smooth deformation as the template configuration moves close to the target.

Clearly, the resulting configuration in the standard small deformation linear vector space PCA approaches fail in the large deformation setting. This is particularly evident for large coefficients γ_j. Consider three landmark configurations $I^{(1)}, I^{(2)}$, and $I^{(3)}$ shown in Figure 17.2 of 2D face images taken from the AAM database [459]. Suppose $\bar{I} = I^{(2)}$, and we take $\phi_1 = I^{(1)} - \bar{I}$ and $\phi_3 = I^{(3)} - \bar{I}$ as two basis elements. Sampling the subspace generated by ϕ_1 and ϕ_3 by generating random coefficients γ_1 and γ_3, and producing

$$\bar{I} + \gamma_1 \phi_1 + \gamma_3 \phi_3,$$

Figure 17.1 2D Face Cartoon landmark LDDMM (see also Plate 42).

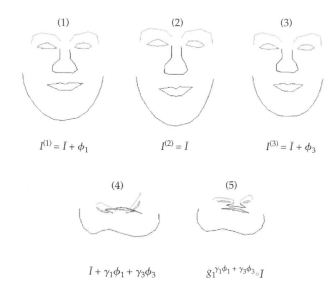

Figure 17.2 Non-linearity artifact in panel 4 overcome by initial momentum approach shown in panel 5. Results taken from Vaillant [284] (see also Plate 43).

we find elements of the subspace. Examine the solution for large coefficients as depicted in panel 4. Notice it no longer represents the "shape" of a face; the mouth appears above the nose and structures run into one another. Now consider the large deformation diffeomorphism PCA. This allows us to overcome the non-linearity of the shape space in a fundamental way. Considering again the same three configurations in Figure 17.2, we compute the initial momentum mapping $\bar{I}$ to I^1 and I^3, respectively. Now, if we again generate random coefficients γ_1 and γ_3, set

$$\hat{\beta}_0 = \gamma_1 \phi_1 + \gamma_3 \phi_3,$$

and apply the evolution Eqns. 17.13, 17.14 to the initial conditions we obtain a natural combination of the two configurations that is guaranteed to be a diffeomorphism of $\bar{I}$. In particular, taking the same coefficients γ_1 and γ_3 which produced the pathological model of panel 4 of Figure 17.2, we obtain panel 5 of Figure 17.2 which clearly retains the structure of a face.

Here are some 3-dimensional (3D) example results from Vaillant, Trouve, and Younes.

Example 17.4 (3D Face Surfaces) Examine statistical results from the landmark matching of Vaillant [284] from Corollary 11.11 and Example 11.12. of Chapter 11. The hippocampi data are part of the morphology testbed for the Biomedical Infomatics Research Network (BIRN, www.nbirn.net).

Shown in Figure 17.3 is PCA via initial momentum of 100 annotated surfaces from the Morphable Faces database [460] of landmarks annotated on the 2D face manifold. The top row shows the mean configuration in frontal and profile view and the subsequent rows show deformation of the mean configuration along the first three principal directions, respectively.

17.4 THE MOMENTUM OF LANDMARKED SHAPE SPACES

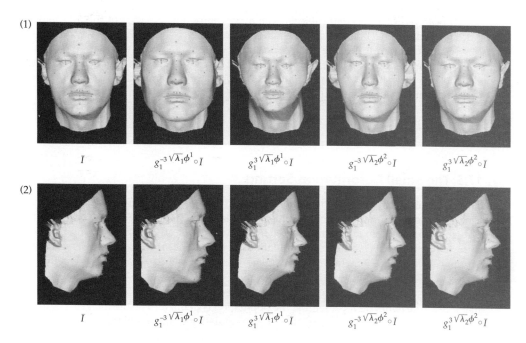

Figure 17.3 First two eigenmodes of 3D PCA applied to the Morphable Faces database of 100 faces. Row 1 shows front views; row 2 shows side views. Column 1 shows template. Columns 2 and 3 show deformation via eigenfunction 1. Columns 4 and 5 show deformation via eigenfunction 2 (see also Plate 44).

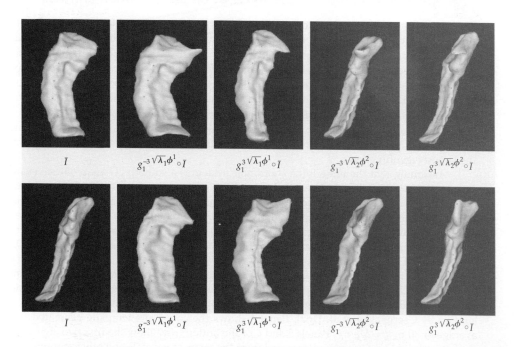

Figure 17.4 First two eigenmodes of 3D PCA applied to hippocampi from Randy Bucker of Washington University. Column 1 shows front and side view. Top row shows deformation via eigenfunction 1. Bottom row shows deformation via eigenfunction 2 (see also Plate 45).

Example 17.5 (3D Hippocampi) Now examine 3D PCA applied to the left hippocampus of 19 normal subjects. The hippocampi data are part of the morphology testbed for the Biomedical Infomatics Research Network (BIRN, www.nbirn.net). Shown in Figure 17.4 are 3D PCA results on the momentum mapping of two hippocampi applied to the left hippocampus of 19 normal subjects. The mean is shown in two views in the top row, and the deformation of the mean along the first three eigenmodes is shown in the subsequent rows.

17.5 The Small Deformation Setting

Diffeomorphisms cannot be added, only composed. Therefore average coordinates and templates generated via averaging of diffeomorphisms is suspect! However, for small deformations a template coordinate system can be constructed from averages of transformations. Empirical estimation of the templates for the various subpopulations of interest use the ideas of *minimum mean-squared estimation* (MMSE) for generating the template. Assume a metric distance based on the quadratic energy $\|(g - \mathrm{id})\|_V^2$. Intuitively, the template yet to be discovered should be defined to be the element $I_\alpha \in \mathcal{I}$ which requires the lowest average energy deformation onto the population of anatomies. The template representing the population is defined to be the image $I \in \mathcal{I}$ minimizing the overall energy of the transformation of the population to the template.

17.6 Small Deformation Gaussian Fields on Surface Submanifolds

The beauty of the general definition of $\{U(x), x \in \mathcal{X}\}$ as a Gaussian random field through its inner product with functions in the Hilbert space is that it immediately generalizes to more general background spaces such as surface submanifolds $M \subset \mathbb{R}^3$. We are interested in 2D manifolds associated with embedded surfaces such as the hippocampus. To quantify the shape of the hippocampus, we use complete orthonormal bases representing the normal deviations of populations of hippocampi. To construct the variability representation, define that the Gaussian random field $U(x), x \in S$ on the domain M is completely specified by its covariance matrix field, mapping $K : M \times M \to \mathbb{R}^3 \times \mathbb{R}^3$. Notice this is a Gaussian field on the submanifold M. The complete orthonormal bases are generated using singular value decomposition (SVD) of the empirical covariances and then deriving the largest eigenfunctions.

Assuming that the underlying coordinate system is a smooth 2D manifold, an example is shown in Figure 17.5 showing a whole MRI–MPRAGE image volume, with a section through the brain delineating the surface of the hippocampus. The left panel shows a section through the brain delineating the surface of the hippocampus; the middle panel shows a rendering of the entire surface bounding the full volume of the hippocampus. Left column panel 2 shows the triangulated graph representing the mean state of the hippocampus. The data are courtesy of Dr. John Csernansky of the Department of Psychiatry at Washington University. Right column four panels show the first four surface harmonics visualized through deformation of the template (panel 2).

Figure 17.6 shows the statistical characterization of the shape of the hippocampus. Panels 1–3 show maps of the initial hippocampus to three in a population of 30 patients, $M_0 \circ g_1, M_0 \circ g_2, M_0 \circ g_3$ studied by Csernansky et al. [162]. Panel 4 shows the composite template generated from the average of 30 maps, $M_0 \circ \bar{g}$. Panels 5 and 6 of the bottom row show the first two eigenshapes of the left and right hippocampus generated from the maps of normals and schizophrenics.

Shown in the bottom right panels of Figure 17.6 are random instances of left and right hippocampi generated via the empirically estimated covariance function for the Gaussian random field on the hippocampus surface.

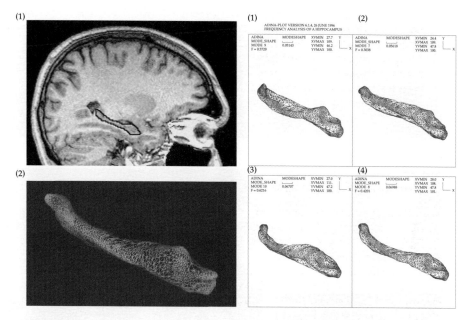

Figure 17.5 Left column panel 1 shows a whole MRI–MPRAGE image volume with a section through the brain delineating the surface of the hippocampus. Left column panel 2 shows the triangulated graph representing the mean state of the hippocampus. The data are courtesy of Dr. John Csernansky of the Department of Psychiatry at Washington University. Right column four panels show the first four surface harmonics visualized through deformation of the template (panel 2). Courtesy Sarang Joshi, Ph.D. thesis [242] (see also Plate 46).

17.7 Disease Testing of Automorphic Pathology

Now examine abnormal anatomies. In formalizing pathologies we shall employ the dichotomy *automorphic pathologies* versus *heteromorphic pathologies* following the terminology in [5]. Automorphic pathologies modify anatomical configurations within the orbit $\mathcal{G} : \mathcal{I}^\alpha \to \mathcal{I}^\alpha$, so that $g : I \in \mathcal{I}^\alpha \mapsto I' = I(g) \in \mathcal{I}^\alpha$, so that the space of anatomies is not exited $\mathcal{I}^\alpha$. This is not true for the heteromorphic pathologies for which the new configuration is generated via a more drastic change of the regular topology, perhaps adding neoplasm or eliminating structures. This requires the expansion to multiple anatomies, $\mathcal{I} = \cup_\alpha \mathcal{I}^\alpha$, with the transformations acting $\mathcal{G} : \mathcal{I}^\alpha \to \mathcal{I}$.

Throughout we shall use the terms *normal/abnormal* in a way that differs from the customary one. Abnormal signifies something exceptional, an anomaly, but not necessarily indicating improper functioning. As well, throughout we shall concentrate solely on change in topological structure. There is of course a second type of automorphic pathology, one corresponding to changes in the textures of the normal textbook pictures $I \in \mathcal{I}$. We shall not explore this here.

17.7.1 Hypothesis Testing on Disease in the Small Noise Limit

Automorphic pathologies though abnormal shall correspond to changes in structure which *preserve the topological structures of the orbit*. If the transformation $g \in \mathcal{G}$ is too far from the identity element in the similarity group so that it falls outside of a confidence set of the null hypothesis, H_0, we can identify this kind of abnormality by estimating the density $p(U)$ and applying the Neyman–Pearson lemma to arrive at an appropriate decision.

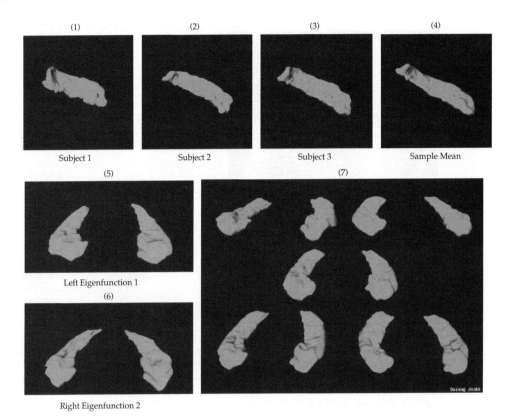

Figure 17.6 Top row: Panels 1–3 show maps of the initial hippocampus to three in a population of 30 patients, $M_0 \circ g_1, M_0 \circ g_2, M_0 \circ g_3$ studied in Csernansky et al. [162]. Panel 4 shows the composite template generated from the average of 30 maps, $M_0 \circ \bar{g}$. Bottom left column shows the first two eigenshapes of the left and right hippocampus generated from a population of maps of normals and schizophrenics. Lower right panels show random instances of left and right hippocampi generated via the empirically estimated covariance function for the Gaussian random field on the hippocampus surface. Taken from Joshi [242] (see also Plate 47).

Under the null (normal) and disease hypotheses let the random deformation have densities $H_0 : \pi_0(U), H_1 : \pi_1(U)$. Assume the image I^D are generated of a particular patient's anatomy, the goal being to diagnose disease. The likelihood function of the measured imaged data I^D expresses the physical characteristics of the sensor(s) used and therefore does not depend upon the choice of prior, thus $f(I^D|U)$ is the same for both hypotheses H_0, H_1. This implies that the density $f_0(\cdot)$ of the observed deformed image I^D under the hypotheses H_0, H_1 becomes

$$H_0 : p_0(I^D) = \int_{\mathcal{G}} \pi_0(U) f(I^D|U) \, dU \, ; \tag{17.16}$$

$$H_1 : p_1(I^D) = \int_{\mathcal{G}} \pi_1(U) f(I^D|U) \, dU. \tag{17.17}$$

The question of how to calculate such integrals has been addressed in [5]. The most powerful test comes out directly from the Neyman–Pearson lemma in terms of the critical region W as

$$W = \left\{ I^D : \frac{p_1(I^D)}{p_0(I^D)} \geq \text{constant} \right\}.$$

Assuming the sensor to be accurate so that the posterior densities $p_0 \propto \pi_0(\cdot) f(I^D|\cdot)$, $p_1 \propto \pi_0(\cdot) f(I^D|\cdot)$ are very peaked, then the vector field U is observed with high accuracy, so that the test can be organized in terms of the vector fields directly. Then the most powerful test effectively becomes

$$W = \left\{ U : \frac{\pi_1(U)}{\pi_0(U)} \geq \text{constant} \right\}.$$

Throughout this chapter we assume that the deformation is highly determined by the data. To carry out disease testing via hypothesis testing we construct the distributions π_0, π_1 empirically under this assumption.

17.7.2 Statistical Testing

To calculate the optimal test statistics the population under study can be grouped into n_0 number of controls and n_1 number of patients. The basis coefficients $\{U_1^0, \ldots, U_{n_0}^0\}$ are random samples from a random process whose mean is $\bar{U}^0$ and covariance Σ, and $\{U_1^1, \ldots, U_{n_1}^1\}$ are random samples from the same random process with mean $\bar{U}^1$ and the same Σ. The sample means and covariances $\hat{U}, \hat{\Sigma}$ formed from each group are calculated as

$$\hat{U}^0 = \frac{1}{n_0} \sum_{i=1}^{n_0} U_i^0 \quad \hat{U}^1 = \frac{1}{n_1} \sum_{j=1}^{n_1} U_j^1 \tag{17.18}$$

and the pooled (common) sample covariance

$$\hat{\Sigma} = \frac{1}{n_0 + n_1 - 2} \left(\sum_{i=1}^{n_0} (U_i^0 - \hat{U}^0)(U_i^0 - \hat{U}^0)^* + \sum_{j=1}^{n_1} (U_j^1 - \hat{U}^1)(U_j^1 - \hat{U}^1)^* \right). \tag{17.19}$$

Then, to establish a hypothesis test of the two groups means assuming an unknown but common covariance, the null hypothesis is

$$\mathcal{H}_0 : \bar{U}^0 = \bar{U}^1, \tag{17.20}$$

and Hotelling's T^2 statistic (for two samples) is

$$T^2 = \frac{n_0 n_1}{n_0 + n_1} \left(\hat{U}^0 - \hat{U}^1 \right)^* \hat{\Sigma}^{-1} \left(\hat{U}^0 - \hat{U}^1 \right). \tag{17.21}$$

The basis coefficients $\{U_1^0, \ldots, U_{n_0}^0\}$ are distributed according to $\mathcal{N}(\bar{U}^0, \Sigma)$, and $\{U_1^1, \ldots, U_{n_1}^1\}$ according to $\mathcal{N}(\bar{U}^1, \Sigma)$. The sample means $\hat{U}^0$ and $\hat{U}^1$ are distributed according to $\mathcal{N}(\bar{U}^0, (1/n_0)\Sigma)$ and $\mathcal{N}(\bar{U}^1, (1/n_1)\Sigma)$, respectively. Consequently, $\sqrt{n_0 n_1/(n_0 + n_1)}(\hat{U}^0 - \hat{U}^1)$ is distributed according to $\mathcal{N}(0, \Sigma)$ under the null hypothesis. Following [461, p. 109], $(n_0 + n_1 - 2)\hat{\Sigma}$ is distributed as $\sum_{i=1}^{n_0+n_1-2} U_i U_i^*$ where U_i is distributed according to $\mathcal{N}(0, \Sigma)$. Thus, T^2 has an F distribution, and the null hypothesis $\mathcal{H}_0$ is rejected with a significance level α if

$$T^2 \geq \frac{(n_0 + n_1 - 2)K}{n_0 + n_1 - K - 1} F^*_{K, n_0+n_1-K-1}(\alpha), \tag{17.22}$$

where $F^*_{K, n_0+n_1-K-1}(\alpha)$ denotes the upper $100\alpha\%$ point of the F_{K, n_0+n_1-K-1} distribution, and K is the total number of basis functions used in calculating the T^2 statistics. Logistic regressions based on χ^2 scores performed on the two groups of coefficients (control and schizophrenia groups) indicate that linear combinations of a subset of the basis functions suffice in describing the difference of the two groups. For each subject, using the subset of basis functions, we calculate the log-likelihood ratio:

$$\Lambda = -\frac{1}{2}\left(U_i - \hat{\bar{U}}^{\text{schiz}}\right)^* \hat{\Sigma}^{-1}\left(U_i - \hat{\bar{U}}^{\text{schiz}}\right) + \frac{1}{2}\left(U_i - \hat{\bar{U}}^{\text{ctrl}}\right)^* \hat{\Sigma}^{-1}\left(U_i - \hat{\bar{U}}^{\text{ctrl}}\right). \quad (17.23)$$

Here, the vector coefficients symbols represent the coefficient vectors with the selected components only (e.g. U_i is the vector coefficient for subject i using only the second, 12th, and 17th components, and similarly, $\hat{\bar{U}}^{\text{ctrl}}$ and $\hat{\bar{U}}^{\text{schiz}}$ are the sample means of the coefficient vectors using the above components for the control and schizophrenia groups, respectively; see example below). The log-likelihood ratios Λ give rise to a classification scheme under two hypotheses: under H_0 (when $\Lambda < 0$) the sufficient statistics is Gaussian distributed with mean $\bar{U}_0$ and variance σ_0^2; under H_1 (when $\Lambda > 0$), mean $\bar{U}_1$ and variance σ_1^2.

Example 17.6 (Schizophrenia) The Csernansky group has examined morphometric differences in schizophrenia. For the study 15 schizophrenia subjects were studied and compared with 15 controls. Shown in Figure 17.7 are results from the schizophrenia study. Wang and Joshi et al. have found that scale and volume are not powerful discriminants of group difference in the two populations—testing based on scale and volume gives no significant discrimination, $p = 0.19$ and $p = 0.27$; however, shape difference is.

Panel 1 shows differences of hippocampal surface patterns between the control and schizophrenia groups visualized as z-scores on the mean surface of the control group. Inward surface deformations due to schizophrenia are visualized in colder colors, outward in warmer colors, and areas which are not deformed in neutral yellow to green colors. Panel 2 shows statistical testing of shape differences: log-likelihood ratios for a linear combination of basis functions {1,3,4,6,10,15}. Multivariate ANOVA indicates significant between-group difference: $p = 0.0028$ ($F = 4.73$, $df = 1,28$). The bottom row shows asymmetry results in schizophrenia. Panel 3 shows difference of hippocampal surface asymmetry patterns between the control and schizophrenia groups visualized as z-scores on the mean flipped right-side surface of the control group. Inward surface deformations due to differences in asymmetry are visualized in colder colors, outward in warmer colors, and areas which are not deformed in neutral yellow to green colors. Panel 4 shows statistical testing of asymmetry pattern differences: log-likelihood ratios for a linear combination of basis functions {2,12,17}. Multivariate ANOVA indicates significant between-group difference in asymmetry: $p = 0.0029$ ($F = 6.03$, $df = 3,26$).

Example 17.7 (Alzheimer's and Aging) The Csernansky group has examined morphometric changes due to normal aging and dementia. For the study 15 younger controls were studied and compared with 18 elderly controls and 18 CDR 0.5 AD patients. Shown in Figure 17.8 are results from the Alzheimer's and normal aging study. Panel 1 shows difference of hippocampal surface patterns between the elderly control (CDR 0) and AD (CDR 0.5) groups visualized as z-scores on the mean surface of the elderly group. Inward surface deformations due to AD are visualized in colder colors, outward in warmer colors, and areas which are not deformed in neutral yellow to green colors. Panel 2 shows statistical testing of shape differences: log-likelihood ratios for a linear combination of basis functions {1,5}. Multivariate ANOVA indicates significant between-group difference: $p = 0.0002$ ($F = 11.4$, $df = 2,33$). The bottom row shows a normal aging comparison. Panel 3 shows difference of hippocampal surface patterns between the younger and the elderly control (CDR 0) groups visualized as z-scores on the mean surface of the younger group. Inward surface deformations due to aging are visualized in colder colors, outward in warmer colors, and areas

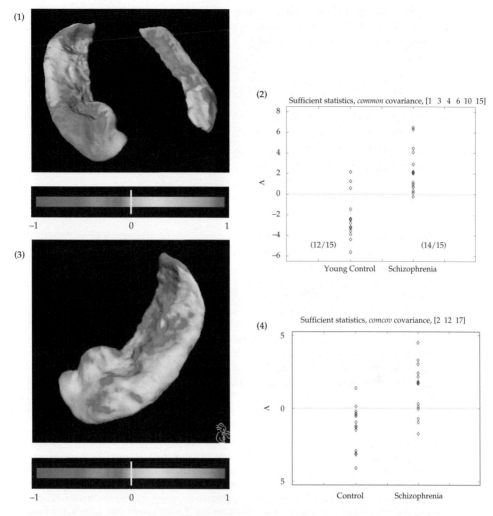

Figure 17.7 Top row: Schizophrenia. Panel 1 shows difference of hippocampal surface patterns between the control and schizophrenia groups visualized as z-scores on the mean surface of the control group. Inward surface deformations due to schizophrenia are visualized in colder colors, outward in warmer colors, and areas which are not deformed in neutral yellow to green colors. Panel 2 shows statistical testing of shape differences: log-likelihood ratios for a linear combination of basis functions {1,3,4,6,10,15}. Multivariate ANOVA indicates significant between-group difference: $p = 0.0028$ ($F = 4.73$, $df = 1, 28$). Bottom row: Asymmetry in Schizophrenia. Panel 3 shows difference of hippocampal surface asymmetry patterns between the control and schizophrenia groups visualized as z-scores on the mean flipped right-side surface of the control group. Inward surface deformations due to differences in asymmetry are visualized in colder colors, outward in warmer colors, and areas which are not deformed in neutral yellow to green colors. Panel 4 shows statistical testing of asymmetry pattern differences: log-likelihood ratios for a linear combination of basis functions {2,12,17}. Multivariate ANOVA indicates significant between-group difference in asymmetry: $p = 0.0029$ ($F = 6.03$, $df = 3, 26$) (see also Plate 48).

which are not deformed in neutral yellow to green colors. Panel 4 shows statistical testing of shape differences: log-likelihood ratios for a linear combination of basis functions {1,2}. Multivariate ANOVA indicates significant between-group difference: $p < 0.0001$ ($F = 348$, $df = 2, 30$).

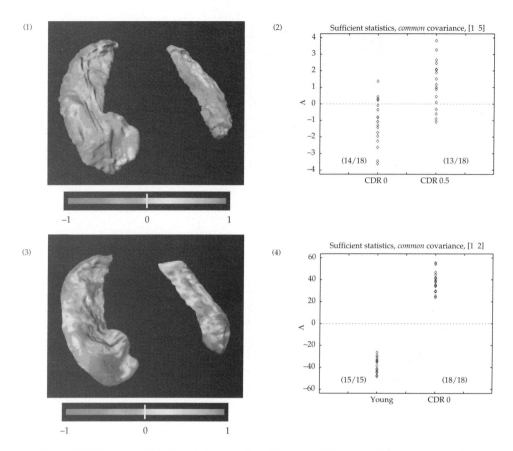

Figure 17.8 Top row: Alzheimer's Disease. Panel 1 shows difference of hippocampal surface patterns between the elderly control (CDR 0) and AD (CDR 0.5) groups visualized as z-scores on the mean surface of the elderly group. Inward surface deformations due to AD are visualized in colder colors, outwards in warmer colors, and areas which are not deformed in neutral yellow to green colors. Panel 2 shows statistical testing of shape differences: log-likelihood ratios for a linear combination of basis functions {1,5}. Multivariate ANOVA indicates significant between-group difference: $p = 0.0002$ ($F = 11.4$, $df = 2, 33$). Bottom row: Normal Aging. Panel 3 shows difference of hippocampal surface patterns between the younger and the elderly control (CDR 0) groups visualized as z-scores on the mean surface of the younger group. Inward surface deformations due to aging are visualized in colder colors, outward in warmer colors, and areas which are not deformed in neutral yellow to green colors. Panel 4 shows statistical testing of shape differences: log-likelihood ratios for a linear combination of basis functions {1,2}. Multivariate ANOVA indicates significant between-group difference: $p < 0.0001$ ($F = 348$, $df = 2, 30$) (see also Plate 49).

Example 17.8 (Alzheimer Progression) The Csernansky group has examined morphometric changes in early AD through two-year follow-up. For the study 18 CDR 0.5 AD subjects were studied and compared with 26 CDR 0 controls using scans collected two years apart for each subject. Shown in Figure 17.9 are results from the Alzheimer's follow-up study. Panel 1 shows deformations from baseline to follow-up for the CDR 0.5 AD group. Inward surface deformations due to AD are visualized in colder colors, outward in warmer colors, and areas which are not deformed in neutral yellow to green colors. Panel 2 shows deformations from baseline to follow-up for the CDR 0 control group. Inward surface deformations due to AD are visualized in colder colors, outward in warmer colors, and areas which are not deformed in neutral yellow to

17.7 DISEASE TESTING OF AUTOMORPHIC PATHOLOGY

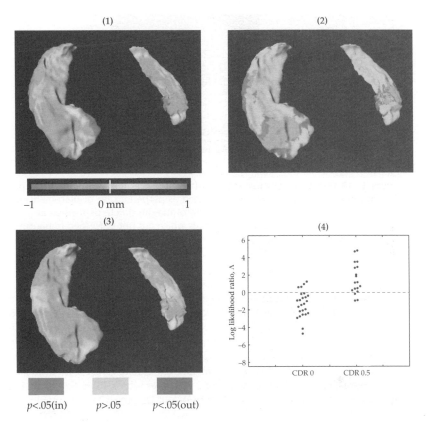

Figure 17.9 AD Progression. Panel 1 shows deformations from baseline to follow-up for the CDR 0.5 AD group. Inward surface deformations due to AD are visualized in colder colors, outwards in warmer colors, and areas which are not deformed in neutral yellow to green colors. Panel 2 shows deformations from baseline to follow-up for the CDR 0 control group. Inward surface deformations due to AD are visualized in colder colors, outward in warmer colors, and areas which are not deformed in neutral yellow to green colors. Panel 3 shows follow-up-versus-baseline "spread" of the between-group inward surface deformation patterns, shown as Wilcoxon's sign rank test map on the CDR 0 mean surface. Areas of significant ($p < 0.05$) inward deformation at baseline of CDR 0.5 group are shown in turquoise color, representing 38% of total hippocampal surface area. By follow-up, areas of significant inward deformation have increased to 47% of total hippocampal surface area. The increased affected areas are shown in purple color. Areas of non-significant surface deformation are shown in green color. Panel 4 shows statistical testing of shape differences: log-likelihood ratios for a linear combination of basis functions {1,2,4,11}. Multivariate ANOVA of the first 12 basis functions indicates significant between-group difference: $p = 0.014$ ($F = 2.66$, $df = 12, 31$) (See also Plate 50).

green colors. Panel 3 shows follow-up-versus-baseline "spread" of the between-group inward surface deformation patterns, shown as Wilcoxon's sign rank test map on the CDR 0 mean surface. Areas of significant ($p < 0.05$) inward deformation at baseline of CDR 0.5 group are shown in turquoise color, representing 38% of total hippocampal surface area. By follow-up, areas of significant inward deformation have increased to 47% of total hippocampal surface area. The increased affected areas are shown in purple color. Areas of non-significant surface deformation are shown in green color. Panel 4 shows statistical testing of shape differences: log-likelihood ratios for a linear combination of basis functions {1,2,4,11}. Multivariate ANOVA of the first 12 basis functions indicates significant between-group difference: $p = 0.014$ ($F = 2.66$, $df = 12, 31$).

510   17 COMPUTATIONAL ANATOMY: HYPOTHESIS TESTING ON DISEASE

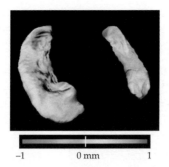

−1   0 mm   1

Figure 17.10 Depression. Panel shows differences of hippocampal surface patterns between the control and depression groups visualized as perpendicular displacements on the mean surface of the control group. Inward surface deformations due to depression are visualized in colder colors, outward in warmer colors, and areas which are not deformed in neutral yellow to green colors. Multivariate ANOVA of the first 10 basis functions indicates significant between-group difference: $p < 0.0001$ ($F = 34.1$, $df = 10, 58$) (see also Plate 51).

Example 17.9 (Depression) Posener et al. [437] have examined morphometric differences in depression. For the study 27 depression subjects were studied and compared with 42 controls. Shown in Figure 17.10 are results from the depression study. The panel shows differences of hippocampal surface patterns between the control and depression groups visualized as perpendicular displacements on the mean surface of the control group. Inward surface deformations due to depression are visualized in colder colors, outward in warmer colors, and areas which are not deformed in neutral yellow to green colors. Multivariate ANOVA of the first 10 basis functions indicates significant between-group difference: $p < 0.0001$ ($F = 34.1$, $df = 10, 58$).

17.8 Distribution Free Testing

Joshi [242] has examined methods based on distribution free testing on shape change between populations. The above mentioned p-values were derived based on the assumptions that the populations *Control* and *Schizophrenics* were Gaussian distributed with different means and common covariance. Joshi uses Fisher's method of randomization to derive a distribution free estimate of the level of significance of the difference. The basis coefficients $\{U_1^0, \ldots, U_{n_0}^0\}$ are random samples from a random process whose mean is $\bar{U}^0$ and covariance Σ, and $\{U_1^1, \ldots, U_{n_1}^1\}$ are random samples from the same random process with mean $\bar{U}^1$ and the same Σ. The sample means formed from each group will be as in Eqns. 17.18 and 17.19. To establish a hypothesis test of the two groups means assuming an unknown but common covariance, the null hypothesis is $\mathcal{H}_0 : \bar{U}^0 = \bar{U}^1$, and Hotelling's T^2 statistic (for two samples) is

$$T^2 = \frac{n_0 n_1}{n_0 + n_1} \left(\hat{\bar{U}}^0 - \hat{\bar{U}}^1 \right)^* \hat{\Sigma}^{-1} \left(\hat{\bar{U}}^0 - \hat{\bar{U}}^1 \right). \tag{17.21}$$

In Fisher's method of randomization, for all permutations of the given two groups, new means and covariances are calculated. Monte Carlo simulations are used to generate a large number of uniformly distributed random permutations (a typical number is 10,000). The collection

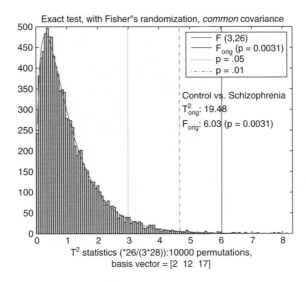

Figure 17.11 Empirical distribution $\hat{F}$ from randomized Fisher's test with 10,000 group permutations, between the group of control and schizophrenia subjects. Basis vectors {2,12,17} are selected. The $p = 0.0031$ value shown is calculated from Eqn. (17.25). Also shown are (i) $\hat{F}(T^2)$ value (solid blue line) of the control-versus-schizophrenia group comparison; (ii) theoretical F-distribution (solid red curve) with (3,26) degrees of freedom superimposed on the empirical distribution; and (iii) $p = 0.05$ (red dotted line) and $p = 0.01$ (red dot–dash line) for reference (see also Plate 52).

of T^2 statistics from each permutation gives rise to an empirical distribution $\hat{F}(\cdot)$ according to (see Eqn. 17.22)

$$F_{K,n_0+n_1-K-1} = \frac{n_0 + n_1 - K - 1}{(n_0 + n_1 - 2)K} T^2. \tag{17.24}$$

The null hypothesis that the two groups have equal means is rejected when

$$p = \int_{T^2}^{\infty} \hat{F}(f)\, df \tag{17.25}$$

falls below a predefined significance level (e.g. 0.05).

Figure 17.11 plots the empirical distribution $\hat{F}$ from randomized Fisher's test with 10,000 group permutations, between the group of control and schizophrenia subjects. Basis vectors {2,12,17} are selected. The $p = .0031$ value shown is calculated from Eqn. 17.25. Also shown are (i) $\hat{F}(T^2)$ value (solid blue line) of the control-versus-schizophrenia group comparison; (ii) theoretical F-distribution (solid red curve) with (3,26) degrees of freedom superimposed on the empirical distribution; and (iii) $p = 0.05$ (red dotted line) and $p = 0.01$ (red dot–dash line) for reference.

Apparently the Gaussian assumption for the coefficient vectors are valid since the empirical distribution of the $\hat{F}$ statistics follows the F-distribution curve.

17.9 Heteromorphic Tumors

Thus far the automorphic pathologies are large deviations of identical topology modifying an anatomical configuration $I \in \mathcal{I}^\alpha$ into another $I' \in \mathcal{I}^\alpha$. Automorphic pathologies never leave the original orbit of anatomies $\mathcal{I}^\alpha$. This is not the case for the heteromorphic pathologies for which a new configuration is generated via a change of the regular topology, perhaps adding neoplasm or eliminating structures.

Dealing with heteromorphic pathologies is a different challenge. New masses due to tumors, for example, introduce deformations which carry the configuration outside of the space $\mathcal{I}^\alpha$, requiring the introduction of a larger sample space $\mathcal{I} = \cup_\alpha \mathcal{I}^\alpha$, with $\mathcal{G} : \mathcal{I}^\alpha \to \mathcal{I}$. To represent this type of heteromorphic pathology, where new mass pushes out the normal tissue, we shall use a deformable template such as in [137] for the mass and then condition the vector field. Introduce the template configuration of the domain $X^{\text{path}} \subset X$ representing the mass and denote its central point by x_{center}. On this template let the random transformations act. To fix ideas let the translation groups act on the background spaces $X^{\text{path}} \subset X$, and use a stochastic difference equation $LU = W$ to describe the normal displacements in $X \setminus X^{\text{path}}$. The normal tissue around x_{center} is displaced by the new mass.

Denote the boundary of the pathology X^{path} as ∂X^{path}, and solve the difference equation in the domain outside of the new mass

$$(LU)(x) = W(x), \quad x \in X \setminus X^{\text{path}},$$

with the boundary conditions $U(x) = x_{\text{center}} - x, x \in \partial X^{\text{path}}$. In other words, the normal material is pushed out from x_{center} and replaced by the tissue in X^{path}.

To represent heteromorphic shape abnormalities assume that the anatomy is enclosed in a bounded rectangular background space $X \subset \mathbb{R}^3$ and use the operator $L = \nabla^2 + c, c < 0$, with Dirichlet boundary conditions on ∂X. Making c negative ensures non-singularity and existence of a solution to the problem below. For simplicity assume the three components of the deformation field $d(x); x \in X$ to be independent; for the coupled case we should replace L by the Navier operator used above.

Let $a \in X$ be a center for an expanding abnormality that expands into a region $X^{\text{path}} \subset X$. For example, X^{path} could be a ball with the center at a but more general shapes will be allowed. The *normal* remainder will be denoted $N = X \setminus X^{\text{path}}$. With the usual representation for the observed image $I^D(x) = I_\alpha[x - U(x)]$ we are led to the following stochastic field equation for the pathology induced displacement with the boundary conditions

$$(\nabla^2 + c)U(x) = W(x), \quad x \in X \setminus X^{\text{path}}, \tag{17.26}$$

$$U(x) = U_A(x) = a - x, \quad x \in \partial X^{\text{path}}, \tag{17.27}$$

$$U(x) = 0, \quad x \in \partial X, \tag{17.28}$$

with W a white noise field. Note that $U(\cdot)$ is not and should not be defined in the abnormal region X^{path}.

Let $G = G(x,y), x,y \in X \setminus X^{\text{path}}$ be the Green's function for the Laplacian ∇^2 in $X \setminus X^{\text{path}}$ with Dirichlet conditions. In $X \setminus X^{\text{path}}$ introduce a function $v(x)$ taking the values $u^{\text{path}}(x)$ on ∂X^{path} and zero on ∂X. For the new field $g = u - v$ we have

$$Lg = Lu - Lv = w - Lv$$

and g satisfies Dirichlet boundary conditions. It follows

$$g(x) = \int_{X \setminus X^{\text{path}}} G(x,y) f(y) \, dy$$

with some function f to be determined. Then

$$Lg(x) = \nabla^2 \int_{X \setminus X^{\text{path}}} G(x,y) f(y) \, dy + c \int_{X \setminus X^{\text{path}}} G(x,y) f(y) \, dy = f(x) + \int_{X^{\text{path}}} G(x,y) f(y) \, dy.$$

17.9 HETEROMORPHIC TUMORS

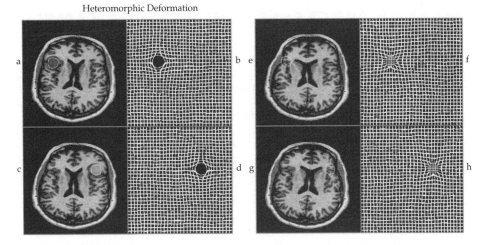

Figure 17.12 Figure showing heteromorphic deformations. The left panel shows an expansion or "push out"; the right panel shows a "push in" (see also Plate 53).

But $Lg = Lu - Lv = w - Lv$ so that we get the integral equation

$$f(x) + \int_{X \setminus X^{\text{path}}} G(x,y)f(y)\,dy = g(x) - Lv(x),$$

and the Fredholm equation should be solved with the Fredholm alternative holding.

The left panel of Figure 17.12 shows an expansion or "push out"; the right panel shows a "push in." Another pathology that may be of interest is when the change in tissue is not a "push out" but "cover," in the sense that the new mass takes the place of some of the normal ones, and invades it without displacing it. This is straightforwardly represented by solving the stochastic difference equation over the whole space X, and then placing a deformed template at some x_{center} covering the rest in that part of X.

18 MARKOV PROCESSES AND RANDOM SAMPLING

ABSTRACT The parameter spaces of natural patterns are so complex that inference must often proceed compositionally, successively building up more and more complex structures, as well as back-tracking, creating simpler structures from more complex versions. Inference is transformational in nature. The philosophical approach studied in this chapter is that the posterior distribution that describes the patterns contains all of the information about the underlying regular structure. Therefore, the transformations of inference are guided via the posterior in the sense that the algorithm for changing the regular structures will correspond to the sample path of a Markov process. The Markov process is constructed to push towards the posterior distribution in which the information about the patterns are stored. This provides the deep connection between the transformational paradigm of regular structure creation, and random sampling algorithms.

The connection to optimal inference is through the posterior distribution. Suppose we choose to generate a minimum-risk estimator; given a risk function $R : \mathcal{I} \times \mathcal{I} \to \mathbb{R}+$ quantifying relative merit between the estimator and the truth, the optimal estimator $I*$ attains the minimum over all estimators $\mathcal{R} = E_{P(dI,dI^D)} R(I, I * (I^D))$. Rarely will these be generated analytically; rather, they must be simulated from the posterior, revealing the remarkable but undisputable fact that our ability to analyze patterns boils down to our ability to synthesize patterns.

Inference based on simulation is of a fundamentally random nature. Are *random algorithms* really necessary? It should be remembered that these probability laws expresses real, for example biological, variability; it is not a mathematical trick designed to achieve certain goals. Granted this, it is clear that the output of an inference *should be* random representing the uncertainty that actually exists in nature. It is only when information is available to such an extent that practically certain inferences can be made that the randomness can be neglected. If so, deterministic approximation can be exploited to increase computational efficiency.

Our random sampling and inference will be constructed from Markov processes characterized through their *generator*, or *backward Kolmogoroff operator*.

18.1 Markov Jump Processes

We will construct Markov processes for inference; there will be two kinds, jumping processes, which can make large moves in state space, and diffusions, which make small moves.

Definition 18.1 Let $\{X_t, t \in T \subseteq \mathbb{R}\}$ be an $\mathcal{X}$-valued (state space $\mathcal{X}$) **Markov process**. Define the **transition law** of the process

$$P(t, x, t + h, dy) = Pr(\omega : X(t + h, \omega) \in dy | X(t) = x), \tag{18.1}$$

with the **Chapman-Kolmogorov equations**

$$P(t, x, t + h, dy) = \int P(t, x, t + r, dz) P(t + r, z, t + h, dy) \quad t \leq r \leq t + h, h \geq 0. \tag{18.2}$$

If the process is in turn homogeneous, then $P(t, x, t + h, dy) = P_h(x, dy)$.

18.1.1 Jump Processes

We shall now focus on a homogeneous Markov process $\{X(t), t \geq 0\}$ on state space $\mathcal{X}$ with the transition law $P_t(x, \cdot)$ (appropriate references include Gikhman and Skorohod [462]).

Definition 18.2 *We shall say $X(t)$ is an $\mathcal{X}$-valued* **homogeneous Markov jump process** *with bounded jump measures $q(x, \cdot) : \mathcal{X} \to [0, \infty)$ if $X(t)$ is continuous in probability for all t, i.e. for all measurable $A \in \mathcal{X}$,*

$$\lim_{s \downarrow t} P(t, x, s, A) = \lim_{\epsilon \downarrow 0} P_\epsilon(x, A) = \delta_x(A), \tag{18.3}$$

with bounded jump measures

$$\lim_{\epsilon \downarrow 0} \frac{P_\epsilon(x, A) - \delta_x(A)}{\epsilon} = q(x, A) \qquad \text{uniformly in } x, A. \tag{18.4}$$

The non-homogeneous setting has $q_t(x, A) = \lim_{\epsilon \downarrow 0}((P(t, x, t+\epsilon, A) - \delta_x(A))/\epsilon)$ with the uniformity in convergence as a function of t as well. Unlike diffusions (studied next), jump processes wander in ϵ time to neighborhoods far from the initial state with probability that goes as ϵ:

$$P_\epsilon(x, A) = \delta_x(A) + q(x, A \setminus \{x\})\epsilon + o(\epsilon).$$

The jump processes of greatest concern herein are conservative, i.e. the sum of the intensity of jumping out of a state and the intensity of staying in that state is zero:

$$q(x, \{x\}) = \lim_{\epsilon \to 0} \frac{1}{\epsilon}(P_t(x, \{x\}) - 1)$$
$$= \lim_{\epsilon \to 0} \left(-\frac{P_t(x, \mathcal{X} \setminus \{x\})}{\epsilon} = -q(x, \mathcal{X} \setminus \{x\}) \right).$$

It is then convenient to define the special notation $q(x) = q(x, \mathcal{X} \setminus \{x\})$ for the rate of jumping out of state x. It is useful to interpret the jump process through its subordinated Markov chain jumping on random times. Define the probability measure $Q(x, \cdot)$ to satisfy

$$q(x) Q(x, dy) = q(x, dy), \tag{18.5}$$

with $\int_{\mathcal{X} \setminus \{x\}} Q(x, dy) = 1$ following from the fact that

$$q(x) = q(x, \mathcal{X} \setminus x) \tag{18.6}$$

$$= q(x) \int_{\mathcal{X} \setminus x} Q(x, dy). \tag{18.7}$$

Example 18.3 (Jump measure for homogeneous Poisson counting process) Let $\{N_t, t \geq 0\}$, on state space $\mathcal{X} = \mathbb{N} = \{0, 1, 2, \dots\}$, be a homogeneous Poisson counting process with intensity λ. Define

$$\Pr\{N_{t+s} = n + 1 | N_t = n\} = \lambda s + o(s)$$
$$\Pr\{N_{t+s} \in A \subset \mathcal{X} | N_t = n\} = o(s), \qquad A \cap \{n, n+1\} = \emptyset.$$

Then the jump measures are $q(n, n+1) = \lambda$, and

$$q(n, n) = \lim_{s \downarrow 0} \frac{1 - P(N_{t+s} = n+1 | N_t = n) - 1 + o(s)}{s} = -\lambda,$$

$$q(n) = q(n, \mathcal{X} \setminus n)$$
$$= \lim_{s \downarrow 0} \frac{P(N_{t+s} = n+1 | N_t = n) + o(s)}{s} = \lambda.$$

Example 18.4 (Inhomogeneous PCP) Let $\{N_t, t \geq 0\}$ on state space $\mathcal{X} = \mathbb{N} = \{0, 1, 2, \dots\}$ be an inhomogeneous Poisson counting process $\{N_t, t \geq 0\}$ with intensity $\lambda_t, t \geq 0$. Define

$$\Pr\{N_{t+s} = n+1 | N_t = n\} = \lambda_t s + o(s)$$
$$\Pr\{N_{t+s} \in A \subset \mathcal{X} | N_t = n\} = o(s), \quad A \cap \{n, n+1\} = \emptyset.$$

Then the jump measures are $q_t(n, n+1) = \lambda_t$, $q_t(n, n) = -\lambda_t$, $q_t(n) = q_t(n, \mathcal{X} \setminus n) = \lambda_t$.

Example 18.5 (Birth–Death Processes) Let $\mathcal{X} = \mathbb{N} = \{0, 1, 2, \dots\}$, and construct a birth–death process $\{N_t, t \geq 0\}$ with birth and death intensities $\lambda_b(n), \lambda_d(n)$. Define

$$\Pr\{N_{t+s} = n+1 | N_t = n\} = \lambda_b(n) s + o(s),$$
$$\Pr\{N_{t+s} = n-1 | N_t = n\} = \lambda_d(n) s + o(s),$$
$$\Pr\{N_{t+s} \in A \subset \mathcal{X} | N_t = n\} = o(s) \quad A \cap \{n-1, n, n+1\} = \emptyset.$$

Then the jump measures are $q(n, n+1) = \lambda_b(n)$, $q(n, n-1) = \lambda_d(n)$, and

$$q(n,n) = \lim_{s \downarrow 0} \frac{1 - P(N_{t+s} = n+1 \cup N_{t+s} = n-1 | N_t = n) - 1 + o(s)}{s}$$
$$= -\bigl(\lambda_b(n) + \lambda_d(n)\bigr),$$
$$q(n) = q(n, \mathcal{X} \setminus n)$$
$$= \lim_{s \downarrow 0} \frac{P(N_{t+s} = n+1 | N_t = n) + o(s)}{s}$$
$$= \lambda_b(n) + \lambda_d(n).$$

18.2 Random Sampling and Stochastic Inference

The *transformations of inference* are guided via the posterior in the sense that the algorithm for changing the regular structures will correspond to the sample path of a Markov process. The dynamics determining the transition measure of the random Markov process of changes will be constructed to push towards the posterior distribution in which the information about the patterns are stored. To illustrate, let us examine the finite state, discrete-time Markov chain case which is more familiar. Suppose we have some probability $\{\pi(x), x \in \mathcal{X}\}$ on finite state space $\mathcal{X}$, and the goal is to draw samples from $\pi(\cdot)$. Examine the following time honored approach. Create an aperiodic Markov chain with stationary transition matrix $Q(i,j), i,j \in \mathcal{X}$ ($\sum_{j \in \mathcal{X}} Q(i,j) = 1$) making the state space $\mathcal{X}$ strongly connected, i.e. $\exists n$ such that for all states $i, j \in \mathcal{X}$ the transition matrix is strictly positive: $Q^n > 0$. Then the Perron–Frobenius theorem is in effect, implying $Q(\cdot, \cdot)$ has a unique left eigenvector with eigenvalue 1. Call this eigenvector $p(\cdot)$. Of course, $Q^n = p\mathbf{1} + O(\alpha^n)$, with $\alpha < 1$; $\mathbf{1}$ is the all 1's row vector.

Now if we can choose the transition dynamics $Q(\cdot, \cdot)$ of the Markov chain properly so that $p(\cdot) = \pi(\cdot)$ the distribution to be sampled from, then we have a process whose transition law converges exponentially quickly to π. It might be expected that averages generated by sampling such a process will converge to their expectation, i.e. some sort of ergodic property should hold. This will generally be the case. So then there are really only two pieces which are needed for the inference to proceed: (i) to choose the Markov dynamics so that $\pi(\cdot)$ on $\mathcal{X}$ is invariant (stationary) for the Markov process, and (ii) to show the dynamics makes $\pi(\cdot)$ on $\mathcal{X}$ the unique invariant measure.

Since we are interested in parameter spaces that have connected components, gradients are natural for searching within these components. Diffusions and stochastic differential equations in the continuous-time Markov process formulation play a fundamental role. In such a setting, the *generator* or *backward Kolmogoroff operator* provides the correct tool for verifying the invariant measure condition, and *Harris recurrence and irreducibility* provides the correct framework for proving uniqueness of the invariant measure.

18.2.1 Stationary or Invariant Measures

This section follows [463] and defines stationarity and the generator.

Definition 18.6 *For a stochastic process $\{X(t), t \geq 0\}$ having values in the state space $\mathcal{X}$ and the transition probabilities $P_t(x, dy)$, then $\pi(dx)$ is* **invariant (stationary)** *if for all $t > 0$,*

$$\int \pi(dx) P_t(x, dy) = \pi(dy).$$

Let $X(t)$ be a Markov process with **generator** *the linear operator A defined by*

$$Af(\cdot) \stackrel{\text{sup-norm}}{=} \lim_{\epsilon \downarrow 0} \frac{\int_{\mathcal{X}} P_\epsilon(\cdot, dy) f(y) - f(\cdot)}{\epsilon}, \quad f(\cdot) \in D, \tag{18.8}$$

$D \subset B$ *the domain on which sup–norm convergence holds, B the supporting Banach space.*

Theorem 18.7 *Let $X(t)$ be a Markov process with* **generator** *the linear operator A of Eqn. 18.8.*
Then $\pi(\cdot)$ is invariant if and only if for all $f \in D$ the domain of A,

$$\int Af(x) \pi(dx) = 0.$$

Proof We first consider the "only if" part. By the definition of invariance, since $\pi(dx)$ is invariant, for all $t \geq 0$,

$$\int \pi(dx) P_t(x, dy) = \pi(dy).$$

Defining $H_t f(\cdot) = \int_{\mathcal{X}} P_t(\cdot, dy) f(y)$, gives

$$0 = \int \int \pi(dx) P_t(x, dy) f(y) - \int \pi(dy) f(y)$$
$$= \int \pi(dx) H_t f(x) - \int \pi(dy) f(y) \tag{18.9}$$
$$= \int \pi(dx) \Big(H_t f(x) - f(x) \Big).$$

This is true for all $t \geq 0$, implying that for all $f \in D$,

$$0 = \lim_{\epsilon \downarrow 0} \frac{1}{\epsilon} \int \pi(dx) \Big(H_\epsilon f(x) - f(x) \Big)$$
$$= \int \pi(dx) Af(x), \tag{18.10}$$

where the sup-norm convergence $(H_\epsilon f - f)/\epsilon \to Af$ is used in bringing the limit inside the integral.

For the "if part" For this we follow [463]. Assume for all $f \in D$, $\int Af(x)\pi(x)dx = 0$. Now it is the case (see Lemma 18.8 below) that for $t > 0$, $\int_0^t H_s f ds \in D$ and

$$H_t f - f = A \int_0^t H_s f \, ds, \quad \forall f \in B. \tag{18.11}$$

Denoting $f_t = \int_0^t H_s f ds \in D$, then

$$\int Af_t(x)\pi(dx) = 0, \tag{18.12}$$

since π is invariant. This implies

$$\int \pi(x)\left(H_t f(x) - f(x)\right) = 0, \quad \forall f \in B. \tag{18.13}$$

Thus $\pi H_t = \pi$, and we have established the sufficiency of the generator condition. To complete the proof, we need the following lemma. □

Lemma 18.8 *If $t > 0$, then $\int_0^t H_s f \, ds \in D$, $\forall f \in B$ and*

$$H_t f - f = A \int_0^t H_s f \, ds, \quad \forall f \in B. \tag{18.14}$$

Proof of lemma:

$$\frac{1}{h}[H_h - I]\int_0^t H_s f \, ds = \frac{1}{h}\int_0^t [H_{s+h}f - H_s f] \, ds$$

$$= \frac{1}{h}\int_h^{t+h} H_s f \, ds - \frac{1}{h}\int_0^t H_s f \, ds$$

$$= \frac{1}{h}\int_h^{t+h} H_s f \, ds - \frac{1}{h}\int_0^h H_s f \, ds, \tag{18.15}$$

which by the fundamental theorem of calculus converges as $h \downarrow 0$ to $H_t f - f$. Apparently $\int H_s f \in D$ and the Theorem proof is completed. □

Example 18.9 Banach Spaces and Domains The generator A is a *linear operator* that is not necessarily defined on all of B, only a subspace $D \subseteq B$. Examples of Banach spaces and domains are as follows. For the jump processes studied the Banach space B will be the bounded measurable functions, with the domain of the jump operator the same, $D = B$. For Feller diffusions on $\mathcal{X}$ a compact state space, (e.g. a closed and bounded subset of $\mathbb{R}^k$ for example), $B = \bar{C}(\mathcal{X})$ bounded continuous functions, and the domain is $D = \bar{C}^\infty(\mathcal{X})$, the bounded, infinitely differentiable functions. For diffusions in $\mathbb{R}^k$ then $B = \hat{C}(\mathbb{R}^k)$, continuous functions vanishing at infinity with $D = \hat{C}^\infty(\mathbb{R}^k)$ the subset that are infinitely differentiable.

18.2.2 Generator for Markov Jump Processes

Suppose we have a homogeneous Markov jump process $\{X(t), t \geq 0\}$ on state space $\mathcal{X}$ with the transition law $P_t(x, \cdot)$. The generator characterizes the process, and defines a necessary condition for the distribution to be stationary.

Theorem 18.10 *Consider the generator (or Kolmogoroff Backward Operator) for the jump process with jump measures $q(x, \cdot) : \mathcal{X} \to [0, \infty)$ defined by $q(x, A) = \lim_{\epsilon \downarrow 0}((P_\epsilon(x, A) - \delta_x(A))/\epsilon)$ (uniformly in x), $A \subset \mathcal{X}$, with bounded continuous intensity $q(x) = q(x, \mathcal{X} \setminus x)$, $q = \sup_{x \in \mathcal{X}} q(x) < \infty$, and probabilities $Q(x, \cdot)$ such that $q(x)Q(x, dy) = q(x, dy)$, then*

$$Af(x) = -q(x)f(x) + \int_\mathcal{X} q(x)Q(x, dy)f(y). \tag{18.16}$$

In particular, the transition law $P_t(x, \cdot)$ of the jump process satisfies the differential equation

$$\frac{\partial P_t}{\partial t}(x, \cdot) = -q(x)P_t(x, \cdot) + \int_\mathcal{X} q(x)Q(x, dy)P_t(y, \cdot), \quad P_0(x, \cdot) = \delta_x(\cdot). \tag{18.17}$$

Proof

$$Af(x) = \lim_{\epsilon \downarrow 0} \frac{\int_\mathcal{X} P_\epsilon(x, dy)f(y) - f(x)}{t}$$

$$= \lim_{\epsilon \downarrow 0} \frac{\int_\mathcal{X} (P_\epsilon(x, dy) - \delta_x(dy) + \delta_x(dy))f(y) - f(x)}{\epsilon}$$

$$= \lim_{\epsilon \downarrow 0} \frac{\int_\mathcal{X} (P_\epsilon(x, dy) - \delta_x(dy))f(y) + f(x) - f(x)}{\epsilon}$$

$$= \int_\mathcal{X} q(x, dy)f(y) \quad \text{using uniform continuity in } x, A$$

$$= -q(x)f(x) + \int_\mathcal{X} q(x)Q(x, dy)f(y)$$

$$= q(x) \int_\mathcal{X} Q(x, dy)(f(y) - f(x)).$$

The domain of the generator A is the Banach space of bounded-measurable functions, since A is a bounded operator, implying that the transition law itself is in the domain and satisfies the differential equation. This follows since $\| \int P_t(x, dy)f(y) \| \leq \| f \| < \infty$ implying that the Banach space supporting the semi-group consists of bounded measurable functions. Since $q = \sup_{x \in \mathcal{X}} q(x) < \infty$, then A is a bounded operator, and for $f \in B$ a bounded measurable function, then $Af \in B$ since

$$\| Af \| = \| q(x) \int_\mathcal{X} Q(x, dy)(f(y) - f(x)) \|$$

$$\leq \| q \| \| f(y) - f(x) \| \leq K \| f \|,$$

implying $P_t(x, \cdot) \in D$, giving the differential equation 18.17. □

Example 18.11 **Poisson process** For a Poisson process with intensity λ, Theorem 18.10 dictates

$$\frac{\partial P_t(n, n)}{\partial t} = -\lambda P_t(n, n) + \sum_{j \neq n} q(n, j) P_t(j, n)$$

$$= -\lambda P_t(n, n)$$

implying the exponential waiting times $P_t(n,n) = e^{-\lambda t}$. The transition probabilities satisfy

$$\frac{\partial P_t(n,m)}{\partial t} = -\lambda P_t(n,m) + \sum_j q(n, n+j) P_t(n+j, m) \qquad (18.18)$$

$$= -\lambda P_t(n,m) + \lambda P_t(n+1, m).$$

18.2.3 Jump Process Simulation

There is a beautiful relationship between Markov processes and Markov chains transitioning on Poisson times. This essentially demonstrates that the invariant distribution condition for the jump Markov process (Eqn. 18.19 below) is not very different than the invariant distribution condition for a Markov chain from classical Monte-Carlo theory corresponding to the left eigenvector having eigenvalue 1. For this we subordinate the Markov process to a Markov chain jumping on Poisson times (e.g. see Cinlar [75]).

Theorem 18.12 *Let $X(t)$ be a Markov process with bounded jump intensity $q(x), x \in \mathcal{X}$, $q = \sup_{x \in \mathcal{X}} q(x) < \infty$, and jump probabilities $Q(x, \cdot)$. Then $\pi(\cdot)$ is invariant for a jump process with jump measures $q(x, \cdot), Q(x, \cdot)$, and jump rates $q(x)$ iff for all $f \in B$ (bounded measurable functions),*

$$\int_{\mathcal{X}} \left(-q(x) f(x) + q(x) \int Q(x, dy) f(y) \right) \pi(dx) = 0. \qquad (18.19)$$

Define the normalized transition probabilities $\tilde{Q}(x, \cdot)$,

$$\tilde{Q}(x, \cdot) = \left(1 - \frac{q(x)}{q}\right) \delta_x(\cdot) + \frac{q(x)}{q} Q(x, \cdot), \qquad \int_{\mathcal{X}} \tilde{Q}(x, dy) = 1. \qquad (18.20)$$

The Markov process is statistically equivalent to a Poisson process jumping on exponential times with intensity q, and with transition probability $\tilde{Q}$. In particular, $\pi(\cdot)$ is the left eigenvector of $\tilde{Q}(\cdot, dy)$ having eigenvalue 1:

$$\pi(dx) = \int \pi(dy) \tilde{Q}(y, dx).$$

Proof The invariant distribution condition 18.19 follows from Theorem 18.7 with $\int Af\pi = 0$. Using Theorem 18.10, $P_t(x, \cdot)$ is in the domain of the generator so that

$$\frac{\partial}{\partial t} P_t(x, \cdot) = \lim_{\epsilon \downarrow 0} \frac{1}{\epsilon} \left(\int_{\mathcal{X}} P_\epsilon(x, dy) P_t(y, \cdot) - P_t(x, \cdot) \right) \qquad (18.21)$$

$$= A P_t(x, \cdot), \quad P_0(x, \cdot) = \delta_x(\cdot). \qquad (18.22)$$

Rewriting the generator A according to

$$Af(x) = \int q(x) Q(x, dy)(f(y) - f(x))$$

$$= q \int (f(y) - f(x)) \tilde{Q}(x, dy)$$

$$= -q f(x) + q \int \tilde{Q}(x, dy) f(y), \qquad (18.23)$$

implies

$$\int_{\mathcal{X}} P_t(x, dy) f(y) = e^{tA} f(x) = e^{-qtI + qt\tilde{Q}} f(x)$$

$$= \sum_{k=0}^{\infty} e^{-qt} \frac{(qt)^k}{k!} \int_{\mathcal{X}} \tilde{Q}^k(x, dy) f(y). \qquad (18.24)$$

The transition law corresponds to a Markov chain jumping on exponential times at rate q. The invariant distribution condition $\int Af(x)\pi(dx) = 0$ gives

$$\int Af(x)\pi(dx) = \int \left(-qf(x) + q \int \tilde{Q}(x, dy) f(y)\right) \pi(dx) \qquad (18.25)$$

$$= \int -f(x) q \pi(dx) + \int f(y) \int q\tilde{Q}(x, dy) \pi(dx) \qquad (18.26)$$

changing variables of integration in the second term

$$= -\int f(x) q \pi(dx) + \int f(x) \int q\tilde{Q}(y, dx) \pi(dy) \qquad (18.27)$$

$$= \int f(x) \left(-q\pi(dx) + \int q\tilde{Q}(y, dx) \pi(dy)\right) = 0. \qquad (18.28)$$

Since this is true for any bounded measurable function $f(x)$,

$$\pi(dx) = \int \pi(dy) \tilde{Q}(y, dx),$$

implying that $\pi(\cdot)$ is the left eigenvector (with eigenvalue 1) of $\tilde{Q}(\cdot, dy)$. □

18.2.4 Metropolis–Hastings Algorithm

We now follow heavily the work of Lanterman [464] and examine the general framework for inference based on the acceptance and rejection of proposals. The principal contribution will be to choose a jumping scheme analogous to the Metropolis-Hastings sampling algorithm [465–467]. At exponentially distributed times, with a *fixed* mean, a candidate x_{prop} is drawn from a *proposal density* $r(x_{\text{old}}, x_{\text{prop}})$; the *acceptance probability* is then computed:

$$\alpha(x_{\text{old}}, x_{\text{prop}}) = \min \left\{ \frac{\pi(x_{\text{prop}}) r(x_{\text{prop}}, x_{\text{old}})}{\pi(x_{\text{old}}) r(x_{\text{old}}, x_{\text{prop}})}, 1 \right\}. \qquad (18.29)$$

The proposal is accepted with probability $\alpha(x_{\text{old}}, x_{\text{prop}})$ and rejected with probability $1 - \alpha(x_{\text{old}}, x_{\text{prop}})$. A wide variety of proposal densities may be used. We will need a characterization of the space of configurations that jump transitions can connect to, so that $r(x_{\text{old}}, x_{\text{prop}}) > 0$ for $x_{\text{prop}} \in \mathcal{J}^1(x_{\text{old}})$ and $r(x_{\text{old}}, x_{\text{prop}}) = 0$ for $x_{\text{prop}} \notin \mathcal{J}^1(x_{\text{old}})$.

Definition 18.13 *A process $\{X_t, t \in T\}$ is said to be **weakly reversible** if $\mathcal{J}(x) = \mathcal{J}^{-1}(x)$ where $\mathcal{J}(x)$ is the set of states that can be reached given that the process started in state x and $\mathcal{J}^{-1}(x)$ is the set of states that can jump into state x:*

$$\mathcal{J}(x) = \{y : y \in \text{support of } Q(x, \cdot)\} \qquad (18.30)$$

$$\mathcal{J}^{-1}(x) = \{y : x \in \text{support of } Q(y, \cdot)\}. \qquad (18.31)$$

Corollary 18.14 *Assume $r(\cdot,\cdot)$ is chosen so the process is weakly reversible, i.e. $\mathcal{J}(x) = \mathcal{J}^{-1}(x)$. The Metropolis-Hastings process for constructing the jump intensities becomes*

$$q(x,dy) = \min\left\{\frac{\pi(y)r(y,x)}{\pi(x)r(x,y)}, 1\right\} r(x,y)dy, \qquad (18.32)$$

and the detailed balance condition of Eqn. 18.19 for the jump process is satisfied.

Proof To see that (18.32) satisfies Eqn. 18.19 of Theorem 18.12. We must satisfy the detailed balance condition $q(x)\pi(x)dx = \int_{\mathcal{X}} q(y,dx)\pi(y)dy$. First define

$$\Omega_{>} = \left\{y \in \mathcal{J}^1(x) : \frac{\pi(y)r(y,x)}{\pi(x)r(x,y)} > 1\right\} \qquad (18.33)$$

$$\Omega_{<} = \left\{y \in \mathcal{J}^1(x) : \frac{\pi(y)r(y,x)}{\pi(x)r(x,y)} < 1\right\}. \qquad (18.34)$$

Substituting the definition of the jump intensity into the left-hand side (LHS) of the detailed balance condition above gives

$$\text{LHS} = \left(\int_{\mathcal{J}^1(x)} \min\left\{\frac{\pi(y)r(y,x)}{\pi(x)r(x,y)}, 1\right\} r(x,y)dy\right) \pi(x)dx \qquad (18.35)$$

$$= \left(\int_{\Omega_{>}} r(x,y)dy\right)\pi(x)dx + \left(\int_{\Omega_{<}} \frac{\pi(y)r(y,x)}{\pi(x)r(x,y)} r(x,y)dy\right)\pi(x)dx. \qquad (18.36)$$

$$= \left(\int_{\Omega_{>}} r(x,y)dy\right)\pi(x)dx + \left(\int_{\Omega_{<}} \pi(y)r(y,x)dy\right) dx. \qquad (18.37)$$

Now define

$$\Omega_{>}^{-1} = \left\{y \in \mathcal{J}^{-1}(x) : \frac{\pi(x)r(x,y)}{\pi(y)r(y,x)} > 1\right\} \qquad (18.38)$$

$$\Omega_{<}^{-1} = \left\{y \in \mathcal{J}^{-1}(x) : \frac{\pi(x)r(x,y)}{\pi(y)r(y,x)} < 1\right\}. \qquad (18.39)$$

Substituting 18.32, the choice for the jump intensity into the right-hand side (RHS) of the detailed balance condition above yields

$$\text{RHS} = \int_{\mathcal{J}^{-1}(x)} \left(\min\left\{\frac{\pi(x)r(x,y)}{\pi(y)r(y,x)}, 1\right\} r(y,x)dx\right) \pi(y)dy. \qquad (18.40)$$

$$= \int_{\Omega_{>}^{-1}} (r(y,x)dx)\pi(y)dy + \int_{\Omega_{<}^{-1}} \left(\frac{\pi(x)r(x,y)}{\pi(y)r(y,x)} r(y,x)dx\right)\pi(y)dy. \qquad (18.41)$$

$$= \int_{\Omega_{>}^{-1}} (r(y,x)dx)\pi(y)dy + \int_{\Omega_{<}^{-1}} (\pi(x)r(x,y)dx)\, dy. \qquad (18.42)$$

The weak reversibility of the jump moves $\mathcal{J}^1(x) = \mathcal{J}^{-1}(x)$ implies that $\Omega_{>}^{-1} = \Omega_{<}$ and $\Omega_{<}^{-1} = \Omega_{>}$, which establishes the equality of the LHS and the RHS, giving detailed balance.

Examples of the general Metropolis–Hastings jump-diffusion scheme described in Section 18.2.4 have appeared throughout the literature. □

18.2.4.1 Metropolis and Heat-Bath

Corollary 18.15 (Metropolis) *In the above, if $r(x_{old}, x_{prop}) = r(r_{prop}, r_{old})$, we have $\alpha(x_{old}, x_{prop}) = \min(\pi(r_{prop})/\pi(r_{old}), 1)$, which corresponds to the traditional Metropolis algorithm [468].*

With $\pi(x) = (e^{-E(x)}/Z), x \in \mathcal{X}$, then

$$q(x, dy) = e^{-(E(y)-E(x))_+} dy, \quad q(x) = \int_{\mathcal{X}\setminus x} e^{-(E(y)-E(x))_+} dy,$$

$$Q(x, dy) = \frac{e^{-(E(y)-E(x))_+} dy}{\int_{\mathcal{X}\setminus x} e^{-(E(y)-E(x))_+} dy}$$

with $(x)_+$ defined to be zero if x is negative or x otherwise, and $\pi(\cdot)$ is an invariant density.

Notice that if we had the means to sample directly from the conditional distributions, then *the Metropolis–Hastings sampler would reduce to the Gibbs sampler or Heat Bath*.

Corollary 18.16 (Heat-Bath) *If $r(x, y)dy = \pi(y)dy$, then $\pi(\cdot)$ is an invariant density.*

In the Bayesian framework, if an informative prior exists that is easy to sample from, then it is beneficial to choose $x_{prop} \in \mathcal{T}^1(x_{old})$, drawing candidates from the prior and accepting and rejecting based on the likelihood. This is the approach of Corollary 2 of Theorems 1 and 2 of [137]. This approach is effective in the ground-to-air tracking study reviewed in a subsequent chapter since drawing from a prior on motion dynamics is fast and effective.

18.3 Diffusion Processes for Simulation

For connected state spaces, we shall use diffusions for simulation.

Definition 18.17 *A real-valued random process $\{X(t), t \in T\}, T \subseteq \mathbb{R}$ is a **diffusion process** if (i) $X(t)$ is Markov, and (ii) $X(t)$ is sample path continuous.*

Dynkin provided a characterization of diffusions, that if $X(t)$ is a real-valued random process with the properties of

(i) right continuity $\forall s \in T : \lim_{t \downarrow s} X(t) = X(s)$;

(ii) $\forall s \in T : \lim_{t \uparrow s} X(t)$ exists; and

(iii) $\forall t \in T : \forall \delta > 0 : P(\{\omega : |X(t+h, \omega) - X(t, \omega)| > \delta\}|X(t) = x) = o(h)$. A sufficient condition for establishing a diffusion follows from the Dynkin property that if $\exists p \geq 3 : \lim_{h \downarrow 0} (1/h) E\{|X(t+h) - X(t)|^p | X(t) = x\} = 0$, then the process is a diffusion. This results from the Markov inequality.

A Poisson process, for example, fails to meet such conditions and it is not a diffusion process. This is obviously true, since a Poisson process fails to be sample-path continuous.

The principal way in which we shall study and characterize diffusions is via their infinitesimal properties, particularly means and variances.

Definition 18.18 *Let $X(t)$ be a **diffusion process**. For every x and $\forall \epsilon > 0$, $\int_{|x-y|>\epsilon} P(t, x, s, dy) = o(s-t)$, and there exists functions $a(t, x)$ and $b(t, x)$ such that for every x,*

infinitesimal mean $\quad a(x, t) = \lim_{h \downarrow 0} \frac{1}{h} E\{X(t+h) - X(t) | X(t) = x\}$ (18.43)

infinitesimal variance $\quad b(x, t) = \lim_{h \downarrow 0} \frac{1}{h} E\{|X(t+h) - X(t)|^2 | X(t) = x\}.$ (18.44)

Example 18.19 (Brownian motion and the Dynkin condition) Brownian motion satisfies the Dynkin condition with $p = 4$. To see this, condition on $W(t) = x$, then the random variable $W(t + h)$ is normally distributed with mean x and variance h, implying

$$\lim_{h\downarrow 0} \frac{1}{h} E\{|W(t+h) - W(t)|^4 | W(t) = x\} = \lim_{h\downarrow 0} \frac{3h^2}{h}$$
$$= 0.$$

Example 18.20 (Infinitesimal means and variances of Brownian motion and diffusions) For a standard Wiener process, $a(x,t) = 0$ and $b(x,t) = 1$. To see this, note $E\{W(t+h) - W(t) | W(t) = x\} = 0$, so the infinitesimal mean is $a(x,t) = 0$. The variance of an increment becomes $E\{|W(t+h) - W(t)|^2 | W(t) = x\} = h$, implying $b(x,t) = 1$.

Let $X(t)$ be defined by the equation

$$X(t) = X_a + \int_a^t m(X(s),s)ds + \int_a^t \sigma(X(s),s)dWs,$$

where $W(t)$ is a Brownian motion process with smoothness properties on the coefficients $m(\cdot), \sigma(\cdot)$. Then $\{X(t), t \in T\}$ is a diffusion (e.g. see [463]), with the infinitesimal mean and variance calculated as follows. Taking conditional expectation, the second integral goes away because increments of X and W are independent and intervals of Brownian motion have zero expected value. The first integral is simplified using the Lipschitz property. From the Lipschitz condition, $|m(X(s),s) - m(x,s)| \leq K|X(s) - x|$, implying

$$\left| \int_t^{t+h} m(X(s),s)ds - \int_t^{t+h} m(x,t)ds \right| \leq \int_t^{t+h} |m(X(s),s) - m(x,s)|dx \quad (18.45)$$

$$\leq K \int_t^{t+h} |X(s) - x|dx \quad (18.46)$$

$$\leq K\sup_{s \in [t,t+h]} |X(s) - x|h. \quad (18.47)$$

By a.s. continuity of the sample paths, $\sup_{s \in [t,t+h]} |X(s) - x| \to 0$ as $h \to 0$, giving

$$a(x,t) = \lim_{h\downarrow 0} \frac{E\{X(t+h) - X(t) | X(t) = x\}}{h} \quad (18.48)$$

$$= \lim_{h\downarrow 0} \frac{E\{\int_t^{t+h} m(X(s),s)ds\}}{h} = m(x,t).$$

For the infinitesimal variance, first compute:

$$(X(t+h) - X(t))^2 = \left(\int_t^{t+h} m(X(s),s)ds\right)^2 + 2\int_t^{t+h} \sigma(X(s),s)dW(s)$$

$$\times \int_t^{t+h} m(X(s),s)ds + \left(\int_t^{t+h} \sigma(X(s),s)dW(s)\right)^2.$$

18.3 DIFFUSION PROCESSES FOR SIMULATION

Taking conditional expectation, the middle term becomes zero obtaining

$$E\{|X(t+h) - X(t)|^2 \mid X(t) = x\}$$
$$= \left(\int_t^{t+h} m(x,s)ds\right)^2 + \sigma^2(x,t) E\left\{\left(\int_t^{t+h} dW(s)\right)^2\right\}$$

The last term above is derived by the independence of $X(t)$ and $W(t)$ at any particular value of t. Dividing by h and taking limits as $h \to 0^+$, the first term goes to zero, and properties of Brownian motion give the infinitesimal variance,

$$b(x,t) = \sigma^2(x,t).$$

Example 18.21 (Langevin stochastic differential equation) Langevin SDEs of the form

$$dX(t) = -\frac{dE(X(t))}{dx} dt + \sigma dW(t),$$

are important with $dE(x)/dx$ satisfying Lipschitz smoothness conditions of (see [213, 463] for examples). The infinitesimal mean and variance just follow from above:

$$a(x,t) = -\frac{dE(x)}{dx}, \quad b(x,t) = \sigma^2.$$

18.3.1 Generators of 1D Diffusions

Examine the diffusion $\{X_t, t \geq 0\}$ in $\mathbb{R}^1$. As discussed in [463], the appropriate setting for diffusions is the Banach space of continuous functions vanishing at infinity, $B = \hat{C}(\mathbb{R})$, with the domain of differentiable functions $D = \hat{C}^\infty(\mathbb{R})$.

Theorem 18.22 *For a real-valued diffusion on the real line $T = \mathbb{R}$, with infinitesimal means and variances $a(x), b(x)$ the generator has domain $D = \hat{C}^\infty(\mathbb{R}) \subset B = \hat{C}(\mathbb{R})$. The generator becomes, for all $f \in \hat{C}^\infty(\mathbb{R})$,*

$$Af(x) = a(x) f'(x) + \frac{1}{2} b(x) f''(x),$$

With $\pi(\cdot)$ invariant iff for all $f \in D = \hat{C}^\infty$,

$$\int_{\mathcal{X}} \left(a(x) f'(x) + \frac{1}{2} b(x) f''(x)\right) \pi(dx) = 0. \tag{18.49}$$

In $\mathbb{R}^n$, $D = \hat{C}^\infty(\mathbb{R}^n)$ with infinitesimal means and variances $a_i(x), b_{ij}(x)$, $i, j = 1, \ldots, n$ and the generator given by

$$Af(x) = \sum_{i=1}^n a_i(x) \frac{\partial}{\partial x_i} f(x) + b_{ij}(x) \frac{\partial^2}{\partial^2 x_i x_j} f(x), \quad f \in \hat{C}^\infty(\mathbb{R}^n). \tag{18.50}$$

Proof Since $f \in \hat{C}^\infty(\mathbb{R})$ vanishes at ∞, use the differentiability to express it in a Taylor series around x:

$$f(y) = f(x) + f'(x)(y-x) + f''(x) \frac{(y-x)^2}{2} + (f''(c) - f''(x)) \frac{(y-x)^2}{2},$$

for some $c = \alpha x + (1-\alpha)y$, $\alpha \in [0,1]$. (See Problem A.15 for further illustration). Therefore, for all $\epsilon > 0$,

$$Af(x) = \lim_{t \to 0} \frac{1}{t} \int P_t(x, dy)(f(y) - f(x))$$

$$= \lim_{t \to 0} \frac{1}{t} \left(\int_{|y-x| \le \epsilon} P_t(x, dy)(f(y) - f(x)) + o(t) \right), \quad f \text{ is bounded, continuous}$$

$$= \lim_{t \to 0} \frac{1}{t} \left(\int_{|y-x| \le \epsilon} P_t(x, dy) \left\{ (y-x)f'(x) + \frac{1}{2}(y-x)^2 f''(x)(1+\alpha_\epsilon) \right\} + o(t) \right)$$

$$= \lim_{t \to 0} \frac{1}{t} \left(f'(x) \int_{|y-x| \le \epsilon} (y-x) P_t(x, dy) + \frac{1}{2} f''(x)(1+\alpha_\epsilon) \right.$$

$$\left. \times \lim_{t \to 0} \frac{1}{t} \int_{|y-x| \le \epsilon} (y-x)^2 P_t(x, dy) + \frac{1}{t} o(t) \right),$$

$$= a(x)f'(x) + \frac{1}{2} b(x) f''(x),$$

where $\alpha_\epsilon = \sup_{|y-x| \le \epsilon} |f''(y) - f''(x)|$, and since $f \in \hat{C}^2$, $\lim_{\epsilon \to 0} \alpha_\epsilon = 0$. □

Example 18.23 (Diffusions and the Wiener process) For a diffusion process, the PDE from Theorem 18.22 is in terms of the infinitesimal means and variances:

$$\frac{\partial P_t}{\partial t}(x, dz) = a(x) \frac{\partial P_t}{\partial x}(x, dz) + \frac{1}{2} b(x) \frac{\partial^2 P_t}{\partial^2 x}(x, dz), \qquad (18.51)$$

with $P_0(x, dz) = \delta_x(dz)$.

For the Wiener process, the transition probability to the set $s \subset \mathbb{R}$ is

$$P_t(x, S) = \int_S \frac{1}{\sqrt{2\pi t}} e^{-(y-x)^2/2t} dy,$$

which as a function of initial state x is twice differentiable and vanishes at $-\infty, \infty$ and is therefore in the domain D of the operator $A = 1/2 f''(x)$.

Let us show that the transition law for a standard Wiener process satisfies the above PDE Eqn. 18.51. For the standard Wiener process, $a(x) = 0, b(x) = 1$, and the generator is $AP_t(x, dz) = (1/2)(\partial^2/\partial x^2) P_t(x, dz)$ with the transition law for a standard Wiener process being $P_t(x, dy) = \left(\frac{1}{\sqrt{2\pi t}} \right) e^{-(y-x)^2/2t} dy$. Therefore, the derivatives become

$$\frac{\partial}{\partial t} P_t(x, dy) = \frac{-1}{2} \frac{1}{(2\pi t)^{3/2}} 2\pi e^{-(y-x)^2/2t} dy$$

$$+ \frac{1}{\sqrt{(2\pi t)}} e^{-(y-x)^2/2t} dy \frac{-(y-x)^2}{2} \left(\frac{-1}{t^2} \right)$$

$$= e^{-(y-x)^2/2t} dy \left(\frac{-1}{2t(2\pi t)^{1/2}} + \frac{(y-x)^2}{2t^2 (2\pi t)^{1/2}} \right), \qquad (18.52)$$

$$\frac{\partial}{\partial x} P_t(x, dy) = \frac{(y-x)}{t} \frac{1}{\sqrt{2\pi t}} e^{-(y-x)^2/2t} dy,$$

$$\frac{\partial}{\partial x^2} P_t(x, dy) = \frac{-1}{t} \frac{1}{\sqrt{2\pi t}} e^{-(y-x)^2/2t} dy + \frac{(y-x)^2}{t^2} e^{-(y-x)^2/2t} dy$$

$$= e^{-\frac{(y-x)^2}{2t}} dy \left(\frac{-1}{t(2\pi t)^{1/2}} + \frac{(y-x)^2}{t^2 (2\pi t)^{1/2}} \right). \qquad (18.53)$$

From Eqns. 18.52, 18.53 it follows that $\frac{\partial}{\partial t} P_t(x, dz) = (1/2)(\partial/\partial^2 x) P_t(x, dz)$.

Example 18.24 Define the S.D.E.

$$dX(t) = -\frac{dE(X(t))}{dx} dt + \sigma dW(t), \quad (18.54)$$

with the drift dE/dx satisfying the Lipschitz condition. This is a Markov process with $a(x) = dE(x)/dx, b(x) = \sigma^2$ and with generator $Af = -(dE(x)/dx)f'(x) + \sigma^2 f''(x)/2$.

18.3.2 Diffusions and SDEs for Sampling

We now examine sampling in the continuum using diffusions. Assume the state-space is the Euclidean space $\mathbb{R}^n$.

Theorem 18.25 Let $\pi(x) = e^{-E(x)}/Z$, $x \in \mathbb{R}^n$. If $X(t)$ is a diffusion with state space $\mathcal{X} = \mathbb{R}^n$ with the properties that the diffusion $X(t)$ on $\mathbb{R}^n$ satisfies the stochastic differential equation

$$dX(t) = -\frac{1}{2}\nabla E(X(t))dt + dW(t) \quad (18.55)$$

with $X(t), \nabla E$ and $W \in \mathbb{R}^n$ denoting the state, gradient and standard vector Brownian motion, respectively, with the gradient ∇E satisfying Lipschitz continuity, then $\pi(\cdot)$ is an invariant density.

Proof Applying A to such f and integrating with respect to μ gives

$$\int_{\mathcal{X}} Af(x)\mu(dx) = \frac{1}{2}\int_{\mathbb{R}^n} \left(-\sum_{i=1}^{n}\frac{\partial E(x)}{\partial x_i}\frac{\partial f(x)}{\partial x_i} + \sum_{i=1}^{n}\frac{\partial^2 f(x)}{\partial x_i^2}\right)\frac{e^{-E(x)}}{Z}dx$$

$$\stackrel{(a)}{=} \left(-\frac{1}{2}\int_{\mathbb{R}^n}\sum_{i=1}^{n}\frac{\partial E(x)}{\partial x_i}\frac{\partial f(x)}{\partial x_i} + \frac{1}{2}\int_{\mathbb{R}^n}\sum_{i=1}^{n}\frac{\partial E(x)}{\partial x_i}\frac{\partial f(x)}{\partial x_i}\right)\frac{e^{-E(x)}}{Z}dx \quad (18.56)$$

$$+ \sum_{i=1}^{n}\frac{e^{-E(x)}}{Z}\frac{\partial f(x)}{\partial x_i}\Big|_{-\infty}^{\infty} = 0, \quad (18.57)$$

where (a) follows from integration by parts of the second derivative term and using the boundary term, which is zero since the derivative on the boundary is zero. $\square$

Example 18.26 Consider a diffusion process which follows the following stochastic differential equation,

$$dX(t) = \frac{-1}{2}\frac{\partial E(X(t))}{\partial X} dt + \sigma \, dW(t), \quad (18.58)$$

with dE/dx satisfying Lipschitz and measurability conditions (e.g. see [213, 463]). The solution is a Markov process, with the generator given by

$$Af(x) = \frac{-1}{2}\frac{\partial E(x)}{\partial x}f'(x) + \frac{1}{2}\sigma^2 f''(x),$$

where $a(x) = (-1/2)(\partial E(x)/\partial x)$ and $b(x) = \sigma^2$. Now clearly

$$\pi(dx) = \frac{e^{-E(x)/\sigma^2}}{Z} dx.$$

is invariant for the Langevin Eqn. 18.58. The infinitesimal means and variances satisfy the Lipschitz conditions, therefore the S.D.E. is a Markov process, hence Theorem 18.7 is in effect:

$$\int Af(x)\pi(dx) = \int_{\mathbb{R}} \left(\frac{-1}{2}\frac{\partial E(x)}{\partial x}f'(x) + \frac{1}{2}\sigma^2 f''(x)\right) \frac{e^{-E(x)/\sigma^2}}{Z} dx$$

$$= \int_{\mathbb{R}} \frac{-1}{2}\frac{\partial E(X(t))}{\partial x}f'(x)\pi(dx) + \int_{\mathbb{R}} \frac{1}{2}f'(x)\frac{e^{-E(x)/\sigma^2}}{Z}\frac{\partial E(x)}{\partial X} dx$$

$$+ \frac{1}{2}f'(x)\frac{e^{-E(x)/\sigma^2}}{Z}\Big|_{-\infty}^{\infty}$$

$$= 0.$$

Notice that $f \in \hat{C}(\mathbb{R})$ vanishing at infinity is used in the boundary condition.

Example 18.27 The Gaussian distribution $\pi(dx) = (1/\sqrt{2\pi\sigma^2})e^{-x^2/2\sigma^2} dx$ is stationary for the Langevin equation

$$dX(t) = -\frac{1}{2}X(t)\, dt + \sigma\, dW(t). \tag{18.59}$$

To see this, the infinitesimal mean and variance of (18.59) is $-(1/2)x$ and σ^2, respectively. The generator of the diffusion becomes

$$Af(x) = -\frac{1}{2}xf'(x) + \frac{1}{2}\sigma^2 f''(x).$$

To show that the Gaussian density is invariant, examine f (with its derivative vanishing at infinity):

$$\int_{-\infty}^{\infty} Af(x)\pi(dx) = \int_{-\infty}^{\infty} \left[-\frac{1}{2}xf'(x) + \frac{1}{2}\sigma^2 f''(x)\right] \frac{1}{\sqrt{2\pi\sigma^2}} e^{-x^2/2\sigma^2} dx$$

$$= \int_{-\infty}^{\infty} \frac{-xf'(x)}{2\sqrt{2\pi\sigma^2}} e^{-x^2/2\sigma^2} dx + \frac{1}{2\sqrt{2\pi\sigma^2}} f'(x)\sigma^2 e^{-x^2/2\sigma^2}\Big|_{-\infty}^{\infty}$$

$$+ \int_{-\infty}^{\infty} \frac{-xf'(x)}{2\sqrt{2\pi\sigma^2}} e^{-x^2/2\sigma^2} dx$$

$$= 0.$$

18.4 Jump-Diffusion Inference on Countable Unions of Spaces

Let us now look at the situation where

$$\mathcal{X} = \cup_k \mathcal{X}(k),$$

with prior probabilities p_k attached to the subconfiguration spaces $\mathcal{X}(k)$. The interpretation of this could be, for example, that k represents the unknown number of objects in the scene or enumerates

18.4 JUMP-DIFFUSION INFERENCE ON COUNTABLE UNIONS OF SPACES

various pathological anatomies. We shall assume for simplicity that each subconfiguration space $\mathcal{X}(k)$ contains a configuration of parameter $x(k) \in \mathcal{X}(k)$. Having defined the configuration space $\mathcal{X} = \cup_k \mathcal{X}(k)$ as the union of spaces over which the inference is to be performed, the crucial part of the problem still remaining is the derivation of the inference algorithm for choosing the graphs and their associated transformations. In other words, *how can we carry out hypothesis formation? The hypothesis space is a countable disconnected union of these connected parameter spaces, with the model order (parametric dimension) a variable to be inferred as well.* Since we take a Bayesian approach, a posterior distribution is constructed sitting on this countable disconnected union of spaces. The parametric representation of the target scene is selected to correspond to the *conditional mean* under this posterior, requiring the generation of expectations over this complex parameter space. Conditional expectations under the posterior are constructed via a Markov process with sample paths moving through the parameter space, satisfying the *ergodic property* that the transition distribution of the Markov process converges to the posterior distribution. This allows for the empirical generation of conditional expectations under the posterior. To accommodate the connected and disconnected nature of the state spaces, the Markov process is forced to satisfy *jump-diffusion dynamics*, i.e. through the connected parts of the parameter space (Lie manifolds) the algorithm searches continuously, with sample paths corresponding to solutions of standard diffusion equations; across the disconnected parts of parameter space, the jump process determines the dynamics. The infinitesimal properties of these jump-diffusion processes are selected so that the Markov process is ergodic in the sense that the transition distribution converges to the posterior.

It is the fundamental difference between diffusions (almost surely continuous sample paths) and jump processes (which make large moves in parameter space in small time) that allows us to explore the very different connected and non-connected nature of hypothesis space.

18.4.1 The Basic Problem

Scenes are assumed to contain a variable number of objects of one or several classes. The variability within each class is described via a template with transformations which deform the template into a large family. These transformations form Lie groups. The objects are described via their associated transformations. Variable numbers of objects are described via finite products of the transformations. The state space describing all possible objects becomes the disjoint union over all finite powers of the transformation spaces. Inference in the Bayesian framework requires a posterior distribution supported on this space.

The basic problem becomes as follows. Identify model k with parameter space $\mathcal{X}(k)$ a connected space of dimension $n(k)$ with prior on model k denoted Pk. Then given are a collection of such spaces with the full hypothesis space $\mathcal{X} = \cup_{k=0}^{\infty} \mathcal{X}(k)$, and posterior distribution μ supported on $\mathcal{X}$ is of the Gibbs type:

$$\mu(\mathcal{A}) = \sum_{k=0}^{\infty} Pk \int_{\mathcal{A} \cap \mathcal{X}(k)} \frac{e^{-H_k(x(k))}}{Z_k} \, dx(k) \qquad (18.60)$$

with normalizer $Z_k = \int_{\mathcal{X}(k)} e^{-H_k(x(k))} \, dx(k)$. Given a measure μ on the countable union of spaces $\mathcal{X} = \cup_{k=0}^{\infty} \mathcal{X}(k)$, the crucial part of the problem is *how shall we carry out hypothesis formation?* We choose to generate conditional expectations with respect to μ empirically, that is by generating a sequence of samples $X(t_1), X(t_2), \ldots$ with the property that $(1/n) \sum_{i=1}^{n} f(X(t_i)) \to \int_{\mathcal{X}} f(x) \mu(dx)$. This we do via the construction of a Markov process $X(t)$ that satisfies *jump-diffusion dynamics* through $\mathcal{X}$ in the sense that (i) on random exponential times the process jumps from one of the countably infinite set of spaces $\mathcal{X}(k), k = 0, 1, \ldots$ to another, and (ii) between jumps it satisfies diffusions of dimension appropriate for that space.

The jump-diffusion is constructed from a set of independent and exponentially distributed times $w_1, w_2, \ldots$, with jump times $t_1, t_2, \ldots$ defined according to $t_i = \inf\{t : \int_{t_{i-1}}^{t} q(X(s))ds \geq w_i\}$, and $t_0 = 0$. The process $X(t)$ within each space $\mathcal{X}(k)$ is a diffusion with infinitesimal drifts $a(x(k)) \in \mathbb{R}^{n(k)}$ and infinitesimal variance matrix $B(x(k))$ (an $n(k) \times n(k)$ matrix). The process moves from one subspace to another on the t_i's with transition measures $q(x, dy)$ defined as above

$$q(x, dy) = \lim_{\epsilon \to 0} \frac{1}{\epsilon}\left(\Pr\{X(t + \epsilon) \in dy | X(t) = x\} - \delta_x(dy)\right), \quad (18.61)$$

with $q(x) = \int_{\mathcal{X} \setminus x} q(x, dy)$, and transition probability measures $Q(x, dy) = q(x, dy)/q(x)$, $\int_{\mathcal{X}} Q(x, dy) = 1$.

For the technical details of the construction of the jump diffusion we rely on adaptation of results from Ethier and Kurtz [463] for characterizing our diffusion processes in the countably infinite number of subspaces. First, we must establish that the addition of the jump and diffusion processes results in a well-defined Markov process. Let $\hat{C}(\mathcal{X})$ be the Banach space of continuous functions on $\mathcal{X}$ vanishing at infinity and $C_c^\infty(\mathcal{X})$ be the space of differentiable functions, compactly supported on $\mathcal{X}$.

Theorem 18.28 *Let $\mu(dx) = \sum_k^{P_k} 1_{\mathcal{X}(k)}(x) \frac{e^{-H_k(x)}}{Z_k} dx$ with the normalizer $Z_k = \int_{\mathcal{X}(k)} e^{-H_k(x)} dx$.*

IF the jump diffusion process $X(t)$ with state space $\mathcal{X} = \bigcup_{k=0}^{\infty} \mathbb{R}^{n(k)}$ has the properties that

(i) the diffusion $X(t)$ within any subspace $\mathbb{R}^{n(k)}$ satisfies the stochastic differential equation

$$dX(t) = -\frac{1}{2}\nabla H_k(X(t))dt + dW_{n(k)}(t) \quad (18.62)$$

where $X(t), \nabla H_k$ and $W_{n(k)} \in \mathbb{R}^{n(k)}$ are the state, gradient and standard vector Brownian motion, respectively, with the gradient ∇H_k satisfying Lipschitz continuity, and

(ii) the jump and transition probability measures $q(x, dy), q(x), Q(x, dy)$ are bounded continuous functions, with the jumps local (finite support) satisfying

$$q(x)\mu(dx) = \int_{\mathcal{X}} q(y)Q(y, dx)\mu(dy), \quad (18.63)$$

then the jump-diffusion process $X(t)$ is a Markov process on $\mathcal{X}$ with unique invariant measure $\mu = \sum_k p_k \mu_k$.

Notice, if the Euclidean spaces are *connected* under the jumps, i.e. for all k, k', /bin/bash: w: command not found finite sequence of jumps carrying the process from $\mathbb{R}^{n(k)}$ to $\mathbb{R}^{n(k')}$, then this implies a recurrence asssuring that μ is the only invariant measure.

Proof The first part of the proof of Theorem 18.28 relies on the fact that the generator or backward Kolmogoroff operator A for the jump-diffusion process characterizes the stationary measure. That is, μ is stationary for $X(t)$ if and only if $\int Af(x)\mu(dx) = 0$ for all f in the *core* of A (p. 239 and the Echeverria theorem, p. 248 of [463]). Now the generator is the superposition of the diffusion and jump generators $A = A^d + A^j$ (diffusion+jump). This follows from Ethier and Kurtz (p. 266, [463]). The core $C_c^2(\mathcal{X})$ is the set of functions $f(x) = \sum_{k=0}^{m} 1_{\mathbb{R}^{n(k)}}(x) f_k(x)$, $m \geq 0$, where $f_k(x) \in C_c^2(\mathbb{R}^{n(k)})$ are

18.4 JUMP-DIFFUSION INFERENCE ON COUNTABLE UNIONS OF SPACES

twice continuously differentiable, compactly supported functions on $\mathbb{R}^{n(k)}$. Applying A to such f and integrating with respect to μ gives

$$\int_{\mathcal{X}} Af(x)\mu(dx) = \frac{1}{2}\sum_{k=0}^{m} \int_{\mathbb{R}^{n(k)}} \left(-\sum_{i=1}^{n(k)} \frac{\partial H_k(x(k))}{\partial x_i} \frac{\partial f_k(x(k))}{\partial x_i} \right. \tag{18.64}$$

$$\left. + \sum_{i=1}^{n(k)} \frac{\partial^2 f_k(x(k))}{\partial x_i^2} \right) \frac{e^{-H_k(x(k))}}{Z} dx(k)$$

$$+ \int_{\mathcal{X}} \mu(dx) q(x) \left(\int_{\mathcal{X}} \bigl(f(y) - f(x)\bigr) Q(x, dy) \right), \tag{18.65}$$

where the first part the standard S.D.E. operator and the second the jump operator. To show $\int Af(x)\mu(dx) = 0$, integrate by parts once the second derivative term in the first part of Eqn. 18.65 and use the second condition (b) of Eqn. 18.63. That completes the first part of the theorem.

Remark: In [471] we show for all $x \in \mathcal{X}$ the associated chain $X(i\Delta), \Delta > 0, i = 0, 1 \ldots$ for each x, $X(i\Delta) = x$, converges in total variation norm to μ. □

19 JUMP DIFFUSION INFERENCE IN COMPLEX SCENES

ABSTRACT This chapter explores random sampling algorithms introduced in for generating conditional expectations in hypothesis spaces in which there is a mixture of discrete, disconnected subsets. Random samples are generated via the direct simulation of a Markov process whose state moves through the hypothesis space with the ergodic property that the transition distribution of the Markov process converges to the posterior distribution. This allows for the empirical generation of conditional expectations under the posterior. To accommodate the connected and disconnected nature of the state spaces, the Markov process is forced to satisfy jump–diffusion dynamics. Through the connected parts of the parameter space (Lie manifolds) the algorithm searches continuously, with sample paths corresponding to solutions of standard diffusion equations. Across the disconnected parts of parameter space the jump process determines the dynamics. The infinitesimal properties of these jump–diffusion processes are selected so that various sample statistics converge to their expectation under the posterior.

We now examine the generation of conditional mean estimates of parameters in the rigid body tracking and recognition scenario. Except under particular sets of assumptions, the posterior distribution will be highly nonlinear in the parameters of the hypothesis space thus precluding the direct closed-form analytic generation of conditional mean estimates. Towards this end, we have taken advantage of the explosion which that occurred over the past years in the statistics community with the introduction of random sampling methods for the empirical generation of estimates from complicated distributions. See for example the reviews [469, 470]. We now explore random sampling algorithms introduced in [137, 303, 471, 472] for generating conditional expectations in hypothesis spaces in which there are a mixture of discrete, disconnected subsets. Random samples are generated via the direct simulation of a Markov process whose state moves through the hypothesis space with the *ergodic property* that the transition distribution of the Markov process converges to the posterior distribution. This allows for the empirical generation of conditional expectations under the posterior. To accommodate the connected and disconnected nature of the state spaces, the Markov process is forced to satisfy *jump–diffusion dynamics*, i.e. through the connected parts of the parameter space (Lie manifolds) the algorithm searches continuously, with sample paths corresponding to solutions of standard diffusion equations; across the disconnected parts of parameter space the jump process determines the dynamics. The infinitesimal properties of these jump–diffusion processes are selected so that various sample statistics converge to their expectation under the posterior.

The original motivation for introducing jump–diffusions in [137, 471] is to accommodate the very different continuous and discrete components of the object discovery process. Given a scene of multiple objects, the problem is to identify the orientation, translation and scale parameters accommodating the variability manifest in the viewing of each object type. For this, the parameter space is sampled using diffusion search, in which the state vector winds continuously through the similarities following gradients of the posterior. The second distinct part of the sampling process corresponds to the target type and number deduction, during which the target types are being discovered, with some subset of the scene only partially "recognized" at any particular time during the process. A jump in hypothesis space corresponds to (i) selecting between different object types, or (ii) hypothesizing a new object in the scene or a "change of mind" via the deletion of an object in the scene. The jump intensities are governed by the posterior density, with the process visiting configurations of higher probability for longer exponential times, and the diffusion equation governing the dynamics between jumps. It is the fundamental difference between diffusions (almost surely continuous sample paths) and jump processes (making large moves in parameter space in small time) that allows us to explore the very different connected and non-connected nature of hypothesis space.

19.1 Recognition of Ground Vehicles

Remote sensing of multiple ground vehicles has been extensively studied using several sensor models including forward looking intra-red (FLIR) and laser radar (LADAR) sensors.

19.1.1 CAD Models and the Parameter Space

To create the basic building blocks of the hypothesized scenes define the primitive generating elements $\mathcal{G}$ made up of disjoint classes $\mathcal{G} = \bigcup_{\alpha \in \mathcal{A}} \mathcal{G}_\alpha \ \alpha \in \mathcal{A}$, the generator indices. The generators studied throughout are airframes and land vehicles. To accommodate variability, apply transformations to the generators from the similarity groups $\mathcal{S}, s : \mathcal{G} \leftrightarrow \mathcal{G}$. These do not change generator type, $\alpha(sg) = \alpha(g)$. Herein, $\mathcal{G}$ are 2-dimensional (2D) manifolds in $\mathbb{R}^3$ (surfaces), and their transformation is the matrix group action. Figure 19.1 shows the basic CAD model used to represent the target classes in several of the experiments to follow. For ground-based, rigid targets, we use translation in the plane and rotation around the $z = x_3$-axis. Then the similarity group $\mathcal{S} : \mathcal{G} \leftrightarrow \mathcal{G}$, consists of the set of rigid motions of the type $s = (O, p) : x \mapsto O(\phi)x + p$, where

$$(O(\phi), p) : \begin{pmatrix} x_1 \\ x_2 \\ x_3 \end{pmatrix} \mapsto \begin{pmatrix} \cos\phi & \sin\phi & 0 \\ -\sin\phi & \cos\phi & 0 \\ 0 & 0 & 1 \end{pmatrix} \begin{pmatrix} x_1 \\ x_2 \\ x_3 \end{pmatrix} + \begin{pmatrix} p_1 \\ p_2 \\ 0 \end{pmatrix}. \tag{19.1}$$

A single ground vehicle in the scene is parameterized by its orientation and position and its class, an element of $\mathbf{SO}(2) \times \mathbb{R}^2 \times \mathcal{A}$. Identifying the similarity transformation of axis-fixed rotation with its Euler representation in the torus $\mathcal{T}$, and translation in the plane $\mathbb{R}^2$, then an m-object parameter vector becomes

$$x(m) = (\phi_1, p_1), (\phi_2, p_2), \cdots \in \mathcal{X}(m) = \left(\mathcal{T} \times \mathbb{R}^2\right)^m. \tag{19.2}$$

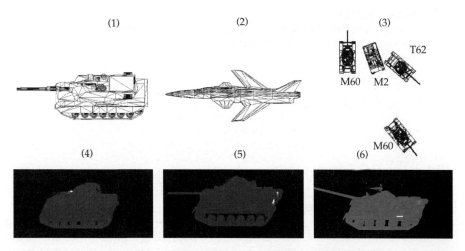

Figure 19.1 Top row: Panels 1 and 2 show the CAD models used for the simulations. Panel 3 shows the top-down and perspective view of the M60, M2, and T62 tank scene. Bottom row: Panels 4, 5, and 6 show the radiant intensities on the CAD models generated via the PRISM simulation package.

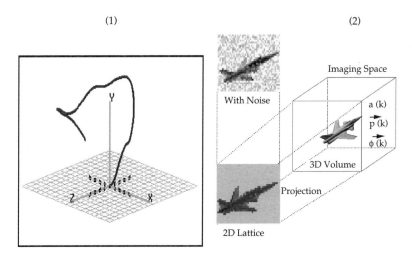

Figure 19.2 Panel 1 shows a single track of $n(1)$ rotations and translations; panel 2 displays the projection system for optical imaging.

The scene of m-targets is depicted in panel 3 of Figure 19.1.

For the tracking and recognition of aircraft, the transformations are rotations and translations acting on $\mathbb{R}^3$. For airframes, the full orthogonal group **SO**(3) is needed (3×3 matrices with determinant 1) of the form $(O, p) : x \mapsto Ox + p$, with the *Euler angle* representation providing natural local coordinates, parameterizing $O(\phi) \in \mathbf{SO}(3)$ via the three angle vector $\phi = (\phi_1, \phi_2, \phi_3)$ of the form

$$O(\phi_1, \phi_2, \phi_3) : \begin{pmatrix} x_1 \\ x_2 \\ x_3 \end{pmatrix} \mapsto \begin{pmatrix} 1 & 0 & 0 \\ 0 & \cos\phi_1 & \sin\phi_1 \\ 0 & -\sin\phi_1 & \cos\phi_1 \end{pmatrix} \begin{pmatrix} \cos\phi_2 & 0 & -\sin\phi_2 \\ 0 & 1 & 0 \\ \sin\phi_2 & 0 & \cos\phi_2 \end{pmatrix}$$
$$\times \begin{pmatrix} \cos\phi_3 & \sin\phi_3 & 0 \\ -\sin\phi_3 & \cos\phi_3 & 0 \\ 0 & 0 & 1 \end{pmatrix} \begin{pmatrix} x_1 \\ x_2 \\ x_3 \end{pmatrix}. \tag{19.3}$$

A single aircraft in the scene is parameterized by its orientation, position and class, and creates a track as it flies through space. Identifying the rotations with the Euler representation, then an m-track configuration has parameter vector in the form

$$x(m) \in \mathcal{X}(m) = \left(\mathcal{T}^3 \times \mathbb{R}^3\right)^{n(m)}. \tag{19.4}$$

An example of a track is shown in panel 1 of Figure 19.2.

19.1.2 The FLIR Sensor Model

The FLIR sensor has a wide field of view relative to object packing dimensions demanding the use of *perspective projection*, which maps according to $(x_1, x_2, x_3) \mapsto (x_1/x_3, x_2/x_3)$. This creates the vanishing point effect in which objects that are further away from the sensor appear closer to the center of the detector. Objects appear skewed depending on where they appear in the image plane. For generating FLIR scenes, standard infra-red radiation models of the targets generated by other groups such as Keweenaw Research Center, Michigan Technological University, are used. The three panels of the bottom row of Figure 19.1 show the infra-red radiance simulated by the PRISM code [299] for an M2, an M60, and a T62. For such modeling, target facets are assumed to radiate known intensities determined from measured data [473] or simulation, or both [474].

For the ground scenario the scene consists of many varied objects, some of which are totally or partially obscuring others, against a highly cluttered background of both natural and man-made origin. Exemplar data are shown in the left column of Figure 19.3 showing images at 3–5 microns (top) and 8–12 microns (bottom). For generating FLIR scenes, the CAD models have superimposed standard infra-red radiation from CAD models of the targets generated using standard software packages from the Keweenaw Research Center, Michigan Technological University. These models are used for generating the 3D radiant scenes. The target facets are assumed to radiate known intensities determined from via radiant model simulation. The resulting measurement is modeled as Poisson with mean field the projection with obscuration and occlusion represented via the transformation $T : I(x(m)) \mapsto TI(x(m))$. If the measurements correspond to a CCD detector, the Poisson model assumes the measured 2D image field $I^D(\cdot)$ is Poisson distributed [31, 31, 32] with mean intensity the convolution of the mean field TI, the projection on the detector plane [32, 317] with the point spread function of the camera. Let the optics in the detector plane have point-spread density $p(\cdot)$. The measured counting process $I^D(y_i)$ at location y_i has mean given by the projection of the temperature signature convolved with the point-spread

$$EI^D(\cdot) = p * TI(\cdot), \tag{19.5}$$

$$H_m(I^D) = -\langle 1, TI(x(m)) \rangle + \langle \log(p * TI(x(m))), I^D \rangle, \tag{19.6}$$

where $*$ denotes the convolution operation.

Panel 3 of Figure 19.3 depicts the mean field TI of the FLIR radiance process projected through the projective geometry with point-spread optics in the detector plane. Panel 4 depicts the

Figure 19.3 Top row: Panels 1 and 2 show 3–5 and 8–12 microns infra-red image of a natural scene obtained during the Grayling I Field Experiment, Smart Weapons Observability Enhancement Joint Test & Evaluation Program. Courtesy Robert Guenther of the Physics Division, Army Research Office. Bottom row: Panel 3 shows an ideal infra-red scene TI through the projective geometry and with Gaussian blur; panel 4 shows the corresponding range mean-field for the LADAR.

19.2 Jump Diffusion for Sampling the Target Recognition Posterior

Random sampling is used to generate samples from the posterior distribution, which is supported on multiple parameter spaces of varying dimensions (varying object number).

19.2.1 The Posterior distribution

The posterior is the product of the prior and data likelihood. The pose angles are taken as uniform around the circle, and uniform in the ground frame over which the observations are made. The prior on the number of objects is taken as Poisson with fixed parameter λ. The posterior potential H_m associated with subspace $\mathcal{X}(m)$ is the superposition of the data likelihood and prior potentials. For an m-object configuration $x(m) \in \mathcal{X}(m) \subset \mathcal{X}$ define the potential $H_m(x), x \in \mathcal{X}(m)$ consisting of the superposition of the log-likelihood and the prior P_m Poisson in object number m with parameter λ. Then, the posterior $\mu(\cdot)$ on $\mathcal{X} = \cup_m \mathcal{X}(m)$ is taken to be of the form

$$\mu(dx) = \sum_{m=0}^{\infty} P_m 1_{\mathcal{X}(m)}(x) \frac{e^{-H_m(x)}}{Z_m} dx, \qquad (19.7)$$

with the m-object potential

$$H_m(x) = H_m(I^D) \quad x(m) \in \mathcal{X}(m) = \left(\mathcal{T} \times \mathbb{R}^2\right)^m, \qquad (19.8)$$

and the normalizer $Z_m = \int_{\mathcal{X}(m)} e^{-H_m(x)} dx$, and dx the Lebesgue measure appropriate for the space $\mathcal{X}(m)$. Let $\pi(\cdot|I^D)$ be the associated density in each of the subspaces. Notice, $\pi(\cdot|I^D)$ does not integrate to one over any of the subspaces, so that

$$\int_{\mathcal{X}(m)} \pi(x|I^D) dx = \frac{\int_{\mathcal{X}(m)} e^{-H_m(x)} dx}{\sum_m \int_{\mathcal{X}(m)} e^{-H_m(x)} dx}.$$

Then, the posterior $\mu(\cdot)$ for tracking on $\mathcal{X} = \cup_m \mathcal{X}(m)$ is similar to that above of the form

$$\mu(dx) = \sum_{m=0}^{\infty} P_m 1_{\mathcal{X}(m)}(x) \frac{e^{-H_m(x)}}{Z_m} dx, \qquad (19.9)$$

with the m-track configuration

$$\mathcal{X}(m) = \left(\mathcal{T}^3 \times \mathbb{R}^3\right)^{n(m)} \times \mathcal{A}^m. \qquad (19.10)$$

19.2.2 The Jump Diffusion Algorithms

A single subspace for one vehicle is $\mathcal{X}_0 = \mathbb{R}^2 \times \mathcal{T}$. The *diffusion* for the space of m ground vehicles flows through the manifold $\mathcal{X}_0^m = \mathbb{R}^{2m} \times \mathcal{T}^m$. Associate the first $2m$ components of the state vector with the flow through $\mathbb{R}^{2m}$, and the last m components with the flow through $\mathcal{T}^m$. Defining $X_t = [X_t^{(1)}, X_t^{(2)}]$, $X_t^{(1)} \in \mathbb{R}^{2m}$, $X_t^{(2)} \in \mathcal{T}^m$, then X_t drifts according to the drift vectors A_m, B_m of dimensions $2m, m$ satisfying the equations

$$X_t^{(1)} = X_{t_0}^{(1)} - \int_{t_0}^{t} A_m(X_s)\,ds + W_t^{(1)}, \tag{19.11}$$

$$X_t^{(2)} = \left[X_{t_0}^{(2)} - \int_{t_0}^{t} B_m(X_s)\,ds + W_t^{(2)} \right]_{\mathrm{mod}\,2\pi}, \tag{19.12}$$

where $W^{(1)}, W^{(2)}$ are standard vector Wiener processes of dimensions $2m, m$, respectively, and where $[\cdot]_{\mathrm{mod}\,2\pi}$ is taken componentwise.

The *jump process* is controlled by the family of simple moves or graph changes $\mathcal{J}$ controlling the movement through the discrete non-connected subspaces. The set $\mathcal{J}$ controls the jump dynamics by moving the process from one subspace to another on the jump times with transition probabilities $Q(x, dy) = q(x, dy)/q(x), \int_\mathcal{X} Q(x, dy) = 1$ and the measures $q(x, dy)$ defined in the standard way [462]:

$$q(x, dy) = \lim_{\epsilon \to 0} \frac{1}{\epsilon} \left(\Pr\{X_{t+\epsilon} \in dy | X_t = x\} - 1_{dy}(x) \right), \quad q(x) = \int_{\mathcal{J}^1(x)} q(x, dy). \tag{19.13}$$

The jumps occur at exponentially distributed times governed by a jump intensity $q(x, dy)$. Between jumps, the process satisfies the stochastic differential Eqns. 19.11 and 19.12 appropriate for that subspace. At each jump, the process moves from state x to a new state y according to the jump transition-probability measure $Q(x, dy) = q(x, dy)/q(x)$.

The algorithm must specify when the process should jump from space to space, and which states within each of the spaces it should jump into, and in which directions it should drift. Proper choice of jump parameters and drifts results in the jump–diffusion process having the special property that μ is invariant for it allowing for empirical generation of conditional expectations under the posterior distribution.

For technical considerations, the jump and diffusion process must satisfy certain regularity conditions to result in a well-defined Markov process (e.g. see [463]). This includes that (i) the drifts be Lipschitz smooth so the diffusion is well defined, and (ii) the jump intensities $q(\cdot)$ and $Q(x, d\cdot)$ are bounded and continuous functions, with the jump transition probability measure $Q(x, d\cdot)$ having finite support (local). To achieve the desired ergodic property, we choose the jump intensities according to the following theorem.

Theorem 19.1 *Assume the drifts and jump parameters satisfy the above regularity conditions. Then,*

1. *if the gradient of the posterior follows the drift given by the gradient of the posterior potential within any of the subspaces,*

$$A_m = \frac{1}{2} \nabla_1 H_m, \quad B_m = \frac{1}{2} \nabla_2 H_m, \tag{19.14}$$

where ∇_1, ∇_2 are the gradients with respect to the $2m$ positions and m orientations, respectively,

2. *the jump intensities satisfy the detailed balance condition*

$$q(x)\pi(x)dx = \int_{\mathcal{J}^{-1}(x)} q(y, dx)\pi(y)\,dy, \tag{19.15}$$

3. *and the family of jump moves is rich enough to permit any subspace to be reached from any other disconnected subspace in a finite number of jump moves,*

THEN, *the posterior distribution $\mu(\cdot)$ is the invariant distribution for the Markov process.*

Proof The proof boils down to showing (i) that μ is invariant for the process, and (ii) verifying that the process is irreducible, and therefore μ is unique. Here, we verify only (i) the stationarity for the jump diffusion Markov process. For the uniqueness part (ii) of the proof, see [137, 303, 472].

The proof of invariance for the diffusion goes as follows. The generator (see Chapter 18, Definition 18.6, Eqn. 18.8) or backward Kolmogoroff operator, for the jump–diffusion process (denote it as $A = A^d + A^j$, diffusion+jump) characterizes the stationary distribution in that μ is stationary for the jump–diffusion if and only if

$$\int Af(x)\mu(dx) = 0 \tag{19.16}$$

for all f in the domain $\mathcal{D}(A)$ of A. This is the Echeverria theorem [463]. Define the set of functions forming the domain of the Euclidean part of the generator A as

$$\left\{ f : f = \sum_{m=0}^{M} 1_{\mathbb{R}^{2m}} f_m, f_m \in \hat{C}^2(\mathbb{R}^{2m}), M \geq 0 \right\}, \tag{19.17}$$

where $\hat{C}^2$ is the set of twice continuously differentiable functions vanishing at ∞. The generator is derived as in Section 18.4.1, Chapter 18. The infinitesimal generator for the diffusion A^d acting on such functions, $f = \sum_{m=0}^{M} 1_{\mathbb{R}^{2m}} f_m$, according to Eqn. 19.16 gives

$$\int_{\mathcal{X}} A^d f(x) \pi(x) \, dx = \sum_{m=0}^{M} \left(-\int_{\mathbb{R}^{2m}} \frac{1}{2} < \nabla_1 H_m(x), \nabla f_m(x) > \frac{e^{-H_m(x)}}{Z} \, dx \right. \tag{19.18}$$

$$\left. + \int_{\mathbb{R}^{2m}} \frac{1}{2} \left(\sum_{i=1}^{2m} \frac{\partial^2 f_m(x)}{\partial x_i^2} \right) \frac{e^{-H_m(x)}}{Z} \, dx \right), \tag{19.19}$$

where $< \cdot, \cdot >$ stands for the vector dot-product and the gradients $\nabla H_m(x), \nabla f_m(x)$ are with respect to the Euclidean positions (elements of $\mathbb{R}^{2m}$). Integration by parts of the second term, with the fact that the functions f_m vanish at the boundary, results in a term that is negative of the first term. Therefore, the posterior π is a stationary density of the diffusion process.

Now extend the domain of the functions to include $\mathcal{T}^m$, so that $\mathcal{X}(m) = \mathbb{R}^{2m} \times \mathcal{T}^m$. Then the generator equation, which must be balanced is identical to that above, just with gradients over all $2m + m$ components with the integral over all functions in the domain

$$\mathcal{D}(A) = \left\{ f : f = \sum_{m=0}^{M} 1_{\mathcal{X}(m)} f_m, f_m \in \hat{C}^2(\mathcal{X}(m)), M \geq 0 \right\}, \tag{19.20}$$

with f_m vanishing at infinity for the $\mathbb{R}^{2m}$ part of the domain, and periodic on the torus $\mathcal{T}^m$ (boundary identified for the torus). Integration by parts yields the same condition as above. In passing to the torus, it must be shown that the infinitesimal generator is identical to the Euclidean case as given above, integrated over the full product space of Euclidean and torul parameters. This follows from the fact that the torus is a group with the form of the generator as derived in [303, 472], and the fact that the usual identification of the torus with the real line wrapped implies it has the same flat geometry (identical tangent space) as the reals.

The jump part of the generator A^j is given by

$$A^j f(x) = q(x) \int_{\mathcal{X}(\mathcal{J}^1(x))} Q(x, dy)(f(y) - f(x)).$$

It must satisfy $\int_{\mathcal{X}} A^j f(x) \mu(dx) = 0$ as well. Computing the adjoint of Eqn. 19.16 gives

$$\int_{\mathcal{X}} A^j f(x) \mu(dx) = \int_{\mathcal{X}} \mu(dx) q(x) \int_{\mathcal{X}(\mathcal{J}^1(x))} Q(x,dy)(f(y) - f(x))$$

$$= \int_{\mathcal{X}} \mu(dx) q(x) \int_{\mathcal{X}(\mathcal{J}^1(x))} Q(x,dy) f(y) - \int_{\mathcal{X}} \mu(dx) q(x) f(x)$$

$$= \int_{\mathcal{X}} f(y) \int_{\mathcal{X}(\mathcal{J}^{-1}(y))} \mu(dx) q(x) Q(x,dy) - \int_{\mathcal{X}} \mu(dx) q(x) f(x). \tag{19.21}$$

Since this is true for all f in the domain of the generator [45], this implies

$$q(x) \pi(x) dx = \int_{\mathcal{X}(\mathcal{J}^{-1}(x))} q(y,dx) \pi(y)\, dy. \tag{19.22}$$

$\square$

19.2.3 Jumps via Gibbs' Sampling

The jump moves are transformations on the models that act by changing one model to another with its resulting scene. For this, identify model m with a member of the countable set $\aleph$. The *simple moves* are drawn probabilistically from the family $\mathcal{J}$, and are applied discontinuously during the dynamics of the algorithm, defining transitions through $\aleph$, $\mathcal{J} : \aleph \to \aleph$. The family of transitions are chosen large enough to act transitively in the sense that given any pair $(m', m'') \in \aleph^2$ it is possible to find a finite chain of transitions that leads from m' to m''. Define the set $\mathcal{J}^1(m)$ of models that can be reached in one move, and $\mathcal{J}^{-1}(m)$ which is the set that can reach m in one move. For the desired ergodic properties of the algorithm, these need to satisfy a weak reversibility: $\mathcal{J}^1 = \mathcal{J}^{-1}$. The set of transformations add, delete, and change the identities of the objects. For ground vehicles $\mathcal{X}_0 = \mathcal{T} \times \mathbb{R}^2$, we have

$$\text{deletion} \quad J_j^d : \mathcal{X}_0^m \times \mathcal{A}^m \to \mathcal{X}_0^{m-1} \times \mathcal{A}^{m-1}, \quad j = 1,\ldots,m, \tag{19.23}$$

$$\text{addition} \quad J^b : \mathcal{X}_0^m \times \mathcal{A}^m \to \mathcal{X}_0^{m+1} \times \mathcal{A}^{m+1}, \tag{19.24}$$

$$\text{identification} \quad J_j^a : \mathcal{X}_0^m \times \mathcal{A}^m \to \mathcal{X}_0^m \times \mathcal{A}^m. \tag{19.25}$$

These are the only transformations of model type that are allowed. To carry the evolution of the state forward from the diffusion the jump measures are delta-dirac measures (placing mass on points in $\mathbb{R}^n$) with respect to the Lebesgue measures in the respective subspaces that the jump transformations move into, i.e. the part of the state that is not being added or deleted remains unchanged after the jump transformation. Let $x(m) \oplus_j y(1) \in \mathcal{X}(m+1)$ stand for the addition of an object $y(1) \in \mathcal{X}_0$ to the jth location in the list of $x(m) \in \mathcal{X}(m)$. Let $x^{d_j}(m) \in \mathcal{X}(m-1)$ denote the deletion of the jth object in the list $x(m) \in \mathcal{X}(m)$. Let $x^{a_j}(m) \in \mathcal{X}(m)$ represent changing the

---

[45] It must be shown that the domain is a separating set, i.e. it is sufficiently large such that it is distribution determining. This is the case for examples we study here.

identity of the jth object in $x(m)$ to a. Define the jump transitions measures to be

$$q(x(m), dy(m+1)) = \sum_{j=1}^{m+1} q^b(x(m), y(m+1)) \delta_{x(m)}(dy^{d_j}(m+1)) dy(1),$$

$$q(x(m), dy(m)) = \sum_{j=1}^{m} \sum_{a \in \mathcal{A}} q^a(x(m), y(m)) \delta_{x(m)}(dy^{a_j}(m)),$$

$$q(x(m), dy(m-1)) = \sum_{j=1}^{m} q^d(x(m), y(m-1)) \delta_{x^{d_j}(m)}(dy(m-1)), \tag{19.26}$$

with total jump intensity

$$q(x(m)) = \int_{\mathcal{X}(\mathcal{J}^1(m))} q(x(m), dy). \tag{19.27}$$

Choosing the jump and diffusion parameters properly makes the invariant probability distribution the posterior. A variety of jump–diffusion processes may be formulated which will result in the desired invariant measure. The different schemes correspond to different choices for the jump intensity $q(x, dy)$. First examine the analog of a Gibbs sampler or heat bath.

Corollary 19.2 *Let the jump intensities satisfy*

$$q(x, dy) = \begin{cases} \pi(y) dy & \text{if } y \in \mathcal{J}^1(x) \\ 0 & \text{otherwise} \end{cases}. \tag{19.28}$$

Then the detailed balance condition Eqn. 19.15 from Theorem 19.1 is satisfied.

Proof The right-hand side (RHS) of Eqn. 19.22 becomes

$$\int_{y \in \mathcal{X}(\mathcal{J}^{-1}(x))} q(y, dx) \pi(y) dy \stackrel{(a)}{=} \int_{y \in \mathcal{X}(\mathcal{J}^{-1}(x))} \pi(x) dx \pi(y) dy \tag{19.29}$$

$$\stackrel{(b)}{=} \int_{y \in \mathcal{X}(\mathcal{J}^1(x))} q(x, dy) \pi(x) dx = q(x) \pi(x) dx, \tag{19.30}$$

with (a,b) following from Eqn. 19.28 and weak reversibility, $\mathcal{J}^1 = \mathcal{J}^{-1}$.

To prove the detailed balance condition Eqn. 19.22, assume that $x \in \mathcal{X}(m)$ for any m, and substitute the jump transitions. The left-hand side (LHS) of Eqn. 19.22 becomes

$$\text{LHS} = \pi(x) dx \int_{\mathcal{X}(\mathcal{J}^1(x(m)))} q(x, dy) \tag{19.31}$$

$$= \pi(x) dx \left(\sum_{j=1}^{m+1} \int_{\mathcal{X}_0} q^b(x, x \oplus_j y(1)) dy(1) + \sum_{j=1}^{m} q^d(x, x^{d_j}) + \sum_{a \in \mathcal{A}} \sum_{j=1}^{m} q^a(x, x^{a_j}) \right) \tag{19.32}$$

$$= \pi(x) dx \left(\sum_{j=1}^{m+1} \int_{\mathcal{X}_0} \pi(x \oplus_j y(1)) dy(1) + \sum_{j=1}^{m} \pi(x^{d_j}) + \sum_{a \in \mathcal{A}} \sum_{j=1}^{m} \pi(x^{a_j}) \right). \tag{19.33}$$

19.2 JUMP DIFFUSION FOR SAMPLING

Now examine the RHS of Eqn. 19.22:

$$\text{RHS} = \sum_{j=1}^{m+1} \int_{\mathcal{X}(m+1)} \pi(y)\,dy q^d(y,x)\delta_{y^{d_j}}(dx) + \sum_{j=1}^{m} \int_{\mathcal{X}(m-1)} \pi(y)\,dy q^b(y,x)\delta_y(dx^{d_j})$$

$$+ \sum_{j=1}^{m} \sum_{a\in\mathcal{A}} \int_{\mathcal{X}(m)} \pi(y)\,dy q^a(y,x)\delta_{y^{a_j}}(dx) \qquad (19.34)$$

$$= \sum_{j=1}^{m+1} \int_{\mathcal{X}_0} \pi(x\oplus_j y(1)) q^d(x\oplus_j y(1),x)\,dy(1)\,dx$$

$$+ \sum_{j=1}^{m} \pi(x^{d_j}) q^b(x^{d_j},x)\,dx + \sum_{j=1}^{m}\sum_{a\in\mathcal{A}} \pi(x^{a_j}) q^a(x^{a_j},x)\,dx \qquad (19.35)$$

$$= \sum_{j=1}^{m+1} \int_{\mathcal{X}_0} \pi(x\oplus_j y(1))\pi(x)\,dx\,dy(1) + \sum_{j=1}^{m} \pi(x)\pi(x^{d_j})\,dx + \sum_{j=1}^{m}\sum_{a\in\mathcal{A}} \pi(x^{a_j})\pi(x)\,dx.$$

Comparing with Eqn. 19.33 shows detailed balance. □

The computational algorithm is as follows.

Algorithm 19.3 *Set the jump number $n = 1$, simulation time $t_0 = 0$, and $X(0) = 0$. Construct a jump–diffusion process $X_t, t \geq 0$ jumping on random times $t_1, t_2, \ldots$ according to the following:*

1. *For $t \geq t_{n-1}$, X_t follows the stochastic integral Eqns. 19.11 and 19.12 in the subspace determined by $X_{t_{n-1}}$ drifting as a diffusion according to Eqns. 19.11 and 19.12 associated with the mth space under the drift vectors A_m, B_m.*

2. *Generate a series of exponential random variables $u_1, u_2, \ldots$ with mean 1, determining the nth jump time determined by the flow and the jump parameters according to*

$$t_n = \inf\{\tau : \int_{t_{n-1}}^{\tau} \int_{\mathcal{X}(\mathcal{J}(X_t))} q(X_t, dy) \geq u_n\}, \qquad (19.36)$$

with X_t following the stochastic differential Eqns. 19.11 and 19.12, in $[t_{n-1}, t_n)$.

3. *At random time t_n, jump from X_{t_n} to an element in $\mathcal{X}(\mathcal{J}^1(X_{t_n}))$ according to the transition probability*

$$Q(X_{t_n}, dy) = \frac{q(X_{t_n}, dy)}{\int_{\mathcal{X}(\mathcal{J}(X_{t_n}))} q(X_{t_n}, dy)}, \qquad y \in \mathcal{X}(\mathcal{J}^1(X_{t_n})). \qquad (19.37)$$

4. *$n \leftarrow n+1$ and return to 1.*

19.2.4 Jumps via Metropolis–Hastings Acceptance/Rejection

This section follows heavily the work of Lanterman [464]. To avoid the complexity of computing the integral in Eqn. 19.27 we shall examine an alternative procedure based on the acceptance and rejection of proposals. The principle contribution will be to choose a jumping scheme analogous to the Metropolis–Hastings sampling algorithm [465–467]. At exponentially distributed times, with *fixed* mean, a candidate x_{prop} is drawn from a *proposal density* $r(x_{\text{old}}, x_{\text{prop}})$; the *acceptance probability* is then computed:

$$\alpha(x_{\text{old}}, x_{\text{prop}}) = \min\left\{\frac{\pi(x_{\text{prop}})r(x_{\text{prop}}, x_{\text{old}})}{\pi(x_{\text{old}})r(x_{\text{old}}, x_{\text{prop}})}, 1\right\}. \tag{19.38}$$

The proposal is accepted with probability $\alpha(x_{\text{old}}, x_{\text{prop}})$ and rejected with probability $1 - \alpha(x_{\text{old}}, x_{\text{prop}})$. A wide variety of proposal densities may be used; of course, $r(x_{\text{old}}, x_{\text{prop}}) > 0$ for $x_{\text{prop}} \in \mathcal{J}^1(x_{\text{old}})$ and $r(x_{\text{old}}, x_{\text{prop}}) = 0$ for $x_{\text{prop}} \notin \mathcal{J}^1(x_{\text{old}})$.

Special cases of the general Metropolis–Hastings jump–diffusion scheme have appeared previously in the literature. If $r(x_{\text{old}}, x_{\text{prop}}) = r(x_{\text{prop}}, x_{\text{old}})$, we have $\alpha(x_{\text{old}}, x_{\text{prop}}) = \min\{\pi(x_{\text{prop}})/\pi(x_{\text{old}}), 1\}$, which corresponds to the traditional Metropolis algorithm [468] in which proposals are drawn from the prior and accepted with the likelihood. When the prior distribution is highly informative this is effective. We shall explore this in the subsequent section. For target detection with a uniform prior on positions and orientations, the prior is not informative; it provides little help in locating new targets or determining target orientation. The approach will be to propose a *birth, death,* or *identity–change* move according to the prior on the number of targets, $p(m)$, and draw the proposal from the posterior density over the space which can be reached via the chosen move type. By sampling from the posterior, the algorithm takes on a "Gibbs within Metropolis" character.

Let $m = \#(x)$ and $\xi = p(m+1) + p(m-1)1_{>0}(m) + p(m)1_{>0}(m)$, where $1_{>0}(m) = 1$ if $m > 0$ and 0 otherwise. Define $r(\cdot, \cdot)$ according to $r(x, y) = 0$ for $y \notin \mathcal{J}^1(x)$, and

$$r(x, x \oplus_j y(1)) = \frac{p(m+1)}{(m+1)\xi} \frac{e^{H(x \oplus_j y(1))}}{\int_{\mathcal{X}_0} e^{H(x \oplus_j y'(1))} dy'(1)}, \tag{19.39}$$

$$r(x, J_j^a x) = \frac{p(m)}{\xi} \frac{e^{H(J_j^a x)}}{\sum_{a' \in \mathcal{A}} e^{H(J_j^{a'} x)}}, \tag{19.40}$$

$$r(x, J_j^d x) = \frac{p(m-1)}{\xi} \frac{e^{H(J_j^d x)}}{\sum_{j'=1}^m e^{H(J_{j'}^d x)}}. \tag{19.41}$$

Algorithm 19.4 *On each jump trial, this choice of $r(\cdot, \cdot)$ corresponds to the following algorithm:*

1. *Draw one of three possible jump choices from the set $\{birth, death, identity - change\}$ according to the distribution $\{p(m+1)/\xi, p(m-1)1_{>0}(m)/\xi, p(m)1_{>0}(m)/\xi\}$.*

2. *If birth, then*
   (a) *Draw $j \in 1...m+1$ uniformly, and draw $y(1)$ from the density*

$$\frac{e^{H(x_{\text{old}} \oplus_j y(1))}}{\int_{\mathcal{X}_0} e^{H(x_{\text{old}} \oplus_j y'(1))} dy'(1)}, \tag{19.42}$$

and accept $x_{\text{prop}} = x_{\text{old}} \oplus_j y(1)$, with probability

$$\min\left\{\frac{(m+1)\int_{\mathcal{X}_0} e^{H(x_{\text{old}} \oplus_j y'(1))} dy'(1)}{\sum_{j'=1}^{m+1} e^{H(J_{j'}^d x_{\text{prop}})}}, 1\right\}. \tag{19.43}$$

3. *Else if identity–change, then*
   (a) Draw $j \in 1...m$ uniformly, and draw identity a from the density

$$\frac{e^{H(J_j^a x_{\text{old}})}}{\sum_{a' \in \mathcal{A}} e^{H(J_j^{a'} x_{\text{old}})}}, \quad (19.44)$$

and accept $x_{\text{prop}} = J_j^a x_{\text{old}}$ unconditionally.

4. *Else if death, then*
   (a) Select $x_{\text{prop}} = J_j^d x_{\text{old}}$ with probability

$$\frac{e^{H(J_j^d x_{\text{old}})}}{\sum_{j'=1}^m e^{H(J_{j'}^d x_{\text{old}})}}, \quad (19.45)$$

and accept x_{prop} with probability

$$\min \left\{ \frac{\sum_{j=1}^m e^{H(J_j^d x_{\text{old}})}}{(m+1) \int_{\mathcal{X}_0} e^{H(x_{\text{prop}} \oplus_j y'(1))} \, dy'(1)}, 1 \right\}. \quad (19.46)$$

19.3 Experimental Results for FLIR and LADAR

19.3.1 Detection and Removal of Objects

To detect ground vehicles, the jump process in the jump diffusion solves the generalized hypothesis test for detection by adding candidates according to their posterior probability. Target detection ideally requires exhaustive evaluation of the posterior over the continuum of $\mathbb{R}^2$, corresponding to the increase in dimension of translation in the plane. The jump parameters associated with a birth of a new target, from $x \mapsto (x \oplus y(1)), y(1) \in \mathbb{R}^2 \times \mathcal{T}$, requires the computation

$$\frac{\pi(x \oplus y(1))}{\int_{\mathbb{R}^2 \times \mathcal{T}} \pi(x \oplus y(1)) \, dy(1)}, \quad y(1) \in \mathbb{R}^2 \times \mathcal{T}. \quad (19.47)$$

This computation is performed by placing a lattice on the position-orientation space. Then $\pi(x \oplus y(1)), y(1) \in \mathcal{L}$ is computed over the lattice, with the jump process adding new detected targets at those locations according to probabilistic selection under the distribution according to Eqn. 19.37.

At times, the algorithm proposes, but then rejects its choices via death moves that remove hypothesized targets. To illustrate, columns 1 and 2 of Figure 19.4 show the death process for removing objects depicting the state of the algorithm beginning with an initial condition in which there are three extra targets. The vehicle in the front represents the most substantial mismatch to the collected data. It is the first target removed by the death process in iteration 2. Further jumps remove the extraneous vehicles in the back.

19.3.2 Identification

Figure 19.4 focuses on the identification component of the algorithm. For a binary alphabet, $\mathcal{A} = \{M60, T62\}$, identification corresponds to the transformation of the scene changing the jth

Figure 19.4 Columns 1 and 2 Death Process: Panel 1 shows an initial condition with extraneous target hypotheses; 2, 3, and 4 show the state of the jump diffusion process at iterations 2, 41, and 65, respectively. Columns 3 and 4 Identification Process: Column 3 shows the match of the M60 to a data from a scene containing an M60 (top) and the match of the T62 to an M60 scene (bottom). Column 4 shows the log-likelihood of the data on the pixel array for the two matches. Brightness means higher relative log-probability. Results taken from Lanterman.

tank in the scene to type $a \in \{M60, T62\}$, requiring the evaluations

$$\frac{\pi(x^{a_j})}{\pi(J_j^{M60}x) + \pi(J_j^{T62}x)}, \quad a \in \{M60, T62\}. \tag{19.48}$$

This solves the generalized hypothesis test by choosing according to the relative probabilities. Shown in column 3 of Figure 19.4 is the M60 (top) and T62 (bottom) superimposed over the data; column 4 shows the difference in log-posterior probabilities (plotted in gray scale). Bright pixels correspond to high evidence of object classification. The relative probability corresponds to an exponentiation of the integral of the brightness pictures. The top choice M60 is far more favorable because of the bright log-probability values.

19.3.3 Pose and Identification

Solving both the pose and identity problems essentially relies on the closeness of the synthesized objects to the actual measured data. Examine airplanes in $\mathbb{R}^3$, with the orientation space of **SO**(3) and translations in $\mathbb{R}^3$ according to $(O, p) : x \mapsto Ox + p$ with

$$(O, p) : \begin{pmatrix} x_1 \\ x_2 \\ x_3 \end{pmatrix} \mapsto \begin{pmatrix} o_{11} & o_{21} & o_{31} \\ o_{12} & o_{22} & o_{32} \\ o_{13} & o_{23} & o_{33} \end{pmatrix} \begin{pmatrix} x_1 \\ x_2 \\ x_3 \end{pmatrix} + \begin{pmatrix} p_1 \\ p_2 \\ p_3 \end{pmatrix}, \tag{19.49}$$

with orthonormal columns for the O-matrix. The recognition of pose and identity is determined by the closeness of the object to the measured data. Consider optical imagery assuming orthogonal projection. The optical imaging measurements are Gaussian random fields whose mean fields are taken to be perspective projection onto the camera. This is depicted in the top row of Figure 19.5 showing a set of imagery of an airplane at different poses. The middle row shows a sequence of two renderings (panels 5 and 7) corresponding to estimated poses as a function of simulation time (left to right). The associated data driving the rotations are shown in panels 6 and 8 via difference images between the actual optical measurement data on the imaging lattice generated from the

19.3 EXPERIMENTAL RESULTS FOR FLIR AND LADAR

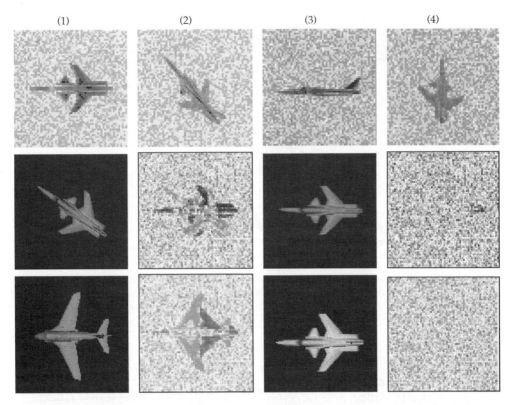

Figure 19.5 Top row shows examples of optical imagery for the airplane. Middle row shows two poses of the target space; panels 5 and 7 show renderings of the target with panels 6 and 8 showing the difference between the optical data synthesized according to the Gaussian model from the target at its estimated pose, subtracted from the true measured data. Bottom row shows identification showing two different identifications. Panels 9 and 11 show renderings of the two different target types; panels 10 and 12 show the difference between the optical data generated from the proposed target at its estimated pose, subtracted from the true measured data. The bottom right panel 12 shows the correctly estimated identity and difference data.

target at its estimated pose subtracted from the true measured data. The rightmost panel shows the most correct pose estimate. Notice the difference image is small and the relative probability preference is proportional to the exponential of the squared differences. Pose is estimated by computing stochastic gradients that follow these probability differences.

For identification of airframes the algorithm jumps between target templates carrying the pose and position parameters with it. The bottom row of Figure 19.5 shows two different target types carrying its pose parameters as the algorithm tries the various target types. The target identity shown in panel 11 is the correct one. Panels 9 and 11 show renderings of the two different target types. Panels 10 and 12 show the difference between the optical data generated from the proposed target at its estimated pose, subtracted from the true measured data. The target identity shown in panel 11 is the correct one. Notice how panel 12 shows difference data, which is small. The identification part of the algorithm checks all the target types. In our experiment, there are three types, $\mathcal{A} = \{1, 2, 3\}$, and the algorithm computes the relative probability of the proposed configuration $x \to x^{a_j}, a \in \{1, 2, 3\}$ choosing according to the probability

$$\frac{\pi(x^{a_j})}{\sum_{a=1,2,3} \pi(x^{a_j})}, \quad a \in \{1, 2, 3\}. \tag{19.50}$$

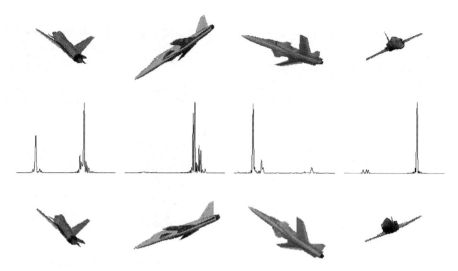

Figure 19.6 Top row shows the target rendered at a sequence of four flight times, time increasing from left to right. Middle row shows the HRR data associated with the target imaged at these orientations. Bottom row shows the estimates of the pose generated at the four flight times. Results taken from O'Sullivan [300].

19.3.4 Identification and recognition via High Resolution Radar (HRR)

Figure 19.6 shows results of target orientation estimation using HRR data. The *HRR* models [308, 475] for target recognition of airframes assume illumination by an S-band radar with a linear FM chirp for the transmitted signal with a center frequency of 3.2 GHz and a bandwidth of 320 MHz. For performing the synthesis in the data likelihood function, the radar signature prediction tool XPATCH is used to generate a library of range profiles. The data generation and imaging model is described in Section 13.3.5 of Chapter 13; the likelihood model is specified via the likelihood Eqn. 13.41. The top row shows the target rendered at a sequence of four flight times, time increasing from left to right. Shown in the middle row are the actual HRR data associated with the target imaged at these orientations. The bottom row shows the estimated pose generated by the algorithm as it diffuses through orientation space searching for the target profile consistent with the HRR data.

19.3.5 The Dynamics of Pose Estimation via the Jump–Diffusion Process

Figures 19.7 and 19.8 show results for FLIR imagery for ground-based targets. Panel 1 of Figure 19.7 shows the data modeled to accommodate the radiant intensity of the scene with superimposed CCD array noise exhibiting photoconversion noise and dead (black) and saturated (white) pixels. Panels 2–6 of Figure 19.7 show successive stages of the jump–diffusion process converging towards the correct estimate of the number of targets and their positions and orientations. The hypothesized targets are manifest as white wireframes superimposed over a noise-free rendering of the true configuration. By iteration 55 (panel 7 of Figure 19.7), the algorithm has discovered the number, positions, and orientations of all of the targets.

19.3 EXPERIMENTAL RESULTS FOR FLIR AND LADAR

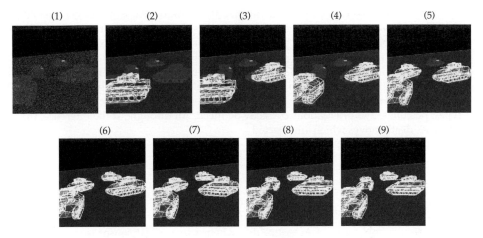

Figure 19.7 FLIR: Panel 1 shows the initial data from a configuration of M2 APC's observed by the FLIR imager. Successive panels 2–9 show a sequence of states of the jump–diffusion process at iterations 1, 10, 12, 25, 27, 34, 55, and 75. Results taken from Lanterman [32].

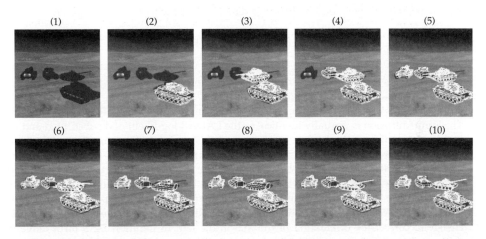

Figure 19.8 FLIR: Panel 1 shows the measured scene for the FLIR imager. Panels 2–10 show iterations 1,3,24,32,34,68,87,88, and 117 of the jump–diffusion process. Results taken from Lanterman [32].

These experiments suggest that operating in the configuration space is exceptionally robust, even with the level of noise seen in panel 2. Between iterations 1 and 10, the algorithm births an object. Due to perspective projection, targets that are closer to the detector appear larger. In iteration 25 (panel 5) the algorithm births a third target, and continues to discover all of the targets.

Interesting states of the sample path of a second FLIR experiment are shown in Figure 19.8. Since the process was started with an empty configuration, the algorithm tries a birth on the first iteration. The M60 on the right appears larger than the other targets since it is closer to the sensor. The algorithm finds it first since it can "explain" the largest amount of data pixels with it. In iteration 3, it mistakes the T62 for an M60. It does this since it has not yet found the adjacent M2 and is trying to explain some of the M2's pixels using the barrel of the M60. This demonstrates the importance of moves that allow changes of type. In iteration 24, the algorithm finds the M2, but the hypothesis is facing the wrong direction. While the diffusions may refine orientation estimates, they are impractical for making large orientation changes, suggesting the necessity of a jump move for making such drastic changes in orientation. Notice the diffusions have found the correct placement of the M60. The remaining M60 is found in iteration 32. The algorithm continues to propose birth moves that are rejected since the data does not support a fifth target.

In iteration 34, the barrel of the incorrectly-guessed M60 is no longer needed to explain the M2 pixels, so a change in identity move swings the hypothesis around, but still incorrectly supposes it to be an M60. It is flipped back the other direction and incorrectly supposed to be an M2 in iteration 68.

Between iterations 68 and 87, the diffusions pull the incorrectly-hypothesized M2 closer to the correct position, so a change in identity move in iteration 88 correctly changes its type to a T62. The correct orientation of the true M2 is found by iteration 117 with the configuration of all of the types, positions, and orientations correctly deduced. All throughout the inference, the algorithm proposes death moves which are rejected.

19.3.6 LADAR Recognition

Similar results hold for the LADAR setting where a laser radar sensor is substituted in place of the FLIR model. The LADAR observes the scene through the effects of *obscuration* and *perspective projection*, in which a point (x, y, z) in 3D space is projected onto the 2D detector according to $(x, y, z) \mapsto (x/z, y/z)$ and discretized into a lattice of points. Following Section 13.3.3, Chapter 13, Eqn. 13.36, the log-likelihood of the range data I^D gives the true range image TI:

$$H_m(I^D) = \sum_i \log\left([1 - Pr_A(y_i)] e^{-1/2\sigma^2 (y_i)(I^D(y_i) + TI(x(m))(y_i))^2} + \frac{Pr_A(y_i)}{\Delta R} \right). \qquad (19.51)$$

The parameters of the model are described in Section 13.3.3, Chapter 13. The idealized LADAR range mean-field is shown in panel 4 of Figure 19.3.

Figure 19.9 shows results for the LADAR similar to that depicted for the FLIR log-likelihood. The jump–diffusion is virtually identical to the LADAR log-likelihood substituted and the Gibbs within Metropolis approach taken for the jump diffusion. Interesting snapshots of the sample path of the algorithm are depicted in the panels. The current hypothesis is shown as a white outline. The algorithm first finds the M2 on the right. Since it is closer to the detector, it takes up more pixels, and the algorithm chooses it since it can "explain" a large portion of the data. By iteration 11 a birth move has discovered the M2 on the left, with the diffusions pulling the M2 on the right to the correct position. In iteration 12, a change in identity move makes a large-scale orientation change to correct the orientation of the M2 on the left. In iteration 24, a birth move finds the

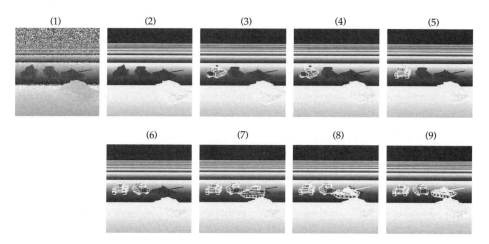

Figure 19.9 Results from LADAR: Panel 1 shows the initial data from a configuration of vehicles observed by the LADAR imager. Panels 2–9 show iterations 1, 3, 11, 12, 24, 32, 38, and 130 of the jump-diffusion process for the LADAR range data. Results taken from Lanterman [32].

19.3 EXPERIMENTAL RESULTS FOR FLIR AND LADAR

M2; also note that the diffusions have refined the position of the leftmost M2. Iteration 32 shows the algorithm birthing an M2 (facing the wrong direction) over the T62. The algorithm changes its mind and switches this incorrect M2 to a T62 (now facing the correct direction) via a change in identity move by iteration 130, and the diffusions have refined the pose. Notice that, in this preliminary experiment, the algorithm incorrectly estimates the orientation of the M2.

19.4 Powerful Prior Dynamics for Airplane Tracking

For airplanes, the basic parameter space is $\mathcal{X}_0 = \mathbb{R}^3 \times \mathbf{SO}(3)$. Our examples employ flight paths which were generated using the flight simulator software. The tracking data results from a narrowband tracking array and a high resolution imaging sensor. The array geometry corresponds to a 64-element cross-array of isotropic sensors located at half-wavelength spacing. The high resolution imaging data consists of Gaussian random fields on a 64×64 imaging lattice. The mean is the projection of the rendered object.

For single objects, the possible jump transformations through parameter space are either addition of a track segment $y \in \mathcal{X}_0$, or deletion of a track segment, or changing the target type. Shown in Figure 19.10 shows the evolution of the random sampling algorithm for estimating the path of a single target. The top row shows the gray track representing the true airplane path

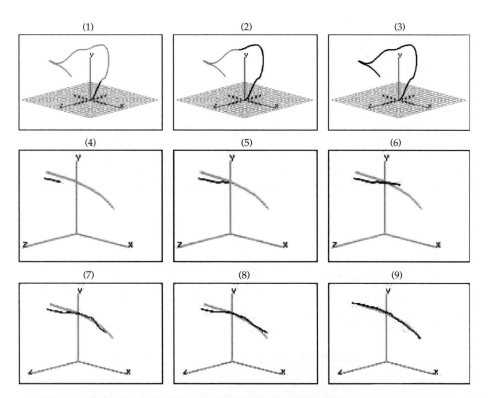

Figure 19.10 Top row: Panel 1 shows the actual track drawn in gray with the mesh representing the ground supporting the observation system in the inertial frame of reference. Panels 2 and 3 display intermediate results from the single-track estimation with the estimates drawn in black. Rows 2 and 3: Sequence of jump moves adding segments to the estimated state from left to right with no diffusion. Bottom two rows: Sequence of panels showing diffusion towards the true track mean.

consisting of 1500 track segments. The estimated track is shown overlapping in black at three different times during the estimation.

The jump diffusion for m airplane tracks has $n(m)$ segments of the diffusion flowing through the orientations and positions $\mathcal{X}_0^{n(m)} = T^{3n(m)} \times \mathbb{R}^{3n(m)}$. The first $3n(m)$ components of the state vector flow through $\mathbb{R}^{3n(m)}$; the last flow through $T^{3n(m)}$ according to $X_t = [X_t^{(1)}, X_t^{(2)}]$, $X_t^{(1)} \in \mathbb{R}^{3n(m)}$, $X_t^{(2)} \in T^{3n(m)}$. The algorithm and its properties are essentially the same as in Theorem 19.2.

19.4.1 The Euler-Equations Inducing the Prior on Airplane Dynamics

Airplane dynamics are straightforwardly expressed using the velocities (v_1, v_2, v_3) projected along the body-frame axes, and the angular velocities (q_1, q_2, q_3). Let $O(\cdot) : [t_0, t] \to \mathbf{SO}(3)$ represent the rotating coordinate system in the body frame satisfying the differential equations

$$\dot{O}_t = O_t Q_t, \quad O_0 = \mathrm{id}, \quad \text{where } Q_t = \begin{pmatrix} 0 & q_{1t} & q_{2t} \\ -q_{1t} & 0 & q_{3t} \\ -q_{2t} & -q_{3t} & 0 \end{pmatrix}, \tag{19.52}$$

where id is the 3×3 identity matrix. The matrix Q_t determines the direction of the flow of $O_t \in \mathbf{SO}(3)$ through the tangent space of the manifold $\mathbf{SO}(3)$. The linear momentum equation, written in body-frame coordinates, becomes $O_t f_t = d/dt \left(O_t v_t \right)$ implying the familiar equation

$$f_t = \dot{v}_t + Q_t v_t. \tag{19.53}$$

The positions in inertial coordinates are related to the velocity in the body reference frame according to $p_t = \int_{t_0}^{t} O_s v_s \, ds + p_{t_0}$.

To induce the prior, assume the forcing function f is a Gaussian white random process; this induces a Gauss–Markov process on the velocities, conditioned on the sequence of Euler angles. Assume the angular and linear velocities are piecewise constant over the time increments $v_n = p_{n+1} - p_n$, $O_{n+1} = O_n e^{Q_n}$, where $Q_n = \begin{pmatrix} 0 & q_{1n} & q_{2n} \\ -q_{1n} & 0 & q_{3n} \\ -q_{2n} & -q_{3n} & 0 \end{pmatrix}$. The discretized equations become $f_n = v_{n+1} - (\mathrm{id} - Q_n) v_n$. Choosing the Gaussian force process to have variance Σ, the velocities are first-order Markov conditioned on the rotations, and the positions are second-order Markov:

$$p(v_n | O_n, O_{n-1}, v_{n-1}) = \det{}^{-1/2}(2\pi \Sigma) e^{-(v_n - (\mathrm{id} - Q_{n-1}) v_{n-1})^* \Sigma^{-1} (v_n - (\mathrm{id} - Q_{n-1}) v_{n-1})}. \tag{19.54}$$

We characterize *rotational motion* through their Euler angle representation $O_1(\phi_1), \ldots, O_n(\phi_n)$ modeling the n Euler angle triples $\phi_1, \ldots, \phi_n$ as a Von–Mises Markov density on the three torus (see [476], for example):

$$p(\phi_n | \phi_{n-1}) = \frac{e^{\sum_{i=1}^{3} \kappa_i \cos(\phi_{i,n} - \phi_{i,n-1})}}{Z(1)}. \tag{19.55}$$

The jump operators for tracks correspond to deletion or removal of the jth track, and addition and deletion of segments to existing tracks with the basic parameter space $\mathcal{X}_0 = T^3 \times \mathbb{R}^3$. For the jump process, let $x(m) \oplus_j y(1)$ represent an $m+1$ plane track formed by adding $y(1) \in \mathcal{X}_0$ to $x(m)$ at the jth location in the list, and $x(m) \oplus_j y$ signify the addition of a segment to the jth track of $x(m), y \in \mathcal{X}_0$. Let $x^{d_j}(m)$ denote the deletion of the jth track in the list $x(m) \in \mathcal{X}(m)$. Let $x^{a_j}(m)$ represent changing the identity of the jth track in $x(m)$ to a.

19.4 POWERFUL PRIOR DYNAMICS FOR AIRPLANE TRACKING

As we have noted in the previous Section 19.2.4, various examples of the general Metropolis–Hastings jump-diffusion scheme can be used depending upon the problem. For selecting model order (object or track number), if an informative prior that is easy to sample from is used, it is beneficial to choose

$$r(x_{\text{old}}, x_{\text{prop}}) = \frac{\pi(\#(x_{\text{prop}}), x_{\text{prop}})}{\int_{\mathcal{J}^1(x_{\text{old}})} \pi(\#(x'), x') \, dx'} \tag{19.56}$$

for $x_{\text{prop}} \in \mathcal{J}^1(x_{\text{old}})$, which corresponds to drawing candidates from the prior and accepting and rejecting them based on the likelihood. This is the approach of Corollary 2 of Theorems 1 and 2 of [137]. This approach is effective in the ground-to-air tracking that we study in this section since drawing from the prior on motion dynamics is fast and effective. For target detection with a uniform prior on positions and orientations, the prior is not informative; it provides little help in locating new targets or determining target orientation.

The jump parameters become

$$q(x(m), dy(m+1)) = \sum_{j=1}^{m+1} q_t^b(x(m), y(m+1)) \delta_{x(m)}(dy^{d_{t_j}}(m+1)) dy(1), \tag{19.57}$$

$$q(x(m), dy(m)) = \sum_{j=1}^{m} q_s^b(x(m), y(m)) \delta_{x(m)}(dy^{d_{s_j}}(m)) dy + \sum_{j=1}^{m} q_s^d(x(m), y(m)) \delta_{x^{d_{s_j}}(m)}(dy(m))$$

$$+ \sum_{j=1}^{m} \sum_{a \in \mathcal{A}} q^a(x(m), y(m)) \delta_{x(m)}(dy^{a_j}(m)), \tag{19.58}$$

$$q(x(m), dy(m-1)) = \sum_{j=1}^{m} q_t^d(x(m), y(m-1)) \delta_{x^{d_{t_j}}(m)}(dy(m-1)). \tag{19.59}$$

Corollary 19.5 (Metropolis jump parameters) *With the parameters of the jump measures chosen as follows, then the detailed balance conditions for the jump process of Theorem 19.2 are satisfied:*

$$q_t^b(x(m), x(m) \oplus_j y(1)) = \frac{1}{5(m+1)} e^{-[L(x(m) \oplus_j y(1)) - L(x(m))]_+} \pi_{\text{prior}}(y(1)), \tag{19.60}$$

$$q_s^b(x(m), x(m) \oplus_j y) = \frac{1}{5m} e^{-[L(x(m) \oplus_j y) - L(x(m))]_+} \pi_{\text{prior}}(y(1)), \tag{19.61}$$

$$q_t^d(x(m), x^{d_{t_j}}(m)) = \frac{1}{5m} \frac{e^{-[L(x^{d_{t_j}}(m)) - L(x(m))]_+}}{Z(1)}, \tag{19.62}$$

$$q_s^d(x(m), x^{d_{s_j}}(m)) = \frac{1}{5m} \frac{e^{-[L(x^{d_{s_j}}(m)) - L(x(m))]_+}}{Z(1)}, \tag{19.63}$$

$$q^a(x(m), x^{a_j}(m)) = \frac{1}{5m} e^{-[L(x^{a_j}(m)) - L(x(m))]_+}. \tag{19.64}$$

The proof follows the Metropolis/Hastings proof.

Figure 19.10 shows magnified views of sections of tracks being estimated by the jump-diffusion algorithm. The series shows successive guesses of the jump process which, continually attempt to add and delete new track segments. On each addition, the new segment is drawn from the prior on flight dynamics, which is parameterized by the track up to that point in time. Hence, the jump algorithm tends to infer track segments that are close to the true track.

The bottom two rows of Figure 19.10 show the diffusion only partly with no addition of track segments. Notice how the state is brought into alignment following the addition of new segments.

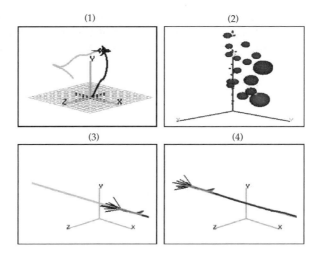

Figure 19.11 Top row: Panel 1 shows the track estimates drawn overlapping in black at several stages of the algorithm. Panel 2 depicts the posterior probability of a target at that position proportional to the size of the sphere drawn. Panels 3 and 4 show candidates from the prior distribution for target path estimation with the high prior probability candidates forming a cone at the end of the track for the algorithm to sample from.

19.4.2 Detection of Airframes

To detect the existence of objects, the jump process queries the Euclidean space based on the tracking data. The jump part of the jump diffusion solves the generalized hypothesis test for detection by adding candidates according to their posterior probability. Target detection ideally requires exhaustive evaluation of the posterior density over the possible positions. The computation is performed by placing a lattice on position space; during initial detection (discovery) there is no orientation. Then $\pi(x \oplus y(1)), y(1) \in \mathbb{R}^3$ is computed on the lattice sites, with the jump process adding new detected targets at those orientations according to probabilistic selection under the distribution according to

$$\frac{\pi(x \oplus y(1))}{\int_{\mathbb{R}^3} \pi(x \oplus y(1))\, dy(1)}. \tag{19.65}$$

The right panel in the top row of Figure 19.11 shows the conditional log-likelihood profile in the azimuth-elevation space of the tracking data vector conditioned on the current estimated configuration as a function of estimation (simulation) time. The algorithm selects points with high probability, and if the data supports target birth, a track is initiated from that point. Shown in the rightmost column are the most likely candidates $y(1)$ on the lattice that have high posterior probability; the probability is depicted via the size of the spherical shells.

19.4.3 Pruning via the Prior distribution

Inducing the prior distributions using the dynamical equations of motion is important in the algorithm. The dynamical equations provide robustness to the estimation procedure and provide an organized strategy for search and pruning of the high-dimensional state spaces. To illustrate

19.4 POWERFUL PRIOR DYNAMICS FOR AIRPLANE TRACKING

this, examine the generalized hypothesis test that the algorithm solves as it generates its family of solutions. Every proposal of a new target during track growth corresponds to the move $x \rightarrow x \oplus y(1)$, $y(1) \in \mathbb{R}^3 \times \mathbf{SO}(3)$. This requires sampling from the distribution according to the calculation of

$$\frac{\pi(x \oplus y(1))}{\int_{\mathbb{R}^3 \times \mathbf{SO}(3)} \pi(x \oplus y(1)) \, dy}. \tag{19.66}$$

This calculation is performed by pruning the state space according to the dynamical equations which determine the distribution.

The bottom row of Figure 19.11 shows the *cone of support* depicting the importance of the dynamics-based prior. To show the support of the prior distribution in phase space, panels 3 and 4 show the 10 highest prior probability candidates placed at the track end for the algorithm to choose from. Panels 3 and 4 correspond to a different time during the inference. The two panels show that if the track vector is close to the true track, the cone of candidates predicts well the future position.

19.5 Deformable Organelles: Mitochondria and Membranes

Now we turn to the most microscopic level of variability, that evident in the ultra-structural and intracellular features of single cells. For this examine the structural features of single cells as studied via electron microscopy. Panel 1 of Figure 19.12 is a sample micrograph illustrating the exquisite variability manifested by the dark closed mitochondria shapes, surrounded by cytoplasm.

19.5.1 The Parameter Space for Contour Models

For representing biological shapes in electron micrographs, the generators become closed curves as in Example 7.29 of Chapter 6. The transformations are scales and rotations $\begin{pmatrix} u_1 & u_2 \\ -u_2 & u_1 \end{pmatrix} =$

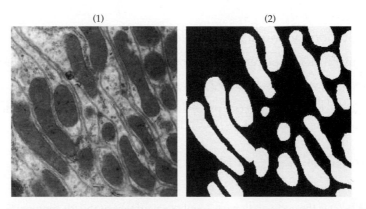

Figure 19.12 Panel 1 shows the micrograph data at 30,000 times magnification with panel 2 showing a hand segmentation into the disjoint partition $\cup_j \mathcal{I}_{g_j}$ and with bounding contours g_j of the two region types, mitochondria (white) and background (black). Data taken from the laboratory of Dr. Jeffrey Saffitz of Washington University.

$\rho \begin{pmatrix} \cos\theta & \sin\theta \\ -\sin\theta & \cos\theta \end{pmatrix}$ acting on the tangent elements of closed circular templates, $g_t, t \in [0,1], g_0 = g_1$. The finite-dimensional transformations of n-similarities arise by assuming piecewise constancy of the transformations over intervals, with the template $g_l^0, l = 0, 1/n, \ldots, 1$ being a polygonal approximation to the circle (see Figure 7.3 of Chapter 6). The n-similarities become the scales and rotations applied to the chords, and adding global translation gives

$$g_{t=\frac{l}{n}} = \begin{pmatrix} u_{00} \\ u_{01} \end{pmatrix} + 2\pi \int_0^{\frac{l}{n}} \begin{pmatrix} u_{1t} & u_{2t} \\ -u_{2t} & u_{1t} \end{pmatrix} \begin{pmatrix} -\sin 2\pi t \\ \cos 2\pi t \end{pmatrix} dt \tag{19.67}$$

$$= \begin{pmatrix} u_{00} \\ u_{01} \end{pmatrix} + \sum_{k=1}^{nl} \begin{pmatrix} u_{1k} & u_{2k} \\ -u_{2k} & u_{1k} \end{pmatrix} \begin{pmatrix} \cos 2\pi k/n - \cos 2\pi(k-1)/n \\ \sin 2\pi k/n - \sin 2\pi(k-1)/n \end{pmatrix}. \tag{19.68}$$

The similarity transformation consists of n-scale/rotations and one global translation. Identifying the transformations with its parametric representation, the parameter space representing closed contours is $2n + 2 - 2$ scale–rotation dimensions and global translation dimensions, since we lose two dimensions for closure. A micrograph of m mitochondria is specified by the scale/rotations and translations, having parameter vectors that are the concatenation of parameters associated with each of the objects:

$$x(m) = (u^{(1)}, u^{(2)}, \ldots u^{(m)}) \in \mathcal{X}(m) = \mathbb{R}^{2nm}, \tag{19.69}$$

where $u^{(i)} = (u_{00}^{(i)}, u_{01}^{(i)}, u_{11}^{(i)}, u_{21}^{(i)}, \ldots)$.

The ideal scene is the set of interiors of the contours $\mathcal{I}_{g_1}, \mathcal{I}_{g_2}, \ldots$ with boundaries given by the contours. This defines a disjoint partition of the background space $\mathbb{Z}^{n^2}$ as the set of simply connected regions $\cup_j \mathcal{I}_{g_j}$. Figure 19.12 is an illustration of the decomposition of the micrograph image into the disjoint partition of simply connected regions. Panel 1 shows the micrograph data with panel 2 showing a hand segmentation into the disjoint partition of connected interiors $\cup_j \mathcal{I}_{g_j}$ of the two region types, mitochondria (white) and background (black).

Gaussian Markov random field models can be used for segmentation of the micrographs; each mitochondria is a conditional Gaussian random field conditioned on the random configuration of the contour.

19.5.2 Stationary Gaussian Contour Model

For the prior stationary processes for the contours are used as in Chapter 9, Example 9.28. Identify the scale–rotation matrices with the 2×1 vectors $u_i = \begin{pmatrix} u_{1i} \\ u_{2i} \end{pmatrix}$. The $u_i, i = 1, \ldots, n$ are modeled as a Gaussian cyclo-stationary, vector process with $2n \times 2n$ block Toeplitz, periodic covariance $K(u^{(j)})$ with 2×2 blocks $K_{k,i}(u^{(j)}) = K_{k-i,0}(u^{(j)})$. The prior on m-contours becomes

$$\pi(x(m) = u^{(1)}, u^{(2)}, \ldots) = \prod_{j=1}^{m} \det^{-\frac{1}{2}} 2\pi K(u^{(j)}) e^{-\frac{1}{2} \sum_{kl} u_k^{(j)*} K(u^{(j)})_{kl}^{-1} u_l^{(j)}}, \tag{19.70}$$

where $K(u^{(j)})$ is the cyclo-stationary Toeplitz covariance of the jth contour. [46]

---

[46] The natural coordinates are the Fourier transform which diagonalizes the covariance into the block independent complex variables $\bar{u}_i = \sum_{k=0}^{n-1} u_k e^{-ji2\pi k/n}$, which are then a realization of an independent Gaussian

Figure 9.1 of Chapter 9 shows examples of eight closed shapes generated from the prior distribution. The mean shape and the covariance spectrum is also represented, demonstrating the typical structure and its range of variation. The covariance was estimated by empirical fitting of the hand-contoured micrographs to generate statistics on the scale–rotation parameters.

The constraint that two objects cannot be superimposed in electron-micrographs is imposed via an intersection penalty with the potential energy with the intersection between adjacent contours $\alpha \sum_{i,j=1}^{m} |\mathcal{I}_{g_i} \cap \mathcal{I}_{g_j}|$. Similar to the ground vehicles example, a Poisson prior on the number of mitochondria can be used, with mean parameter chosen to reflect the micrograph data.

19.5.3 The Electron Micrograph Data Model: Conditional Gaussian Random Fields

We model the micrograph imagery via two compartments, mitochondria interior and cytoplasm exterior. The basic data model builds the electron micrograph imagery as a disjoint partition of Gaussian random fields $I_{g_j}^D, j = 1, \ldots, m$ determined by the random contours $g_1, g_2, \ldots$. Construct the background space $\mathbb{Z}^{n^2}$ as a disjoint partition of simply connected regions $\cup_j \mathcal{I}_{g_j}$, with $\mathcal{I}_{g_j}$ denoting the interior of contour g_j. Model $I_{g_j}^D$ as a Gaussian random field on the subgraph $\mathcal{I}_{g_j} \subset \mathbb{Z}^{n^2}$ associated with the jth mitochondria. Define the cytoplasm image I_{cyt}^D on the complement: I_{cyt}^D on $\mathbb{Z}^{n^2} \setminus \cup_{j=1}^{m} \mathcal{I}_{g_j}$. Then, the measured micrograph images I^D are modeled as realizations of conditionally independent Gibbs random fields, conditioned on the subgraphs specified by the boundaries.

Figure 19.12 shows an illustration of the decomposition of the image into a disjoint partition of simply connected regions. Panel 1 shows the random image I^D; panel 2 shows a partition of the background space into a disjoint partition $\cup_j \mathcal{I}_{g_j}$ depicting the contours in white; panel 3 shows the partition into interiors $\mathcal{I}_g$.

Gaussian Markov random field models can be used for segmentation of the micrographs; each mitochondria is a conditional Gaussian random field conditioned on the random configuration of the contour. Using the asymptotic partition results of Chapter 5 the partition functions of the random contour can be calculated asymptotically and the Bayes posterior required for segmentation can be calculated. Assume the mitochondria are random realizations of the stochastic differential equation

$$L I_i^D = (-\nabla^2 + a) I_i^D = W_i \quad i = (i_1, i_2) \in \mathcal{I}_{g_j}, \tag{19.72}$$

$$\text{with } \nabla^2 I_{i_1,i_2}^D = I_{i_1-1,i_2}^D + I_{i_1+1,i_2}^D + I_{i_1,i_2-1}^D + I_{i_1,i_2+1}^D - 4 I_{i_1,i_2}^D, \tag{19.73}$$

where a is the constant restoring force, with white Gaussian noise variance σ^2. The Laplacian induces nearest-neighbor dependence between pixels on the discrete lattice of points. The three parameters, a, μ and σ, completely specify the model. From Example 8.53 from Chapter 8,

process with 2×2 block covariances

$$\Lambda_i = \begin{pmatrix} \lambda_{11i} & \lambda_{21i}^* \\ \lambda_{21i} & \lambda_{22i} \end{pmatrix} = \sum_{k=0}^{n-1} e^{-j2\pi ik/n} K_{k0}. \tag{19.71}$$

Hermitian symmetry $\bar{u}_{n-i} = \bar{u}_i^*$ implies there are only $2n$ degrees of freedom. The closure constraint on the simple curves ($g_0 = g_1$, see Eqn. 9.52) removes two more degrees of freedom since the highest frequency discrete Fourier transform coefficient satisfies $\bar{u}_{1,n-1} = j\bar{u}_{2,n-1}$, with Hermitian symmetry implying $\bar{u}_{11}^* = j\bar{u}_{21}^*$. Thus the Fourier variables, with closure, require diffusion through $\mathbb{R}^{2n-2}$ for each closed curve.

Table 19.1 Table showing the maximum-likelihood estimates of the parameters μ, noise variance σ^2, and restoring force a estimated from the mitochondria data using the algorithm from Example 5.26 of Chapter 5.

	μ Mean Gray Level	σ^2 Variance of White Noise	a Restoring Force
Mitochondria	58.7	847.3	0.62
Background	130.3	2809.9	0.15

the parameters were estimated using maximum-likelihood to be, for the mitochondria $\mu = 58.7, \sigma^2 = 847, a = 0.62$, and for the background, $\mu = 130, \sigma^2 = 2810, a = 0.15$.

From the asymptotic partition Theorem 5.17 of Chapter 5 (Example 5.26), the spectrum of the 2D Laplacian gives the asymptotic version of the spectrum:

$$\log \det^{\frac{1}{2}} K_{\mathcal{I}_{g_j}} = \frac{|\mathcal{I}_{g_j}|}{4\pi^2} \int_{[-\pi,\pi]^2} \log(2 \sum_{k=1}^{2} (1 - \cos \omega_k) + a) \, d\omega + O(|\partial I(g)|); \tag{19.74}$$

the Gaussian random field density becomes

$$p(I^D|I) = \prod_{j=1}^{m} (2\pi\sigma^2)^{-|\mathcal{I}_{g_j}|/2} \det^{-\frac{1}{2}} K_{\mathcal{I}_{g_j}} e^{-1/2\sigma^2 \|LI_{g_j}^D\|^2} \propto e^{-H_m(I^D)} \quad \text{where} \tag{19.75}$$

$$H_m(I^D) = \sum_{j=1}^{m} \frac{|\mathcal{I}_{g_j}|}{2} \left(\frac{1}{4\pi^2} \int_{[-\pi,\pi]^2} \log 2\pi\sigma^2 (\sum_{k=1}^{2} (2 - 2\cos\omega_k) + a) \, d\omega \right) + \sum_{j=1}^{m} \frac{1}{2\sigma^2} \|LI_{g_j}^D\|^2.$$

Using the asymptotic maximum-likelihood parameter estimation of Chapter 5, Example 5.26 of Theorem 5.17, parameters are estimated from sets of hand-labeled mitochondria and background cytoplasm region types. The parameters are estimated from the micrograph data assuming the Gaussian model with $L = \Delta - a$ having parameters of mean μ, noise power σ^2, and restoring force a. The values estimated from the mitochondria data are shown in Table 19.1.

19.6 Jump–Diffusion for Mitochondria

The posterior potential H_m associated with the subspace $\mathcal{X}(m) = \mathbb{R}^{2nm}$ of m-mitochondria is the superposition of the Gaussian random field potential and the prior on the smoothness of the contours:

$$H_m(x(m)) = \sum_{j=1}^{m} \frac{|\mathcal{I}_{g_j}|}{2} \left(\frac{1}{4\pi^2} \int_{[-\pi,\pi]^2} \log(2\pi\sigma^2 \sum_{k=1}^{2}(2 - 2\cos\omega_k) + a) \, d\omega \right) + \sum_{j=1}^{m} \frac{1}{2\sigma^2} \|LI_{g_j}^D\|^2$$

$$\sum_{j=1}^{m} \left(\log \det^{\frac{1}{2}} 2\pi K(u^{(j)}) + \frac{1}{2} \sum_{kl} u_k^{(j)*} K(u^{(j)})_{kl}^{-1} u_l^{(j)} \right). \tag{19.76}$$

The posterior $\mu(\cdot)$ is similar to the object recognition posterior, exceptitison $\mathcal{X} = \cup_m \mathbb{R}^{2nm}$.

19.6.1 The jump parameters

The jump process has the family of jump moves $\mathcal{J}$ determining which of the jump measures are non-zero, as well as the properties of connectedness and reversibility. We define the graph changes $J \in \mathcal{J}$ to consist of the addition of one object at a time or deletion of one of the existing objects:

$$J_j^b : \mathbb{R}^{2nm} \longrightarrow \mathbb{R}^{2n(m+1)},$$

$$J_j^d : \mathbb{R}^{2nm} \longrightarrow \mathbb{R}^{2n(m-1)},$$

with addition of an object $y(1) \in \mathbb{R}^{2n}$ into the jth location in the list denoted by $x(m) \oplus_j y(1)$, and $x^{d_j}(m) \in \mathbb{R}^{2n(m-1)}$ denoting the jth object removed from the list. There are a total of $j = 1, \ldots, m+1$ birth changes allowed for configurations in $\mathbb{R}^{2nm}$, and m possible death changes, implying that the transition measures from space $\mathbb{R}^{2nm}$ have mass on spaces $\mathbb{R}^{2n(m+1)}, \mathbb{R}^{2n(m-1)}$:

$$q(x(m), dy(m+1)) = \sum_{j=1}^{m+1} q_b(x(m), y(m+1)) \delta_{x(m)}(dy^{d_j}(m+1)) dy(1),$$

$$q(x(m), dy(m-1)) = \sum_{j=1}^{m} q_d(x(m), y(m-1)) \delta_{x^{d_j}(m)}(dy(m-1)). \tag{19.77}$$

The total jump intensity is

$$q(x(m)) = \sum_{j=1}^{m+1} \int_{\mathbb{R}^{2n}} q_b(x(m), x(m) \oplus_j y(1)) \, dy(1) + \sum_{j=1}^{m} q_d(x(m), x^{d_j}(m)). \tag{19.78}$$

The diffusion process satisfies the stochastic differential Eqn. 19.11 with $X_t, t \geq 0$ flowing through all the real subspaces, the mth one $\mathbb{R}^{2nm}$ consisting of m sets of $2n \times 1$ vectors. The drifts are variations of the Gibbs posterior energy with respect to the scale, rotation and translation parameters of each of the m objects. To enforce closure, the stochastic differential equation is performed in the Fourier transform space along the drifts determined by the derivatives $(\partial H_m)/(\partial \bar{u}_{1i}), (\partial H_m)/(\partial \bar{u}_{2i}), i = 1, \ldots, n$, with closure satisfied according to Eqns. 9.52 and 9.53 of Chapter 9. The last two dimensions are constructed from the translation drifts $(\partial H_m(x_0))/(\partial x_0), (\partial H_m(y_0))/(\partial y_0)$.

19.6.2 Computing gradients for the drifts

The drifts are computed as in Chapter 6, Section 8.5 for 1-parameter variations of the deformable models with respect to the translations and rotations. Assume the inside and outside potential models

$$H(g(\gamma)) = \int_{\mathcal{I}_{g(\gamma)}} E_1(x) \, dx + \int_{\mathcal{I} \setminus \mathcal{I}_{g(\gamma)}} E_2(x) \, dx, \tag{19.79}$$

where E_1 is the potential associated with the interior $\mathcal{I}_{g(\gamma)}$ of the contour, and E_2 is the potential in the exterior $\mathcal{I} \setminus \mathcal{I}_{g(\gamma)}$. The gradients are given as in Theorem 8.42. For curves $g_t(\gamma), t \in [0, 1]$

given by

$$g_t(\gamma) = \begin{pmatrix} u_0 \\ u_1 \end{pmatrix} + 2\pi \int_0^t \begin{pmatrix} u_{1l} & -u_{2l} \\ u_{2l} & u_{1l} \end{pmatrix} \begin{pmatrix} -\sin 2\pi l \\ \cos 2\pi l \end{pmatrix} dl \qquad (19.80)$$

with $\gamma \in \{u_{1l}, u_{2l}, x_0, y_0\}$ and continuous potentials E_1, E_2, then

$$\frac{\partial E(g)}{\partial \gamma} = \int_0^1 |J_l(\gamma)|[E_1(g_l) - E_2(g_l)]\, dl, \qquad (19.81)$$

with Jacobian determinants given by

$$|J_l(u_{1t})| = (2\pi)^2 \left(-\sin 2\pi t \frac{\partial g_{yl}}{\partial l} - \cos 2\pi t \frac{\partial g_{xl}}{\partial l} \right) 1_{\geq t}(l), \qquad (19.82)$$

$$|J_l(u_{2t})| = (2\pi)^2 \left(-\cos 2\pi t \frac{\partial g_{yl}}{\partial l} + \sin 2\pi t \frac{\partial g_{xl}}{\partial l} \right) 1_{\geq t}(l), \qquad (19.83)$$

$$|J_l(x_0)| = (2\pi)^2 \frac{\partial g_{yl}}{\partial l}, \qquad |J_l(y_0)| = -(2\pi)^2 \frac{\partial g_{xl}}{\partial l} \qquad (19.84)$$

and the common bottom row of the Jacobian

$$\frac{\partial g_{xl}}{\partial l} = -\sin(2\pi l) u_{1l} - \cos(2\pi l) u_{2l}, \qquad \frac{\partial g_{yl}}{\partial l} = \cos(2\pi l) u_{1l} - \sin(2\pi l) u_{2l}. \qquad (19.85)$$

For computation, there are n unique parameter vectors, $\{u_{1k}, u_{2k}\}_{k=1}^n$ constant over intervals $l \in ((k-1)/n, k/n)$. The continuous curvi-linear integral given by the gradient is computed using the standard discrete approximations of the trapezoid rule.

19.6.3 Jump Diffusion for Mitochondria Detection and Deformation

Figure 19.13 shows results in the electron micrograph data from the jump-diffusion process for segmentation. Panel 1 depicts the underlying mechanism by which the jump algorithm discovers and places new mitochondria candidates. Detection positions for the placement process are shown with the amount of posterior probability depicted via brightness. Shown at every one of the 16 locations depicted are the locations of candidate mitochondria chosen according to the random selection algorithm choosing based on the relative ratios of the locations. The brightness indicates the amount of posterior probability associated with that candidate. This calculation corresponds to computing the posterior probability in configurations according to $(\pi(x \oplus y(1)))/(\sum_{i=1}^{16} \pi(x \oplus y(1)))$, where $\pi(x \oplus y(1))$ has log-probability given by the brightness points shown in Figure 19.13.

Panel 2 depicts the representation of mitochondria resulting from the jump-diffusion algorithm. Gaussian random field models were estimated from the mitochondria and background. The templates are placed in the scene and deformed to maximize the probability that the represented regions in the data are realizations of the appropriate texture model. The vertices of the templates are represented as white squares superimposed on the data.

Table 19.2 quantifies the performance of the pixel-based optimal Bayesian hypothesis testing (left column) compared with the segmentation via the global shape models. The true labels were determined by a manual segmentation. Percentages indicate relative frequency of proper pixel labeling for each model type. Figure 19.14 illustrates the progression of the algorithm through its

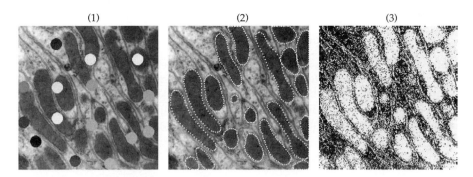

Figure 19.13 Panel 1 shows the jump detections scored via brightness indicating the amount of posterior probability. Panel 2 depicts the representation of mitochondria resulting from the jump-diffusion algorithm. Panel 3 shows a pixel-by-pixel segmentation of the image based on the optimal Bayes hypothesis test for each pixel under the two models.

Table 19.2 Quantifying the performance of the pixel-based optimal Bayesian hypothesis testing (left column) compared with the segmentation via the global shape models. The true labels were determined by segmentation performed manually. Percentages indicate relative frequency of proper pixel labeling for each model type.

	Pixel Based Label (%)	*Global Shape Label(%)*
Mitochondria	88.9	93.2
Background	63.1	93.7
% of Total Pixels Mislabeled	26.2	6.5

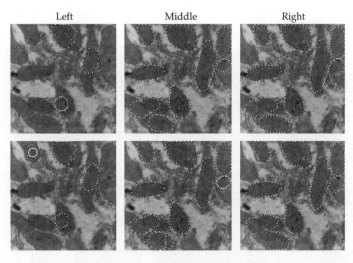

Figure 19.14 Figure showing a birth of a mitochondria (left column), a death of a mitochondria (middle column), and a merge of two mitochondria (right column).

jump moves showing an addition of a mitochondria (left column), a removal of a mitochondria (middle column), and the merging of two mitochondria (right column). The top panel in each column shows the state of the Markov process at the instant before the jump; the bottom panel shows the new configuration after the jump has occurred.

19.6.4 Pseudolikelihood for Deformation

For comparison, one can also examine Besag's pseudolikelihood method for global segmentation. This is a modified Bayes technique that offers an alternative to the purely asymptotic Gaussian random field approach outlined previously. Associate with each region type in the image one of models type θ a MRF field $\{P_{\theta_k}(I_i^D | I_j^D, j \in N_i)\}$, with its neighborhood system $\cup_i N_i$. The asymptotic Bayes likelihood of Eqn. 19.75 in pseudolikelihood is simply the product of the

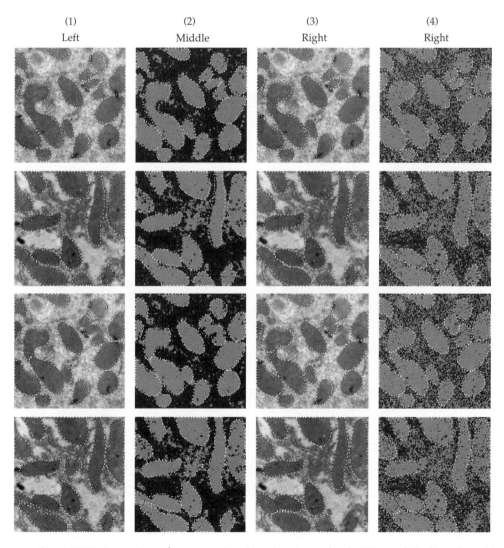

Figure 19.15 Comparison of segmentations based on 4-gray level MRF pseudolikelihood models (left two columns) and the Gaussian asymptotic partition function global Bayes model (right two columns) for two sets of mitochondria (top and bottom rows). Rows 1 and 2 show manual placement of mitochondria templates; rows 3 and 4 show automatic placement by the jump process. Columns 1 and 3 show image data; columns 2 and 4 show pixel-by-pixel segmentation generated by the likelihood ratio test (gray mitochondria, black cytoplasm).

conditional probabilities:

$$\prod_{j=1}^{m} \prod_{i \in \mathbb{Z}^{n^2}} P(I_i^D | I_j^D, j \in N_i). \tag{19.86}$$

The conditional probabilities on the assumed nearest-neighbor are estimated from the set of training micrographs and are based on the local average gray-level feature on 256 gray levels and a 4-gray level texture feature [127, 128]. Estimating the local conditional probabilities for the organelles is straightforward. The Von–Mises estimator for the conditional probabilities are computed by counting the relative frequency of occurrence of $x_i \in \{0, 1, 2, 3\}$ given its neighboring configuration.[47]

Figure 19.15 shows a comparison of segmentations based on the 4-gray level pseudolikelihood model (left column) and global Bayesian model approximating the asymptotic partition functions (right column) for two different sets of mitochondria (top and bottom rows). Rows 1 and 2 show results from manual placement of templates in the image; rows 3 and 4 show automatic placement of templates. Panels 1 and 3 show the vertices of the templates superimposed on the data image; panels 2 and 4 show the vertices superimposed on a pixel-by-pixel segmentation generated by the likelihood ratio test. Pixels that are labeled as mitochondria are colored gray, pixels that are labeled as background are colored black.

Notice how accurately the deformable shapes capture the major mitochondria structures. It also appears, that the extension to the continuum of gray levels along with the asymptotic partition approximation (rather than pseudolikelihood) appears to be slightly more robust at the boundaries of the shapes.

---

[47] We note that the Gaussian potential could also be used with the pseudolikelihood model. However, the Gaussian model affords the opportunity of exploiting the asymptotic partition approximation to the normalizer (see Chapter 5, Theorem 5.17, Eqn. 5.75 on asymptotics of the partition function).

REFERENCES

[1] U. Grenander. A unified approach to pattern analysis. *Adv. Comput.*, 10:175–216, 1970.
[2] U. Grenander. *Pattern Synthesis: Lectures in Pattern Theory*, volume I. Springer-Verlag, New York, 1976.
[3] U. Grenander. *Pattern Analysis: Lectures in Pattern Theory*, volume II. Springer-Verlag, New York, 1978.
[4] U. Grenander. *Regular Structures: Lectures in Pattern Theory*, volume III. Springer-Verlag, New York, 1981.
[5] U. Grenander. *General Pattern Theory*. Oxford University Press, 1994.
[6] C. Shannon. The mathematical theory of communication. *Bell Syst. Tech. J.*, 27:398–403, 1948.
[7] T. Cover and J. Thomas. *Elements of Information Theory*. John Wiley & Sons, New York, 1991.
[8] D. G. Luenberger. *Optimization by Vector Space Methods*. John Wiley and Sons, New York, 1969.
[9] U. Grenander. *Abstract Inference*. John Wiley & Sons, New York, 1981.
[10] H. J. Bierens. *Topics in Advanced Econometrics: Estimation, Testing, and Specification of Cross-Section and Time Series Models*. Cambridge University Press, 1994.
[11] Kjell A. Doksum Peter J. Bickel. *Mathematical statistics: basic ideas and selected topics:* Volume I. Prentice Hall, Englewood Cliffs, New Jersey, 1977.
[12] A. Dempster, N. Laird, and D. Rubin. Maximum likelihood from incomplete data via the em algorithm. *J. R. Stat. Soc.* B, 39(1):1–38, 1977.
[13] I. Csiszar and G. Tusnady. Information geometry and alternating minimization procedures. In *Statistics and Decisions*, pages 205–237, R. Oldenbourg Berlag. Munchen 1984.
[14] B. R. Musicus. *Iterative Algorithms for Optimal Signal Reconstruction and Parameter Identification Given Noisy and Incomplete Data*. M.I.T. Thesis, Cambridge, MA, 1982.
[15] D. J. Thomson. Spectrum estimation and harmonic analysis. *Proc. IEEE*, 70, No. 9:1055–1096, 1982.
[16] B. Friedlander. Lattice methods for spectral estimation. *Proc. IEEE*, 70, No. 9:990–1017, 1982.
[17] R. Kumaresan and D. Tufts. Estimating the angles of arrival of multiple plane waves. *IEEE Trans. Aerospace Electr. Syst*, AES-19, 1983.
[18] M. I. Miller and D. R. Fuhrmann. Maximum likelihood narrow-band direction finding and the EM algorithm. *IEEE Trans Acoust. Speech Signal Proces.*, 38, (9): 560–577, 1990.
[19] E. Becker. *High Resolution NMR: Theory and Chemical Applications*. Academic Press, New York, 1980.
[20] W. P. Aue, E. Bartholdi, and R. R. Ernst. Two-dimensional spectroscopy application to nuclear magnetic resonance. *J. Chem. Phys.*, 64:2229, 1976.
[21] A. Bax. *Two-Dimensional Nuclear Magnetic Resonance in Liquids*. Delft University Press, 1984.
[22] M. I. Miller and A. Greene. Maximum-likelihood estimation for nuclear magnetic resonance spectroscopy. *J. Magn. Reson.*, 83:525–548, 1989.
[23] M. I. Miller, S. C. Chen, D. A. Kuefler, and D. A. D'Avignon. Maximum-likelihood estimation and the em algorithm for 2-d nmr spectroscopy. *J. Magn. Reson.*, Series A 104, 1993.
[24] S. C. Chen, T. J. Schaewe, R. S. Teichman, and M. I. Miller. Parallel algorithms for maximum likelihood nuclear magnetic resonance spectroscopy. *J. Magn. Reson.*, Series A,102:16–23, 1993.
[25] M. I. Miller, D. R. Fuhrmann, J. A. O'Sullivan, and D. L. Snyder. Maximum-likelihood methods for toeplitz covariance estimation and radar imaging. In Simon Haykin, editor, *Advances in Spectrum Estimation*, pages 145–172. Prentice-Hall, Englewood Cliffs, New Jersey, 1990.
[26] P. Moulin, J. A. O'Sullivan, and D. L. Snyder. A method of sieves for multiresolution spectrum estimation and radar imaging. *IEEE Trans. Inform. Theory*, 1992.
[27] M. J. Turmon and M. I. Miller. Maximum-likelihood estimation of complex sinusoids and toeplitz covariances. *IEEE Trans. Acoust., Speech Signal Proces.*, 42(5), 1994.
[28] S. Haykin, V. Kezys, and E. Vertatschitsch. *Maximum Likelihood for Angle-of-Arrival Estimation in Multipath*, Simon Haykin, editor pages 123–144. Prentice-Hall, Englewood cliffs, New Jersey 1990.
[29] S. Kay and L. Marple. Jr. Spectrum analysis- a modern perspective. *Proc. IEEE*, 69, No.11:1380–1418, 1981.
[30] S. Joshi and M. I. Miller. Maximum a Posteriori estimation with Good's roughness for optical sectioning microscopy. *J. Opt. Soc. Am. A*, 10(5):1078–1085, May 1993.
[31] D. L. Snyder, A. M. Hammoud, and R. L. White. Image recovery from data acquired with a charge-coupled-device camera. *J. Opt. Soc. of Am.* A, 10(5):1014–1023, 1993.

[32] A. D. Lanterman, M. I. Miller, and D. L. Snyder. Implementation of jump-diffusion processes for understanding flir scenes. In F.A. Sadjadi, editor, *Automatic Object Recognition V*, vol. 2485, 309–320, SPIE, Orlando, FL, 1995.

[33] S. Joshi. MAP intensity estimation with good's roughness and global shape models for 3D optical sectioning microscopy. Masters Thesis, Department of Electrical Engineering, Sever Institute of Technology, Washington University, St. Louis, MO., 1993.

[34] M. I. Miller, K. B. Larson, J. E. Saffitz, D. L. Snyder, and Jr. L. J. Thomas. Maximum-likelihood estimation applied to electron-microscope autoradiography. *J. Electron Microsc. Tech.*, 2:611–636, 1985.

[35] D. L. Snyder, Jr. L. J. Thomas, and M. M. Ter-Pogossian. A mathematical model for positron emission tomography systems having time-of-flight measurements. *IEEE Trans. Nucl. Sci.*, NS-28:3575–3583, 1981.

[36] L. A. Shepp and Y. Vardi. Maximum-likelihood reconstruction for emission tomography. *IEEE Trans. Med. Imaging*, MI-1:113–121, 1982.

[37] L. A. Shepp, Y. Vardi, J. B. Ra, S. K. Hilal, and Z. H. Cho. Maximum-likelihood with real data. *IEEE Trans. Nucl. Sci.*, NS-31:910–913, 1984.

[38] E. Tanaka. Quantitative image reconstruction with weighted backprojection for single photon emission computed tomography. *J. Comput. Assist. Tomogr.*, 7:692–700, 1983.

[39] M. I. Miller, D. L. Snyder, and T. R. Miller. Maximum likelihood reconstruction for single photon emission computed tomography. *IEEE Trans. Nucl. Sci.*, NS-32, No. 1:769–778, 1985.

[40] S. Geman, K. M. Manbeck, and D. E. McClure. A comprehensive statistical model for single photon emission tomography. In R. Challappa and A. Jain, editors, *Markov Random Field: Theory and Applications*. Academic Press, 1991.

[41] M. I. Miller and D. L. Snyder. The role of likelihood and entropy in incomplete-data problems: Applications to estimating point-process intensities and toeplitz constrained covariances. *Proc. IEEE*, 75, No.7:892–907, July 1987.

[42] D. L. Snyder and M. I. Miller. *Random Point Processes in Time and Space*. Springer-Verlag, 1991.

[43] Y. Vardi, L. A. Shepp, and L. Kaufman. A statistical model for positron emission tomograhy. *J. Am. Stat. Assoc.*, 80:8–35, 1985.

[44] K. Lange and R. Carson. Em reconstruction algorithms for emission and transmission tomography. *J. Comput. Assist. Tomogr.*, 8(2):306–316, 1984.

[45] D. L. Snyder and D. G. Politte. Image reconstruction from list-mode data in an emission tomography system having time-of-flight measurements. *IEEE Trans. Nucl. Sci.*, NS-30:1843–1849, 1983.

[46] M. I. Miller, D. L. Snyder, and T. R. Miller. Maximum-likelihood reconstruction for single-photon emission computed-tomography. *IEEE Trans. Nucl. Sci.*, NS-32:769–778, 1985.

[47] F. S. Gibson and F. Lanni. Diffraction by a circular aperture as a model for three-dimensional optical microscopy. *J. Opt. Soc. Am. A*, 6(9):1357–1367, 1989.

[48] C. Preza, J. M. Ollinger, J. G. McNally, and Jr. L. J. Thomas. Point-spread sensitivity analysis for 3-D fluorescence microscopy. *Proceedings of the SPIE/IS&T's Electronic Imaging: Science and Technology*, 1660, 1992.

[49] D. L. Collins, T. M. Peters, W. Dai, and A. C. Evans. Model-based segmentation of individual brain structures from mri data. In Richard A. Robb, editor, *Visualization in Biomedical Computing 1992*, SPIE 1808, pages 10–23. 1992.

[50] I. Carlbom, D. Terzopoulos, and K. Harris. Computer-assisted registration, segmentation, and 3d reconstruction from images of neuronal tissue sections. *IEEE Trans. Med. Imaging*, 13(2):351–362, 1994.

[51] W. M. Wells, E. I. Grimson, R. Kikinis, and F. A. Jolesz. Adptive segmentation of MRI data. *IEEE Trans. Med. Imaging*, 15(4):429–442, 1996.

[52] W. E. L. Grimson, G. J. Ettinger, T. Kapur, M. E. Leventon, W. M. Wells, and R. Kikinis. Utilizing segmented MRI data in image-guided surgery. *Int. J. Pattern Recogni. Artif. Intell.*, 11(8):1367–1397, 1997.

[53] M. Joshi, J. Cui, K. Doolittle, S. Joshi, D. C. van Essen, L. Wang, and M. I. Miller. Brain segmentation and the generation of cortical surfaces. *NeuroImage*, 9:461–476, 1999.

[54] M. I. Miller, A. Massie, J. T. Ratnanather, K. N. Botteron, and J. G. Csernansky. Bayesian construction of geometrically based cortical thickness metrics. *NeuroImage*, 12:676–687, 2000.

[55] Patrick C. Teo, Guillermo Sapiro, and Brain A. Wandell. Creating connected representations of cortical gray matter for functional MRI visualization. *IEEE Trans. Med. Imaging*, 16(6):852–863, 1997.

[56] Chenyang Xu, Dzung L. Pham, and Jerry L. Prince. Finding the brain cortex using fuzzy segmentation, isosurfaces and deformable surface models. In *XVth Int. Conf. on Info Proc. in Medical Imaging*, June 1997.

[57] A. Chakraborty and J. S. Duncan. Game-theoretic integration for image segmentation. *IEEE Trans. Pattern Anal. Mach. Intell.*, 21(1):12–30, 1999.

[58] G. Schwartz. Estimating the dimension of a model. *Ann. Stat.*, 6:461–464, 1978.

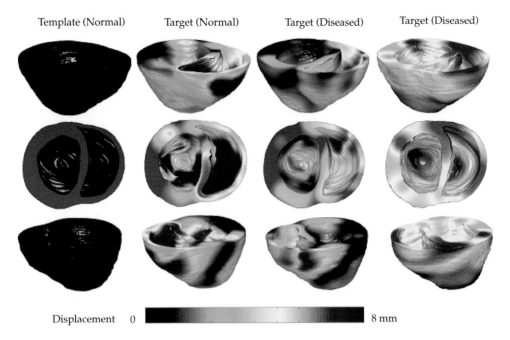

Plate 35 See page 486.

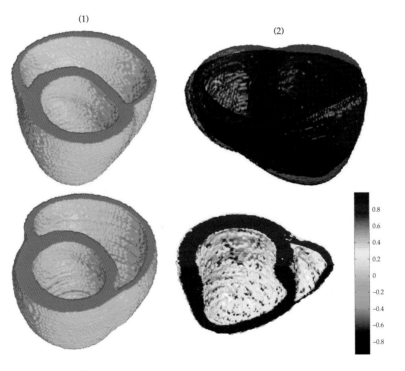

Plate 36 See page 488.

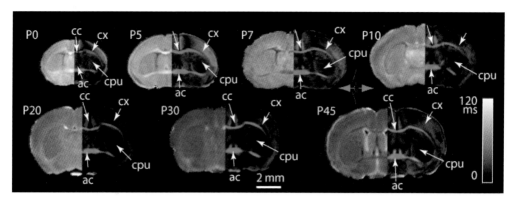

Plate 37 See page 490.

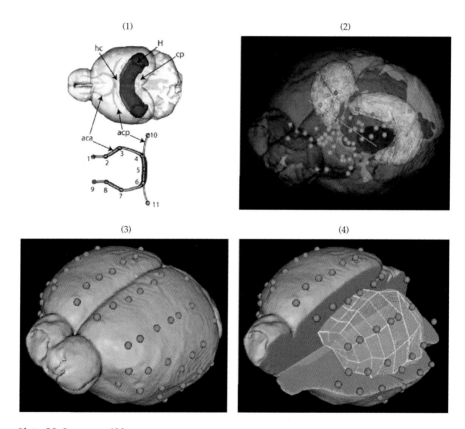

Plate 38 See page 490.

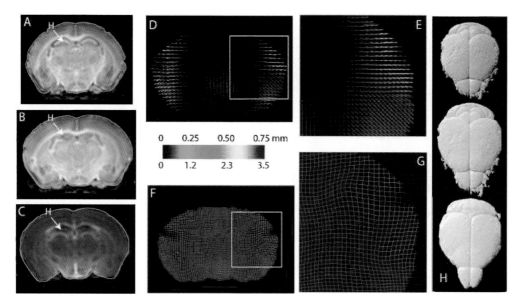

Plate 39 See page 491.

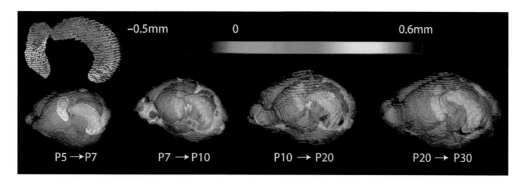

Plate 40 See page 492.

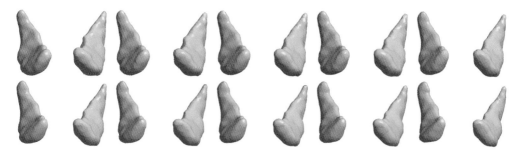

Plate 41 See page 492.

Plate 42 See page 500.

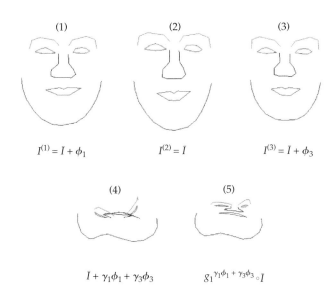

Plate 43 See page 500.

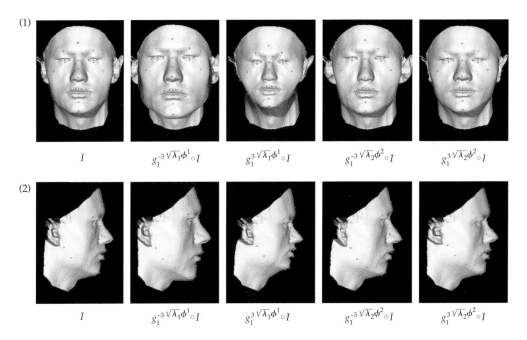

Plate 44 See page 501.

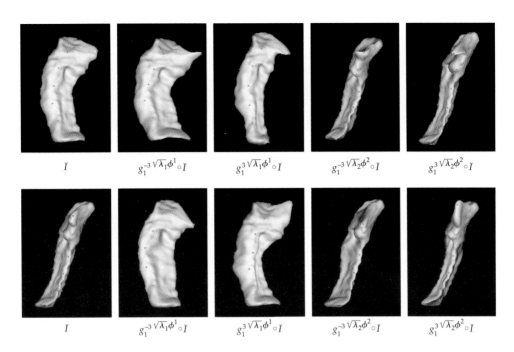

Plate 45 See page 501.

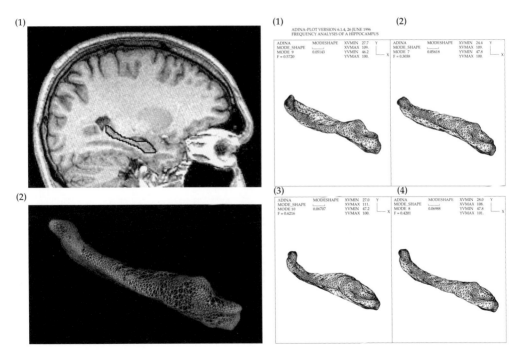

Plate 46 See page 503.

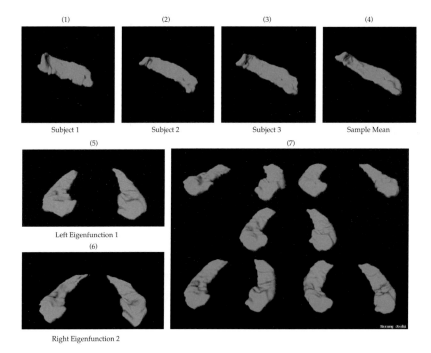

Plate 47 See page 504.

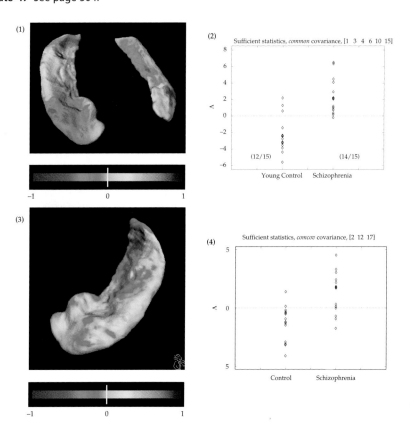

Plate 48 See page 507.

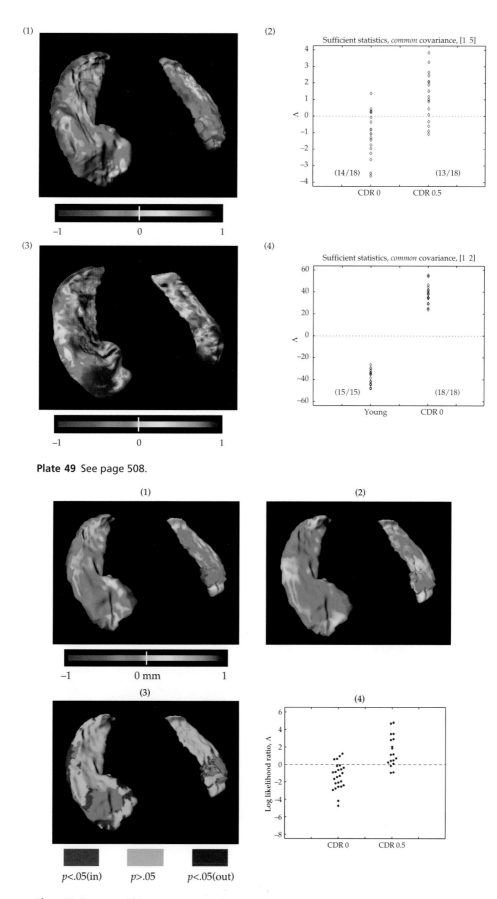

Plate 49 See page 508.

Plate 50 See page 509.

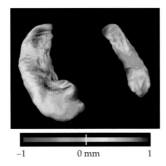

Plate 51 See page 510.

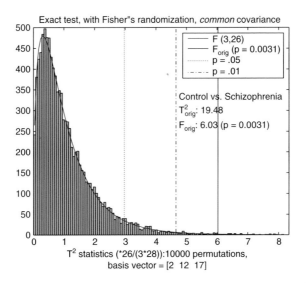

Plate 52 See page 511.

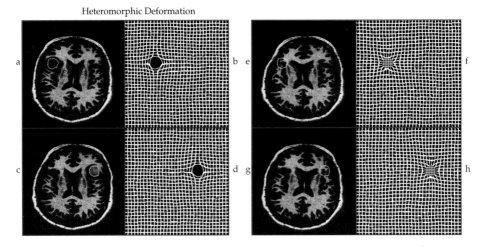

Plate 53 See page 513.

[59] J. Rissanen. A universal prior for integers and estimation by minimum description length. *Ann. Stat.*, 11:416–431, 1983.
[60] B. S. Clarke and A. R. Barron. Information-theoretic asymptotics of bayes method. *IEEE Trans. Inform. Theory*, 36(3), 1990.
[61] G. Polya and G. Szego. *Problems and Theorems in Analysis: Translated by D. Aeppli*. Springer-Verlag, 1976.
[62] A. D. Lanterman. Schwarz, Wallace, and Rissanen: Intertwining themes in theories of model selection. 2000.
[63] E. T. Jaynes. Information theory and statistical mechanics. *Phys. Rev.*, 106:620–630, 1957.
[64] E. T. Jaynes. On the rationale of maximum-entropy methods. *Proc. IEEE*, 70:939–952, 1982.
[65] D. L. Snyder. *Random Point Processes*. John Wiley Sons, New York, 1975.
[66] J. M. Van Campenhout and T. M. Cover. Maximum entropy and conditional probability. *IEEE Trans. Inform. Theory*, IT-27, No. 4:483–489, 1981.
[67] A. P. Dawid. Conditional independence in statistical theory. *J. R. Stat. Soc., Ser. B*, (1):1–31, 1979.
[68] J. Pearl. *From Bayesian Networks to Causal Networks*. 1988.
[69] H. R. Keshavan, J. Barnett, D. Geiger, and T. Verma. Introduction to the special section on probabilistic reasoning. *IEEE Trans. Pattern Anal. Mach. Intell.*, 15, (3):193–195, 1993.
[70] R. G. Cowell, A. P. Dawid, and D. J. Spiegelhalter. Sequential model criticism in probabilistic expert systems. *IEEE Trans. Pattern Anal. Mach. Intell.*, PAMI-15, (3):209–219, 1993.
[71] A. J. Hartemink, D. K. Gifford, T. S. Jaakola, and R. A. Young. Using graphical models and genomic expression data to statistically validate models of genetic regulatory networks. In Altman et al., editor, *Pacific Symposium on Biocomputing 2001*, volume 6, pages 422–433, World Scientific Publishing Company, Singapore, 2001.
[72] T. E. Harris. *The Theory of Branching Processes*. Springer-Verlag, Berlin, 1963.
[73] K. B. Athreya and P. E. Ney. *Branching Processes*. Springer-Verlag, Berlin, 1972.
[74] F. R. Gantmacher. *The Theory of Matrices*. Chelsea Publishing Co., New York, 1959.
[75] Erhan Cinlar. *Introduction to Stochastic Processes*. Prentice-Hall, Englewood Cliffs, 1975.
[76] C. E. Shannon and W. Weaver. *The Mathematical Theory of Communication*. University of Illinois Press, Urbana, IL, 1949.
[77] L. Breiman. *Probability*. Addison-Wesley Publishing Co., Reading, MA, 1968.
[78] U. Grenander. *Probability Measures for Context-Free Languages*. Brown University, Providence, R.I., 1967.
[79] M. I. Miller and J. A. O'Sullivan. Entropies and combinatorics of random branching processes and context-free languagues. *IEEE. Trans. Inform. Theory*, 38, 4, 1992.
[80] Berger and Ye. Entropic aspects random fields. *IEEE. Trans. Info. Theory*, 1990.
[81] N. Chomsky and G. A. Miller. Finite state languages. *Inform. Control*, 1:91–112, 1958.
[82] N. Chomsky. Three models for the description of language. *IRE Trans. Inform. Theory*, IT-2:113–124, 1956.
[83] N. Chomsky. On certain formal properties of grammars. *Inform. Control*, 2:137–167, 1959.
[84] J. E. Hopcroft and J. D. Ullman. *Introduction to Automata Theory, Languages, and Computation*. Addison-Wesley, Reading, MA, 1979.
[85] C. Shannon. Prediction and entropy of printed english. *Bell Syst. Tech. J.*, 30, 1951.
[86] R. B. Banerji. Phrase structure languages, finite machines, and channel capacity. *Inform. Control*, 6:153–162, 1963.
[87] W. Kuich. On the entropy of context-free languages. *Inform. Control*, 16:173–200, 1970.
[88] R. Cameron. Source encoding using syntactic information source models. *IEEE Trans. Inform. Theory*, 34, 4:843–850, 1988.
[89] K. W. Church and R. L. Mercer. Introduction to the special issue on computational linguistics using large corpora. *Comput. Linguist.*, 19(1):1–24, 1993.
[90] K. Church. A stochastic parts program and noun phrase parser for unrestricted text. In ACL, editor, *Proceedings, Second Conference on Applied Natural Language Processing*, pages 136–143, Austin, TX, ACL, 1988.
[91] J. Kupiec. Robust part-of-speech tagging using a hidden markov model. *Comput. Speech Lang.*, 6:225–242, 1992.
[92] P. Brown, V. Della Pietra, P. deSouza, and R. Mercer. Class-based n-gram models of natural language. *Computat Linguistics*, 18(4):467–479, 1992.
[93] K. Lari and S. Young. The estimation of stochastic context-free grammars using the inside-outside algorithm. *Comput Speech Language Process.*, 4:35–56, 1990.
[94] F. Jelinek, J. D. Lafferty, and R. L. Mercer. Basic methods of probabilistic context free grammars. Technical Report RC 16374, IBM, Yorktown Heights, NY, 1990.

[95] F. Jelinek and R. L. Mercer. Interpolated estimation of Markov source parameters from sparse data. In *Proceedings, Workshop on Pattern Recognition in Practice*, pages 381–397, Amsterdam, The Netherlands, 1980.

[96] K. W. Church and W. A. Gale. A comparison of enhanced Good-Turing and deleted estimation methods for estimating probabilities of English bigrams. *Comput Speech Language*, 5, 1991.

[97] L. R. Bahl, F. Jelinek, and R. L. Mercer. A maximum likelihood approach to continuous speech recognition. *IEEE Trans. Pattern Anal Mach. Intell.*, PAMI-5(2):179–190, 1983.

[98] P. Brown, V. Della Pietra, P. V. deSouza J. Lai, and R. L. Mercer. Class-based n-gram models of natural language. *Comput. Linguist.*, 18(4):467–479, 1992.

[99] F. Jelinek. Markov source modeling of text generation. In J. K. Skwirzinski, editor, *The Impact of Processing Techniques on Communication*. Nijhoff, Dordrecht, 1985.

[100] K. Mark, M. I. Miller, U. Grenander, and S. Abney. Parameter estimation for constrained context-free language models. In *1992 DARPA Workshop on Speech and Natural Language*, February 1992.

[101] K. Mark, M. I. Miller, and U. Grenander. Constrained stochastic language models. In Steven Levinson and Larry Shepp, editors, *Image Models (and their Speech Model Cousins)*, pages 131–140. Springer-Verlag, 1995.

[102] T. Cover and R. King. A convergent gambling estimate of the entropy of English. *IEEE Trans. Inform. Theory*, 24(4):413–421, 1978.

[103] P. Brown, S. Della Pietra, V. Della Pietra, J. Lai, and R. Mercer. An estimate of an upper bound for the entropy of English. *Comput. Linguistic*, 18(1):31–40, 1992.

[104] K. E. Mark, M. I. Miller, U. Grenander, and S. Abney. Parameter estimation for constrained context-free language models. In *DARPA Speech and Natural Language Workshop*, Harriman, NY, February 1992.

[105] L. Rabiner. A tutorial on hidden Markov models and selected applications in speech recognition. *Proc. IEEE*, (2):257–285, 1989.

[106] Andrew J. Viterbi. Error bounds for convolutional codes and an asymptotically optimum decoding algorithm. *IEEE Trans. Inform. Theory*, IT-13:260–267, 1967.

[107] D. Forney. The viterbi algorithm. *Proc. IEEE*, 1973.

[108] L. Baum. An inequality and associated maximization technique in statistical estimation of probabilistic functions of a Markov process. *Inequalities*, 3:1–8, 1972.

[109] *Proceedings of the 1997 Large Vocabulary Continuous Speech Recognition Workshop*, 1997. Available at http://www.clsp.jhu.edu/ws97.

[110] J. J. Godfrey, E. C. Holliman, and J. McDaniel. Switchboard: Telephone Speech Corpus for Research and Development. In *IEEE Conference on Acoustics, Speech and Signal Processing*, volume 1, pages 517–520, SanFrancisco, CA, 1992.

[111] H. Hermansky. Perceptual linear predictive (PLP) analysis of speech. *J. Acoust. Soc. Am.*, 87(4):1738–1752, 1990.

[112] V. I. Levenshtein. Binary codes capable of correcting deletions, insertions and reversals. *Sov. Phys. Dokl.*, 10(8):707–710, 1966.

[113] J. Baker. Trainable grammars for speech recognition. In Klatt and Wolf, editors, *Speech Communication Papers for the 97th Meeting of the Acoustical Society of America*, pages 547–550, 1979.

[114] J. Kupiec. A trellis-based algorithm for estimating the parameters of a hidden stochastic context-free grammar. In *DARPA Speech and Natural Language Workshop*, Asilomar, CA, February 1991.

[115] E. Ising. Beitrag sur theorie des ferromagnetismus. *Z. Phys.*, 31:253–258, 1925.

[116] R. L. Dobrushin. Existence of a phase transition in two and three dimensional Ising models. *Theor. Probab. Appl.*, 10:193–213, 1965.

[117] R. L. Dobrushin. The gibbsian random fields for lattices systems with pairwise interactions. *Funct. Anal. Appl.*, 2:292–301, 1968.

[118] R. L. Dobrushin. The problem of uniqueness of a gibbsian random field and the problem of phase transitions. *Funct. Anal. Appl*, 2:302–312, 1968.

[119] R. L. Dobrushin. Prescribing a system of random variables by conditional distributions. *Theoret. Probab. Appl.*, 15:458–486, 1970.

[120] D. Ruelle. *Thermodynamic Formalism, Encyclopedia of Mathematics and Its Applications*, volume 5. Addison Wesley, Reading MA, 1978.

[121] J. Besag. Spatial interaction and statistical analysis of lattice systems (with discussion). *J. R. Stat. Soc.*, B, 36:192–326, 1974.

[122] S. Geman and D. Geman. Stochastic relaxation, Gibbs distributions, and the Bayesian restoration of images. *IEEE Trans. Pattern Anal.*, 6:721–741, November 1984.

[123] R. Kinderman and J. L. Snell. *Markov Random Fields and their Applications*, volume 1. American Mathematical Society- Contemporary Mathematics series, 1980.

[124] D. Geman. Random fields and inverse problems in imaging. *Lect. Notes Math*, 1427:117–193, 1990.
[125] M. I. Miller, B. Roysam, K. Smith, and J. A. O'Sullivan. Representing and computing regular languages on massively parallel networks. *IEEE. Trans. Neural Networ*, 2 (No.1):56–72, 1991.
[126] B. Roysam and M. I. Miller. Combining stochastic and syntactic processing with analog computation methods. *Digit. Signal Process.: Rev. J.*, 2 (2):48–64, 1992.
[127] K. R. Smith and M. I. Miller. *Bayesian Inference of Regular Grammar and Markov Source Models*, pages 388–398. Morgan-Kaufmann, Palo Alto, 1990.
[128] K. R. Smith and M. I. Miller. A Bayesian approach incorporating Rissanen complexity for learning Markov random field texture models. *Proceedings of the Intl. Conference on Acoustics, Speech, and Signal Processing*, 4: (5) pages 2317–2320, April 1990.
[129] M. B. Averintsev. On a method of describing discrete parameter random fields. *Problemy Peredachi Informatsii*, 6:100–109, 1970.
[130] G. R. Grimmett. A theorem about random fields. *Bull. Lond. Math. Soc.*, 5:81–84, 1973.
[131] D. Griffeath. Ergodic theorems for graph interactions. *Adv. Appl. Probab.*, 7:179–194, 1975.
[132] C. J. Preston. *Gibbs States on Countable Sets*. Cambridge University Press, Cambridge, 1974.
[133] D. Mumford. The hierarchical ising model. ??
[134] J. J. Clark and A. L. Yuille. *Data Fusion for Sensory Information Processing Systems*. Kluwer Academic Publishers, Boston, 1990.
[135] D. Mumford. *The Bayesian Rationale for Energy Functionals*. Kluwer Academic, 1994.
[136] A. Blake and Al. Yuille, editors. Active Vision. MIT Press, 1992.
[137] U. Grenander and M. I. Miller. Representations of knowledge in complex systems. *J. R. Stat. Soc. B*, 56(3):549–603, 1994.
[138] Y. Amit and M. I. Miller. Large deviations for coding Markov chains and gibbs random fields. *IEEE. Trans. Inform. Theory*, 39(1):109, 1993.
[139] Y. Amit and M. I. Miller. Large deviations for the asymptotitcs of ziv-lempel codes for 2-d gibbs fields. *IEEE Trans. Inform. Theory*, 38 (4): 1992.
[140] S. C. Zhu, Y. N. Wu, and D. B. Mumford. Minimax entropy principle and its applications to texture modeling. *Neural Comput.*, 9(8):1627–1660, 1997.
[141] J. Daugman. Uncertainty relation for resolution in space, spatial frequency, and orientation optimized by two-dimensional visual cortical filters. *J. Opt. Soc. Am.*, 2(7), 1985.
[142] R. E. Peierls. On Ising's ferromagnet model. *Proc. Camb. Phil. Soc.*, 32:477–481, 1936.
[143] G. R. Grimmett and D. R. Stirzaker. *Probability and Random Processes*, 2nd Ed. Oxford Science Publications, 1995.
[144] In A. Possolo, editor, *Spatial Statistics and Imaging*, volume Lecture notes - monograph series, 20. Institute of Mathematical Statistics, Haywood, California, 1991.
[145] M. Cannon and R. Passmore. Incorporation of background clutter effects into performance modeling of infrared missile seekers. In *Proc. of the 6th Annual Ground Target Modeling and Validation Conf.*, pages 144–151, Houghton, MI, August 1995. U.S. Army TACOM.
[146] *Inequalities 2nd* Ed. Cambridge University Press, Cambridge Mathematical Library, 1988.
[147] F. Klein. *Gesammelte Abhandlungen*. Springer, Berlin, 1971.
[148] W. M. Boothby. *An Introduction to Differentiable Manifolds and Riemannian Geometry*. Academic Press, New York 1986.
[149] M. Artin. *Algebra*. Prentice Hall, Englewood Cliffs, NJ, 1991.
[150] U. Grenander, Y. Chow, and D. Keenan. *HANDS: A Pattern Theoretic Study of Biological Shapes*. Springer-Verlag, New York, 1990.
[151] I. R. Porteous. *Geometric Differentiation*. Cambridge University Press, Cambridge, 1994.
[152] B. O'Neill. *Elementary Differential Geometry*. Academic Press, San Diego, 1966.
[153] N. Khaneja. Statistics and Geometry of Cortical Features. M.S. thesis, Department of Electrical Engineering, Sever Institute of Technology, Washington University, St. Louis, MO, December 1996.
[154] N. Khaneja, U. Grenander, and M. I. Miller. Dynamic programming generation of curves on brain surfaces. *Pattern Anal. Machine Int.*, 20(10):1260–1264, 1998.
[155] D. J. Felleman and D. C. van Essen. Distributed hierarchical processing in the primate cerebral cortex. *Cereb. Cortex*, 1(1):1–47, 1991.
[156] P. M. Thompson, C. Schwartz, R. T. Lin, A. A. Khan, and A. W. Toga. Three-dimensional statistical analysis of sulcal variability in the human brain. *J. Neurosci.*, 16(13):4261–4274, 1996.
[157] B. Hamann. Curvature approximation for triangulated surfaces. In *Computing*, pages 139–153. Springer-Verlag, Berlin, 1993.

[158] S. C. Joshi, J. Wang, M. I. Miller, D. C. van Essen, and U. Grenander. On the differential geometry of the cortical surface. In R. A. Melter, A. Y. Wu, F. L. Bookstein, and W. D. Green, editors, *Proc. SPIE*, volume 2573, pages 304–311. San Diego, CA., 1995.

[159] W. E. Lorensen and H. E. Cline. Marching cubes: A high resolution 3D surface construction algorithm. *comput. Graph.*, 21(4):163–169, 1987.

[160] A. Gueziec and R. Hummel. Exploiting triangulated surface extraction using tetrahedral decomposition. *IEEE Trans Vis. Comput. Graph.*, 1(4):328–334, 1995.

[161] M. Claudio and S. Roberto. Using marching cubes on small machines. *Graph. Model Im. Process.*, 56:182–3, 1994.

[162] J. W. Haller, A. Banerjee, G. E. Christensen, S. Joshi, M. I. Miller, M. W. Vannier, and J. C. Csernansky. Three-dimensional hippocampal volumetry by high dimensional transformation of a neuroanatomical atlas. *Radiol.*, 202(2):504–510, 1997.

[163] J. T. Kent, K. V. Mardia, and J. M. West. Ridge curves and shape analysis. In Monograph, Department of Statistics, University of Leeds., Leeds LS2 9JT, UK, May, 1996.

[164] J. Ph. Thirion and A. Goudon. Computing the differential characteristics of isointensity surfaces. *Comp. Vis. Image Und.*, 61(2):190–202, 1995.

[165] D. C. van Essen and J. H. R. Maunsell. Two-dimensional maps of the cerebral cortex. *J. Comput. Neurol.*, 191:255–281, 1980.

[166] D. C. van Essen and H. Drury. Structural and functional analyses of human cerebral cortex using a surface-based atlas. *J. Neurosci.*, 17:7079–7102, 1997.

[167] D. C. van Essen, H. Drury, S. Joshi, and M. I. Miller. Functional and structural mapping of human cerebral cortex: Solutions are in the surfaces. *Proc. Natl. Acad. Sci.*, 95:788–795, 1998.

[168] D. C. van Essen. Pulling strings to build a better brain: A tension-based theory of morphogenesis and compact wiring in the central nervous sytem. *Nature*, 385:313–318, 1997.

[169] W. Welker. Why does cerebral cortex fissure and fold. *Cerebral Cortex*, 83:3–136, 1990.

[170] J. Ph. Thirion and A. Goudon. The 3D marching lines algorithm. *Graph. Models Image Process.*, 58(6):503–509, 1996.

[171] D. B. Cooper, H. Elliott, F. Cohen, L. Reiss, and P. Symosek. Stochastic boundary estimation and object recognition. In A. Rosenfeld, editor, Image Modeling, pages 63–94. Academic Press, New York, 1981.

[172] J. L. Elion, S. A. Geman, and K. M. Manbeck. Computer recognition of coronary arteries. *J. Am. Coll. Cardiol.*, 17(2): 1991.

[173] A. M. Dale, B. Fischl, and M. I. Sereno. Cortical surface-based analysis - i. segmentation and surface reconstruction. *NeuroImage*, 9(2):179–194, 1999.

[174] B. Fischl, M. I. Sereno, and A. M. Dale. Cortical surface-based analysis - II: Inflation, flattening, and a surface-based coordinate system. *NeuroImage*, 9(2):195–207, 1999.

[175] H. A. Drury, D. C. van Essen, C. H. Anderson, C. H. Lee, T. A. Coogan, and J. W. Lewis. Computerized mappings of the cerebral cortex. a multiresolution flattening method and a surface-based coordinate system. *J. Cognitive Neurosci.*, 8:1–28, 1996.

[176] B. Wandell, S. Chial, and B. Backus. Visualization and measurement of the cortical surface. *J. Cognitive Neurosci.*, 12:739–752, 2000.

[177] S. Angenent, S. Haker, A. Tannenbaum, and R. Kikinis. On the Laplace-Beltrami operator and brain surface flattening. *IEEE Trans. Med. Imaging*, 18(8):700–711, 1999.

[178] M. K. Hurdal, P. L. Bowers, K. Stephenson, D. L. Sumners, K. Rehm, K. Schaper, and D. A. Rottenberg. Quasi-conformally flat mapping the human cerebellum. In C. Taylor and A. Colchester, editors, Lecture Notes in Computer Science, pages 279–286. Springer-Verlag, Berlin, 1999.

[179] M. K. Hurdal, K. Stephenson, P. L. Bowers, D. L. Sumners, and D. A. Rottenberg. Cortical surface flattening: a quasi-conformal approach using circle packings. Submitted, 2002.

[180] C. R. Collins and K. Stephenson. A circle packing algorithm. *Computational Geometry: Theory and Application*, to appear, 2002.

[181] S. G. Krantz. The Riemann mapping theorem. In *Handbook of Complex Analysis*, pages 86–87. Birkhäuser, Boston, MA, 1999.

[182] M. K. Hurdal and K. Stephenson. Cortical cartography using the discrete conformal approach of circle packing. *NeuroImage*, 2004.

[183] J. T. Ratnanather, P. E. Barta, N. A. Honeycutt, N. Lee, H. M. Morris, A. D. Dziorny, M. K. Hurdal, G. D. Pearlson, and M. I. Miller. Dynamic programming generation of boundaries of local coordinatized submanifolds in the neocortex: application to the planum temporale. *NeuroImage*, 2003.

[184] M. Kass, A. Witkin, and D. Terzopolous. Snakes: Active contour models. *Int. J. Comput. Vision*, 1(4): 321–331, 1988.

[185] L. H. Staib and J. S. Duncan. Boundary finding with parametrically deformable models. *IEEE Trans. Pattern Anal. Machine Intell.*, 14:1061–1075, 1992.
[186] Y. Amit and A. Kong. Graphical templates for image matching. *Technical Report: University of Chicago*, (373), 1994.
[187] C. Xu and J. L. Prince. Gradient vector flow: A new external force for snakes. *CVRP*, November 1997.
[188] M. Vaillant and C. Davatzikos. Finding parametric representations of the cortical sulci using an active contour model. *Med. Image Anal.*, 1(4):295–315, 1997.
[189] N. Schultz and K. Conradsen. 2d vector-cycle deformable templates. *Signal Process.*, 71(2):141–153, 1998.
[190] A. Garrido and N. P. De la Blanca. Physically-based active shape models: Initialization and optimization. *Pattern Recogn.*, 31(8):1003–1017, 1998.
[191] C. F. Westin, L. M. Lorigo, O. Faugeras, W. E. L. Grimson, S. Dawson, A. Norbash, and R. Kikinis. Segmentation by adaptive geodesic active contours. In *Medical Image Computing and Computer-Assisted Intervention - Miccai 2000*, volume 1935 of Lecture Notes in Computer Science, pages 266–275. Springer-Verlag, Berlin, 2000.
[192] T. Chen and D. Metaxas. Image segmentation based on the integration of Markov random fields and deformable models. In *Medical Image Computing and Computer-Assisted Intervention - Miccai 2000*, volume 1935 of *Lecture Notes in Computer Science*, pages 256–265. Springer-Verlag, Berlin, 2000.
[193] C. Xu, D. L. Pham, and J. L. Prince. Medical image segmentation using deformable models. In J. M. Fitzpatrick and M. Sonka, editors, SPIE Handbook on Medical Imaging – Volume III: Medical Image Analysis, pages 129–174. SPIE, Bellingham, WA, 2000.
[194] M. Mignotte and J. Meunier. A multiscale optimization approach for the dynamic contour- based boundary detection issue. *Comput. Med. Imag. Graph,* 25(3):265–275, 2001.
[195] A. Yezzi, A. Tsai, and A. Willsky. A fully global approach to image segmentation via coupled curve evolution equations. *J. Vis. Commun. Image R.*, 13(1-2):195–216, 2002.
[196] D. Terzopoulos and D. Metaxas. Dynamic 3d models with local and global deformations: Deformable superquadrics. *IEEE Trans. Pattern. Anal. Mach. Intell.*, 13:703–714, 1991.
[197] I. Cohen, L. Cohen, and N. Ayache. Using deformable surfaces to segment 3-d images and infer differential structures. *Comput. Vis. Graph. Image Process.*, 56(2):242–263, 1992.
[198] M. I. Miller, S. Joshi, D. R. Maffitt, J. G. McNally, and U. Grenander. Mitochondria, membranes and amoebae: 1,2 and 3 dimensional shape models. In K. Mardia, editor, Statistics and Imaging, volume II. Carfax Publishing, Abingdon, Oxon., 1994.
[199] T. F. Cootes, C. J. Taylor, D. H. Cooper, and J. Graham. Active shape models—their training and application. *Comput. Vis. Image Und.*, 61(1):38–59, 1995.
[200] L. H. Staib and J. S. Duncan. Model-based deformable surface finding for medical images. *IEEE Trans. Med. Imag.*, 15(5):1–13, 1996.
[201] J. Montagnat and H. Delingette. Volumetric medical images segmentation using shape constrained deformable models. In *Cvrmed-Mrcas'97*, volume 1205 Lecture Notes in Computer Science, pages 13–22. Springer-Verlag, Berlin, 1997.
[202] J. Montagnat, H. Delingette, and N. Ayache. A review of deformable surfaces: topology, geometry and deformation. *Imag. Vis. Comput.*, 19(14):1023–1040, 2001.
[203] S. Sclaroff and L. F. Liu. Deformable shape detection and description via model-based region grouping. *IEEE Trans. Pattern Anal. Mach. Intell.*, 23(5):475–489, 2001.
[204] F. Farassat. Introduction to generalized functions with applications in aerodynamics and aeroacoustics. In *NASA Technical Paper*, number 3428, 1994.
[205] T. F. Chan and L. A. Vese. Active contours without edges. *IEEE Trans. Image Proc.*, 10(2):266–277, 2001.
[206] G. Strang. *Introduction to Applied Mathematics*. Wellesley Cambridge Press, New York, 1986.
[207] M. C. Delfour and J.-P. Zolesio. Shapes and geometries: Analysis, differential calculus, and optimization. *Adv. Des. Control* SIAM, 4: 2001.
[208] S. Osher and J. A. Sethian. Fronts propagating with curvature-dependent speed: Algorithms based on Hamilton–Jacobi formulations. *J. Comput. Phys.*, 79:12–49, 1988.
[209] J. N. Tsitsiklis. Efficient algorithm for globally optimal trajectories. *IEEE Trans. Automatic Control*, 40(9):1528–1538, 1995.
[210] J. A. Sethian. A fast marching level set method for monotonically advancing fronts. *Proc. Natl. Acad. Sci. USA*, 93:1591–1595, 1996.
[211] H. K. Zhao, T. Chan, B. Merriman, and S. Osher. A variational level set approach to multiphase motion. *J. Comput. Phys.*, 127:179–195, 1996.
[212] J. L. Doob. *Stochastic Processes*. Wiley, New York, 1953.
[213] E. Wong and B. Hajek. *Stochastic Processes in Engineering Systems*. Springer-Verlag, Berlin, 1984.
[214] I. Karatzas and S. E. Shreve. *Brownian Motion and Stochastic Calculus*. Springer-Verlag, Berlin, 1987.

[215] M. Reed and B. Simon. *Functional Analysis I: Revised and Enlarged Edition*. Academic Press, New York, 1980.

[216] J. T. Kent, I. L. Dryden, and C. R. Anderson. Using circulant symmetry to model featureless objects. *Biometrika*, 87(3):527–544, 2000.

[217] N. R. Goodman. Statistical analysis based on a certain multivariate complex gaussian distribution. *Ann. Math. Stat.*, 34(1):152–177, 1963.

[218] G. N. Watson. *A Treatise on the Theory of Bessel Functions*, 2nd ed. Cambridge University Press, Cambridge, England, 1966.

[219] A. M. Yaglom. *Theory of Stationary Random Processes*. Prentice Hall, Englewood Cliffs, NJ 1962.

[220] G. E. Christensen, R. D. Rabbitt, and M. I. Miller. Deformable templates using large deformation kinematics. *IEEE Trans. Image Process.*, 5(10):1435–1447, 1996.

[221] G. E. Christensen, R. D. Rabbitt, and M. I. Miller. 3D brain mapping using a deformable neuroanatomy. *Phys. Med. Biol.*, 39:609–618, 1994.

[222] G. E. Christensen, R. D. Rabbitt, and M. I. Miller. A deformable neuroanatomy textbook based on viscous fluid mechanics. *Proc. 27th Annual Conf. on Information Sciences and Systems, Baltimore, Maryland*, The Johns Hopkins University, pages 211–216. March 1993.

[223] F. L. Bookstein. *The Measurement of Biological Shape and Shape Change*, volume 24. Lecture Notes in Biomathematics, Springer-Verlag, New York, 1978.

[224] F. L. Bookstein. Biometrics, biomathematics and the morphometric synthesis. *Bull. Math. Biol.*, 58(2):313–365, 1996.

[225] Y. Amit, U. Grenander, and M. Piccioni. Structural image restoration through deformable templates. *J. Am. Stat. Assoc.*, 86(414):376–387, 1991.

[226] G. Christensen. Deformable shape models for anatomy. Ph.D. Dissertation, Department of Electrical Engineering, Sever Institute of Technology, Washington University, St. Louis, MO, Aug 1994.

[227] M. Miller, Ayananshu Banerjee, Gary Christensen, Sarang Joshi, Navin Khaneja, U. Grenander, and Larissa Matejic. Statistical methods in computational anatomy. Stat. Methods Med. Res., 6:267–299, 1997.

[228] Giovanni Sansone. *Orthogonal functions*. Interscience Publishers, New York, 1959.

[229] S. Joshi, M. I. Miller, and U. Grenander. On the geometry and shape of brain sub-manifolds. *Int. J. Pattern Recognition Artif. Intell.*, 11(8), 1997.

[230] R. Courant and D. Hilbert. *Methods of Mathematical Physics*, volume 1. Interscience Publishers, New York, 1953.

[231] Muge Bakircioglu, Sarang Joshi, and Michael Miller. Landmark matching on brain surfaces via large deformation diffeomorphisms on the sphere. *Proceedings SPIE Medical Imaging 1999: Image Processing*, July 1999.

[232] J. G. McNally, C. Preza, J. A. Conchello, and L. J. Thomas, Jr. Artifacts in computational optical sectioning microscopy. *J. Opt. Soc. Am. A*, accepted.

[233] K. R. Castleman. *Digital Image Processing*. Prentice-Hall, Englewood Cliffe, NJ, 1979.

[234] D. A. Agard. Optical sectioning microscopy. *Ann. Rev. Biophys. Bioeng.*, 13:191–219, 1984.

[235] C. Preza, M. I. Miller, Jr. L. J. Thomas, and J. G. McNally. A regularized linear method for reconstruction of 3-D microscopic objects from optical sections. *J. Opt. Soc.*, 9(2):219–228, 1992.

[236] A. Pentland and S. Sclaroff. Closed-form solutions for physically based shape modeling and recognition. *IEEE Trans. Pattern Anal. Mach. Intell.*, 13(7):715–729, 1991.

[237] M. N. O. Sadiku. *Numerical Techniques in Electromagnetics*. CRC Press, New York, 1992.

[238] P. P. Silvester and R. L. Ferrari. *Finite Elements for Electrical Engineering*. Cambridge University Press, Cambridge, 1983.

[239] C. A. Brebbia and J. J. Connor. *Fundamentals of Finite Element Technique*. Butterworth, London, 1973.

[240] M. Reed and B. Simon. *Fourier Analysis, Self-Adjointness*. Academic Press, New York 1980.

[241] Y. Amit and M. Piccioni. A nonhomogenous Markov process for the estimation of gaussian random fields with non-linear observations. *Ann. Probab.*, 19:1664–1678, 1991.

[242] S. Joshi. *Large Deformation Diffeomorphisms and Gaussian Random Fields for Statistical Characterization of Brain SubManifolds*. Sever Institute of Technology, Washington University, St. Louis, MO., 1997.

[243] D. L. Snyder and M. I. Miller. The use of sieves to stabilize images produced with the EM algorithms for emission tomography. *IEEE Trans. Nucl. Sci.*, NS-32:3864–3872, 1985.

[244] S. Geman and D. E. McClure. Bayesian image analysis: An application to single photon emission tomography. *Proceedings of the American Statistical Association*, pages 12–18, 1985.

[245] D. L. Snyder, M. I. Miller, Jr. L. J. Thomas, and D. G. Politte. Noise and edge artifacts in maximum-likelihood reconstruction for emission tomography. *IEEE Trans. Med. Imaging*, MI-6(3):228–237, 1987.

[246] M. I. Miller and B. Roysam. Bayesian image reconstruction for emission tomography: Implementation of the EM algorithm and Good's roughness prior on massively parallel processors. *Proc. of the Natl Acad. Sci. USA*, 88:3223–3227, April 1991.

[247] E. S. Chornoboy, C. J. Chen, M. I. Miller, T. R. Miller, and D. L. Snyder. An evaluation of maximum likelihood reconstruction for SPECT. *IEEE Trans. Med. Imaging*, 9(1):99–110, 1990.

[248] T. Hebert and R. Leahy. A generalized EM algorithm for 3-d Bayesian reconstruction from poisson data using Gibbs priors. *IEEE Trans. Med. Imaging*, MI-8(2):194–202, 1989.

[249] P. J. Green. On use of the EM algorithm for penalized likelihood estimation. *J. R. Stat. Soc.*, B, 52(3):443–452, 1990.

[250] K. Lange. Convergence of EM image reconstruction algorithms with Gibbs smoothing. *IEEE Trans. Med. Imaging*, MI-9(4):439–446, 1990.

[251] A. W. McCarthy and M. I. Miller. Maximum likelihood SPECT in clinical computation times using mesh-connected parallel computers. *IEEE Trans. Med. Imaging*, 10(3):426–436, 1991.

[252] G. T. Herman, A. R. De Pierro, and N. Gai. On methods for maximum a posteriori image reconstruction with a normal prior. *J. Visual Commun. Image Representation*, 3(4):316–324, 1992.

[253] C. S. Butler and M. I. Miller. Maximum a posteriori estimation for SPECT using regularization techniques on massively-parallel computers. *IEEE Trans. Med. Imaging*, 12(1):84–89, 1993.

[254] M. I. Miller A. D. Lanterman, D. L. Snyder. Extension of good's roughness for maximum penalized-likelihood estimation for emission tomography. *IEEE Trans. Med. Imaging*.

[255] R. A. Tapia and J. R. Thompson. *NonParametric Probability Density Estimation*. Johns Hopkins University Press, Baltimore,Md., 1978.

[256] I. J. Good. A non-parametric roughness penalty for probability densities. *Nature, London*, 229:29–30, 1971.

[257] I. J. Good and R. A. Gaskins. Nonparametric roughness penalties for probability densities. *Biometrika*, 58(2):255–277, 1971.

[258] I. J. Good and R. A. Gaskins. Density estimation and bump-hunting by the penalized likelihood method exemplified by scattering and meteorite data. *J. Am. Stat. Assoc.*, 75(369):42–73, 1980.

[259] U. Grenander. *Abstract Inference*. John Wiley and Sons, New York, 1981.

[260] T. Poggio, V. Torre, and C. Koch. Computational vision and regularization theory. *Nature*, 317:314–319, 1985.

[261] A. Yuille and N. M. Grzywach. A computational theory for the perception of coherent visual motion. *Nature*, 33(5):71–74, 1988.

[262] E. Levitan and G. T. Herman. A maximum a posteriori probability expectation maximization algorithm for image reconstruction in emission tomography. *IEEE Trans. Med. Imaging*, 6(3):185–192, 1987.

[263] H. Hart and Z. Liang. Bayesian image processing in two dimensions. *IEEE Trans. Med. Imaging*, 6(3):201–208, 1987.

[264] G. T. Herman and D. Odhner. Performance evaluation of an iterative image reconstruction algorithm for positron emission tomography. *IEEE Trans. Med. Imaging*, MI-10:336–346, 1991.

[265] L. S. Joyce and W. L. Root. Precision bounds in superresolution processing. *J. Opt. Soc. Am. A*, 1(2):149–168, 1984.

[266] S. J. Lee, A. Rangarajan, and G. Gindi. Bayesian image reconstruction in spect using higher order mechanical models as priors. *IEEE Trans. Med. Imaging*, 14(4):669–680, 1995.

[267] F. Guichard and J-M Morel. *Image Analysis and Partial Differential Equations*. 2001.

[268] Faisal Beg. Variational and computational methods for flows of diffeomorphisms in image matching and growth in computational anatomy. PhD thesis, The Johns Hopkins University, July 2003.

[269] P. Dupuis, U. Grenander, and M. I. Miller. Variational problems on flows of diffeomorphisms for image matching. *Quart. Appl. Math.*, 56:587–600, 1998.

[270] A Trouvé. Infinite dimensional group action and pattern recognition. *Quar. Appl. Math.*, 1999.

[271] N. Aronszajn. Theory of reproducing kernels. *Trans. Am. Math. Soc.*, 68(3):337–404, 1950.

[272] F. L. Bookstein. *Morphometric Tools for Landmark Data: Geometry and Biology*. Cambridge University Press, 1991.

[273] M. I. Miller, G. E. Christensen, Y. Amit, and U. Grenander. Mathematical textbook of deformable neuroanatomies. *Proc. Natl. Acad. Sci. USA*, 90(24), 1993.

[274] J. C. Gee and R. K. Bajcsy. Elastic matching: Continuum mechanical and probabilistic analysis. In A. W. Toga, editor, Brain Warping. Academic Press, New York, 1999.

[275] P. M. Thompson, J. N. Giedd, R. P. Woods, D. MacDonald, A. C. Evans, and A. W. Toga. Growth patterns in the developing brain detected by using continuum mechanical tensor maps. *Nature*, 404:190–193, 2000.

[276] A. Trouvé. Action de groupe de dimension infinie et reconnaissance de formes. *C. R. Acad. Sci. Paris, Sér. I*, (321):1031–1034, 1995.

[277] A. Trouvé. Diffeomorphisms groups and pattern matching in image analysis. *Int. J. Comput. Vision*, 28:213–221, 1998.

[278] M. I. Miller, A. Trouvé, and L. Younes. On the metrics and Euler-Lagrange equations of computational anatomy. *Ann. Rev. Biomed. Eng.*, 4:375–405, 2002.

[279] V. I. Arnold and B. A. Khesin. Topological methods in hydrodynamics. *Ann. Rev. Fluid Mech.*, 24:145–166, 1992.

[280] D. Mumford. Pattern theory and vision. *Questions Mathématiques En Traitement Du Signal et de L'Image*, chapter 3, pages 7–13. Institut Henri Poincaré, 1998.

[281] M. I. Miller, A. Trouvé, and L. Younes. Geodesic shooting for computational anatomy. *J. Math. Imaging Vision*, 2004. accepted for publication, to appear.

[282] S. Joshi and M. I. Miller. Landmark matching via large deformation diffeomorphisms. *IEEE Trans. Image Process.*, 9(8):1357–1370, 2000.

[283] Grace Wahba. Spline Models for Observational Data, volume Regional Conference Series in Applied Mathematics. SIAM, 1990.

[284] M. Vaillant, M. I. Miller, L. Younes, and A. Trouvé. Statistical analysis of diffeomorphisms via geodesic shooting. *NeuroImage*, 2004.

[285] M. F. Beg, M. I. Miller, A. Trouvé, and L. Younes. Computing metrics via geodesics on flows of diffeomorphisms. *Int. J. Comp. Vision*, 61:139–157, 2004.

[286] M. I. Miller and L. Younes. Group actions, homeomorphisms, and matching: A general framework. *Int. J. Comput. Vision*, 41(1/2):61–84, 2001.

[287] D. G. Kendall. Shape manifolds, procrustean metrics and complex projective spaces. *Bull. Lond. Math. Soc.*, 16:81–121, 1984.

[288] R. Bajcsy, R. Lieberson, and M. Reivich. A computerized system for the elastic matching of deformed radiographic images to idealized atlas images. *J. Comp. Assist. Tomog.*, 7(4):618–625, 1983.

[289] J. Dengler and M. Schmidt. The dynamic pyramid- a model for motion analysis with controlled continuity. *Int. J. Pattern Recognition Arti. Intell.*, 2(2):275–286, 1988.

[290] R. Dann, J. Hoford, S. Kovacic, M. Reivich, and R. Bajcsy. Evaluation of Elastic Matching Systems for Anatomic (CT, MR) and Functional (PET) Cerebral Images. *J. Comp. Assist. Tomog.*, 13(4):603–611, 1989.

[291] R. Bajcsy and S. Kovacic. Multiresolution Elastic Matching. *Comput. Vision, Graphics, Image Process.*, 46:1–21, 1989.

[292] J. Gee, L. L. Briquer, D. R. Haynor, and R. Bajcsy. Matching structural images of the human brain using statistical and geometrical image features. *Visualization in Biomedical Computing*, volume SPIE 2359, pages 191–204, 1994.

[293] R. D. Rabbitt, J. A. Weiss, G. E. Christensen, and M. I. Miller. Mapping of hyperelastic deformable templates using the finite element method. *Presented at the International Symposium on Optical Science, Engineering and Instrumentation*, July 1995.

[294] C. Davatzikos. Spatial transformation and registration of brain images using elastically deformable models. *Comput. Vision Image Understanding*, 66(2):207–222, 1997.

[295] G. E. Christensen. Consistent linear-elastic transformations for image matching. In A. Kuba and M. Samal, editors, *XVIth International Conference on Information Processing in Medical Imaging*, Visegraàd, Hungary, June 1999.

[296] G. E. Christensen, H. J. Johnson, J. W. Haller, M. W. Vannier, and J. L. Marsh. Synthesizing average 3D anatomical shapes using deformable templates. In K.M. Hanson, editor, *Medical Imaging 1999: Image Processing*, Proceedings of SPIE Vol. 3661, pages 574–582, Feb. 1999.

[297] A. W. Toga, P. M. Thompson, and B. A. Payne. Modeling morphometric changes of the brain during development. In R. W. Thatcher, G. R. Lyon, and N. Krasnegor, editors, Developmental Neuroimaging: Mapping the Development of the Brain and Behavior, Academic Press, New York, 1996.

[298] M. I. Miller D. Bitouk and L. Younes. Clutter invariant atr. *IEEE Trans. PAMI*, 27(5), 2005.

[299] *Prism 3.1 User's Manual*. Keweenaw Research Center, Michigan Technological University, 1987.

[300] S. P. Jacobs, J. A. O'Sullivan, M. Faisal A. Srivastava, D. L. Snyder, and M. I. Miller. Automatic target recognition using sequences of hrr radar reflections.

[301] Automatic target recognition using sequences of high resolution radar range-profiles. *IEEE Trans. Aerospace Electron. Syst.*, 2000.

[302] A. Srivastava and U. Grenander. Metrics for target recognition. *Proceedings of the SPIE, Applications of Artificial Neural Networks in Image Processing III*, volume 3307, pages 29–37, January 1998.

[303] Anuj Srivastava, Michael Miller, and Ulf Grenander. Ergodic algorithms on special euclidean groups for atr. *Prog. Syst. Control: Syst. Control Twenty-First Century*, 22:327–350, 1997.

[304] G. H. Golub and C. F. Van Loan. *Matrix Computations*. The Johns Hopkins University Press, Baltimore, 1983.

[305] Cash J. Costello. Medical image registration using the Hilbert-Schmidt estimator. PhD Thesis, Department of Biomedical Engineering, The Johns Hopkins University, 2000.
[306] A. Srivastava. Automated target tracking and recognition using jump diffusion processes. MS Thesis, Washington University, St. Louis, Missouri, December 1993.
[307] M. Loizeaux, A. Srivastava, and M. I. Miller. Estimation of pose and location of ground targets for atr. *Proceedings of the SPIE, Signal Processing, Sensor Fusion and Target Recognition*, volume 3720, 1999.
[308] J. A. O'Sullivan, M. D. DeVore, V. Kedia, and M. I. Miller. Automatic target recognition performance for SAR imagery using conditionally Gaussian model. *IEEE Trans. Aero. Electron. Syst.*, 37(1):91–108, 2001.
[309] M. D. DeVore. Recognition Performance from Synthetic Aperture Radar Imagery Subject to System Resource Constraints. PhD. thesis, Washington University, 2001.
[310] Jr. T. J. Green and J. H. Shapiro. Target detection performance using 3d laser radar images. *Proc. SPIE*, 1471:328–341, 1991.
[311] Jr. T. J. Green and J. H. Shapiro. Maximum-likelihood laser radar range profiling with the expectation-maximization algorithm. *Opt. Eng.*, 31:2343–2354, 1992.
[312] D. Park, J. H. Shapiro, and R. W. Reinhold. Performance analyses for peak-detecting laser radars. *Proc. SPIE*, 663: 38–56, 1986.
[313] J. Shapiro, B. A. Capron, and R. C. Harney. Imaging and target detection with a heterodyne-reception optical radar. *Appl. Opt.*, 20(19):3292–3313, 1981.
[314] Jr. T. J. Green and J. H. Shapiro. Detecting objects in 3d laser radar range images. *Opt. Eng.*, 33:865–873, 1994.
[315] Jr. T. J. Green and J. H. Shapiro. Maximum-likelihood laser radar range profiling with the expectation-maximization algorithm. *Opt. Eng.*, 31:2343–2354, 1992.
[316] J. H. Shapiro. Target reflectivity theory for coherent laser radars. *Appl. Opt.*, 21:3398–3407, 1982.
[317] A. D. Lanterman, M. I. Miller, D. L. Snyder, and W. J. Miceli. The unification of detection, tracking, and recognition for millimeter wave and infrared sensors. In W.J. Miceli, editor, *Radar/Ladar Processing*, volume 2562, pages 150–161. San Diego, CA, 1995.
[318] S. M. Hannon. Detection Processing for Multidimensional Laser Radars. PhD. Thesis, Department of Electrical Engineering and Computer Science, MIT, 1990.
[319] D. R. Wehner. *High Resolution Radar*. Artech House, Dedham, MA., 1987.
[320] H. L. Van Trees. *Detection, Estimation and Modulation Theory, Part III*. John Wiley and Sons, New York, 1971.
[321] D. Bitouk, U. Grenander, M. I. Miller, and P. Tyagi. Fisher information measures for atr in clutter. *Proc. SPIE*, 2001.
[322] J. Kostakis, M. L. Cooper, T. J. Green, M. I. Miller, J. A. O'Sullivan, J. H. Shapiro, and D. L. Snyder. Multispectral active-passive sensor fusion for ground-based target orientation estimation. *Proceedings of the SPIE Conference on Automatic Target Recognition VIII*, 3371(3371):500–507, 1998.
[323] J. Kostakis, M. L. Cooper, T. J. Green, M. I. Miller, J. A. O'Sullivan, J. H. Shapiro, and D. L. Snyder. Multispectral sensor fusion for ground-based ground-based target orientation estimation: Fl(ir, ladar, hrr). *Proceedings of the SPIE Conference on Automatic Target Recognition VIII*, 1999.
[324] J. K. Bounds. The infrared airborne radar sensor suite. MIT Research Laboratory of Electronics Technical Report 610, 1996.
[325] U. Grenander, A. Srivastava, and M. I. Miller. Asymptotic performance analysis of Bayesian target recognition. *IEEE Trans. Inform. Theory*, 46(4):1658–1665, 2000.
[326] M. Lindenbaum. Bounds on shape recognition performance. *IEEE Trans. Pattern Anal. Mach. Intell.*, 17(7):666–680, 1995.
[327] Ben Yen. Target recognition for FLIR and Laser Radar Systems. M.S. Thesis, Department of Electrical Engineering and Computer Science, MIT, Cambridge, MA, 2000.
[328] H. L. Van Trees. *Detection, Estimation and Modulation Theory, Part I*. John Wiley and Sons, New York, 1968.
[329] P. Belhumeur and J. P. Hespanha an D. J. Kriegman. Eigenfaces vs. fisherfaces: Recognition using class specific linear projection. *IEEE Trans. PAMI*, 45–58, 1997.
[330] R. Epstein, P. W. Hallinan, and A. L. Yuille. 5 ± eigenimages suffice: An empirical investigation of low-dimensional lighting models. *Proceedings of IEEE Workshop on Physics-Based Modeling in Computer Vision*, to appear 1995.
[331] A. Yuille. Deformable templates for face recognition. *J. of Cognitive Neurosci.*, 3, 1991.
[332] M. L. Cooper, A. D. Lanterman, S. C. Joshi, and M. I. Miller. Representing the variation of thermodynamic state via principal components analysis. *Proceedings of the Third Workshop On Conventional Weapon ATR*, 481–490, 1996.

[333] M. L. Cooper, U. Grenander, M. I. Miller, and A. Srivastava. Accommodating geometric and thermodynamic variability for forward-looking infrared sensors. *Proceedings of the SPIE, Algorithms for Synthetic Aperture Radar IV*, 3070, 1997.

[334] M. L. Cooper and M. I. Miller. Information measures for object recognition. *Proceedings of the SPIE*, 3370, 1998.

[335] M. Cooper. Information Theoretic Methods for Object Recognition and Image Understanding. PhD. Thesis, Washington University, St. Louis, Missouri, 1999.

[336] A. D. Lanterman. Bayesian inference of thermodynamic state incorporating schwarz-rissanen complexity for infrared target recognition. *Opt. Eng.*, 39(5):1282–1292, 2000.

[337] Andrew Clinton. Splinetree, http://povplace.addr.com, 2000.

[338] J. Huang and D. Mumford. Statistics of natural images and models. 541–547, 1999.

[339] U. Grenander and A. Srivastava. Probability models for clutter in natural images. *IEEE Trans. Pattern Anal. Mach. Intell.*, 23(4):424–429, 2001.

[340] M. L. Cooper and M. I. Miller. Information measures for object recognition accommodating signature variability. *IEEE Trans. Inform. Theory*, 46(5):1896–1907, 2000.

[341] M. I. Miller U. Grenander and A. Srivastava. Hilbert-schmidt lower bounds for estimators on matrix lie groups for atr. *IEEE Trans. on PAMI*, 20(8):790–802, 1998.

[342] A. Srivastava. Inference on Transformation Groups Generating Patterns on Rigid Motions. PhD. Dissertation, Department of Electrical Engineering, Sever Institute of Technology, Washington University, St. Louis, MO, 1996.

[343] L. Shao, A. Hero, W. Rogers, and N. Clinthorne. The mutual information criterion for spect aperture evaluation and design. *IEEE Trans. Med. Imaging*, 8(4):322–36, 1989.

[344] R. G. Gallager. *Information Theory and Reliable Communication*. John Wiley and Sons, Inc., New York, 1968.

[345] T. Berger. *Rate Distortion Theory: A Mathematical Basis for Data Compressions*. Prentice Hill, Inc, Englewood Cliffs, N.J., 1971.

[346] R. E. Blahut. Computation of Channel Capacity and Rate-Distortion Functions. *IEEE Trans. Inform. Theory*, IT-18(4):460–473, 1972.

[347] I. Csiszar. On the Computation of Rate-Distortion Functions. *IEEE Trans. Information Theory*, IT-20: 122–124, 1974.

[348] J. A. O'Sullivan. Alternating minimization algorithms: from blahut-arimoto to expectation-maximization. In A. Vardy, editor, *Codes,Curves, and Signals: Common Threads in Communications*, pages 173–192. Kluwer Academic, Boston, 1998.

[349] U. Grenander and M. I. Miller. Computational anatomy: An emerging discipline. *Quart. Appl. Math.*, 56:617–694, 1998.

[350] J. Feldmar, N. Ayache, and F. Betting. 3D-2D projective registration of free-form curves and surfaces. *Comput Vision Image Und.: CVIU*, 65(3):403–424, 1997.

[351] A. Bartesaghi and G. Sapiro. A system for the generation of curves on 3d brain images. *Hum. Brain Mapp.*, 14:1–15, 2001.

[352] L. M. Lorigo, O. D. Faugeras, W. E. Grimson, R. Keriven, R. Kikinis, A. Nabavi, and C. F. Westin. Curves: curve evolution for vessel segmentation. *Med. Image Anal.*, 5(3):195–206, 2001.

[353] M. E. Rettmann, X. Han, C. Xu, and J. L. Prince. Automated sulcal segmentation using watersheds on the cortical surface. *Neuroimage*, 15(2):329–344, 2002.

[354] A. Cachia, J. F. Mangin, D. Riviere, F. Kherif, N. Boddaert, A. Andrade, D. Papadopoulos-Orfanos, J. B. Poline, I. Bloch, M. Zilbovicius, P. Sonigo, F. Brunelle, and J. Regis. A primal sketch of the cortex mean curvature: a morphogenesis based approach to study the variability of the folding patterns. *IEEE Trans. Med. Imaging*, 22(6):754–765, 2003.

[355] A. Dale and M. Sereno. Improved localization of cortical activity by combining EEG and MEG with MRI cortical surface reconstruction: A linear approach. *J. Cogn. Neurosci.*, 5(2):162–176, 1993.

[356] T. McInerney and D. Terzopoulos. A dynamic finite element surface model for segmentation and tracking in multidimensional medical images with application to cardiac 4d image analysis. *Comp. Med. Imag. Graph.*, 419(1):69–83, 1995.

[357] C. Xu, D. L. Pham, M. E. Rettmann, D. N. Yu, and J. L. Prince. Reconstruction of the human cerebral cortex from magnetic resonance images. *IEEE Trans. Med. Imaging*, 18(6):467–480, 1999.

[358] D. L. Pham, C. Xu, and J. L. Prince. Current methods in medical image segmentation. *Ann. Rev. Biomed. Eng.*, 2:315–337, 2000.

[359] B. Wandell. Computational neuroimaging of human visual cortex. *Ann. Rev. Neurosci.*, 22:145–179, 1999.

[360] X. Zeng, L. H. Staib, R. T. Schultz, and J. S. Duncan. Segmentation and measurement of the cortex from 3-d mr images using coupled-surfaces propagation. *IEEE Trans. Med. Imaging*, 18(10):927–937, 1999.

[361] P. M. Thompson, R. P. Woods, M. S. Mega, and A. W. Toga. Mathematical/computational challenges in creating deformable and probabilistic atlases of the human brain. *Hum. Brain Mapp.*, 9(2):81–92, 2000.

[362] D. C. van Essen, J. W. Lewis, H. A. Drury, N. A. Hadjikhani, R. L. Tootell, M. Bakircioglu, and M. I. Miller. Mapping visual cortex in monkeys and humans using surface-based atlases. *Vision Res.*, 41:1359–1378, 2001.

[363] X. Gu, Y. Wang, T. F. Chan, P. M. Thompson, and S-T Yau. Genus zero surface conformal mapping and its application to brain surface mapping. *IEEE Trans. Med. Imaging*, 23, 2004.

[364] D. Tosun, M. E. Rettmann, X. Han, X. Tao, C. Xu, S. M. Resnick, D. Pham, and J. L. Prince. Cortical surface segmentation and mapping. *NeuroImage*, 2004.

[365] D. C. Van Essen. Surface-based approaches to spatial localization and registration in primate cerebral cortex. *NeuroImage*, 2004.

[366] J-F. Mangin, D. Riviere, A. Cachia, E. Duchesnay, Y. Cointepas, D. Papadopoulus-Orfanos, T. Ochiai, and J. Regis. A framework for studying cortical folding patterns. *NeuroImage*, 2004.

[367] D. L. Collins, C. J. Holmes, Terrence M. Peters, and A. C. Evans. Automatic 3-d model-based neuroanatomical segmentation. *Hum. Brain Mapp.*, 3:190–208, 1995.

[368] R. Kikinis, M. E. Shenton, D. V. Iosifescu, R. W. McCarley, P. Saiviroonporn, H. H. Hokama, A. Robatino, D. Metcalf, C. Wible, C. M. Portas, R. M. Donnino, and F. A. Jolesz. A digital brain atlas for surgical planning, model-driven segmentation, and teaching. *IEEE Trans Visual Comput Graph*, 2(3):232–241, 1996.

[369] P. Teo, G. Sapiro, and B. Wandell. Creating connected representations of cortical gray matter for functional mri visualization. *IEEE Trans. Med. Imaging*, 16:852–863, 1997.

[370] B. Crespo-Facorro, J. J. Kim, N. C. Andreasen, R. Spinks, D. S. O'Leary, H. J. Bockholt, G. Harris, and V. A. Magnotta. Cerebral cortex: a topographic segmentation method using magnetic resonance imaging. *Psychiat. Res.: NeuroImaging*, 100:97–126, 2000.

[371] J. T. Ratnanather, K. N. Botteron, T. Nishino, A. B. Massie, R. M. Lal, S. G. Patel, S. Peddi, R. D. Todd, and M. I. Miller. Validating cortical surface analysis of medical prefrontal cortex. *NeuroImage*, 14:1058–1069, 2001.

[372] D. W. Shattuck, S. R. Sandor-Leahy, K. A. Schaper, D. A. Rottenberg, and R. M. Leahy. Magnetic resonance image tissue classification using a partial volume model. *NeuroImage*, 13:856–876, 2001.

[373] B. Fischl, D. H. Salat, E. Busa, M. Albert, M. Dieterich, C. Haselgrove, A. van der Kouwe, R. Killanay, D. Kennedy, S. Klaveness, A. Montillo, N. Makris, B. Rosen, and A. M. Dale. Whole brain segmentation: Automated labeling of neuroanatomical structures in the human brain. *Neuron*, 33:341–355, 2002.

[374] X. Han, C. Xu, U. Braga-Neto, and J. L. Prince. Topology correction in brain cortex segmentation using a multiscale, graph-based algorithm. *IEEE Trans. Med. Imaging*, 21(2):109–121, 2002.

[375] S. O. Dumoulin, R. D. Hoge, Jr. C. L. Baker, R. F. Hess, R. L. Achtman, and A. C. Evans. Automatic volumetric segmentation of human visual retinotopic cortex. *NeuroImage*, 18(3):576–587, 2003.

[376] B. Fischl, A. van der Kouwe, C. Destrieux, E. Halgren, F. Segonne, D. H. Salat, E. Busa, L. J. Seidman, J. Goldstein, D. Kennedy, V. Caviness, N. Makris, B. Rosen, and A. M. Dale. Automatically parcellating the human cerebral cortex. *Cereb. Cortex*, 14(1):11–22, 2004.

[377] B. Fischl, D. H. Salat, A. J. W. ven der Kouwe, F. Segonne, and A. M. Dale. Sequence-independent segmentation of magnetic resonance images. *NeuroImage*, 2004.

[378] A. Pitiot, H. Delingette, P. M. Thompson, and N. Ayache. Expert knowledge guided segmentation system for brain mri. *NeuroImage*, 2004.

[379] F. L. Bookstein. Landmark methods for forms without landmarks: morphometrics of group differences in outline shape. *Med. Image Anal.*, 1:225–243, 1996.

[380] F. Bookstein. Shape and the information in medical images: A decade of the morphometric synthesis. *Comput. Vision Image Und.*, 66:97118, 1997.

[381] P. R. Andresen, F. L. Bookstein, K. Conradsen, B. K. Ersbll, J. L. Marsh, and S. Kreiborg. Surface-bounded growth modeling applied to human mandibles. *IEEE Trans. Mcd. Imaging*, 19:1053–1063, 2000.

[382] J. C. Gee. On matching brain volumes. *Pattern Recogn.*, 32:99–111, 1999.

[383] J. C. Gee and D. R. Haynor. Numerical methods for high-dimensional warps. In A. W. Toga, editor, *Brain Warping*, pages 101–113. Academic Press, San Diego, CA, 1999.

[384] B. Avants and J. C. Gee. Geodesic estimation for large deformation anatomical shape averaging and interpolation. *NeuroImage*, 23(Supplement 1):S139–S150, 2004.

[385] A. C. Evans, W. Dai, L. Collins, P. Neelin, and S. Marret. Warping of a computerized 3-d atlas to match brain image volumes for quantitative neuroanatomical and functional analysis. *Image Process.*, 1445: 236–246, 1991.

[386] K. J. Friston, C. D. Frith, P. F. Liddle, and R. S. J. Frackowiak. Plastic transformation of pet images. *J. Comp. Assist. Tomog.*, 15:634–639, 1991.

[387] A. W. Toga, P. K. Banerjee, and B. A. Payne. Brain warping and averaging. *J. Cereb. Blood Flow Metab.*, 11:S560, 1991.

[388] D. L. Collins, P. Neelin, T. M. Peters, and A. C. Evans. Automatic 3d intersubject registration of mr volumetric data in standardized talairach space. *J. Comp. Assist. Tomog.*, 192–205, 1994.

[389] K. J. Friston, J. Ashburner, C. D. Frith, J.-B. Poline, J. D. Heather, Liddle, and R. S. J. Frackowiak. Spatial registration and normalization of images. *Hum. Brain Mapp.*, 2:165–189, 1995.

[390] G. E. Christensen, A. A. Kane, J. L. Marsh, and M. W. Vannier. Synthesis of an individualized cranial atlas with dysmorphic shape. *IEEE Proc. Math. Methods Biomed. Image Anal., in press*, 1996.

[391] C. Davatzikos. Spatial normalization of 3-d brain images using deformable models. *J. Comp. Assist. Tomog.*, 20(4):656,665, 1996.

[392] S. K. Kyriacou, C. Davatzikos, S. J. Zinreich, and R. N. Bryan. Modeling brain pathology and tissue deformation using finite element based nonlinear elastic models. *IEEE Trans. Med. Imaging*, 1997. submitted.

[393] D. V. Iosifescu, M. E. Shenton, S. K. Warfield, R. Kikinis, J. Dengler, F. A. Jolesz, and R. W. McCarley. An automated registration algorithm for measuring mri subcortical brain structures. *NeuroImage*, 6:13–25, 1997.

[394] P. M. Thompson and A. W. Toga. Anatomically driven strategies for high-dimensional brain image warping and pathology detection. In A. W. Toga, editor, *Brain Warping*, pages 311–336. Academic Press, San Diego, CA, 1999.

[395] B. Fischl, M. I. Sereno, R. B. H. Tootell, and A. M. Dale. High-resolution intersubject averaging and a coordinate system for the cortical surface. *Hum. Brain Mapp.*, 8(4):272–284, 1999.

[396] S. Warfield, A. Robatino, J. Dengler, F. Jolesz, and R. Kikinis. Nonlinear registration and template-driven segmentation. In *Brain Warping*, pages 67–84. Academic Press, New York 1999.

[397] N. Hata, A. Nabavi, W. M. Wells III, S. K. Warfield, R. Kikinis, P. McL. Black, and F. A. Jolesz. Three-dimensional optical flow method for measurement of volumetric brain deformation from intraoperative mr images. 24(4):531–538, 2000.

[398] A. Guimond, A. Roche, N. Ayache, and J. Meunier. Multimodal Brain Warping Using the Demons Algorithm and Adaptative Intensity Corrections. *IEEE Trans. Med. Imaging*, 20(1):58–69, 2001.

[399] P. Cachier, J.-F. Mangin, X. Pennec, D. Rivière, D. Papadopoulos-Orfanos, J. Régis, and N. Ayache. Multisubject Non-Rigid Registration of Brain MRI using Intensity and Geometric Features. In W. J. Niessen and M. A. Viergever, editors, *4th Int. Conf. on Medical Image Computing and Computer-Assisted Intervention (MICCAI'01)*, volume 2208 of *LNCS*, Utrecht, The Netherlands, pages 734–742, October 2001.

[400] A. Pitiot, A. W. Toga, and P. M. Thompson. Adaptive elastic segmentation of brain mri via shape-model-guided evolutionary programming. *IEEE Trans. Med. Imaging*, 21(8):910–923, 2002.

[401] G. E. Christensen, S. C. Joshi, and M. I. Miller. Volumetric transformation of brain anatomy. *IEEE Trans. Med. Imaging*, 16(6):864–877, 1997.

[402] L. Younes. Computable elastic distances between shapes. *SIAM J. Appl. Math.*, 1998.

[403] J. Glaunes, A. Trouvé, and L. Younes. Diffeomorphic matching of distributions: A new approach for unlabelled point-sets and sub-manifolds matching. Proceeding of *CVPR*, 2004.

[404] A. W. Toga, P. Thompson, and B. A. Payne. *Development Neuroimaging: Mapping the Development of Brain and Behavior, chapter Modeling Morphometric Changes of the Brain During Development*. Academic Press, 1996.

[405] T. Paus, A. Zijdenbos, K. Worsley, D. L. Collins, J. Blumenthal, J. N. Giedd, J. L. Rapoport, and A. C. Evans. Structural maturation of neural pathways in children and adolescents: in vivo study. *Science*, 283:1908–1911, 1999.

[406] E. R. Sowell, B. S. Peterson, P. M. Thompson, S. E. Welcome, A. L. Henkenius, and A. W. Toga. Mapping cortical change across the human life span. *Nat. Neurosci*, 6(3):309–315, 2003.

[407] S. C. Joshi, B. Davis, M. Jomier, and G. Gerig. Unbiased diffeomorphic atlas construction for computational anatomy. *NeuroImage*, 2004.

[408] J. G. Csernansky, L. Wang, S. Joshi, J. P. Miller, M. Gado, D. Kido, D. McKeel, J. C. Morris, and M. I. Miller. Early dat is distinguished from aging by high dimensional mapping of the hippocampus. *Neurology*, 55, 2000.

[409] C. D. Good, I. S. Johnsrude, J. Ashburner, R. N. A. Henson, K. J. Friston, and R. S. J. Frackowiak. A voxel-based morphometric study of ageing in 465 normal adult human brains. *NeuroImage*, 14:21–36, 2001.

[410] J. G. Csernansky, L. Wang, J. Swank, J. Philip Miller, M. Gado, D. Kido, D. McKeel, M. I. Miller, and J. C. Morris. Hippocampal volume and shape predicts cognitive decline in the elderly. *Am. J. Psychiat.*, 2002.

[411] C. D. Good, R. I. Scahill, N. C. Fox, J. Ashburner, K. J. Friston, D. Chan, W. R. Crum, M. N. Rossor, and R. S. Frackowiak. Automatic differentiation of anatomical patterns in the human brain: validation with studies of degenerative dementias. *Neuroimage*, 17(1):29–46, 2002.

[412] J. Gee, L. Ding, Z. Xie, M. Lin, C. DeVita, and M. Grossman. Alzheimer's disease and frontotemporal dementia exhibit distinct atrophy-behavior correlates: a computer-assisted imaging study. *Acad. Radiol*, 10(12):1392–1401, 2003.

[413] P. M. Thompson, K. M. Hayashi, G. de Zubicaray, A. L. Janke, S. E. Rose, J. Semple, D. Herman, M. S. Hong, S. S. Dittmer, D. M. Doddrell, and A. W. Toga. Dynamics of gray matter loss in alzheimer's disease. *J. Neurosci.*, 23(3):994–1005, 2003.

[414] Lei Wang, Jeffrey S. Swank, Irena E. Glick, Mokhtar H. Gado, Michael I. Miller, John C. Morris, and John G. Csernansky. Changes in hippocampal volume and shape across time distinguish dementia of the alzheimer type from healthy aging. *NeuroImage*, 2003. to appear.

[415] M. I. Miller, M. Hosakere, A. R. Barker, C. E. Priebe, N. Lee, J. T. Ratnanather, L. Wang, M. Gado, J. C. Morris, and J. G. Csernansky. Labeled cortical mantle distance maps of the cingulate quantify differences between dementia of the alzheimer type and healthy aging. *Proc. Natl. Acad. Sci. USA*, 100:15172–15177, 2003.

[416] M. Ballmaier, E. R. Sowell, P. M. Thompson, A. Kumar, K. L. Narr, H. Lavretsky, S. E. Welcome, H. DeLuca, and A. W. Toga. Mapping brain size and cortical gray matter changes in elderly depression. *Biol. Psychiat.*, 55(4):382–389, 2004.

[417] M. Ballmaier, A. W. Toga, R. E. Blanton, E. R. Sowell, H. Lavretsky, J. Peterson, D. Pham, and A. Kumar. Anterior cingulate, gyrus rectus, and orbitofrontal abnormalities in elderly depressed patients: an mri-based parcellation of the prefrontal cortex. *Am. J. Psychiat.*, 161(1):99–108, 2004.

[418] M. E. Shenton, R. Kikanis, F. A. Jolesz, S. D. Pollak, M. LeMay, C. G. Wible, H. Hokama, J. Martin, D. Metcalf, M. Coleman, and R. W. McCarley. Abnormalities of the left temporal lobe and thought disorder in schizophrenia. a qualitative magnetic resonance imaging study. *New Engl. J. Med.*, 327(9): 604–612, 1992.

[419] J. L. Rapoport, J. N. Giedd, J. Blumenthal, S. Hamburger, N. Jeffries, T. Fernandez, R. Nicolson, J. Bedwell, M. Lenane, A. Zijdenbos, T. Paus, and A. Evans. Progressive cortical change during adolescence in childhood-onset schizophrenia. a longitudinal magnetic resonance imaging study. *Arch. Gen. Psychiat.*, 56:649–654, 1999.

[420] R. W. McCarley, M. E. Shenton, B. F. O'Donnell, S. F. Faux, R. Kikinis, P. G. Nestor, and F. A. Jolesz. Auditory p300 abnormalities and left posterior superior temporal gyrus volume reduction in schizophrenia. *Arch. Gen. Psychiat.*, 50:190–197, 1993.

[421] P. G. Nestor, M. E. Shenton, R. W. McCarley, J. Haimsen, S. Smith, B. O'Donnell, M. Kimble, R. Kikinis, and F. A. Jolesz. Neuropsychological correlates of mri temporal lobe abnormalities in schizophrenia. *Am. J. Psychiat.*, 150:1849–1855, 1993.

[422] John C. Csernansky, Sarang Joshi, Lei Wang, Mokhtar Gado, J. Philip Miller, Ulf Grenander, and Michael I. Miller. Hippocampal morphometry in schizophrenia by high dimensional brain mapping. *Proc. Natl Acad. Sci.*, 95:11406–11411, 1998.

[423] Lei Wang, Sarang C. Joshi, Michael I. Miller, and John G. Csernansky. Statistical analysis of hippocampal asymmetry in schizophrenia. *NeuroImage*, 14:531–545, 2001.

[424] K. L. Narr, P. M. Thompson, T. Sharma, J. Moussai, R. Blanton, B. Anvar, A. Edris, R. Krupp, J. Rayman, M. Khaledy, and A. W. Toga. Three–dimensional mapping of temporo–limbic regions and the lateral ventricles in schizophrenia: gender effects. *Biol. Psychiat.*, 50:84–97, 2001.

[425] P. M. Thompson, C. Vidal, J. N. Giedd, P. Gochman, J. Blumenthal, R. Nicolson, A. W. Toga, and J. L. Rapoport. Mapping adolescent brain change reveals dynamic wave of accelerated gray matter loss in very early-onset schizophrenia. *Proc. Natl Acad. Sci.*, 98:11650–11655, 2001.

[426] T. D. Cannon, P. M. Thompson, T. G. van Erp, A. W. Toga, V. P. Poutanen, M. Huttunen, J. Lonnqvist, C. G. Standerskjold-Nordenstam, K. L. Narr, M. Khaledy, C. I. Zoumalan, R. Dail, and J. Kaprio. Cortex mapping reveals regionally specific patterns of genetic and disease-specific gray-matter deficits in twins discordant for schizophrenia. *Proc. Natl Acad. Sci. USA*, 99(5):3228–3233, 2002.

[427] G. R. Kuperberg, M. R. Broome, P. K. McGuire, A. S. David, M. Eddy, F. Ozawa, D. Goff, W. C. West, S. C. Williams, A. J. van der Kouwe, D. H. Salat, A. M. Dale, and B. Fischl. Regionally localized thinning of the cerebral cortex in schizophrenia. *Arch. Gen. Psychiat.*, 60(9):878–888, 2003.

[428] R. Tepest, L. Wang, M. I. Miller, P. Falkai, and J. G. Csernansky. Hippocampal deformities in the unaffected siblings of schizophrenia subjects. *Biol. Psychiat.*, 54(11):1234–1240, 2003.

[429] J. G. Csernansky, M. K. Schindler, N. R. Spliner, L. Wang, M. Gado, L. D. Selemon, D. Rastogi-Cruz, P. A. Posener, and M. I. Miller. Abnormalities of thalamic volume and shape in schizophrenia. *Am. J. Psychiat.*, (161):896–902, 2004.

[430] H. E. Hulshoff Pol, H. G. Schnack, R. C. Mandl, W. Cahn, D. L. Collins, A. C. Evans, and R. S. Kahn. Focal white matter density changes in schizophrenia: reduced inter-hemispheric connectivity. *Neuroimage*, 21(1):27–35, 2004.

[431] K. L. Narr, R. M. Bilder, S. Kim, P. M. Thompson, P. Szeszko, D. Robinson, E. Luders, and A. W. Toga. Abnormal gyral complexity in first-episode schizophrenia. *Biol. Psychiat.*, 55(8):859–867, 2004.

[432] K. L. Narr, P. M. Thompson, P. Szeszko, D. Robinson, S. Jang, R. P. Woods, S. Kim, K. M. Hayashi, D. Asunction, A. W. Toga, and R. M. Bilder. Regional specificity of hippocampal volume reductions in first-episode schizophrenia. *Neuroimage*, 21(4):1563–1575, 2004.

[433] D. H. Salat, R. L. Buckner, A. Z. Snyder, D. N. Greve, R. S. Desikan, E. Busa, J. C. Morris, A. M. Dale, and B. Fischl. Thinning of the cerebral cortex in aging. *Cereb. Cortex*, 2004.

[434] E. R. Sowell, P. M. Thompson, S. E. Welcome, A. L. Henkenius, A. W. Toga, and B. S. Peterson. Cortical abnormalities in children and adolescents with attention-deficit hyperactivity disorder. *Lancet*, 362(9397):1699–1707, 2003.

[435] J. G. Levitt, R. E. Blanton, S. Smalley, P. M. Thompson, D. Guthrie, J. T. McCracken, T. Sadoun, L. Heinichen, and A. W. Toga. Cortical sulcal maps in autism. *Cereb. Cortex*, 13(7):728–735, 2003.

[436] C. H. Salmond, M. de Haan, K. J. Friston, D. G. Gadian, and F. Vargha-Khadem. Investigating individual differences in brain abnormalities in autism. *Phil. Trans. R. Soc. Lond. B Biol. Sci.*, 358(1430):405–413, 2003.

[437] J. A. Posener, L. Wang, J. L. Price, M. H. Gado, M. A. Province, M. I. Miller, C. M. Babb, and J. G. Csernansky. High-dimensional mapping of the hippocampus in depression. *Am. J. Psychiat.*, 160(1):83–89, 2003.

[438] L. C. Wiegand, S. K. Warfield, J. J. Levitt, Y. Hirayasu, D. F. Salisbury, S. Heckers, C. C. Dickey, R. Kikinis, F. A. Jolesz, R. W. McCarley, and M. E. Shenton. Prefrontal cortical thickness in first-episode psychosis: a magnetic resonance imaging study. *Biol. Psychiat.*, 55(2):131–140, 2004.

[439] E. R. Sowell, P. M. Thompson, S. N. Mattson, K. D. Tessner, T. L. Jernigan, E. P. Riley, and A. W. Toga. Regional brain shape abnormalities persist into adolescence after heavy prenatal alcohol exposure. *Cereb. Cortex*, 12(8):856–865, 2002.

[440] H. D. Rosas, A. K. Liu, S. Hersch, M. Glessner, R. J. Ferrante, D. H. Salat, A. van der Kouwe, B. G. Jenkins, A. M. Dale, and B. Fischl. Regional and progressive thinning of the cortical ribbon in huntington's disease. *Neurology*, 58(5):695–701, 2002.

[441] J-P. Thirion, S. Prima, G. Subsol, and N. Roberts. Statistical analysis of normal and abnormal dissymmetry in volumetric medical images. *Med. Image Anal.*, 4:111–121, 2000.

[442] David Rey, Gérard Subsol, Hervé Delingette, and Nicholas Ayache. Automatic detection and segmentation of evolving processes in 3D medical images: Application to multiple sclerosis. *Med. Image Anal.*, 2002. to appear.

[443] M. Sailer, B. Fischl, D. Salat, C. Tempelmann, M. A. Schonfeld, E. Busa, N. Bodammer, H. J. Heinze, and A. Dale. Focal thinning of the cerebral cortex in multiple sclerosis. *Brain*, 126(8):1734–44, 2003.

[444] J. G. Csernansky, L. Wang, S. C. Joshi, J. T. Ratnanather, and M. I. Miller. Computational anatomy and neuropsychiatric disease: Probabilistic assessment of variation and statistical inference of group differences, hemispheric asymmetry, and time-dependent change. *NeuroImage*, 2004.

[445] G. E. Christensen, R. D. Rabbitt, M. I. Miller, S. C. Joshi, U. Grenander, and T. A. Coogan. Topological properties of smooth anatomic maps. In *Information Processing in Medical Imaging*, pages 101–112. Kluwer Academic Publishers, 1995.

[446] John W. Haller, Gary E. Christensen, Sarang Joshi, John W. Newcomer, Michael I. Miller, John C. Csernansky, and Michael W. Vannier. Hippocampal MR imaging morphometry by means of general pattern matching. *Radiology*, 199(3):787–791, 1996.

[447] M. Corbetta, E. Akbudak, T. E. Conturo, A. Z. Snyder, J. M. Ollinger, H. A. Drury, M. R. Linenweber, S. E. Petersen, M. E. Raichle, D. C. Van Essen, and G. L. Shulman. A common network of functional areas for attention and eye movements. *Neuron*, 21(4)(1):761–773, 1998.

[448] H. Drury, D. C. van Essen, and J. W. Lewis. Towards probabilistic surface-based atlases of primate cerebral cortex. *J. Neurosci.*, 1999.

[449] Y. Cao, M. I. Miller, R. L. Winslow, and L. Younes. Large deformation metric mapping of vector fields. 2004. in review.

[450] F. M. Beg, P. A. Helm, Elliot McVeigh, M. I. Miller, and R. Winslow. Computational cardiac anatomy using magnetic resonance imaging. *Magn. Reson. Med. (MR)*, 52(5):1167–1174, 2004.

[451] E. T. Ahrens, P. T. Narasimhan, T. Nakada, and R. E. Jacobs. Small animal neuroimaging using magnetic resonance microscopy. *Prog. Nucl. Magn. Reson. Spectrosc.*, 40:275–306, 2002.

[452] J. Zhang, L. J. Richards, P. Yarowsky, H. Huang, P. C. van Zijl, and S. Mori. Three-dimensional anatomical characterization of the developing mouse brain by diffusion tensor microimaging. *Neuroimage*, 20(3):1639–1648, 1053-8119, 2003.

[453] J. Zhang, L. Richards, P. Yarowsky, M. I. Miller, P. van Zijl, and S. Mori. High resolution diffusion tensor imaging in study of postnatal mouse brain development. In *ISMRM 11th Scientific Meeting*, volume 1, Toronto, Canada, 2003.

[454] H. Le and A. Kume. The fréchet mean shape and the shape of means. *Adv. Appl. Probab. (SGSA)*, 32: 101–113, 2000.

[455] R. Bhattacharya and V. Patrangenaru. Nonparametric estimation of location and dispersion on riemannian manifolds. *J. Stat. Plan. Infer.*, 108:23–25, 2002.

[456] K. A. Gallivan, A. Srivastava, and Xiuwen L. Efficient algorithms for inferences on grassmann manifolds. *Proceedings of IEEE Conference on Statistical Signal Processing*, pages 315–318. IEEE, Piscataway NJ; 2003.

[457] P. T. Fletcher, C. Lu, and S. Joshi. Statistics of shape via principal geodesic analysis on lie groups. *Proceedings of CVPR*, pages 95–101. IEEE, Piscataway NJ; 2003.

[458] F. L. Bookstein. *Morphometric Tools for Landmark Data: Geometry and Biology*. Cambridge University Press, Cambridge, 1993.

[459] M. B. Stegmann. Analysis and segmentation of face images using point annotations and linear subspace techniques. Technical report, Informatics and Mathematical Modelling, Technical University of Denmark, DTU, August 2002.

[460] V. Blanz and T. A. Vetter. Morphable model for the synthesis of 3D faces. *SIGGRAPH '99 Conference Proceedings*, pages 187–194, 1999.

[461] T. W. Anderson. *An Introduction to Multivariate Statistical Analysis*. Wiley, Newyork, 1958.

[462] I. I. Gihman and A. V. Skorohod. *Introduction to the Theory of Random Processes*. Saunders, Philadelphia, 1965.

[463] S. N. Ethier and T. G. Kurtz. *Markov Processes*. Wiley, Newyork, 1986.

[464] D. L. Snyder A. D. Lanterman, and M. I. Miller. General metropolis-hasting jump diffusions for automatic target recognition in infrared scenes. *Opt. Eng.*, 36(4):1123–1137, 2000.

[465] W. K. Hastings. Monte carlo sampling methods using markov chains, and their applications. *Biometrika*, 57 (2):97–109, 1970.

[466] P. J. Green. Monte carlo methods: An overview. In invited contribution to *Proceedings of the IMA Conference on Complex Stochastic Systems and Engineering*, 1993.

[467] P. J. Green. Markov chain monte carlo in image analysis. In D. Spiegelhalter W. Gilks, S. Richardson, editor, *Practical Markov Chain Monte Carlo*. Chapman and Hall, to appear 1995.

[468] N. Metropolis, A. Rosenbluth, M. Rosenbluth, A. Teller, and E. Teller. Equation of state calculations by fast computing machines. *J. Phys. Chem.*, 21:1087, 1953.

[469] B. Gidas. Metropolis type monte-carlo simulation algorithms and simulated annealing. *Trends of Contemporary Probability*, 1993. to appear.

[470] J. Besag and P. J. Green. Spatial statistics and bayesian computation. *J. R. Stat. Soc. B*, 55(1):25–38, 1993.

[471] U. Grenander and M. I. Miller. Jump-diffusion processes for abduction and recognition of biological shapes. *Monograph of the Electronic Signals and Systems Research Laboratory*, 1991.

[472] A. Srivastava, U. Grenander, G. R. Jensen, and M. I. Miller. Jump-diffusion markov processes on orthogonal groups for object pose estimation. *J. of Stat. Plan. Infer.*, 103:15–37, April 2002.

[473] C. E. Lucius. Targeting systems characterization facility. *Thermal Imaging* SPIE 636, pages 40–46, 1986.

[474] W. R. Owens. Data-based methodology for infrared signature projection. *Thermal Imaging*, SPIE 636, pages 96–99, 1986.

[475] J. A. O'Sullivan, A. Srivastava, and V. Kedia. Performance analysis of atr from sar imagery. *Proceedings of the SPIE, Algorithms for Synthetic Aperture Radar Imagery V*, April 1998.

[476] K. V. Mardia. *Statistics of Directional Data*. Academic Press, London, 1972.

[477] G. B. Folland. *Real Analysis: Modern Techniques and Their Applications*. John Wiley and Sons, Newyork, 1984.

[478] R. S. Strichartz. *The Way of Analysis*. Jones and Bartlett Publishers, 2000.

[479] R. L. Wheeden and A. Zygmund. *Measure and Integral*. Dekker, Newyork, 1977.

[480] A. Friedman. *Partial Differential Equations of Parabolic Type*. Prentice-Hall, Englewood cliffs, 1964.

[481] D. L. Snyder, A. M. Hammoud, and R. L. White. Image recovery from data acquired with a charge-coupled-device camera. *J. Opt. Soc. Am. A*, 10:1014–1023, 1993.

[482] D. L. Snyder, M. Faisal, A. D. Lanterman, and R. L. White. Implementation of a modified richardson-lucy method for image restoration on a massively parallel computer to compensate for space-variant point-spread of a charge-coupled-device camera. *J. Opt. Soc. Am. A*, 12(12), 1995.

[483] D. L. Snyder, C. W. Helstrom, A. D. Lanterman, M. Faisal, and R. L. White. Compensation for readout noise in ccd images. *J. Opt. Soc. Am. A*, 12:272–283, 1995.

INDEX

Note: page numbers in **bold** refer to definitions and theorems, whilst those in *italics* refer to Figures and Tables.

1-1 constraint languages 75–6
1D Gibbs random fields 119–20
1-single-type branching processes
    extinction probability **63**
    probability generating function **60**
1-type transition processes
    extinction probability **71**
    Markhov chains **54**, 57
        mean 59
2-connection linear graphs 95
2D surface manifolds, eigenfunction models 419–21
2D surfaces **195**
2-type reverse polish branching processes 58

Abelian groups **175**
Abney, Steven, context-free grammar 83
acceptance probability 521, 542
actin-myosin shapes *146–7*
    anisotropic modeling 148
action on configurations **181**
active closed contours 232–4
active contour models, level set methods 237–40
active deformable spheres 228–9, 236–7
active models 226
    active shape models 494
    Gaussian random field models 240–3
    general calculus 229–32
    level set active contour models 237–40
active unclosed snakes and roads 234–6
adding the mean, least-squares estimation 21
addition theorem, spherical harmonics **287**
adjoints, Gaussian processes 124–**5**
affine group **178**
    action on background space 179
affine motions 376–7
aging, hippocampal surface patterns 506–7, *508*
aircraft tracking 157, *158*, *534*, 549–50
    detection of airframes 552
    Euler equations 550–1
    pose and identification 544–5, *546*
    pruning via prior distribution 552–3
all graphs 95, 96
almost-sure continuity **246**
    sample paths **299**, 300–2
alternating minimization, EM sequence 31
Alzheimer's disease, hippocampal surface patterns 506–7, *508*, *509*, 510
A-measurability, random processes **245**, 246
Amit, Y. 275, 301

amoeba
    active deformable surface spheres 290–3
    COSM, EM algorithm reconstruction *33*, 35
    triangulated graph *157*
anatomical manifolds
    automated construction 468–9
    comparison 469
anatomical source models 470–2
    group actions 472–3
anatomical submanifolds, diffeomorphic study 469
anatomy *see* computational anatomy
animal fur
    texture fields *114*
    texture synthesis *136*
anisotropic textures 147–51
    variable orientation 151–3
appendices, website address 4
arc-length of curves **191**
arc-length parameterization 317
arithmetic expression grammar 79–80
Artin, M. 174
assumption of consistency 405
asymptotic approximations of information measures 452–6
asymptotic consistency 406–7
asymptotic covariance, stationary Gaussian random fields 142
asymptotic equipartition (AEP) theorem 65–**6**
    branching processes 72–3
asymptotic maximum likelihood estimators 144
asymptotic normality, maximum-likelihood estimator (MLE) 28–9
asymptotics, Gaussian processes 138–42
a-typical trees 71
autocovariance function
    cyclo-stationary Gaussian processes **135**
    stationary Gaussian processes **133**
automated segmentation, cortex *482–3*
automated target recognition, accommodation of clutter 438
automatic speech recognition *see* speech recognition systems
automorphic pathologies **503**
    disease testing 503–10
auto-regressive image modeling 144–7
    Gaussian processes 127–9
    Green's kernels 268
Aylward, E. 492

background space, action of affine group 179
Bajcsy, R.K. 469

Baker, J. 92
Bakircioglu, M. 289–90
Banach spaces **253**, 519
  generators 518
bandlimited generalized random processes 272
Barron, A.R. and Clark, B.S. 38
Barta, P. 209
Baum-Welch algorithm **89**–90
Bayesian belief networks 52, *53*
Bayesian fusion of information 402–5
Bayesian information criterion 40
Bayesian texture segmentation 110–12
Bayes integration, pose estimation 427–8
Bayes posterior distribution 5–8
Beg, F. 333
  geodesic shooting 353, 354
  landmark matching 342–3
  LDDMM algorithm 476, 477, 479–80
  scalar MRI mapping of heart 486–7
Belhumeur, P. 418, 419
belief networks 52, *53*
Besag, J. 97, 110
  pseudolikelihood method 560–1
bias **24**
bigram language model 76, 81–2
  entropy *86*–7
  generators and configuration 160
bi-harmonic operator 279
binary order
  adequacy 167
  reduction of Markhov random fields 169–70
binary trees 79
binormal vector field **191**–2
Biomedical Informatics Research Network (BIRN) 502
birth-death processes 516
birth moves, Metropolis-Hastings jump-diffusion scheme 542
birth transformations **162**
Bitouk, D. 374, 402, 432, 437, 438
  alternating minimization algorithm 443
  clutter experiments 445
Blahut algorithm 457–9, 460–1, 462
bond function 154, **155**
bonds of generators 154, *155*
bond values, formal languages and grammars 158
Bookstein, F.L. 357, 469
  approximate metric 361, 370
Boothby, W.M. 174, 221
Botteron, K. 204, 212, 213
boundary conditions, Gaussian fields in unit cube 274–5
boundary integral 234
  level set function 238
bounded domains, periodic processes 258–62
brain
  applications of computational anatomy 469
  automated segmentation *482*–3
  caudate, mapping in Huntington's disease 492–3
  cortex
    representation of curves 192–5, 208, *209*, 210
  cortical atlases, metric mapping 483–4
  diffeomorphic studies of submanifolds 469–70, 481–2
  fitting curvatures 202–5
  hippocampus
    3D PCA *501*–2
    diffeomorphic mapping 481, *482*
       LDDMM algorithm 476, *478*
    landmark mapping 344
    MRI-MPRAGE image 502, *503*
    surface patterns in Alzheimer's and aging 506–7, *508*, *509*
    surface patterns in depression *510*
    surface patterns in schizophrenia 506, *507*
  Hoffman phantom 307, *308*
  mouse brain development study 489–92
  MRI registration 397–*8*, *399*
  multiple modality signal registration 429–31
  planar mapping 211–13
  representation of cortical curves 192–5
    dynamic programming 208, *209*, 210
  small deformation vector fields models 280–3
  sulcus, gyrus and geodesic curves on triangulated graphs 205–7
  superior temporal gyrus, use of dynamic programming 208, *209*, 210
  symmetry *180*
  triangulated graphs 156–7
  tumour growth *372*, 373
  whole brain landmark matching 363–5, *364*
brain segmentation, Gaussian mixture modeling 35–8
branching processes
  branching matrix 59–60
  entropy 68–9
  subcritical, critical and supercritical processes 70–1
  extinction probability 63–4
  moment-generating function 60–2
  tree structure 69
  see also multi-type branching processes
Brownian motion 252, 263–4
  almost surely sample path continuity 302
  Dynkin condition 524
  infinitesimal means and variances 524–5
Bucker, R. 501
Burger's equation 338
Burg's maximum-entropy principle, stationary Gaussian processes 143
Butterworth filters 150, 151

CAD models 179, *180*
  3D to 2D projections in Gaussian noise 386
  codebook construction 461–7
calculus, shape activation 229–32
canonical representation 166–7
  directed acyclic graphs 167–9
  Gaussian random fields 170–3
  Markhov random fields 169–70
Cao, Y. 472, 484, 486, 487
capacity, language of strings **77**
Cauchy-Schwartz inequality **18–19**, 317
Center for Language and Speech Processing (CLSP), HMM-based ASR system 91–2
Chan-Vese model 233
  steepest descent direction 238
Chapman-Kolmogorov equations 55, **514**

characteristic function **11**–12
characteristic functional **432**, 433
Chomsky, N., formal languages 74, 158
Chomsky normal form (CNF) grammars
   Inside/Outside algorithm 93
   trellis-based algorithm 94
Christensen, G.E. 275–6, 332, 336, 481, 482
   non-geodesic mapping algorithm 478–80
circle packing 210
circles
   as immersions 221
   local coordinate systems 217
   as manifolds 215
   normal deformation 236
Classical Projection Theorem **19**
clique systems **95**, *96*
   reduction to binary form 169–70
closed contours
   active models 232–4
   directed arcs as generators 185
   transformations 183–4
closure of language 77–8
closure of set **104**, *105*
clutter
   robust deformable templates 438–45
   target detection/identification 445–6
   transported generator model 431–7
CNR (carrier-to-noise ratio) 404–5
   LADAR 393
codebook construction 456–7
   CAD models 460–5
   output symbol distribution 465–7
Collins, C.R. and Stephenson, K. 211
combinatorics, typical trees 73–4
Commanche tank, FLIR imagery 454–6
communications model for image transmission 378–81
compact covariances 310–11
compact operators **254**–5
complete data densities **30**
complete-incomplete data problems 30–1
   principle of maximum entropy 47
complete orthonormal (CON) basis **248**, 255
   finite element generation method 294–6
complexity of models **43**
   model order estimation 38–40
complexity of patterns 1
complex numbers 175
   as Lie group 224
complex random variables 11
complex spherical harmonics 285–6
complex-valued random processes, measurability **245**
computational anatomy (CA) 468–9
   anatomical source model 470–2
      group actions 472–3
   Christensen non-geodesic mapping algorithm 478–80
   data channel model 473–4
   diffeomorphic study of anatomical submanifolds 469–70
   disease testing of automorphic pathology 503–10
   distribution free testing 510–11
   DT-MRI mapping of heart 484–8

extrinsic mapping of surface and volume submanifolds 480–4
Gaussian random fields 495–6
heteromorphic tumours 512–*13*
large deformation diffeomorphic metric mapping (LDDMM) 474–8, 496–8
momentum of landmarked shape spaces 498–502
small deformation Gaussian fields on surface submanifolds 502–3
small deformation setting 502
statistics analysis for shape fields 494–5
vector fields for growth 488–93
computational biology
   graph transformations 165–6
   PDAGs 52–4
   small deformation vector fields models 280–3
computational optical-sectioning microscopy 290, *292*–3
computational time variable, rotational motions 374
conditional distribution **10**
conditional entropy **7**, **8**
   directed acyclic graphs 65
   Markhov chains 66–7
conditional expectation 24
conditionally Gaussian random field models, photometric variability 422–4
conditional mean **22**–3
conditional mean minimum risk estimation 381–5
conditional probability, and maximum entropy 47–8
cone of support, aircraft tracking *552*, 553
configurations, bijection with similarities 184
configurations of patterns 154–*5*, *156*–8
congruence 174, 175
   triangles 180
connected sets, diffeomorphisms 222
conservation of momentum
   geodesics 326, 338–40
conservative jump processes 515
consistency **27**, 405
content of configurations 155
context-free grammars **75**, 83
   generators, configuration and regularity 160–2
   Inside/Outside algorithm 93
   regularity 159
   rewriting rules *159*
   trellis-based algorithm 94
context-free language models 83–4
   entropy *86*, 87
context-sensitive grammars **75**
   generators and configuration 162
   regularity 159
   re-writing rules *159*
continuity **246**
   sample paths **300**, 301–3
   second order processes **249**–50
continuous covariances 254
continuous functions, normed linear space 17
continuous mappings **219**
continuum, representations on 244
contour models, mitochondria 553–6
contours, levels set active models 237–40

convergence **246**
   maximum likelihood estimator 28–9
Cooper, M.L. 424–5, 427, 428
   pose estimation 448–9, 450–2
coordinate charts, manifolds **214**
coordinate frames **195**, 321–3
Corbetta, M. et al. 483, 485
cortex
   fitting curvatures 202–5
   representation of curves 192–5
      dynamic programming 208, *209*, 210
cortical atlases, metric mapping 483–4
cosets **175**
COSM (computational optical sectioning microscopy) 33, 35
Costello, C.J. 397
covariance
   Gaussian fields **123**
   Gaussian fields on sphere 288–9
   Gaussian vector fields **262**
   second order processes **249**, 250
   transported generator model 436
covariance operators 417
covariance trace class 262–3
"cover" pathologies 513
Cover, T. and Thomas, J. 44, 47
Cowell, R.G. et al., belief networks 52, *53*
Cramer decomposition **311**
Cramer-Rao inequality 25–6
crest lines **205**
   brain tissue 206
critical branching processes **60**
   entropy 70–1
cross modality registration 429–31
Csernansky, J.C. 481, 502, 504, 506–9
Csiszar, I. and Tusnady, G., iterative algorithm 31, 47
cumulant generating function **433**–4
CUP CAD model, codebook construction 461–7
curvatures **190**
   fitting to surfaces 198–205
      dynamic programming 207–10
curves **218**
   definitions 190–1
   Frenet representation **191**–5
   ridge curves and crest lines 205
cycles **50**, 52
cyclic graphs *95*, *96*
   directed arcs as generators 185
   generators, configuration and regularity 156
cyclic shift invariant operators 273
cyclomatic number 189
cyclo-stationarity **259**
cyclo-stationary Gaussian processes 134–7
   orthogonal representation 310–11
cyclo-stationary operators 282–4
cylinder
   principal curvatures 199
   shape operator *198*
cytoplasm, image analysis *98*–9, *149*, *151*

data channel model, computational anatomy 473–4
Dawid, A.P. 51
death moves, Metropolis-Hastings jump-diffusion scheme 543, *544*
death transformations **162**
deformable shapes, transport theorem **231**–2
deformable templates **181**, **182**, **347**, 357
deforming closed contours 226–7
deleted interpolation method, probabilities of m-grams 82
dementia, hippocampal surface patterns 506–7, *508*, *509*
Dempster, A., Laird, N. and Rubin, D. 47
   iterative algorithm 31
density of random variables **9**
depression, hippocampal surface patterns *510*
derivative operator, canonical representation 171, *172*
descendents of a site **50**
detectability 408–9
detection 445–6
   aircraft tracking 552
   hypothesis testing 408
   jump processes 543, 558–9
   Metropolis-Hastings jump-diffusion scheme 541–3
detection theory 445
Devore, M.D. 391
Dictyostelium discoideum
   3D motion studies 236
   optical-sectioning microscopy 289
diffeomorphic landmark mapping 343–5
diffeomorphism metric
   for image orbit 349
   N-shapes 361
      approximation for small deformation splines 361–5
diffeomorphisms **219**, 221, 222–3, **332**–3
   anatomic submanifolds 469–70, 480–3
   conditions on generation 333–5
   matrix group actions 223–4
   space of, metric 336–8
diffeomorphism splines, conservation of momentum 340–3
difference operators, Gaussian processes 124, 126–33, *132*
differentiability, second order processes 250–1
differential entropy **8**
differential of mapping 220
differential operators 335–6
   induction of Gaussian processes 271–3
      small deformation vector fields models 280
   in unit cube 274–80
diffusion processes **523**–5
   generators of 1D diffusions 525–7
   stochastic differential equations for sampling 527–8
   see also jump-diffusion inference
diffusion tensor magnetic resonance imaging (DT-MRI) 472, 484, *485*, **487**–8
   mouse brain development study 489–92
dimensional instability 304, *307*
dimension of manifold 214
DIMER graph type 182
Dini's Theorem **257**

directed acyclic graphs (DAGs) *50*
   canonical representation 167–9
   entropies 64–5
   graph transformations *165–6*
   Markhov chains 54–6
   natural language modeling 81–7
   probabilities on (PDAGs) 51–4
directed arcs, as generators of closed contours 185
directed graphs 49
   splitting property **51**
directional derivatives 197, 216–17
direct product groups **175**
discrete graphs *95*, 96
discrete integration, orthogonal group 385
discrete spaces, minimum risk hypothesis testing 12–16
disease testing, automorphic pathologies 503–10
distance between points on sphere 286
distance between sets **105**
distortion (penalty) function **456–7**
distribution free testing, computational anatomy (CA) 510–11
DNA 52
Dobrushin, Lanford, Ruelle equation **116**
Dobrushin, R.L. 97
drift gradients, mitochondria 558–9
drift vectors 537
Drury, H. 483, 485
Dupuis, P. 334
Dijkstra's algorithm 87
dynamic metric growth model 470, 471–2, 488
   growth from landmarked shape spaces 488–93
dynamic programming 88
   generation of curves 207–10
dynamics, incorporation into MMSE estimation 386–8
Dynkin properties, diffusions 523, 524

Echiverria theorem 538
Egevary's representation, cyclo-stationary fields 259
eigenfunction generation, singular value decomposition 418
eigenfunction models
   2D surface manifolds 419–21
   human face 418–19
eigenfunctions of operators **253**–4
eigen-spectrum **253**
   of Brownian motion 263–4
elasticity operator induced stationary priors 313–15
elasticity operators
   small deformation elasticity operator 273, 274–5, 279
   stationary Navier elasticity operator 278–9
elasticity-parameterized Gaussian vector fields, maximum-likelihood estimation 280–2
elastic matching 370
electron-microscopic autoradiography 33, 35
elliptic operators **300**
emission tomography 33, 34
   image deconvolution 307–8
   image reconstruction problems 303, 304
   smoothing 304–5
empirical covariance, photometric variability 416–21
empirical maximum entropy texture coding 113–15

energy density of shapes 229–30
energy function **101**
entropy **7**–8, 453, 454
   asymptotic equipartition (AEP) theorem 65–**6**
   branching processes 68–9
      subcritical, critical and supercritical processes 70–1
   directed acyclic graphs (DAGs) 64–5
   language models 85–7
   Markhov chains 66–8
   stationary Gaussian random fields 142–3
   stationary Gibbs fields 121–2
equilibrium conditions 370
equivalence 405
equivalence classes 174, 175
equivalence relations 175–7
Erlangen program 174
error exponents 15–16
   exponential error exponents theorem **410**–11
error probabilities 407–13
errors, type I and type II **14**
estimators, minimum risk estimation 6–7
Ethier, S.N. and Kurtz, T.G. 530
Euclid, congruence 174
Euclidean group **178**
Euclidean metric 439
   N-shapes 357–9
Euclidean motions, tracking 179
Euclidean space 17, 316
Euler angle representation 534
Euler equations 325–6
   aircraft tracking 550–1
   for photometric and geometric variation 369–70
even spherical harmonics **285**
events **9**
expectation-maximization (EM) algorithm 30, **31**
   Gaussian mixture modeling, brain segmentation 35–8
   Gaussian noise in spectrum estimation 31–3
   image restoration in point processes 33–5
   ML estimation of HMM parameters 89–92
   parameter estimation in natural language models 92–4
expected values **10**
exponential error exponents **410**–11, 413
exponential integral curve **321**
exponential of matrices **320**
exponential maximum entropy density 45
exponential spline smoothing via Good's roughness 305–8
exterior boundary of set **104**
external bonds of configuration
   formal grammars 159
   patterns 155
extinction probability
   branching processes 63–4, **72**
   finite-state Markhov chains **62**–3
extrapolation 6
extrinsic understanding 6
eyelids experiment, Younes 373

fabric
   texture fields *115*
   texture synthesis *136*

face
  3D PCA 500–*1*
  diffeomorphic mapping *481*, 499–*500*
  glasses and eyelids experiments *373*
  illumination variability 418–*19*
  landmark mapping *344*, 345
facial expressions 370
false alarm, probability of (type I error) **14**
false alarm rate **408**
family generation trees 57
family of graph transformations 162
fast marching method 237
field of events 8–**9**
fields, continuity 299
figure eight, as immersion 221
filtered point processes 432
filtering functions, texture representation 113–15
filters, anisotropic 148, 150–1
finite element method, generation of CON 294–6
finite rank operators 254
finite-state grammars **74**, 75
  generators, configuration and regularity 160
  regularity 159
  rewriting rules *159*
finite-state graphs, generators, configuration and regularity 157–8
finite-state languages 76–7
finite-state Markhov chains 54–6
  entropy 66–8
  extinction probability **62**–3
finite volume Gibbs distribution **105**
first order Markhov processes, maximum entropy distribution 45, 46
Fisher information **25**–6, 453–*4*
  monotonicity 403
Fisher information matrix **399**–402
Fisher's method of randomization 510–11
"flattening" of triangulated graphs 210–12
flight path generation 549
FLIR (forward-looking infrared) imaging 33, 393
  asymptotic error exponents *412*, 413
  Commanche database 454–6
  empirical eigenfunction models 419–*21*
  fusion of information 404–5
  ground vehicle recognition 534–5
  mutual information 451–2
  photometric variability *415*
  Poisson projection model 394–5
  pose estimation 425–7, 546–8
  thermal variation *428*
  urban clutter *149*, 151
forest graphs *95*, 96, 190
formal grammars **74**–5
  generators 158–62
formal languages 74–80
Forney, D. 87
Forward/Backward algorithm **90**–1
Fourier transform 313
Frenet representation of curves **191**–2
  in cortex 192–3
Frobenius (Hilbert-Schmidt) distortion 461, 462
Frobenius metric 329–30, 381, **382**

fundamental loops *see* independent loops
fundus beds 206
fundus curve generation 208
fusion of information 402–5

Galton-Watson processes 56, **58**
Gaussian curvature **199**
Gaussian distribution 528
  in generalized sense **271**
Gaussian Markhov random field models 440
  image synthesis 444–5
Gaussian maximum entropy density 45, 46
Gaussian mixture modeling, brain segmentation 35–8
Gaussian noise, spectrum estimation 31–3
Gaussian noise models, deblurring and denoising 308–9
Gaussian priors
  Good's roughness 305–6
  role in image reconstruction 303–4
  and smoothness 304–5
Gaussian processes (fields) **123**–4
  adjoints 124–**6**
  asymptotics 138–41
  characteristic function 11–2
  conditional mean 22–4
  difference operators 124
  induction via difference operators 126–33, *132*
  induction via linear differential operators 271–3
  on sphere 289–93
  stationary Gaussian random fields 133–4
    asymptotic covariance and log-normalizer 142
    entropy 142–3
  in unit cube 274–8
    maximum likelihood estimation 278–80
    small deformation vector fields models 280–3
  see also Gaussian random fields; Gaussian vector fields
Gaussian random fields
  active shapes 240–3
  canonical representation 170–3
  computational anatomy 495–6
  mitochondria 555–6
  on smooth surfaces 293
    Laplace-Beltrami operator with Neumann boundary conditions 293–7
Gaussian vector fields
  Brownian motion 263–4
  mean and covariance **262**–3
  white noise 264
Gauss-Markhov representation, cortical curves 193–5
GEC algorithm 478–80
Gee, J.C. 469
Geman, S. and Geman, D. 97, 107, 108
generalized linear groups **178**
generalized projective sensor transformation **379**
generalized random processes **271**
General Pattern Theory 1
generators 214
  of 1D diffusions 525–7
  of formal languages and grammars 158–62
  group actions **177**
  Markhov jump processes 519–20
  Markhov processes **517**–18

of patterns 154, *155–8*
  canonical representations 166–7
  relationship to similarities 184
  transported generator model 432
gene regulatory networks, *Saccharomyces cerevisiae* 165, 166
genetic regulatory networks, PDAGs 52–4
geodesic connection, normal momentum motion 350–4
geodesic distances, matrix groups 327–9
geodesic evolution equations, landmarks 498–9
geodesic generation 208, *209*
geodesic metric 317–18
geodesics 206
  conservation of momentum 326, 338–40
  in matrix groups 324–6
geodesic shooting *353–4*
Gibbs distribution **101**–4
  Markhov chains 120
  partition function 110–12
  splitting property 104–10
  stationary Gibbs random fields 116–18
    1D 119–20
    entropy 121–2
Gibbs' sampling, jump-diffusion processes 539–41
glasses experiment, Younes 373
Goel, V. 87
Good, I.J. 304
Good's roughness 304, 305
  exponential spline smoothing 306–9
  as Gaussian prior 305–6
Good-Turing method, probabilities of m-grams 82
gradients
  for drifts, jump-diffusion for mitochondria 558–9
  for growth of landmarked shapes 489
grammars
  arithmetic expressions 79–80
  pseudolinear 78–9
grammatical rules 74–5
grammatical transformations 162
Gram-Schmidt orthogonalization 248
graphs
  directed and undirected 49
  representation of regularity 2
graph theory, definitions 189–90
graph transformations 162–3
  directed acyclic graphs (DAGs) 165–6
  multiple object recognition 163–4
  parsing 164–5
Green's formula 239
Green's function 264–5, 267
Green's kernels 267–71, 363
Green's operator 271, 272, 277–8
ground vehicles
  pose estimation 546–8
  recognition 533–5, 539, 543–4
  see also tanks
group actions **176**, **320**–1
  anatomical source models 472–3
  on multiple generators **181**
  on N-shapes **361**
  on set of images 346–7
  on single generator **177**

groups **174**–5
  equivalence relations 175–7
  of geometric and photometric variation **415**
  matrix groups 177–81
  transformations from products of groups 181–4
growth
  equilibrium condition changes 370
  large deformation diffeomorphic metric mapping 475–6
growth connection of dense imagery, normal momentum motion 354–6
growth from landmarked shape spaces 488–93
gyri, curvature 205–7

Haller, J.W. 482
Hamman, B., curvature approximation algorithm 200–1
Hammersley Clifford result 169
Han, X. 240
Harris, T.E. 72
Hartemink, A.J. et al. 53–4
Hart, H. and Liang, Z. 304
heart
  DT-MRI 484–8
  mapping experiments 477
  scalar MRI mapping *486–7*
Heat Bath 523
Helm, P., scalar MRI mapping of heart *486–7*
Herman, G.T. and Odhner, D. 304
Heschl's gyrus *209*, 210, *211*, 212
heteromorphic pathologies 503, 512–13
hidden Markov models (HMM) 87–*8*
  in automatic speech-recognition system 91–2, 99–100
  ML estimation of parameters 89–91
hidden random branching processes, neighbourhood structure 100–1
hidden state sequences, MAP estimators 88
hierarchical directed acyclic graph language model 84–5
  entropy *86*, 87
hierarchical field via line processes 108–10
hierarchy of graphs, multiple object recognition 163–4
high range-resolution radar (HRR) 395–7
  fusion of information 404–5
  target orientation estimation 546
Hilbert-Schmidt distortion 461, 462
  output distribution *465*, *466*
Hilbert-Schmidt estimator 381, **382**, 383, 453–4
  computation for special Euclidean group 384–5
Hilbert-Schmidt metric 381, **382**, *383*
Hilbert-Schmidt theorem **255**
Hilbert spaces **18**–19, **20**, 319–20
  of photometric intensities **416–17**
  of photometric variability 440–1
  random variables 247–*8*
  reproducing kernel Hilbert spaces (RKHS) **265**–6
hippocampus
  3D PCA *501–2*
  diffeomorphic mapping 481, *482*
    LDDMM algorithm 476, *478*
  fitting curvatures 202–3
  landmark mapping *344*
  MRI-MPRAGE image 502, *503*

hippocampus (*Contd.*)
    surface patterns in Alzheimer's and aging 506–7, *508, 509*
    surface patterns in depression *510*
    surface patterns in schizophrenia 506, *507*
Hoffman brain phantom *308*
Holder inequality 17–18
homeomorphisms **219**, 220
homogeneous coordinates 179
homogeneous Markhov jump processes **515**
homogeneous Poisson counting processes, jump measure 515
homogeneous spaces **176**
Huang, J. and Mumford, D. 437
Huntington's disease, mapping of caudate 492–3
hypothesis testing 407–12
    automorphic pathology 505–6
    m-ary multiple hypotheses 412–13
    target detection 445–6

identifiability 408
identification of objects
    communications model for image transmission 378–81
    conditional mean minimum risk estimation 381–5
    hypothesis testing 408
    jump-diffusion processes 543–6
identity-change moves, Metropolis-Hastings jump-diffusion scheme 543
illumination variability, pose estimation 450–2
image deconvolution, time-of-flight PET 307–8
image orbit, diffeomorphism metric 349
image processing
    random Markhov fields 98–9
    segmentation via Ising model 107–8
image recognition algorithms 1
image reconstruction, need for regularization 304
image reconstruction problems 303–4
image restoration in point processes, EM algorithm 33–5
image synthesis, GMRF model 444–5
immersions 221, 226–7
incomplete-data **30**
independence, local coordinate frames 216
independent loops 189–90
independent processes, maximum entropy distribution 45
inexact matching 350–4
inference engine 5
infinite dimensional geometric and photometric variation 370–3
infinitely extended trees **72–4**
infinitesimal mean, diffusion processes **523**, 524
infinitesimal variance, diffusion processes **523**, 524–5
information, monotonicity 447–8
information loss, quantification 449–52
information measures 7–8
    asymptotic approximations 452–6
inhomogeneous Poisson counting processes, jump measure 516
initial momentum 350

Inside/Outside algorithm 92, **93**
integers 175
    equivalence classes 176
integration, and quadratic mean continuity 251–2
interior boundary of set **104**
interior of set **104**, *105*
internal bonds of configuration
    formal grammars 159
    patterns 155
intrinsic mean 495
intrinsic understanding 6
invariance, Markov processes **517**–18
    see also stationarity
invariances 356
invariant metric distances 347–9
Inverse Function Theorem 221
inverse perturbation 350–1
Ising, E. 96
Ising model 103–4
    1D Gibbs random fields 119–20
    low temperature 117–18
    segmentation 107–*8*
isosurface algorithm 202
isotropic Laplacian model, maximum-likelihood estimation 144–7
isotropic processes on sphere 285

Jacobian matrices 220, 221
Jaynes, E.T., principle of maximum entropy **44**, 112
JEEP CAD model, codebook construction 462–6
joint characteristic function 433
joint differential entropy **8**
joint entropy **7**
    directed acyclic graphs **65**
joint measurability **251**–2
joint probability distribution **9**
Joshi, S. 228, 236, 290, 292, 293, 503, 504
    diffeomorphism splines 340–2
    distribution free testing 510
    large deformation landmark mapping 363, *364*
    strongly elliptic operators 301
Joyce, L.S. and Root, W.L. 303
jump-diffusion algorithms 536–9
jump-diffusion inference 532
    aircraft tracking 549–53
    on countable unions of spaces 528–31
    detection and removal of objects 543
    ground vehicle recognition 533–5
    identification of objects 543–6
    mitochondria 556–61
    pose estimation 546–8
    posterior distribution 536
jump parameters, mitochondria 557
jump processes 515–16
    generators 519–20
    Gibbs' sampling 539–41
    Metropolis-Hastings algorithm 521–3, 541–3
jump process simulation 520–1
jump transition measures 540

Kalman filter 6, **21**–2
Karatzas, I. and Shreve, S.E. 299

Karhunen-Loeve Expansion **257–8**, 259, 311, 417
Kendall, D.G., similitude invariant distance 359–60
kernels 265–6, 267
   Green's 267–70
Khaneja, N. 192, 207, 208, 364
Kinderman, R. and Snell, J.L. 97, 117
Klein, F. 174
Kolmogoroff backward operator 519, 538
Kolmogoroff complexity model 174, 184
Kraft inequality 42
Kreigman, D.J. 418, 419
Kuich, W. 78
Kullback Liebler divergence (relative entropy) **8**, 15–6, 114
   information loss 450
Kumar, S. 87
Kupiec, J., trellis-based algorithm 94
kurtosis **434**–6

Lagrangian generation of diffeomorphisms 332
landmarked shape spaces, momentum 498–502
landmark mapping 343–5
landmark matching 342–3
landmark shapes *see N*-shapes
Lange, K. and Carson, R. 35
Langevin stochastic differential equations 525, 528
language **75**
   formal languages 74–80
language modeling
   context-free models 82–4
   EM algorithms for parameter estimation 92–4
   entropy of models 85–7
   hierarchical directed acylic graph model 84–5
   m-gram models 81–2
language synthesis, graph transformations 164–5
Lanterman, D.L. 304, 425, 427, 521, 541, 547
Laplace-Beltrami operator
   CON, smoothing arbitrary function on manifolds 297–9
   Neumann boundary conditions 293–7
Laplacian circulant boundary 277
Laplacian operator 273, 275, 280
   canonical representation 172–3
   on discrete triangulated graph 298–9
   Gaussian fields 130–1
   on sphere 289–93
large deformation diffeomorphic metric mapping (LDDMM) 474–8, 479–80, 496–8
   face *481*, 500–1
   heart 486–8
   mouse brain *491–2*
large deformation landmark mapping, Joshi's algorithm 363–5, *364*
large deformation statistical model **497**
laser radar (LADAR) imaging *392–3*
   asymptotic error exponents *412*, 413
   fusion of information 404–5
   pose estimation 548–9
lattice graphs 95, *96*
   Ising model 103–4
   limitations 244
   magnetic spin models 155–6

Markhov random fields 108–10
leaf model 187–9
least squares estimation 20–2
left group action **320**–1
left invariance 348–9
   photometric transport metric 367–8
left invariant distance, similitudes 359
Legendre functions 285
length, in space of diffeomorphisms 336
length of curve 317
level set methods, active contour models 237–40
Levitan, E. Herman, G.T. 304
Lie algebra 495
Lie group action 223, 224
Lie groups **224**–5
likelihood density **6**
likelihood function, sensors 378
likelihood-ratio testing **14**
limiting entropy rate, stationary Gaussian processes 142
linear differential operators, induction of Gaussian processes 271–3
linear graphs **50**, *95*, 96
   generators, configuration and regularity 156
   tracking of moving bodies 57
linear manifolds 215
linear membrane, road model *236*
linear operator, null space 177
line processes, hierarchical fields 108–10
links 190
Lipschitz property 524
local coordinate frames 216
   of circle and sphere 217
local coordinate patch construction 199
local diffeomorphisms 221
local optimization 323–4
local potential functions **166**
local quadratic charts 200–1
local regularity 155
log-normalizer 110–12
   stationary Gaussian random fields 142
lower neighbourhood *162*
low temperature Ising model 117–18

M2 tanks, Poisson projection model *394*
M60 tank, pose estimation 425–6
macaque cortex, mapping experiment *477*
McNally, J.G. 228, 236
magnetic resonance imaging
   cross modality registration *430*–1
   registration 397–8
magnetic spin models, generators, configuration and regularity 155–6
manifold of *N*-shapes **357**
manifolds **214**–15
   parallelizable **218**
   smooth mappings 219–20
maple leaf model 187–9
MAP (maximum a posteriori) estimation 13–14
   Gaussian case 41–2
   hidden state sequences 88
   multi-modality signature registration **430**

MAP (maximum a posteriori) estimation (Contd.)
  pose **424**–6
  signature and pose, Commanche FLIR 454–6
mappings, definitions 219–20
"Marching Cubes" algorithm 201–2
marginal distributions **9**
Markov chains 54–6
  1D Gibbs random fields 119–20
  entropy 66–8
  extinction probability **62**–3
  m-gram language models 81–2
  neighbourhoods 97
  transition dynamics 516
  unique Gibbs distribution 120
Markov jump processes **515**–16, 537–8
  generators 519–20
Markov jump process simulation 520–1
Markov processes **514**
  generators **517**–18
  jump-diffusion dynamics 529–31
  stationarity (invariance) **517**–18
Markov random fields (MRF) 96–101, **97**
  canonical representation 169–70
Mark, K.
  entropy of bigram mode 86
  language model 84–5
matrix group actions, as diffeomorphisms 223–4
matrix groups 177–81
  geodesics 324–6
  metrics 327–9
maximum entropy
  and conditional probability 47–8
  principle of **44**
maximum entropy models 45–6
  texture representation 112–15
maximum-likelihood estimator (MLE) **26**–7
  anatomical maps 496
  anisotropic textures 148
  asymptotic normality 28–9
  auto-regressive image modeling **144**
  consistency 27–8
  Gaussian mixture modeling, brain segmentation 36–8
  HMM parameters 89–92
  image restoration 33–5
  isotropic Laplacian model 144–7
  sinusoids in Gaussian noise 31–3
mean
  Gaussian fields **123**
  Gaussian vector fields **262**
mean adjustment, least-squares estimation 20
mean branching matrix **59**
mean curvature 199
measurability **245**, 246
measurable events 9
measurable sets 9
medial prefrontal cortex
  fitting curvatures 204–5
  planar maps 212, 213
medical diagnosis, Bayesian belief network 52, 53
medical imaging registration 397–8
Mercer's theorem **255**–7
messenger RNA 53

meteorological photometric variation 420, *421*
method of moments estimation, transported generator model 436–7
Method of Sieves 303
metrics **316**
  between orbits 356–7
  between orbits of special Euclidean group 373–4
  diffeomorphism metric for image orbit 349
  Frobenius metric 329–30
  images under rotational motions 374–6
  induced via photometric and geometric flow 366–8
  in matrix groups 327–9
  N-shapes 357–60
  in $SO(2, 3)$ 330–1
  on space of diffeomorphisms 336–8
metric space norms for clutter 439–42
metric spaces **316**
  for photometric variability 365–6
  of robust deformable templates 415–16
  vector spaces 319–20
Metropolis-Hastings algorithm 521–3
  jump processes 541–3, 551
m-gram language models 76, 81–2
Miller, M.I. 338
minimal geodesics **317**, 318
  Hilbert spaces 319
  space of diffeomorphisms 338
minimum description length (MDL) principle 38, 42–3
minimum energy curves 317, 318
minimum-mean-squared error (MMSE) estimators 16, 381, **382**–4, 502
  conditional mean estimation 22–4
  least-squares estimation 20–2
  projective imagery models
    3D to 1D projections 395–7
    3D to 2D projections in Gaussian noise 385–9
    LADAR imaging 392–3
    medical imaging registration 397–8
    Poisson projection model 393–5
    synthetic aperture radar imaging 389–91
minimum risk estimation, Bayes paradigm 6–7
minimum risk estimator **7**
minimum risk hypothesis testing, discrete spaces 12–16
Minkowski set difference and addition **126**
miss, probability of (type II error) **14**
missed detection rate **408**
mitochondria 137, 476, *479*
  analysis using stationary periodic processes 259–62, *260*
  contour models 553–6
  jump-diffusion inference 556–61
  Markov random field model 98–9
  partition imaging 229, *230*
  segmentation 241–3
  texture parameters 146–7, 148, *149*, *151*
  variability 553
m-memory Markov chains **56**
  canonical representation 168–9
model order estimation 38–40
  minimum description length principle 42–3
model uncertainty, quantification of information loss 449–52

moment-generating function, branching processes 60–2
moments
   determination from characteristic function 11
   nth 10
momentum
   conservation along geodesics 326, 338–40, 474
   conservation for diffeomorphism splines 340–3
   of landmarked shape spaces 498–502
   normal momentum motion
     geodesic connection 350–4
     temporal sequences 354–6
monotonic extension 163
monotonicity
   Fisher information 403
   sensor measurement information gain 447–8
Mori, S. 489
Morphable Faces database 500, 501
mouse brain development study 489–92
moving body systems *158*
   generators, configuration and regularity 157
MRI-MPRAGE image, hippocampus 502, *503*
MSE bounds, incorporation of dynamics 386–8
MSTAR database *389*, 390
mud, texture fields *115*
multiple extension 163
multiple generators, group actions 181
multiple modality signal registration 429–31
multiple object recognition, graph transformations 163–4
multiple sensor fusion 402–5
   error exponent 411
multiple sensors 378
multiplicative numbers, as Lie group 224
multi-type branching processes 56–9
   tree structure 69
   see also branching processes
multivariate Gaussian processes, maximum entropy density 45, 46
multivariate normal distribution 10–**11**
Mumford, D. 107, 108, 338, 418, 431
Musicus, B.R. 49
mutual information **8**, **447**–9
   asymptotic approximations of information measures 452–6
   quantification of information loss 449–52
mutually independent random variables 9

Nadel, S. 397
N-buty alcohol analysis, use of EM algorithm 33
neighbourhood systems **95**, *96*
   hidden Markov chains 99–100
   hidden random branching processes 100–1
   Markov chains 98
   MRF representation of textured images 98–9
   upper and lower neighbourhoods **162**
nested family of graphs, multiple object recognition 163–4
Neymann-Pearson Lemma **14–15**, 503, 504
noise, projective transformations 379–80
noise-equivalent differential temperature (NEΔT) 395
noisy channels 5

non-compact operators 309–10
   orthogonal scale representation 312–15
non-terminating trees **72**–4
normal curvature **199**, *200*
normal deformable surfaces 227–9
normal deformation, circle 236
normal distribution, multivariate 10–**11**
normal form context-free grammars, Inside/Outside algorithm 93
normalized entropy, branching processes 70
normal momentum motion
   geodesic connection 350–4
   temporal sequences 354–6
normal operators **253**
normed linear vector space **17**–18
N-shapes 340, 357
   diffeomorphism metric 361
     approximation for small deformation splines 361–5
   Euclidean metric 357–9
   group actions **361**
   Kendall's similitude invariant distance 359–60
nth moment **10**
nuisance integral 38, 39–40, 41–2, 408, 409
nuisance parameters 407–8
null space of linear operator 177

object recognition
   communications model for image transmission 378–81
   conditional mean minimum risk estimation 381–5
   hypothesis testing 408
   jump-diffusion processes 543–6
Oboukhov's orthogonal representation 286, 288
occipital cortex, fitting curvatures 203–4
occupancy vectors, branching processes 72
odd spherical harmonics **286**
Onsager, L. 96
openGL software 425, 427, 450
open mappings **219**, 220
operational photometric variation 420, *421*
operator norm **253**
operator symmetry 284
operator trace **139**
optical flow, smoothing 305
optical-sectioning microscopy 290, *292*–3
optimal path computation, dynamic programming 207–8
orbits, metric distances 356–7
orbits of group **176**
ordinary differential equations, landmark mapping 343–5
oriented textures 147–53
orthogonal expansions 257–8
orthogonal groups **178**
   discrete integration 385
   MMSE estimator 382–3, 386
orthogonal processes 311
   Cramer decomposition **311**–2
orthogonal scale representation 312–15
orthonormal sequences 248
O'Sullivan, J.A. et al. 389, 390, 391
output symbol distribution 465–7

parallelizable manifolds **218**
parameter estimation 398
  Fisher information 399–402
parents of a site **50**
parity languages 75–6, 77
parsing, graph transformations 164–5
partial ordering 50
partition function **101**
  Gibbs distribution 110–12
partitioning image, conditional probability 229
partitioning of sets 175
part-of-speech tagging 82
paths **50**
Pearl, J. 51
Pearlson, G. 209
Peierls, R.E. 117
penalized estimator **304**
penalized likelihood methods 303, 305–8
penalty (distortion) function **456–7**
Penn Treebank, bigram and trigram data support 82
Pentland, A. and Sclaroff, S. 293
periodic boundary conditions, Gaussian processes 129, 130–1
periodic processes on bounded domains 258–62
Perron-Frobenius theorem **62**, 516
perspective projection 448, 449, 534
PET (positron emission tomography) 33, 35
phase transition, 2D Ising model 117
photometric and geometric flow
  computing infinite dimensional variation 370–3
  Euler equations 369–70
  metrics 366–8
photometric space 416–17
photometric variability 365, 414, *415*
  conditionally Gaussian random field models 422–4
  empirical covariance 416–21
  metric spaces 365–6
  and pose estimation 424–8
phrase structure grammars, generators, configuration and regularity 160–2
piecewise linear roads 235–6
PIE phantom *308*
planar mapping, surface manifolds 210–13
planum temporale
  curvature profiles 298, *299*
  definition via dynamic programming *209*, 210
  planar maps *211*, 212
  surface harmonics 296–7
platelet of a point **206**
Platonic solids 156
Poggio, T. et al. 304
point estimators 6
point processes, image restoration 33–5
Poisson counting processes
  jump measures 515–16
  maximum entropy density 45, 46
Poisson processes 519–20, 523
  characteristic function 12
  image restoration 33–4
Poisson projection model 393–5
polar decomposition 179
pose 378

pose estimation 424–7, 448–9, 544–9
  Bayes integration 427–8
  signature variability 450–2
Posener, J.A. et al. 510
POSET (partially ordered set) graphs 49, 50, 162
positively regularity, matrices **62**, 64
posterior distribution **6**, 536
  and transformations of inference 516
posterior potential, mitochondria 556
power operators **139**
power spectral density
  cyclo-stationary Gaussian processes **135**
  stationary Gaussian processes **133**
Precision Gauges 304
pre-whitening operators 278
Preza, C. et al. 304
principal curvatures **199**
principal curves 205
principle components analysis (PCA) 494, **497–8**
  small deformation PCA versus large deformation PCA 499–502
principle components of landmark momentum **499**
principle of maximum entropy **44**
prior distribution and density **6**
prior induction, aircraft tracking 550–1
PRISM simulator software 37, 420, 425
probabilities on directed acyclic graphs (PDAGs) 51–4
probability distribution function **9**
probability of extinction, Markhov chains **62**–3
probability-generating functions, branching processes 60–1
probability measure **8–9**
probability model-building 43
  maximum entropy models 45–6
  principle of maximum entropy 44
probability spaces **8–9**
Procrustes mean 499
product groups 181–4
projection operators **309**
Projection Theorem **19**
projective imagery models
  MMSE estimators
    3D to 2D projections in Gaussian noise 385–9
    high-resolution radar 395–7
    LADAR imaging 392–3
    medical imaging registration 397–8
    Poisson projection model 393–5
    synthetic aperture radar (SAR) imaging 389–91
  photometric variation 422–3
projective transformations in noise 379–*80*
proofs, website address 4
proposal density 521, 542
pseudolikelihood 110
pseudolikelihood method, Besag 560–1
pseudolinear grammars 78–9
"push in" and "push out" pathologies *513*

Qiu and Bitouk 293, 294–7, 298
quadratic mean continuity **246**
  and integration 251–2
quadratic patch **199**

Rabbitt, R.D. 478
radar cross section (RCS) 395
randomization, Fisher's method 510–11
random processes
    continuity 300
    measurability 245
random variables **9**
    complex 11
    Hilbert space 247–8
    transformation 10
range of interaction **104**
rank of mapping **220**
rate-distortion function calculation, Blahut algorithm 457–9
rate-distortion theory 456–7
    remote rate-distortion problem 459–65
Ratnanather, T. 209, 212
ray traced imagery *437*
real numbers 175
real-valued random variables **9**
recognition 6
recursive estimation, Kalman filtering 21–2
Reed, M. and Simon, B. 253
reflexivity, as equivalence relation 175
regular configurations **155**
regular curve **190**
regular grammar *see* finite-state grammar
regularity
    formal grammars 159
    patterns 155–8
regularity constrained subset of similarities **182**
regularization, role in image reconstruction 304
regular lattices, Ising model 103–4
relative entropy (Kullback Lieber divergence) **8**
remote rate-distortion problem 459–65
reproducing kernel Hilbert spaces (RKHS) **265**–6, 335–6
reproduction alphabet **456**–7
restoration 6
Reverse Polish *58*, 64
reversible transformations 163
re-writing rules, formal grammars 158–*9*
ridge curves **205**
    brain tissue 206
Riemannian manifold 317–18
Riemann Mapping Theorem 210
Riesz-Schauder theorem **255**
right group action **320**–1
risk functions 407
Rissanen, J., minimum description length principle (MDL) 38
road contours, transformations 183
roads, active models 234–6
robust deformable templates 414–**15**
    metric space 415–16
    for natural clutter 438–45
robust multi-modality deformable template **429**
Rodriquez formula 328
rotational motions
    aircraft tracking 550
    metric between images 374–6
rotation determination, variable orientation 152–3
rotation of grid 333

Ruelle, D. 97
run-length languages 75–*6*

Saccharomyces cerevisiae
    gene regulatory networks *165*, 166
    genetic regulatory networks 53–4
Saffitz, J. 230, 243, 553
sample paths, almost-sure continuity **300**, 301–3
sample space **8**, 9
sampling, use of diffusions 527–8
scalar MRI mapping of heart 486–7
scale operator **310**
scale representation, orthogonal 312–15
scale transform 314
scaling 376–7
schizophrenia, hippocampal surface patterns 506, *507*
Schwarz, G. 38
second-order processes **247**–8
    properties 249–51
second-order processes stationary to scale **312**
second-order quadratic charts 200–1
segmentation
    of mitochondria 241–3
    textured images 137–8
    via Ising model 107–*8*
self-adjoint operators **253**, 255
sensing models 379
sensor fusion 402–5
    error exponent 411
sensor measurement information gain, monotonicity 447–8
sensors, likelihood function 378
sensor symmetry 406–7
sentence-language set 161
separability **245**, 246–7
separability program 245
set of equivalence classes **176**
set of equivalences through sensor map **405**
set of images **346**
sets, partitioning by equivalence classes 175
Shannon, C.E. 81
Shannon, C.E. and Weaver, W. 64
Shannon's theory of communications 5
shape functions 293–4
shape operator 196, **197**–8, 199
    ridge curves 205
shape spaces, statistics analysis 494–5
Shapiro, J.H. 393, 404, 413
Shepp, L.A. and Vardi, Y. 34, 35
shift operator **105**, **309**
    sphere 285
signalling models 2
signature variability, pose estimation 450–2
signed distance function 237
similarities, relationship to generators 184
similar triangles 180–1
similitude invariant distance, *N*-shapes 359–60
similitudes **178**
    geodesics 329
singular value decomposition (SVD) 418

sinusoid estimation, generators, configuration and regularity 158
Slutsky's theorem 29
small deformation elasticity operator 274, 275–6
small deformation Gaussian fields on surface submanifolds 502–3
small deformation PCA 499–500
small deformations, template construction 502
small deformation splines, diffeomorphism metric approximation 361–5
small deformation vector field models 280–3
smooth coordinate patch 195
smoothing
 exponential spline smoothing via Good's roughness 306–9
 role in image reconstruction 304
smoothing arbitrary function on manifolds 297–9
smooth manifolds 214
smooth mappings 219–20
smoothness, and Gaussian priors 304–5
smoothness conditions 38–9
 generation of diffeomorphisms 334
smoothness properties 264–5
smooth surfaces, Gaussian random fields 293
smooth vector fields 217–18
snakes 226
 active models 234–6
SNR (signal-to-noise ratio) 404–5
 FLIR 395
 high-resolution radar (HRR) 396
Snyder, D.L., Poisson noise models 34, 393
Snyder, D.L. and Miller, M.I. 306, 307, 432
Snyder, D.L. and Politte, D.G. 35
Sobolev spaces 266
solutions to problems, website address 4
source-channel view 447
source model 379
space of regular configurations 155
spanning trees 189
special Euclidean group (SE) 178
 adding dynamical equations of motion 386–8
 Hilbert Schmidt estimator 384–5
 metrics between orbits 373–4
 MMSE estimator 383–4
special linear groups 178
special orthogonal group 225
spectral density
 cyclo-stationary Gaussian processes 135
 stationary Gaussian processes 133–4
spectrum estimation, Gaussian noise 31–3
speech recognition systems
 hidden Markov models 87, 91–2, 99–100
 trigram model 82
speed of curves 190
SPET (single-photon emission tomography) 33, 35
sphere
 coordinate frames 217
 normal deformation 228–9, 236–7
 shape operator 197–8
 stationary processes 285–9
 Laplacian operator induced Gaussian fields 289–93
spherical harmonics 285–7, 286

addition theorem 288
splitting property
 directed graphs 51, 55
 Gibbs distribution 104–10
squared error 7
squared error metric 439
square-integrable spaces (L^2) 17–18
square-summable spaces (l^2) 17–18
Srivastava, A. 386, 407, 409, 412
 exponential error exponents 410–11, 413
 MMSE estimator theorem 382–3
stabilizers 347–8
stabilizing subgroups 347
standard Brownian motion 263
standard Wiener process 263
static metric mapping model 470–1
stationarity 249
 Markhov processes 517–18
 on sphere 285
stationary Gaussian contour model, mitochondria 554–6
stationary Gaussian random fields 133–4
 asymptotic covariance and log-normalizer 142
 entropy 142–3
stationary Gibbs random fields 116–18
 1D 119–20
 entropy 121–2
stationary Navier elasticity operator 279–80
stationary periodic processes 258–9
 analysis of mitochondria 259–62
stationary processes on sphere 285–9
 Laplacian operator induced Gaussian fields 289–93
statistical communications 2
statistics analysis
 disease testing of automorphic pathology 505–6
 large deformation statistical model 497
 shape spaces 494–5
steepest descent direction, Chan-Vese model 238
Stein's Lemma 15–16
stochastic differential equations (SDEs) 525, 527–8
stochastic formal grammar 75
stochastic languages 75, 160
Stone's representation 313
strongly elliptic operators 301
subcritical branching processes 60
 entropy 70–1
subgroups 174
 of diffeomorphisms 332
submanifolds 222
subspace topology 222
sulci, curvature 205–7
sulcus curves, representation 192–5
supercritical branching processes 60
 entropy 70–1
superior temporal gyrus, use of dynamic programming 208, 209, 210
superposition norm-square 367–8
support set 104
surface harmonics 293
 planum temporale 296–7
surface manifolds, mapping to planar coordinates 210–13

surface normal 196
surfaces
 definitions 195
 fitting curvatures 198–205
  dynamic programming 207–10
 shape operator 196, **197**–8
symmetry
 as equivalence relation 175
 target and object 181
symmetry sets **406**–7
synthesis 6
synthetic aperture radar (SAR) imaging 389–91

T62 tank CAD model, codebook construction 461–5
T-72 tank
 MSTAR data analysis *389–91*
 pose estimation *426*, 427
tangent spaces **195**, 215–17, 218
tangent vectors 215
tangent vectors of surface 196
tanks
 Commanche, FLIR imagery 454–6
 M2 tanks, Poisson projection model *394*
 M60 tank, pose estimation 425–6
 photometric variation *420–1*
 pose estimation 425–7, 546–9
 T62 tank CAD model, codebook construction 461–5
 T-72 tank
  MSTAR data analysis *389–91*
  pose estimation *426*, 427
 thermal variation *428*
 transformations 182
tank/truck discrimination, likelihood ratio 412
Tapia, R.A. and Thompson, J.R. 304
target detection 445–6
 aircraft tracking 552
 hypothesis testing 407
 jump processes 543, 558–9
 Metropolis-Hastings jump-diffusion scheme 541–3
target and object symmetry 181
teacup and teapot data bases 419, *420*
teapot, photometric variability *427–8*
template construction, small deformations 502
templates **181**, **182**
 see also deformable templates; robust deformable templates
temporal sequences, normal momentum motion 354–6
textured images
 random Markhov field representation 98–9
 segmentation 137–8
texture representation
 cyclo-stationary Gaussian modeling 135–7
 maximum entropy models 112–15
textures, anisotropic 147–53
theorems, website address 4
thermal variation, FLIR *428*
Thompson, D'Arcy, "On Growth and Form" 468
thresholding function 7
time-invariance, Markhov chains 54
time-of-flight PET 35
 image deconvolution 307–8

Toeplitz covariance 135, 440
topology **214**
torsion of curves **190**
torus graphs *95*, 96, 129
total jump intensity 540
trace-class property 251
tracking via Euclidean motions 179
transformation of random variables 10
transformations
 application of 2
 groups 174
 translation 218
 see also graph transformations
transition density **6**
transition law 378
transition law of Markhov process **514**
transitive group actions **176**
transitivity as equivalence relation 175
transitivity of transformations **163**
translation 218
transmission tomography 35
transported generator model 431–7, **432**
transport theorem, deformable shapes **231**–2
tree generators, kurtosis 435–6
tree graphs *50*, *95*
tree structure, multi-type branching processes 69
trellis-based algorithm, Kupiec 94
triangle inequality 17–18, 316
 Riemannian manifold 318
triangles, similar and congruent 180
triangulated graphs
 brain tissue 202–3, 206–7
 dynamic programming 207–10
 "flattening" 210–12
 generators, configuration and regularity 156–7
 similarities 186–7
triangulated surfaces, curvature approximation 200–1
trigram language model 76, *82*
 entropy *86*
 generators and configuration 160
Trouvé, A. 334, 338
twigs 190
typical set **66**, **68**
typical trees 71–4

unbiased estimators 24
unclosed contours, vectors as generators 186
unconstrained configurations **155**
understanding 6
undirected graphs 49, 95–6
uniform scale groups **178**
Uniform Weak Law of Large Numbers 27
uniqueness, diffeomorphisms 334–5
unitary operators **309**
unit cube, Gaussian fields 273–7
unit normal curvature vector field **191**–2
unit speed curves **191**
unit tangent vector field **191**–2
univariate normal density 10
upper neighbourhood **162**
urban clutter, image analysis *149*, *151*

Vaillant, M. *344*, 345, 499–500
Van Campenhout, J.M. and Cover, T.M. 48
Van Essen, D. 208, 209, 477, 482, 483, 484
Van Zijl, P. 489
variable orientation, anisotropic textures 151–3
vector fields 191–2, 217–18
vector fields for growth 488–93
vector Lie groups 224
vectors as generators of unclosed contours 186
vector spaces as metric spaces 319–20
vector-valued random processes
 continuity **246**
 measurability **245**
velocity of curve **317**
viscous matching 370
Viterbi algorithm 87, 88–9
V-multi-type (vector) branching processes 57, **58–9**
 entropy 70–1
 extinction probability 63–4, **71**
Von-Mises Markhov density 550
vth probability generating function **60**

waiting times, maximum entropy density 45, 46
weakly reversible processes **521**
white Gaussian processes, on sphere 289
whitening model, maximum entropy 45, 46
white noise 264, 271–2
wide-sense stationarity **249**

wide-sense-stationary uncorrelated-scatter (WSSUS) model 396
Wiener process 263
 almost surely sample path continuity 303
 and diffusions 526–7
 Green's function 270–1
Winslow, R. 477, 484
Wong, E. and Hajek, B. 300

X-29, azimuth-elevation power spectrum *396*
XPATCH simulator 379, 396

Yantis, S. 203–4
yeast (*Saccharomyces cerevisiae*), genetic regulatory networks 53–4
Yen, B. 413
Younes, L. 359, 472
 computing infinite dimensional geometric and photometric variation 370–3
 DT-MRI 484, 486, 487
 glasses and eyelids experiments *373*
Yuille, A. 418
Yuille, A. and Grzywach, N.M. 304

zero level set function 237
Zhang, J. 489
Zhu, S.C. 112, 113, 114